METHODS IN MOLECULAR BIOLOGY

Series Editor
John M. Walker
School of Life and Medical Sciences
University of Hertfordshire
Hatfield, Hertfordshire, AL10 9AB, UK

For further volumes:
http://www.springer.com/series/7651

Calcium-Binding Proteins of the EF-Hand Superfamily

From Basics to Medical Applications

Edited by

Claus W. Heizmann

Division of Clinical Chemistry and Biochemistry, Department of Pediatrics, University of Zürich, Zürich, Switzerland

Humana Press

Editor
Claus W. Heizmann
Division of Clinical Chemistry
and Biochemistry
Department of Pediatrics
University of Zürich
Zürich, Switzerland

ISSN 1064-3745 ISSN 1940-6029 (electronic)
Methods in Molecular Biology
ISBN 978-1-4939-9029-0 ISBN 978-1-4939-9030-6 (eBook)
https://doi.org/10.1007/978-1-4939-9030-6

Library of Congress Control Number: 2018967209

Cover Illustration: The cover image is taken from Kozlyuk et al. (Chapter 18), and it represents the binding of calcium to the S100A12 protein

This Humana Press imprint is published by the registered company Springer Science+Business Media, LLC, part of Springer Nature.
The registered company address is: 233 Spring Street, New York, NY 10013, U.S.A.

Preface

A major direction in medical research leading to clinical applications targets the regulation of intracellular calcium and the various human diseases associated with an altered homeostasis of this global messenger. After entering the cell, Ca^{2+} is reversibly complexed to Ca^{2+}-binding proteins of the EF-hand superfamily which decipher the information carried by Ca^{2+} and pass it on to various targets, receptors, and channels. Many proteins of this family are of major clinical importance in cardiomyopathy, inflammation, allergy, brain disorders, and cancer as predictive biomarkers in modern laboratory medicine and as novel therapeutic drug targets.

One example is the measurement of troponin, the first intracellular Ca^{2+}-sensor protein to be discovered, is nowadays the "gold standard" in the diagnosis of patients with acute coronary syndrome. Parvalbumins were the first Ca^{2+}-binding proteins to have their amino sequence and atomic 3D structure resolved, are selective markers of neurons in the CNS, and are also known to be the major allergens in fish. In addition, parvalbumin gene transfer in an animal model was able to correct diastolic heart failure. Mutations in calmodulin were identified in individuals affected by severe cardiac arrhythmias suggesting that this protein could be a therapeutic target for some cardiac disorders. Inhibition of calcineurin by cyclosporine A or FK506 has been proven to be beneficial in preventing rejection of allogeneic organ transplants, and members of the penta-EF-hand (PEF) protein family (calpains, sorcin, or ALGs) are of clinical relevance in cardiovascular diseases and oncology. S100 proteins constitute the largest subgroup within the EF-hand protein family with intracellular and extracellular, cytokine-like functions. Some members are routinely analyzed in practical laboratory medicine, e.g., S100B is an established biomarker for malignant melanoma and traumatic brain injury, S100A4 is known to be an angiogenic factor inducing tumor progression and metastasis, and S100A8/A9 is involved in inflammatory and autoimmune diseases and several types of cancer. Drug discovery efforts exploiting the identification of selective Ca^{2+}-binding protein inhibitors are underway for specific pharmacological interventions and personalized treatments.

Individual chapters provide step-by-step protocols for studying the many facets of Ca^{2+}-signaling and Ca^{2+}-binding with background information on the principle and application of these techniques. This volume also contains a few chapters with more general reviews on the clinical importance of this protein family.

Part I includes data mining to explore calmodulin bibliography, modern technologies such as deep two-photon and high-resolution imaging, molecular MR imaging methods, measurement of contractility and calcium release in cardiac spheroids, simultaneous recordings of subcellular Ca^{2+}-signals from the cytosol and sarco/endoplasmic reticulum, and designing Ca^{2+}-binding proteins for molecular MR imaging.

Part II includes the expression and purification of calmodulin, calpain, plastins, MICU2, ALG-2, calumenin GCAPs, swiprosin/EFhd2, Ca^{2+}-binding proteins from parasitic worms, and S100 proteins for structural studies (NMR- and X-ray crystallography), other biophysical applications, metal-binding studies, as well as interactions with target proteins and small molecule inhibitors. Other chapters provide detailed protocols of gene transfers of Ca^{2+}-binding proteins to cardiomyocytes, dissecting the role of calcineurin by using Cre-loxP inducible mouse models, analysis of S100 in DNA damage repair, Fluorine-18 labeling of

S100 proteins for positron emission tomography, and tryptophan scanning mutagenesis of EF-hand motifs.

Part III includes the analysis of Ca^{2+}-binding proteins as important diagnostic and prognostic biomarkers in laboratory medicine. This includes high-sensitive troponin assays in the clinical diagnosis of patients with acute coronary syndrome, measurements of S100B as biomarker for head injury and melanoma, S100A7 in psoriasis, S100A8/A9 in myocardial infarction, and S100A12 in fecal samples of children and adults.

All chapters include an introduction to their respective topics, a comprehensive list of the required materials, step-by-step and readily reproducible laboratory protocols, and key tips on troubleshooting to avoid known pitfalls. These methods and techniques are generally applicable to many other areas of basic and applied medical research and laboratory medicine.

This volume is a collection of chapters written by leading experts in the field containing state-of-the-art, lab-based methods and easy-to-follow protocols for daily use, making it interesting for basic and medical researchers, cell and molecular biologists, clinicians, clinical chemists, and the diagnostic industry.

I am very grateful to all the contributors for the opportunity to make this volume possible, to my daughter Melanie Heizmann for her expert secretarial assistance, and to my wife Erika Heizmann for her patience and understanding during the process of editing this book.

Zürich, Switzerland ***Claus W. Heizmann***

About the Editor

Claus W. Heizmann Professor em. of Clinical Biochemistry at the University of Zürich, Switzerland. He received his Diploma in Chemistry from the University of Basel, Switzerland, and his PhD degree in 1970 from the University of Konstanz, Germany. Subsequently, he was trained as a postdoctoral fellow at the University Hospital, Basel, in the laboratory of Dr. Edmond Fischer at the Department of Biochemistry at the University of Washington, Seattle, at the Department of Cell Biology at the Federal Institute of Technology (ETH), Zürich, and was an independent researcher at the Institute of Pharmacology and Biochemistry at the University of Zürich. From 1989 to 2007 he was Director of the Division of Clinical Chemistry and Biochemistry at the Department of Pediatrics, University of Zürich. He was visiting professor at the Mie University, Tsu, Japan, and general secretary of the Swiss Society for Clinical Chemistry (SGKC). His research focuses on "the Structures and Functions of the Ca^{2+}-Binding Proteins of the EF-Hand Superfamily and their Targets in Health and Disease." He was a co-founding, board, and honorary member of the European Calcium Society (ECS) and its first president. He is co-editor of the special issues on calcium published in *Biochimica et Biophysica Acta-Molecular Cell Research*, the author of more than 400 scientific publications, and editor of several books including *Calcium-Binding Proteins and RAGE: From Structural Basics to Clinical Applications* (Vol 963, 2013) and *Calcium-Binding Proteins of the EF-Hand Superfamily: From Basics to Medical Applications* (Vol 1929, 2019).

Contents

Part II The Superfamily of Calcium-Binding Proteins Characterized by the EF-Hand Structural Motif

PART III BIOMARKERS, DIAGNOSTICS AND LABORATORY MEDICINE

Contributors

TAKUYA ACHIHA • *Department of Applied Molecular Biosciences, Graduate School of Bioagricultural Sciences, Nagoya University, Nagoya, Japan*
NOONA AMBARTSUMIAN • *Faculty of Health, Department of Neuroscience, Copenhagen University, Copenhagen, Denmark*
ADDELI BEZ BATTI ANGULSKI • *Department of Integrative Biology and Physiology, University of Minnesota Medical School, Minneapolis, MN, USA*
GOPALA KRISHNA ARADHYAM • *Signal Transduction Laboratory, Department of Biotechnology, Bhupat and Jyoti Mehta School of Biosciences, Indian Institute of Technology Madras, Chennai, India*
RAMONA ASTRAND • *Department of Neurosurgery, Rigshospitalet, Copenhagen University Hospital, Copenhagen, Denmark*
BASKARAN ATHMANATHAN • *Department of Pathology, University of Alabama, Birmingham, AL, USA*
ANTJE BIRKNER • *Institute of Neuroscience, Technical University of Munich, Munich, Germany; Munich Cluster for Systems Neurology (SyNergy) and Center for Integrated Protein Sciences (CIPSM), Munich, Germany*
VITO CALDERONE • *CERM, University of Florence, Sesto Fiorentino, Florence, Italy; Department of Chemistry Ugo Schiff, University of Florence, Sesto Fiorentino, Florence, Italy*
ASIMA CHAKRABORTY • *CSIR-Centre for Cellular and Molecular Biology (CCMB), Hyderabad, India*
WALTER J. CHAZIN • *Department of Biochemistry, Vanderbilt University, Nashville, TN, USA; Center for Structural Biology, Vanderbilt University, Nashville, TN, USA; Department of Chemistry, Vanderbilt University, Nashville, TN, USA*
TAREK CHEHAB • *Centre for Molecular Biology of Inflammation, Institute of Medical Biochemistry, University of Münster, Münster, Germany*
JENS CHRISTOFFERS • *Organic Chemistry, Faculty V, Institut für Chemie, University of Oldenburg, Oldenburg, Germany*
HOUDA COHEN • *Department of Integrative Biology and Physiology, University of Minnesota Medical School, Minneapolis, MN, USA*
NINA CRIADO SANTOS • *Centre for Molecular Biology of Inflammation, Institute of Medical Biochemistry, University of Münster, Münster, Germany; Department of Cell Physiology and Metabolism, University of Geneva, Geneva, Switzerland*
LISA S. CUNDEN • *Department of Chemistry, Massachusetts Institute of Technology, Cambridge, MA, USA*
FABIO D'AMICO • *Department of Biomedical and Biotechnological Sciences, University of Catania, Catania, Italy*
STEVEN M. DAMO • *Department of Biochemistry, Vanderbilt University, Nashville, TN, USA; Center for Structural Biology, Vanderbilt University, Nashville, TN, USA; Department of Life and Physical Sciences, Fisk University, Nashville, TN, USA*
ANDREW S. DAY • *School of Women's and Children's Health, University of New South Wales Medicine, Sydney, Australia; Department of Pediatrics, University of Otago (Christchurch), Christchurch, New Zealand*

BJÖRN-PHILIPP DIERCKS • *The Calcium Signalling Group, Department of Biochemistry and Molecular Cell Biology, University Medical Center Hamburg-Eppendorf, Hamburg, Germany*

REINHARD DUMMER • *Department of Dermatology, University Hospital Zurich, Zurich, Switzerland*

PÉTER ECSÉDI • *Department of Biochemistry, ELTE Eötvös Loránd University, Budapest, Hungary*

MIGUEL FERNANDEZ-TENORIO • *Department of Physiology, University of Bern, Bern, Switzerland*

FRANZISKA FOERTSCH • *Biomolecular Photonics Group, Jena University Hospital, Jena, Germany*

MARCO FRAGAI • *CERM, University of Florence, Sesto Fiorentino, Florence, Italy; Department of Chemistry Ugo Schiff, University of Florence, Sesto Fiorentino, Florence, Italy*

SIMONE FRANCHINI • *Neonatal Intensive Care Unit, Department of Pediatrics, University of Chieti, Chieti, Italy*

ANNA LISA FRAUCHIGER • *Department of Dermatology, University Hospital Zurich, Zurich, Switzerland*

JAN FRIČ • *Center for Translational Medicine, International Clinical Research Center, St. Anne's University Hospital Brno, Brno, Czechia*

BARBARA G. FÜRNROHR • *Division of Molecular Immunology, Department of Internal Medicine III, Nikolaus-Fiebiger-Center, University of Erlangen-Nürnberg, Erlangen, Germany*

VELIA GARCIA • *Center for Structural Biology, Vanderbilt University, Nashville, TN, USA; Department of Chemistry, Vanderbilt University, Nashville, TN, USA*

DIEGO GAZZOLO • *Neonatal Intensive Care Unit, Department of Pediatrics, University of Chieti, Chieti, Italy; Neonatal Intensive Care Unit, Department of Maternal, Fetal and Neonatal Medicine, C. Arrigo Children's Hospital, Alessandria, Italy*

VOLKER GERKE • *Centre for Molecular Biology of Inflammation, Institute of Medical Biochemistry, University of Münster, Münster, Germany*

MARIAGRAZIA GRANATA • *Department of Biomedical and Biotechnological Sciences, University of Catania, Catania, Italy*

BENJAMIN M. M. GRANT • *Department of Medical Biophysics, University of Toronto, Toronto, ON, Canada; Princess Margaret Cancer Centre, University Health Network, Toronto, ON, Canada*

MARIAM GRIGORIAN • *Faculty of Health, Department of Neuroscience, Copenhagen University, Copenhagen, Denmark*

ANDREAS H. GUSE • *The Calcium Signalling Group, Department of Biochemistry and Molecular Cell Biology, University Medical Center Hamburg-Eppendorf, Hamburg, Germany*

CATHLEEN HAASE-KOHN • *Department of Radiopharmaceutical and Chemical Biology, Helmholtz-Zentrum Dresden-Rossendorf, Institute of Radiopharmaceutical Cancer Research, Dresden, Germany*

ROSE C. HADLEY • *Department of Chemistry, Massachusetts Institute of Technology, Cambridge, MA, USA*

JACQUES HAIECH • *CNRS UMR7242 BSC, ESBS, Illkirch Cedex, France*

NAOYA HATANO • *Applied Cell Biology, Graduate School of Interdisciplinary Science and Engineering in Health Systems, Okayama University, Kita-ku, Okayama, Japan*

CLAUS W. HEIZMANN • *Department of Pediatrics, Division of Clinical Chemistry and Biochemistry, University of Zürich, Zürich, Switzerland*
DANIELLE HOF • *Unilabs, Dübendorf, Dübendorf, Switzerland*
ANNA HOLTHENRICH • *Centre for Molecular Biology of Inflammation, Institute of Medical Biochemistry, University of Münster, Münster, Germany*
YUN HYUN HUH • *Cell Logistics and Silver Health Research Center, School of Life Sciences, Gwangju Institute of Science and Technology, Gwangju, Republic of Korea*
OLUWATOSIN Y. IBHAGUI • *Department of Chemistry, Center for Diagnostics and Therapeutics, Georgia State University, Atlanta, GA, USA*
MITSUHIKO IKURA • *Department of Medical Biophysics, University of Toronto, Toronto, ON, Canada; Princess Margaret Cancer Centre, University Health Network, Toronto, ON, Canada*
HIROAKI ISHIDA • *Biochemistry Research Group, Department of Biological Sciences, University of Calgary, Calgary, AB, Canada*
KATHARINE V. JENSEN • *Biochemistry Research Group, Department of Biological Sciences, University of Calgary, Calgary, AB, Canada*
ZONGCHAO JIA • *Department of Biochemical and Molecular Science, Queen University, Kingston, ON, Canada*
NAOKI KANAYAMA • *Applied Cell Biology, Graduate School of Interdisciplinary Science and Engineering in Health Systems, Okayama University, Kita-ku, Okayama, Japan*
RADHIKA KHANDELWAL • *CSIR-Centre for Cellular and Molecular Biology, Hyderabad, India; Academy of Scientific and Innovative Research (AcSIR), Ghaziabad, India*
OLEG KHASSAN • *Biochemistry Research Group, Department of Biological Sciences, University of Calgary, Calgary, AB, Canada*
MARIE-CLAUDE KILHOFFER • *CNRS UMR7242 BSC, ESBS, Illkirch Cedex, France*
UDAY KIRAN • *CSIR-Centre for Cellular and Molecular Biology (CCMB), Hyderabad, India; Academy of Scientific and Innovative Research (AcSIR), Ghaziabad, India*
BENCE KISS • *Department of Biochemistry, ELTE Eötvös Loránd University, Budapest, Hungary*
JÖRG KLINGELHÖFER • *Faculty of Health, Department of Neuroscience, Copenhagen University, Copenhagen, Denmark*
TORSTEN KNIESS • *Department of GMP Radiopharmaceuticals Production, Helmholtz-Zentrum Dresden-Rossendorf, Institute of Radiopharmaceutical Cancer Research, Dresden, Germany*
RYOJI KOBAYASHI • *Sakuragi Psychiatric Hospital, Mima, Tokushima, Japan*
KARL-WILHELM KOCH • *Biochemistry Group, Department of Neuroscience, Faculty VI, University of Oldenburg, Oldenburg, Germany*
RINA KONDO • *Applied Cell Biology, Graduate School of Interdisciplinary Science and Engineering in Health Systems, Okayama University, Kita-ku, Okayama, Japan*
ARTHUR KONNERTH • *Institute of Neuroscience, Technical University of Munich, Munich, Germany; Munich Cluster for Systems Neurology (SyNergy) and Center for Integrated Protein Sciences (CIPSM), Munich, Germany*
NATALIA KOZLYUK • *Department of Biochemistry, Vanderbilt University, Nashville, TN, USA; Center for Structural Biology, Vanderbilt University, Nashville, TN, USA*
DOMINIK KRAUS • *Department of Prosthodontics, Preclinical Education, and Material Sciences, University of Bonn, Bonn, Germany*
MICHAEL R. KREUTZ • *RG Neuroplasticity, Leibniz Institute for Neurobiology, Magdeburg, Germany; Leibniz Group 'Dendritic Organelles and Synaptic Function', University*

Medical Center Hamburg-Eppendorf, Center for Molecular Neurobiology, ZMNH, Hamburg, Germany

GIUSEPPE LAPERGOLA • *Neonatal Intensive Care Unit, Department of Pediatrics, University of Chieti, Chieti, Italy*

MARKUS LAUBE • *Department of Radiopharmaceutical and Chemical Biology, Helmholtz-Zentrum Dresden-Rossendorf, Institute of Radiopharmaceutical Cancer Research, Dresden, Germany*

STEVEN T. LEACH • *School of Women's and Children's Health, University of New South Wales Medicine, Sydney, Australia*

CLAUDIO LUCHINAT • *CERM, University of Florence, Sesto Fiorentino, Florence, Italy; Department of Chemistry Ugo Schiff, University of Florence, Sesto Fiorentino, Florence, Italy*

MASAKI MAGARI • *Applied Cell Biology, Graduate School of Interdisciplinary Science and Engineering in Health Systems, Okayama University, Kita-ku, Okayama, Japan*

MASATOSHI MAKI • *Department of Applied Molecular Biosciences, Graduate School of Bioagricultural Sciences, Nagoya University, Nagoya, Japan*

JOANNA MANGANA • *Department of Dermatology, University Hospital Zurich, Zurich, Switzerland*

CHRISTOPHER B. MARSHALL • *Princess Margaret Cancer Centre, University Health Network, Toronto, ON, Canada*

JUAN MARTÍNEZ-AGUILAR • *Red de Apoyo a la Investigación, Universidad Nacional Autónoma de México, INCMNSZ, Mexico City, Mexico*

RINA MATSUO • *Department of Applied Molecular Biosciences, Graduate School of Bioagricultural Sciences, Nagoya University, Nagoya, Japan*

MARIA CLORINDA MAZZARINO • *Department of Biomedical and Biotechnological Sciences, University of Catania, Catania, Italy*

MARCO MAZZORANA • *Diamond Light Source, Harwell Science and Innovation Campus, Didcot, UK*

CHRISTIAN MELLE • *Biomolecular Photonics Group, Jena University Hospital, Jena, Germany*

JOSEPH M. METZGER • *Department of Integrative Biology and Physiology, University of Minnesota Medical School, Minneapolis, MN, USA*

DIRK MIELENZ • *Division of Molecular immunology, Department of Internal Medicine III, Nikolaus-Fiebiger-Center, University of Erlangen-Nürnberg, Erlangen, Germany*

OLIVIER MIGNEN • *Laboratoire Canalopathies & Signalisation Calcique, INSERM U1227, Université de Bretagne Occidentale (UBO), Brest, France*

ANDREW L. MILLER • *Division of Life Science and State Key Laboratory of Molecular Neuroscience, HKUST, Clear Water Bay, Hong Kong, People's Republic of China*

MARK P. MOLLOY • *Bowel Cancer and Biomarker Research, Kolling Institute, Royal North Shore Hospital, The University of Sydney, St. Leonards, Australia*

ANDREW J. MONTEITH • *Department of Pathology, Microbiology, and Immunology, Vanderbilt University Medical Center, Nashville, TN, USA*

HISAYUKI MORII • *Department of Chemistry, College of Liberal Arts and Sciences, Tokyo Medical and Dental University, Chiba, Japan*

PRABHAKARA REDDY NAGAREDDY • *Department of Pathology, University of Alabama, Birmingham, AL, USA*

MASAYUKI NARA • *Department of Chemistry, College of Liberal Arts and Sciences, Tokyo Medical and Dental University, Chiba, Japan*

CHRISTIN NEUBER • *Department of Radiopharmaceutical and Chemical Biology, Helmholtz-Zentrum Dresden-Rossendorf, Institute of Radiopharmaceutical Cancer Research, Dresden, Germany*

HUNG HUY NGUYEN • *VIB-VUB Centre for Structural Biology (VIB CSB), Vlaams Instituut voor Biotechnologie (VIB), Brussels, Belgium; Structural Biology Brussels (SBB), Department of Biotechnology (DBIT), Vrije Universiteit Brussel (VUB), Brussels, Belgium*

ERNST NIGGLI • *Department of Physiology, University of Bern, Bern, Switzerland*

ELIZABETH M. NOLAN • *Department of Chemistry, Massachusetts Institute of Technology, Cambridge, MA, USA*

SUNIL KIRAN NOOTI • *Department of Pathology, University of Alabama, Birmingham, AL, USA*

LÁSZLÓ NYITRAY • *Department of Biochemistry, ELTE Eötvös Loránd University, Budapest, Hungary*

KRIS PAUWELS • *VIB-VUB Centre for Structural Biology (VIB CSB), Vlaams Instituut voor Biotechnologie (VIB), Brussels, Belgium; Structural Biology Brussels (SBB), Department of Biotechnology (DBIT), Vrije Universiteit Brussel (VUB), Brussels, Belgium*

RÉGINALD PHILIPPE • *Institut National de la Santé et de la Recherche Médicale, U1151, Paris, France*

JENS PIETZSCH • *Department of Radiopharmaceutical and Chemical Biology, Helmholtz-Zentrum Dresden-Rossendorf, Institute of Radiopharmaceutical Cancer Research, Dresden, Germany; Faculty of Chemistry and Food Chemistry, Technische Universität Dresden, Dresden, Germany*

FRANCESCA PLUCHINOTTA • *Department of Pediatric Cardiac Surgery, IRCCS San Donato Milanese Hospital, Milan, Italy*

RAINER PROBSTMEIER • *Neuro- and Tumor Cell Biology Group, Department of Nuclear Medicine, University Hospital Bonn, University of Bonn, Bonn, Germany*

KYOHEI SAKANE • *Applied Cell Biology, Graduate School of Interdisciplinary Science and Engineering in Health Systems, Okayama University, Kita-ku, Okayama, Japan*

MANI SALARIAN • *Department of Chemistry, Center for Diagnostics and Therapeutics, Georgia State University, Atlanta, GA, USA*

DANIEL SCHETELIG • *Department of Computational Neuroscience, University Medical Center Hamburg-Eppendorf, Hamburg, Germany*

ANAND KUMAR SHARMA • *CSIR-Centre for Cellular and Molecular Biology, Hyderabad, India*

YOGENDRA SHARMA • *CSIR-Centre for Cellular and Molecular Biology, Hyderabad, India; Academy of Scientific and Innovative Research (AcSIR), Ghaziabad, India*

GARY S. SHAW • *Department of Biochemistry, The University of Western Ontario, London, ON, Canada*

HIDEKI SHIBATA • *Department of Applied Molecular Biosciences, Graduate School of Bioagricultural Sciences, Nagoya University, Nagoya, Japan*

MÁRTON SIMON • *Department of Biochemistry, ELTE Eötvös Loránd University, Budapest, Hungary*

ERIC P. SKAAR • *Department of Pathology, Microbiology, and Immunology, Vanderbilt University Medical Center, Nashville, TN, USA*

EVANGELIA SKARMOUTSOU • *Department of Biomedical and Biotechnological Sciences, University of Catania, Catania, Italy*

WOO KEUN SONG • *Cell Logistics and Silver Health Research Center, School of Life Sciences, Gwangju Institute of Science and Technology, Gwangju, Republic of Korea*

Thomas Lykke-Møller Sørensen • *Department of Molecular Biology and Genetics, Aarhus University, Aarhus, Denmark*

Gopalkrishna Sreejit • *Department of Pathology, University of Alabama, Birmingham, AL, USA*

Roya Tadayon • *Department of Biochemistry, The University of Western Ontario, London, ON, Canada*

Terunao Takahara • *Department of Applied Molecular Biosciences, Graduate School of Bioagricultural Sciences, Nagoya University, Nagoya, Japan*

Masaru Tanokura • *Department of Applied Biological Chemistry, Graduate School of Agricultural and Life Sciences, University of Tokyo, Tokyo, Japan*

Charlotte M. Thomas • *School of Biological Sciences, Queen's University Belfast, Belfast, UK; Institute for Global Food Security, Queen's University Belfast, Belfast, UK*

Brian R. Thompson • *Department of Integrative Biology and Physiology, University of Minnesota Medical School, Minneapolis, MN, USA*

David J. Timson • *School of Biological Sciences, Queen's University Belfast, Belfast, UK; School of Pharmacy and Biomolecular Sciences, University of Brighton, Brighton, UK*

Hiroshi Tokumitsu • *Applied Cell Biology, Graduate School of Interdisciplinary Science and Engineering in Health Systems, Okayama University, Kita-ku, Okayama, Japan*

Peter Tompa • *VIB-VUB Centre for Structural Biology (VIB CSB), Vlaams Instituut voor Biotechnologie (VIB), Brussels, Belgium; Structural Biology Brussels (SBB), Department of Biotechnology (DBIT), Vrije Universiteit Brussel (VUB), Brussels, Belgium; Institute of Enzymology, Research Centre for Natural Sciences of the Hungarian Academy of Sciences, Budapest, Hungary*

Mitsumasa Tsuchiya • *Emergency and Critical Care Center, Kawasaki Municipal Kawasaki Hospital, Kawasaki, Kanagawa, Japan*

Johan Undén • *Department of Anaesthesia and Intensive Care, Länssjukhuset Halland, Halmstad, Sweden*

Kristen M. Varney • *Center for Biomolecular Therapeutics (CBT), Department of Biochemistry and Molecular Biology, University of Maryland School of Medicine, Baltimore, MD, USA; University of Maryland Marlene and Stewart Greenebaum Comprehensive Cancer Center, Baltimore, MD, USA*

Ravichandran Vignesh • *Signal Transduction Laboratory, Department of Biotechnology, Bhupat and Jyoti Mehta School of Biosciences, Indian Institute of Technology Madras, Chennai, India*

Hans J. Vogel • *Biochemistry Research Group, Department of Biological Sciences, University of Calgary, Calgary, AB, Canada*

Arnold von Eckardstein • *Institute for Clinical Chemistry, University Hospital Zurich, Zürich, Switzerland*

Yuning Wang • *Department of Biochemistry, The University of Western Ontario, London, ON, Canada*

Sarah E. Webb • *Division of Life Science and State Key Laboratory of Molecular Neuroscience, HKUST, Clear Water Bay, Hong Kong, People's Republic of China*

David J. Weber • *Center for Biomolecular Therapeutics (CBT), Department of Biochemistry and Molecular Biology, University of Maryland School of Medicine, Baltimore, MD, USA; University of Maryland Marlene and Stewart Greenebaum Comprehensive Cancer Center, Baltimore, MD, USA*

Matthias Wenghoefer • *Department of Oral and Maxillofacial Plastic Surgery, University of Bonn, Bonn, Germany*

René Werner • *Department of Computational Neuroscience, University Medical Center Hamburg-Eppendorf, Hamburg, Germany*
Paul T. Wilder • *Center for Biomolecular Therapeutics (CBT), Department of Biochemistry and Molecular Biology, University of Maryland School of Medicine, Baltimore, MD, USA; University of Maryland Marlene and Stewart Greenebaum Comprehensive Cancer Center, Baltimore, MD, USA*
Jochen Winter • *Department of Periodontology, Operative and Preventive Dentistry, University of Bonn, Bonn, Germany*
Insa M. A. Wolf • *The Calcium Signalling Group, Department of Biochemistry and Molecular Cell Biology, University Medical Center Hamburg-Eppendorf, Hamburg, Germany*
Andrew G. Woodman • *Biochemistry Research Group, Department of Biological Sciences, University of Calgary, Calgary, AB, Canada*
Wenping Wu • *College of Chemistry, Beijing Normal University, Beijing, China*
Shenghui Xue • *Department of Chemistry, Center for Diagnostics and Therapeutics, Georgia State University, Atlanta, GA, USA; Inlighta Biosciences, Atlanta, GA, USA*
Fuminori Yamaguchi • *Ehime Prefectural University of Health Sciences, Iyo-gun, Ehime, Japan*
Jenny J. Yang • *Department of Chemistry, Center for Diagnostics and Therapeutics, Georgia State University, Atlanta, GA, USA*
Laure Yatime • *Laboratory "Dynamique des Interactions Membranaires Normales et Pathologiques" (DIMNP)—UMR5235, Inserm, CNRS, University of Montpellier, Montpellier Cedex 5, France*
Yu Ra Yeo • *Cell Logistics and Silver Health Research Center, School of Life Sciences, Gwangju Institute of Science and Technology, Gwangju, Republic of Korea*
Teresa Zelante • *Department of Experimental Medicine, University of Perugia, Perugia, Italy*
Wei Zhang • *Department of Applied Molecular Biosciences, Graduate School of Bioagricultural Sciences, Nagoya University, Nagoya, Japan*
Jimin Zheng • *College of Chemistry, Beijing Normal University, Beijing, China*
Christian Zuppinger • *Cardiology Department, DBMR MEM C812, University Hospital Bern, Bern, Switzerland*

Part I

Data Analysis and Modern Technologies to Study Calcium-Binding and Signaling in Cells

Chapter 1

Using Data Mining to Explore Calmodulin Bibliography

Jacques Haiech and Marie-Claude Kilhoffer

Abstract

In this chapter, we present a strategy and the techniques to approach a scientific field from a set of articles gathered from the bibliographic database "Web of Science." The strategy is based on methods developed to analyze social networks. We illustrate its use in studying the calmodulin field. The method allows to structure a huge number of articles when writing a review, to detect the key opinion leaders in a given field, and to locate their own research topic in the landscape of themes deciphered by our own community.

We show that the free software VOSviewer may be used without knowledge in computing science and with a short learning period.

Key words Data mining, Scientometry, Calcium signal, Calmodulin, Social network analysis

1 Introduction

The number of scientific publications has increased exponentially in the last 40 years, and it is becoming almost impossible to perform a bibliographic analysis using a traditional approach by gathering all new articles and analyzing them. Our intelligence needs to be amplified in order to tackle the tsunami of new publications and to get a coherent landscape describing the evolution of a given scientific field.

Such an approach has been developed in scientometry, a discipline that aims at analyzing science by means of quantitative and qualitative indicators. The discipline is linked to the humanities, and therefore, the methods and strategies developed by scientometrists have not yet perfused in biology. The strategy has been used to analyze the emergence of synthetic biology but was performed by a team of sociologists [1].

Electronic supplementary material: The online version of this chapter (https://doi.org/10.1007/978-1-4939-9030-6_1) contains supplementary material, which is available to authorized users.

Claus W. Heizmann (ed.), *Calcium-Binding Proteins of the EF-Hand Superfamily: From Basics to Medical Applications*, Methods in Molecular Biology, vol. 1929, https://doi.org/10.1007/978-1-4939-9030-6_1,

The article aims at presenting some of the strategies and tools used in scientometry and at illustrating how such an approach may be applied to another field. We choose as an example the study of calmodulin.

The number of articles containing calmodulin in the title is higher than 10,000 and over 40,000 when seeking for calmodulin in the abstract or the keywords. As a human being, it is almost impossible to scan this full set of articles and to extract a coherent and logical set of knowledge.

Data mining is a set of techniques allowing the analysis of data from different points of views in order to extract a useful knowledge, mainly by analyzing the relations between the different data or by detecting common patterns.

Such analysis may be performed on a huge set of data, and data mining is therefore often associated with Big Data.

In our case, the number of papers published on calmodulin cannot be considered as enormous, but it is big enough to escape our human abilities. We will use the study of calmodulin to illustrate the capabilities of data mining and the strategies used to extract knowledge from an unstructured set of publications. Such strategies and techniques may be used in any scientific field.

We have realized that the use of this strategy allows students to better approach a new field of knowledge. Moreover, they can take into account the temporal evolution of the field and tackle it via the most significant publications.

2 Materials and Definitions

2.1 Corpus

A corpus is a set of articles and is defined by three parameters: the name of a database, a request, and a date. In our case, we will use either PubMed [2] or Web of Science [3] as bibliographic database (*see* **Note 1**). Both databases allow to download the result of a request under several possible formats (Fig. 1).

2.2 Analyzing and Visualizing

In this paper, we will use VOSviewer version 1.6.8 as the tool to analyze and visualize our data [4].

The use of PubMed files allows to create maps of researchers based on coauthorship links and maps of terms based on co-occurrences in the same publications of keywords either provided by the authors or by the curators of the bibliographic databases.

On top of the previous maps, the use of Web of Science files allows one to easily create maps of scientific publications, scientific journals, researchers, or research organizations based on either co-citation relations (i.e., multiple items being cited by the same

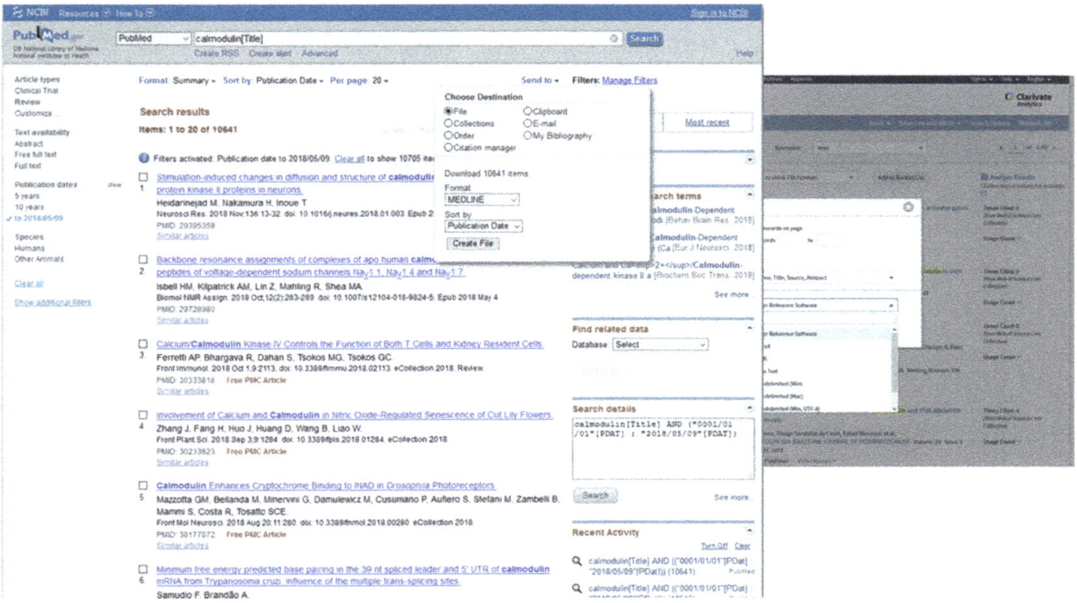

Fig. 1 Corpus creation. To create the corpus, we use either the PubMed database (left panel) or the Web of Science database (right panel) using the simple request "calmodulin" in the title field. The download of the corpus is done in PubMed under the format MEDLINE and in WoS under the format text

publication) or bibliographic coupling relations (i.e., multiple items citing the same publication).

VOSviewer contains an algorithm able to extract terms from titles and abstracts using natural language processing tools. Basically, a term is a suite of adjective(s) and noun(s) finishing by a noun. The algorithm works only on English text. A map of term is built by linking two terms when they are present in the same publication. As we will see in methods and results, each available tool allows to explore an aspect of the gathered corpus.

3 Methods and Results

3.1 The General Strategy

The general strategy in data mining is composed of three steps (Fig. 2):

1. Building the corpus by extracting the data from any source, by loading, and by transforming them including the possibility to cure them.
2. Analyzing the corpus by using tools allowing either to detect common patterns in the corpus or between subsets of the corpus or to detect relationships between specific properties of items of the corpus. In a bibliographic corpus, an item is an article, and properties of an article are authors, title, abstract, year of publication, etc.

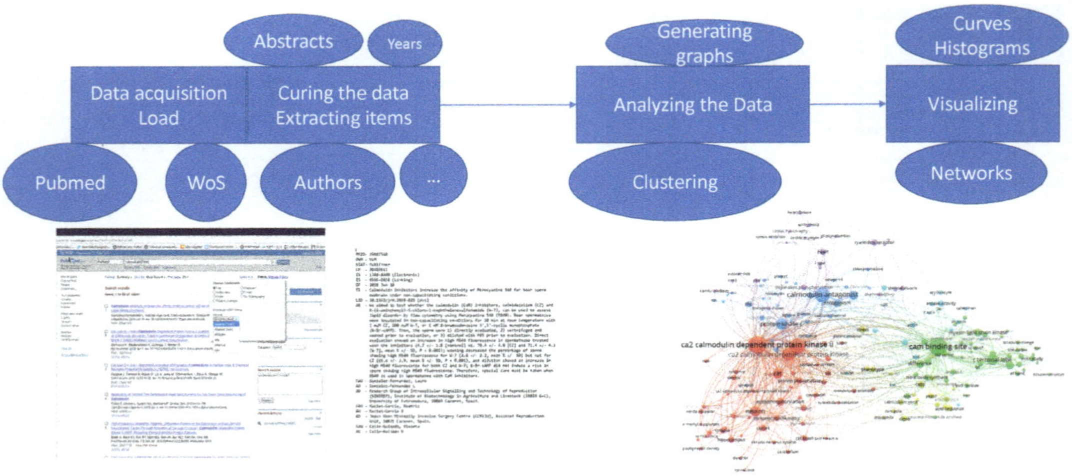

Fig. 2 Scheme illustrating the strategy of data mining when using bibliographic data

3. Visualizing the results from the previous analyses using either tools to represent clusters or graphs.

To illustrate our method, we will tackle the study of calmodulin, the stereotypical calcium-binding protein involved in the modulation of numerous cellular events [5, 6].

To build the corpus, we will use at first a very simple request, namely, looking for "calmodulin" in the title of publications. 10,641 publications were obtained in PubMed and 14,789 publications in Web of Science.

The difference between the two corpora is in part due to the presence of meeting abstracts and notes in the Web of Science database. We obtain 10,389 items when we select only articles and reviews in the Web of Science corpus.

In the following part of the paper, we will use the complete corpus obtained from the Web of Science database (14,789 items) on June 12, 2018.

3.2 What Are the Main Themes and Their Temporal Evolutions in the Calmodulin Field?

To answer this question, after loading the corpus in the VOSviewer software, we use a tool allowing to create a map based on the text from titles and abstracts of the publications. VOSviewer may tag adjectives and nouns, extract terms composed of adjectives and nouns finishing by a noun, and compute a score of relevance for each term. However, to improve the significance of the term set, it is necessary to iteratively build a thesaurus file in order to remove nonsignificant terms from our studied field and to pinpoint the putative synonyms (*see* the thesaurus file as supplemental material S1 and **Note 2**). Such file is a text file with two columns separated by a tabulation. The two columns appear with the headings "label" and "replace by."

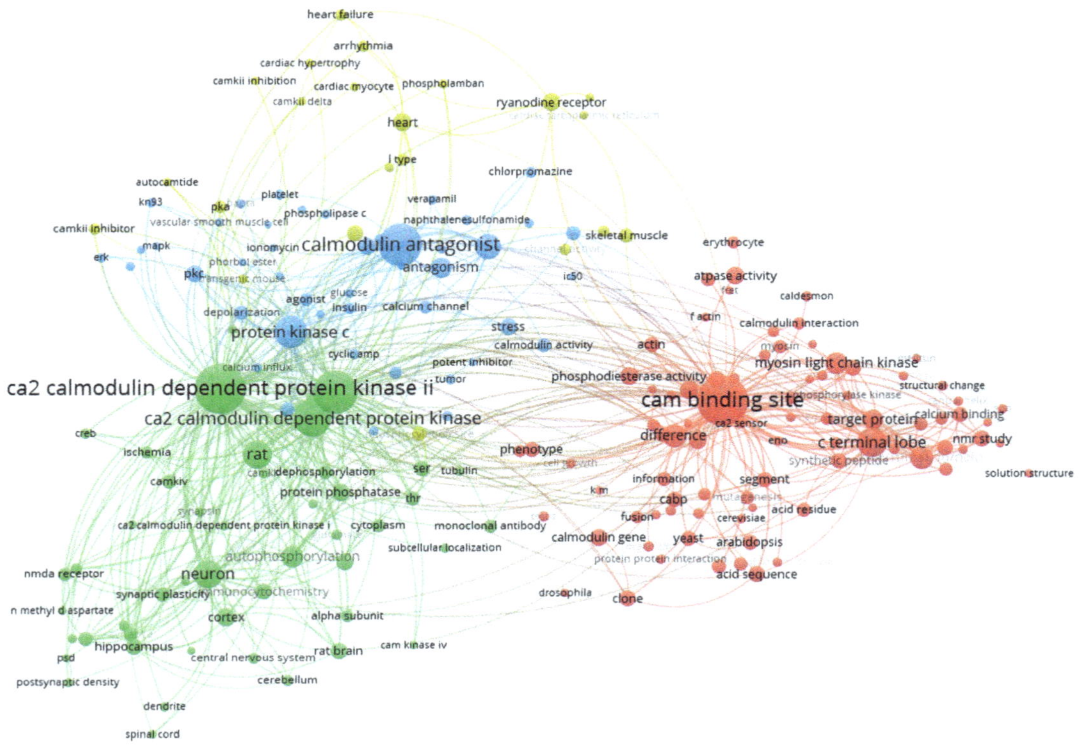

Fig. 3 Term map created with VOSviewer version 1.6.8. The map was created by extracting significant terms from abstract and titles. Links are drawn when two terms appear together in more than 50 items. The 200 most significant terms were kept by using a thesaurus file to remove general terms in the field or by regrouping synonyms (The thesaurus file is given in supplemental materials)

Two terms are linked if they appear in the same publication. The strength of the link is the number of publications where the two terms co-occur. It is a parameter that we set to 50. Then, we choose to select the 200 most relevant terms. VOSviewer may draw the network named the term map. The default parameters given by VOSviewer were used to draw this map (Fig. 3).

To perform the clustering, VOSviewer uses a smart local moving [7] algorithm that we may optimize using two parameters (the two parameters, namely, resolution and min.cluster size, are under the analysis signet). Using resolution equal to 1 and min. cluster size to 10, we obtain four clusters (Fig. 3), one nucleated around calmodulin antagonist (blue color), one around calmodulin-dependent protein kinase II (green color), one around calmodulin-binding site (red color), and one around heart and modulation of calcium channel (yellow color). Cluster 1 is the most ancient and cluster 4 the most recent. By increasing the resolution parameter, the initial clusters are subdivided into smaller ones. For instance, the cluster "calmodulin-binding site" may be split into a cluster dealing with the characterization of calcium sites and another focusing on calmodulin-binding peptides/ targets.

At this point, we have two alternatives:

1. Modifying the corpus to focus on one specific cluster or sub-cluster by adding terms to the initial request.
2. Continuing with the initial corpus to explore other questions.

We will go along with the second part of the alternative.

3.3 How to Find the Main Authors in the Different Topics and the Key Articles to Read?

There are three maps that allow to answer those two questions of the section [8]:

1. The citation map creates a link between two publications when one of them cites the other.
2. The co-citation map links two publications when they are cited by the same article of the corpus (*see* **Note 3**).
3. The bibliographic coupling map creates a link between two articles of the corpus when both cite the same document [9].

Each of those maps may be built with publications or authors as nodes (Fig. 4).

The citation map using authors as nodes (Fig. 4a) has been built keeping the most significant 100 authors. This map pinpoints the five main scientific communities working in the calmodulin field. The first one (in red) does not show a central author nucleating the community although CB Klee is playing a central role, the second one (in green) appears to be organized around fives authors (E Miyamoto, K. Fukunaga, T Yamauchi, TR Soderling, and H Schulman), the third one (in blue) is clearly built around A Means, the fourth one appears to be centralized around H Fujisawa and I Kameshita, and, finally, the fifth one is a small cluster of two authors, Anderson and Wu.

This first map which allows to discriminate five scientific communities also pinpoints five main topics and their evolutions:

1. Cluster 1: This cluster gathers scientists that worked on calmodulin per se, namely, the structure and the calcium-binding properties of calmodulin.
2. Cluster 2: Scientists in this cluster mainly worked on the role of calmodulin kinases and specifically calmodulin kinase II in the brain. The small cluster 5 is a spinoff of this cluster 2. The authors are working on calmodulin kinase II but in the heart.
3. Cluster 3: This cluster is structured around the work of AR Means on calmodulin kinase kinases and, recently, mainly on CaMKK2.

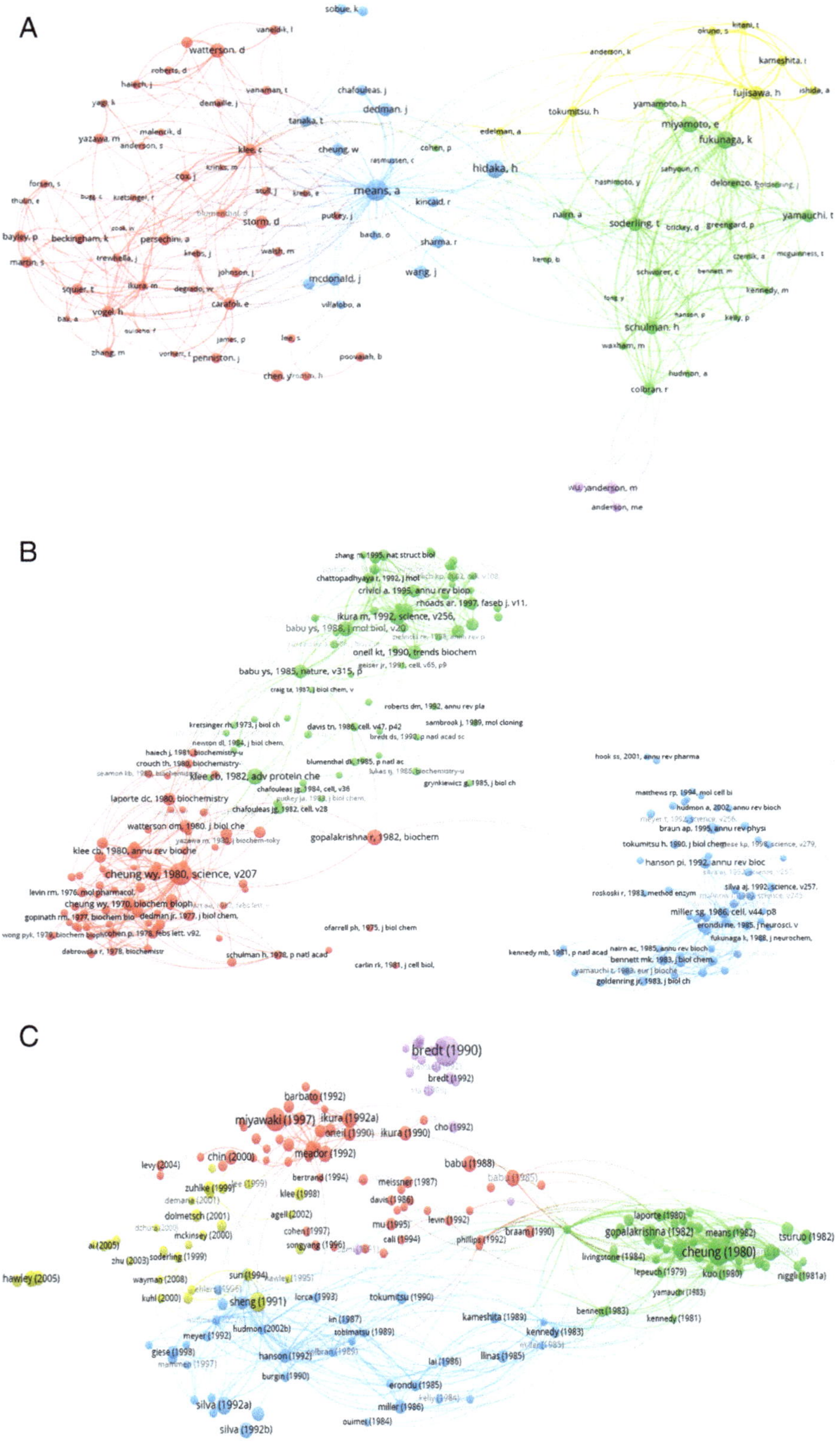

Fig. 4 This figure presents the citation map (panel **a**), the co-citation map (panel **b**), and the bibliographic coupling map (panel **c**) obtained with VOSviewer version 1.6.8 and the parameters discussed in the Subheading 3

4. Cluster 4: The work of the scientists in this cluster mainly deals with the couple calmodulin kinases and the phosphatases that reverse the phosphorylation of the calmodulin kinases.

The co-citation map has been constructed by keeping the 100 most significant articles (Fig. 4b). In this map, we distinguish three clusters:

1. The first cluster presents papers published between 1970 and 1982 dealing with the primary characterization of calmodulin (purification, calcium-binding properties, primary sequence, cloning the calmodulin gene).
2. The second cluster gathers the publications describing mainly the 3D structure of calmodulin and calmodulin complexed with target proteins or peptides, during the period ranging from 1985 to 2003.
3. The third cluster deals with the calmodulin kinases.

Using this tool, it is possible to select for each cluster two publications that are the most cited. We may consider that the selected six papers constitute the fundamental publications to read (Table 1). We may see that those six papers may be considered as an excellent starting point for a newcomer to enter the calmodulin field. Let us notice that this choice of publications has been realized without any "à priori" or implication of knowledge from experts.

Table 1
Fundamental papers in the calmodulin field

Cluster	Publication	References
1	Calmodulin plays a pivotal role in cellular regulation Cheung WY	[10]
1	Calmodulin Klee CB, Crouch TH, Richman PG	[11]
2	Solution structure of a calmodulin-target peptide complex by multidimensional NMR Ikura M, Clore GM, Gronenborn AM, Zhu G, Klee CB, Bax A	[12]
2	Target enzyme recognition by calmodulin: 2.4 A structure of a calmodulin-peptide complex Meador WE, Means AR, Quiocho FA	[13]
3	Neuronal Ca2+/calmodulin-dependent protein kinases Hanson PI, Schulman H	[14]
3	Regulation of brain type II Ca2+/calmodulin-dependent protein kinase by autophosphorylation: a Ca2 + −triggered molecular switch Miller SG, Kennedy MB	[15]

The six papers listed correspond to the two most cited references in each of the three clusters presented in Fig. 4b

The third map is the bibliographic coupling map (Fig. 4c). From a technical point of view, it is the exact opposite of the co-citation map. Co-citation map is a way to represent the "current paradigm" of the domain. It is a retrospective analysis. On the contrary, the bibliographic coupling map seems to be a prospective analysis detecting the trends of the scientific domain. To build this map, we have selected the 200 most cited publications. Keeping the default parameters of VOSviewer for the clustering, we obtained five clusters:

1. The first cluster (in red) integrates work on the structure of calmodulin complexes with target proteins or peptides and work on new calcium dyes using the properties of such complexes. The papers in this cluster were published around the year 2000.
2. The second cluster (in green) gathers work dealing with the biochemical purification and characterization of calmodulin and calmodulin-interacting proteins. Small molecules interacting with calmodulin are used to perturb the system to decipher the molecular mechanism. The work in this cluster peaks around 1980.
3. The third cluster (in blue) is focusing on the CaM kinase II and specifically its role in the brain. Most of the work was done around 1990.
4. The fourth cluster (in yellow) is mainly dealing with CaM kinases in the cardiac tissue and/or the modulation of ion channels. The publications in this cluster are the most recent ones, peaking around 2005.
5. Finally, the fifth cluster (in violet) links most of the work on calmodulin-dependent nitric oxide synthase mainly in endothelial cells, starting around 1990.

These clusters appear to describe the trends in the calmodulin field.

Combining our different maps, we can summarize the state of the art in the calmodulin field, the six publications to read for the beginners in the field and the putative trends (Fig. 5).

Calmodulin has been discovered in 1968 as a calcium-dependent regulator of phosphodiesterase, and the name calmodulin was coined to the protein in 1977. At first, the scientific community aimed at purifying and characterizing the protein and cloning its genes. In parallel, calmodulin-interacting proteins started to be identified, and the molecular mechanism of interaction was investigated. Simultaneously, the interaction of calmodulin with small molecules was studied, mainly in Japan, and triggered a subdomain of investigation on calmodulin antagonists.

The number of calmodulin-interacting proteins rapidly increased up to several hundreds. The 3D structure of calmodulin

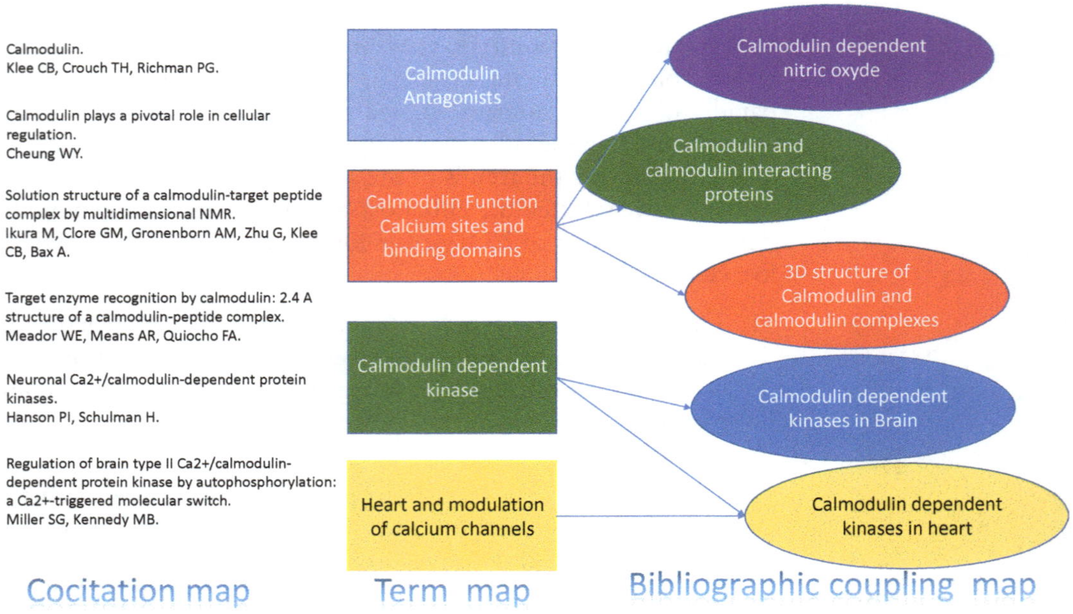

Fig. 5 Scheme summarizing the state of the art in the calmodulin field. The scheme was obtained by combining the results from the different maps presented in Figs. 3 and 4. Colors for the term map and the bibliographic coupling map are the same as those used in the Figs. 3 and 4 for the visualization of the different clusters

was solved first by X-ray crystallography and then by NMR. These results were rapidly followed by the structure of calmodulin complexed to target proteins domains or synthetic calmodulin-interacting peptides.

Among the large family of calmodulin-interacting proteins, the kinase family attracted many scientists. Calmodulin-dependent kinase II became one of the most studied calmodulin targets, and its role was mainly investigated in brain function.

More recently, the discovery of mutations in calmodulin linked to cardiac pathologies prompted the field exploring the role of calmodulin-dependent enzymes in the regulation of cardiac physiology.

Finally, as the oxidative stress appeared to play a specific action in oncogenesis, calmodulin-dependent nitric oxide synthase became a hot topic in the field.

The number of publication was at the acme between 1985 and 1995. Nowadays, as shown in the Fig. 6, the number of publications with the term "calmodulin" in the title is decreasing. We may assume that today, the scientific questions are more global and do not tackle the role of one individual given protein, whatever its importance can be.

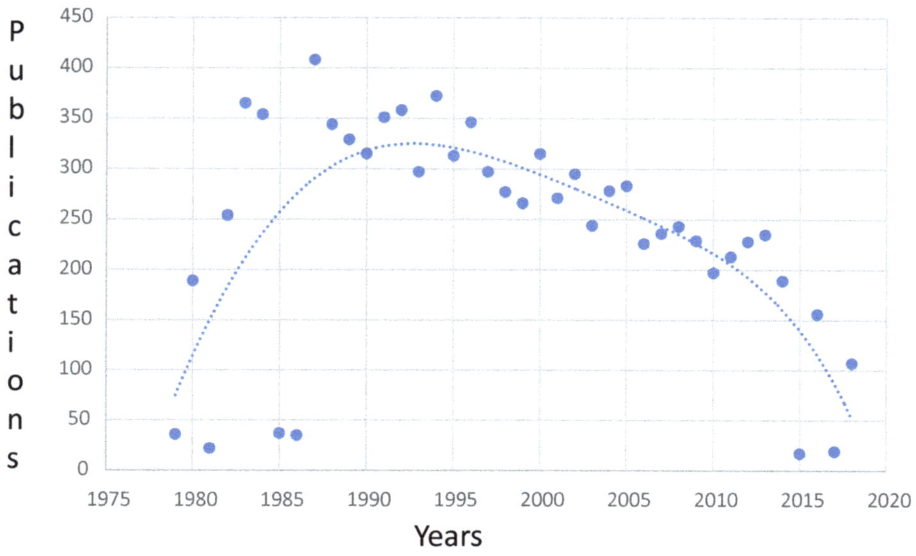

Fig. 6 Number of publications per year with the term "calmodulin" in the title

In summary, in this chapter, we have shown how to use tools initially developed in the field of social network analysis to tackle questions as diverse as:

1. What are the different subtopics in a given scientific field?
2. Who are the key opinion leaders in a given topic?
3. What are the main publications in a given subtopic?
4. How has a scientific field evolved upon time?

We have illustrated the use of such method in the frame of the evolution of the calmodulin field, but this approach may be used in any field and may help answering other questions such the ranking of different scientific institutes in a given scientific domain.

4 Notes

In this section, we will cite some limits of the strategy.

1. The Corpus

   It is important to work with large corpus (few hundred publications at least) in order to get meaningful and robust results.

   On the other hand, it is important to polish the request to get a corpus specific of the studied domain. For instance, using a request seeking for calmodulin in the abstract and not only in the title will gather papers with all kinds of calcium-binding proteins and will not focus on calmodulin.

2. The thesaurus file

   The thesaurus file is important to get a significant network of terms when analyzing the different subtopics of a scientific field.

   The aim is to remove all the terms that are often present in the abstract but that do not bring any meaningful information. For instance, the terms of the request have to be removed or general term such as "biology."

   Building the thesaurus file is an iterative and subjective process. We have noticed that during this iterative process, removing one term may change the geometry of the network. However, the clustering of the terms is not profoundly modified.

3. The co-citation analysis

   The co-citation analysis allows to retrospectively analyze the field. However, when the same type of techniques is used in the field, we end up with technical papers that are not pertinent for a specific field.

   For instance, the paper describing the techniques to make a western blot or to measure the protein concentration must be removed when doing a co-citation analysis.

References

1. Raimbault B, Cointet JP, Joly PB (2016) Mapping the emergence of synthetic biology. PLoS One 11(9):e0161522
2. Vellay SG, Latimer NE, Paillard G (2009) Interactive text mining with Pipeline Pilot: a bibliographic web-based tool for PubMed. Infect Disord Drug Targets 9(3):366–374
3. Li K, Rollins J, Yan E (2018) Web of Science use in published research and review papers 1997-2017: a selective, dynamic, cross-domain, content-based analysis. Scientometrics 115(1):1–20
4. van Eck NJ, Waltman L (2010) Software survey: VOSviewer, a computer program for bibliometric mapping. Scientometrics 84 (2):523–538
5. Berchtold MW, Villalobo A (2014) The many faces of calmodulin in cell proliferation, programmed cell death, autophagy, and cancer. Biochim Biophys Acta 1843(2):398–435
6. Haiech J, Kilhoffer MC, Craig TA et al (1990) Mutant analysis approaches to understanding calcium signal transduction through calmodulin and calmodulin regulated enzymes. Adv Exp Med Biol 269:43–56
7. Waltman L, van Eck NJ (2013) A smart local moving algorithm for large-scale modularity-based community detection. Eur Phys J B 86 (11):471
8. Klavans R, Boyack KW (2017) Which Type of Citation Analysis Generates the Most Accurate Taxonomy of Scientific and Technical Knowledge? J Assoc Inf Sci Technol 68(4):984–998
9. Weinberg BH (1974) Bibliographic coupling: a review. Inf Stor Retr 10(5):189–196
10. Cheung WY (1980) Calmodulin plays a pivotal role in cellular regulation. Science 207 (4426):19–27
11. Klee CB, Crouch TH, Richman PG (1980) Calmodulin. Annu Rev Biochem 49:489–515
12. Ikura M, Clore GM, Gronenborn AM et al (1992) Solution structure of a calmodulin-target peptide complex by multidimensional NMR. Science 256(5057):632–638
13. Meador WE, Means AR, Quiocho FA (1992) Target enzyme recognition by calmodulin: 2.4 A structure of a calmodulin-peptide complex. Science 257(5074):1251–1255
14. Hanson PI, Schulman H (1992) Neuronal Ca2 +/calmodulin-dependent protein kinases. Annu Rev Biochem 61:559–601
15. Miller SG, Kennedy MB (1986) Regulation of brain type II Ca2+/calmodulin-dependent protein kinase by autophosphorylation: a Ca2 +−triggered molecular switch. Cell 44 (6):861–870

Chapter 2

Deep Two-Photon Imaging In Vivo with a Red-Shifted Calcium Indicator

Antje Birkner and Arthur Konnerth

Abstract

Two-photon calcium imaging became in recent years a very popular method for the functional analysis of neural cell populations on a single-cell level in anesthetized or awake behaving animals. Scientific insights about single-cell processing of sensory information but also analyses of higher cognitive functions in healthy or diseased states became thereby feasible. However, two-photon imaging is generally limited to depths of a few hundred micrometers when recording from densely labeled cell populations. Therefore, such recordings are often restricted to the superficial layers 1–3 of the mouse cortex, whereas the deep cell layers 4–6 are hardly accessible with standard two-photon imaging. Here, we provide a protocol for deep two-photon calcium imaging, which allows imaging of neuronal circuits with single-cell resolution in all cortical layers of the mouse primary cortex. This technique can be readily applied to other species. The method includes a reduction of excitation light scattering by the use of a red-shifted calcium indicator and the minimization of background fluorescence by visually guided local application of the fluorescent dye. The technique is similar to previously published protocols for in vivo two-photon calcium imaging with synthetic calcium dyes (Stosiek et al. Proc Natl Acad Sci U S A 100:7319–7324, 2003). Hence, only minor changes of a generic two-photon setup and some adaptations of the experimental procedures are required.

Key words Calcium imaging, Two-photon microscopy, Neuronal activity, Mouse, Cortex, In vivo

1 Introduction

The maximum imaging depth of two-photon microscopy in biological tissue is limited mostly by light scattering, which causes a decreased two-photon excitation efficiency in the focal plane of the excitation laser beam [1]. To compensate for the loss in image brightness in deep structures of the sample, the average laser power needs to be increased, which however leads to an increase of out-of-focus fluorescence excitation as well [2, 3]. As a result, the signal-to-noise ratio of the fluorescence signals progressively degrades when increasing imaging depth. Therefore, imaging of dense cell populations in the mouse cortex in vivo is often restricted to the upper cortical cell layers. The aim of the protocol described here is

Claus W. Heizmann (ed.), *Calcium-Binding Proteins of the EF-Hand Superfamily: From Basics to Medical Applications*, Methods in Molecular Biology, vol. 1929, https://doi.org/10.1007/978-1-4939-9030-6_2,

to describe a method for calcium imaging of all six cortical layers in anesthetized or awake behaving mice. As light scattering is reduced at longer wavelength [4], we use here a red-shifted calcium dye, namely, Cal-590, which is two-photon excited at laser light wavelength of ~1050 nm [5]. Importantly, out-of-focus fluorescence needs to be minimized by restricting the fluorescent dye to a small volume of cells at the desired imaging depth. This reduces the fluorescence in the tissue above the imaging field of view and avoids fluorescent background contamination [5, 6]. The cell labeling technique with calcium indicator Cal-590 AM is based on the previously published method of bulk-loading with the green membrane-permeable acetoxymethyl (AM) ester version of synthetic fluorescent calcium indicators, such as OGB1-AM [7–9]. Once diffused inside the cells, the calcium dye is hydrolyzed by intracellular esterases and trapped inside the cells [7]. Single-cell activity is then reported by calcium-dependent changes in fluorescence. These fluorescence changes are detected with a two-photon microscope by raster scanning the tissue with focused laser light and generating 2D images of the corresponding cell populations. The protocol provides an overview of the necessary hardware modifications of the experimental setup as well as the key experimental procedures required for deep two-photon imaging with the red-shifted calcium dye Cal-590. The hardware modifications include the use of a laser light source emitting light at ~1050 nm. Ideally suited for this purpose are low-cost, robust ytterbium-doped fiber lasers, which do not require noise-intensive water cooling [10]. Furthermore, all optical components of the two-photon microscope need to be suitable for the red-shifted wavelength range (950–1100 nm). Finally, the fluorescence signal can be enhanced, without loss of cellular resolution, by "underfilling" of the objective lens' back aperture with the laser beam [11, 12]. An important factor is the use of highly sensitive detectors, such as hybrid photodetectors. The experimental procedures of animal preparation and in vivo imaging are mostly standard [13, 14], except that the visually guided application of the calcium dye is restricted to the smallest possible volume of brain tissue. By combining these measures, imaging depths up to 900–1000 μm can be readily achieved for imaging neural populations with single-cell resolution in anesthetized and awake behaving animals [5, 6]. We have demonstrated previously that all cortical layers of the mouse brain, including the deepest layer, layer 6, can be accessed with this method with high sensitivity [5, 6]. For all cortical layers, the imaging data was additionally validated by using electrophysiological single-cell in combination with fluorescence recordings [5]. This versatile approach was so far successfully tested in the mouse visual, the auditory, and the barrel cortex, and it can readily be used for the analysis of other brain regions in mice or other species.

2 Materials

2.1 Reagents

1. Mice, C57BL/6 aged 29–39 postnatal days. Animal studies should be performed following the guidelines and regulations of the corresponding authorities.
2. Sodium chloride solution (0.9%).
3. Extracellular buffer solution, containing 150 mM NaCl, 2.5 mM KCl, and 10 mM HEPES.
4. ACSF (artificial cerebrospinal fluid; freshly prepared), containing 125 mM NaCl, 4.5 mM KCl, 1.25 mM NaH_2PO_4, 26 mM $NaHCO_3$, 2 mM $CaCl_2$, 1 mM $MgCl_2$, 20 mM glucose. The solution is bubbled with carbogen to maintain a pH of 7.4.
5. Carbogen (95% oxygen and 5% carbon dioxide).
6. 20% Pluronic F-127 in DMSO (e.g., 2 g Pluronic F-127 in 10 mL DMSO; Thermo Fisher Scientific) (*see* **Note 1**).
7. Fluorescent dye Alexa Fluor 680 (Alexa Fluor 680 C_2 Maleimide, Thermo Fisher Scientific).
8. Membrane-permeable calcium-indicator dye Cal-590 AM (AAT Bioquest) (*see* **Note 2**).
9. Eye cream.
10. Anesthetic agent, here we used Isoflurane (Abbott Laboratories) (*see* **Note 3**).
11. Medical oxygen.
12. Local anesthetic agent (e.g., lidocaine).
13. Analgesic agent (e.g., metamizole).
14. Dental cement to attach the recording chamber to the animal's skull; here we used the single-component dental adhesive OptiBond All-In-One (Kerr) and the viscous dental filling composite Tetric Evoflow (Ivoclar Vivadent), which can be cured with blue light.
15. Low-melting-point agarose (Thermo Fisher Scientific).

2.2 Equipment

2.2.1 Equipment for Animal Preparation

1. Isoflurane vaporizer (e.g., Dräger Vapor 19.3), flow meter (e.g., KOBOLD Messring), and respiration mask for the mouse.
2. Chamber for the mouse for the induction of anesthesia with isoflurane.
3. Heating pad to keep the animal's body temperature constant at around 37 °C.
4. Thermometer to measure the rectal temperature of the mouse.
5. Surgical equipment, including a dissecting microscope (e.g., Olympus), a dental drill (e.g., Meisinger), a stereotaxic

instrument, and forceps as well as scissors (e.g., Fine Science Tools GmbH).

6. Custom-made recording chamber with an outer diameter of 38 mm and a center access opening with a diameter of 7 mm (compare to [13–15]).
7. Dental curing light (in case of light-curable dental cement; e.g., M + W Dental).

2.2.2 Equipment for Fluorescent Dye Preparation

1. Borosilicate glass capillaries (length, 75 mm; outer diameter, 2 mm; e.g., from Hilgenberg GmbH).
2. Pipette puller (e.g., PC-10 from Narishige).
3. Millipore centrifugal filters with pore diameter of 0.45 μm.

2.2.3 Two-Photon Setup

1. Two-photon laser scanning microscope, including a laser light source which provides femtosecond laser pulses at a wavelength of ~1050 nm (Fig. 1) (*see* **Note 4**).
2. Heating pad to keep the body temperature of the mouse at ~37 °C during the experiment.
3. Isoflurane supply system for the mouse, attached to the microscope (including an isoflurane vaporizer, a flow meter, a tubing system, and a respiration mask).
4. Animal monitoring system, including a temperature and breathing rate sensor and corresponding software for an online readout of the parameters (e.g., PowerLab, ADInstruments).
5. Perfusion system, consisting of a pump (e.g., Ismatec), inlet, and outlet tubing and tube heating (custom-made).
6. Motorized micromanipulator to navigate the patch pipette (Luigs and Neumann GmbH).
7. Patch-clamp amplifier (e.g., HEKA) (not mandatory; *see* **Note 5**).
8. Pneumatic drug-ejection system with manometer. The pressure needs to be manually adjustable while imaging the brain tissue with the two-photon microscope.

3 Methods

3.1 Reagent Preparation

1. Prepare fresh ACSF before each experiment, and bubble it with carbogen. For cleaning the craniotomy, the ACSF should be heated up to 37 °C.
2. Dissolve Cal-590 AM in DMSO containing 20% Pluronic F-127 to yield a final concentration of 10 mM. Part of this stock solution can be frozen at −20 °C and used for the next experiment. Dilute the stock solution and Alexa Fluor 680 with

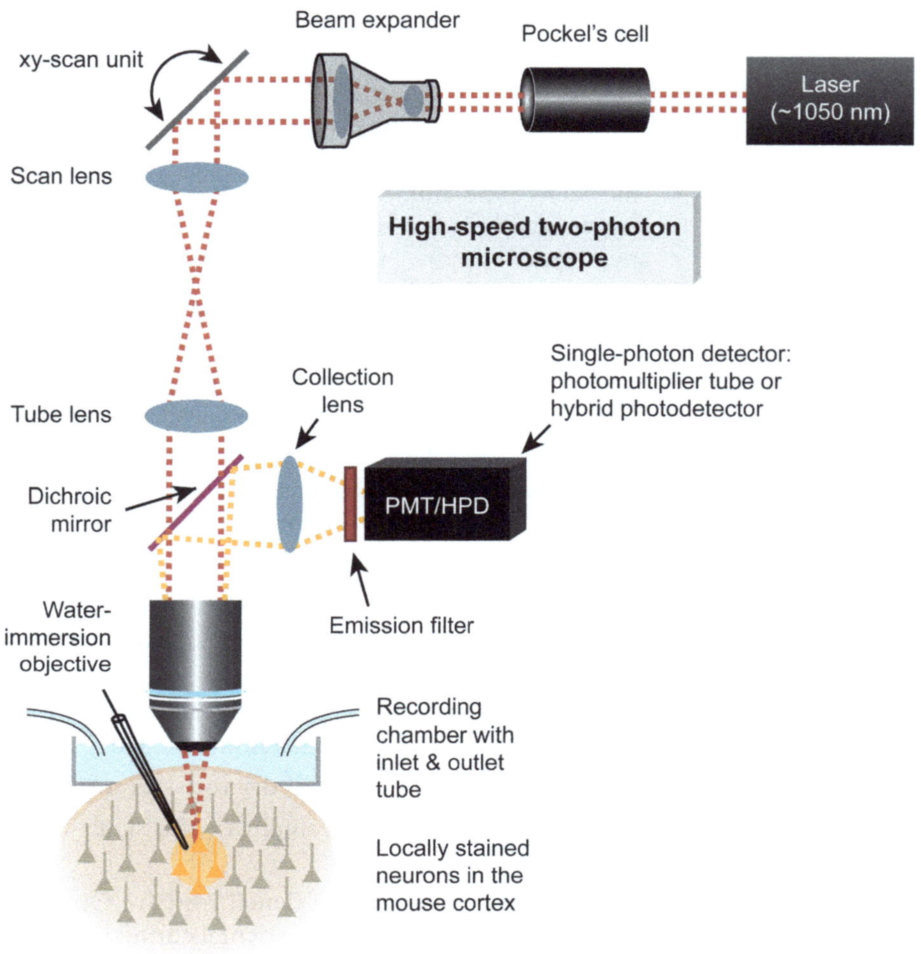

Fig. 1 Scheme of the two-photon laser-scanning microscope for deep imaging. A useful class of laser used for such recordings are ytterbium-doped fiber lasers with <100 fs pulse length. The xy scan unit consists of a resonant (12 kHz) and a standard galvanometric scan mirror, respectively. The optical components, including lenses, filters, and the Pockels cell, were adapted to a wavelength range of 950–1100 nm

extracellular buffer solution (150 mM NaCl, 2.5 mM KCl, and 10 mM HEPES) to final concentrations of 500–1000 μM Cal-590 AM and 10 μM Alexa Fluor 680. Filter the solution if necessary with a Millipore filter.

3. Dissolve 20 mg of agarose powder in 0.9 mL of 0.9% NaCl. Boil the solution just before usage, and cool it down to 38 °C before applying it onto the cortical surface.

3.2 Equipment Setup

1. Start the perfusion system, and adjust the temperature of the ACSF to 37 °C at the inlet tube to the recording chamber (*see* **Note 6**).

2. Start the heating pads for the mouse at the preparation table and at the two-photon microscope to preset a temperature of ~38 °C.
3. Pull micropipettes with a resistance of 3–5 mΩ.

3.3 Animal Preparation

1. Anesthetize the animal with isoflurane (2% (vol/vol) in pure oxygen), and place it on a heating pad in front of the isoflurane/oxygen supply.
2. Monitor the animal's body temperature with a rectal temperature sensor, and keep it at around 37 °C.
3. Cover the animal's eyes with eye cream to prevent them from drying out during anesthesia.
4. Provide analgesia according to the animal care regulations.
5. Check whether the surgical level of anesthesia is reached (e.g., with the pinch withdrawal reflex).
6. Inject the local anesthetic agent subcutaneously above the skull, and wait 10 min until the agent is effective.
7. Remove the skin of an area of ~ 4 cm^2 above the skull with fine scissors.
8. Identify the location of the desired brain area with the stereotactic device, and mark it with a nontoxic pen.
9. Clean the skull with ACSF and dry the skull surface afterward.
10. Carefully remove the periosteum of the skull with a scalpel blade.
11. Apply the single-component dental adhesive OptiBond to the skull, and illuminate it with the dental curing light for 10 s (*see* **Note 7**).
12. Apply the dental filling composite Tetric Evoflow to the skull, but spare a region of at least 5 mm around the desired cortical area.
13. Also apply Tetric Evoflow to the recording chamber, and place it onto the skull.
14. Cure the dental cement with the dental curing light for 60 s (*see* **Note 7**).
15. Perform a small craniotomy with the dental drill and a small-tip steel burr (0.5 mm in diameter) (*see* **Notes 8–10**).
16. Rinse 37 °C warm, fresh ACSF with a plastic pipette onto the craniotomy until superficial bleedings stopped.
17. Transfer the animal to the two-photon microscope, and place it onto the heating pad. Install the perfusion system to circulate fresh ACSF (37 °C).
18. Install the sensors for the animal monitoring, and adjust the temperature and anesthesia level if necessary (*see* **Note 11**).

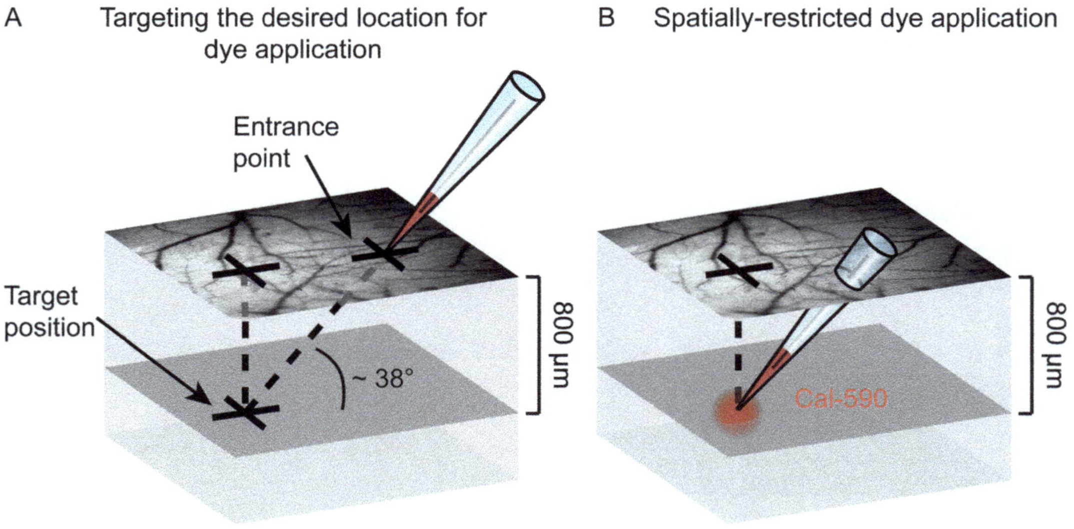

Fig. 2 Targeted local application of the calcium indicator. (**a**) Position of the glass pipette containing the calcium indicator Cal-590 before entering the brain tissue at the calculated point of entrance. (**b**) Position of pipette's tip at the target location and relative size of the stained region (red)

3.4 Spatially Restricted Dye Application

1. Use wide-field illumination and the ocular of the microscope to choose an area within the craniotomy which is free of superficial blood vessels (*see* **Note 12**).
2. The pipette tip needs to be navigated to the vasculature-free region at the desired imaging depth. Therefore, calculate the entrance point of the pipette tip as indicated in Fig. 2 (*see* **Notes 12–15**).
3. Fill a patch pipette with 5 µL of the solution containing Cal-590 and Alexa Fluor 680.
4. Measure the pipette resistance with the patch-clamp amplifier to assure that the pipette is intact and not clogged.
5. Lower the pipette tip, until it touches the cortical surface and until the pipette resistance increases up to about 15 mΩ.
6. Advance the pipette along its axes into the tissue.
7. Visualize the pipette tip with the two-photon microscope, by exciting the Alexa dye with laser light at ~1050 nm (*see* **Note 16**).
8. Once the pipette reached the final location, check the position of the brain surface to assure the correct position in depth of the pipette tip (*see* **Note 14**).
9. Apply slowly increasing pressure to the pipette while imaging the pipette tip within a field of view of 200 µm. Stop the pressure application whenever the field of view is almost filled with fluorescence, yielding a volume of stained cells of a

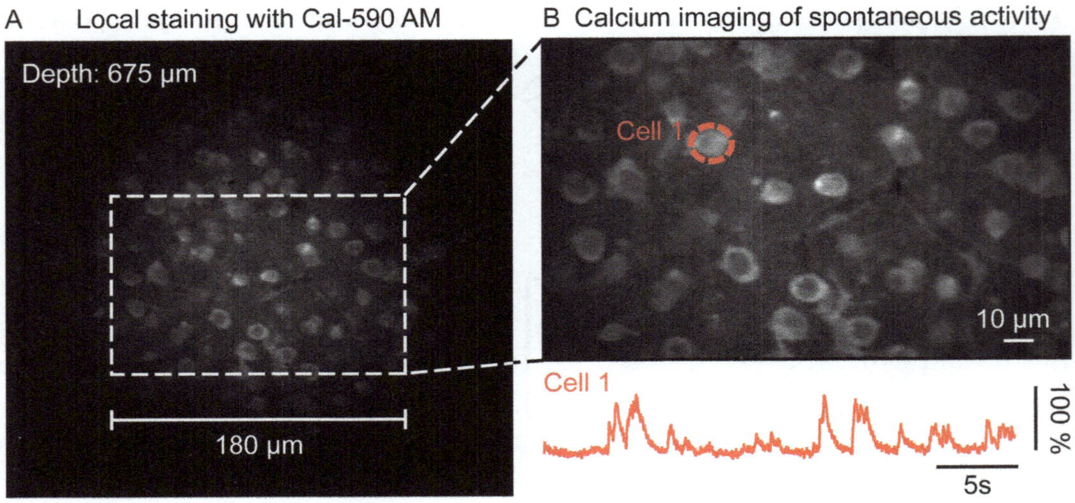

Fig. 3 Two-photon calcium imaging in deep layers of the mouse visual cortex in vivo. (**a**) Spatially restricted staining (<200 μm diameter) with Cal-590 AM at a depth of 675 μm below the cortical surface. (**b**) The recording trace (red) represents the spontaneous action potential-associated activity monitored in the neuron marked as "Cell 1" (reproduced from [6])

diameter of 180–200 μm, as shown in Fig. 3 (reproduced from Birkner et al. 2017 with permission from Elsevier) (*see* **Note 17**).

10. 50–60 min after dye application the cells are well filled with fluorescent dye, and the quality of the staining should be checked under two-photon excitation.
11. Image the labeled region with the two-photon microscope to check whether the staining was successful (*see* **Note 18**). The neurons appear doughnut-shaped, and astrocytes can be identified by their irregular shape.
12. To stabilize the recordings, apply agarose (38 °C) with a plastic pipette onto the dry cortical surface, and wait ~8 min for it to dry before putting back the perfusion. The layer of agarose on top of the brain surface should not exceed a thickness of ~2 mm (*see* **Note 19**).
13. For the two-photon recordings, visualize the cells under the two-photon microscope by slowly increasing the laser power and decreasing the PMT gain, until a good signal-to-noise ratio is obtained (*see* **Note 18**). At a cortical depth between 700 and 900 μm, the required laser power under the objective lens is usually between 60 and 170 mW (*see* [6]) (*see* **Note 20**). Two-photon calcium imaging was performed at a frame rate of 80 Hz to scan 450 × 300 pixels.

4 Notes

1. Pluronic F-127 in DMSO is a strong detergent and can irritate the skin, eyes, and respiratory tract. Wear personal protective equipment.
2. Here, the synthetic calcium dye Cal-590 AM was used for acute in vivo experiments. In principle, these experiments can also be performed with red-shifted genetically encoded indicators, like RCaMPs. The spatial restriction of labeled cell volume can be yielded by targeting specific cell populations at certain cortical depths [16–18] but eventually also by local application of the virus carrying the corresponding RCaMP.
3. Isoflurane is a volatile liquid, and an effective ventilation system is required to prevent the exposure to the experimenter. The safe use and exposure of isoflurane during pregnancy has not been established.
4. Here we used a custom-made two-photon microscope, optimized for deep two-photon imaging with long laser light wavelengths [5, 6]. In principle, a generic two-photon microscope can be used with a laser light source emitting femtosecond laser pulses at a wavelength of ~1050 nm and corresponding optics. The maximum laser power under the objective lens should reach about ~200 mW (depending on the temporal pulse width and pulse repetition rate [6]). For an efficient two-photon excitation at deep cortical layers, the back aperture of the objective lens was slightly "underfilled" by the laser beam [11, 12]. The two-photon microscope used here is schematically shown in Fig. 1.
5. The patch-clamp amplifier is used to measure the pipette resistance during the staining procedure. The readout of the pipette resistance allows recognizing when the pipette touches the brain surface or when the pipette gets clogged. Thus, the use of a patch-clamp amplifier facilitates the staining procedure, but is not mandatory for a successful cell labeling.
6. The ACSF of the perfusion system is heated up to a constant temperature of 37 °C at the surface of the cortex. The temperature of the ACSF within the perfusion system should be kept below 50 °C to prevent calcium precipitation.
7. Whenever the skull surface is illuminated with the dental curing light, cover the target side for two-photon imaging to protect the superficial tissue from potential cellular damage.
8. The craniotomy should be sufficiently large to prevent that fluorescence light is blocked by the skull on its way to the objective lens.

9. The craniotomy is one of the most critical steps because any damage of the brain tissue decreases the image quality. To prevent tissue damage, no pressure should be applied when thinning the skull with the dental drill. Additionally, the drill should not dwell at one position to prevent heating. The skull is thin enough whenever the piece of bone can be removed without applying any force.
10. Do not damage or remove the dura mater to minimize motion artifacts, because the dura stabilizes the tissue.
11. The status of the mouse should be stable during the entire experiment to maintain a good recording quality. Therefore, keep the mouse's breathing rate constant between 90 and 120 BPM and the body temperature at about 37 °C.
12. Only superficial vasculature is visible under wide-field illumination, and deeply located blood vessels remain undetected. The number of blood vessels between recording site and brain surface increases with increasing imaging depth. As a result, some stained regions appear darker than others. Therefore, it is advisable to stain two or three regions and to choose the region with best image quality for the recordings.
13. Adjust the access angle of the pipette to at least 38° to minimize the traveling distance of the pipette inside the tissue. The geometry of the objective lens needs to match the access angle of the pipette. It is advisable to test the geometric configuration and adjust the pipette angle before the experiment.
14. Usually the brain surface is not flat and can be slightly tilted according to the position of the recording chamber. This should be taken into account when calculating the entrance point of the loading pipette. In addition, the final depth of the position of the loading pipette tip should be double-checked just before pressure-injecting the dye.
15. For dye loading, the final depth of the pipette tip should be 20–30 μm below the desired imaging depth, because the best image quality is obtained from the upper half of the stained tissue.
16. Instead of continuously illuminating the brain tissue with laser light while the pipette is navigated toward its final position, it is sufficient to follow the pipette's positions stepwise each 300 μm of advancement to reduce the risk of photodamage.
17. After a traveling distance to a cortical depth between 700 and 900 μm below the brain surface, the pipette tip usually is not clean. Thus, the pressure required for dye ejection, as well as the duration of pressure-injecting the dye, is variable. To estimate the amount of dye ejected, it is crucial to visualize the spread of the Alexa dye. Some experience might be necessary to

estimate the resulting volume and brightness of stained cells based on the amount of ejected Alexa dye. If no dye can be ejected at a pressure level of maximum 700 mbar, then the pipette is clogged and has to be retracted. Sometimes, the pipette tip can be unclogged by retracting the pipette by some microns.

18. Before two-photon imaging the tissue, set the laser power to a minimum and increase the gain of the PMT to avoid photodamage of the illuminated tissue. Start the live imaging and slowly increase the laser power until the cells appear. For the recordings, adjust the laser power and PMT gain to yield a good SNR without detecting any bleaching or photodamage.
19. The application of agarose to the brain surface is a critical step. Agarose of a temperature above 38 °C can damage the tissue and degrade the image quality. However, if the agarose is too cold during the application, it might not form a flat and transparent surface. Moreover, the brain surface needs to be dry (without drying out) before agarose application to avoid a liquid film between the cortical surface and the agarose.
20. The required laser power strongly depends on the imaging depth [1]. For large imaging depths close to 900 μm, we kept the average laser power below 200 mW. Dark images at high laser power values are an indicator for blood vessels covering the recording site. To avoid photodamage, it is advisable to record the data in small periods of time of 30–60 s.

Acknowledgments

This work was supported by the Deutsche Forschungsgemeinschaft (IRTG 1373 and SFB 870) and a European Research Council Advanced Grant, all to A.K.

References

1. Helmchen F, Denk W (2005) Deep tissue two-photon microscopy. Nat Methods 2 (12):932–940. https://doi.org/10.1038/nmeth818
2. Theer P, Hasan MT, Denk W (2003) Two-photon imaging to a depth of 1000 microm in living brains by use of a Ti:Al2O3 regenerative amplifier. Opt Lett 28 (12):1022–1024
3. Ying J, Liu F, Alfano RR (1999) Spatial distribution of two-photon-excited fluorescence in scattering media. Appl Opt 38(1):224–229
4. Shi L, Sordillo LA, Rodriguez-Contreras A, Alfano R (2016) Transmission in near-infrared optical windows for deep brain imaging. J Biophotonics 9(1–2):38–43. https://doi.org/10.1002/jbio.201500192
5. Tischbirek C, Birkner A, Jia H, Sakmann B, Konnerth A (2015) Deep two-photon brain imaging with a red-shifted fluorometric Ca2+ indicator. Proc Natl Acad Sci U S A 112 (36):11377–11382. https://doi.org/10.1073/pnas.1514209112
6. Birkner A, Tischbirek CH, Konnerth A (2017) Improved deep two-photon calcium imaging in vivo. Cell Calcium 64:29–35. https://doi.org/10.1016/j.ceca.2016.12.005

7. Stosiek C, Garaschuk O, Holthoff K, Konnerth A (2003) In vivo two-photon calcium imaging of neuronal networks. Proc Natl Acad Sci U S A 100(12):7319–7324. https://doi.org/10.1073/pnas.1232232100
8. Grienberger C, Konnerth A (2012) Imaging calcium in neurons. Neuron 73(5):862–885. https://doi.org/10.1016/j.neuron.2012.02.011
9. Garaschuk O, Milos RI, Grienberger C, Marandi N, Adelsberger H, Konnerth A (2006) Optical monitoring of brain function in vivo: from neurons to networks. Pflugers Arch 453(3):385–396. https://doi.org/10.1007/s00424-006-0150-x
10. Xu C, Wise FW (2013) Recent advances in fiber lasers for nonlinear microscopy. Nat Photonics 7. https://doi.org/10.1038/nphoton.2013.284
11. Dunn AK, Wallace VP, Coleno M, Berns MW, Tromberg BJ (2000) Influence of optical properties on two-photon fluorescence imaging in turbid samples. Appl Opt 39(7):1194–1201
12. Kondo M, Kobayashi K, Ohkura M, Nakai J, Matsuzaki M (2017) Two-photon calcium imaging of the medial prefrontal cortex and hippocampus without cortical invasion. elife 6:e26839. https://doi.org/10.7554/eLife.26839
13. Garaschuk O, Milos RI, Konnerth A (2006) Targeted bulk-loading of fluorescent indicators for two-photon brain imaging in vivo. Nat Protoc 1(1):380–386. https://doi.org/10.1038/nprot.2006.58
14. Chen X, Leischner U, Varga Z, Jia H, Deca D, Rochefort NL, Konnerth A (2012) LOTOS-based two-photon calcium imaging of dendritic spines in vivo. Nat Protoc 7(10):1818–1829. https://doi.org/10.1038/nprot.2012.106
15. Jia H, Rochefort NL, Chen X, Konnerth A (2011) In vivo two-photon imaging of sensory-evoked dendritic calcium signals in cortical neurons. Nat Protoc 6(1):28–35. https://doi.org/10.1038/nprot.2010.169
16. Pilz GA, Carta S, Stauble A, Ayaz A, Jessberger S, Helmchen F (2016) Functional imaging of dentate granule cells in the adult mouse hippocampus. J Neurosci 36(28):7407–7414. https://doi.org/10.1523/JNEUROSCI.3065-15.2016
17. Bethge P, Carta S, Lorenzo DA, Egolf L, Goniotaki D, Madisen L, Voigt FF, Chen JL, Schneider B, Ohkura M, Nakai J, Zeng H, Aguzzi A, Helmchen F (2017) An R-CaMP1.07 reporter mouse for cell-type-specific expression of a sensitive red fluorescent calcium indicator. PLoS One 12(6):e0179460. https://doi.org/10.1371/journal.pone.0179460
18. Dana H, Mohar B, Sun Y, Narayan S, Gordus A, Hasseman JP, Tsegaye G, Holt GT, Hu A, Walpita D, Patel R, Macklin JJ, Bargmann CI, Ahrens MB, Schreiter ER, Jayaraman V, Looger LL, Svoboda K, Kim DS (2016) Sensitive red protein calcium indicators for imaging neural activity. elife 5:e12727. https://doi.org/10.7554/eLife.12727

Chapter 3

High-Resolution Calcium Imaging Method for Local Calcium Signaling

Björn-Philipp Diercks, René Werner, Daniel Schetelig, Insa M. A. Wolf, and Andreas H. Guse

Abstract

All eukaryotic cells respond to extracellular signals in a physiologically meaningful way. For multicellular organisms, physiologically relevant cooperation is only possible, if cell-cell communication works properly. Here, the extracellular signals must be translated into intracellular signals that ultimately result in cellular responses. This process is termed signal transduction or signaling. Ca^{2+} signaling has been developed in almost all eukaryotic cells. The cellular components used for this highly versatile signaling system are often termed "Ca^{2+} toolbox". Besides Ca^{2+} pumps and Ca^{2+}-binding proteins, the Ca^{2+} channels that are located in the plasma membrane and intracellular membranes and the Ca^{2+}-mobilizing second messengers are major players in shaping the four-dimensional nature of Ca^{2+} signals.

Here, we report on methodological developments to acquire and analyze cellular Ca^{2+} signals with high temporal and spatial resolution with specific focus on (1) photobleaching of Ca^{2+} indicators at high acquisition rate, (2) determination of system noise and spatiotemporal detection limits, and (3) image processing.

Key words Ca^{2+} signaling, Ca^{2+} microdomain cell signaling, High-resolution imaging, Live cell imaging, Deconvolution

1 Introduction

All eukaryotic cells must respond reasonably to extracellular signals. This is true for both uni- and multicellular organisms. While the former must respond to changes in environmental conditions, the latter do so in response to extracellular signaling molecules within multicellular organisms. Physiologically relevant cooperation within such organisms is only possible, if cell-cell communication works properly. Further, the extracellular signals must be translated into signals that can be processed by the intracellular systems of single cells. This process is termed signal transduction or signaling.

Ca^{2+} signaling is the most versatile signaling system developed in eukaryotic cells. Almost all types of eukaryotic cells make use of

Claus W. Heizmann (ed.), *Calcium-Binding Proteins of the EF-Hand Superfamily: From Basics to Medical Applications*, Methods in Molecular Biology, vol. 1929, https://doi.org/10.1007/978-1-4939-9030-6_3,

the high concentration gradient between the extracellular space and the cytosol. Further, many, if not all, cells pump Ca^{2+} into intracellular organelles, thereby creating intracellular Ca^{2+} stores, e.g., the endoplasmic reticulum (ER). The proteins and second messenger molecules involved in generating Ca^{2+} signals are termed "Ca^{2+} toolbox".

The basic components of the "Ca^{2+} toolbox" are:

- Ca^{2+} channels, both located in the plasma membrane and intracellular membranes.
- Ca^{2+} pumps, also both located in the plasma membrane and intracellular membranes.
- Second messenger systems, consisting of:
  - Second messenger forming enzymes, usually controlled by receptors that recognize extracellular signals.
  - Ca^{2+}-mobilizing second messengers, usually small metabolites that bind to and open Ca^{2+} channels.
  - Second messenger-metabolizing enzymes, usually controlled by the concentration of the respective second messenger.
- A small number of Ca^{2+}-binding proteins, located in the cytosol, where they act as Ca^{2+} buffers.

Ca^{2+} signals are four-dimensional since in addition to their three-dimensional spatial distribution, also the temporal dimension needs to be considered. The temporal dimension often relates to oscillatory Ca^{2+} signals with more or less regular oscillations of the amplitude or to biphasic Ca^{2+} signals with a rapid peak phase followed by a sustained plateau phase. While some cell types show regular Ca^{2+} oscillations upon stimulation, e.g., liver cells [1] or RBL cells [2], other cell types show a large variety of signals in the temporal dimension, e.g., T-lymphocytes (*see* supplemental Fig. S7 of [3]).

Regarding spatial Ca^{2+} signaling so far, researchers distinguish between local or global Ca^{2+} signals. Local Ca^{2+} signals are often observed at the initiation of the signaling process, termed as Ca^{2+} microdomains. However, due to the rapid progression from local to global signals and the limited spatiotemporal resolution in live cell imaging, the number of published studies is manageable so far.

This book chapter deals with all the methods needed to acquire and analyze initial local Ca^{2+} signals in T-lymphocytes (*see* Fig. 1), small cells with diameter of about 6 μm, as originally published in [3]. During method development, the following aspects were specifically developed:

- Use of single- vs. dual-wavelength Ca^{2+} indicators.
- Use of excitation vs. emission shift Ca^{2+} indicators.
- Photobleaching of Ca^{2+} indicators at high acquisition rate.

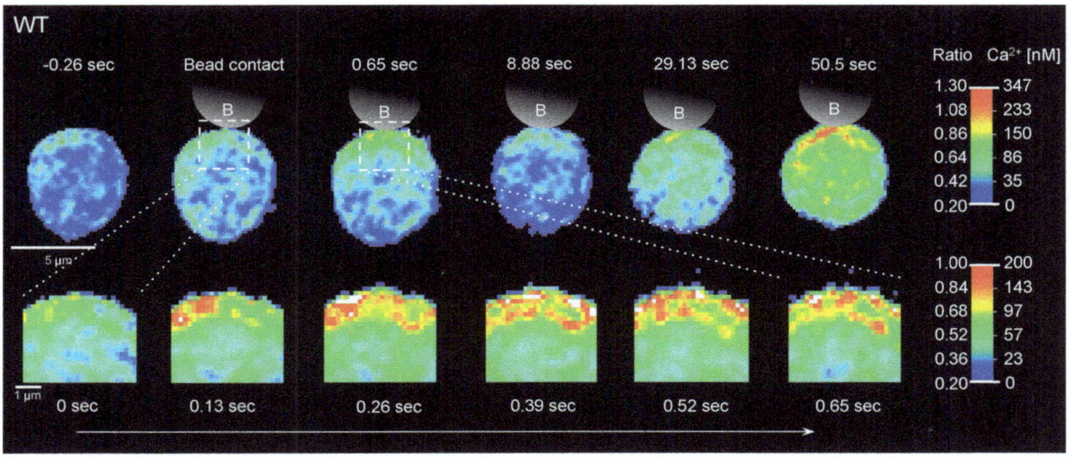

Fig. 1 Framescan of a WT T-lymphocyte stimulated via the T-cell receptor/CD3 complex and CD28. Initial Ca^{2+} microdomains in a WT T-lymphocyte upon stimulation of CD3 and CD28 (bead contact indicated schematically) shown by Framescan (top row) and a magnified region (3 × 3 μm) near the bead contact site (bottom row). In this presentation, five sequential images were merged to generate an average image to decrease noise

- Determination of system noise and spatiotemporal detection limits.
- Image post-processing and analysis.

2 Materials

2.1 Buffer and Media

ACK buffer: 4.3 g ammonium chloride, 0.5 g $KHCO_3$, 0.0186 g Na_2-EDTA in 400 mL H_2O, pH 7.2–7.4.

Ca^{2+}-free buffer: 140 mM NaCl, 5 mM KCl, 1 mM $MgSO_4$, 20 mM HEPES (pH 7.4).

1 mM NaH_2PO_4, 5 mM glucose.

Ca^{2+} buffer: 140 mM NaCl, 5 mM KCl, 1 mM $MgSO_4$, 1 mM $CaCl_2$, 20 mM HEPES, 1 mM NaH_2PO_4, 5 mM glucose, pH 7.4.

PBS-T: PBS + 0.1% Tween 20.

Isolation/culture medium: RPMI, 10% fetal calf serum (FCS), 1% penicillin/streptomycin.

2.2 Calcium Indicators

Fluo-4-AM and Fura-Red-AM are dissolved in DMSO, divided into aliquots, and stored at −20 °C until required for use (*see* **Note 1**). Only use each aliquot once.

2.3 Antibodies

0.5 mg/mL Anti-CD3.

0.5 mg/mL Anti-CD28.

0.5 mg/mL Anti-IgG (control).

2.4 Filter Set

Fluo-4/Fura-Red excitation 480/40 nm, beam splitter 495, Fluo-4 emission 542/50, Fura-Red emission 650/57 (all in nanometers).

2.5 Further Materials

5 mg/mL BSA, 0.1 mg/mL Poly-L-Lysine, 200 mM EGTA, 1.41 mM ionomycin. Store aliquots at −20 °C.

Silicon grease.

Protein G magnetic beads.

Aluminum foil.

3 Methods

Carry out all procedures at room temperature unless otherwise specified.

3.1 Isolation of Primary T-Lymphocytes

1. Lymph nodes and spleen are dissected from mice and kept in cold medium until isolation (max. time 2 days).
2. Grind lymph nodes and spleen in isolation medium with a plunger taken from a syringe and a cell strainer (40 μm nylon). Transfer to 50 mL tube. Keep cells on ice.
3. Centrifuge, 300 × *g*, for 5 min, 4 °C.
4. Discard supernatant and re-suspended pellet in 5-mL cold ACK buffer (for erythrocyte lysis). Incubate 3–5 min.
5. Stop lysis by adding 25 mL EH medium.
6. Centrifuge, 300 × *g*, for 5 min, 4 °C.
7. Discard supernatant and re-suspended pellet in 2 mL PBS. Transfer 100 μL in a small tube to control efficiency of T-lymphocyte isolation (contains B- and T-cells).
8. T-lymphocytes are enriched by negative selection, e.g., using *EasySep™ Mouse T Cell Enrichment Kit* according to manufacturer's instructions.
9. To control the efficiency of T-lymphocyte enrichment, cells are stained for TCRβ (T-lymphocytes) and CD45R (B-lymphocytes) by FACS.
10. Count enriched T-lymphocytes (*see* **Note 2**).

3.2 Loading of T-Lymphocytes with Fluo-4 and Fura-Red

1. Centrifuge two to six Mio cells for 5 min, 300 × *g*.
2. Keep lights down and wrap Ca^{2+} indicators and tubes in aluminum foil. Ca^{2+} indicators are dissolved in DMSO. Dilute Ca^{2+} indicators in medium, 500 μL per cell pellet, e.g., 10 μM Fluo-4 and 20 μM Fura-Red.

3. Discard supernatant, and re-suspended pellet with the medium containing your desired indicator.
4. Keep in the dark and incubate for 20 min. Then, add 2 mL medium and incubate for further 30 min.
5. Centrifuge for 5 min, 300 × *g*.
6. Discard supernatant and wash with 2 mL Ca^{2+} buffer.
7. Centrifuge for 5 min, 300 × *g*.
8. Discard supernatant and re-suspended pellet in Ca^{2+} buffer (*see* **Note 3**).
9. Let cells rest for 20 min in the dark (AM will be cleaved intracellularly, thereby trapping the indicator inside the cell).
10. Store cells on ice water and in the dark until measurement.

3.3 Coating Protein G Magnetic Beads with Antibodies

1. Mix the protein magnetic bead suspension gently with a pipette, and transfer 12.5 μL to a new reaction tube (*see* **Note 4**).
2. Place the tube into a magnetic stand, allow the beads to migrate to the magnet, and remove the storage buffer with a pipette.
3. Wash the beads by adding 500 μL PBS-T and vigorously vortexing for 10 s. Return the tube into the magnetic stand and remove the buffer.
4. Re-suspend the beads in 5 μL anti-CD3, 5 μL anti-CD28 (or 10 μL anti-IgG as control), and 7.5 μL PBS-T.
5. Incubate for 30–60 min with continuous mixing.
6. Wash three times with 500 μL PBS-T using the magnetic stand.
7. Wash once with 500 μL Ca^{2+} buffer using the magnetic stand.
8. Re-suspend beads in 500 μL Ca^{2+} buffer (*see* **Note 5**).

3.4 Preparing Slides for Imaging

1. Add 5 μL BSA per slide.
2. Spread the BSA homogeneously until no smears are visible.
3. Let them dry for 10–20 min.
4. Afterwards add 5 μL Poly-L-Lysine and spread it homogeneously.
5. Prepare chamber slides by mounting rubber O-rings onto slides with silicon grease (*see* **Note 6**).

3.5 Measurement of Ca^{2+} Microdomains

1. Technical setup:

   Slides are mounted onto a widefield microscope using a 100-fold magnification.

   A xenon lamp is used as a light source, and frames are acquired with an electron-multiplying charge-coupled device (EMCCD) camera. Images are acquired in 14-bit mode with twofold binning and a dual view to split the emission wavelengths (*see* **Note 7**).

2. Transfer 10 μL of T-lymphocyte suspension in the center of the prepared slide. Let the cells adhere approx. 5 min.
3. Add carefully 90 μL of calcium buffer to the cells.
4. Mount slides onto the microscope, acquire a transmitted light image, and start measurement (*see* **Note 8**).
5. Measure the basal cell activity for 1 min (1 frame per 5 s).
6. Change to maximum frame rate (e.g., 40 fps) for 3 min.
7. Directly add 5–10 μL of coated beads carefully with a pipette to the cells (*see* **Note 9**).
8. Acquire a transmitted light image after the measurement.

3.6 Image Post-processing

Use, for instance, MATLAB and its Image Processing and Optimization Toolboxes for image post-processing. The individual steps are:

1. Import the recorded fluorescence microscopy data as spatio-temporal image sequences Fluo-4(x, y, t) = Fluo-$4_t(x, y)$ and Fura-Red(x, y, t) = Fura-Red$_t(x, y)$ into MATLAB; (x, y) denotes the location of image pixels and t the temporal image frames. For import of the original .ics files into MATLAB, use, e.g., the Bio-Formats package from The Open Microscopy Environment [4].
2. Import transmitted light images that were acquired directly before and after recording of the fluorescence microscopy data.
3. For each cell that is intended to be processed, manually select a rectangular region of interest (ROI).
4. Compensate for photobleaching by additive frame-by-frame bleaching correction (*see* **Note 10**). Since bleaching of Fluo-4 can be considered negligible compared to Fura-Red [5], only bleaching characteristics of Fura-Red are taken into account. First, approximate the temporal development of the frame-specific mean signal intensity of the Fura-Red ROIs by a biexponential decay curve $h_{\text{Fura-Red}}(t) = c_1 e^{-\lambda_1 t} + c_2 e^{-\lambda_2 t}$ [6]; fit $h_{\text{Fura - Red}}$ to the image data in a least-squares sense (MATLAB function `lscurvefit`). Pixel-based bleaching correction then reads Fura-Red$^{\text{corr}}(x, y, t)$ = Fura-Red$_t(x, y)$+$(h_{\text{Fura-Red}}(t = 0) - h_{\text{Fura-Red}}(t))$.
5. Deconvolve the Fura-Red^{corr} and Fluo-4 ROI image sequences. We apply the Lucy-Richardson (LR) algorithm (`deconvlucy`, max. number of iterations = 4) independently to each frame. LR requires a point spread function (PSF), which in our case is analytically generated using a scalar diffraction-limited model [7, 8]. Respective MATLAB source

code can be found at MATLAB Central File Exchange [9]. Application of different deconvolution algorithms and/or a measured PSF is possible (*see* **Note 11**).

6. Spatially align the ROI image sequences by registration of the first frames of the Fluo-4 and Fura-Red image sequences (`imregister`; transformation type: translation; metric mean-squared intensity error), and apply the resulting translation to all frames of Fura-Red^{corr} image sequence.
7. Remove image background by semiautomated thresholding. The transmitted light image acquired at data recording beginning is semi-transparently overlayed to the first frame of the Fluo-4 image sequence and an intensity threshold manually selected such that the remaining visible area in the Fluo-4 image optimally matches the cell shape in the light microscopy data. The same procedure is repeated for the last frame and the transmitted light image acquired after recording the fluorescence microscopy image sequences. For the Fluo-4 frames in between, a frame-specific threshold is defined by linear interpolation of the two aforementioned thresholds. The Fura-Red^{corr} images are handled analogously.
8. After background removal, compute a ratio image sequence by frame-by-frame pixel-wise division of the Fluo-4 by the Fura-Red^{corr} image intensities for all none-background pixels of the ROI.

3.7 Calibration of $[Ca^{2+}]_i$

Calibration is based on [6] for ratiometric dyes. Since an additive frame-by-frame bleaching correction is applied for Fura-Red, mean Fura-Red fluorescence remains stable throughout each measurement and is included in the calibration as Fura-Red$_{t0}$ (initial mean Fura-Red fluorescence). Calculation of $[Ca^{2+}]_i$ is thus determined in situ by R_{min} [Fluo-4$_{min}$/Fura-Red$_{t0}$ after EGTA chelation] and R_{max} [Fluo-4$_{max}$/Fura-Red$_{t0}$] in single-cell measurements as described below. The K_d of Fluo-4 and Fura-Red is 408 ± 12 nM as published in [3].

1. Cells and slides are prepared as described above.
2. Prior to applying cells on the slide, 100 μL of cells are disrupted using 3 μM ionomycin (incubate 5–10 min). Pipette cells on the slide and let adhere. Due to the Ca^{2+} gradient, Ca^{2+} will penetrate, and the maximum Ca^{2+}-fluorescence (**Fluo-4$_{max}$**) may be determined by using the standard acquisition setup.
3. Minimum Ca^{2+}-fluorescence (**Fluo-4$_{min}$**) is achieved by chelating Ca^{2+} with 4 mM EGTA. Add 3 μM ionomycin and incubate for 5–10 min (*see* **Note 12**).
4. The fluorescence of Fura-Red$_{t0}$ can be obtained at the start of every measurement.

3.8 Discrimination Between Signal and Noise in Fluo-4/Fura-Red Ratioing Experiments

One of the primary challenges with imaging data is the discrimination between noise and signals. Noise is defined, among other definitions, as "unwanted disturbance superimposed upon a useful signal, which tends to obscure the signal's information content" [10].

1. Dilute 10 μM Fluo-4 and 20 μM Fura-Red in a cell-free Ca^{2+}-EGTA buffer ($[Ca^{2+}] = 100$ nM).
2. Measure with the same settings as for the bead addition.
3. Post-process the images as explained above, but skip the background removal, since the whole image can be used for analysis purposes.
4. Select a period of the resulting image sequence with stable mean intensity of the individual image frames over time.
5. In the first frame of the period, randomly sample a larger number (e.g., 1000) of regions of interest (ROIs) with or slightly larger than the size of the spatial imaging resolution and the expected minimum local Ca^{2+} signal size (*see* **Note 13**).
6. For each frame, determine a θ_{noise} threshold such that for all ROIs, the minimum intensity of the ROI pixels is above the frame-specific mean intensity of the entire image frame plus θ_{noise}.
7. Select the largest frame-specific θ_{noise} as final threshold to discriminate between signal and noise (*see* **Note 14**).

3.9 Identification of Subcellular Ca^{2+} Microdomains

Subcellular Ca^{2+} microdomains are defined as small, compact connected sets of pixels with high $[Ca^{2+}]_i$ values. They are identified independently for each cell and on a frame-by-frame basis, pursuing the following steps:

1. Compute the frame-specific mean $[Ca^{2+}]_i$ value, written as $\mu_t([Ca^{2+}]_i)$ with t denoting the frames of the image sequence, by averaging the $[Ca^{2+}]_i$ values of all inner cell pixels at frame t.
2. For each frame t, determine all connected components $C_{\text{cell}}^{t,k}$, $k = 1,2,\ldots,$ of cell pixels x such that $\forall k \forall x \in C_{\text{cell}}^{t,k} : [Ca^{2+}]_i(x,t) > \mu_t\left([Ca^{2+}]_i\right) + \theta_{\text{noise}}$ with θ_{noise} as introduced in Subheading 3.8 in order to discriminate between signal and noise.
3. For each connected component $C_{\text{cell}}^{t,k}$ check the following conditions:

   3.1 Minimum and maximum pixel set size: Does $C_{\text{cell}}^{t,k}$ consist of at least 6 and not more than 20 pixels?

   3.2 Pixel set circularity: With circularity of a pixel set Ω being defined by $f_{\Omega}^{\text{circ}} = (4\pi A_{\Omega})/P_{\Omega}^2$ with A_{Ω} as area and P_{Ω} as perimeter of Ω (i.e., f_{Ω}^{circ} values are between 0 and 1, with $f_{\Omega}^{\text{circ}} = 1$ for a circle), is f_{Ω}^{circ} for $\Omega = C_{\text{cell}}^{t,k}$ larger than 0.5? Component area and perimeter can be assessed by the `regionprops` function of the MATLAB Image Processing Toolbox.

If connected components $C_{\text{cell}}^{t,k}$ fulfill conditions 3.1 and 3.2, they are interpreted as Ca^{2+} microdomains. Ca^{2+} values for individual microdomains usually correspond to the maximum $[Ca^{2+}]_i$ value of all pixels $x \in C_{\text{cell}}^{t,k}$.

3.10 Dartboard Projection: Compartment-Wise Analysis of Ca^{2+} Microdomains

Correlation of initial Ca^{2+} microdomains to the applied stimulus and tracking the spatiotemporal progression from local to global signals for a larger number of cells and cell groups further require normalization of cells, i.e., analysis of the Ca^{2+} in a common spatial cell coordinate system. Under the assumption of approximately circular cell shape in the 2D images and neglecting cell motion and deformation during image acquisition (*see* **Note 15**), we employ a "dartboard projection" approach. The following steps are applied to each cell and the corresponding image sequence:

1. Segment the cell in the first frame of the calibrated Ca^{2+} image sequence: Edge detection and binarization of the edge image by threshold (MATLAB function `edge`, e.g., using the Sobel operator), dilatation of resulting binary image (`imdilate`), and morphological closing of residual holes (`imfill`). An example MATLAB script can be found at [11].
2. Apply circular Hough transform to the cell segmentation image (`imfindcircles`). Estimate the transform radius range based on the size of the segmented cell and the respective ROI (e.g., radii between 0.5 and 0.2 times the maximum cell ROI side length).
3. Generate the "dartboard": Divide the found circle into the desired number of annuli and annuli segments (standardized cell compartments) of same area. The number of annuli and annuli segments depends on the exact analysis goal.
4. For each frame, assign the identified Ca^{2+} microdomains to annuli segments based on the spatial location of the microdomain center of mass.
5. Rotate the resulting dartboard such that the spatial area of stimulation (bead contact zone) corresponds to a pre-defined dartboard area that is consistent for all cells.

After cell normalization, cell-specific compartment-wise information about the Ca^{2+} microdomains can be aggregated. The principle of the dartboard projection and an application for group-wise Ca^{2+} signal analysis are illustrated in Fig. 2.

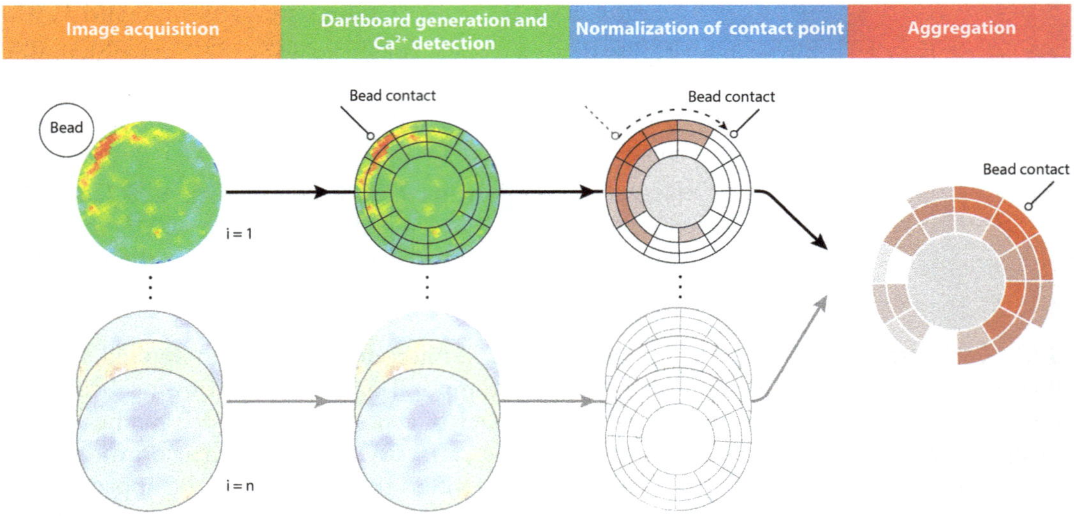

Fig. 2 Generation of a dartboard projection with a WT T-lymphocyte. (**a**) Summarizes the dartboard workflow. First a dartboard is projected on every T-lymphocyte. All T-lymphocytes are normalized on bead contact site and time. The mean number of Ca^{2+} microdomains is analyzed and depicted in a dartboard. (**b**) Primary WT T-lymphocytes were stimulated by an anti-CD3/anti-CD28 coated beads. Bead contact site is indicated as black arrow. Initial Ca^{2+} microdomains were detected in the first 15 s (left column) or as indicated (right columns) after bead contact. The dartboard shows mean values of the number of Ca^{2+} microdomains in all T-lymphocytes. Max. value is 1.7 (dark red) as indicated

4 Notes

1. Best use brown reaction tubes to aliquot Fluo-4 and Fura-Red. Our stock solution for Fluo-4 is 1 mM and for Fura-Red is 4 mM.
2. Between 3 and 12 Mio T-lymphocytes can be isolated from one mouse using spleen and lymph nodes. The freshly isolated T-lymphocytes can be cultivated overnight but not any longer without any stimulation. Therefore, it is best to isolate and image the T-lymphocytes on the same day.
3. Re-suspend one Mio T-lymphocyte in 80 μL Ca^{2+} buffer.
4. For a good bead-cell interaction, the beads should be similar in size compared to the cells. We use magnetic beads with a diameter of 10 μm for primary T-lymphocytes.
5. It is critical for the imaging that you have the right concentration of beads you add to the cells, because each T-lymphocyte should only be activated by one bead. We usually add 5–10 μL of the coated bead solution to every slide. If the whole field of view is full of beads, you should dilute the bead solution, or if there are no beads in the field of view, you should add more bead solution or concentrate it.

6. We find that it is best to coat each slides fresh, but it is also possible to prepare them 1 day in advance.
7. The exposure time has to be evaluated for every microscope since it is dependent on the light source, objective, filters, and the camera. We use an exposure time of 25 ms (40 fps). Instead of a dual view, also a dual cam can be used to split the emission wavelengths.
8. It is essential to acquire a transmitted light image before and after the measurement to control where the beads have contact with T-lymphocytes.
9. This is the crucial step. We add the beads close to the light point on the slide and move the pipette slightly during the addition. It probably takes a while to add the beads at the right position with the right angle without moving the slide.
10. Additive bleaching correction refers to elevation of pixel intensity levels based on a frame-specific (but not pixel-specific!) correction term. In contrast, multiplicative correction refers to a multiplication of the pixel intensities by a correction term. Due to noise amplification introduced by multiplicative correction, we prefer the more conservative additive bleaching correction strategy.
11. Deconvolution and bleaching correction significantly affect image appearance and signal-to-noise ratio (*see* Fig. 3). Hence, the applied algorithms directly influence quantitative evaluation of local Ca^{2+} signals and especially measures like number of microdomains per cell and time period or maximum Ca^{2+} concentration within a microdomain [12]. It should, therefore, be noted that replacing deconvolution or bleaching correction algorithms by alternative methods necessarily requires re-evaluation of noise levels. The dependence of quantitative Ca^{2+} microdomain measures on applied post-processing algorithms makes it furthermore hard to compare respective numbers across different publications.
12. The calibration has to be done on the same microscope with the exact same settings as for the measurements.
13. For our microscope with two times binning, the calculated spatial resolution is 0.368 μm (3 × 3 pixel), but the resolution has to be evaluated specifically for every system.
14. The discrimination between signal and noise has to be reevaluated for every system.
15. Violation of the assumptions of circular cell shape and negligible cell motion/deformation during image acquisition affects accuracy of the analysis results. In our experiments, the

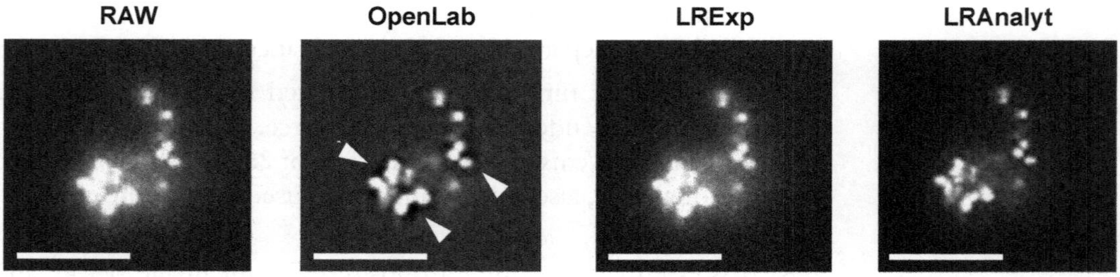

Fig. 3 Influence of different deconvolution approaches on image quality. Jurkat T-lymphocytes were loaded with 75 nM LysoTracker® for 1 h at 37 °C. Imaging was performed using a Leica IRBE2 microscope with 100× magnification, a Sutter DG-4 as light source, and an electron-multiplying charge-coupled device camera. Arrowheads indicate dark artifacts. Scale bar represents 10 μm. *RAW* unprocessed, *LR* Lucy-Richardson deconvolution, *Exp* experimentally acquired PSF, *Analyt* analytically computed PSF

assumptions were appropriate for primary T-lymphocytes. Yet, for other cell lines like Jurkat T-cells, they are questionable, and approaches for motion/deformation compensation and cell shape normalization have to be implemented [13].

Acknowledgment

The authors acknowledge financial support by Deutsche Forschungsgemeinschaft (SFB1328 projects A01 (to AHG), A02 (to IMAW and RW), grants GU 360/15-1 and 15-2 (AHG)), Joachim-Herz-Stiftung, Infectophysics Consortium, project 04, and Landesforschungsförderung Hamburg ReAd Me (project 1, IMAW and AHG).

References

1. Berrie CP, Cobbold PH (1995) Both activators and inhibitors of protein kinase C promote the inhibition of phenylephrine-induced $[Ca^{2+}]_i$ oscillations in single intact rat hepatocytes. Cell Calcium 18:232–244
2. Barker SA, Lujan D, Wilson BS (1999) Multiple roles for PI 3-kinase in the regulation of PLCgamma activity and Ca^{2+} mobilization in antigen-stimulated mast cells. J Leukoc Biol 65:321–329
3. Wolf IMA, Diercks B-P, Gattkowski E et al (2015) Frontrunners of T cell activation: Initial, localized Ca^{2+} signals mediated by NAADP and the type 1 ryanodine receptor. Sci Signal 8:ra102
4. Linkert M, Rueden CT, Allan C et al (2010) Metadata matters: access to image data in the real world. J Cell Biol 189:777–782
5. Thomas D, Tovey SC, Collins TJ et al (2000) A comparison of fluorescent Ca^{2+} indicator properties and their use in measuring elementary and global Ca^{2+} signals. Cell Calcium 28:213–223
6. Grynkiewicz G, Poenie M, Tsien RY (1985) A new generation of Ca^{2+} indicators with greatly improved fluorescence properties. J Biol Chem 260:3440–3450
7. Stokseth PA (1969) Properties of a defocused optical system. J Opt Soc Am 59:1314–1321
8. Pankajakshan P, Kam Z, Dieterlen A et al (2010) Point-spread function model for fluorescence MACROscopy imaging. In: Matthews MB (ed) Signals, systems and computers (ASILOMAR), 2010 Conference Record of the Forty Fourth Asilomar Conference on, Pacific Grove, November 2010, IEEE, pp 1364–1368

9. Pankajakshan P (2011) Widefield fluorescence microscope point-spread function. https://de.mathworks.com/matlabcentral/fileexchange/31945-widefield-fluorescence-microscope-point-spread-function. Accessed 20 May 2018
10. Dyer SA (2004) Wiley survey of instrumentation and measurement. John Wiley & Sons, Hoboken, pp 759–808
11. Detecting a cell using image segmentation. https://mathworks.com/help/images/examples/detecting-a-cell-using-image-segmentation.html. Accessed 20 May 2018
12. Schetelig D, Wolf IMA, Diercks B-P et al (2015) A modular framework for post-processing and analysis of fluorescence microscopy image sequences of subcellular calcium dynamics. In: Handels H, Deserno TM, Meinzer H-P et al (eds) Bildverarbeitung für die Medizin 2015. Springer, Berlin Heidelberg, pp 401–406
13. Antoni S-T, Ismail OMF, Schetelig D et al (2017) Systematic analysis of jurkat T-cell deformation in fluorescence microscopy data. In: Maier-Hein K, Deserno TM, Handels H et al (eds) Bildverarbeitung für die Medizin 2017. Springer, Berlin Heidelberg, pp 275–280

Chapter 4

Measurement of Contractility and Calcium Release in Cardiac Spheroids

Christian Zuppinger

Abstract

There is a need for organotypic in vitro models that resemble the native tissue in functionality and tissue architecture for disease models and drug development. To this end, many 3D culture formats have been developed over time. Among the most often used type is the scaffold-free multicellular aggregate, also called spheroid, that forms by self-assembly. However, working with 3D cultures can be challenging because single cells are not as accessible as in 2D cultures and standard lab procedures must be adapted or replaced altogether. This chapter describes methods to create cardiac spheroids consisting of human iPSC-derived cardiomyocytes and cardiac fibroblasts and how to measure contractility or calcium signals using quantitative video analysis and confocal microscopy. Emphasis is on the particular challenges that 3D cultures pose and on affordable methods that do not require specialized equipment.

Key words 3D culture, Cardiomyocytes, Calcium, Contractility, Beating, Spheroid, Microtissue, Video, Image processing, Confocal microscopy

1 Introduction

Cell culture systems that consist of aggregated cells or matrices populated by cells are commonly called "3D culture" in contrast to the classic culture in the form of a monolayer on plastic or glass surfaces in bottles and culture dishes. Many different formats of 3D cultures and cell culture technologies have been developed and evaluated in the last decade. Additionally, sophisticated microfluidic systems and bioreactors have been developed together with improved sensor technology [1]. In this text, we will cover the scaffold-free cardiac spheroid (also called multicellular aggregate, spheroidal microtissue, or cardiosphere) that forms by self-assembly without the addition of hydrogels or matrix proteins.

Electronic Supplementary Material: The online version of this chapter (https://doi.org/10.1007/978-1-4939-9030-6_4) contains supplementary material, which is available to authorized users.

Claus W. Heizmann (ed.), *Calcium-Binding Proteins of the EF-Hand Superfamily: From Basics to Medical Applications*, Methods in Molecular Biology, vol. 1929, https://doi.org/10.1007/978-1-4939-9030-6_4,

Spheroids can be made in many ways, floating in rotation systems where aggregates form in nonuniform sizes, by the hanging drop method (gravity-assisted cell assembly), in soft gels, and on nonadhesive surfaces such as coated multi-well plates or agarose molds. Advantages and disadvantages of different 3D cell culture models in the cardiovascular field, as well as comparisons with standard 2D cultures, have been reviewed elsewhere [2]. Several lab techniques have already been successfully used with cardiac spheroids such as viability assays using fluorescent dyes or luminescence reporters, classic histology and immunolabeling, or protein biochemistry and analysis of gene expression [3–5]. Nevertheless, the multicellular nature of 3D cultures can be a serious challenge for applying standard laboratory techniques that rely on rapid diffusion of dyes, optical transparency, or complete tissue lysis. Although these cellular aggregates are small, in fact hardly visible to the naked eye, and handled only by pipetting (*see* Fig. 1), spheroids of several hundred micrometers may still be too large and compact to allow complete penetrance of reagents in fixed material or diffusion into living microtissues. Therefore, some assays for live cells that make use of diffusing fluorescent dyes, such as calcium-sensitive dyes or

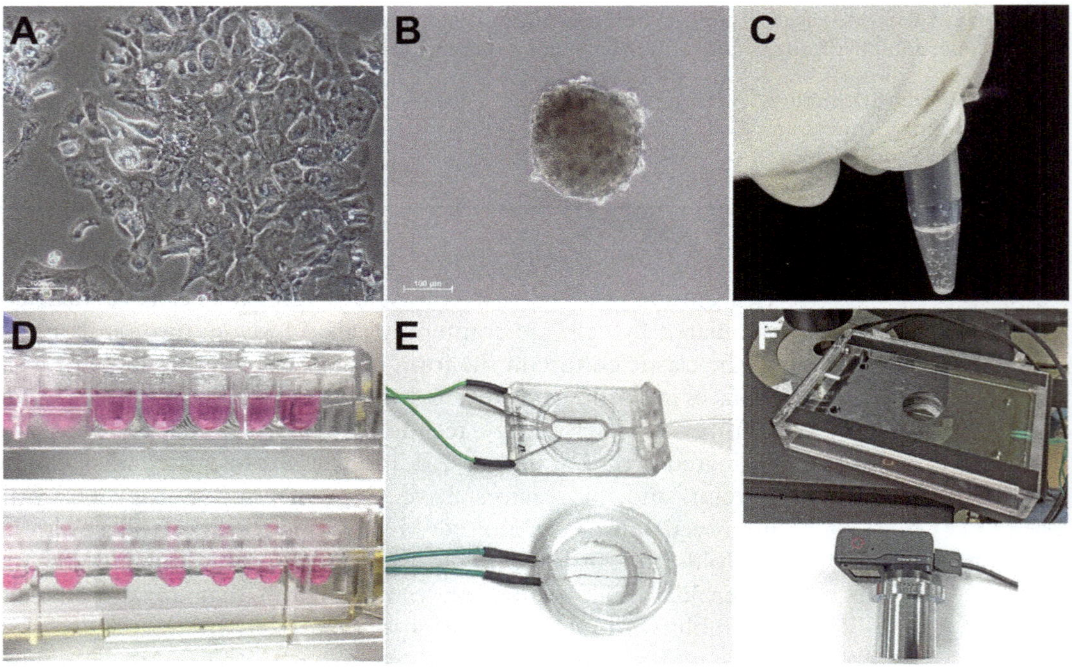

Fig. 1 Working with spheroids. Panel (**a**) shows a standard 2D culture of hiPSC-CMs. Panel (**b**) shows a cardiac spheroid after harvesting. Panel (**c**) shows a test tube with collected spheroids for size comparison. In panel (**d**) two methods for making microtissues are shown: U-shaped cell-repellent wells (above) and the InSphero hanging drop system (below). Panel (**e**) shows two types of field pacing chambers: a commercially available perfusion chamber (above) and a self-made culture dish (below). Panel (**f**) shows an ibidi warming chamber for 35 mm dishes (above) and the GoPro camera with adapter tube (below)

reporters used in viability assays, may capture only the activity of the outer layers of a spheroid. Unlike with monolayer-cultured cells, these microtissues are nearly opaque, and cell borders, nuclei, or other cellular features are not distinguishable in bright-field microscopy. Even by confocal or light-sheet microscopy, it may not be possible to obtain high-resolution data from the entire breadth of a spheroid (own observation). This is due to optical properties of the tissue and dyes but also to practical limitations like the small working distances of higher aperture lenses and the shape of sample holders. Furthermore, there is an upper limit for spheroid size which is set by hypoxia and cell death in the core of the avascular microtissues. Studies in tumor tissue have shown that cells experience normoxic conditions within a distance of 10–100 μm from a blood vessel [6].

In contrast to viability end-point assays, the measurement of parameters such contractility and calcium transients provides an assessment of cardiac organotypic features that are relevant for the phenotypic testing for cardiotoxicity and drug efficacy. Assessment of calcium transients provides information about mechanisms which are fundamentally involved in the excitation-contraction machinery of the cardiomyocytes and may be linked to arrhythmias in patients. The availability of low-cost consumer cameras delivering comparably high frame rates and free open-source software for video image analysis opens new opportunities for research groups which have limited access to specialized or costly equipment. As a label-free method, video measurement can be repeated multiple times during treatment. Additionally, drugs could be used that take longer time to show effects such as hypertrophy of muscle cells or fibrotic changes. The measurement of contractility demonstrates the health status of the microtissues and provides insight regarding drug effects. Spontaneous beating activity in cardiac microtissues is well visible by microscope inspection and usually begins a few days after starting the aggregation process, and it can persist up to several months in culture [3].

2 Materials

2.1 Production and Culture of Cardiac Spheroids

1. Cells: Human iPSC-derived cardiomyocytes (hiPSC-CMs) are from Cellular Dynamics International (Madison, WI, USA) or from Ncardia (Cologne, Germany) and primary human cardiac fibroblasts (HCF) from Sciencell (Carlsbad, CA, USA) or PromoCell (Heidelberg, Germany).
2. Cell culture medium: hiPSC-CMs are delivered with medium from the same manufacturer (*see* **Note 1**). HCF can be expanded on fibronectin-coated cell culture flasks with

"Dulbecco's Modified Eagle Medium" (DMEM) low glucose and supplemented with 10% fetal calf serum and antibiotics.

3. Additional cell culture reagents: Accumax (Merck, Darmstadt, Germany) for detachment of hiPSC-CM, if the cardiomyocytes are shipped in a culture flask or a pre-plating step is used (*see* **Note 2**). Fibronectin from human plasma (Merck) for coating culture flasks for the expansion of HCF.
4. Nonstandard cell culture lab equipment: Programmable multi-pipettor ViaFlo-300 (INTEGRA Biosciences, Zizers, Switzerland) for medium exchange and spheroid seeding and transfer steps. Glass-bottom culture dishes with a no zero coverslip insert for calcium imaging with oil or water immersion lenses, either in a format with larger surface (ibidi, GmbH, Martinsried, Germany) or microwell dishes with a round 7 mm diameter wells (MatTek Corp., Ashland, USA). For pipette tips, low-retention tips are generally preferred as spheroids may stick inside the tip when transferred.
5. Cell culture material for the production of spheroids: The InSphero (Schlieren, Switzerland) system consisting of GravityPLUS™ hanging drop assembly and the 96-well GravityTRAP™ spheroid culturing plates with conical wells are currently available from PerkinElmer (PerkinElmer, Schwerzenbach, Switzerland). Ready-to-use 96-well plates with U-shaped wells and ultra-low attachment/cell-repellent surface are available from Greiner (Greiner Bio-One Intl., Kremsmünster, Austria) (*see* Fig. 1d) (*see* **Note 3**).

2.2 Equipment and Reagents for Contractility Measurement

1. Camera and microscope: A modified GoPro HeroBlack 6 camera (Back-Bone Gear Inc., Kanata, Canada) without lenses but with a C-mount adapter ring added is used to record short movie sequences of 5–10 s at a speed of 240 frames per second (i.e., one full image every 4.2 ms) at 1920 × 1080 pixel size. The stored movies are of MP4 type and using the H.265 codec format. The camera is equipped with a microSDXC-type UHS U3 storage card with 64GB capacity (SanDisk). An additional adapter tube may be necessary for attachment to the microscope port. The inverted microscope is an Eclipse TE2000-U (Nikon, Tokyo, Japan) with a Nikon Plan Fluo 10×/0.3 phase-contrast lens. A compact heating chamber with warmed lid is directly placed on the stage of the microscope and connected to a temperature controller with separated settings for chamber and lid (ibidi, GmbH, Martinsried, Germany) (*see* Fig. 1f).
2. Electrical field pacing of microtissues is performed either using a perfusion chamber (Warner Instruments, Hamden, USA) or a self-made dish with two parallel platinum wire electrodes (*see* Fig. 1e). The pacing chamber is attached to a MyoPacer

(IonOptix, Boston, USA) set to a frequency of 3 Hz, 10 Volts, bipolar pulses of 4 ms duration (*see* **Note 4**).

3. Software for movie processing and video analysis: For converting GoPro movies to standard H.264 format and reducing frame size in order to limit the amount of data, the free software "HandBrake" (https://handbrake.fr) is used (*see* **Note 5**). "Prism Video" from NCH Software http://www.nchsoftware.com/prism is used to export single image files from videos. A macro for the image processing software NIH-ImageJ (National Institutes of Health, USA) named "Musclemotion 1.0" by Sala et al. [7] is used to extract contractility features from the video sequences. This open access software is available from GitHub https://github.com/l-sala/MUSCLEMOTION.

2.3 Equipment and Reagents for Calcium Measurements

1. Reagents for calcium measurements: Spheroids are stained and measured in standard cell culture medium. The calcium-sensitive fluorescent dye Calbryte520 AM (AAT Bioquest, Sunnyvale, USA) is solved as a 5 milliM stock solution in anhydrous DMSO and stored at −20 °C until use. To avoid motion artifacts, the specific inhibitor of myosin ATPase 10 microM blebbistatin (Merck) is used in the staining and recording medium (*see* **Note 6**). Optionally, 0.1% Pluronic F-127 surfactant (Thermo Fisher Scientific) and 1.25 milliM probenecid (Merck) are added to the staining medium (*see* **Note** 7).

2. Equipment for measuring calcium signals: Confocal microscope Zeiss LSM710 with a Zeiss EC Plan-Neofluar 40×/1.3 oil immersion lens. Fluorescence excitation and filters are the same as used for FITC fluorochromes. The microscope is equipped with an "Incubator XLmulti S1" hood that heats lenses, the stage, and surrounding air to 37 °C. Software for image acquisition is ZEN2.3 SP3 running on a HPZ800 workstation. Line scans of calcium transients can be analyzed in detail using a software such as Igor Pro (WaveMetrics, Lake Oswego, USA). Videos of calcium cycling in spheroids (*see* Video 1) were recorded using the Nikon Eclipse microscope with a Nikon60×/1.4 Plan Apo oil immersion lens and a Digital Sight DS-Ri1 camera at a resolution of 320 × 256 pixels at 25 frames per second and in 2 × 2 binning mode (*see* **Note 8**).

3 Methods

3.1 Production and Culture of Cardiac Spheroids

1. Expand HCF by culturing the cells on fibronectin-coated dishes or flasks in DMEM, and detach the cells when they are still at a sub-confluent stage. Process cryopreserved hiPSC-CM according to the protocol supplied by the manufacturer. Detach cells from culture flasks, and count viable cell density for each cell type. Then dilute each solution with culture medium so that together a mixture of cardiomyocytes and fibroblasts at a ratio of 4:1 results. If the hanging drop method is used, then the total amount of cells in a volume of 40 microliter, i.e., one droplet, determines the size of the spheroid. A size between 3000 and 5000 cells per spheroid is recommended. Carefully mix the cell solution in a reservoir for multi-pipettes, and remix between seeding steps. Set the multi-pipette to repeatedly dispend 40 microliters at slow speed, then fill the 96 GravityPlus from the top with the cell solution. Put on the lid and place the entire assembly in an incubator for 3–4 days. If you use nonadhesive wells, deliver the appropriate amount of cell solution in wells of a pre-wetted spheroid plate, gently shake the plate, and incubate.
2. Harvest spheroids into a pre-wetted GravityTrap plate or spheroid plate, and inspect the formed microtissues and their spontaneous contraction activity. Make sure that all the spheroids have approximatively similar size before an experiment is started.
3. On the day of the measurement, collect the desired group of spheroids in a test tube, and let them sediment or gently centrifuge. Then discard excess medium and use a 1 mL pipette to place the spheroids in the center of a culture dish or pacing chamber for recording.

3.2 Measurement of Contractility

1. Find the spheroids in the microscope using the 10× phase-contrast lens, and wait at least 10 min for warm-up, sedimentation of the spheroids, and stabilization of the system (*see* **Note 9**).
2. Turn on the GoPro camera, and switch to the video mode that records 240 frames per second at 1920 × 1080 pixels. Make sure that the spheroid is in frame and in focus, and then record a constant quantity of video for each spheroid. Since this camera is recording sound, a short annotation about the experiment may be dictated as the soundtrack of the movie.
3. After a recording session, the camera is connected to a computer via USB cable for the transfer of the data. The video may now be converted and image dimensions reduced for faster processing and storage.

4. Load video images as a stack into ImageJ. In ImageJ, areas outside of the spheroid may easily be trimmed off using the "crop" command, and unwanted parts of neighboring spheroids can be deleted from the image by applying a polygon selection and setting the contrast to pure white in the entire stack using the contrast and brightness command. Save the trimmed stack to disk.
5. The "Musclemotion" macro is installed as a button in the ImageJ control bar using the "install macro" command. Upon start of the macro, the user is asked to choose a directory for the results and the location of the stack of images, as well as some analysis settings. The software outputs several images of recognized motion signals over time including the result of the peak-finding mechanism (*see* Fig. 2). Finally, a table of numeric results is presented that shows the values for each recognized peak and the following parameters: contraction duration; time to peak; relaxation time; time to 90, 50, and 10% of the peak and baseline; peak amplitude; contraction amplitude; and peak-to-peak time (*see* [7] for more information about use and algorithms of this software).
6. Inspect the numerical results, and discard peaks that contain the value zero in any of the parameters. If necessary, restart the macro and adjust analysis parameters.
7. An example for an experiment, a comparison of the spontaneous contractility of spheroids with and without HCF, is shown in Fig. 3.

3.3 Measurement of Calcium Signals in Cardiac Spheroids

1. On the day of the measurement, collect spheroids in a test tube, and sediment or gently centrifuge. Discard excess medium. Prepare cardiomyocyte medium or a buffer of choice with 5 microM Calbryte520 AM, and incubate for 1 h at 37 °C in the dark.
2. Incubate another 15 min at room temperature in the dark.
3. Remove the staining solution, and replace it with fresh medium without dye but containing 10 microM blebbistatin.
4. Transfer the spheroids to a glass-bottom culture dish or pacing chamber, and wait 10 min for warm-up. Protect the spheroids from light while the microscope is readied for the measurement.
5. Find the spheroids in the field of view using epifluorescence with FITC filters and a 10× lens for overview, and then switch to a 40× oil immersion lens for high-resolution measurement. When the microtissue is in focus, apply recording parameters using the 488 nm line of the argon laser to adjust focus, gain, and background settings (*see* Fig. 5a).

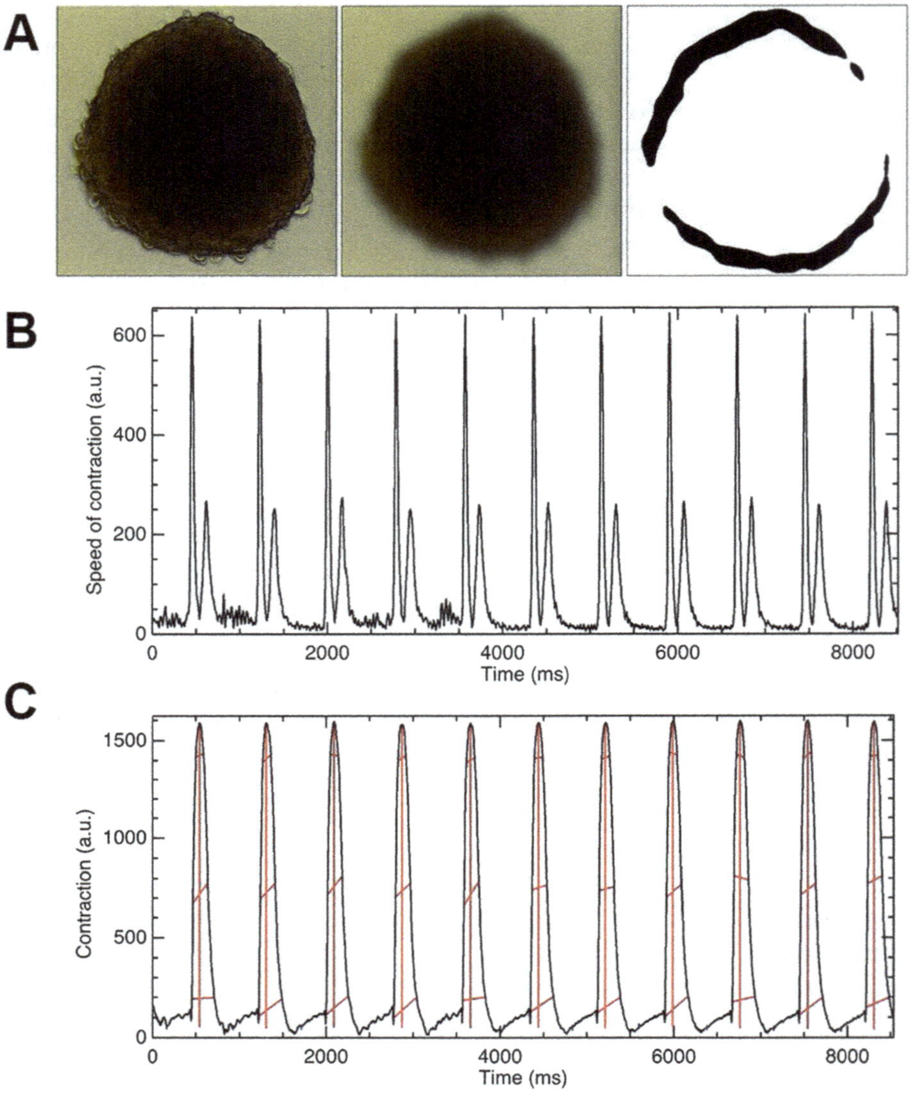

Fig. 2 Output of the "Musclemotion" macro. Panel (**a**) shows three views of the stack of images in ImageJ that the macro produces with the original stack on the left, blurred version in the middle, and a depiction of the moving areas of the movie on the right. Panel (**b**) shows detected motion in the movie shown as double peaks. Panel (**c**) shows the representation of recognized contraction peaks with labels for 10, 50, and 90% of transient duration

6. Take a first XY image as an overview of the entire spheroid. This image is too slow to measure single calcium transients, but it provides a "shutter" map of active areas in the microtissue and helps to set the region of interest for a line scan (Fig. 5b).
7. Change the acquisition mode from "frame" to "line," and check "time series" in the experiments section of the ZEN software (*see* Fig. 4). Enter the length of each time step

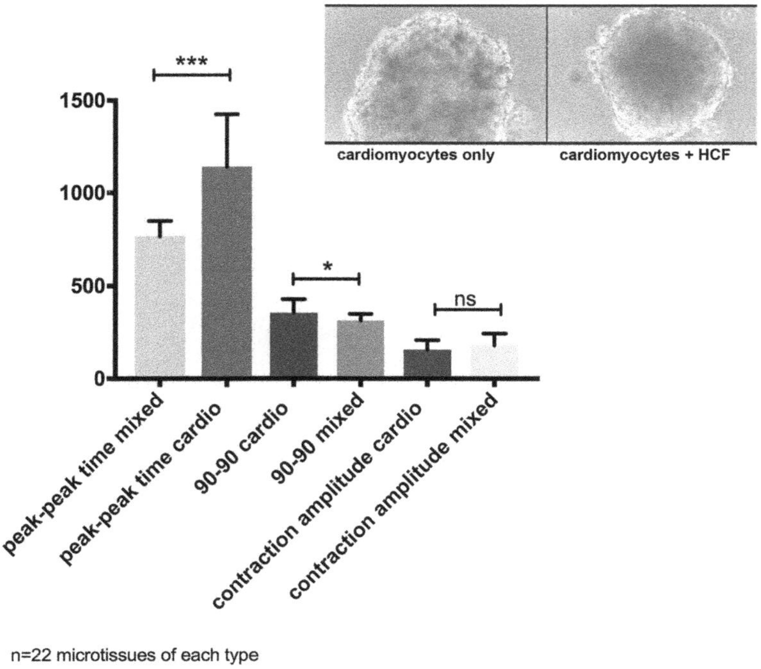

Fig. 3 Example of an experiment performed using quantitative video analysis. Spheroids made of hiPSC-CMs without and with HCF (morphology is shown in the inset) were cultured for 1 month, and 22 spheroids of each type were measured. The mixed spheroids showed significantly shorter peak-to-peak times, i.e., a faster rate of spontaneous contractions

(1 millisecond in this example), and adjust acquisition parameters so that a relatively low-noise signal is produced. The duration of one line scan is displayed. Set the acquisition parameters so that the scan speed does not exceed the length of one time step. In this way, the timing of the resulting YT strip is defined.

8. Select a short line segment and start scanning (*see* **Note 10**).
9. Use an image processing software for further analysis of the calcium signals. Profile plots are a common way of presenting the data [4]. Kinetics of the calcium transient can be studied using curve fitting methods with programs such as Igor Pro (Fig. 5c, d).

4 Notes

1. For co-culture with HCF, it is necessary to test the cardiomyocyte medium first on HCF 2D standard cultures.
2. Accumax is used undiluted and is deactivated by temperature, so detachment of cells works best at room temperature.

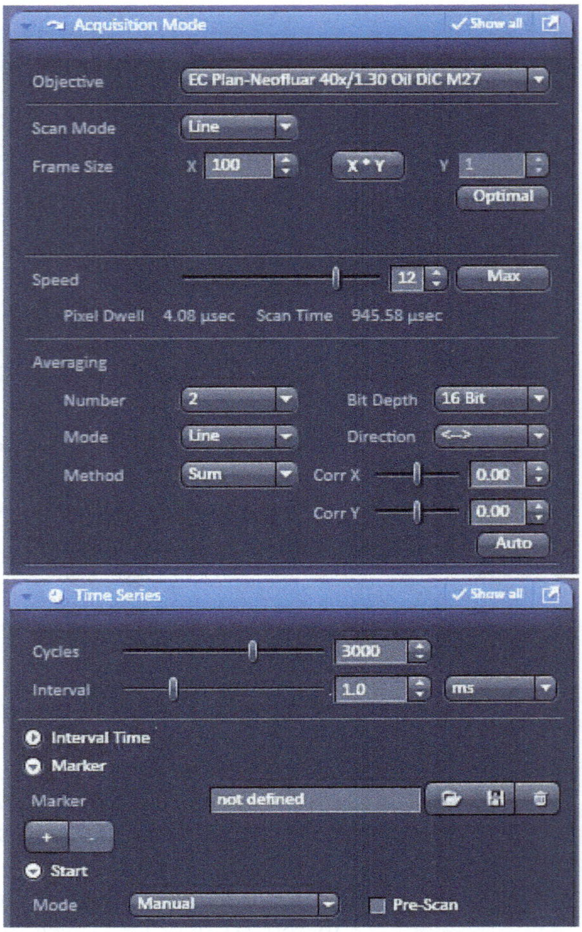

Fig. 4 Settings for line scans and time series in the ZEN software

3. The hanging drop method has a higher assembling power than the nonadhesive wells that sometimes lead to a remaining halo of non-integrated cells (own observation). Both methods are useful for the production of cardiac spheroids made of hiPSC-CM and cardiac fibroblasts. Spheroids made only of hiPSC-CM often form looser aggregates and tend to break up, especially when less than 5000 cells are used.
4. The spontaneous contraction rate of untreated cardiac spheroids usually is below 3 Hz at 37 °C. Therefore, setting the pacing to 3 Hz reveals if the spheroid is following the field pacing. Pacing rates up to 5 Hz may be tolerated by the cardiac spheroids (*see* Video 2).
5. Settings in the program "Handshake": take care that in the tab "Video" the framerate is set to "constant framerate" and under "Quality" it is set at "constant quality, RF=0."

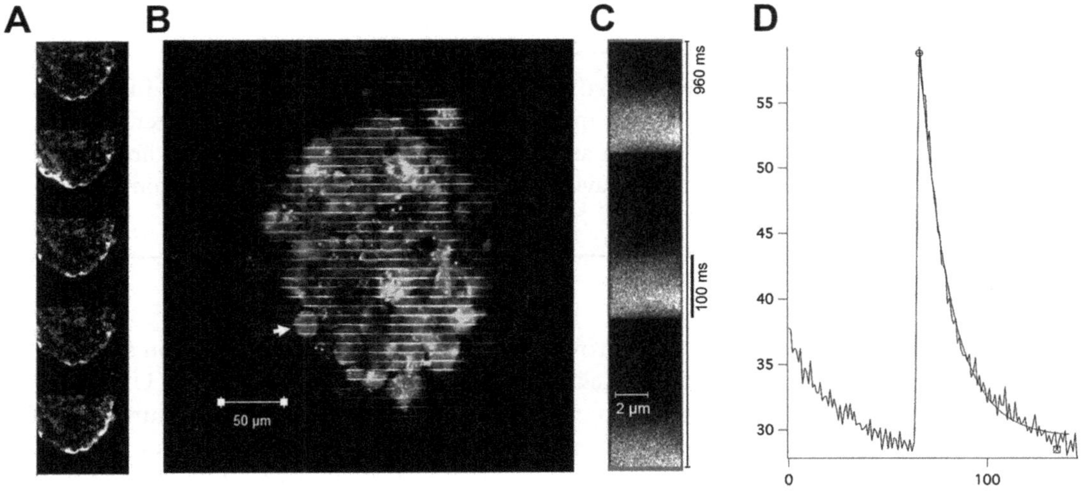

Fig. 5 Measurements of calcium transients using confocal microscopy. Panel (**a**) shows several frames of a time series recorded in XY mode every 200 ms. The images demonstrate calcium cycling activity in different regions of the spheroid. The calcium-sensitive dye apparently did not reach the inner part of the microtissue. Panel (**b**) shows a slow XY scan resulting in a "shutter" pattern as the brightness of the cells changes with each transient. The arrow down left points at the single cell that produced the line scan in panel (**c**). Panel (**c**) shows a narrow line selection recorded for a total of 960 ms. Panel (**d**) shows an example of transient analysis in Igor Pro where a profile plot was obtained from a smoothed line scan and a curve fitting is applied to the transient in order to compare the kinetics of reuptake of calcium into the sarcoplasmic reticulum

6. Blebbistatin is an electromechanical uncoupler which stops the motion of the spheroids but has little effect on calcium cycling in contrast to other inhibitors that have been used for this purpose [8].
7. These compounds are added to the medium to improve penetrance of the dye into the microtissue and to reduce leakage of the de-esterified indicator from the cells. Both Pluronic F-127 and probenecid are not easy to solve in culture medium and may need warming and stirring. It is recommended to prepare stock solutions in DMSO that are kept at room temperature until use. However, stocks of the dye Calbryte520 in DMSO have to be stored frozen and anhydrous until use.
8. The use of a confocal microscope is required when measuring calcium signals in spheroids at high resolution, because the signal can be restricted to a small volume, even to single cells, which would not be possible otherwise. A standard microscope can be instrumental for capturing the activity of an entire spheroid as a nearly synchronously contracting tissue.
9. Take care that the spheroid is not drifting through the field of view during video recording as this disturbs the video analysis. If drifting is a problem, let the spheroids slightly attach to the bottom of the culture by coating it first with gelatin or

fibronectin, and keep the dish in the incubator for some hours up to overnight.

10. While the voxel size and filter settings are stored in the *.LSM and other image formats, the specific parameters of a time series such as the duration of a single line scan are not automatically saved with the file and have to be written down.

Acknowledgment

Funding was provided by the Swiss Heart Foundation to C. Z. We acknowledge technical support by Patrick Furer (University of Bern) for help with making the field pacing culture dish. We thank Sarah Longnus (Cardiac Surgery Department, University Hospital Bern) for support and discussions. This study was performed with the support of the Microscopy Imaging Center (MIC).

References

1. Bürgel SC, Diener L, Frey O et al (2016) Automated, multiplexed electrical impedance spectroscopy platform for continuous monitoring of microtissue spheroids. Anal Chem 88:10876–10883. https://doi.org/10.1021/acs.analchem.6b01410
2. Zuppinger C (2016) 3D culture for cardiac cells. Biochim Biophys Acta 1863:1873–1881. https://doi.org/10.1016/j.bbamcr.2015.11.036
3. Kelm JM, Ehler E, Nielsen LK et al (2004) Design of artificial myocardial microtissues. Tissue Eng 10:201–214. https://doi.org/10.1089/107632704322791853
4. Beauchamp P, Moritz W, Kelm JM et al (2015) Development and characterization of a scaffold-free 3D spheroid model of iPSC-derived human cardiomyocytes. Tissue Eng Part C Methods 21:852–861. https://doi.org/10.1089/ten.TEC.2014.0376
5. Garzoni LR, Rossi MID, de Barros APDN et al (2009) Dissecting coronary angiogenesis: 3D co-culture of cardiomyocytes with endothelial or mesenchymal cells. Exp Cell Res 315:3406–3418. https://doi.org/10.1016/j.yexcr.2009.09.016
6. Yeom CJ, Goto Y, Zhu Y et al (2012) Microenvironments and cellular characteristics in the micro tumor cords of malignant solid tumors. Int J Mol Sci 13:13949–13965. https://doi.org/10.3390/ijms131113949
7. Sala L, van Meer BJ, Tertoolen LT et al (2017) MUSCLEMOTION: a versatile open software tool to quantify cardiomyocyte and cardiac muscle contraction in vitro and in vivo. Circ Res 122(3):e5–e16. https://doi.org/10.1161/CIRCRESAHA.117.312067
8. Fedorov VV, Lozinsky IT, Sosunov EA et al (2007) Application of blebbistatin as an excitation-contraction uncoupler for electrophysiologic study of rat and rabbit hearts. Heart Rhythm 4:619–626. https://doi.org/10.1016/j.hrthm.2006.12.047

Chapter 5

Simultaneous Recording of Subcellular Ca^{2+} Signals from the Cytosol and Sarco/Endoplasmic Reticulum: Compartmentalized Dye Loading, Imaging, and Analysis

Ernst Niggli and Miguel Fernandez-Tenorio

Abstract

An increase in the cytosolic Ca^{2+} concentration triggers the contraction in cardiomyocytes. In these cells sarcoplasmic reticulum (SR) is the major source of Ca^{2+}, and the release from this store is mediated by the ryanodine receptors (RyRs). These receptors are regulated by cytosolic and intra-SR $[Ca^{2+}]$. The cytosolic Ca^{2+} regulation is well established, but there are some limitations to determine indirectly the intra-SR Ca^{2+} concentration and understand its role in the RyRs regulation. Therefore, the interest to directly measure the free intra-SR Ca^{2+} concentration ($[Ca^{2+}]_{SR}$) has led to the application of a low-affinity Ca^{2+} indicator (Fluo-5N AM) to follow changes of $[Ca^{2+}]_{SR}$ in cardiomyocytes. However the loading of this AM-ester dye into the SR has remained a challenge in freshly isolated mouse cardiomyocytes. Here, we describe an optimized protocol to measure changes of $[Ca^{2+}]_{SR}$ in mouse cardiomyocytes using fluorescent Ca^{2+} indicators and confocal microscopy. The application of this protocol allows to evaluate directly intra-SR Ca^{2+} in real time in various mouse models of cardiac disease, including transgenic animals harboring mutants of RyRs or other Ca^{2+} signaling proteins.

Key words Calcium signaling, Cardiomyocytes, Fluo-5N, Confocal calcium imaging, Cell permeabilization, Sarcoplasmic reticulum, Excitation-contraction coupling

1 Introduction

In the heart Ca^{2+} ions play an important role and are responsible for the muscle contraction. In cardiomyocytes during excitation-contraction (EC) coupling, the depolarization of the plasma membrane first activates L-type Ca^{2+} channels, and the Ca^{2+} entry subsequently triggers Ca^{2+} release from sarcoplasmic reticulum (SR) through ryanodine receptors (RyRs). This mechanism is called Ca^{2+}-induced Ca^{2+} release (CICR) and generates the transient cytosolic Ca^{2+} increase, which is necessary for the cell contraction. These changes in the cytosolic Ca^{2+} concentration can be measured and imaged using well-established intensometric fluorescent Ca^{2+}

Claus W. Heizmann (ed.), *Calcium-Binding Proteins of the EF-Hand Superfamily: From Basics to Medical Applications*, Methods in Molecular Biology, vol. 1929, https://doi.org/10.1007/978-1-4939-9030-6_5,

indicators (e.g., Fluo-3, Fluo-4, Rhod-1, Rhod-2 [1, 2]) or with ratiometric Ca^{2+} dyes (such as Fura-2 or Indo-1 [3, 4]). Abnormal function of the SR Ca^{2+} release mechanism has been demonstrated in a wide range of pathologies (e.g., hypertrophy, cardiomyopathies, heart failure, channelopathies of the RyRs, various forms of cardiac arrhythmias) [5–11]. Posttranslational protein modifications of the RyRs (e.g., phosphorylation, oxidation) or RyR mutations can render these channels abnormally Ca^{2+} sensitive and increase the propensity for spontaneous arrhythmogenic Ca^{2+} releases (e.g., Ca^{2+} waves). In cardiomyocytes, the SR plays an important role as the predominant Ca^{2+} store. The Ca^{2+} concentration in the SR ($[Ca^{2+}]_{SR}$) itself can also modulate RyRs, in addition to the cytosolic activation of the channels occurring with every heartbeat. Diastolic Ca^{2+} release, such as the mentioned Ca^{2+} waves can, for example, be induced by SR Ca^{2+} overload, which is thought to make the RyRs more sensitive for Ca^{2+}. The intra-SR Ca^{2+} concentration can be sensed via an allosteric signal involving calsequestrin, triadin, and junctin [12] or directly recognized by the RyRs [13, 14]. Since any dysfunctions of elements that participate in the SR Ca^{2+} uptake/release balance (e.g., SERCA and RyR, for review *see* [6]) can alter $[Ca^{2+}]_{SR}$ and provoke cardiac Ca^{2+} signaling pathologies, direct spatially and temporally resolved measurements of free $[Ca^{2+}]_{SR}$ have become much more desirable over the last decade. Several indirect methods have been proposed to estimate experimentally the intra-SR Ca^{2+} content based on the amount of Ca^{2+} released from the SR (e.g., measuring the amplitude of a cytosolic Ca^{2+} transient during the application of high concentrations of caffeine, an agonist of RyRs, or by integrating the inward current resulting from the cytosolic Ca^{2+} extrusion mediated by the electrogenic Na^{+}/Ca^{2+} exchanger during that Ca^{2+} transient [15–18]). These methods have considerable limitations. They are static measurements of one point in time, give no spatial and temporal information, and can be affected by Ca^{2+} release events occurring just before the caffeine application. Over the last few years, several chemical low-affinity Ca^{2+} indicators have been developed (e.g., mag-Fluo-4 [19–21], Fluo-5N [9, 22–24]) to directly quantify changes of $[Ca^{2+}]_{SR}$, which occur in the high micromolar range. Fluo-5N has been used to image SR Ca^{2+} in single cardiomyocytes isolated from rabbits [23, 24], dog hearts [9], and mice [25]. Unfortunately, for several years the targeting of the indicators to the SR of mouse cardiomyocytes by the AM-ester loading technique has remained particularly challenging. This is especially unfortunate because of the wide availability of mouse models for cardiac diseases. In this chapter, we introduce an optimized protocol to measure SR Ca^{2+} signals in freshly isolated mouse cardiomyocytes using Fluo-5N AM to load the SR with a Ca^{2+} indicator. One of the main problems in the delivery of the Ca^{2+} dye to the SR is that this has to diffuse across the cytosol and can be de-esterified by cytosolic esterases and remain entrapped in this cell compartment. Another potential problem is the dye uptake by mitochondria, which occupy a large fraction of the cell volume. This dye contamination will generate a detectable fluorescence signal arising from the cytosolic Ca^{2+} transient that interferes with the decreasing intra-SR Ca^{2+} signal, despite the very low Ca^{2+} affinity of Fluo-5N. Therefore, it is advantageous to remove the residual dye contamination from the cytosol. For this purpose we propose three different techniques: removal by dialysis via patch-clamp electrode, irreversible permeabilization with saponin, and reversible permeabilization with streptolysin O. With all these approaches, the direct imaging of SR Ca^{2+} concentrations and cytosolic Ca^{2+} signals in a wide range of mouse models of cardiac disease will allow us to improve the understanding of these diseases.

2 Materials

All solutions must be prepared with ultrapure water (deionized water with a resistance of 18 MΩ cm^{-1} at 25 °C) to avoid contamination with Ca^{2+}. Prepare solutions and reagents at room temperature (unless other conditions are indicated).

2.1 Cell Isolation

1. Ca^{2+}-free modified Tyrode's solution (in mM): 140 NaCl, 1.1 $MgCl_2$, 5.4 KCl, 1 NaH_2PO_4, 5 HEPES, 10 glucose, pH 7.4. Adjust pH with NaOH. Store at 4 °C.
2. Heparin. Prepare stock at 2000 U/mL. Dissolve 66.67 mg of heparin (150 U/mg) in 5 mL of sterilized saline (0.9% NaCl). Store at 4 °C.
3. Enzymatic solution. Protease type XIV (0.21 U/mL, Sigma). Prepare a stock with 105 U/mL, split in 100 μL aliquots, and store them at −20 °C. Collagenase type II (160 U/mL, Worthington, Switzerland). Weight the amount necessary for 50 mL of Ca^{2+}-free modified Tyrode's solution. Dissolve collagenase powder in modified Tyrode's solution, inverting the tube several times, and add 100 μL of protease stock. Avoid the formation of foam. Keep at RT.
4. Use these solutions on the Langendorff system (*see* Fig. 1).

2.2 Ca^{2+} Fluorescent Dyes

2.2.1 Fluo-5N AM: Intra-SR Ca^{2+} Dye with Green Fluorescence Emission

1. Pluronic acid solution (20% w/v). Prepare a stock mixing 0.2 g of Pluronic acid F-127 (Biotium) and 1 mL of anhydrous DMSO. Sonicate the mixture until Pluronic acid is completely dissolved. Split in small aliquots and store at −20 °C. Sonicate frozen aliquots again after defrosting, before use.
2. BSA stock (12% w/v). Dissolve 120 mg of BSA in 1 mL of modified Tyrode's solution (250 μM of Ca^{2+}).

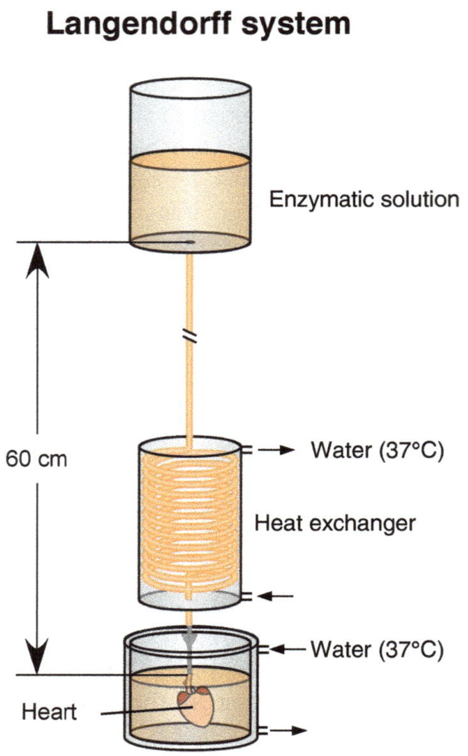

Fig. 1 Scheme of Langendorff system. Solution flow is controlled by gravity. The temperature is adjusted in the heat exchanger to 37 °C by an external heating circulator

3. Prepare a Fluo-5N AM stock solution (44.3 μM) dissolving 50 μg of Fluo-5N AM (Invitrogen) in 10 μL of Pluronic acid solution (20% w/v). Sonicate the mixture during 10 min to obtain a homogeneous solution. Add 40 μL of BSA stock (12%) to 950 μL of modified Tyrode's solution (250 μM of Ca^{2+}) to obtain a final BSA concentration of 0.5%. Mix this solution and the Fluo-5N AM-Pluronic acid solution mixture until you obtain a homogenous solution (*see* **Note 1**). Separate in aliquots and store at −20 °C until experiment day. Use them within 1 week after preparation (*see* Fig. 2a).

2.2.2 Rhod-2 Salt: Cytosolic Ca^{2+} Dye with Red Fluorescence Emission

1. Prepare a stock of Rhod-2 (5 mM). Dissolve 1 mg of Rhod-2 tripotassium salt (Biotium) in 201 μL of deionized water. Make small aliquots and store at −20 °C.

2.3 Patch-Clamp Technique

1. Patch pipette internal solution (in mM): 120 Cs-Asp, 20 TEA-Cl, 5 K_2ATP, 8 NaCl, 5.6 $MgCl_2$, and 20 HEPES, pH 7.2; adjust with CsOH (Free Mg 0.74 mM). Keep at −20 °C. On the day of the experiment, add 10 μL of Rhod-2 (5 mM stock) to 1 mL of patch pipette internal

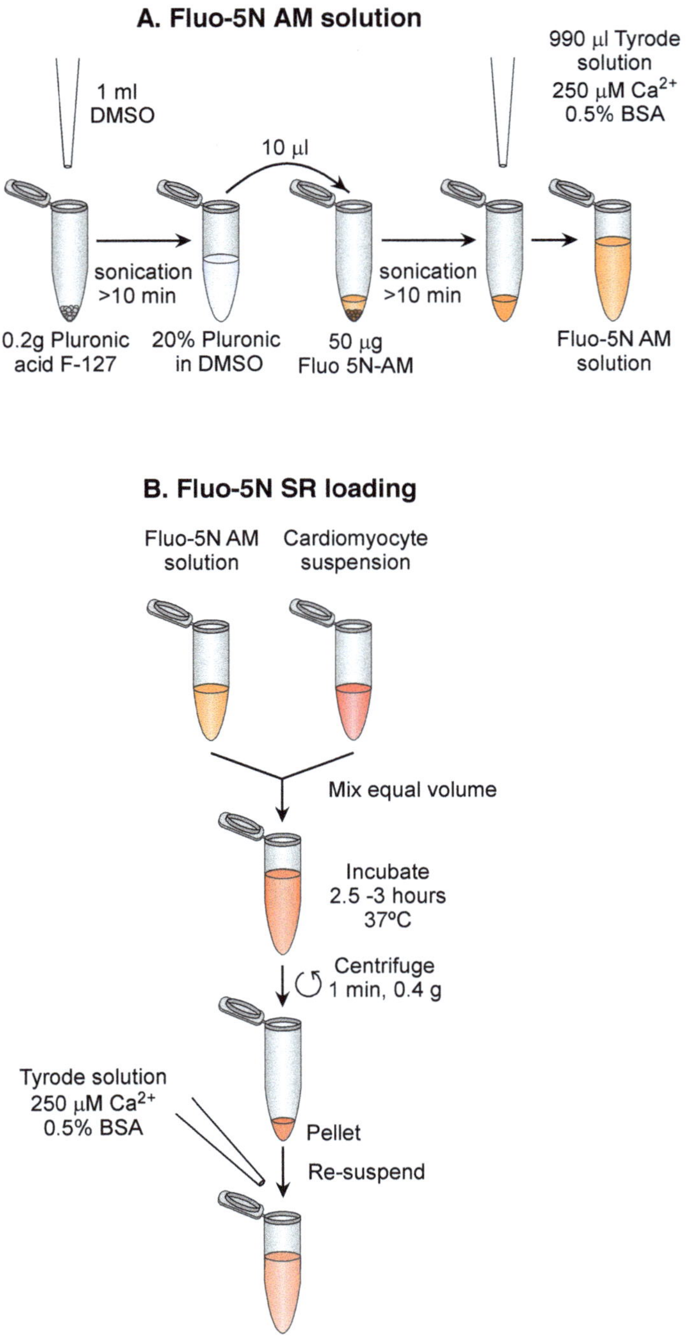

Fig. 2 Protocols to prepare stock of Fluo-5N AM (**a**) and to load Fluo-5N into the SR (**b**). For step-by-step details, *see* text

solution to get a final concentration of 50 μM Rhod-2. Filter this solution using a 0.2 μm mesh. Keep on ice during the experiment.

2. Prepare patch-clamp pipettes from glass of your choice with a tip diameter of approximately 1 μm (pipette resistance between 2 and 2.5 MΩ). Filamented glass makes filling the tip easier.

2.4 Permeabilization with Saponin

1. Prepare Ca^{2+}-free modified Tyrode's solution supplemented with 100 μM of EGTA to complex residual Ca^{2+}. Add 3.8 mg of EGTA to 100 mL of Ca^{2+}-free modified Tyrode's solution. pH 7.4, continuously adjust pH with NaOH while dissolving EGTA.
2. Saponin-free internal solution (in mM): 120 K-Asp, 3 K_2ATP, 3 $MgCl_2$, 0.1 EGTA, 10 phosphocreatine, 10 HEPES, 5 U/mL creatine phosphokinase, pH 7.2, adjust with KOH. Filter this solution using a mesh of 0.4 μm and store at −20 °C.
3. Saponin-containing permeabilization solution (in mM): 100 K-Asp, 20 KCl, 3.7 $MgCl_2$, 1 EGTA, 10 HEPES and 0.005% saponin, pH 7.2, adjust with KOH.
4. Final saponin-free internal solution (50–100 nM free Ca^{2+}). $[Ca^{2+}]_{free}$ can be either calculated [26] or better be adjusted by measuring the Ca^2 fluorescence signal of the final solution, using a ratiometric Ca^{2+} dye (e.g., Indo-1) and a fluorescence spectrofluorometer (e.g., NanoDrop), after taking as a reference a calibrated $[Ca^{2+}]$ curve (*see* **Note 2**). After $[Ca^{2+}]$ adjustment dextran 40 K, 4% can be added to increase the density of the solution and reduce lattice swelling and the extrusion of small proteins. Use 10 μM of blebbistatin or 10 mM of BDM to prevent contraction and reduce motion artifacts during the experiments.

2.5 Reversible Permeabilization with Streptolysin O (SLO)

1. Prepare a stock of SLO (8000 U/mL, with a specific activity of 41.28 U/mL). Dissolve the corresponding amount of SLO in cold deionized water. Split it in aliquots of 25 μL and store at −20 °C until the experimental day.
2. Make a stock of reducing agent dithiothreitol (DTT; 200 mM). Add 30.9 mg of DTT to 1 mL of water. Store at −20 °C.
3. Permeabilization solution (in mM): 100 K-Asp, 20 KCl, 3.7 $MgCl_2$, 1 EGTA and 10 HEPES, pH 7.2, adjust with KOH.

3 Methods

3.1 Cell Isolation

1. Inject 4 μL of heparin i.p. per gram of mouse body weight to avoid blood clot formation during surgery.

2. After 10 min, euthanize the mouse by an approved method.
3. The heart should be quickly removed and placed in cold Ca^{2+}-free modified Tyrode's solution.
4. Remove remainder of other tissues surrounding the aorta (in ice).
5. Cannulate the aorta and perfuse slowly with 1 mL of a Ca^{2+}-free solution supplemented with 100 U/mL of heparin.
6. Mount the cannula on a Langendorff system (*see* **Note 3**).
7. In order to remove the rest of the blood from the myocardium, perfuse with Ca^{2+}-free solution (about 10 mL or for 2 min) at 37 °C, non-recirculating.
8. Perfuse with enzymatic solution for 15 min (at 37 °C, recirculating). Keep the heart bathed in enzymatic solution all the time (*see* Fig. 1).
9. Separate ventricles and place in a petri dish with 5 mL of Ca^{2+}-free modified Tyrode's solution at 37 °C, and mince in small pieces.
10. Place the small tissue pieces into fresh Ca^{2+}-free modified Tyrode's solution.
11. Gently shake the pieces to mechanically isolate the cells.
12. Remove tissue pieces by filtering with a mesh of 100 μm pore diameter.
13. Slowly raise the Ca^{2+} concentration from nominally 0 Ca^{2+} to 250 μM within 1 h (4 × 12.5 μM, 4 × 25 μM, and 2 × 50 μM).
14. Store the cell suspension at room temperature (21–22 °C) until use.

For more details for mouse cardiomyocyte isolation, *see* [27].

3.2 Fluo-5N AM Loading

1. Slowly thaw (at 4 °C) the aliquot of Fluo-5N AM solution. Vortex and spin the solution to the tip of the vial.
2. Check cell survival after external Ca^{2+} increase. If more than 50% of the cells show a hypercontracted popcorn shape, take 1 mL of cell suspension, and leave cells to decant for 30 min. Then remove carefully 900 μL of supernatant, and add 400 μL of fresh modified Tyrode's solution (250 μM Ca^{2+}) supplemented with BSA (0.5%) (*see* **Note 4**).
3. Mix an equal volume of myocyte suspension and Fluo-5N AM solution. Avoid damage to the cells (*see* **Note 5**).
4. Incubate mixture during 2.5–3 h in an incubator at 37 °C without CO_2 (*see* **Note 6**).
5. Centrifuge at 0.4 × *g* during 1 min, and resuspend the pellet in fresh modified Tyrode's solution (250 μM Ca^{2+}). Keep at RT and protect from light until use in the experiment (*see* Fig. 2b).

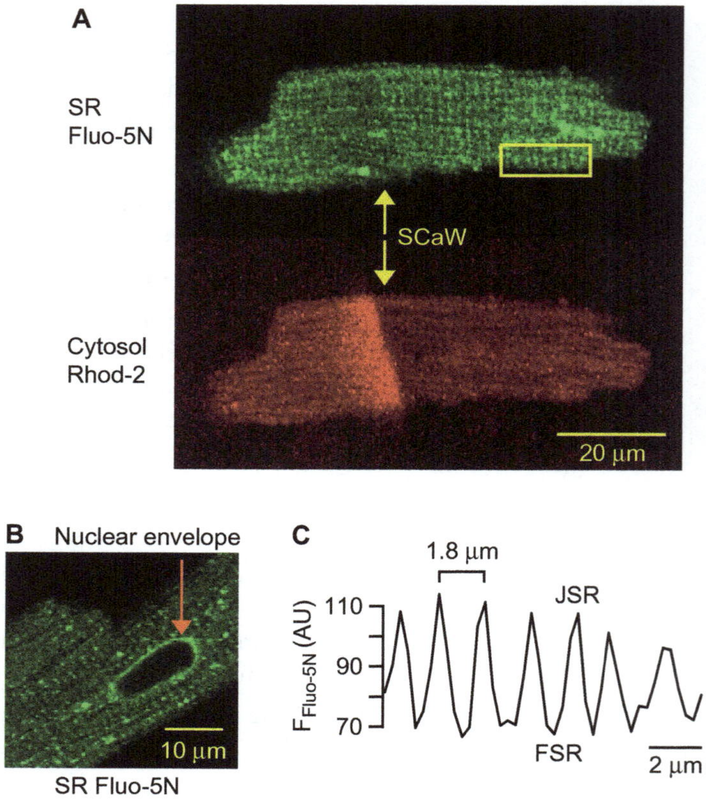

Fig. 3 Fluo-5N cellular distribution. (**a**) 2D images show intra-SR Fluo-5N (green) and cytosolic Rhod-2 (red) distribution in a permeabilized mouse ventricular myocyte. Arrows indicate a spontaneous Ca^{2+} wave (SCaW). (**b**) Fluo-5N is also loading into the nuclear envelope showing a bright ring surrounding the nucleus. (**c**) Pattern of Fluo-5N intensity measured along the longitudinal cell axis (yellow box in panel **a**)

Cells with a good Fluo-5N loading present a pattern of fluorescence signal typical of SR, showing bright transversal lines separated by approximately 1.8 μm and a halo in the nuclear envelope surrounding the nucleus (*see* Fig. 3).

3.3 Removal of Cytosolic Fluo-5N AM Contamination

The cytosol contains nonspecific esterases that can cleave the lipophilic blocking groups from Fluo-5N AM and entrap it in this compartment. Since the residual dye can bind to the cytosolic Ca^{2+} during the Ca^{2+} release from the SR, the resulting increase of the cytosolic fluorescence signal can obscure the decreasing signal in the SR (*see* Fig. 4 and [25]). Therefore, it is necessary to remove this Fluo-5N contamination from the cytosol. Three techniques can be used to remove the dye contamination from the cytosol: cell dialysis using a patch clamp, cell permeabilization with saponin, or reversible cell permeabilization with SLO.

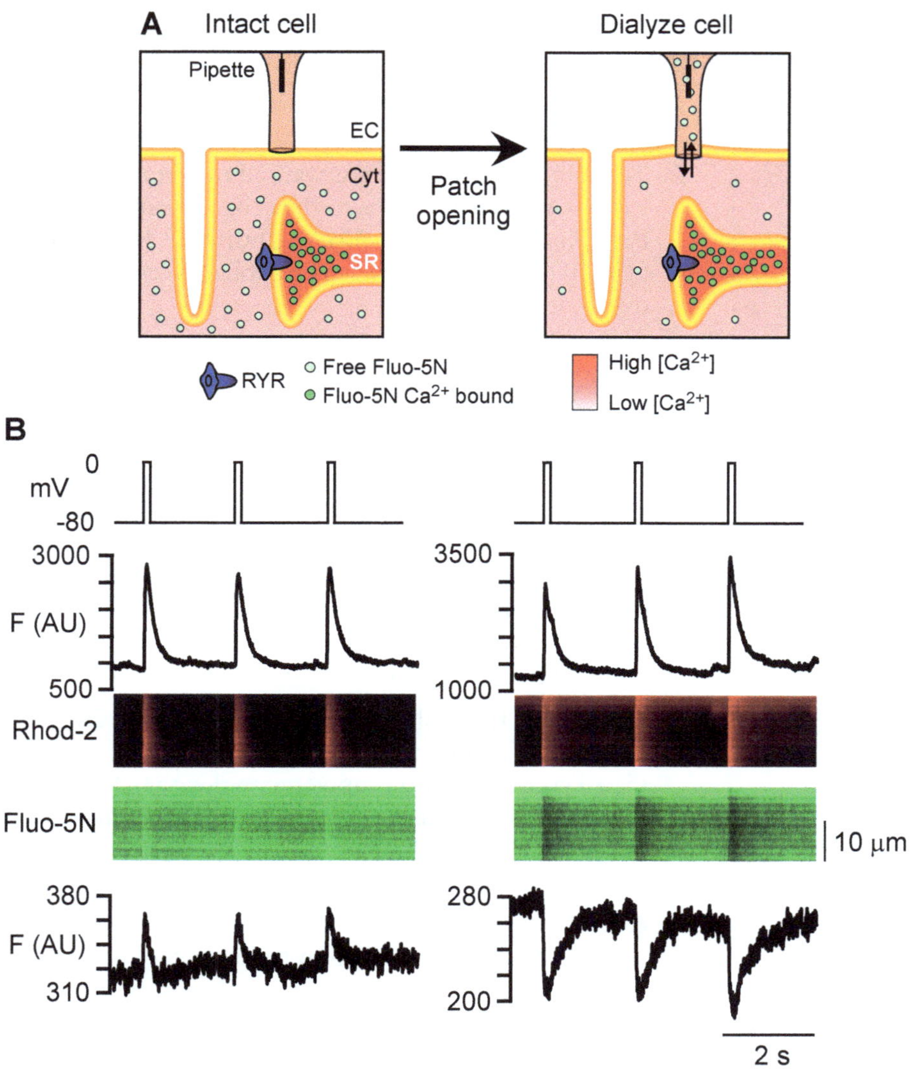

Fig. 4 Patch clamp in the whole-cell configuration to remove Fluo-5N contamination from the cytosol. (**a**) Scheme illustrates how cell dialysis via a patch pipette can reduce the Fluo-5N entrapped into the cytosol. (**b**) Representative line-scan images and fluorescence traces of cytosolic Ca^{2+} (Rhod-2 in red) and SR Ca^{2+} (Fluo-5N in green) during voltage-clamp depolarizations immediately after patch opening (left panel) and after 10 min of cell dialysis (right panel). Note early overlap and later disappearance of the contaminating cytosolic Ca^{2+} signal

3.3.1 Cell Dialysis Using Patch Clamp

The patch-clamp technique, in its whole-cell configuration, in addition to control the membrane voltage also permits us to exchange solution between the cytosol and the patch pipette, typically used to load the cells with non-permeable chemical compounds, peptides, antibodies, etc. This feature can also be used to dialyze Fluo-5N out and minimize any residual dye in the cytosol (*see* Fig. 4a). A large diameter of the patch pipette tip is an advantage for this procedure. For details regarding the patch-clamp technique, *see* [28].

1. Once the gigaseal is open, dialyze the cell for at least 10 min (*see* **Note 7**).
2. To verify that Fluo-5N was washed out from the cytosol, apply a train of depolarizations to the cell, and simultaneously record the Fluo-5N signal. Also a puff of high concentration (10–20 mM) of caffeine can be applied to evaluate the cytosolic dye contamination.
3. If you observe a transient increase of Fluo-5N fluorescence (*see* Fig. 4b left panel), the cell has still Fluo-5N contamination, and longer dialysis is required until the cytosolic Fluo-5N signal disappears completely (*see* Fig. 4b right panel).

3.3.2 Cell Permeabilization with Saponin

For intracellular Ca^{2+} signals not requiring electrical activity and a functioning cell membrane (e.g., for the analysis of Ca^{2+} waves and Ca^{2+} sparks), the sarcolemma can be permeabilized with saponin to remove the residual cytosolic Fluo-5N by simple diffusion. Saponin binds to cholesterol molecules in the sarcolemma generating micelles and pores in the membrane (*see* Fig. 5a, b).

1. Centrifuge the cells at 0.4 × *g* for 1 min, 24 °C. Remove carefully the supernatant, and add 100 μL of Ca^{2+}-free modified Tyrode's solution supplemented with 0.1 mM of EGTA to bind the Ca^{2+} from the external solution. Resuspend the pellet (*see* **Note 8**).
2. Centrifuge the cells at 0.4 × *g* for 1 min, 24 °C. Remove carefully the supernatant and resuspend the pellet in 100 μL saponin-free internal solution to lower the high concentration of Na^{+}, and place the cells in a solution which more closely corresponds to the intracellular medium before opening pores in the sarcolemma.
3. Centrifuge the cells at 0.4 × *g* for 1 min, 24 °C. Remove carefully the supernatant, and resuspend the pellet in 100 μL saponin-containing permeabilization solution (0.005% of saponin). Incubate for up to 30 s at RT.
4. Centrifuge the cells at 0.4 × *g* for 1 min, 24 °C. Rapidly but carefully remove the supernatant, and resuspend the pellet in final saponin-free internal solution (50–100 nM of Ca^{2+}).

Verify that there is no cytosolic spike in the Fluo-5N signal before the SR depletion (*see* Fig. 5c).

3.3.3 Reversible Cell Permeabilization with Streptolysin O

Membrane permeabilization is an efficient method to remove the dye contamination from the cytosol, but for studies of SR Ca^{2+} release signals modulated by electrical cell activity or by receptors located in the sarcolemma, the integrity of this membrane is crucial.

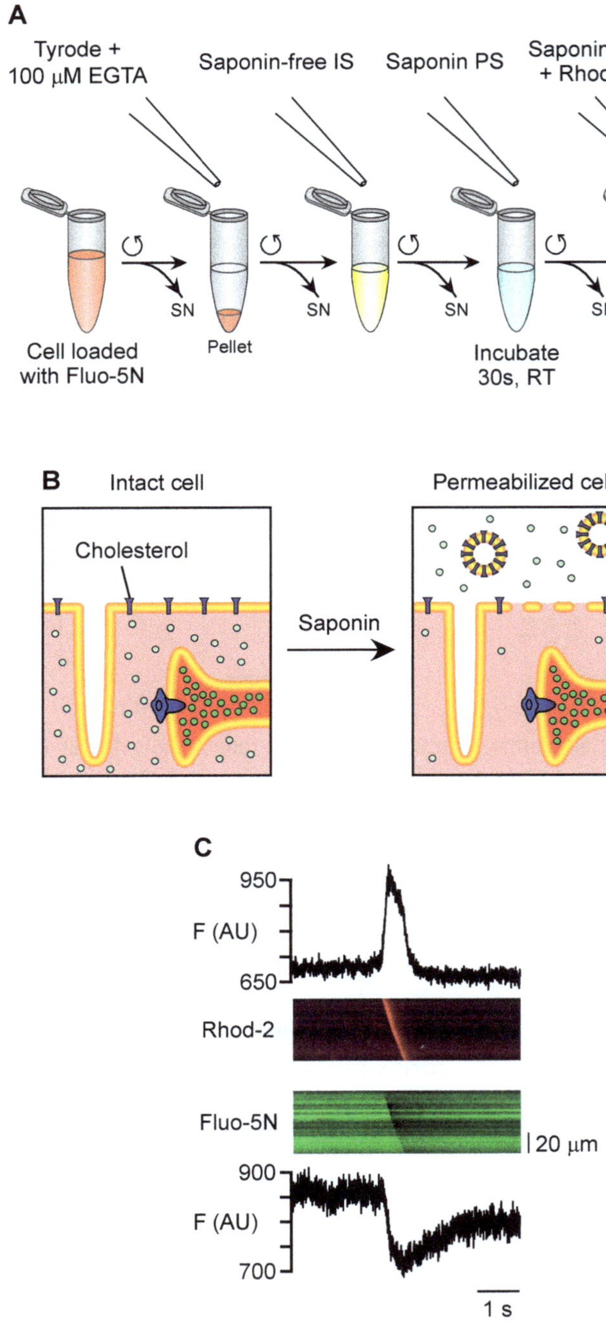

Fig. 5 Permanent permeabilization with saponin completely removes cytosolic Fluo-5N contamination. (**a**) Protocol for cardiomyocyte permeabilization with saponin. (**b**) Saponin binds to cholesterol in the plasma membrane generating pores. (**c**) Representative line-scan (cytosolic Ca^{2+} and SR Ca^{2+} signals recorded simultaneously) and traces of a spontaneous wave. ↺ = centrifuge 1 min at $0.4 \times g$. *SN* supernatant, *IS* internal solution, *PS* permeabilization solution

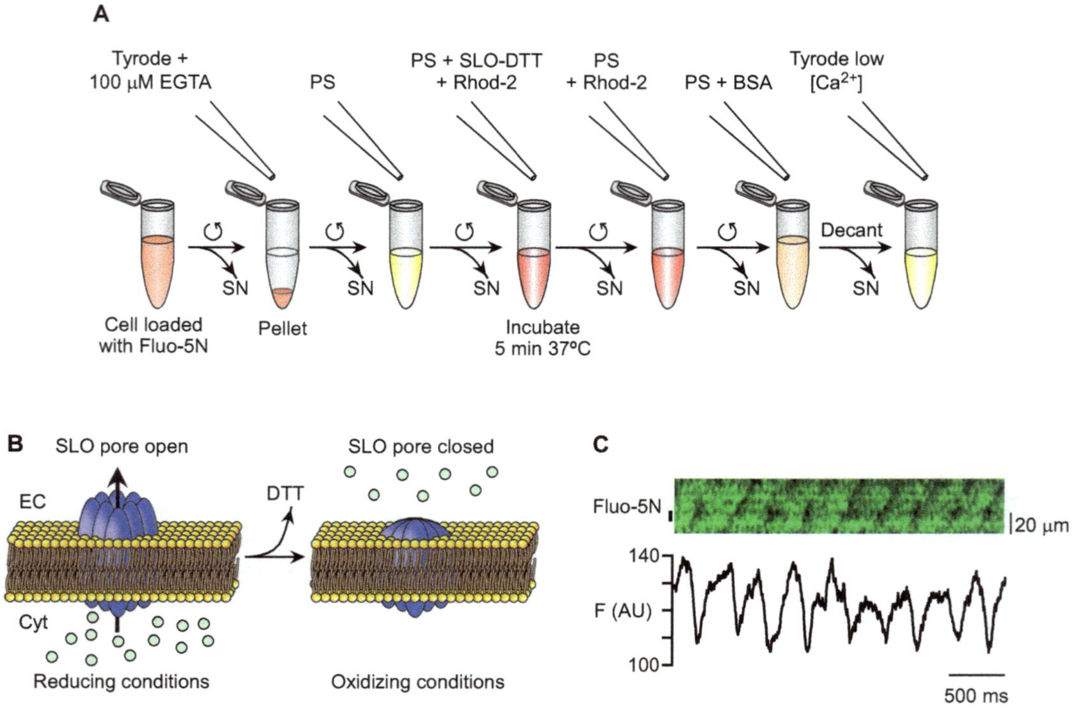

Fig. 6 Reversible permeabilization with SLO to reduce the cytosolic Fluo-5N contamination maintaining the properties of the plasma membrane. (**a**) Protocol for cardiomyocyte reversible permeabilization using SLO. (**b**) Contaminating Fluo-5N can diffuse out of the cell though the temporal plasma membrane pores created by SLO under reducing conditions. (**c**) Spontaneous SR Ca^{2+} waves recorded in 6 mM extracellular Ca^{2+}. The absence of fluorescence spikes at the beginning of Ca^{2+} waves suggests successful removal of cytosolic Fluo-5N contamination. ↻ = centrifuge 1 min at 0.4 × *g*. *SN* supernatant, *PS* permeabilization solution

Therefore, the reversible permeabilization approach using SLO would be a good solution for these measurements. SLO is a hemolytic toxin, which binds to sarcolemmal cholesterol molecules and forms pores. But these pores in the sarcolemma form only under reducing conditions. Therefore, they can be transient and reversible (*see* Fig. 6a, b) [29].

1. Activation of SLO. Mix one volume of SLO (8000 U/mL) and two of the reducing agent DTT (200 mM) (ratio of 1:2). Incubate 2 h at RT. Add 15 μL of the mixture to 400 μL of permeabilization solution. Final concentration of DTT is 5 mM and SLO 0.1 U/mL.
2. Centrifuge the cells at 0.4 × *g* for 1 min, 24 °C. Remove carefully the Tyrode's solution containing Ca^{2+}, and add 200 μL of Ca^{2+}-free modified Tyrode's solution supplemented with 100 μM of EGTA.
3. Centrifuge the cells at 0.4 × *g* for 1 min, 24 °C. Carefully remove the supernatant, and add 100 μL of permeabilization solution supplemented with 1 mM EGTA.

4. Centrifuge the cells at 0.4 × *g* for 1 min, 24 °C. Carefully remove the supernatant, and resuspend the pellet in permeabilization solution with 0.2 U/mL of streptolysin O, 5 mM of DTT, and 25 μM of Rhod-2 salt. Incubate for 5 min at 37 °C.
5. Centrifuge the cells at 0.4 × *g* for 1 min, 24 °C. Rapidly but carefully remove the supernatant, and resuspend the pellet in permeabilization solution (1 mM EGTA and 25 μM of Rhod-2 salt).
6. Centrifuge the cells at 0.4 × *g* for 1 min, 24 °C. Remove the supernatant, and resuspend the pellet in permeabilization solution (1 mM EGTA, BSA 5%). Decant the cells.
7. Remove the supernatant, and resuspend the pellet Ca^{2+}-free modified Tyrode's solution.
8. Increase slowly the $[Ca^{2+}]_0$ in several steps until it reaches 1.8 mM (*see* **Note 9**).

Figure 6c shows a confocal line-scan image of SR Ca^{2+} signal recorded in the presence of high $[Ca^{2+}]$ (6 mM) in the extracellular solution to induce SR Ca^{2+} overload and spontaneous Ca^{2+} waves. The absence of a fluorescence overshoot at the beginning of waves suggests successful removal of residual cytosolic Fluo-5N.

3.4 Ca^{2+} Imaging

Ca^{2+} release signals in cardiomyocytes are relatively short events (20–500 ms) with rapid kinetics. Therefore a fast image recording system is necessary. Some subcellular Ca^{2+} events can be highly localized (e.g., spark/blink pairs); thence an excellent and near diffraction-limited spatial resolution is also required for the recording of these events. Laser-scanning confocal microscopes offer a good compromise between temporal and spatial resolutions, particularly in the line-scan mode. In the line-scan mode, one individual line is repeatedly and rapidly scanned at a high frequency, typically at around 500 Hz. The resulting image then contains one spatial and one temporal dimension (e.g., *see* Fig. 4). Below you can find some tips and recommendations to obtain a good Ca^{2+} image recording:

1. Simultaneous recording of the cytosolic and SR signals helps to identify Ca^{2+} events. Fluo-5N can be excited at 488 nm and Rhod-2 at 560 nm (*see* **Note 10**). To minimize spectral cross talk and bleed-through, Fluo-5N fluorescence (i.e., the intra-SR signal) is best detected between 515 and 560 nm and Rhod-2 > 585 nm (*see* **Note 11**).
2. Reduce the laser power as much as possible to minimize photobleaching of Fluo-5N. The amount of Fluo-5N entrapped in the SR is very small. We employ GaAsP PMTs with improved quantum yield to minimize the excitation power requirements.

3. Sampling times of 4 μs per pixel (512 pixels per line) result in a good balance between temporal resolution, signal-to-noise ratio due to photon statistics, and dye photobleaching.
4. Avoid recording longer than 10 s from the same line to reduce Fluo-5N photobleaching.

3.5 Ca^{2+} Image Analysis

For Ca^{2+} image processing and signal analysis, the freely available ImageJ software can be used [30].

1. Background fluorescence (and other offsets) should be removed from all images. For intact cells choose a ROI outside of the cells, measure the mean fluorescence intensity, and subtract that value from all images of the same cell (*see* **Note 12**). For permeabilized cells the background fluorescence subtraction for Fluo-5N can be carried out like for intact cells. For Rhod-2 the background fluorescence has to be measured in a solution without Ca^{2+} before the experiment, because the final internal solution has a $[Ca^{2+}]$ of 100 nM.
2. To study the dynamics of Ca^{2+} events, select a ROI, and make a profile along the time axis.
3. The Fluo-5N signal normally presents a slow decay since this dye is photobleached, particularly during long excitation periods. To correct this phenomenon, the signal trace can be fitted with a single exponential function (avoid Ca^{2+} events). After this, divide the raw signal by the fitting curve.
4. Since Fluo-5N and Rhod-2 are not ratiometric dyes, it is necessary to normalize the signal to compare between experiments. For Rhod-2, the signal can be expressed as $\Delta F/F_0$, where F is the signal in each instant and F_0 is the fluorescence baseline measured in a period without Ca^{2+} events (*see* Fig. 7a). In the case of Fluo-5N, the minimal fluorescence value during caffeine application can be used to normalize the SR signal for variable Fluo-5N loading, since the absolute fluorescence level reached in caffeine mainly depends on the amount of Fluo-5N entrapped in the SR (*see* **Note 13**). With Fluo-5N the Ca^{2+} signal can be expressed as $\Delta F/F_{\mathrm{Caff}}$ (*see* Fig. 7b for details).
5. For detection of SR Ca^{2+} blinks, make a self-ratio (*see* **Note 14**) to the Fluo-5N image, taking a ROI without Ca^{2+} events as reference. This removes the striation pattern arising from the sarcomeric dye distribution. After this procedure, smooth the image to improve the signal-to-noise ratio (*see* Fig. 8a).
6. When spontaneous Ca^{2+} waves are recorded to study the dynamics of Ca^{2+} release, the line-scan images can be de-skewed (*see* **Note 15**) and aligned before extracting the profile (*see* Fig. 8b). De-skewing restores the correct time course of the Ca^{2+} signals.

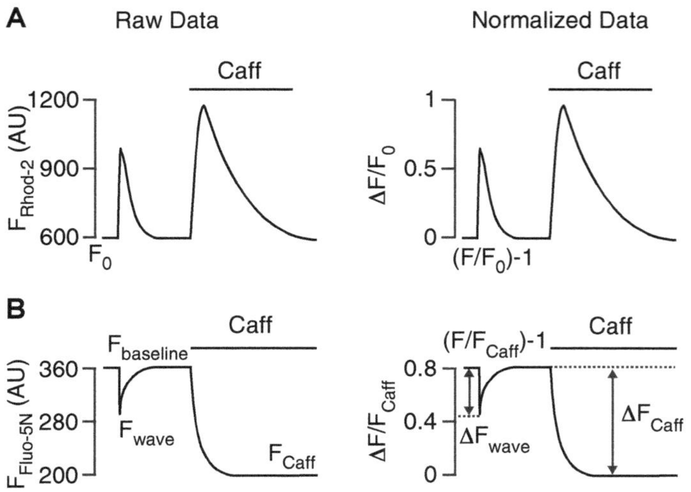

Fig. 7 Normalization of cytosolic (**a**) and intra-SR (**b**) Ca^{2+} signals. (**a**) Cytosolic fluorescence signal (left) is normalized with respect to F_0, which is the resting fluorescence signal before Ca^{2+} events and expressed as $\Delta F/F_0$ (right). (**b**) In the case of intra-SR Ca^{2+} signal, the reference for normalization of the fluorescence (left) is the fluorescence level during application of a high concentration of caffeine (10 mM), and the signal is expressed as $\Delta F/F_{Caff}$ (right)

4 Notes

1. Avoid foam formation during mixing since this impairs subsequent Fluo-5N loading into the SR.
2. We use a Ca^{2+} calibration buffer kit (Molecular Probes/ Thermo Fisher) to derive the Ca^{2+} calibration curve in the range 0–39 μM. Increase Ca^{2+} concentration slowly and ensure that there are no precipitations. A ratiometric Ca^{2+} dye must be used to compare between samples.
3. Before mounting the heart in the Langendorff system, remove all bubbles from the perfusion system since these can block small coronary arteries, impairing the retrograde perfusion and the enzymatic tissue digestion.
4. Lysed cells release cytosolic esterases to the extracellular medium, and these can de-esterify Fluo-5N AM before it can enter in the cells, thereby decreasing the efficiency of dye loading.
5. To avoid damage the myocytes, use pipette tips with a tip diameter of 2–3 mm. For example, cut a 200 μL tip and fire polish to remove sharp borders.

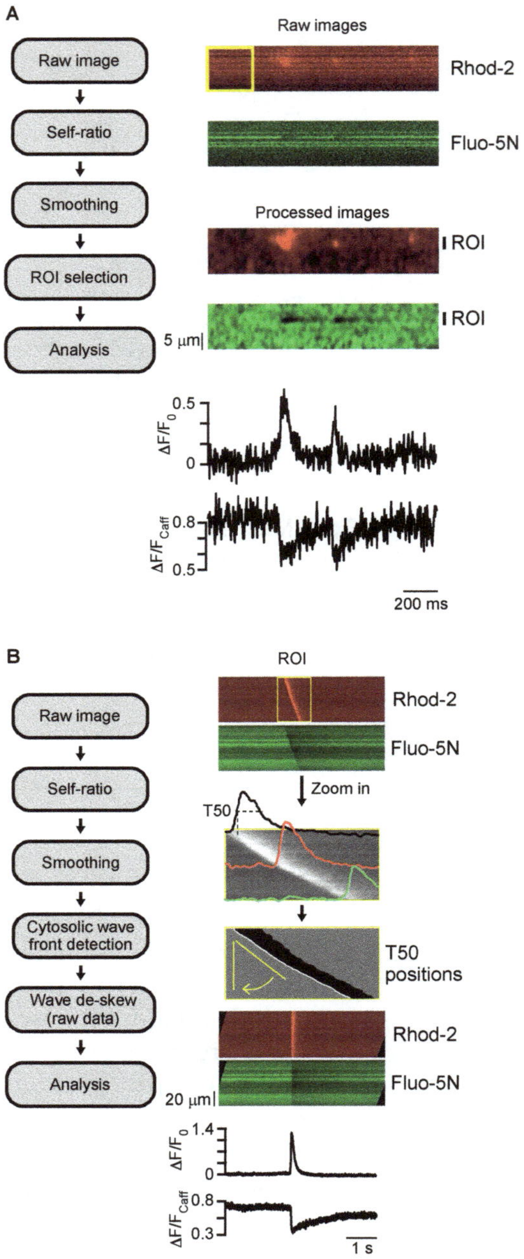

Fig. 8 Ca^{2+} image processing. (**a**) For better detection of local intra-SR Ca^{2+} signals (blinks), raw images are processed by removing the striated pattern corresponding to SR dye distribution (by performing a line-wise self-ratio) and by reducing the noise (by 3 × 3 pixels smoothing). The left panel shows the image processing flowchart. Right panel shows cytosolic and intra-SR signals before and after image processing. Yellow box represents the ROI of resting fluorescence selected as the reference for the self-ratio procedure. Lower panel corresponds to cytosolic and intra-SR Ca^{2+} traces obtained from the raw

6. Don't use an incubator with 5% of CO_2 since the pH of the solution can change.
7. To favor the exchange between cytosol and pipette solution, keep the patch well open and switch between positive and negative pipette pressure several times.
8. If the Ca^{2+} concentration in the initial solution is higher than 250 μM, it is recommended to repeat **step 1** to be sure that Ca^{2+} is lowered in the solution before the cells are permeabilized.
9. High $[Ca^{2+}]_0$ can be applied to check the integrity of the sarcolemma. If the membrane is damaged, cytosolic $[Ca^{2+}]$ will increase dramatically to toxic levels. Long exposition to this high Ca^{2+} will activate hypercontraction and the mechanisms of death cell.
10. Rhod-2 can also be excited with a 488-nm laser with lower excitation efficiency (5–10% of maximal absorption).
11. To reduce the interfering signal from Fluo-5N in the Rhod-2 channel, you can use an emission long-pass filter (585 nm) that allows to record only signals with a wavelength higher than 585 nm.
12. If during an experiment the application of some drug interferes with the fluorescence, a new background measurement is necessary in the presence of the drug.
13. When converting the Fluo-5N signals into absolute intra-SR Ca^{2+} concentrations, a small quench of the Fluo-5N fluorescence by caffeine would need to be considered.
14. Self-ratio. For each x-pixel, determine a mean of n t-lines (avoid Ca^{2+} events), and divide each t-line by that mean.
15. Ca^{2+} wave de-skew. We have developed a macro, running under ImageJ, based on the detection of the time when cytosolic Ca^{2+} signal reaches 50% of the Ca^{2+} wave amplitude (T50). This is done for each x-pixel along the scanned line. Subsequently, the macro shifts and synchronizes each individual pixel row, according to the T50 position to align them to the same time. Since the signal-to-noise ratio of the Fluo-5N signal from the SR is smaller than that of the cytosolic Rhod-2 signal, the cytosolic T50 positions can be taken as the reference, also to de-skew the Fluo-5N line-scan image.

Fig. 8 (continued) data and ROI selected. (**b**) (left panel) Flowchart summarizing the processing of the Ca^{2+} images to study the global dynamics of SR Ca^{2+} release during a Ca^{2+} wave. Images are de-skewed using a macro based on the detection of the 50% gray-scale threshold (T50) of the cytosolic Ca^{2+} wave (white pixels in the middle right panel). De-skewing restores the real time course of the Ca^{2+}signals during the wave. Compare with traces in 5C

Acknowledgment

This work was supported by a grant from the Swiss National Science Foundation (grant 156375) and the Microscopy Imaging Center (MIC) of the University of Bern.

References

1. Kao JP, Harootunian AT, Tsien RY (1989) Photochemically generated cytosolic calcium pulses and their detection by fluo-3. J Biol Chem 264:8179–8184
2. Minta A, Kao JP, Tsien RY (1989) Fluorescent indicators for cytosolic calcium based on rhodamine and fluorescein chromophores. J Biol Chem 264:8171–8178
3. Grynkiewicz G, Poenie M, Tsien RY (1985) A new generation of Ca^{2+} indicators with greatly improved fluorescence properties. J Biol Chem 260:3440–3450
4. Peeters GA, Hlady V, Bridge JH, Barry WH (1987) Simultaneous measurement of calcium transients and motion in cultured heart cells. Am J Phys 253:H1400–H1408
5. Molkentin JD, Lu JR, Antos CL, Markham B, Richardson J, Robbins J, Grant SR, Olson EN (1998) A calcineurin-dependent transcriptional pathway for cardiac hypertrophy. Cell 93:215–228
6. Bers DM, Eisner DA, Valdivia HH (2003) Sarcoplasmic reticulum Ca^{2+} and heart failure: roles of diastolic leak and Ca^{2+} transport. Circ Res 93:487–490
7. Lindner M, Erdmann E, Beuckelmann DJ (1998) Calcium content of the sarcoplasmic reticulum in isolated ventricular myocytes from patients with terminal heart failure. J Mol Cell Cardiol 30:743–749
8. Hobai IA, O'Rourke B (2001) Decreased sarcoplasmic reticulum calcium content is responsible for defective excitation-contraction coupling in canine heart failure. Circulation 103:1577–1584
9. Kubalova Z, Terentyev D, Viatchenko-Karpinski S, Nishijima Y, Györke I, Terentyeva R, da Cuñha DNQ, Sridhar A, Feldman DS, Hamlin RL et al (2005) Abnormal intrastore calcium signaling in chronic heart failure. Proc Natl Acad Sci U S A 102:14104–14109
10. Benkusky NA, Farrell EF, Valdivia HH (2004) Ryanodine receptor channelopathies. Biochem Bioph Res Comm 322:1280–1285
11. Betzenhauser MJ, Marks AR (2010) Ryanodine receptor channelopathies. Pflugers Arch - Eur J Physiol 460:467–480
12. Györke I, Hester N, Jones LR, Györke S (2004) The role of calsequestrin, triadin, and junctin in conferring cardiac ryanodine receptor responsiveness to luminal calcium. Biophys J 86:2121–2128
13. Chen W, Wang R, Chen B, Zhong X, Kong H, Bai Y, Zhou Q, Xie C, Zhang J, Guo A et al (2014) The ryanodine receptor store-sensing gate controls Ca^{2+} waves and Ca^{2+}-triggered arrhythmias. Nat Med 20:184–192
14. Jiang D (2005) Enhanced store overload-induced Ca^{2+} release and channel sensitivity to luminal Ca^{2+} activation are common defects of RyR2 mutations linked to ventricular tachycardia and sudden death. Circ Res 97:1173–1181
15. Callewaert G, Cleemann L, Morad M (1989) Caffeine-induced Ca^{2+} release activates Ca^{2+} extrusion via Na^{+}-Ca^{2+} exchanger in cardiac myocytes. Am J Phys 257:C147–C152
16. Varro A, Negretti N, Hester SB, Eisner DA (1993) An estimate of the calcium content of the sarcoplasmic reticulum in rat ventricular myocytes. Pflugers Arch - Eur J Physiol 423:158–160
17. Negretti N, Varro A, Eisner DA (1995) Estimate of net calcium fluxes and sarcoplasmic reticulum calcium content during systole in rat ventricular myocytes. J Physiol 486:581–591
18. Díaz ME, Trafford AW, O'Neill SC, Eisner DA (1997) Measurement of sarcoplasmic reticulum Ca^{2+} content and sarcolemmal Ca^{2+} fluxes in isolated rat ventricular myocytes during spontaneous Ca^{2+} release. J Physiol 501:3–16
19. Shmigol AV, Eisner DA, Wray S (2001) Simultaneous measurements of changes in sarcoplasmic reticulum and cytosolic $[Ca^{2+}]$ in rat uterine smooth muscle cells. J Physiol 531:707–713
20. Valverde CA, Kornyeyev D, Ferreiro M, Petrosky AD, Mattiazzi A, Escobar AL (2010) Transient Ca^{2+} depletion of the sarcoplasmic reticulum at the onset of reperfusion. Cardiovasc Res 85:671–680
21. Kornyeyev D, Reyes M, Escobar AL (2010) Luminal Ca^{2+} content regulates intracellular Ca^{2+} release in subepicardial myocytes of intact

beating mouse hearts: effect of exogenous buffers. Am J Physiol Heart Circ Physiol 298: H2138–H2153

22. Kabbara AA, Allen DG (2001) The use of the indicator fluo-5N to measure sarcoplasmic reticulum calcium in single muscle fibres of the cane toad. J Physiol 534:87–97
23. Shannon TR, Guo T, Bers DM (2003) Ca^{2+} scraps: local depletions of free [Ca^{2+}] in cardiac sarcoplasmic reticulum during contractions leave substantial Ca^{2+} reserve. Circ Res 93:40–45
24. Brochet DXP, Yang D, Di Maio A, Lederer WJ, Franzini-Armstrong C, Cheng H (2005) Ca^{2+} blinks: rapid nanoscopic store calcium signaling. Proc Natl Acad Sci U S A 102:3099–3104
25. Fernández-Tenorio M, Niggli E (2016) Real-time intra-store confocal Ca^{2+} imaging in isolated mouse cardiomyocytes. Cell Calcium 60:331–340
26. Bers DM, Patton CW, Nuccitelli R (2010) A practical guide to the preparation of Ca^{2+} buffers. In: Whitaker M (ed) Calcium in Living Cells. Methods in Cell Biology, vol 99. Elsevier Inc, Amsterdam, pp 1–26
27. Louch WE, Sheehan KA, Wolska BM (2011) Methods in cardiomyocyte isolation, culture, and gene transfer. J Mol Cell Cardiol:1–11
28. Molleman A (2003) Patch clamping: an introductory guide to patch clamp electrophysiology. John Wiley & Sons, Chichester, UK
29. Fawcett JM, Harrison SM, Orchard CH (1998) A method for reversible permeabilization of isolated rat ventricular myocytes. Exp Physiol 83:293–303
30. Schneider CA, Rasband WS, Eliceiri KW (2012) NIH Image to ImageJ: 25 years of image analysis. Nat Meth 9:671–675

Chapter 6

The Use of Complementary Luminescent and Fluorescent Techniques for Imaging Ca^{2+} Signaling Events During the Early Development of Zebrafish (*Danio rerio*)

Sarah E. Webb and Andrew L. Miller

Abstract

We have visualized many of the Ca^{2+} signaling events that occur during the early stages of zebrafish development using complementary luminescent and fluorescent imaging techniques. We initially microinject embryos with the luminescent Ca^{2+} reporter, *f*-holo-aequorin, and using a custom-designed luminescent imaging system, we can obtain pan-embryonic visual information continually for up to the first ~24 h postfertilization (hpf). Once we know approximately when and where to look for these Ca^{2+} signaling events within a complex developing embryo, we then repeat the experiment using a fluorescent Ca^{2+} reporter such as calcium green-1 dextran and use confocal laser scanning microscopy to provide time-lapse series of higher-resolution images. These protocols allow us to identify the specific cell types and even the particular subcellular domain (e.g., nucleus or cytoplasm) generating the Ca^{2+} signal. Here, we outline the techniques we use to precisely microinject *f*-holo-aequorin or calcium green-1 dextran into embryos without affecting their viability or development. We also describe how to inject specific regions of early embryos in order to load localized embryonic domains with a particular Ca^{2+} reporter. These same techniques can also be used to introduce other membrane-impermeable reagents into embryos, including Ca^{2+} channel antagonists, Ca^{2+} chelators, fluorescent dyes, RNA, and DNA.

Key words Luminescent and fluorescent Ca^{2+} reporters, Microinjection, Holo-aequorin, Calcium green-1 dextran, Zebrafish embryos

1 Introduction

Accumulating evidence indicates that Ca^{2+} ions play a key role in the signal transduction pathways that orchestrate embryonic development [1–9]. Indeed, at the earliest stages of embryogenesis, distinct Ca^{2+} signals have been shown to accompany (and in a number of cases initiate and/or regulate) specific developmental events [8, 10–18]. As development proceeds, however, and the embryonic cells divide and start to differentiate, the Ca^{2+} signals become more complex [19–25].

Claus W. Heizmann (ed.), *Calcium-Binding Proteins of the EF-Hand Superfamily: From Basics to Medical Applications*, Methods in Molecular Biology, vol. 1929, https://doi.org/10.1007/978-1-4939-9030-6_6,

The temporal and/or spatial patterns of Ca^{2+} signals generated during embryogenesis are commonly recorded via the use of intracellular fluorescent or bioluminescent Ca^{2+} reporters. A number of fluorescent Ca^{2+} indicators are commercially available from companies such as Thermo Fisher Scientific (Waltham, MA, USA). These include single-wavelength Ca^{2+} reporters, such as fluo-3, fluo-4, and calcium green-1 dextran; dual excitation/single emission reporters, for example, fura-2; and single excitation/dual emission reporters, such as Indo-1. There are also a number of luminescent Ca^{2+} reporters, including aequorin [26], obelin [27, 28], clytin [29], and mitrocomin [30, 31]; however, of these, aequorin is the most commonly utilized. Aequorin (also called holo-aequorin) comprises a 21-kDa protein (apoaequorin), along with a ~400 Da hydrophobic prosthetic group (coelenterazine), and oxygen.

Holo-aequorin and coelenterazine were initially isolated from *Aequorea* spp. jellyfish [26, 32], but various analogues of coelenterazine (including native, *f*-, *h*-, and *hcp*-) have been synthesized chemically since the late 1970s [33–36], and apoaequorin has been produced using molecular cloning methods since the mid-1980s [37–39]. For many years, a number of recombinant semisynthetic holo-aequorins (including native, *f*-, *h*-, and *hcp*-holo-aequorin), which exhibited varying sensitivities to Ca^{2+}, could be obtained directly from Prof. Osamu Shimomura who synthesized them at the Marine Biological Laboratory (Woods Hole, MA, USA) and provided them via a company that he established (The Photoprotein Laboratory, Falmouth, MA, USA). However, since Prof. Shimomura's death, it is now difficult to obtain these different types of holo-aequorin, although native holo-aequorin can be obtained commercially from companies such as NanoLight Technology (Pinetop, AZ, USA). Furthermore, native coelenterazine and a number of the synthetic analogues are available commercially from various companies including NanoLight Technology, Thermo Fisher, Sigma Aldrich, and Promega. In addition, in recent years an increasing number of fluorescent and luminescent genetically encoded Ca^{2+} indicators (GECI) have been developed, which can be expressed in developing zebrafish embryos in a ubiquitous manner or else targeted to specific tissues or cells [25, 40–44]. Most recently, two GECI lines of zebrafish have been developed, which express GCaMP6s ubiquitously (via the use of the *βactin2* or *ubi* promoters) from the earliest stages of development [44]. Many of the Ca^{2+} signals observed in early zebrafish embryos via the microinjection of holo-aequorin and calcium green-1 dextran have now been confirmed using these GCaMP6s transgenic lines [44].

Here, we describe techniques that are routinely used in our laboratory for microinjecting zebrafish embryos (normally at the 1-cell stage) with Ca^{2+} reporters. We use a complementary approach and use *f*-holo-aequorin and calcium green-1 dextran in conjunction with luminescence imaging (via a specialized custom-designed EMCCD-based imaging system) and time-lapse laser scanning confocal imaging, respectively. The former provides long-duration but relatively low-resolution information about Ca^{2+} signaling events that occur during embryogenesis. This method is especially powerful for showing large-scale pan-embryonic Ca^{2+} signaling events. Once we know when and approximately where a Ca^{2+} signal is generated, then we repeat the experiments using calcium green-1 dextran to confirm the signals observed with holo-aequorin and provide higher-resolution information about the type and specific location of the cells that are generating the signals.

We have used this complementary luminescent/fluorescent Ca^{2+} reporter approach to help characterize the Ca^{2+} signals generated during a number of the early developmental events that occur during the first ~24 h postfertilization (hpf) of zebrafish embryogenesis. For example, we identified both fast short-range and slow long-range Ca^{2+} waves that propagate around the external yolk syncytial layer (E-YSL) in zebrafish embryos during the late blastula and early gastrula stages [45] (Fig. 1A). In this case, *f*-holo-aequorin or calcium green-1 dextran was microinjected into the upper part of the yolk of embryos at the 128-cell stage rather than at the 1-cell stage in order to preferentially load the reporter into the E-YSL as it formed (Fig. 1Aa). As calcium green-1 dextran is a single-wavelength dye, we co-injected it with rhodamine B dextran to correct for differences in reporter distribution as well as to identify any Ca^{2+}-independent changes in fluorescence that might occur due to cells moving out of the selected viewing plane [8, 45]. Pan-embryonic and shorter-range Ca^{2+} signals were initially identified via imaging the *f*-holo-aequorin-generated luminescence with a custom-designed EMCCD-based luminescence imaging microscope system, after which time-lapse confocal imaging with calcium green-1 dextran was used to confirm that the Ca^{2+} signals were real and demonstrate that they were generated in both the nuclei and cytoplasm of the E-YSL.

We have also used this complementary luminescent/fluorescent Ca^{2+} reporter approach to investigate Ca^{2+} signals generated during the formation of the enveloping layer of cells, which takes place in the blastula period [8, 46]; during epiboly in the gastrula period [19, 47] (Fig. 1B); during somitogenesis in the segmentation period [24] (Fig. 1C); and during slow muscle development in the mid-segmentation/early pharyngula periods [25] (Fig. 1D). In all of these examples, *f*-holo-aequorin or calcium green-1 dextran was microinjected into the upper part of the yolk of zebrafish embryos at the 1-cell stage (Fig. 1Ba); the reporter was then transferred into the blastodisc via a combination of diffusion and

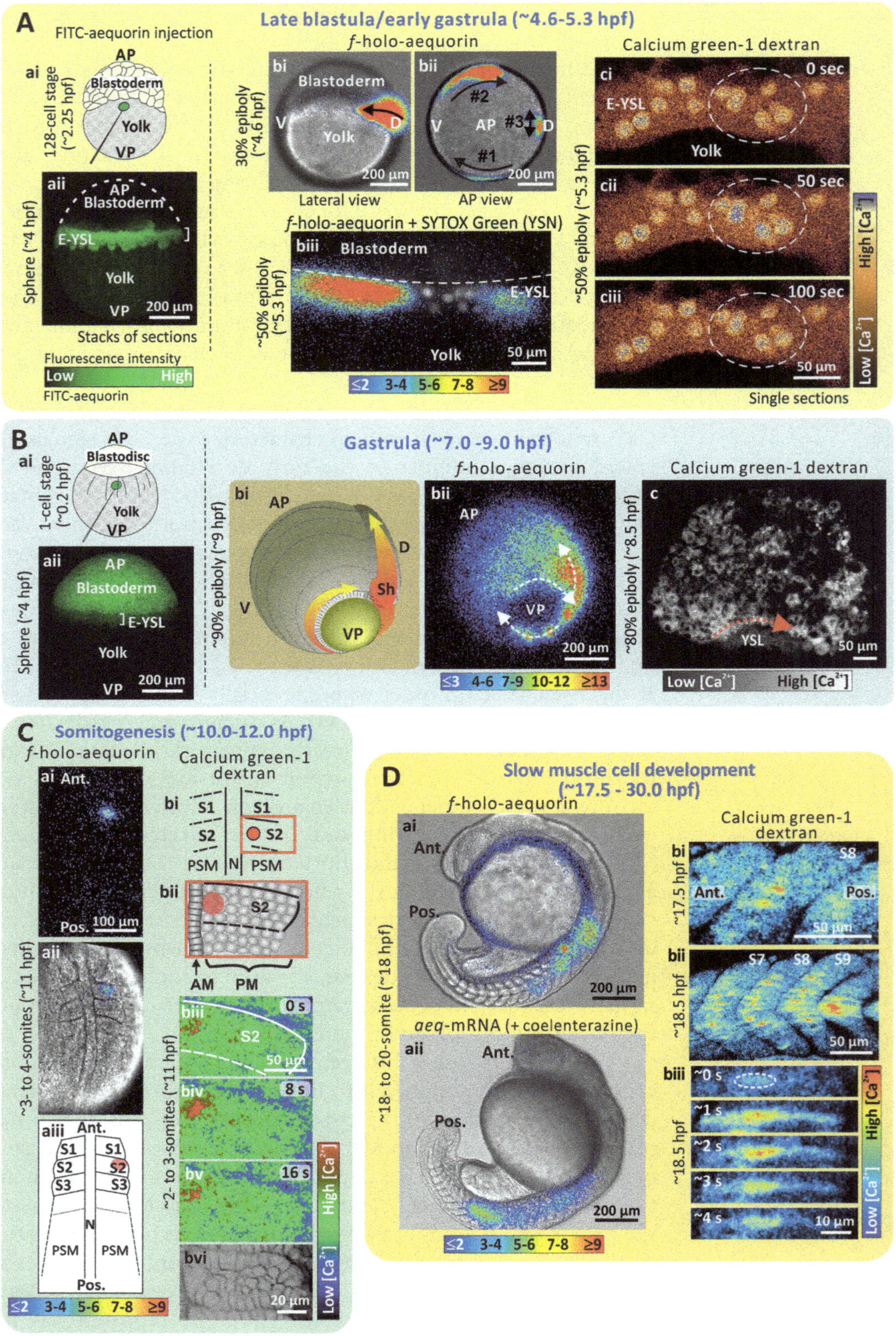

Fig. 1 Representative examples of several of the Ca^{2+} signaling events that have been identified during the early development of zebrafish embryos, using a complementary luminescence and fluorescence Ca^{2+}

ooplasmic segregation [16, 48, 49], after which it was partitioned into all the subsequent cell progenies following successive rounds of cell division. Using this 1-cell stage microinjection protocol, *f*-holo-aequorin-generated Ca^{2+} signals can be imaged continually for approximately the first 24 h of zebrafish development. After this, the reporter is all used up, although it can be reconstituted by incubating embryos in solutions containing membrane-permeable coelenterazine [25, 50]. Calcium green-1 dextran does not get used up in the same way that *f*-holo-aequorin does; however, we have not yet attempted to image Ca^{2+} signals beyond ~24 hpf of zebrafish development [25].

Fig. 1 (continued) imaging approach via microinjection with *f*-holo-aequorin or calcium green-1 dextran, respectively. Images are shown (**A**) during the late blastula/early gastrula period (i.e., from ~4.6 h post-fertilization (hpf) to ~5.3 hpf); (**B**) during the late gastrula period (i.e., ~7.0 hpf to ~9.0 hpf); and during (**C**) somitogenesis (~10 hpf – ~12 hpf) and (**D**) slow muscle development (~17.5 hpf – ~30 hpf) in the segmentation and early pharyngula periods. (**A**) If embryos are microinjected at the 128-cell stage, then by the sphere stage, the injectate is localized in the external yolk syncytial layer (E-YSL). (**B–D**) In contrast, if embryos are injected at the 1-cell stage, then the injectate becomes distributed throughout all the cells of the embryo. (**Aaii**, **Baii**) The distribution of injectate after the 128-cell or 1-cell stage injection of FITC-aequorin is shown. Representative (**Ab**, **Bb**, **Ca**, **Da**) *f*-holo-aequorin-generated luminescent and (**Ac**, **Bc**, **Cb**, **Db**) the complementary calcium green-1 dextran-generated fluorescent images are shown. The luminescent images represent (**Ab**) 10 s, (**Bbii**) 120 s, (**Cai**, **Caii**) 20 s, and (**Dai**, **Daii**) 60 s of accumulated light, respectively. (**Aai**, **Bai**, **Bbi**, **Caiii**, **Cbi**, **Cbii**) The schematics show the orientation of the embryos and (when appropriate) the location of the Ca^{2+} signals generated (in red/yellow) in these examples. AP, animal pole; VP, vegetal pole; E-YSL, external yolk syncytial layer; D, dorsal; V, ventral; Sh, shield; Ant., anterior; Pos., posterior; S, somite; N, notochord; PSM, presomitic mesoderm; AM, adaxial mesoderm; PM, paraxial mesoderm. The color scale in **Aa** (for **Aaii** and **Baii**) indicates fluorescence intensity of FITC-aequorin; the color scales in panels **Ab**, **Bbii**, **Ca**, and **Da** indicate luminescence flux in photons/pixel; and the color scales in panels **Ac**, **Bc**, **Cb**, and **Db** indicate $[Ca^{2+}]$, determined from the fluorescence intensity. Panels **Aa–Ac** and **Ba** are reproduced with permission, from "Characterization of Ca^{2+} signaling in the external yolk syncytial layer during the late blastula and early gastrula periods of zebrafish development" by Yuen MYF, Webb SE, Chan CM, Thisse B, Thisse C, Miller AL, 2013. Biochimica Biophysica Acta 1833: 1641–1656. Elsevier ©. Panel **Bb** is reproduced with permission, from "Ca^{2+} signaling and early embryonic patterning during the blastula and gastrula periods of zebrafish and *Xenopus* development" by Webb SE, Miller AL, 2006. Biochimica Biophysica Acta 1763: 1192–1208, Elsevier ©. Panel **Bc** is reproduced with permission, from "Imaging intercellular calcium waves during late epiboly in intact zebrafish embryos" by Webb SE, Miller AL, 2003. Zygote 11: 175–182, Cambridge University Press ©. Panel **C** is reproduced with permission, from "Visualization of stochastic Ca^{2+} signals in the formed somites during the early segmentation period in intact, normally developing zebrafish embryos" by Leung CF, Miller AL, Korzh V, Cong SW, Sleptsova-Friedrich I, Webb SE, 2009. Development, Growth & Differentiation 51: 617–637, Japanese Society of Developmental Biologists ©. Panel **Da** is reproduced with permission, from "Transient expression of apoaequorin in zebrafish embryos: extending the ability to image calcium transients during later stages of development" by Cheung CY, Webb SE, Meng A, Miller AL, 2006. International Journal of Developmental Biology 50: 561–569, UPV/EHU Press ©. Panel **Db** is reproduced with permission, from "Visualization, characterization, and modulation of calcium signaling during the development of slow muscle cells in intact zebrafish embryos" by Cheung CY, Webb SE, Love DR, Miller AL, 2011. International Journal of Developmental Biology 55: 153–174, UPV/EHU Press ©

In addition to microinjecting *f*-holo-aequorin and calcium green-1 dextran into zebrafish embryos, we have used the same methodology to inject *f*-holo-aequorin into other large embryos, such as those of *Oryzias latipes* (medaka), *Puntius conchonius* (rosy barb), *Fundulus heteroclitus* (mummichog), and *Xenopus laevis* (African clawed frog) [51]. Furthermore, as well as Ca^{2+} reporters, we routinely use this microinjection protocol to introduce other membrane-impermeable reagents into embryos. These include FITC-aequorin (a fluorescent form of aequorin, which does not generate luminescence in the presence of Ca^{2+}; Fig. 1Aaii, Baii) [45]; the nuclear dyes, SYTOX Green and Hoechst 33258 [45]; Ca^{2+} channel antagonists and inhibitors such as heparin, de-N-sulfate heparin, ryanodine, and ruthenium red [52]; the Ca^{2+} chelator, 5,5′-dibromo BAPTA [15, 45, 52]; and mRNA and DNA [9, 45, 50, 53, 54]. For example, we have microinjected *apoaequorin* mRNA (*aeq*-mRNA) into embryos and demonstrated that following incubation with *f*-coelenterazine to reconstitute active *f*-holo-aequorin, the Ca^{2+} signals generated in the trunk at the ~18–20 somite stage are similar to those generated following injection of *f*-holo-aequorin alone [50] (compare Fig. 1Daii with Fig. 1Dai). Most recently, this microinjection technique has been used to inject *tpcn2* morpholino oligonucleotides or *tpcn2* gRNA and nls-zCas9-nls mRNA into zebrafish embryos to knock down and knock out two-pore channel 2 (TPC2), respectively [9, 53]. Thus, once this technique is established in a laboratory, it has a versatile range of uses when investigating developmental events in large embryos such as those of zebrafish.

2 Materials

2.1 Microelectrode Preparation (Fig. 2)

1. Borosilicate glass capillaries (1.0 mm diameter, with filling fiber; e.g., World Precision Instruments, LLC., Sarasota, FL, USA).
2. Horizontal puller (e.g., PN-3, Narishige Scientific Instrument Laboratory, Tokyo, Japan).
3. Upright binocular microscope (e.g., Galen III, Cambridge Instruments, Leica, Wetzlar, Germany) with eyepiece reticle.
4. Micromanipulator (e.g., MM-33 N, Narishige Scientific Instrument Laboratory).
5. Microgrinder (e.g., EG-40, Narishige Scientific Instrument Laboratory).
6. Microscope slides (×2, glass).
7. Reusable adhesive putty (e.g., Blu-Tack).
8. Petri dishes (100 × 20 mm).

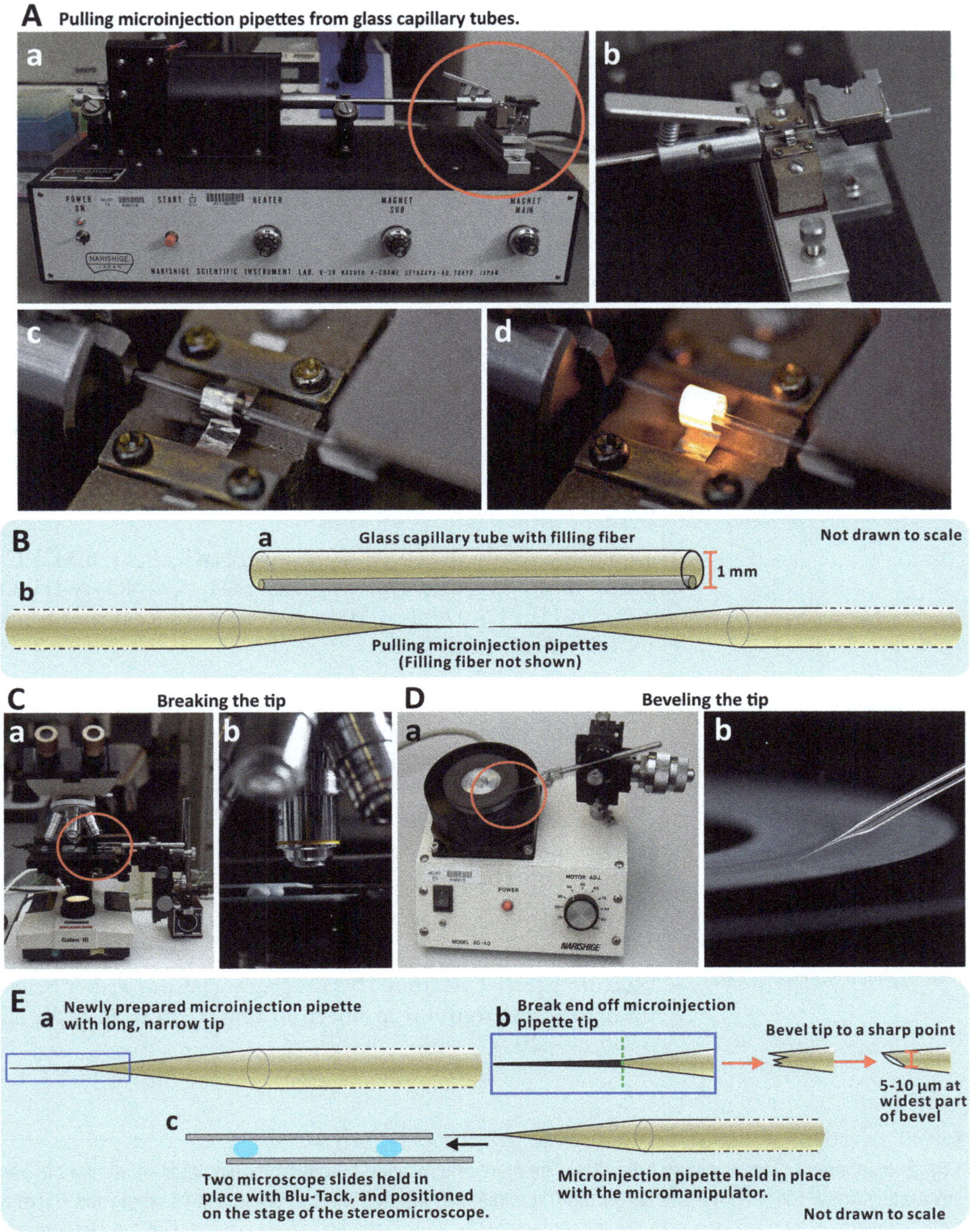

Fig. 2 Microinjection pipette preparation. (**A**) Pulling microinjection pipettes from glass capillary tubes. (**Aa**) The Narishige horizontal puller: the red circle shows where the glass capillary is being held in the jaws (on the left) and clamp (on the right). (**Ab**) Closeup view of the capillary held in the jaws and clamp. (**Ac**, **Ad**) Closeup views of the glass capillary and the heating filament, (**Ac**) before and (**Ad**) during the pulling process when the latter heats up to melt the glass. (**B**) Schematic diagram to show how microinjection pipettes are pulled from a

2.2 Microinjection/Imaging Chamber Preparation

1. Glass bottom microwell dishes (35-mm dish with 14 mm microwell and No. 1.5 cover glass; e.g., MatTek Corp., Ashland, MA, USA).
2. Petri dishes (35 × 10 mm).
3. Diamond-tipped glass cutter.
4. Borosilicate glass capillaries (*see* Subheading 2.1, **item 1**) cut into sections of ~2–3 mm and ~10 mm in length (with the diamond knife).
5. Microwave.
6. Heat-resistant gloves (for use with the microwave).
7. Glass Pasteur pipettes (150 mm).
8. Parafilm.
9. Agarose: 1% solution in 1× Danieau's solution. Heat gently with microwave until the agarose powder goes into solution with gentle swirling. Use heat-resistant gloves to prevent burns.
10. Danieau's solution (10×): 17.4 mM NaCl, 0.21 mM KCl, 0.12 mM $MgSO_4.7H_2O$, 0.18 mM $Ca(NO_3)_2.4H_2O$, 1.5 mM HEPES. Dilute 1:10 to give a 1× solution, and adjust to pH 7.2 just prior to use.

2.3 Calcium Reporters

1. Native holo-aequorin (e.g., NanoLight Technology, Pinetop, AZ, USA): 1% solution in Ca^{2+}-free Milli-Q water. Store 5 nL aliquots at −80 °C in 0.5 mL microcentrifuge tubes that have the lid removed and are themselves held inside 1.5 mL microcentrifuge tubes. These 0.5 mL and 1.5 mL tubes and the pipette tips used for aliquoting holo-aequorin should first be washed with 100 μM EGTA.
2. Ethylene glycol tetraacetic acid (EGTA): 100 μM solution in Milli-Q water.
3. Calcium green-1 dextran 10 kDa (e.g., Thermo Fisher Scientific): 400 μM solution in injection buffer. Store in 5 μL aliquots at −20 °C.

Fig. 2 (continued) glass capillary tube. (**Ca**) The stereomicroscope and micromanipulator setup used to very accurately break the long narrow tip off the microinjection pipette; the red circle shows where the microinjection pipette is held in place with the micromanipulator so that the tip can be viewed with the stereomicroscope. (**Cb**) Closeup view of the microinjection pipette (held in the micromanipulator) next to the edge of the microscope slide setup (on the stage of the stereomicroscope), which is used to break the tip. (**Da**) The Narishige microgrinder used to bevel the microinjection pipette tip to a sharp point. (**Db**) Close-up of the microinjection pipette tip being beveled on the grinding wheel. (**E**) Schematic diagram to show the long narrow tip of the microinjection pipette being broken off and then the resulting wider tip being beveled to a sharp point. The region bounded by the blue rectangle in (**Ea**) is shown enlarged in (**Eb**)

4. Rhodamine B dextran 10 kDa (e.g., Thermo Fisher Scientific): 400 μM solution in injection buffer. Store in 5 μL aliquots at −20 °C.
5. Injection buffer: 150 mM KCl, 5 mM HEPES in Milli-Q water, pH 7.2. Filter prior to use.
6. Filters: Minisart, 0.2 μm pore size, single use, sterile (e.g., Sartorius Stedim Biotech SA, Goettingen, Germany).
7. Benchtop centrifuge.
8. Microcentrifuge tubes (0.6 mL and 1.5 mL).
9. Gilson Pipetman (P10, P100, and P1000) and associated tips.

2.4 Zebrafish Embryos

1. Danieau's solution: *see* Subheading 2.2, **item 10**.
2. Protease from *Streptomyces griseus*: Dissolve 100 mg protease in 20 mL 1× Danieau's solution. Store 1 mL aliquots at −20 °C until required.
3. Diamond-tipped glass cutter.
4. "Wide-bore" glass Pasteur pipettes (150 mm): Cut ~4.5–5 cm from the narrow end of a glass Pasteur pipette using a diamond-tipped glass cutter, and then flame polish the cut end to smooth any rough edges.
5. Petri dishes (60 × 15 mm and 100 × 20 mm).
6. Tweezers (fine or superfine points, No. 0 or 5; e.g., Regine, Switzerland).

2.5 Microinjection of Holo-aequorin (Fig. 3)

1. Bottom illumination stereomicroscope (e.g., Stemi SV6, Carl Zeiss Inc., Jena, Germany).
2. Micromanipulator (e.g., MMJ, Carl Zeiss Inc.).
3. Pressure injection system (e.g., PLI-188, Medical Systems Corp., Nikon, Tokyo, Japan).
4. Benchtop centrifuge: *see* Subheading 2.3, **item 7**.
5. Gilson Pipetman (P10) and associated tips.
6. Microelectrodes: *see* Subheading 2.1.
7. Microinjection chambers: *see* Subheading 2.2.
8. EGTA (100 μM): *see* Subheading 2.3, **item 2**.
9. Eyelash brush (prepared by attaching a human eyelash to the narrow end of a glass Pasteur pipette).
10. Permanent marker pen.
11. Polypropylene tube with lid (50 mL).
12. Tissues.
13. EGTA-washed vegetable oil (e.g., Wesson soybean oil): Add equal volumes of 100 μM EGTA (*see* Subheading 2.3, **item 2**)

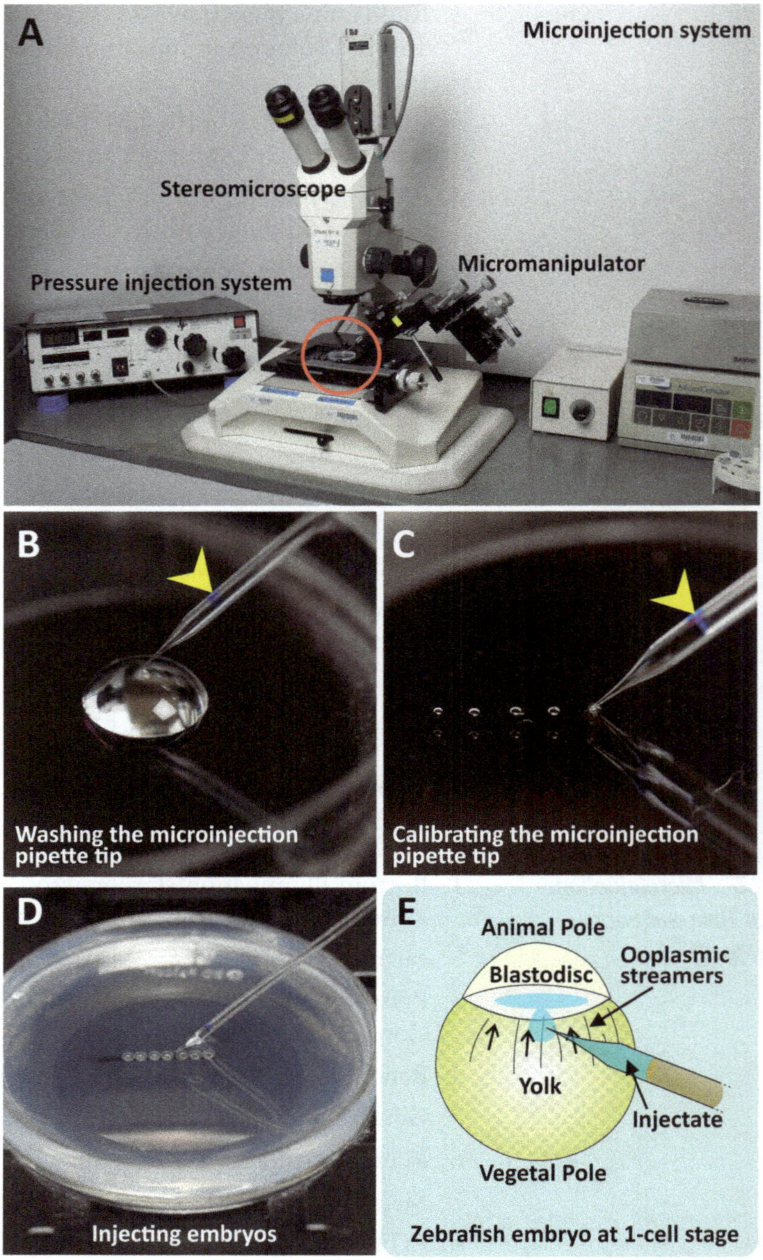

Fig. 3 Embryo microinjection. (**A**) The Stemi SV6 stereomicroscope, Zeiss MMJ micromanipulator, and PLI-188 Medical Systems Corp. pressure injection system used to microinject zebrafish embryos with luminescent or fluorescent Ca^{2+} reporters. The region of the stereomicroscope stage bounded by the red circle is enlarged in panels (**B–D**). (**B**) Washing and (**C**) calibrating the microinjection pipette tip. The yellow arrowheads show the pen mark, which indicates the upper limit of microinjection pipette tip washing with EGTA, which is required for the holo-aequorin injections. (**D**) Injecting holo-aequorin or calcium green-1 dextran into zebrafish embryos. (**E**) Schematic illustration to show the location of microinjection into the upper region of the yolk of zebrafish embryos at the 1-cell stage

and vegetable oil to a 50 mL polypropylene tube with lid. Shake vigorously to mix completely, and then allow the mixture to sit over the next few days, and settle into two layers. Store the upper oil layer in an EGTA-washed and air-dried lid of a 35-mm Petri dish. Store the Petri dish lid containing the oil in a 100-mm Petri dish, and keep this covered with the lid to prevent dust contamination. EGTA-washed oil should be prepared and transferred to the Petri dish lid at least 1 week before holo-aequorin injection. Once prepared it can be used for all the experiments conducted over a period of several weeks.

14. Holo-aequorin: *see* Subheading 2.3, **item 1**.

2.6 Microinjection of Calcium Green-1 Dextran (and Rhodamine B Dextran)

1. Bottom illumination stereomicroscope: *see* Subheading 2.5, **item 1**.
2. Micromanipulator: *see* Subheading 2.5, **item 2**.
3. Pressure injection system: *see* Subheading 2.5, **item 3**.
4. Microelectrodes: *see* Subheading 2.1.
5. Microinjection chambers: *see* Subheading 2.2.
6. Vegetable oil (e.g., Wesson soybean oil).
7. Eyelash brush: *see* Subheading 2.5, **item 9**.
8. Calcium green-1 dextran and rhodamine B dextran: *see* Subheading 2.3, **items 3** and **4**.

2.7 Ca^{2+} Measurements

1. Imaging chambers: *see* Subheading 2.2.
2. Photomultiplier tube (PMT) system (e.g., Science Wares Inc., East Falmouth, MA): To record temporal holo-aequorin-generated Ca^{2+} signaling information.
3. Luminescence imaging microscope (LIM) system (e.g., Science Wares Inc.): To record temporal and spatial holo-aequorin-generated Ca^{2+} signaling information.
4. Confocal microscope equipped with 20×, 40×, and 63× objective lenses and Argon (blue) and DPSS (light green) lasers suitable for 488 nm and 561 nm excitation, respectively, for recording temporal and spatial fluorescent Ca^{2+} signals.
5. Triton X-100: 1% solution in 1× Danieau's solution (*see* Subheading 2.2, **item 10**).

3 Methods

Carry out all procedures at room temperature unless otherwise stated. To transfer embryos between dishes, it is important to use the wide-bore glass Pasteur pipettes to prevent damage.

3.1 Microinjection Pipette Preparation (Fig. 2)

1. Prepare microinjection pipettes from the borosilicate glass capillary tubes (with filling fiber) using the horizontal puller (Fig. 2A). Adjust the magnet (main and sub) and heater settings of the puller so that microinjection pipettes are pulled with an extended taper as illustrated in Fig. 2B.
2. Gently break off the narrow point of each micropipette as follows. The micropipette is loaded into the micromanipulator, and the tip is visualized via the stereomicroscope (Fig. 2Ca). On the stage of the stereomicroscope are two microscope slides, one held in place above the other with small balls of reusable adhesive putty, which are placed in each corner (Fig. 2Cb, Ec). The edge of the microscope slide is brought into focus via adjusting the stereomicroscope focus control, and the tip of the micropipette is brought into focus via the micromanipulator. The micropipette is then gently nudged against the edge of the microscope slide to break the tip (Fig. 2C, E).
3. Bevel the resulting tip to a 30° angle with the microgrinder (Fig. 2D, E). If the tip is touching the grinding wheel of the microgrinder correctly, then it should take <1 min to bevel the tip. The diameter of the sharpened tips can be checked with the stereomicroscope using the eyepiece reticle, and only accept those that are ~5–10 μm at the widest part of the bevel.
4. Microinjection pipettes should be prepared in advance, and they can be stored in a 90-mm Petri dish and kept in place along a strip of reusable adhesive putty. It is important that the lid is kept on the Petri dish at all times to prevent dust from contaminating the micropipettes.

3.2 Injection Chamber Preparation

1. Heat the solid 1% agarose in the microwave until it has melted (i.e., ~3–5 min on high).
2. Cut an intact borosilicate glass capillary into small pieces with the glass cutter. These pieces of capillary will be used to prepare grooves in the agarose, and they can be stored in a small container (e.g., a 35-mm Petri dish) and reused. The grooves are used to hold embryos in place during microinjection. Thus, capillaries should be cut into sections of ~2–3 mm in length for preparing the grooves to hold single dechorionated embryos and sections of ~10 mm in length for the grooves to hold a row of several intact (i.e., still with the chorion) embryos (Fig. 3D).
3. Using a glass Pasteur pipette, transfer molten agarose into a 35-mm Petri dish lid, and place one or more of the capillary sections lengthwise into the center of the dish ensuring that they are immersed by the agarose by only ~50–60% of the depth.

4. Fine (or superfine) tweezers can be used to adjust the location of the capillary sections in the agarose until it sets to ensure that they are in the correct location (*see* **Note 1**).
5. When the agarose is cooled and set (within just a few min), carefully remove the capillary sections to form the grooves.
6. The injection chambers can be prepared a few days before the experiment and stored at 4 °C; two chambers can be stored together (open rims apposed) and sealed with Parafilm to prevent the agarose from drying out.

3.3 Imaging Chamber Preparation

1. Use the agarose and short glass capillaries described above—*see* Subheading 3.2.
2. For these chambers, transfer the molten agarose into the central, circular depression of glass bottom microwell dishes.
3. In this case, the short agarose grooves are used to hold a single embryo during imaging.
4. For these chambers, make sure that the glass capillary pieces touch the glass coverslip and that enough agarose is poured so that it forms a flat layer that covers the circular depression entirely without bulging upward or overflowing. In this way, the grooves that are formed are the correct depth to hold an embryo in place for imaging, and embryos are close enough to the coverslip to allow for effective focusing when imaging with an inverted microscope.
5. Like the injection chambers, the imaging chambers can be prepared a few days in advance of the experiment; simply cover with the lid, and then seal around the edge with Parafilm to prevent the agarose from drying out.

3.4 Embryo Dechorionation

Embryos can be dechorionated (i.e., the external membrane—chorion—can be removed), to enhance their optical clarity during imaging, as follows:

1. Incubate newly fertilized embryos for ~1–2 min with 2 mL protease solution in a 35-mm Petri dish.
2. Gently swirl the dish to ensure that all embryos are exposed to the enzyme, and observe their morphology closely.
3. When the embryos move freely inside the chorion and/or the three layers that make up the chorion [55] start to separate, quickly transfer them into a 100-mm Petri dish containing Danieau's solution (*see* **Note 2**).
4. Rinse them several times with fresh Danieau's solution to dilute, and thus stop the action of the enzyme. By this stage, the protease-digested chorions should have ruptured, and the Danieau's washes will result in them becoming separated from the main embryo.

5. Once dechorionated, embryos should be maintained and stored in a Petri dish coated with agarose as they readily stick to (and lyse on) an uncoated plastic surface (*see* **Note 3**).

3.5 Microinjection of Holo-aequorin (Fig. 3)

3.5.1 Washing and Calibrating the Microinjection Pipette

1. All the glass- and plasticware associated with the holo-aequorin injections should be prewashed for 1–2 min with 100 μM EGTA to chelate any contaminating Ca^{2+} on their inner surfaces.
2. This includes the following: (1) the lid of a 35-mm Petri dish, where the small droplet of holo-aequorin will be placed prior to being front-loaded into the microinjection pipette; (2) the Gilson pipette tip used to transfer the holo-aequorin aliquot from its storage Eppendorf tube to the Petri dish lid; and (3) the microinjection pipette itself.
3. Regarding (1), fill the lid of a 35-mm Petri dish with EGTA, incubate for 1–2 min, and then empty and allow to air-dry.
4. Regarding (2), *see* Subheading 3.5.2, **step 3**.
5. Regarding (3), place two 50-μL drops of EGTA side by side on a second 35-mm Petri dish lid (Fig. 3B). Fix a beveled microinjection pipette into the micromanipulation pipette holder, and then use the pressure injection system to inject air through the empty microinjection pipette and into the first 50-μL drop of EGTA. This will clean the tip of the microinjection pipette of any fragments of glass that might remain following the beveling process.
6. Using the stereomicroscope, micromanipulator and pressure injection system described in Subheading 2.5 (Fig. 3A), front-fill the microinjection pipette with EGTA from the second 50-μL droplet (Fig. 3B), and mark the side of the microinjection pipette with a permanent marker to indicate the EGTA level and hence the region washed (*see* yellow arrowheads; Fig. 3B, C).
7. Calibrate microinjection pipettes before (and after) each injection by injecting droplets of EGTA under EGTA-washed vegetable oil (*see* Subheading 2.5, **item 13**; Fig. 3C).
8. Measure the diameter of the droplets with a micrometer-calibrated eyepiece reticle [56] (*see* **Notes 4** and **5**).
9. During calibration and subsequent injection, maintain the injection pressure at ~15 pounds per square inch (PSI or 105 kPa), and change the injection time in order to fine-tune the volume of injectate (*see* **Note 6**).
10. Calculate the volume of the droplets injected into the oil with the following equation: Volume = $3/4\pi r^3$, where "r" is the radius of the spherical droplet.

11. Injectate volumes of ~1–8 nL are typically microinjected into early zebrafish embryos.
12. Once the microinjection pipette is calibrated, store the tip under EGTA-washed oil whenever it is not in use, to prevent it from getting blocked due to dust and/or the evaporation and crystallization of the injectate solutions.

3.5.2 Filling the Microinjection Pipette with Holo-aequorin and Microinjecting the Embryos

1. Remove a 5-nL aliquot of holo-aequorin from the −80 °C freezer, and while it is thawing, use a tissue to remove any frost that might be on the outside of the tube. This helps to prevent the holo-aequorin from being contaminated with (Ca^{2+}-containing) water.
2. Once thawed, centrifuge the holo-aequorin at ~14,000 rpm (~16,800 × *g*) for ~5 min with the benchtop centrifuge. This step is required to pellet any crystals of holo-aequorin that might have formed, which might result in the microinjection pipette getting blocked.
3. Use a P10 Gilson Pipetman, and wash a P10 pipette tip with 10 μL EGTA by repeated (~5–10 times) filling and emptying. Keep the pipette tip in place, and take care that it doesn't touch anything, and then adjust the Pipetman to collect a volume of 0.5 μL.
4. Carefully remove 0.5 μL holo-aequorin supernatant from the microcentrifuge tube, and place it on the EGTA-washed 35-mm Petri dish lid surface (*see* Subheading 3.5.1, **step 3**).
5. Using the stereomicroscope, micromanipulator, and pressure injection system described in Subheading 3.5.1, **step 6** (Fig. 3A), empty the microinjection pipette of the prewashing EGTA, and immediately fill it with holo-aequorin. The volume of holo-aequorin loaded into the microinjection pipette must be kept below the volume of EGTA used for the prewash.
6. Replace the 35-mm EGTA-washed Petri dish lid with an injection chamber (*see* Subheading 3.2) containing embryos held in agarose grooves on the stereomicroscope stage (Fig. 3D).
7. If intact (i.e., still inside the chorion) embryos are to be microinjected, then a few can be lined up to fill the long grooves of the injection chamber (*see* Subheading 3.2, **step 2**), and they can be gently rolled with tweezers so that they are all lying in the same orientation.
8. If dechorionated embryos are to be microinjected, then just one embryo should be placed in each of the short grooves of the injection chamber (*see* Subheading 3.2, **step 2**). As dechorionated embryos are very delicate, they should be manipulated as little as possible. Use an eyelash brush to gently roll the embryo to the required reorientation.

9. Inject holo-aequorin into the upper region of the yolk of embryos at the 1-cell stage (Fig. 3E). The Ca^{2+} reporter gets transferred into the blastodisc by a combination of diffusion and ooplasmic streaming, with the latter starting at ~15 min postfertilization (mpf) [16, 48, 49] (*see* **Notes 7** and **8**).

3.6 Luminescence Data Acquisition

1. After microinjection, transfer the holo-aequorin-loaded embryos into an imaging chamber—*see* Subheading 3.3.
2. Maintain embryos at ~28.5 °C until they reach the stage of development under investigation.
3. Ensure that the embryos are in the correct orientation.
4. Transfer the imaging chamber containing the holo-aequorin-injected embryos to the PMT or luminescence imaging microscope (LIM) system for data acquisition [15, 56, 57].

3.7 Holo-aequorin "Burnout"

1. At the end of each experiment, suspend data acquisition.
2. Add ~100 μL 1% Triton X-100 to the imaging chamber, and resume data acquisition. This detergent permeabilizes the plasma membrane in embryos, which exposes any remaining unspent cytoplasmic holo-aequorin to extracellular Ca^{2+} in the medium [24].
3. This is especially useful if during data acquisition, the level of luminescence in an embryo is suspiciously low, as it helps to distinguish between a genuine inhibition of the endogenous Ca^{2+} signals and an insufficiency of active holo-aequorin.

3.8 Microinjection of Calcium Green-1 Dextran

1. Prepare microinjection pipettes, as described in Subheading 3.1.
2. Prepare injection and imaging chambers, as described in Subheadings 3.2 and 3.3, respectively.
3. If required, dechorionate embryos, as described in Subheading 3.4.
4. Microinjection pipette preparation. For calcium green-1 dextran injections, glassware and plasticware do not need to be prewashed with EGTA. However, the tip of microinjection pipettes should still be cleaned prior to use to remove any glass fragments; thus, inject air though the microinjection pipette into a droplet of injection buffer.
5. Mix a 1:1 ratio (~1 nL of each) of calcium green-1 dextran and rhodamine B dextran on the lid of a 35-mm Petri dish, and front-load into the microinjection pipette (*see* **Note 9**).
6. Calibrate microinjection pipettes before (and after) each injection by injecting droplets of this mixture under vegetable oil.
7. The remaining calibration steps are as described in Subheading 3.5.1, **steps 8–11**.

8. Following calibration, store the microinjection tip under vegetable oil to prevent it from getting blocked.
9. Microinject intact or dechorionated embryos with the calcium green-1 dextran/rhodamine B dextran mixture as described in Subheading 3.5.2, **steps 7–9**.
10. Immediately transfer the injected embryos to the dark at ~28.5 °C, and maintain in these conditions until they develop to the stage of interest.

3.9 Fluorescence Data Acquisition

1. Transfer the imaging chamber containing the calcium green-1 dextran/rhodamine B dextran injected embryos to the confocal laser scanning microscope.
2. Acquire calcium green-dextran fluorescence with 488 nm excitation and 500–550 nm detection and rhodamine B dextran fluorescence with 561 nm excitation and 570–700 nm detection, to monitor Ca^{2+} signaling and Ca^{2+}-independent changes in fluorescence, respectively.
3. Acquire time-lapse images in the order of ~160 images or more, every 100 s.

4 Notes

1. If the pieces of glass capillary clump together in the molten agarose, this is likely due to the agarose being too hot. Allow the agarose to cool slightly (i.e., for ~1 min), and then use tweezers to move the pieces of capillary to the required location of the injection or imaging chamber.
2. If the embryos don't show any obvious indications that the chorions are being lysed even after being incubated in protease solution for >2 min, continue with the Danieau's wash step, as described. If the chorions are not removed during this wash step, then repeat the dechorionation procedure using a new aliquot of protease solution. If the chorions are still not removed, then the protease might have become inactive, so it is necessary to purchase a new batch.
3. If embryos lyse during (or shortly following) the dechorionation procedure, then this indicates that they have been incubated in protease for too long. Instead of relying on a fixed incubation time, use the changing morphology of the embryo inside the chorion and/or the morphology of the chorion itself to dictate when to start the Danieau's wash steps. Thus transfer embryos into Danieau's solution when they start to float freely inside (and contact the wall of) the chorion and/or when separation of the three layers of the chorion occurs.
4. If the EGTA droplets rapidly move out of the field of view, making calibration difficult, the EGTA-washed oil should be

maintained overnight in the 35-mm Petri dish lid before use. This phenomenon usually occurs when the EGTA-washed oil is first transferred to the Petri dish lid or if oil is spilt and the bottom of the lid containing the oil is wiped with a tissue.

5. If a stream of small EGTA droplets is injected into the EGTA-washed oil instead of just one single drop, this might be because the filling fiber is misaligned in the microinjection pipette. Thus, either attempt to adjust the position of the filling fiber by increasing the injection duration to generate large droplets before continuing with the calibration as usual or else switch to a new microinjection pipette.
6. If the microinjection pipette cannot maintain a stable injectate level, make small adjustments to the balance pressure on the pressure injection system.
7. If microinjection of the holo-aequorin into embryos is unsuccessful, inject droplets under EGTA-washed vegetable oil to check if the microinjection pipette is blocked. if it is blocked, then attempt to unblock the tip by injecting into a droplet of fresh EGTA, and increase the injection duration. If the blockage continues, then prepare a new microinjection pipette.
8. Embryos can be injected with holo-aequorin and SYTOX Green (e.g., Thermo Fisher Scientific). Prepare SYTOX Green as a 5 mM stock solution in DMSO and then dilute with injection buffer to ~2.5 μM just prior to use. Embryos should be injected with holo-aequorin first (as described) and then with SYTOX Green (~1 nL). SYTOX Green fluorescence can be visualized using 488 nm excitation and 500–550 nm detection wavelengths.
9. Calcium green-1 dextran and rhodamine B dextran can be co-injected with a nuclear stain such as Hoechst 33258 (at 1.6 mM; e.g., Thermo Fisher Scientific). Prepare a mixture of Hoechst 33258, calcium green-1 dextran, and rhodamine B dextran at a ratio of 2:9:9 by volume, respectively, just prior to microinjection. Hoechst 33258 fluorescence can be visualized using multiphoton excitation microscopy using 790 nm excitation and 410–510 nm detection wavelengths.

Acknowledgments

This work was supported by the Hong Kong Research Grants Council (RGC) General Research Fund awards 16101714 and 16100115 and the ANR/RGC joint research scheme award A-HKUST601/13. We also acknowledge funding support from the Hong Kong Innovation and Technology Commission (ITCPD/17-9). We thank Andrew Ho for helping us to photograph the equipment shown in Figs. 2 and 3.

References

1. Jaffe LF (1999) Organization of early development by calcium patterns. BioEssays 21:657–667
2. Berridge MJ, Lipp P, Bootman MD (2000) The versatility and universality of calcium signaling. Nature Rev Mol Cell Biol 1:11–21
3. Webb SE, Miller AL (2003) Calcium signaling during embryonic development. Nature Rev Mol Cell Biol 4:539–551
4. Porter GA Jr, Makuck RF, Rivkees SA (2003) Intracellular calcium plays an essential role in cardiac development. Dev Dyn 227:280–290
5. Whitaker M (2006) Calcium at fertilization and in early development. Physiol Rev 86:25–88
6. Slusarski DC, Pelegri F (2007) Calcium signaling in vertebrate embryonic patterning and morphogenesis. Dev Biol 307:1–13
7. Rosenberg SS, Spitzer NC (2011) Calcium signaling in neuronal development. Cold Spring Harb Perspec Biol 3(10):a004259. https://doi.org/10.1101/cshperspect.a004259
8. Zhang J, Webb SE, Ma LH, Chan CM, Miller AL (2011) Necessary role for intracellular Ca^{2+} transients in initiating the apical-basolateral thinning of enveloping layer cells during the early blastula period of zebrafish development. Develop Growth Diff 53:679–696
9. Kelu JJ, Webb SE, Parrington J, Galione A, Miller AL (2017) Ca^{2+} release via two-pore channel type 2 (TPC2) is required for slow muscle cell myofibrillogenesis and myotomal patterning in intact zebrafish embryos. Dev Biol 425:109–129
10. Brownlee C, Dale B (1990) Temporal and spatial correlation of fertilization current, calcium waves and cytoplasmic contraction in eggs of *Ciona intestinalis*. Proc R Soc Lond B 239:321–328
11. Fluck RA, Miller AL, Jaffe LF (1991) Slow calcium waves accompany cytokinesis in medaka fish eggs. J Cell Biol 115:1259–1265
12. Homa ST, Carroll J, Swann K (1993) Fertilization and early embryology: the role of calcium in mammalian oocyte maturation and egg activation. Hum Reprod 8:1274–1281
13. Swann K, McDougall A, Whitaker M (1994) Calcium signalling at fertilization. J Mar Biol Assoc UK 74:3–16
14. Chang DC, Meng C (1995) A localized elevation of cytosolic free calcium is associated with cytokinesis in the zebrafish embryo. J Cell Biol 131:1539–1545
15. Webb SE, Lee KW, Karplus E, Miller AL (1997) Localized calcium transients accompany furrow positioning, propagation, and deepening during the early cleavage period of zebrafish embryos. Dev Biol 192:78–92
16. Leung CF, Webb SE, Miller AL (1998) Calcium transients accompany ooplasmic segregation in zebrafish embryos. Develop Growth Differ 40:313–326
17. Créton R, Kreiling JA, Jaffe LF (2000) Presence and roles of calcium gradients along the dorsal-ventral axis in *Drosophila* embryos. Dev Biol 217:375–385
18. Whitaker M (2008) Calcium signalling in early embryos. Phil Trans Royal Soc B 363:1401–1418
19. Gilland E, Miller AL, Karplus E, Baker R, Webb SE (1999) Imaging of multicellular large-scale rhythmic calcium waves during zebrafish gastrulation. Proc Natl Acad Sci U S A 96:157–161
20. Leclerc C, Webb SE, Daguzan C, Moreau M, Miller AL (2000) Imaging patterns of calcium transients during neural induction in *Xenopus laevis* embryos. J Cell Sci 113:3519–3529
21. Tada M, Concha ML (2001) Vertebrate gastrulation: calcium waves orchestrate cell movements. Curr Biol 11:R470–R472
22. Wallingford JB, Ewald AJ, Harland RM, Fraser SE (2001) Calcium signaling during convergent extension in *Xenopus*. Curr Biol 11:652–661
23. Rogers KL, Picaud S, Roncali E, Boisgard R, Colasant C, Stinnakre J, Tavatian B, Brûlet P (2007) Non-invasive *in vivo* imaging of calcium signaling in mice. PLoS One 2:e974
24. Leung CF, Miller AL, Korzh V, Cong SW, Sleptsova-Friedrich I, Webb SE (2009) Visualization of stochastic Ca^{2+} signals in the formed somites during the early segmentation period in intact, normally developing zebrafish embryos. Develop Growth Differ 51:617–637
25. Cheung CY, Webb SE, Love DR, Miller AL (2011) Visualization, characterization and modulation of calcium signaling during the development of slow muscle cells in intact zebrafish embryos. Int J Dev Biol 55:153–174
26. Shimomura O, Johnson FH, Saiga Y (1962) Extraction, purification and properties of aequorin, a bioluminescent protein from the luminous hydromedusan, *Aequorea*. J Cell Comp Physiol 59:223–239
27. Campbell AK (1974) Extraction, partial purification and properties of obelin, the calcium-activated luminescent protein from the hydroid *Obelia geniculata*. Biochem J 143:411–418
28. Markova SV, Vysotski ES, Blinks JR, Burakova LP, Wang B-C, Lee J (2002) Obelin from the

bioluminescent marine hydroid *Obelia geniculata*: cloning, expression, and comparison of some properties with those of other Ca^{2+}-regulated photoproteins. Biochemist 41:2227–2236

29. Inouye S, Sahara Y (2007) Expression, purification and characterization of a photoprotein, clytin, from *Clytia gregarium*. Protein Expr Purif 53:384–389
30. Inouye S, Sahara Y (2009) Expression and purification of the calcium binding photoprotein mitrocomin using ZZ-domain as a soluble partner in *E. coli* cells. Protein Expr Purif 66:52–57
31. Burakova LP, Natashin PV, Markova SV, Eremeeva EV, Malikova NP, Cheng C, Liu ZJ, Vysotski ES (2016) Mitrocomin from the jellyfish *Mitrocoma cellularia* with deleted C-terminal tyrosine reveals a higher bioluminescence activity compared to wild type photoprotein. J Photochem Photobiol B 162:286–297
32. Shimomura O, Johnson FH (1975) Chemical nature of bioluminescence systems in coelenterates. Proc Natl Acad Sci U S A 72:1546–1549
33. Shimomura O, Johnson FH (1978) Peroxidized coelenterazine, the active group in the photoprotein aequorin. Proc Natl Acad Sci U S A 75:2611–2615
34. Shimomura O, Musicki B, Kishi Y (1988) Semi-synthetic aequorin: an improved tool for the measurement of calcium ion concentration. Biochem J 251:405–410
35. Shimomura O, Musicki B, Kishi Y (1989) Semi-synthetic aequorins with improved sensitivity to Ca^{2+} ions. Biochem J 261:913–920
36. Shimomura O, Inouye S, Musicki B, Kishi Y (1990) Recombinant aequorin and recombinant semi-synthetic aequorin. Biochem J 270:309–312
37. Inouye S, Noguchi M, Sakari Y, Takagi Y, Miyata T, Iwanaga S, Miyata T, Tsuji FI (1985) Cloning and sequence analysis of cDNA for the luminescent protein aequorin. Proc Natl Acad Sci U S A 82:3154–3158
38. Prasher D, McCann RO, Cormier MJ (1985) Cloning and expression of the cDNA coding for aequorin, a bioluminescent calcium-binding protein. Biochem Biophys Res Comm 126:1259–1268
39. Shimomura O, Inouye S (1999) The *in situ* regeneration and extraction of recombinant aequorin from *Escherichia coli* cells and the purification of extracted aequorin. Prot Express Purific 16:91–95
40. Higashijima S, Masino MA, Mandel G, Fetcho JR (2003) Imaging neuronal activity during zebrafish behaviour with a genetically encoded calcium indicator. J Neurophysiol 90:3987–3997
41. Ashworth R, Brennan C (2005) Use of transgenic zebrafish reporter lines to study calcium signalling in development. Brief Funct Gen Proteo 4:186–193
42. Horikawa K, Yamada Y, Matsuda T, Kobayashi K, Hashimoto M, Matsu-ura T, Miyawaki A, Michikawa T, Mikoshiba K, Nagai T (2010) Spontaneous network activity visualized by ultrasensitive Ca^{2+} indicators, yellow Cameleon-Nano. Nature Meth 7:729–732
43. Muto A, Kawakami K (2011) Imaging functional neural circuits in zebrafish with a new GCaMP and the Gal4FF-UAS system. Commun Integr Biol 4:566–568
44. Chen J, Xia L, Bruchas MR, Solnica-Krezel L (2017) Imaging early embryonic calcium activity with GCaMP6s transgenic zebrafish. Dev Biol 430:385–396
45. Yuen MYF, Webb SE, Chan CM, Thisse B, Thisse C, Miller AL (2013) Characterization of Ca^{2+} signaling in the external yolk syncytial layer during the late blastula and early gastrula periods of zebrafish development. Biochim Biophys Acta 1833:1641–1656
46. Ma LH, Webb SE, Chan CM, Zhang J, Miller AL (2009) Establishment of a transitory dorsal-biased window of localized Ca^{2+} signaling in the superficial epithelium following the mid-blastula transition in zebrafish embryos. Dev Biol 327:143–157
47. Webb SE, Miller AL (2003) Imaging intercellular calcium waves during late epiboly in intact zebrafish embryos. Zygote 11:175–182
48. Roosen-Runge EC (1938) On the early development-bipolar differentiation and cleavage of the zebrafish, *Brachydanio rerio*. Biol Bull 75:119–133
49. Beams HW, Kessel RG, Shim CY, Tung HN (1985) Scanning electron microscope studies on blastodisc formation in zebrafish, *Brachydanio rerio*. J Morphol 189:41–49
50. Cheung CY, Webb SE, Meng A, Miller AL (2006) Transient expression of apoaequorin in zebrafish embryos: extending the ability to image calcium transients during later stages of development. Int J Dev Biol 50:561–569
51. Webb SE, Miller AL (2007) Ca^{2+} signalling during embryonic cytokinesis in animal systems. In: Krebs J, Michalak M (eds) Calcium: a matter of life and death, vol 17. Elsevier, Amsterdam, Netherlands, pp 443–468
52. Lee KW, Webb SE, Miller AL (2003) Ca^{2+} released via IP_3 receptors is required for furrow

deepening during cytokinesis in zebrafish embryos. Int J Dev Biol 47:411–421

53. Kelu JJ, Chan HLH, Webb SE, Cheng AHH, Ruas M, Parrington J, Galione A, Miller AL (2015) Two-pore channel 2 activity is required for slow muscle cell-generated Ca^{2+} signaling during myogenesis in intact zebrafish. Int J Dev Biol 59:313–325
54. Kelu JJ, Webb SE, Galione A, Miller AL (2018) TPC2-mediated Ca^{2+} signalling is required for the establishment of synchronized activity in developing zebrafish primary motor neurons. Dev Biol 438(1):57–68
55. Bonsignorio D, Perego L, Del Giacco L, Cotelli F (1996) Structure and macromolecular composition of the zebrafish egg chorion. Zygote 4:101–108
56. Miller AL, Karplus E, Jaffe LF (1994) Imaging Ca^{2+} with aequorin using a photon imaging detector. In: Nuccitelli R (ed) Methods in cell biology: a practical guide to the study of calcium in living cells, vol 40. Academic Press Inc., San Diego, CA, pp 305–338
57. Webb SE, Rogers KL, Karplus E, Miller AL (2010) The use of aequorins to record and visualize Ca^{2+} dynamics: from subcellular microdomains to whole organisms. In: Whitaker M (ed) Methods in cell biology: calcium in living cells, vol 99. Academic Press Inc., San Diego, CA, pp 263–300

Chapter 7

Cellular Ca^{2+}-Responding Nanoluciferase Reporter Gene System Directed by Tandemly Repeated Pseudo-palindromic NFAT-Response Elements

Wei Zhang, Terunao Takahara, Takuya Achiha, Hideki Shibata, and Masatoshi Maki

Abstract

Luciferase reporter gene systems based on the NFAT-response element (RE) have been used to monitor intracellular Ca^{2+} elevation. However, Ca^{2+} mobilization agent (e.g., ionomycin) alone is not adequate to activate the currently often employed reporter gene that contains the NFAT-RE found in the *IL2* promoter. In addition to activation of NFAT through the Ca^{2+}-calmodulin/calcineurin pathway, activation of AP-1 as a partner transcription factor is essential for the *IL2*-based NFAT-RE system. Here, we describe a detailed method for the recently developed new reporter gene system containing the NFAT-RE from the *IL8* promoter. This system enables us to monitor endpoint effects of Ca^{2+}-mobilizing agonists independent of AP-1 activation.

Key words Calcineurin, Calcium signaling, Calmodulin, Carbachol, Luminescence, NanoLuc, NFAT, Reporter gene assay, Transcription factor

1 Introduction

NFATs are transcription factors involved in immune system regulation, development, cancer progression, and apoptosis. Among the five NFAT paralogues (NFAT1–5), NFAT1–4 are present in the cytoplasm as hyperphosphorylated forms in the basal state. Subsequent to intracellular Ca^{2+} mobilization, they are dephosphorylated through the calmodulin/calcineurin pathway and translocate to the nucleus to promote transcription of specific genes [1]. NFATs recognize specific DNA sequences [the NFAT-response elements (REs)] in the promoter region of the target genes. Reporter gene systems in cell-based assays have been developed to analyze *cis*-elements for transcription factors that regulate target gene expressions. Cells are transfected with the reporter gene expression plasmids containing response elements, a minimal

Claus W. Heizmann (ed.), *Calcium-Binding Proteins of the EF-Hand Superfamily: From Basics to Medical Applications*, Methods in Molecular Biology, vol. 1929, https://doi.org/10.1007/978-1-4939-9030-6_7,

promoter, and a reporter gene. Chloramphenicol acetyltransferase (CAT), β-galactosidase, secretory alkaline phosphatase (SEAP), and luciferases have been used as enzymatic reporters [2]. Luciferases catalyze the oxidation of the substrates like coelenterate luciferin (coelenterazine) in reactions that generate bioluminescent signals. Various types of luciferases are found in different organisms such as firefly *Photinus pyralis* (Fluc) [3, 4], sea pansy *Renilla reniformis* (Rluc) [5, 6], mesopelagic copepod *Gaussia princeps* (Gluc) [7], and deep-sea luminous shrimp *Oplophorus gracilirostris* (nanoluciferase; NanoLuc; Nluc) [8]. These luciferases are used separately or concomitantly to monitor biological processes including intracellular signal transductions and gene expressions [9].

To monitor intracellular Ca^{2+} elevation, a dual-luciferase assay system has been designed, in which NFAT-RE derived from the *IL2* promoter (e.g., 5′-GGAGGAAAAACTGTTTCATACAGAAGGC GT-3′; GenBank accession number, DQ904462.1) is used to promote transcription of a reporter gene [10]. However, intracellular Ca^{2+} mobilization alone (e.g., ionomycin) is not adequate to activate reporter genes driven by this type of NFAT-response elements, and AP-1 activators in addition to ionomycin or other Ca^{2+}-mobilizing reagents are needed. Both NFAT monomer and AP-1 (Fos/-Jun heterodimer) are essential to bind to this non-palindromic NFAT-RE sequence for transcriptional activation [11]. Addition of the AP-1 activator (e.g., PMA, phorbol 12-myristate 13-acetate, also named TPA, 12-O-tetradecanoyl-phorbol-13-acetate; a protein kinase C (PKC) activator) often complicates the obtained results due to the potential Ca^{2+}- and NFAT-independent reactions [12] (Fig. 1). On the other hand, the NFAT-RE in the *IL8* promoter has a pseudo-palindromic sequence (5′-GGAATTTCC-3′) that is recognized by a dimer form of NFAT [13].

In this chapter, we describe a protocol of the newly developed NanoLuc reporter assay system [14], which is directed by a tandemly repeated pseudo-palindromic NFAT-RE from the *IL8* promoter (Fig. 2). We co-express NanoLuc [*NlucP*, experimental reporter gene driven by the 9 tandem repeats of *IL8* NFAT-RE (9× IL8 NFAT-RE) and a minimal synthetic promoter], Fluc (*luc2*, transfection efficiency control reporter gene driven by the *PGK* promoter), and NFAT in HEK293 cells. After stimulation with a Ca^{2+}-mobilizing reagent, luciferase activities of NanoLuc and Fluc are sequentially measured using a dual-luciferase reagent kit. The ratio of NanoLuc to Fluc (Nluc/Fluc) is expressed as normalized relative luciferase activity (RLA). Although the efficiency of transcriptional activation (expressed as fold stimulation) by ionomycin is low, the 9× IL8 NFAT-RE reporter system does not require PMA (Fig. 3). Stimulation with carbachol—an acetylcholine receptor agonist known to mobilize Ca^{2+} in HEK293 cells [15, 16]—is sufficient to monitor the transcriptional activation of the Nluc reporter gene expression in the presence of exogenously expressed

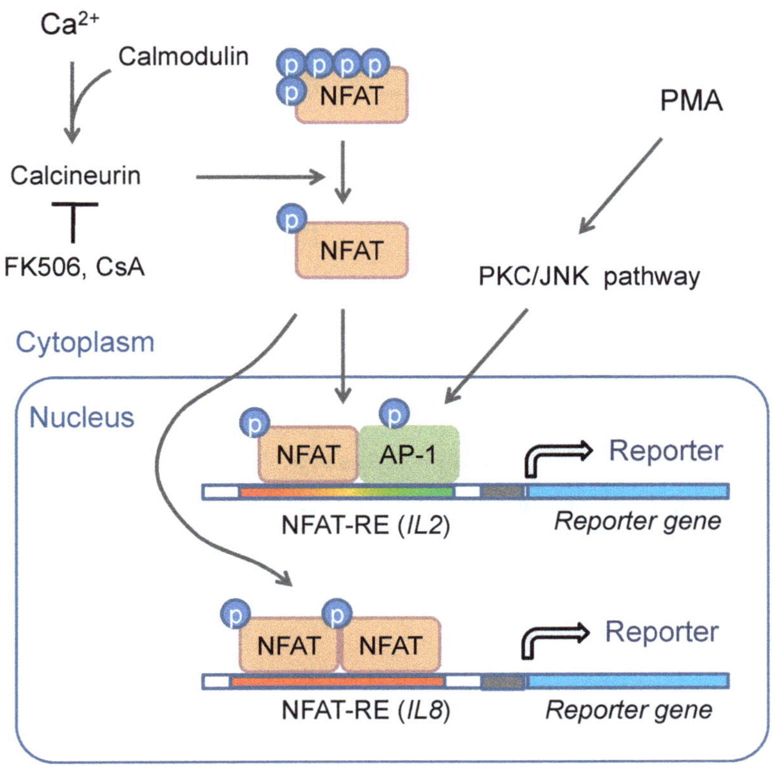

Fig. 1 NFAT-response element (RE)-directed nanoluciferase reporter system. Under the resting cell condition, inactive NFAT is localized in the cytoplasm in a hyperphosphorylated conformation. Subsequent to cell stimulation and intracellular Ca^{2+} elevation, NFAT is dephosphorylated through Ca^{2+}-calmodulin/calcineurin pathway and translocates to the nucleus to regulate gene expression. NFAT binds either to a non-palindromic composite NFAT-RE (*IL2*) by cooperating with partner transcription factors such as AP-1 or to a pseudo-palindromic NFAT-RE (*IL8*) as a dimer. Immunosuppressants FK506 and cyclosporine A (CsA) suppress the NFAT activation by inhibiting calcineurin

NFATs [14]. Since NFAT4 shows the highest stimulating efficiency among NFAT paralogues (Fig. 4), we describe a protocol of the reporter assay under exogenous expression of NFAT4 using carbachol as a Ca^{2+}-mobilizing reagent, and results of exemplar reporter assays are presented.

2 Materials

2.1 Transfection

1. Plasmids: pCMV-3× HA-A/NFAT4, pNL3.2 [*NlucP*/9× IL8 NFAT-RE/minP] [14], pGL4.53 [*luc2*/PGK] (Promega) (*see* **Note 1**).

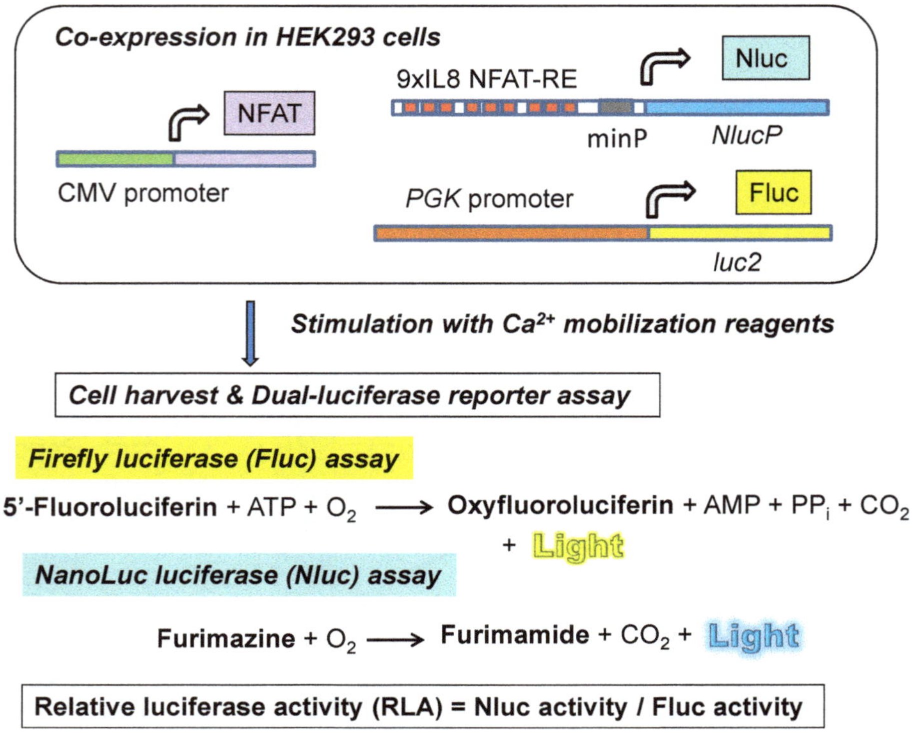

Fig. 2 Dual-Luciferase Reporter Assay System. 9× IL8 NFAT-RE and minimal promoter-directed NanoLuc luciferase (Nluc), *PGK* promoter-directed firefly luciferase (Fluc), and CMV promoter-directed NFAT are transiently co-expressed in HEK293 cells, followed by stimulation with Ca^{2+} mobilization reagents. After appropriate stimulation time (~6 h), cells are lysed and subjected to measurement of luminescent signals as luciferase activities. A kit of Nano-Glo® Dual-Luciferase Reporter Assay System (Promega) has an advantage of measuring two different luciferase activities sequentially using the same sample. First, the Fluc activity is initiated by mixing an aliquot of cell lysate with Fluc assay reagent containing substrate (5'-fluoroluciferin). Next, the Nluc assay reagent is added to the same sample to quench firefly luciferase activity and provide the substrate (furimazine) to measure Nluc activity. The ratio of Nluc activity to Fluc activity (Nluc/Fluc) is calculated to normalize the variation in transfection efficiency and sample handling and expressed as relative luciferase activity (RLA)

2. Mammalian culture cells for DNA transfection: Grow human embryonic kidney (HEK) 293 cells (RIKEN BioResource Center Cell Bank; also available from American Type Culture Collection) in Dulbecco's Modified Eagle's Medium (DMEM) supplemented with 0.584 mg/mL glutamine, 5% fetal bovine serum (FBS) at 37 °C under humidified air containing 5% CO_2.
3. Mammalian cell culture 24-well plates (Iwaki Science Products, Tokyo, Japan).
4. FuGENE 6 Transfection Reagent (Promega). Stored at 4 °C. Warm up to room temperature before use.

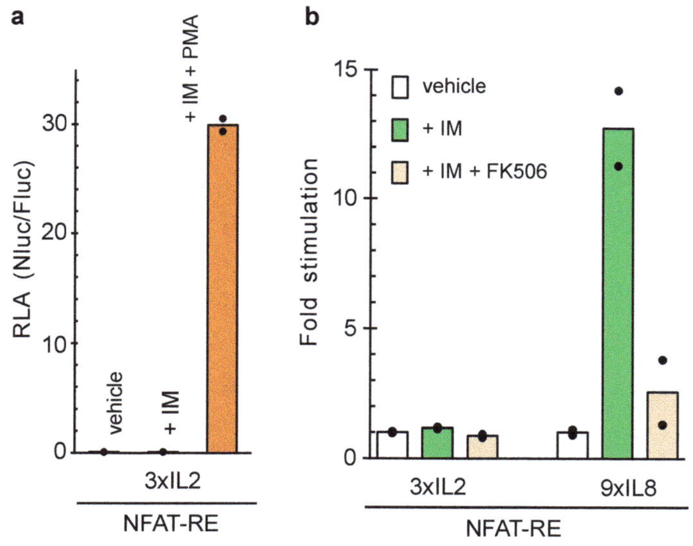

Fig. 3 Comparison of luciferase reporter activities between the NFAT-REs from the *IL2* promoter and the *IL8* promoter. (**a**) HEK293 cells were co-transfected in duplicates with the expression plasmids for murine NFAT1 (pcDNA3-mNFAT1) (1 ng) [14], Fluc (transfection efficiency control pGL4.53 [*luc2*/PGK] (20 ng)), and 3× IL2 NFAT-RE-directed Nluc pNL3.2 [*NlucP*/3× IL2 NFAT-RE/minP] (50 ng). The reporter assays were performed by stimulation of the transfected HEK293 cells with ionomycin (IM, 1 μM; vehicle, 0.007% ethanol) and PMA (10 ng/mL; vehicle, 0.01% DMSO) for 6 h. Cell lysates were used to measure luminescent signals of Nluc and Fluc using a Nano-Glo® Dual-Luciferase Reporter Assay System. Luminescence was measured with a single-tube type luminometer (ATTO luminescencer-MCA AB-2250). The ratio of Nluc to Fluc (Nluc/Fluc) is expressed as normalized relative luciferase activity (RLA). Dots and bars represent individual and averaged RLA values, respectively, obtained from the duplicate transfection assays. (**b**) The activities of 9× IL8 NFAT-RE- and 3× IL2 NFAT-RE-directed luciferase reporters were analyzed by stimulation with ionomycin (IM, 1 μM; vehicle, 0.007% ethanol) for 6 h with or without pre-treatment with FK506 (10 μM; vehicle, 0.16% ethanol) for 1 h. The measurement of luminescent signals was performed as described above. Fold stimulation represents the ratio of RLA values obtained by stimulation to those obtained without stimulation (the resting condition). Dots and bars represent individual and averaged fold stimulation values, respectively, obtained from the duplicate transfection assays

5. Opti-MEM™ Reduced Serum Medium (Thermo Fisher Scientific). Stored at 4 °C. Warm up to room temperature before use.

2.2 Stimulation

1. Carbachol (CCh): Dissolve carbachol (Santa Cruz Biotechnology) with water (100 mM stock solution), and store at −20 °C. Dilute to 1.14 mM with culture medium immediately before use (*see* **Note 2**).

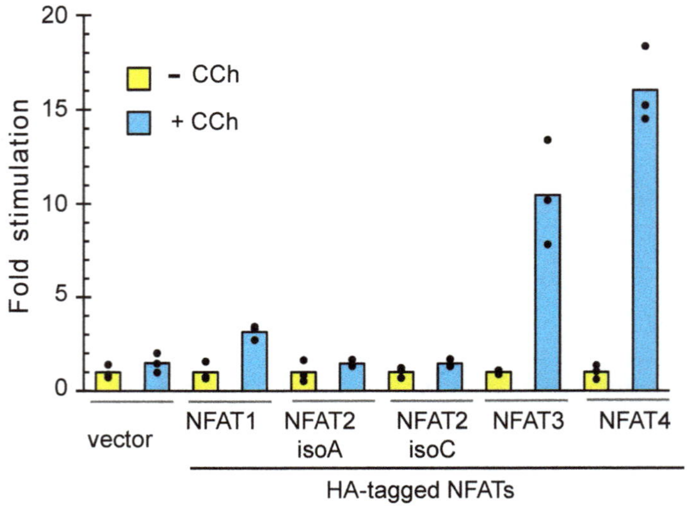

Fig. 4 Effects of carbachol (CCh) on different NFAT isoforms using the 9× IL8 NFAT-RE luciferase reporter. HEK293 cells were co-transfected in triplicates with the reporter plasmids and expression plasmid (1 ng) for HA-tagged human NFAT1, NFAT2 isoform A (isoA), NFAT2 isoform C (isoC), NFAT3, or NFAT4. The DNA-transfected cells were treated with or without CCh (100 μM) for 6 h, and then the reporter assays were performed. The ratio of Nluc to Fluc (Nluc/Fluc) is expressed as normalized relative luciferase activity (RLA). Fold stimulation represents the ratio of RLA values obtained by stimulation to those obtained without stimulation (the resting condition). Dots and bars represent individual and averaged fold stimulation values, respectively, obtained from the triplicate transfection assays. Luminescence was measured with a single-tube type luminometer (ATTO luminescencer-MCA AB-2250). Data were taken from Zhang et al. (ref. 14) and expressed as fold stimulation instead of RLA (Nluc/Fluc)

2.3 Cell Harvest

1. Phosphate-buffered saline (PBS): 137 mM NaCl, 2.7 mM KCl, 8 mM Na_2HPO_4, and 1.5 mM KH_2PO_4 (pH 7.4). Dilute 10× PBS stock solution (stored at room temperature) with water. Prepare at the time of use, or store at 4 °C for less than 1 month.
2. Passive lysis buffer (PLB): Dilute 5× PLB (Promega) with water.

2.4 Dual-Luciferase Reporter Assay

2.4.1 Signal Stability Analysis of Fluc and Nluc

1. Fluc assay reagent: Dissolve 1 vial of lyophilized ONE-Glo™ EX Luciferase Assay Substrate with 10 mL of ONE-Glo™ EX Luciferase Assay Buffer. Dispense 1 mL each of the firefly luciferase assay reagent into 1.5 mL light-shielding microtubes, and store at −20 °C. Thaw and take aliquots. Warm up to room temperature before use. All reagents described above are contained in the Nano-Glo® Dual-Luciferase® Reporter Assay System (Promega).

2. Nluc assay reagent: Dispense NanoDLR™ Stop & Glo® Buffer 1 mL each into 1.5 mL microtubes, and store at −20 °C. Thaw and take aliquots to dilute 100× NanoDLR™ Stop & Glo® Substrate (furimazine included). Warm up to room temperature before use. All reagents described above are contained in the Nano-Glo® Dual-Luciferase® Reporter Assay System (Promega).
3. Special equipment: A photon-counting type microplate reader (Monochromator Multimode Microplate Reader-Berthold technologies Mithras^2 LB 943 or equivalent).
4. White 384-well plates for luminescence measurement (Greiner Bio-One, Frickenhausen, Germany).

2.4.2 Comparison of Different Types of Luminometers

1. Special equipment: a personal single-tube type luminometer (ATTO luminescencer-MCA AB-2250 or equivalent).
2. 75 × 12 mm clear polystyrene (PS) tubes (4.5 mL, Sarstedt, Nümbrecht, Germany).

3 Methods

3.1 Transfection

1. Seed 2.5 × 10^4 HEK293 cells in 24-well plates in 0.5 mL of the medium 1 day before transfection.
2. Dilute pCMV-3× HA-A/NFAT4 (1 ng), pNL3.2 [*NlucP/*9× IL8 NFAT-RE/minP] (50 ng), pGL4.53 [*luc2*/PGK] (20 ng), with Opti-MEM™ Reduced Serum Medium to a final volume of 19.7 μL per one transfection sample. Prepare the plasmid DNA premix enough for multiple transfections.
3. Add 0.3 μL of FuGENE 6 Transfection Reagent per 1 plasmid DNA premix (19.7 μL).

   Prepare the transfection cocktail enough for multiple transfections.
4. Gently mix the transfection cocktail, and incubate at room temperature for 30 min.
5. Add 20 μL of the transfection cocktail to each well of the seeded cells in 0.5 mL of the medium.

3.2 Stimulation

1. Prepare 50 μL of 1.14 mM carbachol (CCh) solution per well by dilution from 100 mM CCh stock solution with culture medium.
2. After approximately 18 h post-transfection, stimulate cells with 100 μM CCh by addition of 50 μL of 1.14 mM CCh solution into cells (up to a final volume of 570 μL) (*see* **Note 3**).
3. Incubate the stimulated cells for 6 h in the CO_2 incubator (*see* **Note 4**).

3.3 Cell Harvest

1. Remove the medium with an aspirator, and rinse cells twice with 200 μL of PBS per well.
2. Add 50 μL of PLB into each well to harvest cells, and transfer the cell lysates of each well to an empty 0.6 mL tube.
3. Freeze the cell lysates at −20 °C for 30 min (*see* **Note 5**).
4. Thaw the cell lysates at room temperature for luciferase assay (*see* **Note 6**).

3.4 Dual-Luciferase Reporter Assay

3.4.1 Signal Stability Analysis of Fluc and Nluc

1. Prepare reagents to measure luciferase activities of Fluc and Nluc.
2. Take an aliquot of stimulated cell lysates, and make different dilutions with PLB for luciferase assays using a multimode microplate reader (Mithras^2 LB 943) with a luminescence mode.
3. Program the MikroWin 2010 software connected with the plate reader Mithras^2 LB 943 at the following settings:

   [Temperature] 25 °C; [Plate Parameter] Plate type, 384; Measurement order, By rows; [Shake] Duration, 5.0 s; Speed, Slow; Diameter, 1.0 mm; Type, Linear; [Measurement] Counting time, 1.0 s; Cycle time, 300 s; Total time, 3600 s; Repeated Operation; HiSense mode (380–650 nm); Counter position: Top.
4. Add 7.5 μL of the Fluc assay reagent and 3.75 μL of the cell lysates into a well of a 384-well plate (*see* **Note 7**).
5. Load the plate into the plate reader to start the first measurement of Fluc activities (*see* **Note 8**).
6. Take the plate out and add 7.5 μL of the Nluc assay reagent into the same well, and then load the plate into the plate reader again to start the measurement of Nluc activities.
7. Record the luminescent signals of Fluc and Nluc (Fig. 5) (*see* **Note 9**).

3.4.2 Comparison of Luminescence Units Measured with Different Types of Luminometers

1. Prepare reagents to measure luciferase activities of Fluc and Nluc.
2. Analyze the same cell lysates (Subheading 3.4.1, **step 2**) for comparison of luciferase activities measured with different types of luminometers (Subheading 3.4.2, **steps 3–4** for the plate reader luminometer and Subheading 3.4.2, **steps 5–8** for the single-tube type luminometer).
3. Program the plate reader Mithras^2 LB 943 with luminescence measurement mode as shown in Subheading 3.4.1, **step 3** except for [Shake] Duration, 1 s, and without recording the luminescence signals every 5 min for 1 h. Proceed to

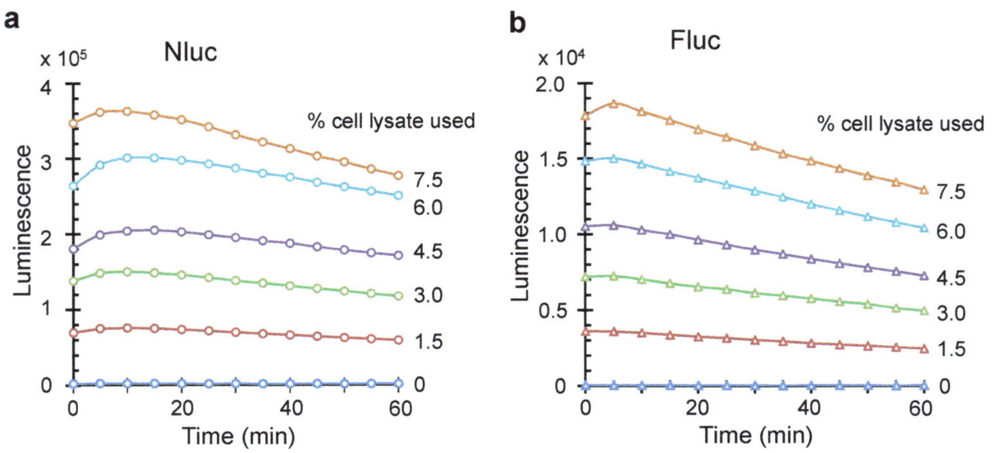

Fig. 5 Stability assessment of luminescent signals produced by Fluc and Nluc reactions. HEK293 cells were co-transfected with the reporter plasmids and an NFAT4 expression plasmid. The DNA-transfected cells were treated with CCh (100 μM) for 6 h, and cell lysates (50 μL) were diluted with PLB into different lysate concentrations to assess dose dependency of the luciferase activities (% cell lysate used are 0, 1.5, 3.0, 4.5, 6.0, and 7.5). The luminescent signals of Nluc (**a**) and Fluc (**b**)) in 3.75 μL of diluted samples were collected by Monochromator Multimode Microplate Reader (Berthold technologies Mithras^2 LB 943, a plate reader luminometer) at every 5 min for 1 h after manually mixing reaction reagents. Luminescent signals gradually decreased, but the slopes were smaller when using smaller amounts of the luciferase-containing lysates (*see* **Note 9**)

Subheading 3.4.1, **steps 4–6** for luciferase assays of Nluc (Fig. 6a) and Fluc (Fig. 6b).

4. Record the luminescent signals of Nluc and Fluc measured with the plate reader. The calculated ratio of Nluc to Fluc (Nluc/Fluc) is expressed as normalized relative luciferase activity (RLA) (Fig. 6c).
5. Add 7.5 μL of the Fluc assay reagent and 3.75 μL of the cell lysates into a luminometer tube, and then vortex the reaction mixture gently.
6. Place the tube in the luminometer tube holder of ATTO luminescencer-MCA AB-2250, and then start the measurement of Fluc activities at the following setting: delay, 10 s; integrated time of analysis, 10 s.
7. After measuring luminescence of Fluc activity, remove the luminometer tube, and add 7.5 μL of the Nluc assay reagent into the same tube; vortex the reaction mixture gently. Place the tube in the luminometer tube holder, and then measure luminescence of Nluc reaction at the following setting: delay, 60 s; integrated time of analysis, 10 s.
8. Record the reading of activities of Fluc and Nluc from ATTO luminescencer. The calculated ratio of Nluc to Fluc (Nluc/

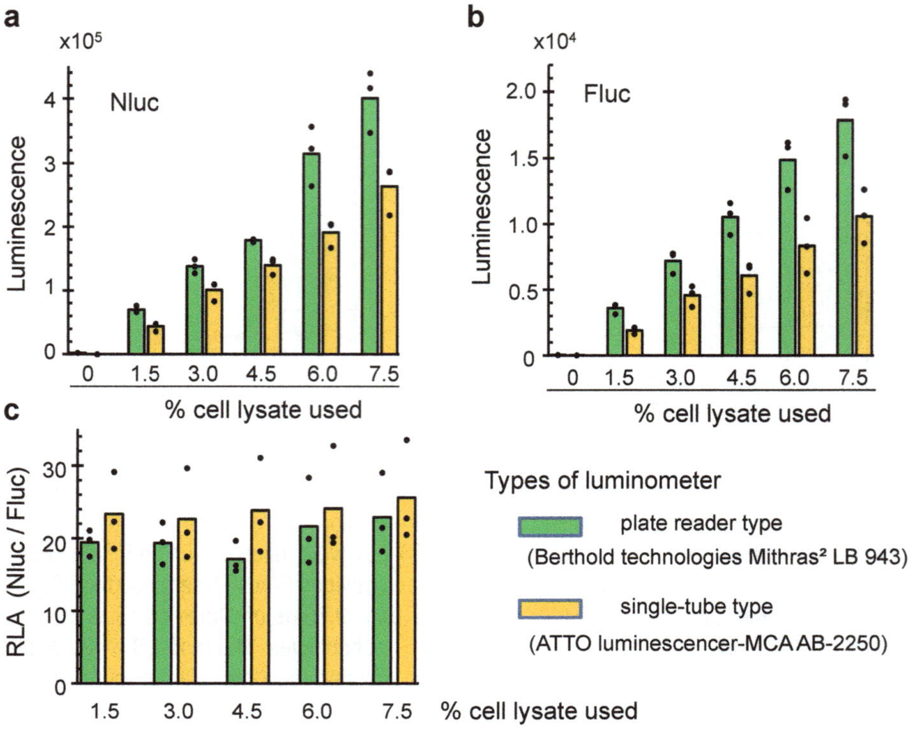

Fig. 6 Comparison between different types of luminometers on the measurement of luciferase activities. The luminescent signals of Nluc (**a**) and Fluc (**b**) collected by different types of luminometers were compared between ATTO luminescencer (MCA AB-2250, a personal single-tube type luminometer) and Monochromator Multimode Microplate Reader (Berthold technologies Mithras^2 LB 943, a plate reader luminometer). The DNA-transfected HEK293 cells were treated with CCh (100 μM) for 6 h, and cell lysates were diluted with PLB into different lysate concentrations to assess dose dependency of the luciferase activities (% cell lysate used are 0, 1.5, 3.0, 4.5, 6.0, and 7.5). Dots and bars represent individual and averaged luminescent signal values, respectively, obtained from the triplicate transfection assays. (**c**) The ratio of Nluc to Fluc (Nluc/Fluc) is expressed as normalized relative luciferase activity (RLA) (*see* **Note 10**)

Fluc) is expressed as normalized relative luciferase activity (RLA).

9. Data read from ATTO luminescencer are calculated as the average of 10 s-analysis (counts/s) to compare with data read from the plate reader Mithras^2 LB 943. (Fig. 6a–c).

4 Notes

1. Expression plasmid of the *PGK* promoter-driven *luc2* (Fluc) is used to normalize experimental variability of transfection efficiency and sample handling. Rluc cannot be used for this purpose due to the similarity of the luciferase reaction of Nluc (substrate, coelenterazine, and its analogue; no requirement of

ATP/Mg^{2+}). Herpes simplex virus (HSV)-1 *TK* promoter is frequently used for expression of the control luciferase for normalization. However, caution is required because the *TK* promoter is modulated by cellular factors. PMA, an activator of the PKC/JNK pathway, activates the HSV-1 *TK* promoter in HEK293 cells that express adenovirus E1A protein [17].

2. In the case of using other Ca^{2+} mobilization reagents, effects of vehicles used to dissolve reagents need to be assessed. Chemicals used to stimulate cells are also reported to influence luciferase activities [18, 19].
3. Dose-dependent CCh-stimulated reporter gene activities are shown in Fig. 7. An optimum concentration of CCh stimulation in this study is 100 μM.
4. Time course of Nluc reporter gene expression by CCh stimulation is shown in Fig. 8. An optimum time of CCh stimulation in this study is 6 h. Optimum conditions for other Ca^{2+} mobilizing agonists may need to be tested.
5. Passive lysis buffer (PLB) will passively lyse cells without the requirement of a freeze-thaw cycle. However, lysis efficiency is dependent upon the cell types and needs to be determined. *See* Luciferase Assay System Technical Bulletin #TB281 provided from Promega.
6. Luciferase activity is temperature dependent (Reference: Nano-Glo® Dual-Luciferase® Reporter Assay System technical manual). Samples and reagents should be pre-warmed to room temperature before use.
7. An appropriate starting volume of samples needs to be determined depending on plate types and well sizes. The 96-well plate needs more volume to cover the well's bottom surface.
8. Alternatively, with an autoinjection mode, the luciferase reaction can be started by injecting the Fluc reagent to each well after loading the sample-containing plate to the plate reader. However, due to the large dead space volume of the injection syringe and tubing (~1.6 mL), manual mixing of reagents is recommended to save reagents when assaying a small number of samples (7.5 μL reagent/sample).
9. A glow-type luciferase assay kit allows stable luminescent signal emission. These results indicate that starting luciferase reactions manually (e.g., 24 samples) on a bench and recording with the plate reader instantly (1 s for each sample) cause no serious problem of differences in the total reaction time among samples.
10. Luminescent signals of both Nluc and Fluc gradually decreased probably due to consumption of the substrate, but the slopes were smaller by using smaller amounts of the luciferase-

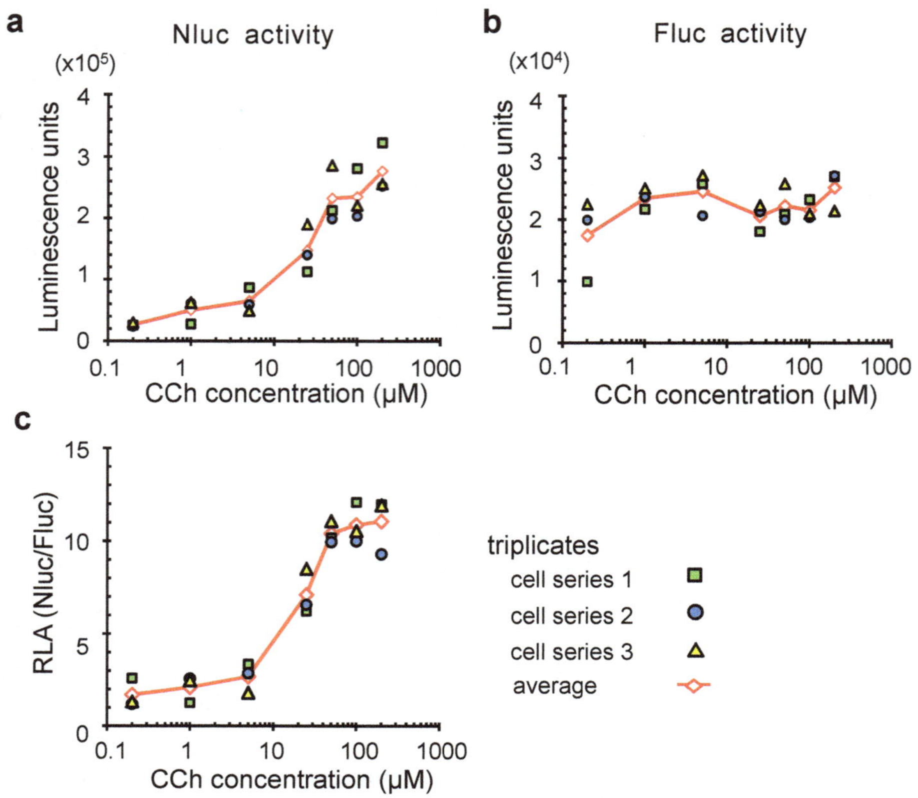

Fig. 7 Dose-dependent activation of Nluc reporter gene expression by carbachol (CCh). The DNA-transfected HEK293 cells were unstimulated or stimulated with different concentrations of CCh (0.2–200 μM) for 6 h. Cells were lysed and subjected to measurement of (**a**) Nluc and (**b**) Fluc luciferase activities. Calculated relative luciferase activities (RLA, Nluc/Fluc) are shown in (**c**) for normalization of experimental variance in triplicate cell series 1–3. Individual and the averaged values in the triplicate assays are shown and marked with different symbols as indicated. Data were taken from Zhang et al. (ref. 14), and figures were modified. A graph in semi-logarithmic scale of CCh concentrations is presented. Luminescence was measured with a microplate reader (Berthold technologies Mithras[2] LB 943)

containing lysates. When reactions of a series of samples are to be started by manual mixing on the bench (~15 s/sample) and measured all at once with the plate reader (1 s/sample), the net reaction time becomes variable from sample to sample. Higher dilution rates are recommended to reduce the differences in luminescent signals due to variation of the reaction time. Differences in the amounts of used cell lysates seem smaller in the RLA (Nluc/Fluc) values calculated from the data obtained from the single-tube type luminometer. After each luciferase reaction was started by addition of reagents at a fixed time interval (100 s), the luminescence was measured one by one to keep the reaction time constant. A large variation among

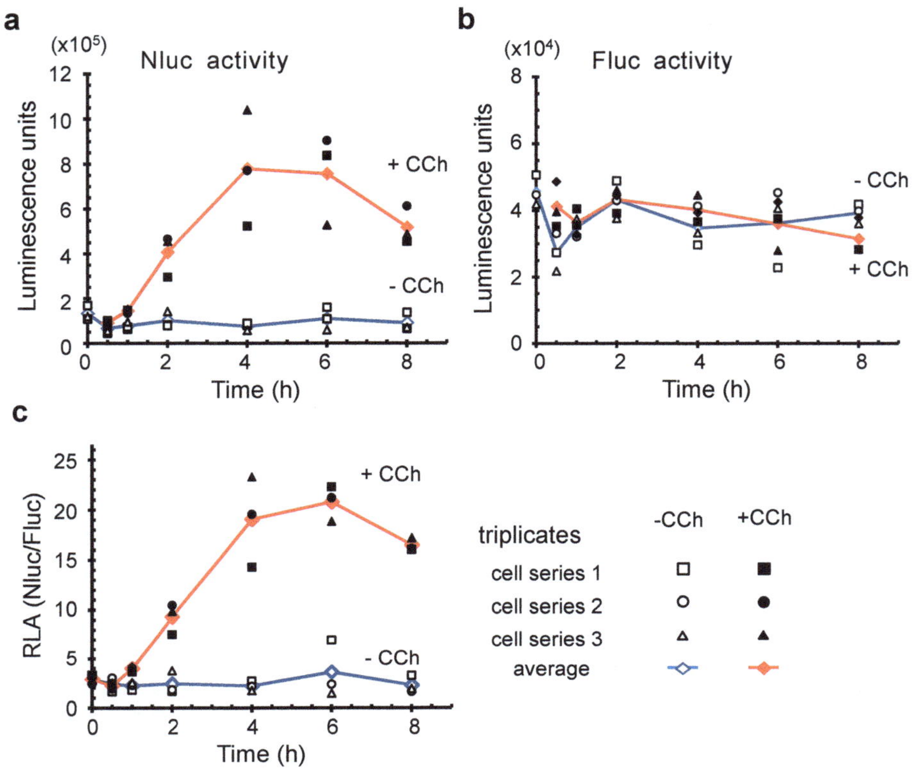

Fig. 8 Time course of the Nluc reporter gene expression by carbachol (CCh) stimulation. The DNA-transfected cells were stimulated with or without carbachol (CCh, 100 μM; vehicle, medium), and culture was continued for a different period of time (0, 0.5, 1, 2, 4, 6, and 8 h) in triplicates. Cells were harvested at indicated time and subjected to measurement of (**a**) Nluc and (**b**) Fluc luciferase activities. Calculated relative luciferase activities (RLA, Nluc/Fluc) are shown in (**c**) for normalization of experimental variance in triplicate cell series 1–3. Individual and the averaged values in the triplicate assays are shown and marked with different symbols as indicated: unfilled symbols (−CCh); filled symbols (+CCh); a line graph, average. Data were taken from Zhang et al. (ref. 14), and figures were modified. Luminescence was measured with a single-tube type luminometer (ATTO luminescencer-MCA AB-2250)

three-cell series might be caused by pipetting handling. Increase of the reaction volume might reduce the variance.

Acknowledgments

This work was supported by JSPS KAKENHI Grant Number 17H03803 (Grant-in-Aid for Scientific Research (B) to M.M.). We thank current and past members of the Laboratory of Molecular and Cellular Regulation for valuable suggestions and discussion.

References

1. Hogan PG, Chen L, Nardone J, Rao A (2003) Transcriptional regulation by calcium, calcineurin, and NFAT. Genes Dev 17:2205–2232. https://doi.org/10.1101/gad.1102703
2. Naylor LH (1999) Reporter gene technology: the future looks bright. Biochem Pharmacol 58:749–757
3. Williams TM, Burlein JE, Ogden S, Kricka LJ, Kant JA (1989) Advantages of firefly luciferase as a reporter gene: application to the interleukin-2 gene promoter. Anal Biochem 176:28–32
4. Ignowski JM, Schaffer DV (2004) Kinetic analysis and modeling of firefly luciferase as a quantitative reporter gene in live mammalian cells. Biotechnol Bioeng 86:827–834. https://doi.org/10.1002/bit.20059
5. Srikantha T, Klapach A, Lorenz WW, Tsai LK, Laughlin LA, Gorman JA, Soll DR (1996) The sea pansy *Renilla reniformis* luciferase serves as a sensitive bioluminescent reporter for differential gene expression in *Candida albicans*. J Bacteriol 178:121–129
6. Bhaumik S, Lewis XZ, Gambhir SS (2004) Optical imaging of *Renilla* luciferase, synthetic *Renilla* luciferase, and firefly luciferase reporter gene expression in living mice. J Biomed Opt 9:578–586. https://doi.org/10.1117/1.1647546
7. Remy I, Michnick SW (2006) A highly sensitive protein-protein interaction assay based on *Gaussia* luciferase. Nat Methods 3:977–979. https://doi.org/10.1038/nmeth979
8. Hall MP, Unch J, Binkowski BF, Valley MP, Butler BL, Wood MG, Otto P, Zimmerman K, Vidugiris G, Machleidt T, Robers MB, Benink HA, Eggers CT, Slater MR, Meisenheimer PL, Klaubert DH, Fan F, Encell LP, Wood KV (2012) Engineered luciferase reporter from a deep sea shrimp utilizing a novel imidazopyrazinone substrate. ACS Chem Biol 7:1848–1857. https://doi.org/10.1021/cb3002478
9. England CG, Ehlerding EB, Cai W (2016) NanoLuc: a small luciferase is brightening up the field of bioluminescence. Bioconjug Chem 27:1175–1187. https://doi.org/10.1021/acs.bioconjchem.6b00112
10. Weaver JR, Good K, Walters RD, Kugel JF, Goodrich JA (2007) Characterization of the sequence and architectural constraints of the regulatory and core regions of the human interleukin-2 promoter. Mol Immunol 44:2813–2819. https://doi.org/10.1016/j.molimm.2007.01.027
11. Macian F, Lopez-Rodriguez C, Rao A (2001) Partners in transcription: NFAT and AP-1. Oncogene 20:2476–2489. https://doi.org/10.1038/sj.onc.1204386
12. Pfeifhofer C, Kofler K, Gruber T, Tabrizi NG, Lutz C, Maly K, Leitges M, Baier G (2003) Protein kinase C theta affects Ca^{2+} mobilization and NFAT cell activation in primary mouse T cells. J Exp Med 197:1525–1535. https://doi.org/10.1084/jem.20020234
13. Jin L, Sliz P, Chen L, Macian F, Rao A, Hogan PG, Harrison SC (2003) An asymmetric NFAT1 dimer on a pseudo-palindromic κB-like DNA site. Nat Struct Biol 10:807–811. https://doi.org/10.1038/nsb975
14. Zhang W, Takahara T, Achiha T, Shibata H, Maki M (2018) Nanoluciferase reporter gene system directed by tandemly repeated pseudo-palindromic NFAT-response elements facilitates analysis of biological endpoint effects of cellular Ca^{2+} mobilization. Int J Mol Sci 19(2): E605. https://doi.org/10.3390/ijms19020605
15. Luo D, Broad LM, Bird GS, Putney JW Jr (2001) Signaling pathways underlying muscarinic receptor-induced $[Ca^{2+}]_i$ oscillations in HEK293 cells. J Biol Chem 276:5613–5621. https://doi.org/10.1074/jbc.M007524200
16. Vetter I, Lewis RJ (2010) Characterization of endogenous calcium responses in neuronal cell lines. Biochem Pharmacol 79:908–920. https://doi.org/10.1016/j.bcp.2009.10.020
17. Shifera AS, Hardin JA (2009) PMA induces expression from the herpes simplex virus thymidine kinase promoter via the activation of JNK and ERK in the presence of adenoviral E1A proteins. Arch Biochem Biophys 490:145–157. https://doi.org/10.1016/j.abb.2009.08.013
18. Shifera AS, Hardin JA (2010) Factors modulating expression of Renilla luciferase from control plasmids used in luciferase reporter gene assays. Anal Biochem 396:167–172. https://doi.org/10.1016/j.ab.2009.09.043
19. Auld DS, Inglese J (2016) Interferences with luciferase reporter enzymes. In: Sittampalam GS, Coussens NP, Brimacombe K, Grossman A, Arkin M, Auld D, Austin C,

Baell J, Bejcek B, TDY C, Dahlin JL, Devanaryan V, Foley TL, Glicksman M, Hall MD, Hass JV, Inglese J, Iversen PW, Kahl SD, Kales SC, Lal-Nag M, Li Z, McGee J, McManus O, Riss T, Trask OJ Jr, Weidner JR, Xia M, Xu X (eds) Assay guidance manual [Internet]. Eli Lilly & Company and the National Center for Advancing Translational Sciences, Bethesda, MD

Chapter 8

Designing Calcium-Binding Proteins for Molecular MR Imaging

Mani Salarian, Shenghui Xue, Oluwatosin Y. Ibhagui, and Jenny J. Yang

Abstract

Early diagnosis, noninvasive detection, and staging of various diseases, remain one of the major clinical barriers to effective medical treatment and prevention of disease progression toward major clinical consequences. Molecular imaging technologies play an indispensable role in the clinical field in overcoming these major barriers. The increasing application of imaging techniques and agents in early detection of different diseases such as cancer has resulted in improved treatment response and clinical patient management. In this chapter we will first introduce criteria for the design and engineering of calcium-binding protein (CaBP) parvalbumin as a *pro*tein Gd-MRI *c*ontrast *a*gent (ProCA) with unprecedented metal selectivity for Gd^{3+} over physiological metal ions. We will then discuss the further development of targeted MRI contrast agent for molecular imaging of PSMA biomarker for early detection of prostate cancer.

Key words CaBP, Parvalbumin, Molecular imaging, MRI, ProCA, Contrast agent, PSMA

1 Introduction

Magnetic resonance imaging (MRI) is a noninvasive imaging modality providing high-resolution, 3D images of anatomic structures, as well as functional and physiological information about tissues in vivo. MRI has the capacity to detect abnormalities in deep tissues and provide whole-body imaging and, therefore, has been recognized as one of the primary diagnostic imaging techniques [1, 2]. Exogenous MRI contrast agents are usually applied to enhance the contrast between pathological and normal tissues by changing the longitudinal and transverse (i.e., T1 and T2) relaxation times of water protons.

Gd^{3+} chelators have high magnetic moments, asymmetric electronic ground states, and potential for increased MRI contrast; therefore, they are among the most frequently used MRI contrast agents [3, 4]. Relaxivity, defined as the capability of the agent to change the relaxation time of water, is dependent on several factors

Claus W. Heizmann (ed.), *Calcium-Binding Proteins of the EF-Hand Superfamily: From Basics to Medical Applications*, Methods in Molecular Biology, vol. 1929, https://doi.org/10.1007/978-1-4939-9030-6_8,

such as the number of water molecules in the coordination shell, the exchange rate of the coordinated water with the bulk water, and the rotational correlation time τ_R of the molecule [5, 6].

There are several important criteria for an MRI contrast agent to have: first, they should have high relaxivity for high contrast-to-noise ratio (CNR) and dose efficiency; second, good thermodynamic and chemical stability and especially good metal selectivity for the Gd^{3+} over other physiological metal ions, to avoid the release of toxic Gd^{3+}; third, adequate vascular tissue retention time for optimized imaging time; and fourth, timely excretion from the body.

To date, the most widely used MRI contrast agent in diagnostic imaging is Gd-DTPA (diethylene triamine pentaacetic acid) or related derivatives such as Gd-DTPA-BMA (bismethylamide), which are based on small-molecule Gd^{3+} chelators. These small molecular Gd^{3+} chelators have longitudinal and transverse proton relaxivities, r_1 and r_2, less than 5 $mM^{-1}\ s^{-1}$ at clinical and high field strength, which are much lower than the theoretically maximal value (>100 $mM^{-1}\ s^{-1}$), so they therefore lack sufficient sensitivity [7]. Moreover, these small-molecule contrast agents demonstrate very short blood circulation (less than 30 min) and tissue retention time, limiting their MRI applications [8]. To increase correlation time, τ_R, and therefore their sensitivity, small chelators have been covalently or non-covalently linked to macromolecules such as linear polymers [9], dendrimers [10, 11], carbohydrates [12], proteins [13, 14], viral capsids [15], and liposomes [16]. However, conjugation yields limited improvement due to challenges associated with internal mobility, restricted water-exchange rates, limited applications to imaging biomarkers on the blood vessel because of large size and low tissue penetration, and incomplete elimination from the body resulting in Gd^{3+} toxicity.

Calcium-binding proteins (CaBPs) have been an attractive target in molecular imaging studies since they represent the important role of calcium as a second messenger in different cellular signaling pathways. For instance, fluorescent calcium sensors are widely applied in optical imaging at cellular and cell population levels. Calcium-sensitive dyes combined with laser scanning microscopy have been recently used to monitor neural network activities in small brain areas [17], as well as characterizing patterns of interaction between cells in vertebrate embryos [18]. Until now, however, only very limited cases have reported the application of engineered CaBPs as molecular magnetic resonance imaging (MRI) Gd^{3+} contrast agents, possibly due to challenges associated with the engineering and development of protein-based MRI contrast agents and designing suitable Gd^{3+}-binding sites in these proteins.

We have pioneered the development of a novel class of MRI contrast agents using protein design and engineering. Proteins were investigated as a suitable biomolecule for contrast agent development based on the fact that approximately 1/3 of all proteins are metalloproteins with diversified metal selectivity and biodistribution, and they are biocompatible. Previous studies have also applied shorter peptides, and increases in relaxivity were observed when Gd^{3+} binds to calcium-binding peptides [19] or proteins such as concanavalin A and bovine serum albumin (BSA). However, application of these short EF-hand peptides or proteins as MRI contrast agents has been limited due to their weak metal-binding affinity for Gd^{3+} (K_d ~100 μM for Eu^{3+}) and lack of dynamic flexibility [19, 20]. Jasanoff and coworkers previously reported calmodulin (CaM) modified superparamagnetic iron oxide (SPIO) nanoparticles as Ca^{2+}-responsive MRI T2 agents, but these constructs lacked the ability to penetrate cell membranes, and effective imaging was limited by the dark contrast associated with the iron oxide nanoparticles [21].

Taking a different approach, work in our laboratory has focused on engineering Gd^{3+}-binding sites in CaBPs to develop Gd^{3+} MRI contrast agents [22–28]. In this chapter, we first report our detailed analysis of metal-binding sites in various natural CaBPs to gain insight into coordination chemistry and metal selectivity. Then, we describe our approach to designing and engineering protein-based contrast agents by creating high coordination Gd^{3+}-binding sites in a stable Ca^{2+}-binding protein using amino acid residues and water molecules as metal coordinating ligands and thereby transforming EF-hand motifs into Gd^{3+}-binding motifs, resulting in MRI contrast agents for molecular imaging of cancer and their biomarkers. The design of Gd^{3+}-binding proteins based on CaBPs for molecular imaging described in this chapter largely relies on the concept of the EF-hand calcium-binding protein (α-parvalbumin), and that will be the focus of the chapter. The protocol will further summarize our methods for biophysical characterization of the engineered protein and development of different metal-binding assays. The designed protein demonstrates a high selectivity for Gd^{3+} over physiological metal ions such as Ca^{2+}, Zn^{2+}, and Mg^{2+}. Moreover, it shows a 20-fold increase in longitudinal and transverse relaxation rate values over clinical small-molecule contrast agents, such as Gd-DTPA. In addition, these contrast agents, designed from CaBPs, have stronger contrast enhancement and much longer blood retention time than Gd-DTPA in mice. With superior in vivo properties, the protein contrast agent ProCA has shown great promise as a molecular imaging probe to target disease biomarkers and extending applications of magnetic resonance imaging (MRI) (Fig. 1).

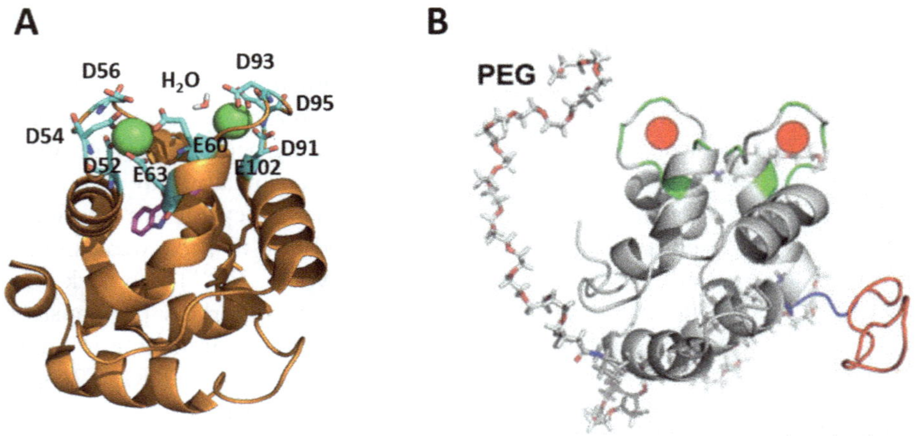

Fig. 1 (**a**) Model structure of ProCA32 based on α-parvalbumin. (**b**) Model structure of PEGylated PSMA-targeted ProCA32 for the molecular imaging of PSMA in prostate cancer

2 Materials

2.1 Bioinformatic Analysis of Calcium-Binding Sites Based on MUG Algorithm

1. Protein sequence search from NCBI: http://www.ncbi.nlm.nih.gov/protein/.
2. MUltiple Geometries (MUG) algorithm to accurately predict CaBPs with multiple-binding sites that undergo local conformational change or side-chain rotations upon Ca^{2+}-binding: http://chemistry.gsu.edu/faculty/Yang/Calciomics.htm [29].
3. Java program for MUG implementation: http://java.sun.com/.
4. PyMOL, used for molecular visualization: http://pymol.sourceforge.net/.

2.2 Molecular Cloning Reagents

1. Primer phosphorylation: T4 Polynucleotide Kinase (PNK) and PNK buffer (New England Biolabs). ATP (Sigma).
2. PCR: KOD hot start DNA polymerase and buffer (Novagen). dNTP (Novagen). $MgSO_4$ (Novagen).
3. Ligation: T4 DNA ligase and ligation buffer (New England Biolabs). ATP (Sigma).
4. Transformation: Competent cells: DH5α and BL21(DE3) pLysS (Invitrogen). LB broth media and LB agar (EMD Chemicals). Ampicillin (Inalco S.p.A, Milano, Italy).
5. Molecular cloning kits: QIAprep Spin Miniprep Kit and QIAquick Gel Extraction Kit (Qiagen).

2.3 Protein Expression and Purification

1. Protein expression for α-parvalbumin (ProCA32):$6#Escherichia coli BL21(DE3)pLysS cell strain. Isopropyl β-D-1-thiogalactopyranoside (IPTG) stock solution: 1 M IPTG in

ddH_2O, sterilized by injection through a 0.2 μm syringe filter. UV-VIS spectrophotometer (Shimadzu).

2. Protein purification and preparation of α-parvalbumin (ProCA32):
   Sonication and French press. Lysis buffer: 10 mM HEPES, pH 7.2, with 100 μM PMSF and 1 μL of benzonuclease (Novagen). DNA precipitation with 3% (wt/vol) streptomycin sulfate. HiTrap Q column (GE Healthcare). Chelex-100 for metal removal. FPLC (GE Healthcare). ICP-OES (Agilent Technologies). PEGylation buffer: 10 mM HEPES at pH 7.2. PEGylation reagents: trimethyl-PEG-NHS Ester reagent, TMS (PEG) 12 (Thermo Scientific). PEGylation quench buffer: 100 mM Tris–HCl, pH = 8.0.

2.4 Instruments for Biophysical Characterization of Engineered Proteins

QuantaMaster™40 fluorometer (PTI, Birmingham, NJ). Jasco-810 circular dichroism spectropolarimeter (JASCO, Easton, MD). Relaxometer at 1.47 T (Bruker). Stelar Spinmaster Fast Field Cycling NMR spectrometer FFC-2000 (Bruker). ICP-OES (Agilent Technologies).

2.5 Fluorescent-Based Metal-Binding Assays

Metal-binding affinity assays:

10 μM ProCA32, 5 mM EGTA or 5 mM DTPA, 100 mM KCl, 50 mM HEPES at pH 7.2., 10 mM or 20 μM stock solution of $CaCl_2$ or $TbCl_3$.

2.6 Water Coordination Number by Terbium Lifetime Luminescence

10 μM of Tb^{3+}-DTPA, Tb^{3+}-EDTA, Tb^{3+}-NTA, and Tb^{3+}-H_2O, and 10 μM of Tb^{3+}-ProCA in H_2O and D_2O.

2.7 Transmetallation Studies of ProCA32

50 or 100 μM ProCA, 100 μM $GdCl_3$, 100 μM $ZnCl_2$, and 1.2 mM PO_4^{3-} buffer.

3 Methods

3.1 Design of Gd^{3+}-Binding Sites in α-Parvalbumin

Run MUG algorithm with three major steps:

1. In **step 1**, for any protein with determined holo (bound) structure, locate clusters of oxygen atoms in close proximity to each other in the three-dimensional structure, and treat each cluster as a potential ligand group.
2. In **step 2**, recognize a point CC (calcium center) for the insertion of a calcium ion using a grid algorithm for each oxygen cluster.
3. In the final step, sequentially filter and apply various restrictions to the structure of the (cluster, CC) pair. After passing all filters,

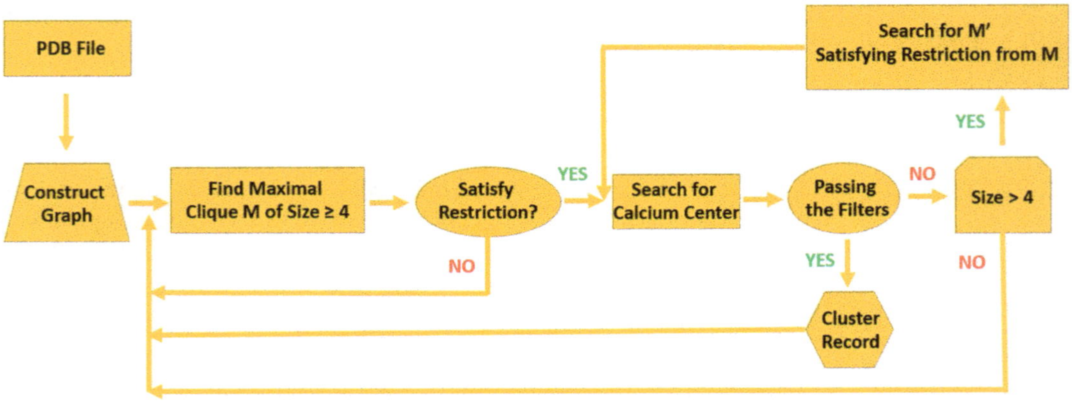

Fig. 2 Flow chart of MUG algorithm for prediction of Gd^{3+}- and Ca^{2+}-binding sites in CaBPs. Restriction: a cluster must have at least four oxygen atoms of specified types where at least n ($n = 2, 3, 4$) oxygen atoms are from amino acids. Both the type and n are the input parameters

the cluster can be a predicted ligand group, and CC is the predicted calcium position [29–32]. This process is summarized in Fig. 2.

4. Develop the protein-based MRI contrast agent (ProCA32) based on rat α-parvalbumin. Parvalbumin has two functional Ca^{2+}-binding sites (CD and EF sites).
5. Achieve the mutations of S56D and F103 W in α-parvalbumin by site-directed mutagenesis.
6. Generate the S56D mutation in the CD site of the protein, and on the amino acid serine, which directly coordinates Ca^{2+} via its main chain oxygen. The replacement of aspartic acid introduces more negative charge in the coordination sphere, increasing the selectivity of Gd^{3+} over Ca^{2+}. In this contrast agent, the affinity of Gd^{3+} was significantly increased. Introduce the F103 W mutation in the EF site for ease of determination of metal-binding affinities by Luminescence Resonance Energy Transfer (LRET).

3.2 ProCA32 Expression and Purification

1. Transform the correct plasmid encoding the engineered protein into BL21(DE3)pLysS competent cells for protein expression. Positive colonies were selected on ampicillin-resistant LB agar plates.
2. Inoculate a single colony into 10 mL LB-amp medium, and shake the culture overnight at 200–250 rpm at 37 °C.
3. Transfer the overnight culture to 1 L fresh LB-amp medium, and allow the culture to grow until the OD_{600} reaches 0.6–0.8.
4. Induce protein expression with 0.3 mM IPTG, and allow the bacteria to grow for another 3–4 h at 200 rpm, 37 °C.
5. Harvest cells by centrifugation at 5000 × *g* for 20 min, and collect the cell pellets in a 50 mL centrifuge tube.

6. Store the cell pellets at −80 °C for future use.
7. Sonicate the cell pellet in 10 mM HEPES buffer, pH 7.2, 100 μM PMSF, and 1 μL of benzonuclease, and then pass it through a French pressure cell.
8. Incubate the supernatant of the cell lysate at 85 °C for 10 min, cool to 4 °C in an ice-water bath, and centrifuge (41,100 × *g*, 20 min, 4 °C).
9. Then precipitate the unwanted components, such as DNA, by addition of 3% (wt/vol) streptomycin sulfate.
10. After dialysis at 4 °C for 24 h against 10 mM HEPES at pH 8.0, purify the protein mixture using a HiTrap Q column.
11. After removing unbound proteins with three column volumes of HEPES at pH 8.0, elute the column with a 0–1 M NaCl gradient in 10 mM HEPES at pH 8.0. Remove metals in ProCA32 by Chelex-100, and analyze metal content in ProCA32 by ICP-OES.

3.3 Structural Analysis of ProCA32

1. Secondary structure analysis by circular dichroism spectroscopy:

   Acquire the circular dichroism spectrum of the engineered protein with a Jasco-810 spectropolarimeter at ambient temperature using a quartz cell of 1-mm path length. Obtain all spectra as the average of at least 8 scans with a scan rate of 50 nm/min.

   (a) Prepare 300 mL of the engineered protein sample with a final concentration of 10–20 mM in 10 mM Tris–HCl at pH 7.4. Transfer ~250 mL to a 1-mm-path-length quartz cuvette.

   (b) Record CD spectrum in the far-UV (190–260 nm) range. Adjust the protein concentration until the CD signals are between −10 and −20 mdeg.

   (c) Record and use the CD signals from the buffer for background subtraction.

2. NMRD studies of ProCA32:

   (a) Conduct magnetic relaxation dispersion experiments over the magnetic field range corresponding to proton Larmor frequencies from 0.01 to 40 MHz at 25 °C.

   (b) Regulate the temperature using the Stelar temperature controller to set airflow temperature over the sample.

   (c) Measure relaxation rate constants with an automated profile-acquisition procedure using a 20-MHz polarization field and 16 different relaxation delays at each relaxation field, and detect the free induction decay at 15.8 MHz following each field cycle.

3.4 ^{17}O NMR Study of ProCA32

Measure the water-exchange rates of ProCA using variable temperature ^{17}O NMR.

(a) Prepare 10 μM ProCA containing 20 mM Gd^{3+} in 2% (vol/vol) ^{17}O water with 10 mM HEPES buffer at pH 7.0.

(b) Load this solution, or 2% (vol/vol) ^{17}O water without protein, in a spherical NMR bulb, and then insert it into a 5-mm NMR sample tube filled with 400 μL of ddH_2O containing 3% (vol/vol) D_2O.

(c) Collect the variable temperature Fourier transform ^{17}O NMR at 54.24 MHz using Bruker 400-MHz NMR spectrometer equipped with broadband probe. Measure the line widths of ^{17}O at different temperatures ranging from 25 to 50 °C.

(d) Calculate the reduced transverse relaxation rate ($1/T_{2r}$) by Eqs. (1) and (2):

$$\frac{1}{T_{2r}} = \frac{\pi}{P_m} \times (\Delta_{\text{ProCA}} - \Delta_{\text{solvent}}) \tag{1}$$

$$P_m = \frac{q \times [\text{Gd}^{3+}]}{[\text{H}_2\text{O}]} \tag{2}$$

where P_m is the molar fraction of solvent in the exchanging site compared with bulk water, which you can determine by Eq. (2).

Δ_{ProCA} is the half-width of the ^{17}O NMR signal of solvent in the presence of 10 mM ProCA. Δ_{solvent} denotes the half-width of the ^{17}O NMR signal of solvent without contrast agent, while q is the water coordination number of ProCA32-P40, [Gd^{3+}] is the concentration of Gd^{3+}, and [H_2O] is the water concentration.

(e) Fit the water-exchange rate by Eqs. (3)–(5):

$$\frac{1}{T_{2r}} = \frac{1}{\tau_m}\left(\frac{\Delta\omega_m^2}{\tau_m^{-2} + \Delta\omega_m^2}\right) \tag{3}$$

$$\frac{1}{\tau_m} = k_{\text{ex}} = \frac{k_{\text{ex}}^{298} T}{298.15} \exp\left[\frac{\Delta H^{\ddagger}}{R}\left(\frac{1}{298.15} - \frac{1}{T}\right)\right] \tag{4}$$

$$\Delta\omega_m = \frac{A}{T} \tag{5}$$

where $k_{ex}{}^{298}$ is the water molecule exchange rate at 298.15 K, T is the temperature in Kelvin, $\Delta H^{\ddagger}$ is the enthalpy of activation, R is the universal gas constant, and $\Delta\omega_m$ is the difference between the resonance frequency of ^{17}O nuclei of inner-sphere water and bulk water. Assume that temperature dependence of $\Delta\omega_m$ is the simple reciprocal function of A/T (Eq. 5) [33], where you can determine A as a parameter in the treatment of the line broadening data.

3.5 Metal-Binding Affinity Assays

1. Excitation and emission wavelength setup for metal-binding experiments: $6#Monitor tryptophan (Trp) fluorescence using excitation at 280 nm and emission between 290 and 370 nm with 2–4 nm bandpasses. Acquire Tb^{3+} LRET data by excitation of tryptophan at 280 nm and emission from 500 to 650 nm.
2. Ca^{2+}-binding affinity determination:

   To calculate Ca^{2+}-binding affinity to ProCA, use a calcium-EGTA buffer system, which contains 50 mM HEPES, 100 mM NaCl, and 5 mM EGTA at pH 7.2. Then, titrate calcium chloride at different concentrations into the buffer system, and calculate the free calcium concentration according to Tsein's method [34] (K_d of EGTA is 1.51 × 10^{-7} M) using Eq. (6). Calculate the Ca^{2+}-binding affinity for ProCA ($K_{d_{\mathrm{Ca,ProCA}}}$) by fitting Trp fluorescence change over various concentrations of free Ca^{2+} using Eq. (7):

$$\left[\mathrm{Ca}^{2+}\right]_{\mathrm{free}} = K_{d_{\mathrm{Ca,EGTA}}} \times \frac{[\mathrm{Ca-EGTA}]}{[\mathrm{EGTA}]_{\mathrm{free}}} \tag{6}$$

$$f = \frac{\left[\mathrm{Ca}^{2+}\right]_{\mathrm{free}}^{n}}{K_{d_{\mathrm{Ca,ProCA}}^{n}} + \left[\mathrm{Ca}^{2+}\right]_{\mathrm{free}}^{n}} \tag{7}$$

   where $[Ca^{2+}]_{free}$ is the free Ca^{2+} concentration at each titration point, $K_{d_{\mathrm{Ca,EGTA}}}$ is the dissociation constant of EGTA for Ca^{2+}, [Ca-EGTA] denotes the concentration of total Ca-EGTA complex at each titration point, f represents fractional change of fluorescent signal, and $[\mathrm{EGTA}]_{\mathrm{free}}$ is the concentration of EGTA at each titration point.
3. Tb^{3+}-binding affinity determination:

   Obtain Tb^{3+}-binding affinity for ProCA using a Tb^{3+}-DTPA buffer system. Add 30 μM ProCA into the Tb^{3+}-DTPA buffer system, which contains 5 mM DTPA, 50 mM HEPES, and 100 mM NaCl at pH 7.2. Then, titrate $TbCl_3$ into the buffer, and calculate the free Tb^{3+} concentration by Eq. (8). Excite ProCA at 280 nm, and observe the Tb^{3+}-LRET signal at emission wavelengths between 500 and 650 nm. Calculate the Tb^{3+}-binding affinity for ProCA32 ($K_{d_{\mathrm{Tb,DTPA}}}$) by fitting the Tb^{3+}-LRET signal change over various concentrations of free Tb^{3+} using Eq. (9):

$$\left[\mathrm{Tb}^{3+}\right]_{\mathrm{free}} = K_{d_{\mathrm{Tb,DTPA}}} \times \frac{[\mathrm{Tb-DTPA}]}{[\mathrm{DTPA}]} \tag{8}$$

$$f = \frac{\left[\mathrm{Tb}^{3+}\right]_{\mathrm{free}}^{n}}{K_{d_{\mathrm{Tb,ProCA}}^{n}} + \left[\mathrm{Tb}^{3+}\right]_{\mathrm{free}}^{n}} \tag{9}$$

where $[Tb^{3+}]_{free}$ denotes the free Tb^{3+} concentration calculated from the buffer system, $K_{d_{Tb,DTPA}}$ is the dissociation constant between Tb^{3+} and DTPA, [Tb-DTPA] represents the concentration of Tb-DTPA complex, $[DTPA]_{free}$ is the free DTPA concentration in the buffer, f is the fractional change, and $K_{d_{Tb,ProCA}}$ is the dissociation constant between Tb^{3+} and ProCA32.

3.6 Relaxivity Studies of ProCA32

Measure the relaxation times (T1 and T2) using a Bruker relaxometer at 1.47 T and MRI scanner at 4.7 and 7 T. Prepare the contrast agents at different concentrations in 50 mM HEPES and 100 mM NaCl, pH 7.4. Calculate the relaxivities by Eq. (10):

$$r_i = \frac{\frac{1}{T_{\text{isample}}} - \frac{1}{T_{\text{ibuffer}}}}{[\text{Gd}]_T} \quad i = 1, 2, \tag{10}$$

where r_i denotes the per Gd^{3+} relaxivity, T_{isample} is the relaxation time of water after adding ProCA, T_{ibuffer} is the relaxation time of water without adding contrast agents, and $[\text{Gd}]_T$ is the total Gd^{3+} concentration in the tested solution (Fig. 3).

3.7 Determination of Water Coordination Number of ProCA

Determine the number of coordination water molecules in the inner sphere of Gd^{3+}-ProCA by the difference in Tb^{3+} luminescence decay between H_2O and D_2O [35]. Measure the Tb^{3+} lifetime using a fluorescence spectrophotometer.

(a) Prepare Tb^{3+}-ProCA complexes in H_2O and D_2O, separately.

(b) After excitation at 265 mm, record Tb^{3+} emission decay over time at 545 nm.

(c) Fit luminescence decay using monoexponential decay.

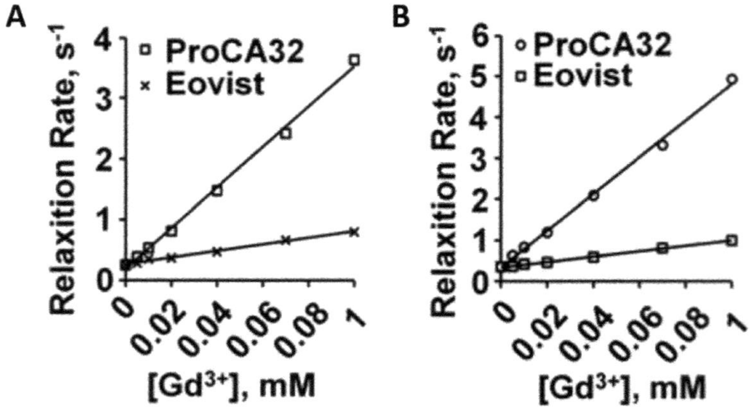

Fig. 3 (**a**) Changes in R_1 relaxation rate over different concentrations of Gd^{3+} in ProCA32 and clinical contrast agent, Eovist at 1.5 T and 37 °C. (**b**) Changes in R_2 relaxation rate over different concentrations of Gd^{3+} in ProCA32 and Eovist at 1.5 T and 37 °C

(d) Establish a standard curve of water number (denoted q) over ΔK_{obs} (the difference of the decay constant between H_2O and D_2O) using well-characterized chelators, such as Tb^{3+}-DTPA ($n = 1$), Tb^{3+}- EDTA ($n = 3$ when [EGTA]:[Tb^{3+}] = 1:1), Tb^{3+}-NTA, and Tb^{3+} in aqueous solution ($n = 9$).

(e) Calculate the water number of Tb^{3+}-ProCAs by fitting ΔK_{obs} into the standard curve. Obtain the water numbers of Gd-DTPA and Eovist from published studies [36].

3.8 Transmetallation Assay

Consider the transmetallation of Gd^{3+} chelators by physiological ions such as Fe^{3+}, Cu^{2+}, and Zn^{2+} in the development of the contrast agents. Under physiological conditions, Zn^{2+} and PO_4^{3-} are the major molecules to compete with contrast agents for Gd^{3+}. Measure the relaxation rate changes of ProCA32 loaded with different Gd^{3+} ratios in phosphate buffer supplemented with $ZnCl_2$ for different lengths of time. Due to the high relaxivity, reduce both the Gd^{3+} concentration and chelator concentrations to 25–50 times in this study. To better compare the transmetallation properties of ProCA32 with clinical MRI contrast agents, reduce the concentrations of Zn^{2+} and PO_4^{3-} by 25 times.

3.9 Molecular Imaging of PSMA Using Targeted ProCA32

For molecular imaging of PSMA in prostate cancer, a contrast agent should have targeting moieties with affinity at least in the μM range. Moreover, it must have high relaxivity in order to produce significant contrast differences in the tumor site before and after injection of contrast agents. Another important criterion is strong Gd^{3+} affinity and high Gd^{3+} selectivity over physiological metal ions such as Zn^{2+} and Ca^{2+} to prevent free Gd^{3+} from being released, which could result in nephrogenic systemic fibrosis. Additionally, because PSMA is mainly expressed on prostate tumor cells that are far away from the blood vessels, the contrast agent must have strong tumor penetration. Develop the PSMA-targeted ProCA32 by engineering a PSMA-targeting peptide at the C-terminal of ProCA32 (Fig. 1b). Insert a flexible linker between ProCA32 and the targeting peptide to prevent the protein from interfering with the targeting capability of the peptide. Addition of the targeting moiety at the C-terminal of ProCA32 has little effect on the expression and folding of the protein variants. PEGylate the protein to improve its in vivo properties [23]. The incorporation of a PSMA-targeting peptide to ProCA32 with a flexible linker equips ProCA32 with PSMA-targeting capacity while maintaining high metal affinity and selectivity, as well as high relaxivities.

3.10 Parameters of Molecular Imaging with MRI

Collect MR images of the mice on both 4.7 and 7.0 T Varian MR animal scanners. Acquire different MR pulse sequences such as T1-weighted, T2-weighted, and T2 map images. Acquire T1-weighted images using a 2D spin-echo pulse sequence with

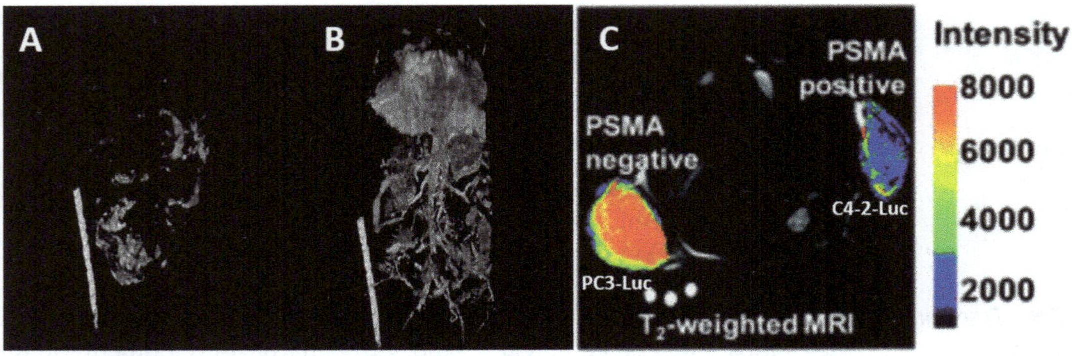

Fig. 4 Three-dimensional MRI of mouse (**a**) before and (**b**) after injection of ProCA32. The bright line next to the mouse is a tube filled with the contrast agent as a reference. (**c**) MR imaging of PSMA in xenografted mice tumors by PSMA-targeted ProCA32 (ProCA32.PSMA). T2-weighted MRI of the mice implanted with both PC3-Luc and C4–2-Luc tumors after injection of ProCA32.PSMA were collected. The PSMA-positive tumor, C4–2-Luc, exhibited decreased MRI signal intensity at 30 min–48 h postinjection of ProCA32.PSMA, while the PSMA-negative tumor, PC3-Luc, did not show any significant MRI signal change

the following parameters: TR/TE, 400/11 ms; FOV, 4 cm × 4 cm with a matrix size of 512 × 512; slice thickness, 1 mm; and 23 slices. Acquire T2-weighted images using a 2D RARE pulse sequence with the following parameters: TR/TE, 4.4 s/56 ms; FOV, 4 cm × 4 cm with a matrix size of 256 × 256; and slice thickness of 1 mm. Collect The T2 map before the contrast agent injection and at different time points after IV injection of 5 mM ProCA32 or ProCA32.PSMA using the MSME sequence with the following parameters: TR = 3 s, different length of TE, matrix = 128 × 128, FOV = 4 cm × 3 cm, and slice thickness = 1 mm. Generate the final T2-map images by using an ImageJ plugin MRI T2 calculator (*see* Fig. 4).

4 Notes

1. To ensure that the protein loaded to FPLC is completely apo- or Ca^{2+}-free, incubate the supernatant with 20 mM EGTA for a few hours at 4 °C, and then inject into an FPLC for column binding [22, 23].
2. When different fractions of FPLC are eluting the protein, you should take 20 μL of samples, and mix it with 120 μL of 1× Bio-Rad Protein Assay solution. If the solution turns blue, the eluent should include the purified protein.
3. To monitor the protein expression and purification process, you are advised to collect samples at each step and analyze them with SDS-PAGE. For example, we always keep 11 samples for SDS-PAGE analysis: cells before induction, cells after induction (1.5 and 3 h), cell pellets after sonication,

supernatant before binding and after 85 °C incubation, supernatant after streptomycin/dialysis, fractions after column binding, flow through, and eluent [22, 23].

4. Snap-freeze the cell pellets with liquid nitrogen and store at −80 °C for long-term use. It is strongly suggested to proceed to purification steps in the same day to minimize degradation and/or unfolding of engineered proteins.
5. For α-parvalbumin purification, the pH of 8.0 is important, and it is optimized since it helps the binding of the protein to the HiTrap Q column [22].
6. The 15 min interval between each titration point is essential for reaching full equilibrium between metals and protein in all metal-binding assays. It is important to observe the gradual increase or decrease of luminescence signal until it reaches saturation [22, 23].

Acknowledgments

We appreciate the critical review by Dr. Michael Kirberger and previous works by Drs. Fan Pu, Jingjuan Qiao, and Jie Jiang. This work is supported in part by grants R42 CA183376, R41CA177034, R41AA112713, EB007268, 1R01GM081749 from National Health Institute to Jenny J. Yang. This work was also supported by the Molecular Basis of Disease (MBD) Fellowship to M. Salarian and Center for Diagnostics and Therapeutics (CDT) Fellowship to O. Y. Ibhagui.

References

1. Tyszka JM, Fraser SE, Jacobs RE (2005) Magnetic resonance microscopy: recent advances and applications. Curr Opin Biotechnol 16 (1):93–99. https://doi.org/10.1016/j.copbio.2004.11.004
2. Lippard SJ (2006) The inorganic side of chemical biology. Nat Chem Biol 2(10):504–507. https://doi.org/10.1038/nchembio1006-504
3. Lauffer RB, Vincent AC, Padmanabhan S, Villringer A, Saini S, Elmaleh DR, Brady TJ (1987) Hepatobiliary MR contrast agents: 5-substituted iron-EHPG derivatives. Magn Reson Med 4(6):582–590
4. Aime S, Barge A, Cabella C, Crich SG, Gianolio E (2004) Targeting cells with MR imaging probes based on paramagnetic Gd(III) chelates. Curr Pharm Biotechnol 5(6):509–518
5. Burai L, Scopelliti R, Toth E (2002) EuII-cryptate with optimal water exchange and electronic relaxation: a synthon for potential pO2 responsive macromolecular MRI contrast agents. Chem Commun (Camb) 20:2366–2367
6. Geraldes CF, Sherry AD, Cacheris WP, Kuan KT, Brown RD 3rd, Koenig SH, Spiller M (1988) Number of inner-sphere water molecules in Gd^{3+} and Eu^{3+} complexes of DTPA-amide and -ester conjugates. Magn Reson Med 8(2):191–199
7. Caravan P (2006) Strategies for increasing the sensitivity of gadolinium based MRI contrast agents. Chem Soc Rev 35(6):512–523. https://doi.org/10.1039/b510982p
8. Weinmann HJ, Press WR, Gries H (1990) Tolerance of extracellular contrast agents for magnetic resonance imaging. Investig Radiol 25 (Suppl 1):S49–S50
9. Opsahl LR, Uzgiris EE, Vera DR (1995) Tumor imaging with a macromolecular

paramagnetic contrast agent: gadopentetate dimeglumine-polylysine. Acad Radiol 2 (9):762–767

10. Langereis S, de Lussanet QG, van Genderen MH, Meijer EW, Beets-Tan RG, Griffioen AW, van Engelshoven JM, Backes WH (2006) Evaluation of Gd(III)DTPA-terminated poly (propylene imine) dendrimers as contrast agents for MR imaging. NMR Biomed 19 (1):133–141. https://doi.org/10.1002/nbm.1015
11. Bryant LH Jr, Brechbiel MW, Wu C, Bulte JW, Herynek V, Frank JA (1999) Synthesis and relaxometry of high-generation (G = 5, 7, 9, and 10) PAMAM dendrimer-DOTA-gadolinium chelates. J Magn Reson Imaging 9 (2):348–352
12. Sirlin CB, Vera DR, Corbeil JA, Caballero MB, Buxton RB, Mattrey RF (2004) Gadolinium-DTPA-dextran: a macromolecular MR blood pool contrast agent. Acad Radiol 11 (12):1361–1369. https://doi.org/10.1016/j.acra.2004.11.016
13. Lanza GM, Winter PM, Caruthers SD, Morawski AM, Schmieder AH, Crowder KC, Wickline SA (2004) Magnetic resonance molecular imaging with nanoparticles. J Nucl Cardiol 11 (6):733–743
14. Lanza GM, Winter P, Caruthers S, Schmeider A, Crowder K, Morawski A, Zhang H, Scott MJ, Wickline SA (2004) Novel paramagnetic contrast agents for molecular imaging and targeted drug delivery. Curr Pharm Biotechnol 5(6):495–507
15. Anderson EA, Isaacman S, Peabody DS, Wang EY, Canary JW, Kirshenbaum K (2006) Viral nanoparticles donning a paramagnetic coat: conjugation of MRI contrast agents to the MS2 capsid. Nano Lett 6(6):1160–1164. https://doi.org/10.1021/nl060378g
16. Strijkers GJ, Mulder WJ, van Heeswijk RB, Frederik PM, Bomans P, Magusin PC, Nicolay K (2005) Relaxivity of liposomal paramagnetic MRI contrast agents. MAGMA 18 (4):186–192. https://doi.org/10.1007/s10334-005-0111-y
17. Stosiek C, Garaschuk O, Holthoff K, Konnerth A (2003) In vivo two-photon calcium imaging of neuronal networks. Proc Natl Acad Sci U S A 100(12):7319–7324. https://doi.org/10.1073/pnas.1232232100
18. Wallingford JB, Ewald AJ, Harland RM, Fraser SE (2001) Calcium signaling during convergent extension in Xenopus. Curr Biol 11 (9):652–661
19. Caravan P, Greenwood JM, Welch JT, Franklin SJ (2003) Gadolinium-binding helix-turn-helix peptides: DNA-dependent MRI contrast agents. Chem Commun (Camb) 20:2574–2575
20. Kim Y, Welch JT, Lindstrom KM, Franklin SJ (2001) Chimeric HTH motifs based on EF-hands. J Biol Inorg Chem 6(2):173–181
21. Atanasijevic T, Shusteff M, Fam P, Jasanoff A (2006) Calcium-sensitive MRI contrast agents based on superparamagnetic iron oxide nanoparticles and calmodulin. Proc Natl Acad Sci U S A 103(40):14707–14712. https://doi.org/10.1073/pnas.0606749103
22. Xue S, Yang H, Qiao J, Pu F, Jiang J, Hubbard K, Hekmatyar K, Langley J, Salarian M, Long RC, Bryant RG, Hu XP, Grossniklaus HE, Liu ZR, Yang JJ (2015) Protein MRI contrast agent with unprecedented metal selectivity and sensitivity for liver cancer imaging. Proc Natl Acad Sci U S A 112 (21):6607–6612. https://doi.org/10.1073/pnas.1423021112
23. Pu F, Salarian M, Xue S, Qiao J, Feng J, Tan S, Patel A, Li X, Mamouni K, Hekmatyar K, Zou J, Wu D, Yang JJ (2016) Prostate-specific membrane antigen targeted protein contrast agents for molecular imaging of prostate cancer by MRI. Nanoscale 8(25):12668–12682. https://doi.org/10.1039/c5nr09071g
24. Pu F, Qiao J, Xue S, Yang H, Patel A, Wei L, Hekmatyar K, Salarian M, Grossniklaus HE, Liu ZR, Yang JJ (2015) GRPR-targeted protein contrast agents for molecular imaging of receptor expression in cancers by MRI. Sci Rep 5:16214. https://doi.org/10.1038/srep16214
25. Xue S, Qiao J, Jiang J, Hubbard K, White N, Wei L, Li S, Liu ZR, Yang JJ (2014) Design of ProCAs (protein-based Gd(3+) MRI contrast agents) with high dose efficiency and capability for molecular imaging of cancer biomarkers. Med Res Rev 34(5):1070–1099. https://doi.org/10.1002/med.21313
26. Qiao J, Xue S, Pu F, White N, Jiang J, Liu ZR, Yang JJ (2014) Molecular imaging of EGFR/HER2 cancer biomarkers by protein MRI contrast agents. J Biol Inorg Chem 19 (2):259–270. https://doi.org/10.1007/s00775-013-1076-3
27. Li S, Jiang J, Zou J, Qiao J, Xue S, Wei L, Long R, Wang L, Castiblanco A, White N, Ngo J, Mao H, Liu ZR, Yang JJ (2012) PEGylation of protein-based MRI contrast agents improves relaxivities and biocompatibilities. J Inorg Biochem 107(1):111–118. https://doi.org/10.1016/j.jinorgbio.2011.11.004
28. Pu F, Xue S, Qiao J, Patel A, Yang JJ (2016) Towards the Molecular Imaging of Prostate Cancer Biomarkers Using Protein-based MRI

Contrast Agents. Curr Protein Pept Sci 17 (6):519–533

29. Wang X, Kirberger M, Qiu F, Chen G, Yang JJ (2009) Towards predicting Ca^{2+}–binding sites with different coordination numbers in proteins with atomic resolution. Proteins 75 (4):787–798. https://doi.org/10.1002/prot.22285
30. Kirberger M, Wang X, Deng H, Yang W, Chen G, Yang JJ (2008) Statistical analysis of structural characteristics of protein Ca^{2+}–binding sites. J Biol Inorg Chem 13 (7):1169–1181. https://doi.org/10.1007/s00775-008-0402-7
31. Wang X, Zhao K, Kirberger M, Wong H, Chen G, Yang JJ (2010) Analysis and prediction of calcium-binding pockets from apo-protein structures exhibiting calcium-induced localized conformational changes. Protein Sci 19(6):1180–1190. https://doi.org/10.1002/pro.394
32. Zhou Y, Yang W, Kirberger M, Lee HW, Ayalasomayajula G, Yang JJ (2006) Prediction of EF-hand calcium-binding proteins and analysis of bacterial EF-hand proteins. Proteins 65(3):643–655. https://doi.org/10.1002/prot.21139
33. Dees A, Zahl A, Puchta R, Hommes NJ, Heinemann FW, Ivanovic-Burmazovic I (2007) Water exchange on seven-coordinate Mn (II) complexes with macrocyclic pentadentate ligands: insight in the mechanism of Mn (II) SOD mimetics. Inorg Chem 46 (7):2459–2470. https://doi.org/10.1021/ic061852o
34. Grynkiewicz G, Poenie M, Tsien RY (1985) A new generation of Ca^{2+} indicators with greatly improved fluorescence properties. J Biol Chem 260(6):3440–3450
35. Sudnick DR, Horrocks WD Jr (1979) Lanthanide ion probes of structure in biology. Environmentally sensitive fine structure in laser-induced terbium(III) luminescence. Biochim Biophys Acta 578(1):135–144
36. Caravan P, Ellison JJ, McMurry TJ, Lauffer RB (1999) Gadolinium(III) chelates as MRI contrast agents: structure, dynamics, and applications. Chem Rev 99(9):2293–2352

Chapter 9

Coordination to Divalent Cations by Calcium-Binding Proteins

Masayuki Nara, Hisayuki Morii, and Masaru Tanokura

Abstract

Fourier-transform infrared spectroscopy (FTIR) is a powerful tool for examining the metal coordination of the side chain COO^- groups of Glu and Asp on Ca^{2+}-binding proteins in solution. The behavior of COO^- symmetric stretch can be investigated by using protein samples in H_2O solution. However, it is difficult to obtain information about the behavior of the COO^- antisymmetric stretch in H_2O solution, because the COO^- antisymmetric stretching band overlaps with the amide II band. Therefore, to obtain reliable infrared spectra in the region of COO^- antisymmetric stretch, exchangeable protons in the protein should be completely deuterated by incubating the apoprotein dissolved in D_2O under mild heating conditions.

Key words FTIR, Coordination structure, Carboxylate group, Synthetic peptide analogue, HD exchange

1 Introduction

Fourier-transform infrared spectroscopy (FTIR) is a powerful tool for examining the metal coordination of the side chain COO^- groups of Glu and Asp on Ca^{2+}-binding proteins in solution, which play important role in Ca^{2+}-mediated functions [1–6]. Infrared studies have been done on the relationship between COO^- stretching frequencies and coordination types for acetate salts [7–9]. In general, the carboxylate (COO^-) groups can coordinate to metal ions in three modes (Fig. 1): "unidentate" (or "monodentate"), "bidentate" (or "chelating"), and "bridging" (or "bridging bidentate") modes [7, 8, 10]. As a special case of the bridging mode, the pseudo-bridging coordination mode features a water molecule replacing one of the two ligands in the bridging coordination [7, 10]. The frequency of the COO^- antisymmetric stretch of the unidentate species is higher than that of the ionic (metal-free) species, which is in turn higher than that of the bidentate species. The reverse is the case for the COO^- symmetric

Claus W. Heizmann (ed.), *Calcium-Binding Proteins of the EF-Hand Superfamily: From Basics to Medical Applications*, Methods in Molecular Biology, vol. 1929, https://doi.org/10.1007/978-1-4939-9030-6_9,

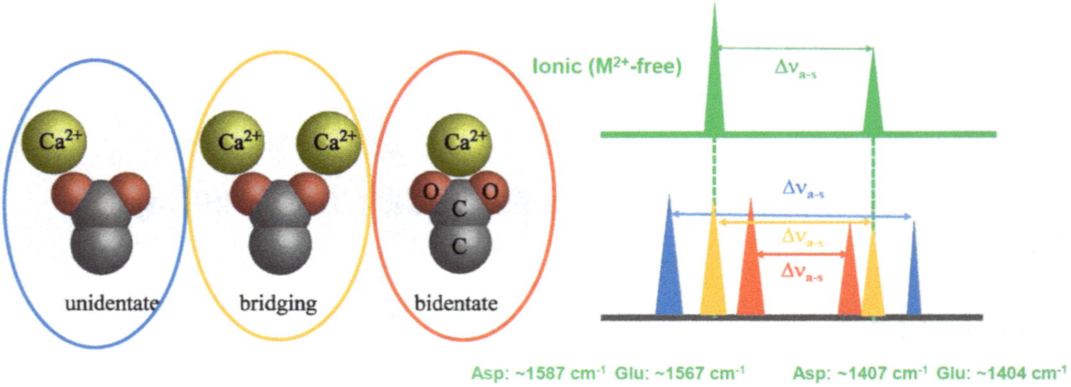

Fig. 1 General tendency in the relationship between $\Delta\nu_{a-s}$ and the coordination types of the COO^- groups to metal ions

stretch. As a result, the $\Delta\nu_{a-s}$ values for unidentate, bridging, bidentate, and ionic species are in the following order:

$$\Delta\nu_{\text{a-s}}(\text{unidentate}) > \Delta\nu_{\text{a-s}}(\text{ionic}) \sim \Delta\nu_{\text{a-s}}(\text{bridging}) > \Delta\nu_{\text{a-s}}(\text{bidentate}),$$

where $\Delta\nu_{\text{a-s}}$ (ionic) is approximately 180 cm^{-1} and 160 cm^{-1} for β-COO^- group of Asp and γ-COO^- group of Glu, respectively [1, 2] (*see* **Note 1**).

The behavior of COO^- symmetric stretch can be investigated by using protein samples in H_2O solution, where the handling of HD exchange in sample solution is not necessary. However, it is difficult to obtain information about the behavior of the COO^- antisymmetric stretch in H_2O solution, because the COO^- antisymmetric stretching band overlaps with the amide II band. Therefore, to obtain reliable infrared spectra in the region of COO^- antisymmetric stretch, exchangeable protons in the protein should be completely deuterated by incubating the apoprotein dissolved in D_2O under mild heating conditions (e.g., for 60 min at 60 °C). If it is difficult to exchange H for D completely for amide groups of membrane proteins and biological systems, it is suitable to analyze the COO^- symmetric stretch alone. As an alternative method, complete HD exchange can be carried out at room temperature without heating by using synthetic peptide analogues corresponding to the Ca^{2+}-binding domain of a complicated protein, since it is easy to exchange H for D completely for amide groups of a shorter peptide.

2 Materials

Prepare all solutions using ultrapure water (prepared by purified deionized water, to attain of 18 MΩ-cm at 25 °C) and analytical grade reagents. Prepare and store all reagents at room temperature.

2.1 Protein Solutions

1. HEPES: Purchase from Dojindo Molecular Technologies, Inc.
2. EGTA and EDTA: Purchase from Dojindo Molecular Technologies, Inc.
3. KCl, $CaCl_2$, $MgCl_2$, and NaOH: Purchase from FUJIFILM Wako Pure Chemical Co.
4. D_2O: Purchase from Sigma-Aldrich.
5. Buffer solution: 40 mM HEPES-NaOD, pD 7.4. Weigh 0.19 g of HEPES and transfer to the clinder. Add D_2O to a volume of 10 mL. Mix and adjust pD with 1 M NaOD. Make up to 20 mL with D_2O at 4 °C.
6. Ca^{2+}-binding protein: Prepare M^{2+}-free protein powder as described in the method.

2.2 Synthesis of Peptides

1. Support resin for elongation of peptidyl chain: Fmoc-NH-SAL-PEG resin (4-[(2,4-Dimethoxyphenyl)-*N*-(9-fluorenylmethoxycarbonyl)aminomethyl]phenoxyacetamido polyethylene glycol-grafted polystyrene resin, Watanabe Chem.).
2. Fmoc amino acids: *N*-(9-fluorenylmethoxycarbonyl)amino acids, containing side chain protection with *tert*-butyl (tBu) group for Ser, Thr, and Tyr; trityl (triphenyl methyl, Trt) group for His, Asn, Gln, and Cys; *tert*-butoxy (OtBu) group for Asp and Glu; *tert*-butoxycarbonyl (Boc) group for Lys and Trp; 2,2,4,6,7-pentamethyldihydrobenzofuran-5-sulfonyl (Pbf) group for Arg.
3. Reagents for amide-coupling: 1-[bis(dimethylamino)methyliumyl]- 1H-benzotriazole-3-oxide hexafluorophosphate (HBTU), 1-hydroxy-1H-benzotriazole (HOBt), and *N*-methylmorpholine (NMM), with 1:1:2 equivalents for Fmoc amino acid in *N*,*N*-dimethylformamide (DMF) as solvent.
4. Reagents for removal of Fmoc group in each stepwise elongation: 2% 1,8-diazabicyclo[5.4.0]undec-7-ene (DBU) and 2% piperidine in DMF containing 2.5% HOBt.
5. Operation of elongation: automatically with a peptide synthesizer, Shimadzu PSSM-8 system, or manually with reaction vessels of syringe barrel equipped with filter.
6. Reagents for cleavage from supporting resin and deprotection of side chains: a mixture of 90% trifluoroacetic acid, 4% ethanedithiol, 4% triisopropylsilane, and 2% water.
7. Isolation of crude synthetic peptide: Precipitate with diethyl ether from cleavage mixture.

2.3 Purification of Peptides

1. Reverse-phase HPLC: Shimadzu LC-6A and Tosoh SC-8020 with FC-8010.
2. Column as stationary phase: YMC-Pack Pro C18 S-3.

3. Solvent for injection, DMF containing 4% lithium chloride.
4. Eluents, water containing 0.1% trifluoroacetic acid and acetonitrile in a gradient-elution mode.
5. Detection, UV absorption at 220 nm.
6. Isolation of purified peptide: Remove solvent from fractionated peptide solution by evaporation in vacuo.
7. Identification by mass spectrometry: Shimadzu QP-8000 in an electron spray ionization manner (ESI-MS). Preparation of TFA-free peptide, gel filtration (size-exclusion chromatography) with PD-10 (GE-Healthcare) desalting column by elution with a buffer solution of 50 mM $(NH_4)_2CO_3$ and 150 mM KCl (pH 8.5).

2.4 FTIR Measurements

FTIR instrument: Spectrum-100 Fourier-transform infrared spectrometer (Perkin-Elmer) equipped with an ATR unit and TGS and MCT detectors.

1. Diamond/ZnSe 1-reflection top–plate (Perkin-Elmer).
2. Barium fluoride (BaF_2) plates: Purchase from Pier Optics Co., Ltd., Japan.
3. Calcium fluoride (CaF_2) plates: Purchase from Pier Optics Co., Ltd., Japan.

3 Methods

1. Removal of Ca^{2+}: Prepare M^{2+}-free calcium protein by chromatography, followed by lyophilization. Remove contaminating Ca^{2+} by the treatment with trichloroacetic acid (TCA) [11].
2. Exchange for amide protons: To obtain reliable infrared spectra in the regions of the COO^- antisymmetric stretching vibration and the amide I′ band, deuterate exchangeable amide protons in the proteins completely by incubating the apoproteins dissolved in D_2O at about 60 °C for at least 60 min (*see* **Note 2**). Freeze-dry the solution after cooling at room temperature.
3. Protein sample solution for FTIR measurements: Dissolve the powder of a deuterated apoprotein in D_2O containing 40 mM HEPES-NaOD, 5 mM EGTA, and 0.1 M KCl (pD 7.4) in order to obtain apo samples for FTIR measurements. Dissolve the powder of a deuterated apo peptide in D_2O containing 40 mM HEPES-NaOD and 0.1 M KCl in addition to 0.1 M $CaCl_2$ (pD 7.4) in order to obtain the sample of M^{2+}-bound protein. The concentrations of protein for FTIR measurements are usually about 1–2 mM.
4. Preparation of synthetic peptide: Carry out the preparation of peptides by the usual solid-phase method using the Fmoc-

strategy employing the HBTU-HOBt-NMM (1:1:2) reagent system for coupling and DBU-piperidine-HOBt (2, 2, 2.5% in DMF) for deblocking Fmoc groups [12]. Operate every synthetic cycle automatically with a peptide synthesizer. Use the resin Fmos-NH-SAL-PEG as a solid-phase support to provide the peptides with amide-type C-termini. Cleave the elongated peptide chains from the resin with a trifluoroacetate (TFA)-EDT-TIPS-water (90:4:4:2) mixed solution. Purify the peptides by reversed-phase HPLC and analyze by ESI-MS with QP-8000. Eliminate TFA in advance by using PD10 desalting column (*see* **Note 3**).

5. Peptide sample for IR measurements: Prepare samples for FTIR measurements by dissolving the freeze/dried apo peptides in D_2O, and incubate at higher temperature for 30–60 min, long enough to ensure complete exchange of any amide protons for deuterons. Dissolve the powder of deuterated apo peptide in D_2O containing 40 mM HEPES-NaOD, 5 mM EGTA (or EDTA), and 0.1 M KCl (pD 7.4) in order to obtain apo samples for FTIR measurements. Dissolve the powder of deuterated apo peptide in D_2O containing 40 mM HEPES-NaOD and 0.1 M KCl in addition to 0.1 M $CaCl_2$ (pD 7.4) in order to obtain the sample of M^{2+}-loaded peptide. The concentrations of peptide for FTIR measurements are usually about 10 mM (*see* **Note 4**).
6. Transmission infrared measurements: Carry out FTIR measurements at 25 °C on a commercially available Fourier-transform infrared spectrometer equipped with a triglycine sulfate (TGS) detector at a resolution of 2 cm^{-1}. Pump nitrogen gas or dry air constantly into the sample room of spectrometer to minimize water vapor, which absorbs in the spectral region of interest. Average interferograms from 200 to 500 scans to obtain one spectrum.
7. Place about 12 μL of sample solution between two BaF_2 plates (*see* **Note 5**) separated by a 0.012 mm thick Mylar® spacer. Seal the gap between the two BaF_2 plates with aluminum tape to suppress the evaporation of water (Fig. 2). Measure the FTIR spectrum of the solvent (water) in the same way.
8. ATR-FTIR measurements: Perform ATR-FTIR measurements at 25 °C on a FTIR spectrometer equipped with a universal ATR unit and a liquid nitrogen-cooled mercury cadmium tellurium (MCT) semiconductor-detector at 2 or 4 cm^{-1}.

   Place a sample solution on a Diamond/ZnSe 1-reflection top-plate (about 1.5 mm in diameter) (*see* **Note 6**). The sampling depth of the ATR method was approximately 1–2 μm over the range of 2000–1000 cm^{-1}. Average interferograms from 200 to 500 scans to obtain one spectrum.

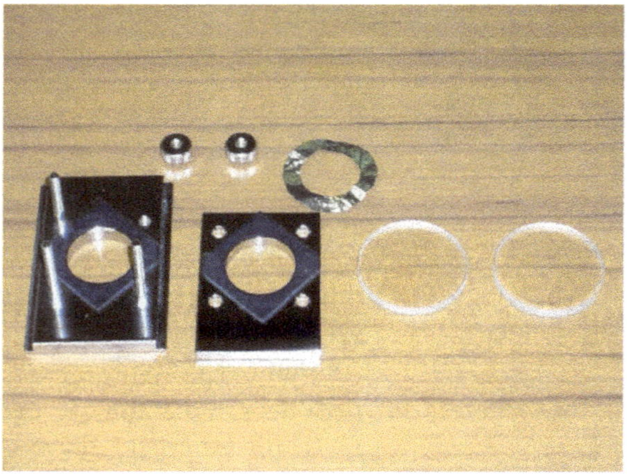

Fig. 2 Window cells and cell holder for transmittance measurements

9. Data analysis: Remove the slight contributions from water vapor in each spectrum completely by subtracting the water vapor spectrum (*see* **Note 7**). To eliminate the contributions of D_2O from the spectrum of the sample solution, subtract the spectrum of the solvent from the spectrum of the sample solution after multiplying by an appropriate factor. The criterion on the determination of the factor is the absorbance of the D-O-D bending vibrational band at about 1205 cm^{-1}(*see* **Note 8**). Calculate second-derivative spectra by using a software such as IGOR PRO or GRAMS. If necessary, calculate the difference spectra between Ca^{2+}-bound and apo forms by using the band at 1515 cm^{-1}, which due to the CC stretching mode of tyrosine ring as an internal standard, because the band is not related to the Ca^{2+} ligand directly (*see* **Note 9**). The downshift of the COO^- antisymmetric stretching mode from 1565 cm^{-1} to 1555–1540 cm^{-1} upon Ca^{2+} binding is a commonly observed feature of FTIR spectra for EF-hand proteins (*see* **Note 10**).

4 Notes

1. The band positions of COO^- stretches due to β-COO^- group of Asp and γ-COO^- group of Glu are as follows: Asp $\nu_{as}(COO^-)$ 1584 cm^{-1}, $\nu_s(COO^-)$ 1402 cm^{-1} and Glu $\nu_{as}(COO^-)$ 1567 cm^{-1}, and $\nu_s(COO^-)$ 1407 cm^{-1} in D_2O solution [13].
2. 2.No NH proton signals are observed in the ^1H nuclear magnetic resonance spectrum of completely deuterated Ca^{2+}-binding protein dissolved in D_2O.
3. Contaminated TFA shows a sharp band at 1674 cm^{-1} in the second-derivative spectra. As far as our interest is focused on

the COO^- stretching vibrational region, the contaminated TFA is not a problem in the study. However, in order to examine the secondary structure by using the amide I′ band, it is necessary to remove TFA completely.

4. Aggregation is observed for some solutions of synthetic peptide analogues. Examine FTIR spectra for peptide samples in wet film under D_2O atmosphere, when the absorbance of peptide was too weak in solution to analyze the amide I′ and COO^- stretching bands in detail.
5. Calcium fluoride (CaF_2) and barium fluoride (BaF_2) are available for transmittance FTIR experiments in an aqueous solution. From our experience, the solubility of CaF_2 in water may be left out of consideration, although the solubility of Ca^{2+} from the CaF_2 windows cannot be excluded.
6. After measuring an ATR-FTIR spectrum, a drop of sample is collected by using a micropipette.
7. Imperfect elimination of bands due to atmospheric water vapor disturbs the analyses of second-derivative spectra, because the bands of vapor are overlapped with those of sample. In particular, be careful of atmospheric water vapor for ATR-FTIR measurements, because the spectra of sample and reference are obtained by the single beam, although the dual beams are available for transmission infrared measurements.
8. To eliminate the contributions of H_2O from the spectrum of the sample solution, subtract the spectrum of the solvent from the spectrum of the sample solution after multiplying by an appropriate factor, so that the spectral line in the region of 2600–1800 cm^{-1} approached zero.
9. Photosensitive Ca^{2+} chelator will be available for obtaining direct information regarding structural changes induced in protein molecule upon Ca^{2+} binding [14].
10. As an example of Ca^{2+}-binding proteins, Fig. 3 shows the FTIR absorbance and second-derivative spectra of apo, Mg^{2+}-bound and Ca^{2+}-bound vertebrate troponin C in D_2O solutions across the range of 1800–1300 cm^{-1}. The band at 1550 cm^{-1} is due to side chain Glu COO^- groups coordinated to the Ca^{2+} ion in the bidentate mode. This band is not observed in the apo and Mg^{2+}-bound states. The band at 1600 cm^{-1} for the Mg^{2+}-bound state indicates the COO^- groups of side chain Asp interact with Mg^{2+} in the pseudo-bridging mode. It is noted that strict "unidentate mode" does not exist in solution, and therefore, unidentate mode should be pseudo-bridging.

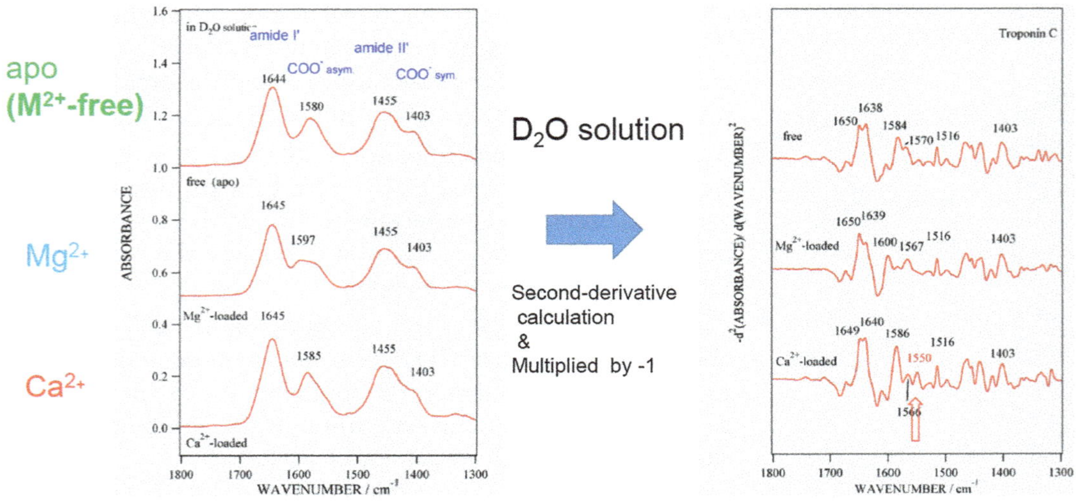

Fig. 3 FTIR absorbance spectra and FTIR second-derivative spectra of M^{2+}-free, Mg^{2+}-bound, and Ca^{2+}-bound vertebrate troponin C in solutions containing 40 mM HEPES-NaOD (pD 7.4) and 100 mM KCl

References

1. Nara M, Tanokura M (2008) Infrared spectroscopic study of the metal-coordinations structures of calcium-binding proteins. Biochem Biophys Res Comm 369:225–239
2. Nara M, Morii H, Tanokura M (2013) Coordination to divalent cations by calcium-binding proteins studied by FTIR spectroscopy. Biochim Biophys Acta Biomembr 1828:2319–2327
3. Nara M, Morii H, Tanokura M (2013) Infrared study of synthetic peptide analogues of the calcium-binding site III of troponin C: the role of helix F of an EF-hand motif. Biopolymers 99:42–347
4. Suzuki N, Imai LF, Kato Y, Nagata K, Ohashi Y, Kuchitsu K, Tanokura M, Sakamoto A, Nara M, Nakano M, Yonezawa N (2013) Coordination structures of Mg^{2+} and Ca^{2+} in three types of tobacco calmodulins in solution: Fourier transform infrared spectroscopic studies of side chain COO^- Groups. Biopolymers 99:472–483
5. Nara M, Yumoto F, Kagi H, Tanokura M (2008) Infrared spectroscopic study of the binding of divalent cations to Akazara scallop troponin C: the effect of the methylene side chain of glutamate residues. Biopolymers 89:595–599
6. Yumoto F, Nara M, Kagi H, Iwasaki W, Ojima T, Nishita K, Nagata K, Tanokura M (2001) Coordination structures of Ca^{2+} and Mg^{2+} in Akazara scallop troponin C in solution. FTIR spectroscopy of side-chain COO^- groups. Eur J Biochem 268:6284–6290
7. Deacon GB, Phillips RJ (1980) Relationships between the carbon-oxygen stretching frequencies of carboxylato complexes and the type of carboxylate coordination. Coord Chem Rev 33:227–250
8. Nakamoto K (1997) Complexes of alkoxides, alcohols, ethers, ketones, aldehydes, esters and carboxylic acids. In: Nakamoto K (ed) Infrared and Raman spectra of inorganic and coordination compounds part B, 5th edn. Wiley, New York, pp 57–62
9. Tackett JE (1989) FT-IR characterization of metal acetates in aqueous-solution. Appl Spectrosc 43:483–499
10. Nara M, Torii H, Tasumi M (1996) Correlation between the vibrational frequencies of the carboxylate group and the types of its coordination to a metal ion: an ab initio molecular orbital study. J Phys Chem 100:19812–19817
11. Tanokura M, Yamada K (1984) Heat capacity and entropy changes of calmodulin induced by calcium binding. J Biochem 95:643–649
12. Shaw GS (2001) Synthetic calcium-binding peptides. In: Hans HJ (ed) Methods in molecular biology, vol 173. Humana, Totowa, NJ, pp 175–181
13. Barth A (2000) The infrared absorption of amino acid side chains. Prog Biophys Mol Biol 74:141–173
14. Nara M, Tanokura M, Tasumi M (1994) UV-laser induced changes in the infrared spectrum of calmodulin in the presence of a photosensitive Ca^{2+} chelator. J Mol Struct 324:107–112

Chapter 10

Chaperoning Against Amyloid Aggregation: Monitoring In Vitro and In Vivo

Ravichandran Vignesh and Gopala Krishna Aradhyam

Abstract

Protein aggregation and inclusion body formation have been a key causal phenomenon behind a majority of neurodegenerative disorders. Various approaches aimed at preventing the formation/elimination of protein aggregates are being developed to control these diseases. Molecular chaperones are a class of protein that not only direct the functionally relevant fold of the protein but also perform quality control against stress, misfolding/aggregation. Genes that encode molecular chaperones are induced and expressed in response to extreme stress conditions to "salvage" the cell by the "unfolded protein response" (UPR) signaling pathway. Here we describe in detail the various in vitro and in vivo assays involved in identifying the chaperone activity of proteins using human calnuc as a model protein. Calnuc is a Golgi resident, calcium-binding protein, identified as chaperone protein and is reported to protect the cells against the cytotoxicity caused by amyloidosis and ER stress. Calnuc is also reported to regulate $G_{\alpha i}$ activity and inflammation apart from the role of chaperoning against amyloid proteins.

Key words Protein aggregation, Molecular chaperones, Misfolding, Calnuc, Calcium, Stress

1 Introduction

Protein aggregation is the main reason for a growing number of neurodegenerative diseases across the globe. Aggregation and amyloid formation of proteins happens because of a shift in dynamic equilibrium between growing insoluble oligomers and soluble monomer, leading to formation of fibrils and plaque. These amyloid fibrils/plaques accumulate in extracellular spaces, prompting cytotoxicity or cell death, thereby causing multiple neurodegenerative disorders [1–3]. Under these conditions, certain class of proteins, called chaperones, protect the cell by interacting with partially folded aggregates, high-order oligomers, and fibrils and rescue the cells against fibrillar toxicity. They also maintain protein homeostasis and enable the de novo protein folding/structure intact under such abnormal conditions [4].

Claus W. Heizmann (ed.), *Calcium-Binding Proteins of the EF-Hand Superfamily: From Basics to Medical Applications*, Methods in Molecular Biology, vol. 1929, https://doi.org/10.1007/978-1-4939-9030-6_10,

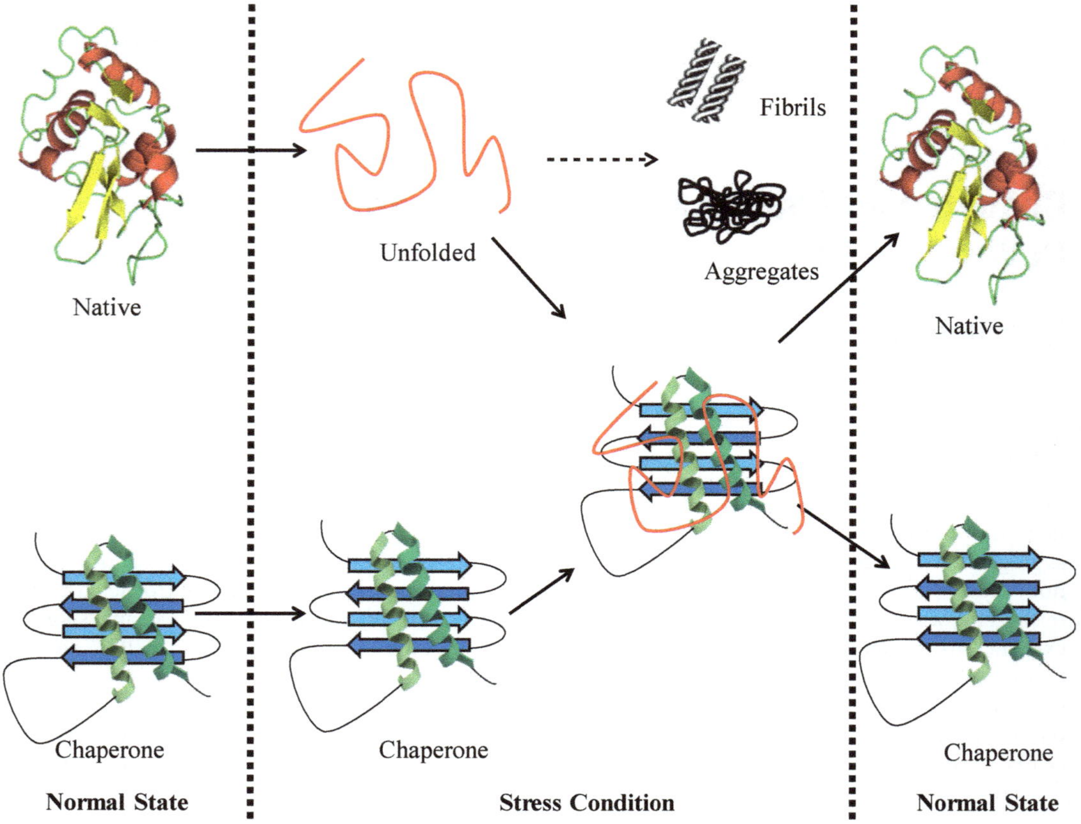

Fig. 1 Protein aggregation and chaperone action. Stress conditions like unfolded protein response (UPR) cause accumulation of mis–/unfolded proteins. The conditions also favor the formation of aggregates, fibrils, and plaques. Chaperones interact with them in energy-dependent/energy-independent manner assisting them to attain a correct fold

These set of proteins, which aid in proper folding of, other, susceptible proteins and protect them from misfolding/aggregation, are known as molecular chaperones (Fig. 1). Chaperones have high affinity toward non-native, unfolded/partially folded proteins and prevent the formation of non-native interactions, in an ATP-independent/−dependent manner [5, 6]. Heat shock proteins, protein disulfide isomerase (PDI), and glucose-related proteins 78 and 94 (GRP78, GRP94) are classic examples of group of molecular chaperones [7, 8].

Molecular chaperones also protect model substrate proteins (those that are susceptible to aggregation) against thermal and chemical stress without compromising their biological activity and prevent or dissociate fibril formation, thereby protecting the cells against the fibril-induced cytotoxicity. These substrate proteins demonstrate diverse quaternary structures, folding rates, functions, and irreversible side reactions during folding/refolding process [9]. Typical in vitro chaperone assay consists of stabilization of functional fold of the protein against thermal- and/or chemical-induced aggregation and prevents aggregation during unfolding-refolding process [10]. In vivo chaperoning assays also utilize luciferase or green fluorescent protein as model substrates toward heat stress to cells [11, 12]. Ex vivo protection against the fibrillar species aims at analyzing the cytotoxic ability of fibrils in the presence and absence of chaperones [13]. This screen, to identify the anti-amyloidogenic function of chaperone activity of proteins, could lead to identifying lead answers toward treating neurodegenerative disorders.

2 Materials

All the buffers are prepared in decalcified, deionized ultrapure water and stored at room temperature unless mentioned. Cell culture buffers and media are stored at 4 °C. All the reagents purchased were of analytical grade.

2.1 Subcloning and Expression Vectors

1. pTYB12 (New England Biolabs) with human calnuc (without signal sequence) or any gene of interest in bacterial expression vector (chaperone of your interest) (*see* **Note 1**).
2. pDsRed2-N1 vector with the chaperone of your interest as an insert (for co-localization as Aβ(1–42) is tagged with GFP).
3. pEGFP-C2 vector with Aβ(1–42) gene as insert [obtained from Prof. Seongman Kang, School of Life Sciences and Biotechnology, Korea University, Seoul].
4. Restriction enzymes and DNA ligase (for subcloning, if necessary).
5. pGL3 vector (any appropriate vector used for luciferase assay) and pcDNA6 V5HisA with calnuc as insert (or your gene of interest).
6. *E. coli* DH5α (for cloning) and BL21DE3 (for overexpression) strains.

2.2 Overexpression and Purification

1. Overexpression and purification of calnuc/Aβ(1–42) is not explained here (as it is not a main objective of the study) (*see* **Notes 2** and **3**).

2.3 Substrate Proteins for Monitoring Aggregation/Fibrillation/Enzyme Activity

2.3.1 Aggregation Experiments

1. Catalase.
2. Alcohol dehydrogenase (ADH).
3. Malate dehydrogenase (MDH) (*see* **Note 4**).

2.3.2 Enzyme Activity

1. Citrate synthase (CS).
2. Glutathione S-transferase (GST).
3. Malate dehydrogenase (MDH).

2.3.3 Fibrillation Studies

1. Insulin.
2. Amyloid β(1–42).
3. Hen egg-white lysozyme.

2.4 Thermal Stress, Enzyme Assay, and Spectroscopic Measurements

1. Assay buffer (thermal stress/enzyme assay): 20 mM HEPES pH 7.5, 50 mM NaCl.
2. Enzyme assay for citrate synthase: 0.5 μM citrate synthase, 20 μM 5,5′-dithiobis-(2-nitrobenzoic acid) (DTNB, also called as Ellman's reagent), 50 μM oxaloacetate, 10 μM acetyl Co-A (*see* **Note 5**).
3. Enzyme assay for malate dehydrogenase: 50 μM oxaloacetate, 13 μM NADH.
4. Enzyme assay for GST: 4 mM reduced glutathione, 0.4 mM 1-chloro-2,4-dinitrobenzene (CDNB), 5 μg of GST.
5. Ultraviolet-visible spectrophotometer with a thermostatic cell holder assembly (or circulating water bath), micro-magnetic beads, and quartz cuvette (*see* **Note 6**).

2.5 Recovery of Enzymatic Activity

1. Unfolding buffer: 20 mM HEPES pH 7.5, 40 mM NaCl, 20 mM DTT, 6 M guanidine hydrochloride (GdmCl) (*see* **Note 7**).
2. Refolding buffer: 20 mM HEPES (pH 7.5), 40 mM NaCl, and 10 mM DTT.

2.6 Fibrillation

1. Insulin fibrils: 25 mM HCl and 100 mM NaCl (pH 1.5).
2. Lysozyme fibrils: 25 mM HCl (pH 1.6).
3. Water bath or dry bath (*see* **Note 8**).

2.7 Thioflavin T Assay

1. Thioflavin T stock: 1 mM ThT is prepared in 95% (w/v) methanol, prepared just before use protected from light (*see* **Note 9**).
2. Spectrofluorimeter connected to thermostat or temperature-controlled circulating water bath.
3. Micro-magnetic beads and fluorescence cuvette.

2.8 Electron Microscopy

1. HR-SEM: Initiate fibrils in the presence/absence of calnuc, glass slides, and gold sputtering.
2. TEM: Prepare fibrils in the presence/absence of calnuc, Formvar, and carbon-coated nickel grids (*see* **Note 10**).

2.9 Maintenance of HEK293T and IMR32 Cell Lines

1. Dulbecco's Modified Eagle Medium (DMEM).
2. Serum and antibiotics (antibiotic-antimycotic).
3. Trypsin with 5 mM EDTA.
4. Phosphate-buffered saline (1× PBS): 137 mM NaCl, 2.7 mM KCl, 10 mM Na_2HPO_4, 1.8 mM KH_2PO_4. Dissolve all salts in water; adjust pH to 7.2–7.4, and autoclave.
5. Sterile vertical laminar air flow chamber.
6. CO_2 incubator.
7. Inverted bright-field microscope.

2.10 Cell Toxicity Assays

1. MTT stock: Dissolve 5 mg/mL of MTT in sterile 1× PBS, serum-free media, DMSO, UV-visible spectrophotometer/plate reader (*see* **Note 11**).
2. LDH leakage: 0.3 mM sodium pyruvate, 0.5 mM NADH, and UV-visible spectrophotometer/plate reader.
3. DAPI staining: Ice cold PBS, 4% formaldehyde, 0.1% Triton X-100, 1000× DAPI, and fluorescence microscope.
4. 1000× DAPI: 0.2 mg of DAPI in sterile H_2O and store in dark at 4 °C.

2.11 Luciferase Assay

1. 20 μg/mL cycloheximide (protein synthesis inhibitor, to prevent synthesis of heat shock proteins).
2. Temperature-controlled water bath.
3. Lysis buffer: 100 mM phosphate (pH 7.8), 10 μM DTT, and 1% Triton X-100.
4. Luciferase substrate buffer: 100 mM Tris-acetate pH 7.8, 10 mM magnesium acetate, 1 mM EDTA, 2 mM ATP, and 1 mM D-luciferin (*see* **Note 12**).
5. 2× β-gal assay buffer: 200 mM potassium phosphate pH 7.3, 100 mM β-mercaptoethanol, 2 mM $MgCl_2$, 1.33 mg/mL of ortho-Nitrophenyl-β-galactoside (ONPG).
6. Luminometer.

2.12 Transfection of HEK293T/IMR32 Cells

1. Transfection reagent (pick any transfection reagent and follow the protocol as per the manufacturers' instruction).
2. Serum-free DMEM media.
3. Fluorescence microscope imaging system (for studying co-localization).

3 Methods

3.1 Overexpression of Human Recombinant Calnuc

1. Calnuc is overexpressed and purified as described (this is not described in detail as the protocol will depend on the protein of interest) [14].

3.2 Cloning into Mammalian Expression Vectors

1. Human calnuc from pTYB12 is amplified with certain adapter sequences (for restriction digestion), is cleaved with restriction enzymes, and is ligated into the correspondingly linearized mammalian expression vectors (pcDNA6 V5HisA, pDsREd2-N1).
2. Confirm the subcloning by colony PCR and nucleotide sequencing with appropriate primers.

3.3 Overexpression and Purification of Aβ1–42 Peptide

1. Aβ(1–42) is subcloned into pGEX-6-P1 vector and overexpressed and purified as per the protocol described [15].

3.4 In Vitro Chaperone Assays

Prevention of thermal-induced aggregation of model substrate proteins is a primary assay in studying the chaperone activity.

3.4.1 Prevention of Thermal Aggregation

1. Dissolve the 1 mg/mL substrate proteins in appropriate buffer.
2. Incubate the proteins in room temperature for 5 min before initiating aggregation.
3. Set the temperature of the UV-visible spectrophotometer thermostatic cell assembly to 50 °C.
4. After incubation, add the samples to the quartz cuvette and record time-dependent absorbance at 360 nm for 20 min, with constant stirring (*see* **Note 13**).
5. Repeat the abovementioned steps in the presence of chaperone protein (1–2 μM of calnuc) by incubating with the substrate proteins in room temperature for 5 min before inducing aggregation (Fig. 2).

3.4.2 Protection of Biological Activity During Thermal Stress

The enzymatic assays for various substrates are studied spectrophotometrically using appropriate substrates and corresponding absorbance measurements (*see* **Note 14**) [16].

1. Assay for malate dehydrogenase:
   (a) Malate dehydrogenase catalyzes the conversion of oxaloacetate to malic acid with NADH as coenzyme (Fig. 3a).
   (b) Set up the enzymatic reaction with 0.2 μM of MDH, 50 μM of oxaloacetate, and 13 μM of NADH.
   (c) Adjust the temperature of the spectrophotomer cell assembly to 50 °C and monitor the absorbance at 340 nm with constant stirring.

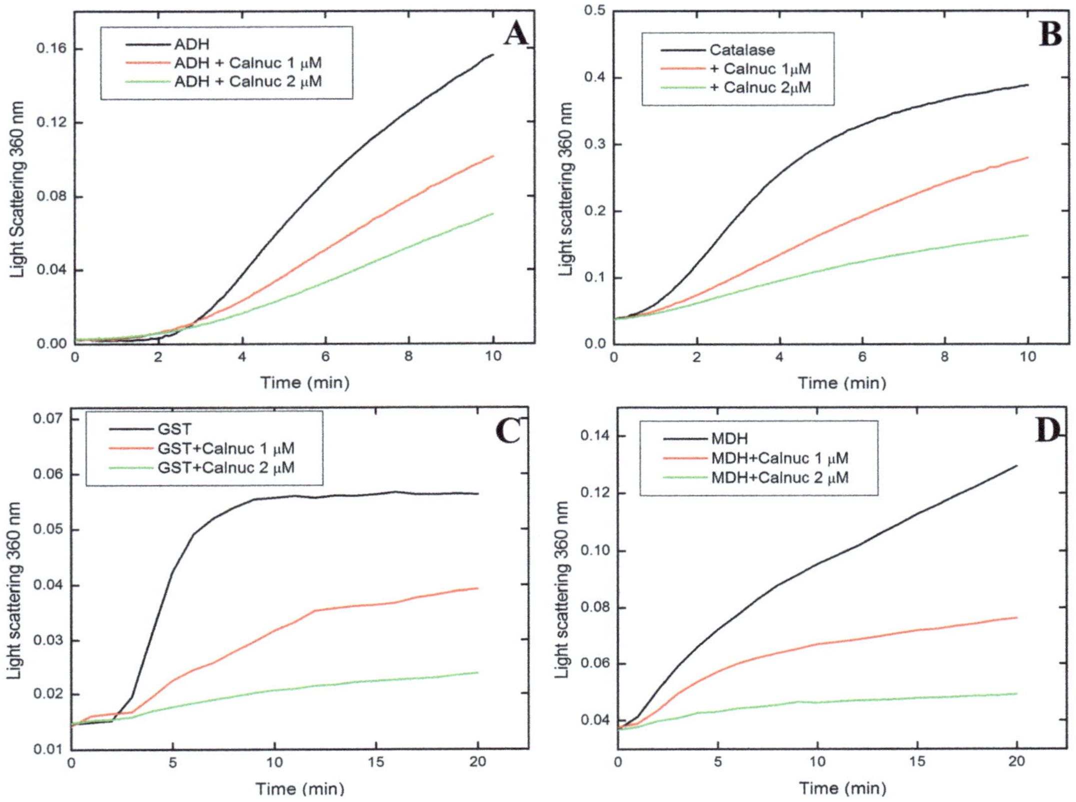

Fig. 2 Calnuc, as a model chaperone protein, protects the thermal-induced aggregation of substrate proteins like (**a**) alcohol dehydrogenase, (**b**) catalase, (**c**) glutathione S-transferase, and (**d**) malate dehydrogenase, at 50 °C, in a dose-dependent manner

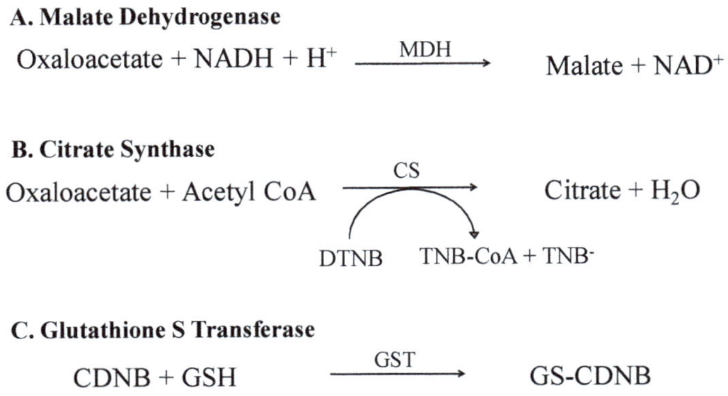

Fig. 3 Mechanism of various enzyme assays for the substrate proteins observed in this study using the synthetic substrates

2. Assay for citrate synthase:
   (a) Citrate synthase aids in the conversion of oxaloacetate to citrate, liberated free thiol group reacts with Ellman's reagent forming TNB (that demonstrates absorbance at 412 nm) (Fig. 3b).
   (b) Set up the enzymatic reaction with citrate synthase (0.5 μM), oxaloacetate (50 μM), acetyl Co-A (10 μM), and DTNB (20 μM) in the assay buffer.
   (c) Adjust the temperature of the spectrophotomer cell assembly to 50 °C and monitor the absorbance at 412 nm with constant stirring.
3. Assay for glutathione S-transferase:
   (a) GST catalyzes the conjugation of 1-chloro, 2,4-dinitrobenzene (CDNB) with reduced glutathione (GSH) (Fig. 3c).
   (b) Set up the reaction by adding 4 mM reduced glutathione (GSH), 0.4 mM CDNB, and 5 μg of GST.
   (c) Adjust the temperature of the spectrophotomer cell assembly to 50 °C and monitor the absorbance at 360 nm with constant stirring.
4. Repeat the abovementioned reaction steps by incubating the MDH/CS/GST with 2 μM of calnuc at room temperature for 5 min before inducing thermal stress (Fig. 4a–c).

3.4.3 Recovery of Residual Activity of MDH

1. Add 6 μM of MDH in unfolding buffer at 25 °C for 1 h.
2. Dilute the mixture 100 times in refolding buffer.
3. Set up the enzyme assay (mentioned in Subheading 3.4.2.1) with the refolded protein at different refolded time intervals.
4. Calculate the relative enzyme activity with respect to MDH (native form) as 100% (Fig. 4d).

3.4.4 Fibrillation

1. Prepare insulin (1 mg/mL) in fibrillation buffer in 37 °C for 1 week or 50 °C for 24 h.
2. Prepare lysozyme fibrils by dissolving 2 mg/mL of lysozyme in acidic buffer 50 °C for 10 days.
3. Fibril starts appearing as visible white fibers.
4. Incubate 10 μM Aβ(1–42) in assay buffer (20 mM HEPES pH 7.5, 50 mM NaCl) for 72 h at 37 °C (Fig. 5).
5. Prepare additional aliquot of the fibrils in the presence of 5 μM of calnuc. Only calnuc is also incubated under similar conditions (as negative control) (*see* **Note 15**).

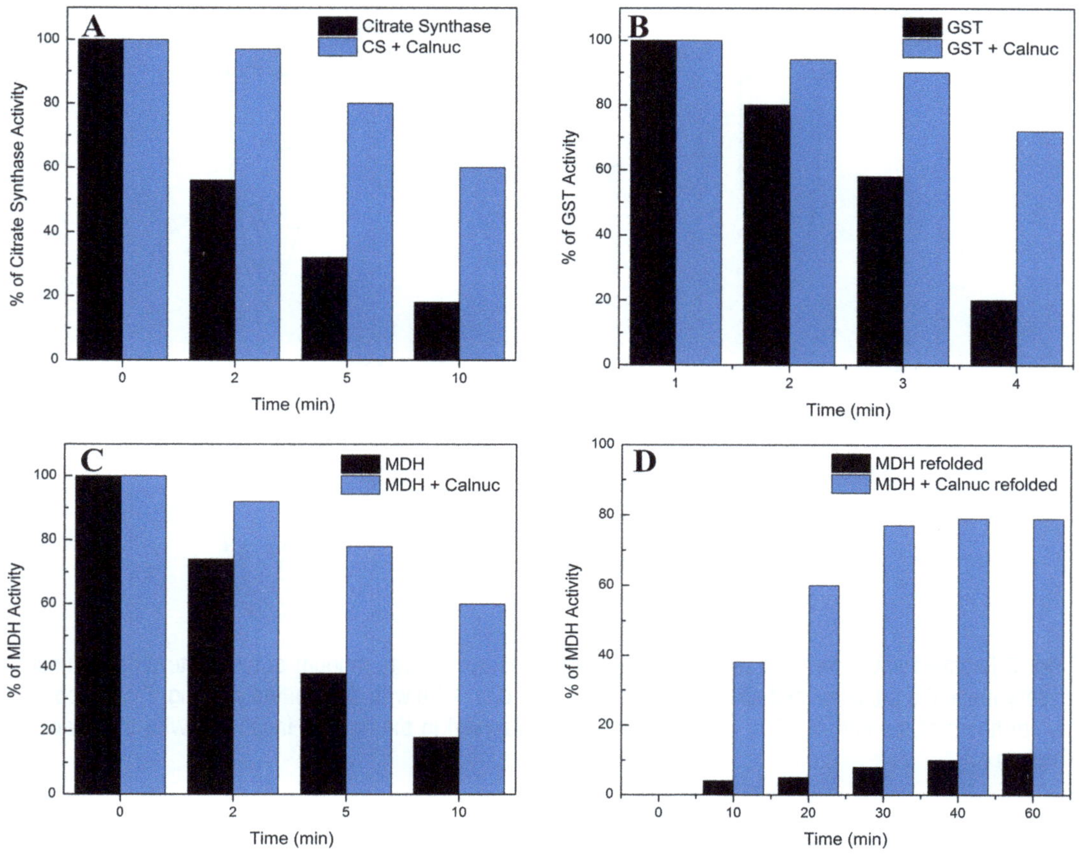

Fig. 4 Chaperone proteins can protect the biological activity of the substrate proteins during thermal stress. Enzyme assays of (**a**) citrate synthase, (**b**) GST, and (**c**) MDH are studied in the presence and absence of chaperone at 50 °C. Chaperone proteins also protect the activity of MDH when subjected to chemical stress followed by refolding the protein (**d**)

3.4.5 Thioflavin T (ThT) Assay

Thioflavin T is a benzothiazole derivative dye that interacts with the cross β-fibril core. This binding causes a tremendous increase in ThT quantum yield, and this property is used to study the fibrillation process [17].

1. Add 10 μM of ThT (from freshly prepared stock) to the (insulin/lysozyme) fibrils with constant stirring.
2. Set up the excitation at 440 nm, and monitor fluorescence emission intensity between 460 and 600 nm with excitation and emission slit widths set to 5 nm.
3. Repeat the fluorescence measurements for fibrils formed in the presence of calnuc and negative control (Fig. 6a, c, d).
4. Thermal aggregation or fibrillation assays can be performed in the presence of cofactors (test protein) or ATP or H_2O_2.

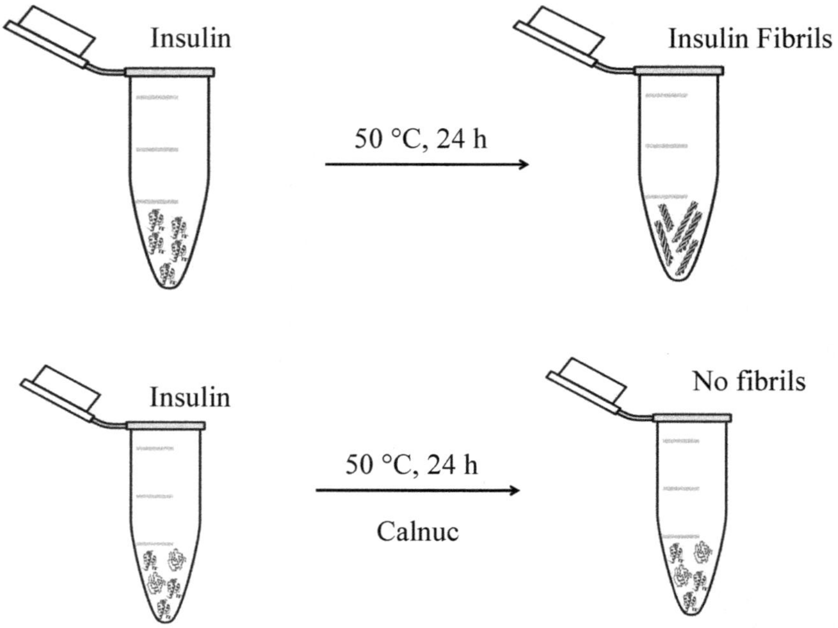

Fig. 5 Reaction setup for fibrillation includes heating the appropriate amount of substrate proteins (insulin/lysozyme) in the presence and absence of calnuc at 50 °C for 24 h with appropriate shaking. Fibrils appear as white fibers at the end of 24 h. No white fibrils are observed in the sample incubated with the chaperone protein

5. Time-dependent ThT assay:
   (a) Monitor time-dependent ThT fluorescence by incubating the freshly prepared insulin/lysozyme in acidic buffer.
   (b) Excite the sample at 440 nm and monitor emission intensity at 480 nm.
   (c) Set the temperature of circulating water bath to 50 °C with constant stirring of the reaction mixture.
   (d) Repeat the experiment with other conditions (Subheading 3.4.4.5) (Fig. 6b).

3.4.6 Electron Microscopy

Electron microscopy is an important tool in visualization of fibril formation and knowing their morphology and size [18].

1. Set up fibrillation (50–100 μL) as the following:
   (a) 1 mg/mL of insulin
   (b) 1 mg/mL of insulin and 0.5 mg/mL of calnuc
   (c) 0.5 mg/mL of calnuc.
2. After the time period (as mentioned in Subheading 3.4.5), prepare slides for electron microscopy.

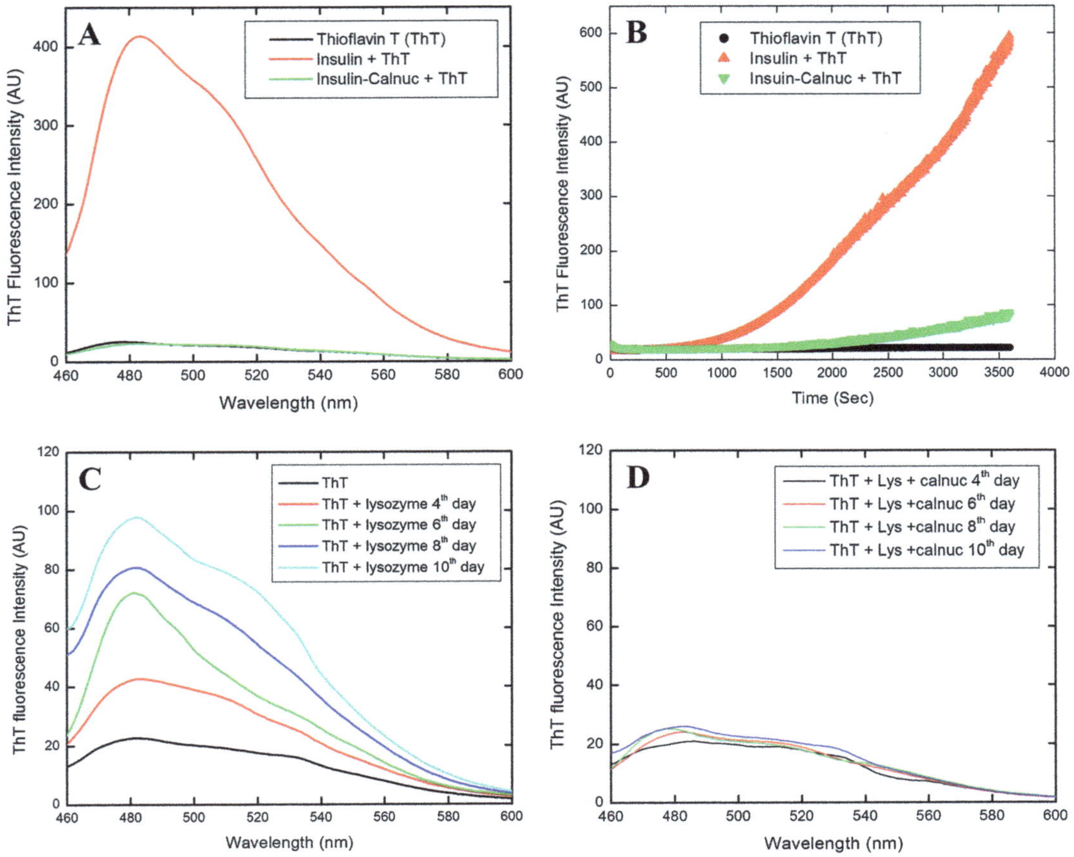

Fig. 6 Chaperone protein prevents the formation of fibrils in insulin (**a**, **b**) and lysozyme (**c**, **d**). ThT fluorescence of insulin fibrils is prevented by the action of chaperone protein as observed in steady state fluorescence (**a**) and time-dependent measurements (**b**). Chaperone protein also prevents fibrillation if lysozyme is studied at various time points, in the absence (**c**) and presence (**d**) of chaperone

3. For SEM, add the samples drop by drop onto a glass slide under vacuum for drying. Sputter with gold before imaging (Fig. 7a, b).
4. For TEM, add the samples onto a carbon-coated nickel grid and dry it under sterile conditions. Load the grids and image (Fig. 7c, d).

3.5 Ex Vivo Chaperone Assay (Cytotoxicity Assays)

Amyloid fibrils are known to induce apoptotic cell death in various cell lines/model systems [19]. Effect of chaperones in preventing fibril-induced cytotoxicity validates the identification of potent chaperone (*see* **Notes 16** and **17**).

3.5.1 MTT Assay

1. Maintain IMR32 cells as described [20].
2. Seed 3000–5000 IMR32 cells in each well of a 96-well plate, with 200 μL of DMEM media supplemented with antibiotics and 10% serum.

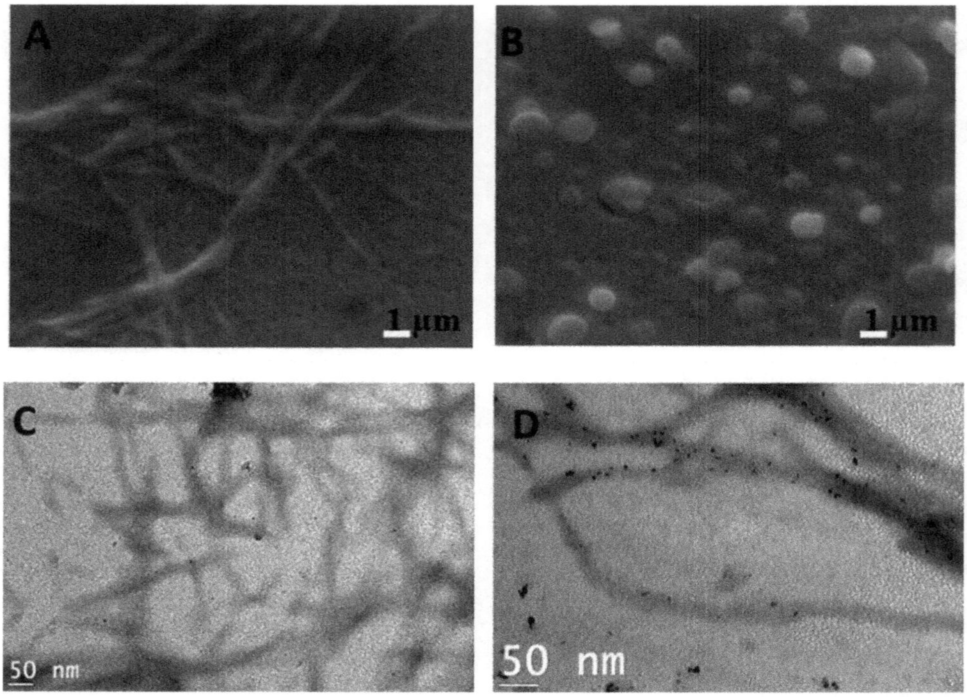

Fig. 7 SEM images show the morphology of insulin fibrils in the absence (**a**) and presence (**b**) of chaperone protein. (**c**) and (**d**) are HR-TEM images of insulin fibrils formed under same conditions without chaperone and with chaperone incubation, respectively

3. After 24 h of seeding, add 10 μM of fibrils (Insulin/Aβ(1–42)) to the wells.
4. Add different concentrations of calnuc (1–5 μM) in different wells. Untreated cells (without fibril/calnuc treatment) are used as control.
5. After 24 h of treatment, aspirate the media without dislodging the cells.
6. Add 100 μL of serum-free media and 10 μL of MTT (5 mg/mL stock), and incubate in the dark at 37 °C for 4 h.
7. Observe the formation of purple coloration (Fig. 8) (*see* **Note 11**).
8. Dissolve the reduced MTT (formazan) crystals by adding 200 μL of DMSO by gentle pipetting.
9. Measure absorbance at 590 nm in a plate reader.
10. Calculate % viability by taking the untreated cells (control) as 100% viable (Fig. 9, red bars).

3.5.2 LDH Leakage Assay

1. Measurement of LDH in culture media denotes the loss in cell viability (Fig. 8b).

A. MTT Assay

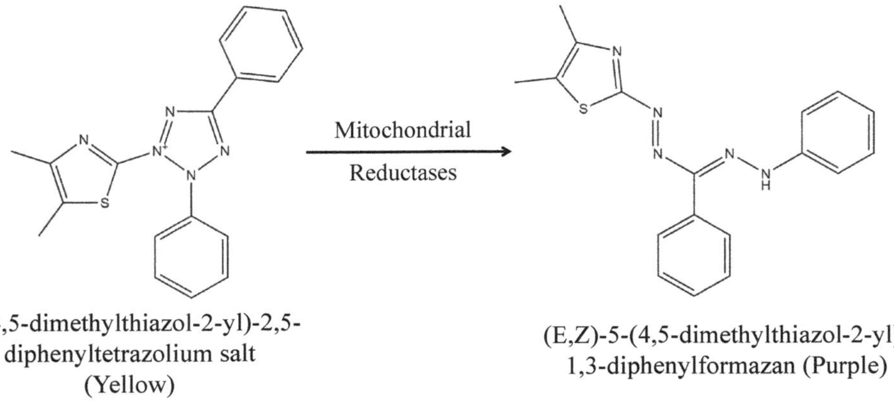

B. LDH leakage assay

$$\text{Pyruvate} + \text{NADH} + \text{H}^+ \xrightarrow{\text{LDH}} \text{Lactate} + \text{NAD}^+$$

Fig. 8 Reaction mechanism of MTT and LDH leakage assay to measure the cytotoxicity of the fibrils. MTT (tetrazolium salt) is reduced to formazan (**a**) by active mitochondrial reductases (yellow to purple color), indicating viability of the cell. Leakage of lactate dehydrogenase from the apoptotic cells is measured spectrophotometrically using pyruvate (**b**) as substrate

2. Seed 2×10^5 cells/well in a 6-well plate and grow them at 37 °C at 5% CO_2.
3. After 24 h of seeding, treat the cells with 5 μM of insulin fibrils or 10 μM of Aβ(1–42) in a well.
4. Mix 5 μM of calnuc and appropriate concentration of fibrils, and add in another well of the 6-well plates.
5. Treat 5 μM of calnuc only in one well. Untreated cells are taken as control.
6. Aspirate 500 μL of culture media after 24 h of treatment to a fresh tube.
7. Add 0.18 mM of NADH and 0.5 mM of sodium pyruvate to initiate the reaction at 25 °C.
8. Measure the time-dependent absorbance at 340 nm (decrease in NADH absorbance as it dissociates to form NAD^+).
9. Calculate % LDH activity by taking untreated cells as 100% activity (Fig. 9, blue bars).

3.5.3 DNA Fragmentation

1. Fragmented nuclear DNA denotes loss in cell viability.
2. Culture 2×10^5 cells/well in 6-well plates for 24 h at 37 °C.
3. Well A: Add 5 μM of insulin fibrils or 10 μM of Aβ(1–42).

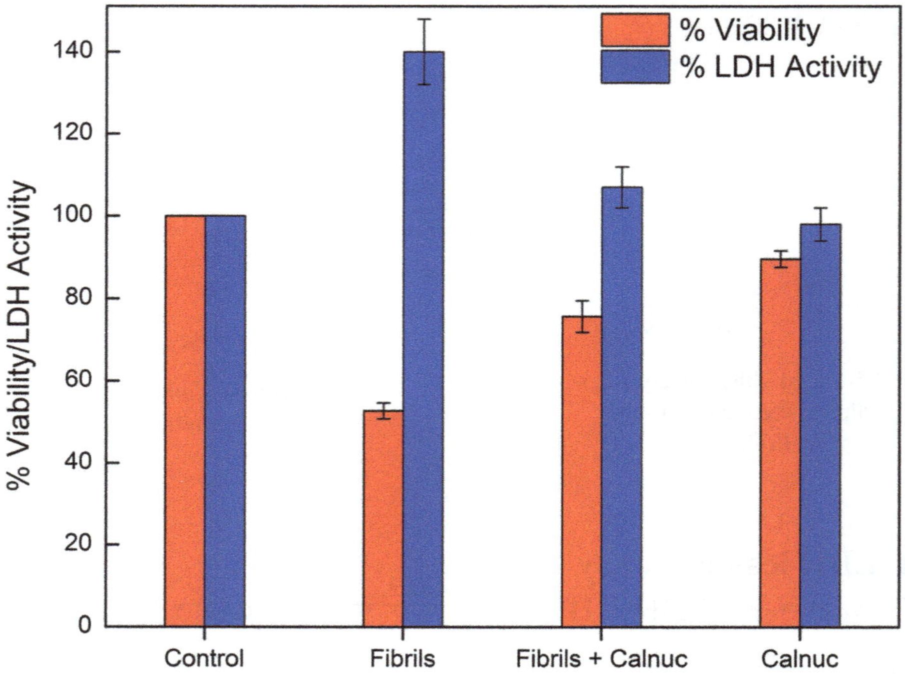

Fig. 9 Cytotoxicity assays. (**a**) % viability of cells (untreated control cells as 100% viable) revealed that chaperone protein protects the cell against the cytotoxicity of the fibrillar species. (**b**) % LDH activity showed that the leakage of cellular LDH is prevented by the action of chaperone protein

(a) Well B: Add 5 μM of calnuc and 5 μM of insulin fibrils or 10 μM of Aβ(1–42), after proper mixing.

(b) Well C: Add appropriate volume of serum-free DMEM (positive control).

(c) Well D: No treatment (control).

4. After 48 h of treatment, remove the media and wash the wells, thrice with ice cold sterile-filtered 1× PBS.
5. Fix the cells in 4% formaldehyde for 5–10 min, and wash with ice cold sterile-filtered 1× PBS.
6. Permeabilize the cells by treating the fixed cells with 0.1% Triton X-100 for 20–30 s.
7. Wash again with 1× PBS.
8. Add 1× DAPI to cells, incubate for 20 min in dark at room temperature, and wash with PBS to remove excess unbound stain.
9. Visualize the nucleus (blue colored) under fluorescence microscope (Fig. 10). Appearance of patches of blue color denotes fragmented nuclei.

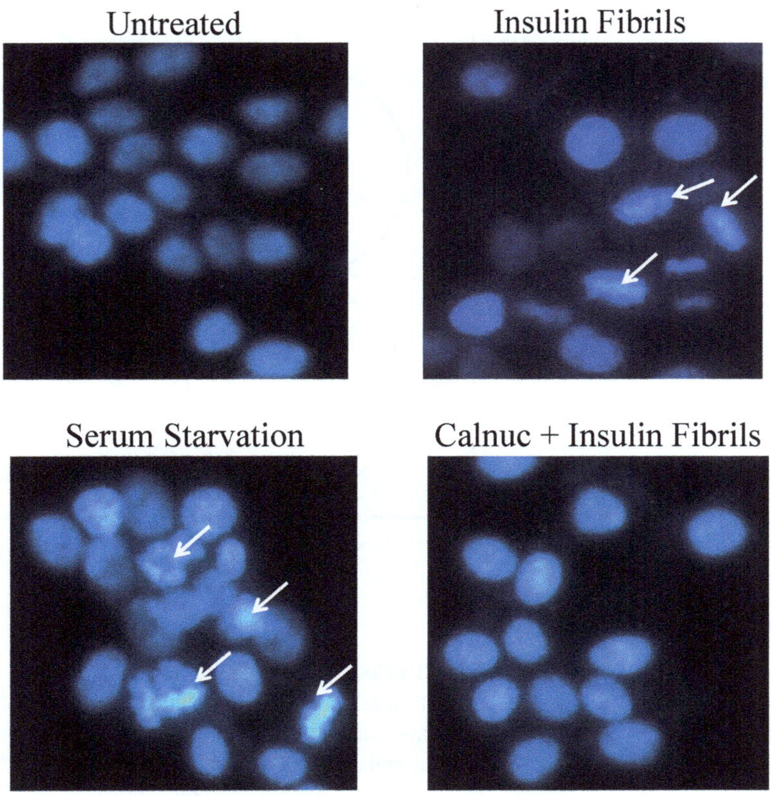

Fig. 10 Nuclear DNA fragmentation revealed by DAPI staining showed that nuclei remain intact in control (**a**), fragmented in serum starved (**b**), and fibril treated (**c**). Fragmentation is not observed when chaperone is added along with fibrils (**d**). White arrows indicate fragmented nuclei

3.6 *In Vivo* Chaperone Assay

3.6.1 Luciferase Assay

1. Grow HEK293T cells to 70% confluency in 35 mm dish (Fig. 11).
2. Do transfection as follows (as per the standard protocol):
   (a) Set A: Well 1–3 transfect pGL3 only.
   (b) Set B: Well 1–3 transfect pGL3 and pcDNA6 V5HisA calnuc.
3. After 24 h of transfection, treat the cells with 20 μg/mL cycloheximide for 30 min.
4. Posttreatment, Plates 2 and 3 of Set A and B are subjected to heat shock at 45 °C for 15 min.
5. Plate 1A and 1B (control) and Plate 2A and 2B (heat shock) are used for assay.
6. Allow the Plates 3A and 3B to cool to room temperature and used for assay.

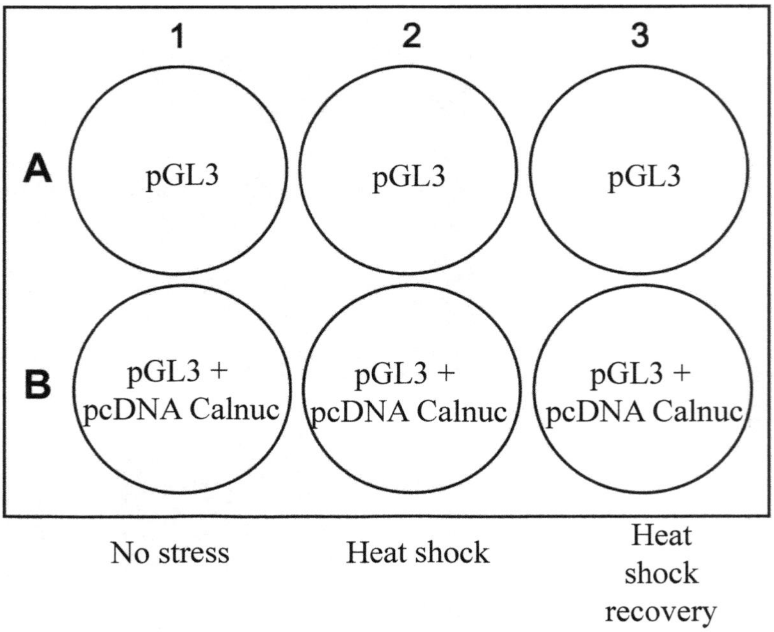

Fig. 11 Experimental setup for in vivo chaperone assay using luciferase reporter system. pGL3.0 expresses luciferase in mammalian system, and the expressed cells are subjected to heat shock and one followed by heat shock recovery. Control cells are not subjected to heat shock. The experiment is repeated by co-transfecting with chaperone expressing plasmid

7. Wash the cells twice with ice cold sterile-filtered 1× PBS.
8. Lyse the cells in 200 μL of lysis buffer for 45 min in ice.
9. Centrifuge at 10,000 × *g* for 10 min and collect the supernatant.
10. Take 50 μL of lysate on to RIA vials, and measure the luciferase activity in a luminometer (instruments add the substrate and water automatically, or else add manually and measure the reading) (Fig. 12).
11. Mix 50 μL of lysate and 50 μL of 2× β-gal assay buffer in a 96-well plate.
12. Incubate at 37 °C for 20 min. You will visualize yellow color.
13. Measure absorbance at 420 nm gives the β-gal activity.
14. Normalize the luciferase reading with either β-galactosidase activity.

3.7 Co-localization of Calnuc and Aβ (1–42)

1. Grow HEK293T cells to 70% confluency in six 35 mm plate.
2. Co-transfect GFP-Aβ(1–42) and DsRed-calnuc to HEK293T cells.

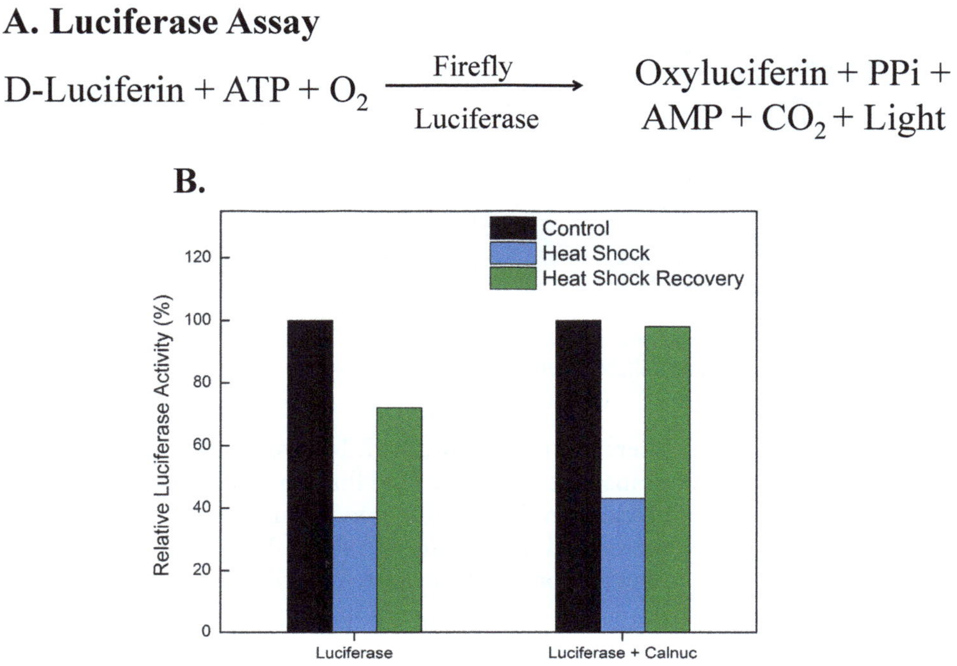

Fig. 12 (**a**) Mechanism of luciferase activity where light is emitted when luciferin is oxidized to form oxyluciferin and luminescence. The emitted light is measured in luminometer, and it corresponds to the activity of the enzyme. (**b**) Luciferase assay showed that heat shock embarked the loss in enzymatic activity and is recovered partially. When luciferase is co-expressed with chaperone, the activity is protected and recovered completely upon recovery from thermal shock

3. After 24 h of transfection, image the cells under a fluorescence microscope.
4. Observe GFP-Aβ(1–42) fluorescence by exciting with blue light and capture the image.
5. Observe DsRed-calnuc fluorescence by exciting with green light and capture the image.
6. Merge both of the captured images in an image editing software. Appearances of yellow spots indicated the regions of co-localization (Fig. 13).
7. Intracellular interactions can also be studied using other suitable methods (*see* **Note 18**).

4 Notes

1. Calnuc is used as a "chaperone protein." The concentration used in this study has been characterized by trying various concentrations. In case of a novel protein, various concentrations can be tried. Also calcium-dependent chaperone

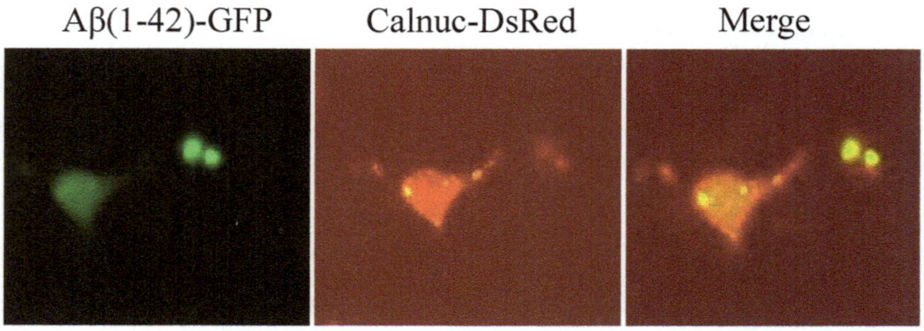

Fig. 13 Co-localization of GFP-Aβ(1–42) and DsRed-calnuc (yellow spots are observed when the images are merged together) showed the intracellular interactions of these proteins

interaction can be studied. In case of calnuc, calcium is known to inactive the chaperone function, so all assays are performed in calcium-free buffer conditions. Hence all the assays are performed in calcium-free form. Various stoichiometric ratios of chaperone and substrate protein have to be tried.

2. In case of purification, all buffers used are ice cold, and purification was done in cold room. All the buffers were made in autoclaved deionized water. Decalcification of water can be made by using chelex beads or any other appropriate methods. Phosphate buffers are not ideal for studying calcium-binding proteins as they tend to precipitate calcium.
3. The users may purchase the purified peptide for their study or overexpress the peptide as chimera, as per the protocol described.
4. The widely used model substrate proteins for chaperone assay are citrate synthase, malate dehydrogenase, luciferase, GFP, and RuBisCO. Non-aggregating proteins (any thermostable) can be used a negative control (optional). Apart from insulin/lysozyme/Aβ(1–42), other fibril-forming proteins can also be studied, viz., α-synuclein, $\beta_2$microglobulin, or serum amyloid component. Few chaperone proteins have tendency in chaperoning toward only specific substrates.
5. DTNB is poorly soluble and gives an emulsion. Extended stirring at ice cold conditions is required. Look for the solubility of the small molecules before preparing the stock.
6. All spectroscopic measurements are repeated at least thrice to check for reproducibility. During spectroscopic measurements, a constant stirring is advised to avoid settling of aggregates. Also, aggregation of target protein is to be tested as a control.
7. For enzymatic assays, mix the reactants first and add the enzyme and immediately start the absorbance measurements. The time interval between adding and starting the measurements affect the reading. Also prepare the enzyme solutions

freshly before use. We used HEPES because the pK_a doesn't vary on temperature. Tris has to be avoided as its pK_a varies with temperature.

8. Water bath gives proper heat exchange toward fibril formation than dry bath. Also occasionally mix the tubes by gentle inverting. Repeated mixing or vigorous mixing should be avoided.
9. ThT has to be prepared freshly in methanol/ethanol solutions in amber tubes to avoid light exposure.
10. Electron microscopic images are highly essential, to show the fibril formation and chaperoning potential of the target protein. Fibrils appear as long fibers, and smooth surfaces indicate the native conditions.
11. In case of MTT assay, the reagent is to be stored in dark-colored tubes or wrapped in aluminum foil. The assay plates are also to be wrapped in aluminum foil till measurements. During assay, the color of MTT changes from yellow to purple denotes the function of active mitochondrial reductases.
12. Luciferase assay substrates are to be freshly prepared, and the stock D-luciferin can be prepared and stored in −20 °C. β-gal assay is essential for normalization as transfection control. The expression of calnuc/chaperone is confirmed by Western blotting. All the substrates must be stored in amber tubes or wrapped with aluminum foil to avoid light exposure (during sample preparation), and the substrates are added in the dark to the reaction mixture.
13. Instead of absorbance measurements, aggregation can be monitored by fluorescence measurements by monitoring excitation and emission of the samples at 360 or 400 nm.
14. Apart from prevention of aggregation, the chaperones also should restore biological activity during stress and unfolding. Thus we performed enzymatic assays at elevated temperatures and refolded protein in the presence/absence of test protein.
15. All the assay/reaction conditions should consist of aggregation-prone proteins, aggregation-prone proteins + calnuc, and calnuc only, unless mentioned in the corresponding sections.
16. All cell-based assays must be replicated at least thrice, and mean and standard error of three experiments is used in graphs.
17. Cytotoxicity assay denotes ex vivo potential and rescuing effect. Any cytotoxic assays can be performed like trypan blue exclusion assay, mitochondrial membrane potential or annexin binding, etc.
18. Co-localization assay is performed to validate the in vivo interaction between the fibrillating species and target protein. Apart from co-localization, pulldown assay can also be used to validate the in vivo interaction.

References

1. Reiss AB, Arain HA, Stecker MM, Siegart NM, Kasselman LJ (2018) Amyloid toxicity in Alzheimer's disease. Rev Neurosci 29 (6):613–627. https://doi.org/10.1515/revneuro-2017-0063
2. Eisele YS, Monteiro C, Fearns C, Encalada SE, Wiseman RL, Powers ET, Kelly JW (2015) Targeting protein aggregation for the treatment of degenerative diseases. Nat Rev Drug Discov 14:759–780. https://doi.org/10.1038/nrd4593
3. Ross CA, Poirier MA (2004) Protein aggregation and neurodegenerative disease. Nat Med 10(Suppl):S10–S17. https://doi.org/10.1038/nm1066
4. Liberek K, Lewandowska A, Zietkiewicz S (2008) Chaperones in control of protein disaggregation. EMBO J 27:328–335. https://doi.org/10.1038/sj.emboj.7601970
5. Stull F, Koldewey P, Humes JR, Radford SE, Bardwell JCA (2016) Substrate protein folds while it is bound to the ATP-independent chaperone Spy. Nat Struct Mol Biol 23:53–58. https://doi.org/10.1038/nsmb.3133
6. Beissinger M, Buchner J (1998) How chaperones fold proteins. Biol Chem 379:245
7. Schroder M, Kaufman RJ (2005) ER stress and the unfolded protein response. Mutat Res 569:29–63. https://doi.org/10.1016/j.mrfmmm.2004.06.056
8. Liu CY, Kaufman RJ (2003) The unfolded protein response. J Cell Sci 116:1861–1862. https://doi.org/10.1242/jcs.00408
9. Haslbeck M, Buchner J (2015) Assays to characterize molecular chaperone function in vitro. Methods Mol Biol 1292:39–51. https://doi.org/10.1007/978-1-4939-2522-3_3
10. Glover JR, Lindquist S (1998) Hsp104, Hsp70, and Hsp40: a novel chaperone system that rescues previously aggregated proteins. Cell 94:73–82
11. Nollen EA, Brunsting JF, Roelofsen H, Weber LA, Kampinga HH (1999) In vivo chaperone activity of heat shock protein 70 and thermotolerance. Mol Cell Biol 19:2069–2079
12. Mymrikov EV, Daake M, Richter B, Haslbeck M, Buchner J (2017) The chaperone activity and substrate spectrum of human small heat shock proteins. J Biol Chem 292:672–684. https://doi.org/10.1074/jbc.M116.760413
13. Dehle FC, Ecroyd H, IF M, Carver JA (2010) alphaB-Crystallin inhibits the cell toxicity associated with amyloid fibril formation by kappa-casein and the amyloid-beta peptide. Cell Stress Chaperones 15:1013–1026. https://doi.org/10.1007/s12192-010-0212-z
14. Kanuru M, Raman R, Aradhyam GK (2013) Serine protease activity of calnuc: regulation by Zn2+ and G proteins. J Biol Chem 288:1762–1773. https://doi.org/10.1074/jbc.M112.382846
15. Long F, Cho W, Ishii Y (2011) Expression and purification of 15N- and 13C-isotope labeled 40-residue human Alzheimer's beta-amyloid peptide for NMR-based structural analysis. Protein Expr Purif 79:16–24. https://doi.org/10.1016/j.pep.2011.05.012
16. Bergmeyer HU (1974) Methods of enzymatic analysis. Verlag Chemie; Academic Press, Weinheim
17. Lindberg DJ, Wenger A, Sundin E, Wesen E, Westerlund F, Esbjorner EK (2017) Binding of Thioflavin-T to Amyloid Fibrils Leads to Fluorescence Self-Quenching and Fibril Compaction. Biochemistry 56:2170–2174. https://doi.org/10.1021/acs.biochem.7b00035
18. Gras SL, Waddington LJ, Goldie KN (2011) Transmission electron microscopy of amyloid fibrils. Methods Mol Biol 752:197–214. https://doi.org/10.1007/978-1-60327-223-0_13
19. El-Agnaf OM, Jakes R, Curran MD, Middleton D, Ingenito R, Bianchi E, Pessi A, Neill D, Wallace A (1998) Aggregates from mutant and wild-type alpha-synuclein proteins and NAC peptide induce apoptotic cell death in human neuroblastoma cells by formation of beta-sheet and amyloid-like filaments. FEBS Lett 440:71–75
20. Kanuru M, Aradhyam GK (2017) Chaperone-like Activity of Calnuc Prevents Amyloid Aggregation. Biochemistry 56:149–159. https://doi.org/10.1021/acs.biochem.6b00660

Part II

The Superfamily of Calcium-Binding Proteins Characterized by the EF-Hand Structural Motif

Chapter 11

Ca^{2+}-Binding Proteins of the EF-Hand Superfamily: Diagnostic and Prognostic Biomarkers and Novel Therapeutic Targets

Claus W. Heizmann

Abstract

A multitude of Ca^{2+}-sensor proteins containing the specific Ca^{2+}-binding motif (helix-loop-helix, called EF-hand) are of major clinical relevance in a many human diseases. Measurements of troponin, the first intracellular Ca-sensor protein to be discovered, is nowadays the "gold standard" in the diagnosis of patients with acute coronary syndrome (ACS). Mutations have been identified in calmodulin and linked to inherited ventricular tachycardia and in patients affected by severe cardiac arrhythmias. Parvalbumin, when introduced into the diseased heart by gene therapy to increase contraction and relaxation speed, is considered to be a novel therapeutic strategy to combat heart failure. S100 proteins, the largest subgroup with the EF-hand protein family, are closely associated with cardiovascular diseases, various types of cancer, inflammation, and autoimmune pathologies. The intention of this review is to summarize the clinical importance of this protein family and their use as biomarkers and potential drug targets, which could help to improve the diagnosis of human diseases and identification of more selective therapeutic interventions.

Key words Calcium-binding, EF-hand, Cardiomyopathy, Cancer, Inflammation, Allergy, Biomarkers, Diagnostic, Clinical chemistry, Laboratory medicine, Drug targets

1 Introduction

Calcium regulates multiple intracellular activities. After entering the cell, Ca^{2+} is reversibly complexed to specific Ca^{2+}-binding proteins which decipher the information carried by Ca^{2+} and passes it on to various targets regulating a myriad of biological functions. Proteins of this family are of major clinical relevance particularly in the field of heart diseases [1–10, 19, 20], cancer [11–18], inflammation [21–23], and allergies [24–26]. These calcium sensors contain highly specific Ca^{2+}-binding sites, a characteristic helix-loop-helix structure called the EF-hand, with one Ca^{2+} binding in the loop [27]. This highly conserved motif is found to date in a

Claus W. Heizmann (ed.), *Calcium-Binding Proteins of the EF-Hand Superfamily: From Basics to Medical Applications*, Methods in Molecular Biology, vol. 1929, https://doi.org/10.1007/978-1-4939-9030-6_11,

great number of Ca^{2+}- binding proteins grouped into distinct subfamilies [28–40]. Humans alone may express up to 100 EF-hand containing proteins, spanning a myriad of Ca^{2+}-regulated biological processes. Protein structures, their metal-binding properties, their interactions with numerous targets and receptors, and their signaling pathways and biological functions have been summarized in many reviews [28–45], special issues on calcium [46, 47], and books [48–54].

The intention of this survey is to summarize and list the clinical importance of members of this protein family as diagnostic and predictive tools in cardiology (Table 1), cancer (Table 2), and inflammation and allergies (Table 3) helping to improve the diagnosis of human diseases, leading to more selective therapeutic interventions for a future personalized medicine.

2 Clinical Relevance of Ca^{2+}-Binding Proteins in Cardiology

2.1 Troponin

The first EF-hand protein identified was troponin, which regulates muscle contraction [235]. The cardiac troponin complex is composed of the three subunits: the actomyosin ATPase inhibitory troponin I (cTnI), the tropomyosin-binding troponin T (cTnT), and the Ca^{2+}- sensing troponin C (cTnC). The heart-specific isoforms of TnT and TnI were found to be organ-specific biomarkers of cardiac necrosis when released into the blood. In contrast, the two isoforms of TNC, the "fast skeletal" TnC and the "slow skeletal-cardiac" TnC [236], are not cardiac-specific and therefore of no diagnostic use in cardiology. At the present, highly sensitive fully automated immunoassays for blood cTnT and cTnI are the "gold standard" for the diagnosis of acute myocardial infarction (AMI) [61–63]. A critical review of the different commercial available assays, their standardization, and analytical evaluation is available [64] and summarized on the website of the International Federation of Clinical Chemistry and Laboratory Medicine (IFCC). Genetic investigations of cardiomyopathy have revealed significant changes in the Ca^{2+}-regulatory mechanisms, in which cardiac troponin plays a key role. More than a hundred mutations have been reported in all three subunits of cardiac troponin associated with inherited hypertrophic (HCM), dilated (DCM), and restrictive cardiomyopathies [55, 59, 60]. Most mutations have been found in cTnT and cTnI, with only a few in cTnC in HCM and DCM [4, 56–58, 65]. Functional studies have indicated that some of these mutations directly affect the Ca^{2+}-sensitivity and Ca^{2+}-affinity of cTnC and normal heart functioning. Troponin modulating drugs are being developed for future treatment of cardiomyopathy and heart failure [58].

2.2 Calmodulin

Calmodulin, discovered in 1970 [237–239], interacts, activates, and regulates more than 100 different proteins (enzymes as well as structural elements) in a Ca^{2+}-dependent manner in different cellular compartments [30, 240–242]. In recent years, calmodulin was found to be associated with a number of human diseases including heart failure [6–8, 240, 243]. Mutations in calmodulin linked to dominantly inherited ventricular tachycardia (CPVT) and in individuals affected by severe cardiac arrhythmias [68–70] have been identified. These results show that cardiac contraction is most sensitive to calmodulin integrity and that calmodulin could be a therapeutic target for some cardiac disorders.

2.3 Parvalbumin

Parvalbumin was the first Ca^{2+}-binding protein to have its amino acid sequence [244] and atomic 3D structure resolved [27]. On this basis, Kretsinger developed the concept of the Ca^{2+}-binding, EF-hand structural motif [27, 28]. Parvalbumin functions as a Ca^{2+}-buffer, enhancing relaxation of fast-contracting skeletal muscles in small mammals (e.g., rodents), but is absent from slow-contracting/relaxing skeletal muscle and heart of humans [245–247]. Based on these findings, it was suggested that parvalbumin could be therapeutically introduced into the human heart to speed up relaxation after each heartbeat in diastolic heart failure (HF) [5, 75]. Transgenic de novo expression of parvalbumin enhanced relaxation in rat hearts in vivo, without affecting the time course of force development or contraction amplitude. Based on these results, it was suggested that ectopic expression of parvalbumin in the myocardium of patients with diastolic dysfunction may be a new therapeutic strategy helping to improve Ca^{2+}-handling and cardiac performance.

2.4 S100 Proteins

S100 proteins constitute the largest subgroup within the EF-hand protein family. S100 proteins are involved in many cellular activities [33–40, 248–252]. S100 proteins are small, acidic proteins that are mostly in the form of homo- or heterodimers but can also form tetramers and hexamers [253, 254]. S100 proteins display a relatively large range of Ca^{2+}-affinity and also bind zinc and copper [34–36, 40] at positions distinct from the Ca^{2+}-binding sites, resulting in modulation of their activities. Interestingly, of the 20 genes encoding S100 proteins identified so far, 16 are clustered on a region of chromosome 1q21 that is prone to molecular rearrangements in cancer [255]. Based on these findings, a S100 nomenclature was introduced and approved by the HUGO nomenclature committee [251, 256, 257]. Besides their distinct intracellular functions, certain S100 proteins are secreted from cells and exert cytokine-like functions through the activation of cell surface receptors such as the receptor for advanced glycation end products (RAGE; [18, 254, 258–262]) or Toll-like receptor 4 (TLR-4; [263, 264]). Several members of the S100 protein family have received increasing attention due to their close association with human diseases including heart failure [265] and are established biomarkers in routine chemistry laboratories for diagnosis and monitoring, supporting the clinical management of those diseases (Table 1).

One example is S100A1, predominantly expressed in the heart and skeletal muscle [266–268]. The roles of S100 A1 in vivo were revealed by using a mouse model carrying a targeted deletion of the S100A1 protein-coding sequence [269]. Mutant and heterozygous mice failed to augment left ventricular pressure, contraction, and relaxation rates in response to beta-adrenergic stimulation, whereas at rest cardiac function was normal. After myocardial infarction, S100A1$^{-/-}$ mice showed more severely impaired contractility, relaxation, and fractional shortening compared to WT mice [19]. We have investigated the role of S100A1 in the electrical activity of the heart by using S100A1 KO mice that were generated by gene-trap mutagenesis [20] and have shown that S100A1 deficiency resulted in a prolonged ventricular repolarization in response to sympathetic activation [270]. Studies of Ca^{2+}-signaling in mouse cardiomyocytes with the ablated S100A1 protein indicated that the chronical absence of this protein results in enhanced L-type Ca^{2+}-channel activity combined with a blunted SR Ca^{2+}-release amplification [271]. This may have implications in a variety of cardiac pathologies where abnormal RyR Ca^{2+} sensitivity or reduced S100A1 level have been described [77]. Furthermore, it was found that S100A1 relocates in association with ischemic myocardial injuries during cardiopulmonary bypass, and it was proposed that serial measurements of S100A1 serum levels in patients undergoing heart surgery may be relevant for the diagnosis of ischemic injury of the heart [272]. A S100A1-specific Sandwich ELISA test was established to measure S100A1 plasma levels in patients with signs of acute myocardial ischemia. The concentration time course was distinct from the "classical" biomarkers, CK, CKMB, and troponin I, showing an early rise and fast decline in plasma after the ischemic event. S100A1 may be included in combinatorial measurements as an early diagnostic biomarker for ischemic coronary diseases [273]. Decreased cardiac expression of S00A1 is characteristic of heart failure. Therefore, cardiac adenoviral S100A1 gene transfer to failing cardiomyocytes of rodents was applied, resulting in the normalization of S100A1 protein levels, contractile functions, and Ca^{2+}-homeostasis [79, 80]. It was proposed that S100A1 gene transfer to restore S100A1 protein levels in the failing myocardium may be a future therapeutic strategy to support the injured heart and to reverse or prevent heart failure [81, 82]. Besides S100A1, also S100A2 has the potential to differentially influence Ca^{2+}-cycling and contractility of cardiomyocytes [83]. Therefore, S100A2 and S100A4, in addition to parvalbumin and S100A1, may be further candidates for a differentiated heart failure therapy promoting survival of cardiac myocytes [83, 274]. Furthermore, S100A1, together with troponins and other components, is considered to be a potential biomarker in the diagnosis of infective endocarditis [85].

2.5 Myosin Light Chains

Myosin light chains bind to the neck region of myosin and belong to the EF-hand Ca^{2+}-binding protein family [275]. Myosin light chain-2 (MYL2, also named MLC-2) exists as three isoforms in mammalian muscles, encoded by three distinct MLC-2 genes: MCL-2f, fast-twitch skeletal muscle isoform; MLC-2v, cardiac ventricular and slow-twitch skeletal muscle form; and MLC-2a, cardiac atrial form. Genetic loss-of-function studies of the cardiac forms in mice have demonstrated their key roles in cardiac function during early embryogenesis. In adults, MLC-2v phosphorylation regulates cardiac myosin cycling kinetics and normal cardiac function. Dephosphorylation of MLC-2v has been implicated in dilated cardiomyopathy and heart failure [90].

2.6 Calcineurin

Calcineurin (Cn), a serine-threonine protein phosphatase, is a heterodimer consisting of two subunits: CnA and CnB (containing four EF-hands each). On activation, calmodulin (four EF-hands) binds as a third subunit [9, 10, 276–279]. The role of Cn in the activation of T cells was first reported in studies of the immunosuppressive compounds cyclosporine and FK506, both Cn inhibitors, which can prevent cardiac hypertrophy and dilated myopathy in rodent models [91]. Constitutive overexpression of activated Cn leads to cardiac hypertrophy in mice, whereas genetic deletion of the stress-responsive isoform of Cn (CnB beta) inhibits the development of myocardial hypertrophy [92, 93]. Cn levels, when measured in left ventricular samples of heart failure individuals, indicate that Cn may play a critical role in the progression of human heart failure [94, 95].

Recently, the EF-hand-containing CnB homologous protein 3 (CHP3), abundantly expressed in the heart, was suggested to be a novel negative regulator of cardiomyocyte hypertrophy [280].

2.7 CIB

CIB is a Ca^{2+}-binding protein associated with cardiac hypertrophy [96–98]. It was originally found to bind to integrin alpha IIb in platelets, inhibiting the activation of the alpha IIb beta3 integrin complex [97]. CIB1 is a 22-kDa protein with four EF-hand domains, but only the third and fourth C-terminal domains bind calcium at physiological concentrations [98]. CIB1 interacts with many signaling proteins including CnB. Ablation of CIB1 was found to protect the mouse heart from hypertrophy and the development of heart failure, suggesting that CIB1 could be a potential target for future therapeutic strategies [96, 98, 99].

2.8 Penta-EF-Hand (PEF) Protein Family

Calpains are a family of cysteine proteinases activated by Ca^{2+} [281–283]. Members of this penta-EF-hand (PEF) protein family include sorcin, grancalcin, ALG-2, and peflin and are involved in many biological activities [283–285]. The human m-calpain is composed of a large 78–80 kDa catalytic subunit and a small 29-kDa regulatory subunit; each subunit contains five EF-hand motifs. Calpain activity is tightly controlled by its endogenous inhibitor calpastatin. Overactivation of calpain and mutations of calpain, abrogating calpain activity, have been associated with human diseases including cardiac and cerebral ischemia [100]. The role of the calpain/calpastatin system in regulating the complex signaling pathways in the normal heart is important to understand the underlying mechanisms of cardiovascular diseases [101–103].

Table 1
Clinical relevance of EF-hand calcium-binding proteins in cardiology

Proteins	Disease association	References
Cardiac troponin		
TN-C	Mutations, hypertrophic cardiomyopathy	[55–58]
TN-T	Mutations, cardiomyopathies	[59, 60]
	Biomarker, myocardial infarction	[61–64]
TN-I	Mutations, cardiomyopathies (incl. peripartum)	[59, 60, 65, 66]
	Biomarker, autoantibodies, myocardial infarction	[61–64, 67]
Calmodulin	Mutations, tachycardia, sudden cardiac death	[6, 7, 68–72]
	Arrhythmia, cardiac arrest in infants	[73, 74]
Parvalbumin	Gene therapy to enhance cardiac performance	[5, 75]
S100 proteins		
S100B	Left ventricular remodeling after myocardial Infarction	[76]
S100A1	Altered expression in cardiomyopathy, hypertension	[77, 78]
	Heart failure gene therapy	[79–82]
S100A2/A6	Gene transfer to increase contractility, relaxation	[83]
S100A6	Myocardial ischemia-reperfusion	[84]
S100A11	Potential biomarker for infective endocarditis	[85]
S100A12	Renal and cardiovascular diseases	[86]
S100A9, A12	Acute coronary syndrome	[87]
DAMPs[a]	Cardiac injury, myocarditis	[88, 89]
MLC-2	Dephosphorylation has been implicated in dilated cardiomyopathy and heart failure	[90]
Calcineurin	Cardiac hypertrophy, heart failure	[91–95]
CIB1	Cardiac hypertrophy	[96–99]
Calpains	Overactivation, mutations, cardiovascular diseases	[100–103]
Sorcin	Mutations, cardiac hypertrophy	[104–108]

[a]DAMPs, damage-associated molecular patterns include S100 proteins [88]

Sorcin belongs to the PEF family of proteins [104, 286]. These proteins undergo Ca^{2+}-induced conformational changes which enable them to translocate from the cytoplasm to membranes and to interact with a variety of target proteins, thereby regulating diverse signaling processes. In the heart, sorcin modulates key regulatory processes such as the calcium-induced Ca^{2+}-release and excitation-contraction coupling [105]. A natural F112 L sorcin mutant was found to be associated with hypertrophic cardiomyopathy and hypertension in two unrelated families [106]. This F112 L variant was unable to translocate to the membrane and showed a decrease in both Ca^{2+}-affinity and interaction with target proteins (e.g., annexin II, ryanodine receptor, the cardiac L-type Ca channel, and the cardiac Na^{+}-Ca^{2+}-exchanger), resulting in an alteration of E-C coupling in the heart [105–108].

3 Clinical Relevance of Ca^{2+}-Binding Proteins in Oncology

Calcium remodeling of Ca^{2+}-homeostasis and dysregulation of Ca^{2+}-signaling in cells is closely associated with an uncontrolled cell multiplication, an increase in cell migration, angiogenesis, invasion, and cancer progression [287]. The complex changes in Ca^{2+} concentrations in cells are regulated by Ca^{2+} channels and pumps, exchangers, and the superfamily of Ca^{2+}-binding proteins characterized by the EF-hand structural motif. This chapter will focus and list (Table 2) the Ca^{2+}-binding proteins important in cancer progression, diagnosis, and treatment, highlighting their use as biomarkers and as therapeutic targets.

3.1 S100 Proteins

S100 proteins, the largest subfamily of Ca^{2+}-binding proteins, interact with many protein targets, cell surface receptors (RAGE and TLR-4), and channels, reflecting their broad range of intra- and extracellular, cytokine-like functions [33–40, 248–259]. Despite the high degree of structural similarity, their biological functions and contribution to tumorigenic processes—within each type of cancer—are very specific. Most S100 genes are located in a tight gene cluster on human chromosome 1q21. Within this chromosomal region, several rearrangements and deletions occur in tumors linked to a dysregulation of S100 gene expression [120, 251, 255–257]. The role of S100 proteins in human cancer, diagnosis, and treatment [12–15] as well as drug discovery efforts exploiting the identification of selective S100 inhibitors (already in clinical trials in patients with cancer [11]) has been reviewed [13, 120, 260, 287–289].

S100B and its role as a ligand of the cell surface receptor RAGE in melanoma have been recently discussed [111–114, 117, 290]. Besides the already established prognostic factors (primary tumor thickness, ulceration, mitotic activity, and the presence of tumor infiltrating lymphocytes) and a list of new promising biomarkers identified from tissue microarrays [113], S100B is considered to be the best-studied melanoma biomarker incorporated into the American Joint Committee on Cancer (AJCC) melanoma staging system. S100B is routinely analyzed in practical laboratory medicine, and elevated S100B levels closely correlate with a poor survival rate [113, 115]. Developing inhibitors against S100B as an antagonist [11, 116, 118] to block the interaction with RAGE is a promising goal for future pharmaceutical interventions and cancer treatment.

S100A1, normally expressed in the myocardium [248, 251, 266–269], has recently been found to be highly upregulated in ovarian cancer tissues when compared with fallopian and normal ovarian epithelium tissues [119]. The S100A1 expression is associated with lymph node metastasis and tumor grade stages, indicating an important role in the malignancy of ovarian cancer and also the value of S100A1 in the detection of this condition.

S100A2-A6. A prognostic significance of S100A2 in laryngeal squamous cell carcinoma has been found, allowing discrimination of high- and low-risk patients in the lymph node-negative subgroup for a better adjusted therapy [121]. S100A2, differentially expressed in normal human bronchial epithelial cell versus tumorigenic cells, has been found to be a marker for early stages of lung carcinogenesis [122, 123] also in combination with S100A6 [124]. Different expressions of S100 A2-A6 proteins were distinguished between distinct clinical entities in both WHO grade II and III astrocytic [126] and other tumors of the brain [127].

S100A3 is a unique member of the S100 protein family with a low affinity for Ca^{2+} but high affinity for Zn^{2+} ions [34, 40, 291–293], deposited in the mature endocuticle or medulla of keratinized hair [294]. Recently, S100A3 was also found to be associated with hepatocellular carcinoma [128], various types of astrocytic tumors [126], and gastric cancers [129].

S100A4 (metastasin), a key player in tumor progression and metastasis, is by far the best-studied and prominent member of the S100 protein family [11–15, 130–132]. The importance of S100A4 and its selective expression in cellular invasion and metastasis, is exemplified by its presence in a great variety of different types of cancer, as listed in Table 2. S100A4 can also act via an extracellular route, as an angiogenic factor inducing tumor progression. Inhibiting the process of tumor angiogenesis is a possibility by blocking either S100A4 secretion or its extracellular binding to targets/receptors. The receptor for advanced glycation end products (RAGE) is one of its interaction partners. The interaction of S100A4 with RAGE mediates S100A4-induced colorectal cell motility [133]. Further study of S100A4 and RAGE as two independent prognostic biomarkers of cancer metastasis and as a drug targets is a promising aim, with potential to improve clinical management, therapeutic outcome, and survival of patients in the future.

S100A7 (psoriasin) and the homologous protein S100A15 (koebnerisin) were first described in psoriatric keratinocytes and were then found to be implicated in a range of intra- and extracellular functions [143, 199, 295, 296]. Despite their homology, S100A7 and S100A15 are distinct in in their expression and binding to different receptors in a zinc-dependent (S100A7 to RAGE) and zinc-independent (S100A15 to a G-protein-coupled receptor) manner, respectively. Dysregulation of these proteins is associated with several malignant tumors including, e.g., epithelial carcinogenesis (S100A7/A15; [143]), osteosarcoma [144], ovarian cancer (S100A7; [145]), cervical cancer (S100A7; [146]), and pancreatic cancer (S100A7; [147]). Targeting the S100A7/A15-receptor axis is suggested as a new targeted strategy for preventing the invasion and metastasis of these types of cancer. Another therapeutic strategy is using neutralizing monoclonal antibodies against S100A7 for cancer treatment [148].

S100A8/A9 proteins (calprotectin/Mrp8 and Mrp 14) are involved in many types of cancer, inflammatory, and autoimmune diseases [203, 263, 264, 297]. Their expression and proposed cytokine-like functions in inflammation and cancer indicate key roles in inflammation-associated cancer [23]. These include colorectal cancer [149], chronic lymphocytic leukemia (CLL; [150]), acute myeloid leukemia (AML; [159, 298]), squamous cell carcinomas (SCCs; [134, 151]), breast cancer [152], laryngeal carcinoma [153], hepatocellular carcinoma [154], head and neck cancer [155], bladder cancer [156], thyroid carcinoma [157] as well as melanoma [158]. These proteins (listed in Table 2) represent promising marker genes to evaluate the risk potential of various types of cancer in molecular pathology.

S100A11 was found to be highly expressed in ovarian cancer cells, and it was suggested that S100A11 may become important for the diagnosis and treatment of ovarian cancer [161]. A further study showed that S100A11 is able to promote human pancreatic cancer cell proliferation through upregulation of the Pi3K/AKT signaling pathway opening the possibility for a targeted therapy of this cancer type [162]. Proteins were identified in various metastatic cancers, and this included S100A11 and its target protein annexin II. These data suggested a new role of the S100A11-annexin complex in plasma membrane repair of tumor cells and as targets for treating metastatic cancers [163].

S100A13 was identified by screening expressed sequence tag databases followed by its biochemical characterization and cellular expression in human and mice [223, 299, 300]. A multiplexed, targeted mass spectrometry analysis identified several S100 proteins (including S100A13) in thyroid tumors, suggesting that S100A13 may be a novel candidate for papillary carcinomas (PTC; [164]). Another study showed an association between S100A13 and the high mobility group A (HMGA1) in the modulation of thyroid cancer growth and invasion [165]. A proteomics analysis of cutaneous malignant melanoma (CMM) revealed an association between the expression of S100A13 and chemotherapy resistance [16]. CMM is commonly unresponsive to standard dacarbazine (DTIC) and temozolomide (TMZ) chemotherapy, and the use of S100A13 as a predictive marker of therapy response would be of great clinical help.

S100A14 was first identified by suppression subtractive hybridization and shown to be overexpressed in several cancer tissues [17]. Subsequent studies showed that S100A14 regulates proliferation, migration, and invasion of human cervical cancer cells [166], induces differentiation, and suppresses metastasis in gastric cancer [167] and is co-expressed together with S100A16 in human breast cancer correlating with a poor prognosis and promoting cancer cell invasion [168, 169]. S100A16 was found to be a prognostic marker and therapeutic target for colorectal cancer [170].

S100P possesses intracellular functions and extracellular functions by binding to cell surface receptor RAGE activating downstream signaling pathways [171–175]. S100P is expressed in many pancreatic tumors [176] and is associated with tumor growth and invasion. S100P is an important protein that contributes to the aggressiveness of pancreatic cancer. The antiallergic drug cromolyn binds to S100P preventing RAGE activation inhibiting pancreatic tumor growth and metastasis in two animal models. Cytotoxic treatment using cromolyn in combination with gemcitabine or applying a small RAGE antagonistic peptide appears to be a promising aim for further investigations to diagnose and treat pancreatic and possibly other types of cancer [171–174], malignant melanoma [177], breast cancer [178], and nasopharyngeal carcinoma [179].

S100-fused-type proteins (e.g., hornerin, filaggrin, trichohyalin, repetin) encode multidomain proteins with an S100-like domain comprised of two different Ca^{2+}-binding loops fused to a larger repetitive peptide via a spacer sequence. These proteins are classified as a separate protein (SFTP) family. Interestingly, most S100 genes are clustered together with the SFTP genes in a region of human chromosome 1q21 known as the epidermal differentiation complex (EDC; [251, 256, 257, 301–305]). S100 and SFTP genes are associated with the maturation of the skin, and skin disorders, but also with cancer. Recently, the SFTP hornerin was identified as a regulator of tumor vascularity of a human pancreatic ductal carcinoma specimen using phage display-based functional proteomics [180].

3.2 Sorcin

Sorcin is mainly expressed in the heart and brain but is overexpressed in many cancer cells as a result of the increased drug resistance to several chemotherapeutic agents [104, 107, 181, 182]. Multidrug resistance (MDR) is one of the major causes of treatment failure in pediatric acute leukemia (ALL), one of the most common types of cancer in children. It was found that overexpression of sorcin is a prognostic marker for multidrug-resistant pediatric ALL, correlating with upregulated MDRl/P-glycoprotein (P-gp) expression in these patients [183]. Sorcin may be a biomarker of MDR and therapeutic target for reversing multidrug resistance in ALL and other types of cancers, e.g., gastric cancer [184], colorectal cancer [185], and breast cancer [186].

3.3 CIB1

CIB1 was identified as an important regulator of cardiac hypertrophy, but recent studies have indicated a more diverse role in calcium signaling pathways and cellular processes [96–99, 187]. Several studies have investigated the role of CIB1 in a number of other human diseases including cancer, suggesting that CIB1 may be a novel drug target in human tumors.

3.4 Swiprosin-1

Swiprosin-1, an actin-binding protein also known as EF-hand domain containing 2 (**EFHD2**), is overexpressed in various human cancer tissues, especially at highly invasive stages of malignant melanoma. Swiprosin-1 stimulates cancer invasion and metastasis through activating the Rho family of small GTPases suggesting that these proteins could be considered as a potential therapeutic target to prevent cancer invasion and metastasis [189]. The expression of **EFHD2** was found to correlate with postsurgical recurrence of cancer in patients with stage I adenocarcinoma, and the multivariate Cox regression analysis revealed that EFHD2 expression was an independent clinical predictor for this cancer type [190].

3.5 Tescalcin

Tescalcin is a Ca^{2+}- and Mg^{2+}-binding protein with one consensus EF-hand and three additional domains with EF-hand homology. Tescalcin shares sequence homology with the calcineurin-B homologous protein3 (CHP3), can inhibit the phosphatase activity of calcineurin A [306], and has emerged as a regulator of cell differentiation and growth [307]. Tescalcin is overexpressed in colorectal cancer (CRC) contributing to the invasive and metastatic activity. Patients with CRC showed increasing serum levels of CRC [191, 308]. The protein is therefore considered to be a potential diagnostic marker and oncotarget in CRC.

3.6 LETM1

LETM1, a leucine zipper/EF-hand containing transmembrane protein, is located in the inner membrane of mitochondria. Elevated levels of expression have been correlated with carcinogenesis and tumor progression of breast cancer [192, 193] and were found to be an independent marker for a poor prognosis for patients with this type of cancer.

Table 2
Clinical relevance of EF-hand calcium-binding proteins in oncology

Proteins	Disease association	References
Calmodulin	Autophagy in cancer, glioblastoma invasion, HRAS- and NRAS-driven cancers	[109, 110]
S100 proteins		
S100B	Melanoma metastasis, breast cancer	[111–118]
S100B/A1	S100B as antagonist of S100A1-RAGE V interaction for potential cancer treatment	[116]
S100A1	Ovarian cancer	[119]
S100A2	Lung carcinogenesis, breast cancer, tumor suppression	[120–125]
S100A2-A6	Classification of brain tumors	[126, 127]
S100A3	Hepatocellular carcinoma, anticancer effect of sodium cantharidinate, astrocytic tumors, gastric cancer	[126, 128, 129]
S100A4	Metastasin colon, rectal, gastric, ovarian, breast cancers, endometrioid carcinoma, melanoma, radioresistance of pancreatic cancer cell, brain metastases and radiation response, leukoplakias and oral squamous cell carcinomas anticancer lymphocyte cytotoxicity	[13–15, 130–142]
S100A7	Psoriasin, ovarian, cervical, pancreatic cancers	[143–148]
S100A8/A9/calprotectin	Inflammation-associated cancer, breast, calprotectin colorectal cancers, CLL[f], AML[g], cutaneous squamous, laryngeal, hepatocellular carcinoma, head, neck, bladder cancer	[23, 134, 149–159]
S100A11	Breast, ovarian, pancreatic cancers	[161–163]
S100A13	Thyroid tumors, cutaneous malignant melanoma	[16, 164, 165, 223]
S100A14	Cervical, breast, gastric cancers	[17, 166–169]
S100A16	Colorectal, breast cancers	[168, 170]
S100 P	Breast cancer, pancreatic ductal adenocarcinoma, nasopharyngeal carcinoma	[171–179]
S100 fused-type protein/ hornerin (SFTP family; EDC[a])	Pancreatic cancer	[180]
Sorcin	Multidrug resistance, gastric, colorectal, breast cancers, childhood ALL[b]	[181–186]
CIB1[c]	Tumor progression	[187, 188]
Swiprosin-1	Malignant melanoma	[189]
EFHD2[e]	Stage I lung adenocarcinoma	[190]
Tescalin	Colorectal cancer	[191]
LETM1[d]	Breast, bladder cancers	[192–194]

[a]Epidermal differentiation complex
[b]Acute lymphoblastic leukemia
[c]Calcium and integrin-binding protein
[d]Leucine zipper/EF-hand containing transmembrane-1, a mitochondrial inner membrane protein
[e]EF-hand domain-containing protein D2
[f]Chronic lymphocytic leukemia
[g]Acute myeloid leukemia

4 Clinical Relevance of Ca^{2+}-Binding Proteins in Inflammation and Allergies

The disease association of these proteins is listed in Table 3.

4.1 Calpains

Calpains, a family of cysteine proteases [281–283], are involved in the process of several inflammation-associated diseases, including the regulation of inflammatory mediators, and may be considered as therapeutic targets for these diseases [160].

4.2 Calcineurin

Calcineurin is a serine-threonine protein phosphatase [9, 10, 276–279]. Recent results indicate that calcineurin-NFAT signaling is essential for infection control, and calcineurin inhibitors used in translation medicine (e.g., cyclosporine A and tacrolimus) are now being tested in a range of inflammatory conditions and autoimmune pathologies [195].

4.3 S100

S100 proteins interact with a number of target proteins/receptors and have intra- as well as extracellular functions [33–40, 248–259].

S100B was reported to be correlated with active vitiligo depigmentation [196, 197]. Vitiligo is the most common depigmenting disorder, caused by immune destruction of melanocytes. Serum levels of S100B were increased in patients with active nonsegmental vitiligo and associated with disease progression. These findings imply that S100B may be biomarker for this disease and a new target for treatment.

S100A4/metastasin is implicated in metastatic tumor progression (Table 2) and chronic inflammation (Table 3). S100A4 induced the expression of acute-phase responsive proteins, serum amyloid A (SAA) 1 and SAA3 proteins, and cytokines in an organ-specific manner. SAA proteins in connection with S100A4 serve as a link between inflammation and tumor progression [198]. S100A4 was also identified as a candidate gene in allergy by using a translational strategy. Treatment with an anti-S100A4 antibody resulted in decreased signs of allergy in a mouse model and in allergen-challenged T cells from allergic patients. This analytical strategy may be generally applicable to identify diagnostic and therapeutic candidate genes in allergy [24].

S100A7/psoriasin is localized in epithelial cells and regulates cell proliferation and differentiation. An increase of S100A7 expression is found in response to inflammatory stimuli, such as in psoriasis (a chronic inflammatory autoimmune-mediated skin disease). This suggests that S100A7 may be a therapeutic target in psoriasis [200].

An overexpression of S100A7 was found in the plaques of patients with psoriasis vulgaris with joint inflammation, but not in patients without joint involvement. This showed differences in S100A7 expression depending on joint involvement [201].

S100A7 expression is an important regulator of keratinocyte migration, upregulated in wound tissues and considered as a potential biomarker for human wound healing [202].

S100A8 and S100A9 proteins are characterized by a unique expression pattern with strong prevalence in cells of myeloid origin. These proteins were first discovered in body fluids of children with cystic fibrosis [309]. S100A8 /A9 expression is low in healthy people, whereas in inflammation, specific cell populations release homo- or hetero-complexes depending on the phase and type of inflammation. These proteins are associated with a number of chronic inflammatory diseases listed in Table 3. S100A8 and S100A9 (calprotectin) have been detected in serum and synovial fluid with juvenile idiopathic arthritis (JIA) and can be used (in combination with S100A12) to determine the relapse risk in patients with JIA after stopping anti-inflammatory treatment [204, 310]. The usefulness and validity of serum calprotectin level in childhood rheumatic diseases including inflammatory bowel diseases, cystic fibrosis, and psoriatic arthritis have been shown to correlate with—or even be superior to—other laboratory variables of inflammation such as C-reactive protein (CRP) and erythrocyte sedimentation rate (ESR) [206–216].

Sensitive and automated ELISA calprotectin (S100A8/A9) assays are routinely used in clinical chemistry laboratories to diagnose the two main forms of inflammatory bowel diseases, Crohn's disease and Ulcerative colitis. High-performance liquid chromatography coupled with electrospray ionization mass spectrometry (HPLC-ESI-MS) was able to identify and quantify S100 proteins and their isoforms (S100A7, two isoforms; S100A8/A9, four isoforms; S100A11 and S100A12, two isoforms) in human newborn (and even preterm) saliva with only a few ul analyte available. This noninvasive methodology allowed a simultaneous analysis of several S100 proteins and their posttranslational modifications [311, 312]. As a consequence of the various pro-inflammatory properties of S100 proteins, strategies targeting these molecules are an option for anti-inflammatory therapies, by in vivo administration of S100 antibodies or by therapeutic drug targeting to inhibit the release of these cytokine-like molecules at sites of inflammation. S100A11 (calgizzarin) was associated with rheumatoid and osteoarthritis and is an additional biomarker [224] to S100A/A9 and S100A12 for a more tailored diagnostic of inflammatory diseases. S100A12/calgranulin C is associated with the Kawasaki disease [225] and Mooren's ulcer [226] as well as atherosclerosis [230]. In addition, S100A12 seems to be more specific for inflammatory bowel disease in school-aged children and adolescents when compared to established stool markers such as S100A8/A9 and lactoferrin [227, 228]. A further study evaluated the impact of several S100 proteins (A7, A8, A9, and A12) in the chronic skin disease, psoriasis, and found that S100A12 showed the closest association with disease activity and therapeutic response and might therefore provide an additional and valuable biomarker for psoriasis [229].

4.4 Parvalbumin

Parvalbumin and its isoforms have been found to be a major fish allergen in humans. Approximately 70% of fish allergies, mediated by parvalbumin IgE antibodies, lead to respiratory and gastrointestinal symptoms [25, 26, 231, 233, 234]. A FAST (food allergy-specific immunotherapy) project was initiated to develop a safe and effective subcutaneous immunotherapy for fish allergy, using recombinant hypoallergenic carp parvalbumin, Cyp C 1. These results have now warranted first-in-man immunotherapy studies to further develop this new vaccine for a safe and effective therapy for fish allergy [232].

Table 3
Clinical relevance of EF-hand calcium-binding proteins in inflammatory, autoimmune diseases, and allergies

Proteins	Disease association	References
Calpain	Inflammation-associated diseases	[160]
Calcineurin	Inflammation and autoimmune pathologies	[195]
S100 proteins		
S100B	Vitiligo	[196, 197]
S100A4	Link between inflammation and tumor progression	[198]
S100 A7	Psoriasis; wound healing	[199–202]
S100 A8/A9[a]	Rheumatoid arthritis, juvenile idiopathic rheumatoid arthritis, pediatric inflammatory bowel disease, psoriatric arthritis, gastrointestinal inflammation, cystic fibrosis, hyperinflammation in neonates, Crohn's disease, inflammatory lung disorders, autoinflammation, cutaneous wound healing	[21, 22, 203–222]
S100A11[c]	Rheumatoid arthritis	[224]
S100A12[b]	Kawasaki disease, Mooren's ulcer, Crohn's disease inflammatory bowel disease (IBD), psoriasis, vascular inflammation, and atherosclerosis	[207, 219, 225–230]
Parvalbumin	Fish allergy	[25, 26, 231–234]
S100A4	Allergies, seasonal allergic rhinitis	[24]

[a]Calprotectin, DAMPs, alarmins
[b]Calgranulin C
[c]Calgizzarin

5 Concluding Remarks

The intention of this review is to summarize the importance of the Ca^{2+}-binding proteins and their use as novel biomarkers and drug targets to support practical laboratory medicine in the diagnosis and monitoring of disease progression for selective therapeutic interventions. At the end of the nineteenth century, Sidney Ringer obtained the first hint that Ca^{2+} was involved in the regulation of contraction in isolated rat heart [313]; in the 1960s troponin was discovered by Setsuro Ebashi [235] as the first intracellular calcium sensor protein, regulating muscle contraction. Parvalbumin was found in fish and amphibians by Hamoir et al. [314] and Pechère et al. [315], calmodulin is discovered by Cheung [237] and the Kakiuchi group [238] as an activator of the brain cyclic nucleotide phosphodiesterase, and the S100 proteins were first described by Moore [316] as small acidic proteins in a 100% saturated ammonium solution of a brain extract. Today, troponin immunoassays are the "gold standard" in Clinical Chemistry and Laboratory Medicine for the diagnosis of acute myocardial infarction [61–64]. Many mutations associated with heart failure have been reported in all three subunits of cardiac troponin [3, 4] and in calmodulin [6–8], making them potential therapeutic targets for cardiac disorders. Parvalbumin [5] and S100A1 [79] gene transfers to the failing heart are suggested to be a new therapeutic strategy helping to restore Ca^{2+}-handling and cardiac performance in patients. Members of the S100 protein family are routinely used in Practical Laboratory Medicine to analyze and monitor disease conditions, e.g., melanoma (S100B; [317]), brain injury (S100B; [111]), and inflammatory and autoimmune diseases (S100A8/A9/calprotectin; Crohn's disease; [203, 220, 221, 297]). Covalent small molecule inhibitors of S100B have been developed that inhibit S10B-p53 complex and restore active p53 in malignant melanoma [317]. A similar approach was to target the estrogen receptor/S100 interface by a short peptide derived from S100P in order to reduce regression of tumor growth receptor-positive breast cancer [318]. Some S100 proteins are secreted into the microenvironment activating cell surface receptors (e.g., RAGE, Toll-like receptors (TLRs), scavenger receptors, and G-protein-coupled receptors) where they inhibit extracellular interactions with drugs such as sodium cantharidinate, tasquinimod, oxyclozanide, cromolyn, pentamidine, or blocking antibodies. They are considered as promising therapeutic candidates to fight cancer in the future [128, 319–321]. So far, only a small number of this largest family of Ca^{2+}-binding proteins (with over 200 human genes coding for proteins containing the EF-hand Ca^{2+}-binding motif) have been explored for their clinical relevance, and this does not include their intra- and extracellular targets/surface receptors or posttranslational modification (e.g., S100 proteins: phosphorylation, myristoylation, or sumoylation; [3, 260]. A method of choice for a simultaneous analysis of Ca^{2+}-binding proteins and their modifications in very small amounts of body fluids (e.g., S100 proteins in the saliva of preterm human newborns) is the HPLC-electrospray-ionization-MS technology [311, 312] and the selected reaction monitoring mass spectrometry (SRM-MS) applied for the concurrent and specific quantification analysis of S100 proteins in cell lines and tumor tissues [164, 322]. A new and sophisticated bioanalytical technology is the focal molography (a next-generation biosensor that visualizes specific biomolecular interactions in real time) allowing label-free detection of trace amounts of all kinds of biomolecules in biological samples [323]. This technique is known for its simplicity and robustness and may have a future impact on the development of point-of-care diagnostic devices applied for analyses of the Ca^{2+}-binding proteins of human diseases.

Acknowledgment

I would like to thank Drs Anthea Rowlerson and Dean Malencik for critical reading of the manuscript and Melanie Heizmann for secretarial assistance.

References

1. Schaub MC, Heizmann CW (2008) Calcium, troponin, calmodulin, S100 proteins: from myocardial basics to new therapeutic strategies. Biochem Biophys Res Commun 369:247–264
2. Wehrens XH, Marks AR (2004) Novel therapeutic approaches for heart failure by normalizing calcium cycling. Nat Rev Discov 3:565–573
3. Landstrom AP, Ackerman MJ (2012) Beyond the cardiac myofilament: hypertrophic cardiomyopathy-associated mutations in genes that encode calcium-handling proteins. Curr Mol Med 12:507–518
4. Veltri T, Landim-Vieira M, Parvatiyar MS et al (2017) Hypertrophic cardiomyopathy cardiac troponin C mutations differentially affect slow and cardiac muscle regulation. Front Physiol 8:221. https://doi.org/10.3389/fphys.2017.00221
5. Wang W, Barnabei MS, Asp ML et al (2013) Non-canonical EF-hand motif strategically delays Ca^{2+} buffering to enhance cardiac performance. Nat Med 19:305–314
6. Sondergaard MT, Sorensen AB, Skov LL et al (2015) Calmodulin mutations causing catecholaminergic polymorphic ventricular tachycardia confer opposing functional and biophysical molecular changes. FEBS J 282:803–816
7. Crotti L, Johnson CN, Graf E et al (2013) Calmodulin mutations associated with recurrent cardiac arrest in infants clinical perspectives. Circulation 127:1009–1017
8. Piazza M, Taiakina V, Dieckmann T, Guillemette JG (2017) Structural consequences of calmodulin EF-hand mutations. Biochemistry 56:944–956
9. Wolska BM (2009) Calcineurin and cardiac function: is more or less better for the heart? Am J Physiol Heart Circ Physiol 297: H1576–H1577
10. Olson EN, Molkentin JD (1999) Prevention of cardiac hypertrophy by calcineurin inhibition. Hope or hype? Circ Res 84:623–632
11. Bresnick AR, Weber DJ, Zimmer DB (2015) S100 proteins in cancer. Nat Rev Cancer 15:96–109
12. Gross SR, Sin CG, Barraclough R, Rudland PS (2014) Joining S100 proteins and migration: for better or worse, in sickness and in health. Cell Mol Life Sci 71:1551–1579

13. Stein U, Burock S, Herrmann P et al (2011) Diagnostic and prognostic value of metastasis inducer S100A4 transcripts in plasma of colon, rectal, and gastric cancer patients. J Mol Diagn 13:189–198
14. Grigorian M, Ambartsumian N, Lukanidin E (2008) Metastasis-inducing S100A4 protein: implications in non-malignant human pathologies. Curr Mol Med 8:492–496
15. Mishra SK, Siddique HR, Saleem M (2012) S100A4 calcium-binding protein is a key player in tumor progression and metastasis: preclinical and clinical evidence. Cancer Metastasis Rev 31:163–172
16. Azimi A, Pernemalm M, Frostvik Stolt M et al (2014) Proteomics analysis of melanoma metastasis: association between S100A13 expression and chemotherapy resistance. Br J Cancer 110:2489–2495
17. Pietas A, Schlüns K, Marenholz I et al (2002) Molecular cloning and characterization of the human *S100A14* gene encoding a novel member of the S100 family. Genomics 79:513–522
18. Ramasamy R, Shekhtmann A, Schmidt AM (2016) The multiple faces of RAGE-opportunities for therapeutic intervention in aging and chronic disease. Expert Opin Ther Targets 4:431–446
19. Most P, Seifert H, Gao E et al (2006) Cardiac S100A1 protein levels determine contractile performance and propensity towards heart failure after myocardial infarction. Circulation 114:1258–1268
20. Ackermann GE, Marenholz I, Wolfer DP et al (2006) S100A1-deficient male mice exhibit increased exploratory activity and reduced anxiety-related responses. Biochem Biophys Acta Mol Cell Biol 1763:1307–1319
21. Kopec-Medrek M, Widuchowska M, Kucharz EJ (2016) Calprotectin in rheumatic diseases: a review. Reumatologia 54:306–309
22. Ometto F, Friso L, Astorri D et al (2017) Calprotectin in rheumatic diseases. Exp Biol Med (Maywood) 242:859–873
23. Gebhardt C, Nèmeth J, Angel P, Hess J (2006) S100A8 and S100A9 in inflammation and cancer. Biochem Pharmacol 72:1622–1631
24. Bruhn S, Fang Y, Barrenäs F et al (2014) A generally applicable translational strategy identifies S100A4 as a candidate gene in allergy. Sci Transl Med 6:218ra4. https://doi.org/10.1126/scitranslmed.3007410
25. Swoboda I, Bugajska-Schretter A, Linhart B, Verdino P, Keller W et al (2007) A recombinant hyperallergenic parvalbumin mutant for immunotherapy of IgE-mediated fish allergy. J Immunol 178:6290–6296
26. Freidl R, Gstoettner A, Baranyi U, Swoboda I, Stolz F, Focke-Tejkl M et al (2017) Blocking antibodies induced by immunization with a hyperallergenic parvalbumin mutant reduce allergic symptoms in a mouse model of fish allergy. J Allergy Clin Immunol 139:1897–1905
27. Kretsinger RH, Nockolds CE (1973) Carp muscle calcium-binding protein. II. Structure determination and general description. J Biol Chem 248:3313–3326
28. Kawasaki H, Kretsinger RH (2017) Structural and functional diversity of EF-hand proteins: evolutionary perspectives. Protein Sci 26:1898–1920
29. Haynes LP, McCue HV, Burgoyne RD (2012) Evolution and functional diversity of calcium-binding proteins (CaBP's). Front Mol Neurosci 5:1–13
30. Krebs J, Heizmann CW (2007) Calcium-binding proteins and the EF-hand principle. In: Krebs J, Michalak M (eds) Calcium: a matter of life or death. Elsevier, Amsterdam, pp 51–93
31. Bunick CG, Nelson MR, Mangahas S et al (2004) Designing sequence to control protein function in an EF-hand protein. J Am Soc 126:5990–5998
32. Chazin WJ (2011) Relating form and function of EF-hand calcium binding proteins. Accounts Chem Res 44:171–179
33. Donato R, Cannon BR, Sorci G et al (2013) functions of S100 proteins. Curr Mol Med 13:24–57
34. Fritz G, Heizmann CW (2004) 3D structures of the calcium and zinc binding S100 proteins. In: Messerschmidt A, Bode W, Cygler M (eds) Handbook of metalloproteins, vol 3. Wiley, Chichester, pp 529–540
35. Leclerc E, Heizmann CW (2011) The importance of Ca^{2+}/Zn^{2+}-signaling S100 proteins and their receptor RAGE in translational medicine: impact on diagnostics and therapy in human disorders. Front Biosci S3:1232–1262
36. Cavalier MC, Melville Z, Aligholizaden E, Raman EPYW et al (2016) Novel protein-inhibitor interactions in site 3 of Ca^{2+}-bound S100B as discovered by X-ray crystallography. Acta Cristallogr D Struct Biol 72:753–760
37. Xiao Y, Shaw GS, Konermann L (2017) Calcium-mediated control of S100 proteins: allosteric communication via agitator/signal blocking mechanism. J Am Chem Soc 139:11460–11470
38. Heizmann CW (2005) The importance of calcium-binding proteins in childhood diseases. J Pediatr 147:731–738

39. Fritz G, Botelho HM, Morozova-Roche LA, Gomes CM (2010) Natural and amyloid self-assembly of S100 proteins: structural basis of functional diversity. FEBS J 277:4578–4590
40. Unno M, Kawasaki T, Takahara H et al (2011) Refined crystal structures of human Ca^{2+}/Zn^{2+}-binding S100A3 protein characterized by two disulfide bridges. J Mol Biol 408:477–490
41. Fernandez-Fernandez MR, Veprintsev DB, Fersht AR (2005) Proteins of the S100 family regulate the oligomerization of p53 tumor suppressor. Proc Natl Acad Sci U S A 102:4735–4740
42. Moravkova P, Kohoutova D, Reichrt S et al (2016) Role of S100 proteins in colorectal carcinogenesis. Gastroenterol Res Pract 2016:2632703. https://doi.org/10.1155/2016/2632703
43. Yen MC, Huang YC, Kan JY et al (2018) S100B expression in breast cancer as a predictive marker for cancer metastasis. J Oncol 52 (2):433–440
44. Ismail TM, Bennett D, Platt-Higgins AM et al (2016) S100A4 elevation empowers expression of metastasis effector molecules in human breast cancer. Cancer Res 77(3):780–789. https://doi.org/10.1158/0008-5472.CAN-16-1802
45. Dukhanina EA, Lukyanova TI, Dukhanin AS, Georgieva SG (2017) The role of S100A4 protein in the anticancer cytotoxicity: its presence is required on the surface of CD4+CD25+PGRPs+S100A4+ lymphocyte and undesirable on the surface of target cells. Cell Cycle 17(4):479–485. https://doi.org/10.1080/15384101.2017.1415678
46. Donato R, Heizmann CW (2010) S100B protein in the nervous system and cardiovascular apparatus in normal and pathological conditions. Cardiovasc Psych Neurol 2010:929712
47. Haiech J, Heizmann CW, Krebs J (2015 and 2017) Special issues on calcium: Biochim Biophys Acta: Mol Cell Res 1853:1919–2182 and 1864: 839–1131
48. Heizmann CW (ed) (1991) Novel calcium-binding proteins: Fundamentals and clinical implications. Springer, Heidelberg
49. Carafoli E, Brini M (eds) (2007) Calcium signaling and disease: molecular pathology of calcium, Subcellular biochemstry, vol 45. Springer, Heidelberg
50. Krebs J, Michalak M (eds) (2007) Calcium: a matter of life or death, New comprehensive biochemistry, vol 41. Elsevier, Amsterdam
51. Heizmann CW (ed) (2013) Calcium-binding proteins and RAGE: From structural basics to clinical applications, Methods in molecular biology, vol 963. Springer Protocols, Humana, New York
52. Kretsinger RH, Uversky VN, Permyakov EA (eds) (2013) Encyclopedia metalloproteins. Springer, Berlin
53. Krebs J (ed) (2018) Membrane dynamics and calcium signaling, Advances in experimental medicine and biology, vol 981. Springer, Heidelberg
54. Heizmann CW (ed) (2019) Calcium-binding proteins of the EF-hand superfamily: from basics to medical applications, Methods in molecular biology, vol 1929. Springer Protocols, Humana, New York (in press)
55. Pinto JR, Parvatiyar MS, Jones MA et al (2009) A functional and structural study of troponin C mutations related to hypertrophic cardiomyopathy. J Biol Chem 284:19090–19100
56. Landstrom AP, Parvatiyar MS, Pinto JR et al (2008) Molecular and functional characterization of novel hypertrophic cardiomyopathy susceptibility mutations in TNNC1-encoded troponin C. J Mol Cell Cardiol 45:281–288
57. Parvatiyar MS, Landstrom AP, Figueiredo-Freitas C et al (2012) A mutation in TNNC1-encoded cardiac troponin C, TNNC1-A31S, predisposes to hypertrophic cardiomyopathy and ventricular fibrillation. J Biol Chem 287:31845–31855
58. Li MX, Hwang PM (2015) Structure and function of cardiac troponin C (TNNC1): implications for heart failure, cardiomyopathies, and troponin modulating drugs. Gene 571:153–166
59. Watkins H, Ashrafian H, Redwood C (2011) Inherited cardiomyopathies. N Engl J Med 364:1643–1656
60. Lu QW, Wu YY, Morimoto S (2013) Inherited cardiomyopathies caused by troponin mutations. J Geriatr Cardiol 10:91–101
61. Hof D, Klingenberg R, von Eckardstein A (2013) Sensible use of high-sensitive troponin assays. In: Heizmann CW (ed) Calcium-binding proteins and RAGE: from structural basics to clinical applications, Methods in molecular biology, vol 963. Springer-Protocols, Humana, New York, pp 385–406
62. White HD, Thygesen K, Alpert JS, Jaffe AS (2014) Clinical implications of the third universal definition of myocardial infarction. Heart 100:424–432
63. Korff S, Katus HA, Giannitsis E (2006) Differential diagnosis of elevated troponins. Heart 92:987–993
64. Roongsritong C, Warraich I, Bradley C (2004) Common causes of troponin elevations in the absence of acute myocardial

infarction: incidence and clinical significance. Chest 125:1877–1884

65. Hwang PM, Cai F, Pineda-Sanabria SE, Corson DC, Sykes BD (2014) The cardiac-specific N-terminal region of troponin I positions the regulatory domain of troponin C. Proc Natl Acad Sci 111:14412–14417
66. Li Y, Zhu G, Paolocci N et al (2017) Heart failure-related hyperphosphorylation in the cardiac Troponin I C terminus has divergent effects on cardiac function in vivo. Circ Heart Fail 10(9):e003850. https://doi.org/10.1161/CIRCHEARTFAILURE117.003850
67. Haghikia A, Kaya Z, Schwab J et al (2015) Evidence of autoantibodies against cardiac troponin I and sarcomeric myosin in peripartum cardiomyopathy. Basic Res Cardiol 110 (6):60. https://doi.org/10.1007/1007/s00395-015-0517-2
68. Nyegaard M, Overgaard MT, Sondergaard MT et al (2012) Mutations in calmodulin cause ventricular tachycardia and sudden cardiac death. Am J Human Gen 91:703–712
69. Limpitikul WB, Dick IE, Joshi-Mukherjee R et al (2014) Calmodulin mutations associated with long QT syndrome prevent inactivation of cardiac L-type Ca^{2+} currents and promote proarrhythmic behavior in ventricular myocytes. J Mol Cell Cardiol 74:115–124
70. Hwang HS, Nitu FR, Yang Y et al (2014) Divergent regulation of ryanodine receptor 2 calcium release channel by arrhythmogenic human calmodulin missense mutations. Circ Res 114:1114–1124
71. Sondergaard MT, Tian X, Lia Y et al (2015) Arrhythmogenic calmodulin mutations affect the activation and termination of the cardiac ryanodine receptor-mediated Ca^{2+} release. J Biol Chem 290:26151–26162
72. Jimènez-Jaimez J, Palomino-Doza P, Ortega A et al (2016) Calmodulin 2 mutation N98S is associated with unexplained cardiac arrest in infants due to low clinical penetrance electric disorders. PLoS One 11(4):e0153851. https://doi.org/10.1371/journal.pone.01533851
73. Crotti L, Johnson CN, Graf E, De Ferrari GM, Cueno BF et al (2013) Calmodulin mutations associated with recurrent cardiac arrest in infants. Circulation 127:1009–1017
74. Makita N, Yagihara N, Crotti L et al (2014) Novel calmodulin mutations associated with congenital arrhythmia susceptibility. Circ Cardiovasc Genet 7:466–474
75. Coutu P, Bennett CN, Favre EG, Day SM, Metzger JM (2004) Parvalbumin corrects slowed relaxation in adult cardiac myocytes expressing hypertrophic cardiomyopathy-linked alpha-tropomyosin mutations. Circ Res 94:1235–1241
76. Tsoporis JN, Marks A, Haddad A, Dawood F, Liu PP, Parker TG (2005) S100B expression modulates left ventricular remodeling after myocardial infarction in mice. Circulation 111:598–606
77. Remppis A, Greten T, Schäfer BW et al (1996) Altered expression of the Ca^{2+}-binding protein S100A1 in human cardiomyopathy. Biochim Biophys Acta 1313:253–257
78. Ehlermann P, Remppis A, Guddat O et al (2000) Right ventricular upregulation of the Ca^{2+}-binding protein S100A1 in chronic pulmonary hypertension. Biochim Biophys Acta 1500:249–255
79. Most P, Pleger ST, Völkers M et al (2004) Cardial adenoviral S100A1 gene delivery rescues failing myocardium. J Clin Invest 114:1550–1563
80. Most P, Raake P, Weber C, Katus HA, Pleger ST (2013) S100A1 gene therapy in small and large animals. In: Heizmann CW (ed) Calcium-binding proteins and RAGE: from structural basics to clinical applications, Methods in molecular biology, vol 963. Springer-Protocols, Humana, New York, pp 407–420
81. Belmonte SL, Margulies KB, Blaxall BC (2011) S100A1: another step towards therapeutic development for heart failure. J Am Coll Cardiol 58:974–976
82. Pleger ST, Most P, Boucher M et al (2007) Stable myocardial-specific AAV6-S100A1 gene therapy results in chronic functional heart failure rescue. Circulation 115:2506–2515
83. Wang W, Asp ML, Guerrero-Serna G, Metzger JM (2014) Differential effects of S100 proteins A2 and A6 on cardiac Ca^{2+} cycling and contractile performance. J Mol Cell Cardiol 72:117–125
84. Mofid A, Newman NS, Lee PJ et al (2017) Cardiac overexpression of S100A6 attenuates cardiomyocyte apoptosis and reduces infarct size after myocardial ischemia-reperfusion. J Am Heart Assoc 6(2):e004738. https://doi.org/10.1161/JAHA116.004738
85. Snipsoyr MG, Ludvigsen M, Petersen E, Wiggers H, Honoré B (2016) A systematic review of biomarkers in the diagnosis of infective endocarditis. Int J Cardiol 202:564–570
86. Nazari A, Khorramdelazad H, Hassanshahi G et al (2017) S100A12 in renal and cardiovascular diseases. Life Sci 191:253–258
87. Buyukterzi Z, Can U, Alpaydin S et al (2017) Enhanced S100A9 and S100A12 expression

in acute coronary syndrome. Biomark Med 11:229–237

88. Turner NA (2016) Inflammatory and fibrotic responses of cardiac fibroblasts to myocardial damage associated molecular patterns (DAMPs). J Mol Cell Cardiol 94:189–200
89. Müller I, Vogl T, Pappritz K et al (2017) Pathogenic role of the damage-associated molecular patterns S100A8 and S100A9 in Coxsackievirus B3-induced myocarditis. Circ Heart Fail 10(11):e004125
90. Sheikh F, Lyon RC, Chen J (2015) Functions of myosin light chain-2 (MLC-2) in cardiac muscle and disease. Gene 569:14–20; Corrigendum: 571 (2015) 151
91. Sussman MA, Lim HW, Gude N et al (1998) Prevention of cardiac hypertrophy in mice by calcineurin inhibition. Science 281:1690–1693
92. Molkentin JD, Lu JR, Antos CL et al (1998) A calcineurin-dependent transcriptional pathway for cardiac hypertrophy. Cell 93:215–228
93. Bueno OF, Wilkins BJ, Tymitz KM et al (2002) Impaired cardiac hypertrophic response in calcineurin Abeta-deficient mice. Proc Natl Acad Sci U S A 99:4586–4591
94. Bueno OF, van Rooij E, Molkentin JD et al (2002) Calcineurin and hypertrophic heart disease: novel insights and remaining questions. Cardiovasc Res 53:806–821
95. Tsao L, Neville C, Musaro A, McCullagh KJ, Rosenthal N (2000) Revisiting calcineurin and human heart failure. Nat Med 6:2–3
96. Heineke J (2013) Screening for novel calcium-binding proteins that regulate cardiac hypertrophy: CIB1 as an example. In: Heizmann CW (ed) Calcium-binding proteins and RAGE: from structural basics to clinical applications, Methods in molecular biology, vol 963. Springer-Protocols, Humana, New York, pp 279–301
97. Naik UP, Patel PM, Parise IV (1997) Identification of a novel calcium-binding protein that interacts with the integrin alpha IIb cytoplasmic domain. J Biol Chem 272:4651–4654
98. Gentry HR, Singer AU, Betts I et al (2005) Structural and biochemical characterization of CIB1 delineates a new family of EF-hand containing proteins. J Biol Chem 280:8407–8415
99. Heineke J, Auger-Messier M, Correll RN et al (2010) CIB1 is a regulator of pathological cardiac hypertrophy. Nat Med 16:872–879
100. Huang Y, Wang KKW (2001) The calpain family and human disease. Trends Mol Med 7:355–362
101. Latavernier E, Perez J, Bellocq A et al (2008) Targeting the calpain/calpastatin system as a new strategy to prevent cardiovascular remodeling in angiotensin II-induced hypertension. Circ Res 102:720–728
102. Nishida K, Yamaguchi O, Otsu K (2015) Degradation system in heart failure. J Mol Cell Cardiol 84:212–222
103. Bompotis GC, Deftereos S, Angelidis C et al (2016) Altered calcium handling in reperfusion injury. Med Chem 12:114–130
104. Ilari A, Fiorillo A, Poser E et al (2015) Structural basis of sorcin-mediated calcium-dependent signal transduction. Sci Rep 5:16828. https://doi.org/10.1038/srep/6828
105. Collins LP, Meyers MB, Zhang J et al (2007) Expression of a sorcin missense mutation in the heart modulates excitation-contraction coupling. FASEB J 21:475–487
106. Franceschini S, Ilari A, Verzili D et al (2008) Molecular basis for the impaired function of the natural F112L sorcin mutant: X-ray crystal structure, calcium affinity, and interaction with annexin VII and the ryanodine receptor. FASEB J 22:295–306
107. Zamparelli C, Macquaide N, Colotti G et al (2010) Activation of the cardiac Na(+)-Ca(2+) exchanger by sorcin via the interaction of the Ca^{2+}-binding domains. J Mol Cell Cardiol 49:132–141
108. Fowler MR, Colotti G, Chiancone E et al (2008) Sorcin modulates cardiac L-type Ca^{2+}-current by functional interaction with the alpha1C subunit in rabbits. Exp Physiol 93:1233–1238
109. Li T, Yi L, Hai L et al (2018) The interactome and spatial redistribution feature of Ca^{2+}receptor protein calmodulin reveals a novel role in invadopodia-mediated invasion. Cell Death Dis 9:292. https://doi.org/10.1038/s1419-017-02553-7
110. Nussinov R, Zhang M, Tsai CJ, Jang H (2017) Calmodulin and IQGAP1 activation of PI3K*a* and Akt in KRAS, HRAS and NRAS-driven cancers. Biochim Biophys Acta 1864(6 Pt B):2304–2314. https://doi.org/10.1016/j.bbadis.2017.10032
111. Astrand R, Unden J, Romner B (2013) Clinical use of the calcium-binding S100B protein. In: Heizmann CW (ed) Calcium-binding proteins and RAGE: from structural basics to clinical applications, Methods in molecular biology, vol 963. Springer-Protocols, Humana, New York, pp 373–384
112. Harpio R, Einarsson R (2004) S100 proteins as cancer biomarkers with focus on S100B in malignant melanoma. Clin Biochem 37:512–518
113. Gogas H, Eggermont AMM, Hauschild A et al (2009) Biomarkers in melanoma. Ann Oncol 20

(Suppl 6):vi8–vi13. https://doi.org/10.1093/annonc/mdp251

114. Leclerc E (2011) The roles of S100 proteins and RAGE in melanoma. In: Tanaka Y (ed) Breakthrough in melanoma research, vol 1. InTech, Croatia, pp 331–356
115. Livingstone E, Krajewski C, Eigentler TK, Windemuth-Kieselbach C et al (2015) Prospective evaluation of follow-up in melanoma patients in Germany—results of a multicentre and longitudinal study. Eur J Cancer 51:653–667
116. Khan MI, Su YK, Zou J, Yang LW, Chou RH, Yu C (2018) S100B as an antagonist to block the interaction between S100A1 and the RAGE V domain. PLoS One 13(2):e0190545
117. Tandler N, Mosch B, Pietzsch J (2012) Protein and non-protein biomarkers in melanoma: a critical update. Amino Acids 43:2203–2230
118. Syed DN, Aljohani A, Waseem D, Mukhtar H (2018) Ousting RAGE in melanoma: a viable therapeutic target? Semin Cancer Biol 49:20–28
119. Tian T, Li X, Hua Z et al (2017) S100A1 promotes cell proliferation and migration and is associated with lymph node metastasis in ovarian cancer. Discov Med 23:235–245
120. Goh JY, Feng M, Wang W et al (2017) Chromosome 1q21.3 amplification is a trackable biomarker and actionable target for breast cancer recurrence. Nat Med 23:1319–1330
121. Lauriola L, Michetti F, Maggiano N et al (2000) Prognostic significance of the Ca^{2+}-binding protein S100A2 in laryngeal squamous-cell carcinoma. Int J Cancer 89:345–349
122. Diederichs S, Bulk E, Steffen B, Ji P, Tickenbrock L et al (2004) S100 family members and trypsinogens are predictors of distant metastasis and survival in early-stage non-small cell lung cancer. Cancer Res 64:5564–5569
123. Zhang Y, Wang H, Wang J et al (2015) Global analysis of chromosome 7 genes among patients with lung adenocarcinoma, squamous carcinoma, large-cell carcinoma, small-cell carcinoma, or non-cancer. Cancer Metastasis Rev 34(2):333–345
124. Wang T, Liang Y, Thakur A et al (2016) Diagnostic significance of S100A2 and S100A6 levels in sera of patients with non-small cell lung cancer. Tumour Biol 37 (2):2299–2304
125. Wicki R, Franz C, Scholl FA et al (1997) Repression of the candidate tumor suppressor gene S100A2 in breast cancer is mediated by site-specific hypermethylation. Cell Calcium 22:243–254
126. Camby I, LeFranc F, Titeca G et al (2000) Differential expression of S100 calcium-binding proteins characterizes distinct clinical entities in both WHO grade II and III astrocytic tumors. Neuropathol Appl Neurobiol 26:76–90
127. Camby I, Nagy N, Lopes MB et al (1999) Supratentorial pilocytic astrocytomas, astrocytomas, anaplastic astrocytomas and glioblastomas are characterized by differential expression of S100 proteins. Brain Pathol 9:1–19
128. Tao R, Wang ZF, Qui W et al (2017) Role of S100A3 in human hepatocellular carcinoma and the anticancer effect of sodium cantharidinate. Exp Ther Med 13:2812–2818
129. Liu J, Li GL, Dong HW et al (2008) In silico analysis and verification of S100 gene expression in gastric cancer. BMC Cancer 8:261. https://doi.org/10.1186/1471-2407-8-261
130. Ismail TM, Zhang S, Fernig DG et al (2010) Self-association of calcium binding protein, S100A4 and metastasis. J Biol Chem 285:914–922
131. Kim EJ, Helfman DM (2003) Characterization of the metastasis-associated protein, S100A4. Roles of calcium binding and dimerization in cellular localization and interaction with myosin. J Biol Chem 278:30063–30073
132. Fei F, Qu J, Zhang M, Li Y, Zhang S (2017) S100A4 in cancer progression and metastasis: a systematic review. Oncotarget 8:73219–73239
133. Dahlmann M, Okhrimenko A, Marcinkowski P, Osterland M et al (2014) RAGE mediates S100A4-induced cell motility via MAPK/ERK and hypoxia signaling and is a prognostic biomarker for human colorectal cancer metastasis. Oncotarget 5:3220–3233
134. Reckenbeil J, Kraus D, Probsmeier R et al (2016) Cellular distribution and gene expression pattern of metastasin (S100A4), calgranulin A (S100A8), and calgranulin B (S100A9) in oral lesions as marker for molecular pathology. Cancer Investig 34:246–254
135. Lv Y, Niu Z, Guo X et al (2018) Serum S100 calcium binding protein A4 (S100A4, metastasin) as a diagnostic and prognostic biomarkers in epithelial ovarian cancer. Br J Biomed Sci 75(2):88–91. https://doi.org/10.1080/09674845.2017.1394052
136. Tahara S, Nojima S, Ohshima K et al (2016) S100A4 accelerates the proliferation and invasion of endometrioid carcinoma and is associated with the 'MELF 'pattern. Cancer Sci 107(9):1345–1352
137. Egeland EV, Boye K, Park D et al (2017) Prognostic significance of S100A4-expression and

subcellular localization in early-stage breast cancer. Breast Cancer Res Treat 162:127–137
138. Dahlmann M, Kobelt D, Walther W et al (2016) S100A4 in cancer metastasis: Wnt signaling-driven interventions for metastasis restriction. Cancer (Basel) 8(6):E59. https://doi.org/10.3390/cancers8060059
139. Herwig N, Belter B, Pietzsch J (2016) Extracellular S100A4 affects endothelial cell integrity and stimulates transmigration of A375 melanoma cells. Biochem Biophys Res Commun 477(4):963–969
140. Kozono S, Ohuchida K, Ohtsuka T et al (2013) S100A4 mRNA expression level is a predictor of radioresistance of pancreatic cancer cells. Oncol Rep 30(4):1601–1608
141. Mudduluru G, Ilm K, Fuchs S, Stein U (2017) Epigenetic silencing of miR-520c leads to induced S100A4 expression and its mediated colorectal cancer progression. Oncotarget 8(13):21081–21094
142. Zakaria R, Platt-Higgins A, Rathi N et al (2016) Metastasis-inducing proteins are widely expressed in human brain metastases and associated with intracranial progression and radiation response. Br J Cancer 114 (10):1101–1108
143. Hattinger E, Zwicker S, Ruzicka T et al (2013) Opposing functions of psoriasin (S100A7) and koebnerisin (S100A15) in epithelial carcinogenesis. Curr Opin Pharmacol 13(4):588–594
144. Kataoka K, Ono T, Murata H et al (2012) S100A7 promotes the migration and invasion of osteosarcoma cells via the receptor for advanced glycation end products. Oncol Lett 3:1149–1153
145. Lin M, Xia B, Qin L, Chen H, Lou G (2018) S100A7 regulates ovarian cancer cell metastasis and chemoresistance through MAPK signaling and is targeted by miR-330-5p. DNA Cell Biol 37(5):491–500. https://doi.org/10.1089/dna.2017.3953
146. Tian T, Li X, Hua Z et al (2017) S100A7 promotes the migration, invasion and metastasis of human cervical cancer cells through epithelial-mesenchymal transition. Oncotarget 8(15):24964–24977. https://doi.org/10.18632/oncotarget.15329
147. Liu Y, Bunston C, Hodson N et al (2017) Psoriasin promotes invasion, aggregation and survival of pancreatic cancer cells: association with disease progression. Int J Oncol 50:1491–1500
148. Padilla L, Dakhel S, Adan J et al (2017) S100A7: from mechanism to cancer therapy. Oncogene 36:6749–6761
149. Turville J, Aghahoseini A, Sivarajasingham N, Abbas K et al (2016) Faecal calprotectin in patients with suspected colorectal cancer: a diagnostic accuracy study. Br J Gen Pract 66: e499–e506
150. Prieto D, Sotelo N, Seija N et al (2017) S100A9 protein in exosomes from chronic lymphocytic leukemia cells promote NF-*k*B activity during disease progression. Blood 130:777–788
151. Shin JM, Chang IK, Lee YH et al (2016) Potential role of S100A8 in cutaneous squamous cell carcinoma differentiation. Ann Dermatol 28:179–185
152. Gunaldi M, Okuturlar Y, Gedikbasi A et al (2015) Diagnostic importance of S100A9 and S100A12 in breast cancer. Biomed Pharmacother 76:52–56
153. Topuz MF, Binnetoglu A, Yumusakhuylu AC et al (2017) Circulating calprotectin as a biomarker of laryngeal carcinoma. Eur Arch Otorhinolaryngol 274:2499–2504
154. Huang CH, Kuo CJ, Liang SS et al (2015) Onco-proteomics identifies urinary S100A9 and GRN as potential combinatorial biomarkers for early diagnosis of hepatocellular carcinoma. Biochim Biophys Acta Clin 3:205–213
155. Argyris PP, Slama ZM, Ross KF, Khammanivong A, Herzberg MC (2018) Calprotectin and the initiation and progression of head and neck cancer. J Dent Res 97(6):674–682. https://doi.org/10.1177/0022034518756330
156. Yasar O, Akcay T, Obek C, Turegun FA (2017) Significance of S100A8, S100A9 and calprotectin levels in bladder cancer. Scand J Clin Lab Invest 77(6):437–441. https://doi.org/10.1080/00365513.2017.1336567
157. Reeb AN, Li W, Sewell W (2014) S100A8 is a novel therapeutic target for anaplastic thyroid carcinoma. J Clin Endocinol Metab 25: jc20142988
158. Hibino T, Sakaguchi M, Miyamoto S, Yamamoto M et al (2012) S100A9 is a novel ligand of EMMPRIN that promotes melanoma metastasis. Cancer Res 73(1):172–183. https://doi.org/10.1158/0008-5472.CAN-11_3843
159. Tamburini J (2017) S100 proteins in AML: differentiation and beyond. Blood 129:1893–1894
160. Ji J, Su L, Liu Z (2016) Critical role of calpain in inflammation. Biomed Rep 5(6):647–652
161. Liu Y, Han X, Gao B (2015) Knockdown of S100A11 expression suppresses ovarian cancer cell growth and invasion. Exp Ther Med 9:1460–1464

162. Xiao M, Li T, Ji Y et al (2018) S100A11 promotes human pancreatic cancer PANC-1 cell proliferation and is involved in the PI3K/AKT signaling pathway. Oncol Lett 15:175–182
163. Jaiswal JK, Nylandsted J (2015) S100 and annexin proteins identify cell membrane damage as the Achilles heel of metastatic cancer cells. Cell Cycle 14:502–509
164. Martinez-Aguilar J, Clifton-Bligh R, Molloy MP (2015) A multiplexed, targeted mass spectrometry assay of the S100 protein family uncovers the isoform-specific expression in thyroid tumours. BMC Cancer 15:199. https://doi.org/10.1186/s12885-015-1217-x
165. Zhong J, Liu C, Chen YJ et al (2016) The association between S100A13 and HMGA1 in the modulation of thyroid cancer proliferation and invasion. J Transl Med 14:80. https://doi.org/10.1186/s12967-016-0824-x
166. Wang X, Yang J, Qian J et al (2015) S100A14, a mediator of epithelial-mesenchymal transition, regulates proliferation, migration and invasion of human cervical cancer cells. Am J Cancer Res 5:1484–1495
167. Zhu M, Wang H, Cui J et al (2017) Calcium-binding protein S100A14 induces differentiation and suppresses metastasis in gastric cancer. Cell Death Dis 8:e2938. https://doi.org/10.1038/cddis.2017.297
168. Tanaka M, Ichikawa-Tomikawa N, Shishito N et al (2015) Co-expression of S100A14 and S100A16 correlates with poor prognosis in human breast cancer and promotes cancer cell invasion. BMC Cancer 15:53. https://doi.org/10.1186/s12885-015-1059-6
169. Xu C, Chen H, Wang X et al (2014) S100A14, a member of the EF-hand calcium-binding proteins, is overexpressed in breast cancer and acts as a modulator of HER2 signaling. J Biol Chem 289:827–837
170. Sun X, Wang T, Zhang C, Ning K et al (2018) S100A16 is a prognostic marker for colorectal cancer. J Surg Oncol 117:275–283
171. Arumugam T, Ramachandran V, Gomez SB et al (2012) S100P-derived RAGE antagonistic peptide reduces tumor growth and metastasis. Clin Cancer Res 18(16):4356–4364
172. Arumugam T, Ramachandran V, Maxwell D et al (2013) Designing and developing S100P inhibitor 5-methyl cromolyn (C5OH) for pancreatic cancer therapy. Mol Cancer Ther 12(5):654–662
173. Arumungam T, Ramachandran V, Logsdon CD (2006) Effect of cromolyn on S100P interactions with RAGE and pancreatic cancer growth and invasion in mouse models. J Natl Cancer Inst 98:1806–1818
174. Penumutchu SR, Chou RH, Yu C (2014) Interaction between S100P and the anti-allergy drug cromolyn. Biochem Biophys Res Commun 454:404–409
175. Namba T, Homan T, Nishimura T et al (2009) Up-regulation of S100P expression by non-steroidal anti-inflammatory drugs and its role in anti-tumorigenic effects. J Biol Chem 284:41584167
176. Matsunaga T, Ohtsuka T, Asano K et al (2017) S100P in duodenal fluid is a useful diagnostic marker for pancreatic ductal adenocarcinoma. Pancreas 46:1288–1295
177. Zhu L, Ito T, Nakahara T et al (2013) Upregulation of S100P, receptor for advanced glycation end products and ezrin in malignant melanoma. J Dermatol 40:973–979
178. Peng C, Chen H, Wallwiener M et al (2016) Plasma S100P level as a novel prognostic marker of metastatic breast cancer. Breast Cancer Res Treat 157:329–338
179. Liu Y, Wang C, Shan X et al (2017) S100P is associated with proliferation and migration in nasopharyngeal carcinoma. Oncol Lett 14:525–532
180. Gutknecht MF, Seaman ME, Ning B et al (2017) Identification of the S100 fused-type protein hornerin as a regulator of tumor vascularity. Nat Commun 8(1):552. https://doi.org/10.1038/s41467-017-00488-6
181. Colotti G, Poser E, Fiorillo A et al (2014) Sorcin, a calcium binding protein involved in the multidrug resistance mechanisms in cancer cells. Molecules 19:13976–13989
182. Genovese I, Fiorillo A, Ilari A et al (2017) Binding of doxorubicin to sorcin impairs cell death and increases drug resistance in cancer cells. Cell Death Dis 8(7):e2950. https://doi.org/10.1038/cddis.2017.34
183. Dabaghi M, Rahgozar S, Moshtaghian J et al (2016) Overexpression of sorcin is a prognostic biomarker for multidrug-resistant pediatric acute lymphoblastic leukemia and correlates with upregulated MDR1/P-gp. Genet Test Mol Biomarkers 20:516–521
184. Tuo H, Shu F, She S et al (2017) Sorcin induces gastric cancer cell migration and invasion contributing to STAT3 activation. Oncotarget 8(61):104258–104271
185. Tong W, Sun D, Wang Q, Suo J (2015) Sorcin enhances metastasis and promotes epithelial-to-mesenchymal transition of colorectal cancer. Cell Biochem Biophys 72(2):453–459
186. Gong Z, Sun P, Chu H et al (2014) Overexpression of sorcin in multidrug-resistant human breast cancer. Oncol Lett 8(6):2393–2398

187. Leisner TM, Freeman TC, Black JL, Parise LV (2016) CIB1: a small protein with big ambitions. FASEB J 30(8):2640–2650
188. Wang X, Peng X, Zhang X et al (2017) The emerging roles of CIB1 in cancer. Cell Physiol Biochem 43(4):1413–1424
189. Hu YH, Oh S, Yeo YR et al (2015) Swiprosin-1 stimulates cancer invasion and metastasis by increasing the Rho family of GTPase signaling. Oncotarget 6(15):13060–13071
190. Fan CC, Cheng WC, Huang YC et al (2017) EFHD2 promotes epithelial-to-mesenchymal transition and correlates with postsurgical recurrence of stage I lung carcinoma. Sci Rep 7(1):14617. https://doi.org/10.1038/s41598-017-15186-y
191. Kang J, Kang YH, Oh BM et al (2016) Tescalcin expression contributes to invasive and metastatic activity in colorectal cancer. Tumor Biol 37(10):13843–13853
192. Li N, Zheng Y, Xuang C et al (2015) LETM1 overexpression is correlated with the clinical features and survival outcome of breast cancer. Int J Clin Exp Pathol 8 (10):12893–12900
193. Wang CA, Liu Q, Chen Y et al (2015) Clinical implications of leucine zipper/EF hand-containing transmembrane-1 overexpression in the prognosis of the triple-negative breast cancer. Exp Mol Pathol 98(2):254–259
194. Huang B, Zhang J, Zhang X et al (2017) Suppression of LETM1 by siRNA inhibits cell proliferation and invasion of bladder cancer cells. Oncol Rep 38(5):2935–2940
195. Bendickova K, Tidu F, Fric J (2017) Calcineurin-NFAT signalling in myeloid leucocytes: new prospects and pitfalls in immunosuppressive therapy. EMBO Mol Med 9(8):990–999. https://doi.org/10.15252/emmm.201707698
196. Birlea SA (2017) S100B: correlation with active vitiligo depigmentation. J Invest Dermatol 137(7):1408–1410
197. Speeckaert R, Voet S, Hoste E, van Geel N (2017) S100B is a potential activity marker in nonsegmental vitiligo. J Invest Dermatol 137 (7):1445–1453
198. Hansen MT, Forst B, Cremers N et al (2014) A link between inflammation and metastasis: serum amyloid A1 and A3 induce metastasis, and are targets of metastasis-inducing S100A4. Oncogene 34(4):424–435. https://doi.org/10.1038/onc.2013.568
199. Jia J, Duan Q, Guo J, Zheng Y (2014) Psoriasin, a multifunctional player in different diseases. Curr Protein Pept Sci 15:836–842
200. D'Amico F, Skarmoutsou E, Granata M et al (2016) S100A7: a rAMPing up AMP molecule in psoriasis. Cytokine Growth Factor Rev 32:97–104
201. Cubillos S, Norgauer J (2016) Low vitamin D-modulated calcium-regulating proteins in psoriasis vulgaris plaques: S100A7 overexpression depends on joint involvement. Int J Mol Med 38(4):1083–1092
202. Rangaraj A, Ye L, Sanders AJ et al (2017) Molecular and cellular impact of psoriasin (S100A7) on the healing of human wounds. Exp Ther Med 13(5):2151–2160
203. Vogl T, Stratis A, Wixler V et al (2018) Autoinhibitory regulation of S100A8/S100A9 alarmin activity locally restricts sterile imflammation. J Clin Invest 28(5):1852–1866. https://doi.org/10.1172/JCI89867
204. Rothmund F, Gerss J, Ruperto N et al (2013) Validation of relapse risk biomarkers for routine use in patients with juvenile idiopathic arthritis. Arthritis Care Res (Hoboken) 66 (6):949–955. https://doi.org/10.1002/acr.22248
205. Nefla M, Holzinger D, Berenbaum F, Jacques C (2016) The danger from within: alarmins in arthritis. Nat Rev Rheumatol 12(11):669–683
206. Mariani A, Marsili M, Nozzi M et al (2014) Serum calprotectin: review of its usefulness and validity in paediatric rheumatic diseases. Clin Exp Rheumatol 33(1):109–114
207. Leach ST, Yang Z, Messina I, Song C et al (2007) Serum and mucosal S100 proteins, calprotectin (S100A8/S100A9) and S100A12, are elevated at diagnosis in children with inflammatory bowel disease. Scand J Gastro 42(11):1321–1331
208. Malickova K, Kalousova M, Fucikova T et al (2010) Anti-inflammatory effect of biological treatment in patients with inflammatory bowel diseases: calprotectin and IL-6 changes do not correspond to sRAGE changes. Scand J Clin Lab Invest 70(49):294–299
209. Lopez RN, Leach ST, Lemberg DA et al (2016) Faecal biomarkers in inflammatory bowel disease. J Gastroenterol Hepatol 32 (3):577–582. https://doi.org/10.1111/jgh.13611
210. Rosso C, Caviglia GP, Pellicano R (2016) The usefulness of fecal calprotectin determination in pediatric intestinal diseases. Minerva Pediatr 68(6):478–486
211. Foell D, Wulffraat N, Wedderburn LR et al (2010) Methotrexate withdrawal at 6 vs 12 month in juvenile idiopathic arthritis in remission: a randomized clinical trial. JAMA 303 (13):1266–1273

212. Schonthaler HB, Guinea-Viniegran J, Wculek SK et al (2013) S100A8-S100A9 protein complex mediates psoriasis by regulating the expression of complement factor C3. Immunity 39(6):1171–1181
213. Kostakis ID, Cholidou KG, Vaiopoulos AG et al (2013) Fecal calprotectin in pediatric inflammatory bowel disease: a systematic review. Dig Dis Sci 58(2):309–319
214. Kane D, Roth J, Frosch M et al (2003) Increased perivascular synovial membrane expression of myeloid-related proteins in psoriatic arthritis. Arthritis Rheum 48:1676–1685
215. Obry A, Lequerré T, Hardouin J et al (2014) Identification of S100A9 as biomarker of responsiveness to the methotrexate/etanercept combination in rheumatoid arthritis using a proteomic approach. PLoS One 9 (12):e115800
216. Garg M, Leach ST, Coffey MJ et al (2017) Age-dependent variation of fecal calprotectin in cystic fibrosis and healthy children. J Cystic Fibrosis 16(5):631–636
217. Ulas T, Pirr S, Fehlhaber B et al (2017) S100-alarmin-induced innate immune programming protects newborn infants from sepsis. Nat Immunol 18(6):622–632. https://doi.org/10.1038/ni.3745
218. Hiroshima Y, Hsu K, Tedla N et al (2017) S100A8/A9 and S100A9 reduce acute lung injury. Immunol Cell Biol 95(5):461–472
219. Boschetti G, Garnero P, Moussata D et al (2015) Accuracies of serum and fecal S100 proteins (calprotectin and calgranulin C) to predict the response to TNF antagonists in patients with Crohn's disease. Inflamm Bowel Dis 21(2):331–336
220. Holzinger D, Kessel C, Omenetti A, Gattorno M (2015) From bench to bedside and back again: translational research in autoinflammation. Nat Rev Rheumatol 11 (10):573–585. https://doi.org/10.1038/nrrheum.2015
221. Kessel C, Holzinger D, Foell D (2012) Phagocyte-derived S100 proteins in autoinflammation: putative role in pathogenesis and usefulness as biomarkers. Clin Immunol 147(3):229–241. https://doi.org/10.1016/j.clim.2012.11.008
222. Taverna D, Pollins AC, Sindona G et al (2015) Imaging mass spectrometry for assessing cutaneous wound healing: analysis of pressure ulcers. J Proteome Res 14(2):986–996
223. Wicki R, Schäfer BW, Erne P, Heizmann CW (1996) Characterization of the human and mouse cDNAs coding for S100A13, a new member of the S100 protein family. Biochem Biophys Res Commun 227:594–599
224. Andrés Cerezo L, Sumova B, Prajzlerova K et al (2017) Calgizzarin (S100A11): a novel inflammatory mediator associated with disease activity of rheumatoid arthritis. Arthritis Res Ther 19(1):79. https://doi.org/10.1186/s13075-017-1288-y
225. Foell D, Ichida F, Vogl T et al (2003) S100A12 (EN-RAGE) in monitoring Kawasaki disease. Lancet 361:1270–1272
226. Gottsch JD, Liu SH, Minkovitz JB et al (1995) Autoimmunity to a cornea-associated stromal antigen in patients with Mooren's ulcer. Ophthalmol Vis Sci 36:1541–1547
227. Heida A, Kobold ACM, Wagenmakers L et al (2017) Reference values of fecal calgranulin C (S100A12) in school aged children and adolescents. Clin Chem Lab Med 56(1):126–131. https://doi.org/10.1515/cclm-2017-0152
228. Heida A, van de Vijver E, Muller Kobold A, van Rheenen P (2017) Selecting children with suspected inflammatory bowel disease for endoscopy with the calgranulin C or calprotectin stool test. Protocol of the CACATU study. BMJ Open 7(5):e015636
229. Wilsmann-Theis D, Wagenpfeil J, Holzinger D et al (2015) Among the S100 proteins, S100A12 is the most significant marker for psoriasis disease activity. J Eur Acad Dermatol Venerol 30(7):1165–1170. https://doi.org/10.1111/jdv.1326910
230. Farokhzadian J, Mangolian Shahrbabaki P, Bagheri V (2017) S100A12-CD 36 axis: a novel player in the pathogenesis of atherosclerosis? Cytokine 17:30211–30219. https://doi.org/10.1016/j.cyto.2017.07.010
231. Elsayed S, Apold J, Holen E et al (1991) The structural requirements of epitopes with IgE binding capacity demonstrated by three major allergens from fish, egg and tree pollen. Scand J Clin Lab Invest Suppl 204:17–31
232. Zuidmeer-Jonggejan L, Huber H, Swoboda I et al (2015) Development of a hypoallergenic recombinant parvalbumin for first-in-man subcutaneous immuntherapy of fish allergy. Int Arch Allergy Immunol 166(1):41–51
233. Ruethers T, Raith M, Sharp MF et al (2018) Characterization of Ras k 1 a novel major allergen in Indian mackerel and identification of parvalbumin as the major fish allergen in 33 Asia-Pacific fish species. Clin Exp Allergy 48 (4):452–463
234. Kuehn A, Swoboda I, Arumugam K et al (2014) Fish allergens at a glance: variable allergenicity of parvalbumin, the major fish

allergens. Front Immunol 5:179. https://doi.org/10.3389/fimm.2014.00179
235. Ebashi S, Ebashi F, Kodama A (1967) Troponin as the Ca^{2+}-receptive protein in the contractile system. J Biochem 62:137–138
236. Jin JP, Zang Z, Bautista JA (2008) Isoform diversity, regulation, and functional adaptation of troponin and calponin. Crit Rev Eukariot Gene Expr 18:93–124
237. Cheung WY (1970) Cyclic 3′,5′-nucleodide phosphodiesterase. Demonstration of an activator. Biochem Biophys Res Commun 38:533–538
238. Kakiuchi S, Yamazaki R (1970) Calcium-dependent phosphodiesterase activity and its activator factor (PAF) from brain. Biochem Biophys Res Commun 41:1104–1110
239. Teo TS, Wang JH (1973) Mechanism of activation of a cyclic adenosine 3′,5′-monophosphate phosphodiesterase from bovine heart by calcium ions. Identification of the protein activator as a Ca^{2+}-binding protein. J Biol Chem 248:5950–5955
240. Sorensen AB, Sondergaard MT, Overgaard MT (2013) Calmodulin in a heartbeat. FEBS J 280:5511–5532
241. Kursula P (2014) The many structural faces of calmodulin: a multitasking molecular jackknife. Amino Acids 46:2295–2304
242. Marshall CB, Nishikawa T, Osawa M et al (2015) Calmodulin and STIM proteins: two major calcium sensors in the cytoplasm and endoplasmic reticulum. Biochem Biophys Res Commun 460:5–21
243. Ikeda S, He A, Kong SW et al (2009) Micro RNA-1 negatively regulates expression of the hypertrophy-associated calmodulin and Maf2a genes. Mol Cell Biol 29:193–204
244. Coffee CJ, Bradshaw RA (1973) Carp muscle calcium-binding protein I. Characterization of the tryptic peptides and the complete amino acid sequence of component B. J Biol Chem 248:3305–3312
245. Celio MR, Heizmann CW (1982) Calcium-binding protein parvalbumin is associated with fast contracting muscle fibres. Nature 297:504–506
246. Heizmann CW, Berchtold MW, Rowlerson AM (1982) Correlation of parvalbumin concentration with relaxation speed in mammalian muscles. Proc Natl Acad Sci U S A 79:7243–7247
247. Stuhlfauth I, Reininghaus J, Jockusch H, Heizmann CW (1984) Calcium-binding protein, parvalbumin is reduced in mutant mammalian muscle with abnormal contractile properties. Proc Natl Acad Sci U S A 81:4814–4818
248. Schäfer BW, Heizmann CW (1996) The S100 family of EF-hand calcium-binding proteins: functions and pathology. Trends Biochem Sci 21:134–140
249. Zimmer DB, Wright-Sadosky P, Weber DJ (2003) Molecular mechanisms of S100-target interactions. Microsc Res Tech 60:552–559
250. Heizmann CW, Fritz G (2009) The family of S100 cell signaling proteins. In: Bradshaw RA, Dennis EA (eds) The Handbook Cell Signalling, 2nd edn. Academic, Oxford, pp 983–994
251. Marenholz I, Heizmann CW, Fritz G (2004) S100 proteins in mouse and man: from evolution to function and pathology (including an update of the nomenclature). Biochem Biophys Res Commun 322:111–1122
252. Hermann A, Donato R, Weiger TM, Chazin WJ (2012) S100 calcium binding proteins and ion channels. Front Pharmacol 3:67. https://doi.org/10.3389/fphar.2012.00067
253. Moroz OV, Antson AA, Dodson EJ et al (2002) The structure of S100A12 in a hexameric form and its proposed role in receptor signaling. Acta Cristallogr D Biol Crystallogr 58:407–413
254. Ostendorp T, Leclerc E, Galichet A et al (2007) Structural and functional insights into RAGE activation by multimeric S100B. EMBO J 26:3868–3878
255. Engelkamp D, Schäfer BW, Mattei MG, Erne P, Heizmann CW (1993) Six S100 genes are clustered on human chromosome 1q21: identification of two genes coding for the two previously unreported calcium-binding proteins S100D and S100E. Proc Natl Acad Sci U S A 90:6547–6551
256. Schäfer BW, Wicki R, Engelkamp D, Mattei MG, Heizmann CW (1995) Isolation of a YAC clone covering a cluster of nine S100 genes on human chromosome 1q21: rationale for a new nomenclature of the S100 calcium-binding protein family. Genomics 25:638–643
257. Marenholz I, Lovering RC, Heizmann CW (2006) An update of the S100 nomenclature. Biochem Biophys Acta Mol Cell Res 1763:1282–1283
258. Koch M, Chitayat S, Dattilo BM et al (2010) Structural basis for ligand recognition and activation of RAGE. Structure 18:1342–1352
259. Rouhiainen A, Kuja-Panula J, Tumova S, Rauvala H (2013) RAGE-mediated cell signaling. In: Heizmann CW (ed) Calcium-binding proteins and RAGE: from structural basics to clinical applications, Methods in molecular biology, vol 963. Springer Protocols, Humana, New York, pp 239–263

260. Leclerc E, Vetter S (2015) The role of S100 proteins and their receptor RAGE in pancreatic cancer. Biochim Biophys Acta 1852:2706–2711
261. Jensen LJ, Flyvbjerg A, Bjerre M (2015) Soluble receptor for advanced glycation end product: a biomarker for acute coronary syndrome. Biomed Res Int 2015:815942. https://doi.org/10.1155/2015/815942
262. Junarek J, Ray R, Banach M, Rai V (2015) Receptor for advanced glycation end-products in neurodegenerative diseases. Rev Neurosci 26:691–698
263. Hiratsuka S, Watanabe A, Sakurai Y et al (2008) The S100A8-serum amyloid A3-TLR-4 paracrine cascade establishes a pre-metastatic phase. Nat Cell Biol 10:1349–1355
264. Ehrchen JM, Sunderkötter C, Foell D, Vogl T, Roth J (2009) The endogenous Toll-like receptor 4 agonist S100A8/S100A9 (calprotectin) as innate amplifier of infection, autoimmunity, and cancer. J Leukoc Biol 86:557–566
265. Imbalzano E, Mandraffino G, Casciaro M et al (2016) Pathophysiological mechanism and therapeutic role of S100 proteins in cardiac failure: a systematic review. Heart Fail Rev 21 (5):463–473
266. Wright NT, Cannon BR, Zimmer DB, Weber DJ (2009) S100A1: structure, function, and therapeutic potential. Curr Chem Biol 3:138–145
267. Duarte-Costa S, Castro-Ferreira R, Neves JS, Leite-Moreira AF (2014) S100A1: a major player in cardiovascular performance. Physiol Res 63:669–681
268. Desjardins JF, Teichert-Kuliszewska K, Parker T (2010) S100A1: a pluripotent regulator of cardiac and vascular function. Can J Cardiol 26(Suppl. A):9A–12A
269. Du XJ, Cole TJ, Tenis N et al (2002) Impaired cardiac contractility response to hemodynamic stress in S100A1 deficient mice. Mol Cell Biol 22:2821–2829
270. Ackermann GE, Domenighetti AA, Deten A et al (2008) S100A1 deficiency results in prolonged ventricular repolarization in response to sympathetic activation. Gen Physiol Biophys 27:127–142
271. Gusev K, Ackermann GE, Heizmann CW, Niggli E (2009) Ca^{2+} signaling in mouse cardiomyocytes with ablated S100A1 protein. Gen Physiol Biophys 28:371–383
272. Brett W, Mandinova A, Remppis A et al (2001) Translocation of S100A1 calcium-binding protein during heart surgery. Biochem Biophys Res Commun 284:698–703
273. Kiewitz R, Acklin C, Minder E et al (2000) S100A1, a new marker for acute myocardial ischemia. Biochem Biophys Res Commun 274:865–871
274. Schneider M, Kostin S, Strom CC et al (2007) S100A4 is upregulated in injured myocardium and promotes growth and survival of cardiac myocytes. Cardiovasc Res 75:40–50
275. Grabarek Z (2006) Structural basis for diversity of the EF-hand calcium-binding proteins. J Mol Biol 359:509–525
276. Klee CB, Crouch TH, Krinks MH (1979) Calcineurin: a calcium and calmodulin-binding protein of the nervous system. Proc Natl Acad Sci U S A 76:6270–6273
277. Klee CB, Ren H, Wang X (1998) Regulation of the calmodulin-stimulated protein phosphatase, calcineurin. J Biol Chem 273:13367–13370
278. Feske S, Rao A, Hogan PG (2007) The Ca^{2+}-calcineurin-NAFT signaling pathway in: Krebs J, Michalak M (eds) Calcium: a matter of life or death. Elsevier B.V, Amsterdam, pp 363–401
279. Medyouf H, Ghysdael J (2008) The calcineurin/NFAT signaling pathway. Cell Cycle 7:297–303
280. Kobayashi S, Nakamura TY, Wakabayashi S (2015) Calcineurin B homologous protein 3 negatively regulates cardiomyocyte hypertrophy via inhibition of glycogen synthase kinase 3 phosphorylation. J Mol Cell Cardiol 84:133–142
281. Khorchid A, Ikura M (2002) How calpain is activated by calcium. Nat Struct Biol 9:239–241
282. Zatz M, Starling A (2005) Calpains and disease. New Engl J Med 352:2413–2423
283. Campbell RL, Davies PL (2012) Structure-function relationships in calpains. Biochem J 447:335–351
284. Osuki K, Shibata H, Maki M (2013) Biochemical and immunological detection of physical interactions between penta-EF-hand protein ALG-2 and its binding partners. In: Heizmann CW (ed) Calcium-binding proteins and RAGE: from structural basics to clinical applications, Methods in molecular biology, vol 963. Springer-Protocols, Humana, New York, pp 187–200
285. Lei HY, Zhou XL, Ruan ZR et al (2015) Calpain cleaves most components in the multiple aminoacyl-tRNA synthetase complex and affects their functions. J Biol Chem 290:26314–26127
286. Maki M, Kitaura Y, Satoh H, Ohkouchi S, Shibata H (2002) Structures, functions and molecular evolution of the penta-EF-hand

Ca^{2+}-binding proteins. Biochim Biophys Acta 1600:51–60

287. Monteith GR, Prevarskaya N, Roberts-Thomson SJ (2017) The calcium-cancer signalling nexus. Nat Rev Cancer 17(6):367–380
288. Chen H, Xu C, Jin Q, Liu Z (2014) S100 family in human cancer. Am J Cancer Res 4:89–115
289. Wang T, Huo X, Chong Z et al (2018) A review of S100 protein family in lung cancer. Clin Chim Acta 476:54–59
290. Unden J, Romner B (2009) A new objective method for CT triage after minor head injury: serum S100B. Scand J Clin Lab Invest 69:13–17
291. Föhr UG, Heizmann CW, Engelkamp D, Schäfer BW, Cox JA (1995) Purification and cation binding properties of the recombinant S100 calcium-binding A3, an EF-hand motif protein with a high affinity for zinc. J Biol Chem 270:21056–21061
292. Fritz G, Heizmann CW, Kroneck PM (1998) Probing the structure of the human Ca^{2+}-and Zn^{2+}-binding protein S100A3: spectroscopic investigation of its transition metal ion complexes, and three-dimensional structural model. Biochim Biophys Acta 1448:264–276
293. Fritz G, Mittl PR, Vasak M, Grütter MG, Heizmann CW (2002) The crystal structure of the metal-free human EF-hand protein S100A3 at 1.7-A resolution. J Biol Chem 277:33092–33098
294. Kizawa K, Unno M, Heizmann CW, Takahara H (2014) Importance of citrullination on hair protein molecular assembly during trichocytic differentiation. In: Nicholas AP, Bhattacharya SK (eds) Protein deimination in human health and disease. Springer, New York, pp 129–148
295. Wolf R, Mascia F, Dharamsi A et al (2010) Gene from a psoriasis susceptibility locus primes the skin for inflammation. Sci Transl Med 2(61):61ra90
296. Murray JI, Tonkin ML, Whiting AL et al (2012) Structural characterization of S100A15 reveals a novel zinc coordination site among S100 proteins and altered surface chemistry with functional implications for receptor binding. BMC Struct Biol 12:16. https://doi.org/10.1186/1472-6807-12-16
297. Loser K, Vogl T, Voskort M et al (2010) The Toll-like receptor 4 ligands Mrp8 and Mrp14 are crucial in the development of autoreactive $CD8^{+}$T cells. Nat Med 16:713–718
298. Laouedj M, Tardif MR, Gil L et al (2017) S100A9 induces differentiation of acute myeloid leukemia cells through TLR4. Blood 129:1980–1990
299. Ridinger K, Schäfer BW, Durussel I, Cox JA, Heizmann CW (2000) S100A13: biochemical characterization and subcellular localization in different cell lines. J Biol Chem 275:8686–8694
300. Hsieh HL, Schäfer BW, Heizmann CW (2002) S100A13 and S100A6 exhibit distinct translocation pathways in endothelial cells. J Cell Sci 115:3149–3158
301. Marenholz I, Volz A, Ziegler A et al (1996) Gene analysis of the epidermal differentiation complex (EDC) on human chromosome 1q21: chromosomal orientation, new markers, and a 6-mB YAC contig. Genomics 37:295–302
302. Kizawa K, Takahara H, Unno M, Heizmann CW (2011) S100 and S100 fused-type protein families in epidermal maturation with special focus on S100A3 in mammalian hair cuticles. Biochimie 93:2038–2047
303. Mlitz V, Strasser B, Jaeger K et al (2014) Trichohyalin-like proteins have evolutionary conserved roles in the morphogenesis of skin appendages. J Invest Dermatol 134(11):2685–2692
304. Bunick CG, Presland RB, Lawrence OT et al (2015) Crystal structure of human profilaggrin S100 domain and identification of target proteins annexin II, stratifin, and HSP27. J Invest Dermatol 135(7):1801–1809
305. .Mlitz V, Hussain T, Tschachler E, Eckhart L (2017) Filaggrin has evolved from an, S100 fused-type protein (SFTP) gene present in a common ancestor of amphibians and mammals. Exp Dermatol 26(10):955–957. https://doi.org/10.1111/exd.13317
306. Gutierrez-Ford C, Levay K, Gomes AV et al (2003) Characterization of tescalcin, a novel EF-hand protein with a single Ca^{2+}–binding site: metal binding properties, localization in tissues and cells, and effect on calcineurin. Biochemistry 42:14553–14565
307. Kolobynina KG, Solovyova VV, Levay K, Rizanov AA, Slepak VZ (2016) Emerging roles of the single EF-hand Ca^{2+} sensor tescalcin in the regulation of gene expression, cell growth and differentiation. J Cell Sci 129(19):3533–3540
308. Kang YH, Han SR, Kim JT et al (2014) The EF-hand calcium-binding protein tescalcin is a potential oncotarget in colorectal cancer. Oncotarget 5(8):2149–2160
309. Edgeworth J, Gorman M, Benett R et al (1991) Identification of p8,14 as a highly abundant heterodimeric calcium binding protein complex of myeloid cells. J Biol Chem 266(12):7706–7713

310. Roth J, Vogl T, Sorg C, Sunderkotter C (2003) Phagocyte-specific S100 proteins: a novel group of proinflammatory molecules. Trends Immunol 24:155–158
311. Castagnola M, Inzitari R, Fanali C et al (2011) The surprising composition of the salivary proteome of preterm human newborn. Mol Cell Proteomics 10:M110.003467
312. Castagnola M, Cabras T, Iavarone F et al (2013) Detection of Ca^{2+}-binding S100 proteins in human saliva by HPLC-ESI-MS. In: Heizmann CW (ed) Calcium-Binding Proteins and RAGE: from structural basics to clinical applications, Methods in molecular biology, vol 963. Springer Protocols, Humana, New York, pp 357–371
313. Ringer A (1883) A further contribution regarding the influence of different constituents of the blood on the concentration of the heart. J Physiol 4:29–43
314. Hamoir G, Focant B, Distèche M (1972) Proteinic criteria of differentiation of white, cardiac and various red muscles in carp. Comp Biochem Physiol B 41(4):665–674
315. Pechère JF, Demaille J, Capony JP (1971) Muscular parvalbumins: preparative and analytical methods of general applicability. Biochim Biophys Acta 236(2):391–408
316. Moore BW (1965) A soluble protein characteristic of the nervous system. Biochem Biophys Res Commun 189:739–744
317. Cavalier MC, Pierce AD, Wilder PT et al (2014) Covalent small molecule inhibitors of Ca(2+)-bound S100B. Biochemistry 53 (42):6628–6640
318. Lee DH, Asare BK, Rajnarayanan RV (2016) Discovery at the interface: towards novel antiproliferative agents targeting human estrogen receptor/S100 interactions. Cell Cycle 15 (20):2806–2818
319. Padilla L, Dakhel S, Hernadez JL (2014) S100 to receptor for advanced glycation end-products binding assay: looking for inhibitors. Biochem Biophys Res Comm 446:404–409
320. Grum-Schwensen B, Klingelhöfer J, Beck M (2015) S100A4-neutralizing antibody suppresses spontaneous tumor progression, pre-metastatic niche formation and alters T-cell polarization balance. BMC Cancer 15:44. https://doi.org/10.1186/s12885-015-1034-2
321. Cho CC, Chou RH, Yu C (2016) Pentamidine blocks the interaction between mutant S100A5 and RAGE V domain and inhibits the RAGE signaling pathway. Biochem Biophys Commun 477(2):188–194
322. Martínez-Aguilar J, Molloy MP (2019) Targeted mass spectrometry of S100 proteins. In: Heizmann CW (ed) Calcium-binding proteins of the EF-hand superfamily: from basics to medical applications, Methods in molecular biology, vol 1929. Springer Protocols, Humana, New York (in press)
323. Gatterdam V, Frutiger A, Stengele KP et al (2017) Focal molography is a new method for the in situ analysis of molecular interactions in biological samples. Nat Nanotechnol 12 (11):1089–1095

Chapter 12

Gene Transfer of Calcium-Binding Proteins into Adult Cardiac Myocytes

Brian R. Thompson, Houda Cohen, Addeli Bez Batti Angulski, and Joseph M. Metzger

Abstract

Heart failure is the leading cause of combined morbidity and mortality in the USA with 50% of cases being diastolic heart failure. Diastolic heart failure results from poor myocardial relaxation and inadequate filling of the left ventricular chamber caused in part by calcium-handling dysregulation. In this chapter we describe methods to investigate new approaches of novel human Ca^{2+} binding protein motifs to restore normal Ca^{2+} handling function to diseased myocardium. Gene transfer of parvalbumin into adult cardiac myocytes has been studied as a potential therapeutic, specifically as a strategic Ca^{2+} buffer to correct cardiac mechanical dysfunction in disease. This chapter provides protocols for studying wild-type parvalbumin isoforms and parvalbumins with strategically designed EF-hand motifs in adult cardiac myocytes via acute adenoviral gene transfer. These protocols have been used extensively to optimize parvalbumin function as a potential therapeutic for failing heart muscle.

Key words Parvalbumin, Calcium, Adult cardiac myocyte, Contractility, Calcium imaging, Gene transfer

1 Introduction

Heart failure is the leading cause of combined morbidity and mortality in this country with an estimated five million Americans affected and up to 600,000 new cases per year [CDC/AHA statistics]. Heart failure can be subdivided clinically into systolic and diastolic heart failure. Systolic heart failure is a progressive and largely intractable clinical syndrome characterized by an overall decline in the pumping function of the heart. Diastolic heart failure is just as common as systolic heart failure and results from poor myocardial relaxation and inadequate filling of the left ventricular chamber. Diastolic heart failure is increasing in prevalence with the majority of patients being women and the elderly. There is no cure for heart failure except heart transplantation, which is woefully inadequate for the millions of heart

Claus W. Heizmann (ed.), *Calcium-Binding Proteins of the EF-Hand Superfamily: From Basics to Medical Applications*, Methods in Molecular Biology, vol. 1929, https://doi.org/10.1007/978-1-4939-9030-6_12,

failure patients. Thus there is an urgent need to discover and implement new approaches to treat this disease.

There is extensive scientific literature demonstrating myocardial intracellular Ca^{2+} dysregulation as a key element underlying defective heart performance in failure [1]. We and others have therefore used this foundation as a starting point to investigate new approaches of novel human Ca^{2+} binding protein motifs to restore normal Ca^{2+} handling function to diseased myocardium. The growing burden of cardiovascular disease in this country and throughout the world necessitates unwavering vigor directed at the mechanistic basis of disease with the goal of identifying new therapeutic targets and effective treatment modalities. The focus here on novel physiologically optimized EF-hand Ca^{2+} binding motifs offers an exciting new therapeutic approach for the diseased and failing heart: specifically, strategic Ca^{2+} buffering to correct cardiac mechanical dysfunction in disease. We agree with the premise that accelerating Ca^{2+} transient decay would be corrective for diastolic dysfunction and would be beneficial to heart performance in the long term, providing excessive SR Ca^{2+} load is avoided. We have designed novel EF-hand Ca^{2+} binding motifs that facilitate fast Ca^{2+} transient decay while maintaining Ca^{2+} transient peak amplitude [2].

Parvalbumin is a cytosolic calcium buffer endogenously expressed in glycolytic skeletal muscle [3]. Due to parvalbumin's affinities for Mg^{+2} and Ca^{+2} and the cytosolic contents of these cations, parvalbumin almost exclusively binds Mg^{+2} at rest [4]. Upon calcium-induced calcium release from the sarcoplasmic reticulum, parvalbumin binds Ca^{+2} which allows contraction to occur and facilitates faster relaxation. Gene transfer of parvalbumin into adult cardiac myocytes has been studied as a potential therapeutic in diastolic heart failure where there is slow Ca^{+2} reuptake and relaxation [5–7]. This chapter provides protocols for studying parvalbumin and mutants in adult cardiac myocytes via adenoviral gene transfer [1]. These protocols have been used extensively to study parvalbumin and mutants to optimize parvalbumin function for cardiac myocytes and the heart [2, 8–10].

2 Materials

2.1 Isolation of Adult Cardiac Myocytes and Adenoviral-Mediated Gene Transfer

2.1.1 Instruments and Perfusion Apparatus

1. Gather surgical instruments and glassware (Fig. 1). Clean thoroughly and autoclave.
2. Surgical instruments: Large scissors, small scissors, small hemostat, large forceps, needle nose forceps, curved forceps, and 14 gauge luer lock compatible cannula.
3. Wide-bore glass Pasteur pipettes siliconized (sigma, SL2) for trituration.
4. 4–0 surgical silk.

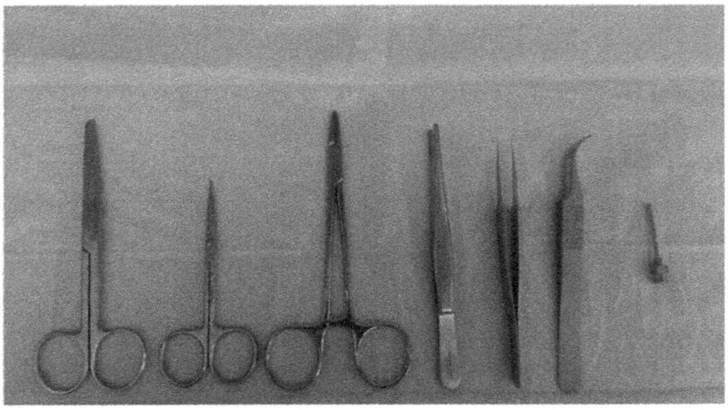

Fig. 1 Surgical instruments: Large scissors, small scissors, small hemostat, large forceps, needle nose forceps, curved forceps, and 14 gauge luer lock compatible cannula

5. The perfusion apparatus used for adult cardiac myocyte isolation (Fig. 2) consists of two solution reservoirs placed above and connected to a double-barreled warming coil with changeover stopcock (Harvard "Baker" perfusion). Circulating 37 °C water bath should be turned on ahead of time to warm up the perfusion apparatus.
6. The reservoirs of Ca^{2+}-free and Ca^{2+}+KHP solutions are placed at around 65 cm from the bench level and oxygenated.
7. The perfusion apparatus is cleaned three times with 70% ethanol then rinsed three times with sterile ddH_2O before and after each rat preparation.

2.1.2 Solutions

1. Ca^{2+}-free Krebs-Henseleit buffer (Ca^{2+}-free KHB): 118 mM NaCl, 4.8 mM KCl, 1.2 mM KH^2PO^4, 2 mM $MgSO_4$, 25 mM HEPES, and 11 mM glucose.
2. Ca^{2+} Krebs-Henseleit buffer (Ca^{2+}-KHB): add 1.0 mM $CaCl^2$ to Ca^{2+}-free KHB (*see* **Note 1**).
3. Ca^{2+}-free KHB with collagenase: 30 mg of type 2 collagenase from Worthington to solve in 80 mL Ca^{2+}-free KHB during Ca^{2+}-free KHB retrograde perfusion (*see* **Note 2**).
4. Ca^{2+}-KHB with 2% bovine serum albumin: solve 30 mg BSA in 15 mL Ca^{2+} KHP.
5. Culture media: M199 (M3769-1 L from Sigma) + 1× Pen Strep.
6. Plating media: supplement M199 culture media with 5% heat-inactivated fetal bovine serum.
7. Filter sterilize all solutions using a 0.2 μm filter (*see* **Note 3**).

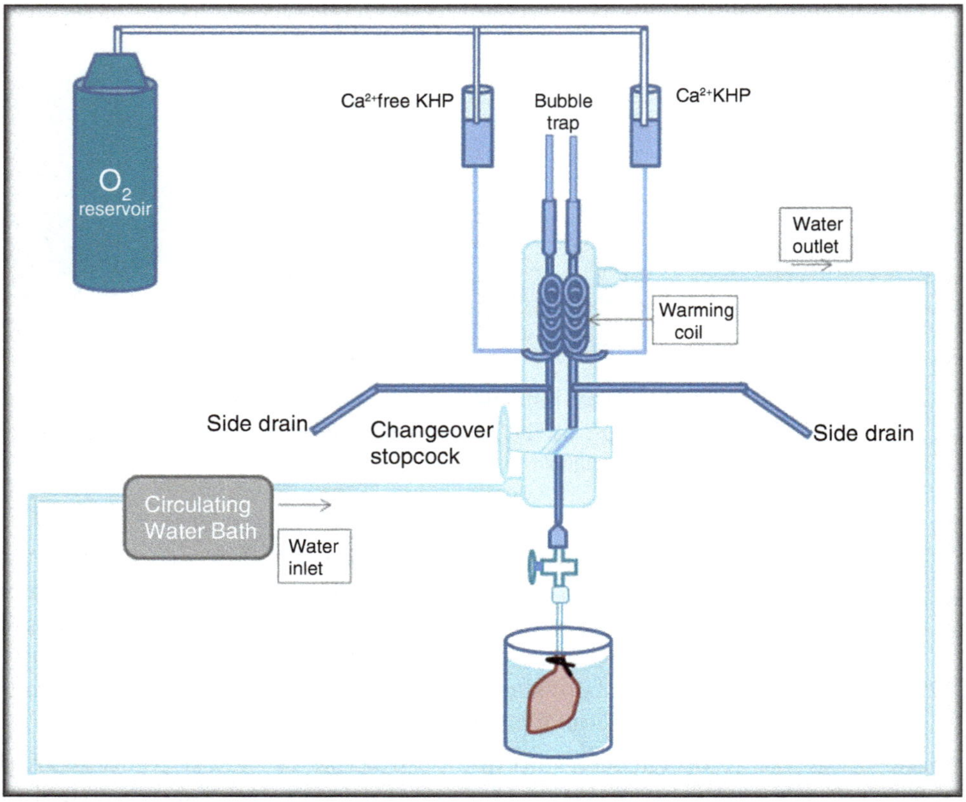

Fig. 2 Heart perfusion apparatus

2.2 Analysis of Calcium-Handling Protein Expression and Localization

2.2.1 Western Blot Solutions

1. Lysis buffer (RIPA buffer): 50 mM Tris, pH 8.0; 150 mM NaCl; 0.1% Triton X-100; 0.5% sodium deoxycholate; 0.1% sodium dodecyl sulfate (SDS); protease inhibitors tablet.
2. Loading buffer: 62.5 mM Tris–HCl, pH 6.8; 2% SDS; 25% glycerol; 0.01% bromophenol blue.
3. Running buffer (Tris/glycine/SDS): 25 mM Tris, pH 8.3; 192 mM glycine; 0.1% SDS.
4. Tris-buffered saline with Tween 20 (TBST) buffer: 20 mM Tris, pH 7.5; 150 mM NaCl, 0.1% Tween 20.
5. Transfer buffer: 25 mM Tris base, 190 mM glycine, 20% methanol.

2.2.2 Immunofluorescence Solutions

1. Phosphate-buffered saline (PBS): 137 mM NaCl; 2.7 mM KCl; 8 mM Na_2HPO_4; and 2 mM KH_2PO_4.
2. 3% paraformaldehyde in PBS: 3 g of paraformaldehyde to 100 mL of PBS.
3. Blocking buffer: 1× PBS containing 20% normal goat serum (NGS) and 0.5% Triton X-100.
4. Antibody dilution buffer: 1× PBS containing 2% NGS and 0.5% TX-100.

2.3 Cellular Functional Analysis

2.3.1 Solutions

1. Tyrode's solution: 140 mM NaCl, 0.5 mM $MgCl_2$, 0.33 mM NaH_2PO_4, 5 mM Hepes, 5.5 mM glucose, 1.8 mM $CaCl_2$, 5 mM KCl, pH to 7.4.
2. Tyrode's solution no Na^+, no Ca^{+2}: Same as Tyrode's solution with no $CaCl_2$ added and LiCl replaces NaCl.
3. Tyrode's solution no Na^+, no Ca^{+2} with caffeine: Same as Tyrode's solution no Na^+, no Ca^{+2} with addition of 20 mM caffeine.

2.3.2 Equipment

1. Ionoptix contractility and calcium imaging system.
2. Myocam.
3. HyperSwitch filter box.
4. Xenon arc lamp.
5. Fluorescence Systems Interface.
6. Fura-2 filter cubes and PMT.
7. Inverted microscope with 40× objective.
8. MyoPacer.
9. Heated stage adapter temperature control unit.
10. Stimulation chamber.
11. Motorized stage and recall software.
12. IonWizard analysis software.
13. Perfusion apparatus and vacuum pump.

3 Methods

3.1 Adult Rat Ventricular Cardiac Myocytes Isolation

The following protocol includes heart extraction and retrograde perfusion using a modified Langendorff apparatus for an initial digestion. The ventricles are then minced, and cardiomyocytes are separated by gentle trituration. Ca^{2+} is gradually added to make the isolated cardiomyocytes Ca^{2+}-tolerant prior to plating.

3.1.1 Terminal Thoracotomy and Heart Excision

1. Adult female Sprague Dawley rat (225–275 g) is initially injected intraperitoneally with 0.3 cc of heparin (1000 IU/mL). After 10–15 min, anesthetize rat with 0.7 cc of sodium pentobarbital (diluted to 50 mg/mL). Note: use isoflurane to shortly anesthetize the rat and deliver the IP injection.
2. For all subsequent steps, a surgical mask and sterile gloves should be worn to prevent potential bacterial and mycobacterial contamination. Sterile surgical instruments and glassware should be used throughout the isolation procedure. *See* Fig. 3 for a timeline.

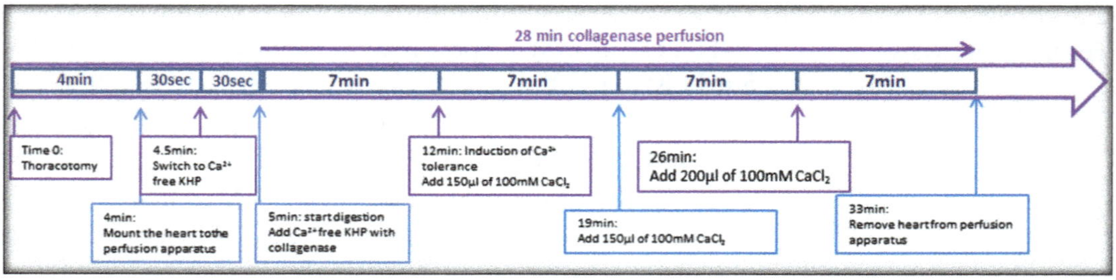

Fig. 3 Heart perfusion timeline

3. Once the rat becomes unresponsive (30–60 s after pentobarbital injection), spray 70% ethanol over the chest and abdomen.
4. Cut open the thorax. Rapidly and carefully pull the heart upward using curved forceps, and excise above the aortic arch.
5. Place the heart in ice-cold 10 mL Ca^{2+}-free KHP solution (in 100 mm dish placed on ice bucket), trim excess tissue, and mount the heart on the cannula. Tie surgical silk twice to secure the aorta to the cannula. The tip of the cannula should be above the aortic valve. If the cannula tip is placed into the left ventricle, the coronary perfusion won't be successful.
6. Slowly flush the heart with Ca^{2+}-KHP in the cannula-mounted syringe. The heart will resume beating. Note: the cannula is fixed to syringe filled with Ca^{2+}-KHP and flushed prior to mounting the heart to avoid formation of air bubbles.

3.1.2 Retrograde Perfusion and Enzymatic Digestion

1. Start a slow drip of oxygenated Ca^{2+}-KHP, and then transfer the cannulated heart to the perfusion apparatus. Note: watch for bubbles; make sure there are no bubbles in the running perfusion apparatus before mounting the cannulated heart.
2. Increase the flow rate and wait for the solution to run clear (about 30 s). In the meantime, open the side valve to dispose of unoxygenated Ca^{2+}-free KHP solution setting in tubing.
3. Switch the stopcock to open oxygenated Ca^{2+}-free KHP, and allow the heart to perfuse until the ventricles stop contracting (less than 1 min). Simultaneously, dissolve 30 mg of collagenase type 2 from Worthington in 80 mL Ca^{2+}-free KHP by swirling.
4. Once the heart stops contracting, drain as much Ca^{2+}-free KHP through the side valve as possible, and then add the 80 mL of Ca^{2+}-free KHP with collagenase solution to the Ca^{2+}-free side. Adjust O_2 flow to avoid foam formation. At the end of the procedure, the final volume of collagenase solution should be close to 80 mL.
5. Once the collagenase solution is added, place 50 mL beaker under the heart, so it is immersed in collagenase solution, and

start the timer. Flow rate should be approximately 10 mL/min. Recirculate the collagenase solution during the remaining time of the procedure.

6. After 7 min of perfusion with the collagenase solution, add 150 μL of 100 mM $CaCl_2$ to the collagenase solution (final $[Ca^{2+}] = 0.1875$ mM). Let the heart perfuse for 7 more min, and add 150 μL of 100 mM $CaCl_2$ to the collagenase solution (final $[Ca^{2+}] = 0.375$ mM). After another 7 min, add 150 μL of 100 mM $CaCl_2$ to the collagenase solution (final $[Ca^{2+}] = 0.625$ mM), and continue the perfusion for an additional 7 min for a total of 28 min of perfusion with collagenase solution. Note: it is important to not perfuse the heart for more than 30 min with the collagenase solution as this tends to decrease the final yield (*see* **Note 4**).
7. Remove the heart and cannula from perfusion apparatus, and collect all the collagenase solution from the perfusion apparatus in 250 mL Erlenmeyer flask.

3.1.3 Trituration and Induction of Ca^{2+} Tolerance

1. Transfer the heart to a plastic petri dish with 2–3 mL of collagenase solution. Using small scissors and needle nose forceps, remove the atria and aorta. Cut ventricles into 1–2 mm pieces, and gently transfer into sterile 50 mL beaker with 15–20 mL collagenase solution (*see* **Note 5**).
2. Gently swirl the tissue suspension at room temperature, then pour off supernatant (fraction 1), and add 15–20 mL of collagenase solution recovered in the Erlenmeyer flask.
3. Centrifuge fraction one tube at 40 × *g* for 15 s, recover supernatant in a new 50 mL conical, and discard fraction one cell pellet as cell viability is usually very low (around 1–10%).
4. Using wide-bore siliconized Pasteur pipette, gently triturate the minced tissue for 2–3 min. Recover supernatant in 15 mL conical labeled fraction 2, add 15–20 mL collagenase solution to the minced tissue, and place on the shaker.
5. Centrifuge fraction two tube at 40 × *g* for 15 s, recover supernatant in the 50 mL conical, and discard fraction two cell pellet (*see* **Note 6**).
6. Gently triturate the minced tissue for 2–3 min. Recover supernatant in 15 mL conical labeled fraction 3, add 15–20 mL collagenase solution to the minced tissue, and place on the shaker (*see* **Note 7**).
7. Centrifuge fraction three tube at 40 × *g* for 15 s; recover supernatant in the 50 mL conical. Resuspend cell pellet in 1–2 mL of Ca^{2+}-KHP with 2% BSA.
8. Assess cell viability by looking at a drop of the cell suspension under the microscope. If 70% or more of the cells from the

fraction are viable (rod shaped), place the tube on the shaker and save the fraction.

9. Repeat step 6, 7, and 8 to obtain fraction 4–15 (*see* **Note 8**).
10. Combine the different fractions, and bring to a final volume of Ca^{2+}-KHP with 2% BSA to 10 mL in a 15 mL conical with a sterile plastic transfer pipette (wide mouth).
11. Increase Ca^{2+} concentration to 1.75 mM by adding 3–25 μl aliquots of 100 mM $CaCl^2$ at 5 min intervals. Incubate tube on the shaker at room temperature in between Ca^{2+} additions.
12. After final Ca^{2+} addition and incubation, centrifuge at 40 × *g* for 15 s. In the tissue culture hood, aspirate supernatant, and gently resuspend cells in 10 mL of warm (37 °C) M199 media supplemented with 5% FBS.
13. Use hemocytometer to count live cells in the homogenous cell suspension; make sure cells are well resuspended and not settling out.

3.2 Primary Culture and Adenovirus-Mediated Gene Transfer

3.2.1 Cultureware Preparation and Plating

1. In cell culture hood, place 1 × 25 mm^2 coverslip per well (6well plate). Coat each coverslip with 100 μL laminin (40 μg/mL from Life Technologies), and leave uncovered under UV for 10 min to sterilize. Cover plates and allow laminin to set for an additional 10 min prior to plating.
2. Adjust cell concentration to 1 × 10^5 myocytes/mL in M199 culture media supplemented with 5% FBS.
3. Remove excess laminin from coverslips by aspirating.
4. Using aerosol-resistant tips, carefully plate 200 μL cell suspension per coverslip (2 × 10^4 rod-shaped cells/coverslip).
5. Place plate in incubator carefully so as to not disturb cells (Fig. 4). Incubate for 2 h at 37 °C (*see* **Note 9**).

3.2.2 Transduction: Gene Transfer

1. Just before the 2 h incubation is over, make the desired dilutions of the adenovirus which will be used for gene transfer. For parvalbumin use Flag-tagged versions for easy detection. Typical multiplicities of infection (MOI) for adult cardiac myocytes are between 100 and 500 without toxicity. This should be empirically determined for each virus. The virus should be thawed and kept on ice (do not allow it to warm to room temp). Resuspend the virus before diluting. Pipette up and down 20–25 times.
2. In a microcentrifuge tube, mix the virus with serum-free M199 media. A total volume of 200 μL per coverslip will be used. Mix well to get an even distribution of the virus.
3. At the end of the 2 h incubation, gently aspirate the plating media by carefully placing the Pasteur pipette tip at the edge of

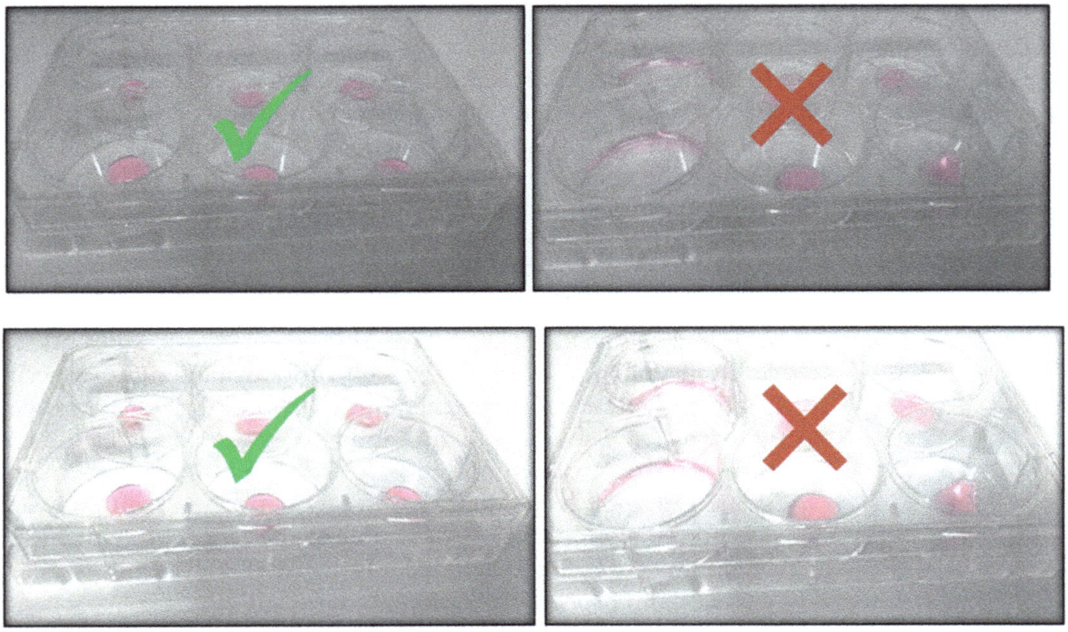

Fig. 4 Cell plating: the surface tension should maintain the droplet on the coverslip. Make sure to not disturb the droplet when moving the plate to the incubator to avoid cells dispersion and to obtain a good yield of live cells

the bubble. Many of the unattached myocytes (dead) will be aspirated away.

4. Using an aerosol-resistant pipette tip, add 200 μL of virus dilution to the coverslip, taking care that it stays on the coverslip. Carefully return the cells to the incubator for 1 h without disturbing the bubble.
5. After 1 h, add 2 mL M199 culture media to each well (without removing the virus-containing media). Try to aim delivery away from coverslip so that myocytes are not dislodged. Return the dish to the incubator.
6. 24 h later, aspirate the old media and replace with M199 culture media. Media should be changed every 2–3 days thereafter. Serum-free media is used for cardiomyocyte primary culture to minimize dedifferentiation of cardiomyocytes. Using the protocol described earlier, ventricular cardiomyocytes can be maintained for up to 7 days without detectable changes in contractile protein stoichiometry and isoform composition.

Parvalbumin, a cytosolic calcium-binding protein, can be characterized using both Western blot analysis and immunofluorescence staining. Western blot is a widely used technique to detected levels of expression of parvalbumin protein. Through this technique it is

3.3 Analysis of Calcium-Handling Protein Expression and Localization

possible to measure the parvalbumin expression levels in biological tissues such as fast-contracting muscles, where levels are highest. Immunofluorescence is used to determine the localization and relative expression and allows visualization of the distribution of parvalbumin in the sample. Together, both techniques, which will be described next, determine the expression pattern and subcellular location of parvalbumin in biological samples.

3.3.1 Western Blot Analysis

Western blotting (WB), also called immunoblotting, is a common and well-accepted procedure to determine levels of protein expression in a cell or tissue extracted. This technique measures protein levels in a biological sample through antibody binding to a specific protein of interest. In this technique, previously separated proteins are transferred to a membrane, and antibodies specific to the epitopes on the separated proteins are then used to probe the membrane. The precise binding that occurs between an antibody and the protein epitope allows detection of the target protein. In this section we are going to describe how to use WB technique to detect parvalbumin protein expression.

3.3.2 SDS-PAGE

The first step on a SDS-PAGE procedure is the choice of which gel to use. The choice of which gel to use depends on several factors including the number of samples, the size of proteins you are looking at, and if you are attempting to separate proteins of similar molecular weights. The differences between the distinct types of gel are primarily the pH at which the acrylamide is polymerized and the pH at which is run. It is possible to make your own gel, or precast gels also can be used. The most common precast gel used in our lab is the Criterion (Tris-Glycine eXtended). Gel percentage of 12% is ideal for small proteins, such as parvalbumin.

1. After resuspending the sample in a lysis buffer, leave on ice for 30 min to 1 h.
2. Boil the sample for 2 min, and after that spin in microcentrifuge at maximum speed for 2 min.
3. Remove supernatant and transfer to a new tube. Take a small volume of lysate to perform a protein quantification assay. Use Bio-Rad BCA assay to determine the protein concentration for each sample lysate. This assay is sensitive and tolerates the presence of detergents.
4. To a volume of protein sample (cell or tissue lysate), add equal volume of loading buffer.
5. To reduce and denature, boil each sample at 95 °C for 5 min. Centrifuge at 16,000 × *g* for 5 min. These samples can be stored at −80 °C or can be used to procedure with gel electrophoresis.

6. Load equal amounts of protein into the wells of the SDS-PAGE gel, along with molecular weight marker. Load 50 μg of total protein from cell lysate or tissue homogenate.
7. Run the gel for 1–2 h at 100 V.

3.3.3 Transfer

The membrane used for transfer can be either nitrocellulose or PVDF (polyvinylidene difluoride). If you are going to use PVDF, activate this membrane with methanol for 1 min, and rinse with transfer buffer before use.

1. In preparation for transferring proteins, preincubate the filter paper, sponges, and membrane (either PVDF or nitrocellulose) in transfer buffer.
2. When gel is finished running, remove cartridge, and crack open sides using metal spatula.
3. Remove the gel from the plates, and place it into the transfer buffer for at least 15 min (*see* **Note 10**).
4. Building the transfer sandwich: Place a sponge (thin abrasive pads) topped by a piece of filter paper on to the black side of the folder. Next, carefully place the gel (pre-equilibrated in transfer buffer) on top of this. In a similar method, apply the membrane to the gel. It is particularly critical that air bubble is not introduced at this junction. Next, place a second piece of filter paper on top of the membrane. At this step, firmly roll out any air bubbles. Finally, place the second sponge on top of the assembly. Close the folder and secure the locking mechanism; it should be tight.
5. Place the sandwich into the transfer tank and fill with transfer buffer over the top. For tank transfers, it is important that the transfer buffer be stirred during the transfer.
6. Run at constant current (amps) at 350 mA for 1–5 h or at 250 mA overnight.

3.3.4 Antibodies

1. Immunodetection is performed by initially blocking nonspecific binding sites with Tris-buffered saline with Tween 20 (TBST) containing 5% (w/v) nonfat dry milk for 1 h at room temperature.
2. The membrane is then incubated in TBST + 5% milk containing the primary monoclonal antibody (for Flag-tagged parvalbumin you can use anti-Flag M2 antibody) overnight at 4 °C and then washed three times in TBST + 5% milk for 5 min each.
3. Incubate the membrane for 1 h at room temperature in TBST + 5% milk containing a secondary antibody conjugated to IR 680 or 800. Wash the membrane three times with TBST + 5% milk for 5 min each.

4. Antibody binding is detected by Licor Odyssey Imaging. Use image analysis software to read the band intensity of the loading control proteins. Anti-beta actin can be used as a loading control for parvalbumin detection by WB.
5. Use the loading control protein levels to normalize parvalbumin levels.

3.3.5 Immunofluorescence

Immunofluorescence (IF) is a powerful technique that combines the use of the specificity of antibodies to their antigen with optical microscopy imaging to visualize proteins within fixed cell or tissue samples. Through IF, parvalbumin can be detected using a two-step approach with unlabeled primary antibody followed by fluorophore-conjugated secondary antibody (called indirect detection). This technique can be useful to gain insight into parvalbumin localization allowing the identification of co-localization as well as changes in subcellular localization.

Fixation and Permeabilization

1. Fix cells in coverslips with 3% paraformaldehyde in phosphate-buffered saline (PBS) for 10 min. Fixation is usually performed on ice to ensure better preservation and morphology, but fresh, cold fixative added during the fixation protocol works well.
2. Following the fixation, wash samples three times in ice-cold PBS for 5 min each.
3. For the permeabilization step, incubate the cells with 0.1% of TX-100 in PBS for 10 min.
4. Prepare humid chamber for blocking and antibody incubation steps. Line staining tray (or any other sealable plastic box) with gauze, saturate gauze with dH_2O, and then place cottonless swabs on top of gauze.
5. Drain off excess 1× PBS and transfer the coverslips to humid chamber.
6. For the blocking step, add ~200 μL of PBS containing 20% normal goat serum (NGS) and 0.5% Triton X-100 (TX-100) on each coverslip. Incubate for 30 min at room temperature. The PBS-TX + NGS should be first filtered through a 0.22 μm syringe filter.
7. Gently aspirate off serum. Do not wash in 1× PBS after this step, and do not allow sections to dry out at any point.

3.3.6 Antibodies

1. Apply primary antibody (anti-Flag M2 antibody for Flag-parvalbumin) diluted in PBS containing 2% NGS and 0.5% TX-100. Incubate at room temperature for 1.5 h. Always centrifuge Ab for this and all subsequent Abs for 1 min to remove any protein complexes or bacteria.

2. Wash samples three times in PBS containing 0.5% TX-100 for 5 min each.
3. Drain off excess liquid and place samples in humid chamber. Block nonspecific secondary antibody binding by incubating samples in PBS containing 20% NGS and 0.5% TX-100 for 30 min. Usually use 20% NGS unless having trouble with obtaining a positive stain.
4. Gently aspirate off serum. Incubate samples with secondary Ab (Alexa Fluor conjugated to goat anti-mouse IgG) diluted in PBS containing 2% NGS and 0.5% TX-100 for 1 h.
5. Wash samples three times in PBS containing 0.5% TX-100 for 5 min each.
6. If wanting to detect another protein with a second mouse primary antibody, follow the steps below. If only staining with one mouse primary, proceed to step 15.
7. Drain off excess liquid and place samples in humid chamber. Incubate samples in goat anti-mouse IgG overnight at 4 °C to neutralize unreacted sites on the first primary antibody and Alexa Fluor-conjugated secondary Ab.
8. Wash samples three times in PBS containing 0.5% TX-100 for 5 min each.
9. Drain off excess liquid and place samples in humid chamber. Incubate samples with goat anti-mouse Fab fragments for 1.5 h at room temperature. Fab fragments are diluted 1:20 in PBS-TX with 2% NGS. These steps eliminate any recognition by the second set of primary and secondary antibodies.
10. Wash samples three times in PBS containing 0.5% TX-100 for 5 min each.
11. Block samples with 20% NGS in 1× PBS containing 0.5% TX-100 for 30 min at room temperature.
12. Drain off excess liquid and place samples in humid chamber. Add the second primary antibody diluted in PBS containing 2% NGS and 0.5% TX-100 for 1 h at room temperature. Wash three times in PBS containing 0.5% TX-100 for 5 min each.
13. Block samples in PBS containing 20% NGS and 0.5% TX-100 for 30 min at room temperature.
14. Apply second secondary antibody (Alexa Fluor-conjugated to goat anti-mouse IgG) diluted in PBS containing 2% NGS + 0.5% TX-100 for 1 h at room temperature.
15. Wash samples three times in PBS. Drain off excess liquid, and mount coverslip with a drop of mounting medium. Seal coverslip with nail polish to prevent drying and movement under microscope. Samples are ready for microscopic analysis.

3.4 Cellular Functional Analysis

Functional characterization of parvalbumin-transduced adult cardiac myocytes combines sarcomere length dynamics and calcium transient analysis using the Ionoptix contractility and calcium imaging system. Individual cardiac myocytes, 1–6 days post transduction, are analyzed for alterations in both contractility amplitude and kinetics and both calcium transient amplitude and kinetics. This analysis allows one to understand the impact of parvalbumin on excitation-contraction coupling.

3.4.1 Sarcomere Length Dynamics

1. Mount coverslips to temperature-controlled stage.
2. Apply a small amount of vacuum grease to the bottom of the stimulation chamber in a circle at the edge of where the coverslip will fit.
3. With a forceps remove one coverslip from the culture media, and quickly dry the bottom with a Kimwipe.
4. Gently attach to the bottom of the stimulation chamber with cell side up, and then press down on a flat surface to firmly attach.
5. Place in the heated stage and fill the chamber with warm Tyrode's solution.
6. With a forceps remove excess vacuum grease from the chambers edges.
7. Add temperature probe to the solution and monitor temperature until 37 °C.
8. Visualize the cells and start stimulation (usually 0.2 Hz to 1 Hz) (*see* **Note 11**).

3.4.2 Collection of Sarcomere Dynamics

1. Once the myocytes are at temperature and are stimulated, open IonWizard software and collect new experiment. Make sure the experiment includes sarcomere length detection.
2. Find a contracting myocyte through the microscope, and switch the light path to the MyoCam (*see* **Note 12**).
3. Position the myocyte in the video window so that the striations are vertical in the window.
4. Drag the pink sarcomere length detection box on the myocyte, and position to get good signal. The more of the myocyte that is included, the higher is the signal to noise.
5. Verify the resting sarcomere length (healthy cells usually have sarcomere lengths in excess of 1.7 μm).
6. Start the experiment, and verify that the contraction is smooth and returning to baseline lengths. Collect at least ten contractions and then stop the experiment, and save the file for analysis.

7. If studying a drug such as isoproterenol, mark the location of the cell in the recall software to be able to recall that cell, and collect pairwise plus and minus drug (*see* **Note 13**).
8. Collect 10–15 cells per coverslip from 3 to 4 independent rat preps for statistical power.

3.4.3 Analysis in IonWizard 6.0

1. Open the file in IonWizard 6.0.
2. Average the contractions to one contraction under operations. The more contractions analyzed the better is the signal to noise in the average.
3. Under marks add a monotonic transient. Pay attention to the offset (how prior to the stimulation mark you want to analyze for baseline) and duration (make sure the duration is long enough to return 100% to baseline) (*see* **Note 14**).
4. Under operations select monotonic transient analysis. A fifth-order polynomial fit results with five colors fit over the average contraction. Make sure the five colors fit well with each other and with the curve otherwise change your offset or duration.
5. Export the trace and resulting data to graphing program for compiling all the myocytes collected (*see* **Note 15**).

3.4.4 Calcium Transient Analysis and SR Calcium Load

1. Reconstitute Fura-2 AM powder in DMSO to a final concentration of 1 mM. Use fresh DMSO for consistent results. Avoid repeat freeze-thaw cycles by aliquoting single day use amounts.
2. Add 1 μL Fura-2 AM to 1 mL of M199 in a six-well dish for a final concentration of 1 μM. Add 1 mL of M199 to two more wells of the dish for subsequent wash steps.
3. Using forceps transfer a coverslip of cells to the Fura-2 AM M199 well. Leave at room temperature in the dark for 10 min.
4. Transfer the coverslip from the Fura-2 AM well to a M199 only well for 5 min, and repeat again to the next well for 5 more minutes.
5. Load the coverslip onto the stimulation chamber as above, and fill with Tyrode's solution.
6. Wait for an additional 5–10 min for de-esterification of the Fura-2 AM. The de-esterification time is 20 min total after removing from the Fura-2 AM well.
7. Verify correct temperature and start stimulating as above.

3.4.5 Collection of Calcium Transients

1. Collection of calcium data is similar to the above-described collection of sarcomere length dynamics with the exceptions below.

2. Make sure the experiment is set for calcium and sarcomere length (*see* **Note 16**).
3. After positioning a myocyte in the video window as above, reduce the window size to just smaller than the myocyte using the shutter on the MyoCam. Avoid cells that are adjacent to each other as the fluorescence can bleed through. Position the sarcomere length box to measure simultaneously.
4. Start the experiment, and monitor windows for the numerator, denominator, and the ratio as well as sarcomere length changes.
5. Again collect ten or more transients for analysis. The Fura-2 ratio is usually noisier than the sarcomere length curve so analyzing more transients can clean this up.
6. After collecting data from the cell, move to a portion of the coverslip nearby that is empty of cells, and collect background fluorescence for 5 s. This will be used in the analysis. Make sure this signal has no changes with stimulation and is truly background.

3.4.6 Caffeine Perfusion for SR Load Functional Assessment

1. Load cells with Fura-2 AM as above and mount to the stimulation chamber as above.
2. As cells are loading, set up a perfusion apparatus with flow control and in-line temperature control. Three 60 mL syringes work well for the three solutions that will be used. Fill one with Tyrode's; one with Tyrode's no Ca^{+2}, no Na^{+}; and one with Tyrode's no Ca^{+2}, no Na^{+} plus caffeine.
3. Once coverslip is mounted, perfuse with Tyrode's solution, and allow media to come to temperature.
4. Start pacing and set your window in IonWizard to bracket a cell as above.
5. Start collection (this is continuous collection of data throughout the experiment) of 10 contractions while monitoring the 340/380 nm ratio with Tyrode's.
6. Stop pacing and switch perfusion to Tyrode's no Ca^{+2}, no Na^{+} for 20 s.
7. Switch perfusion to Tyrode's no Ca^{+2}, no Na^{+} plus caffeine for 8 s.
8. Switch perfusion back to Tyrode's no Ca^{+2}, no Na^{+} to wash out caffeine until 340/380 nm ratio is back to baseline.
9. Switch to normal Tyrode's, and pace again to make sure the cell survived and is functioning normally (*see* **Note 17**).

3.4.7 Analysis

1. Analysis for calcium transients is as for sarcomere dynamics except for the background subtraction. Once averaged and monotonic transient analysis is complete, go to operations

and click constants. Highlight background option, and enter the numerator background value and the denominator background value from the background file collected. This is done for every file.

2. Analysis for caffeine transients is manually entered marks to surround the caffeine peak by going to mark options in IonWizard. This will allow monotonic transient analysis of the caffeine peak as above.
3. Similar outputs to graph as sarcomere length dynamics are generally used.

3.5 Summary

The protocols above have been used extensively to study parvalbumin in adult cardiac myocytes [2, 5, 8–10]. The protocols allow one to study structure function of parvalbumin through mutagenesis and live cell function. In this way optimized parvalbumin mutants have been developed and will continue to be developed [9, 10]. By following these protocols, one can engineer adult cardiac myocytes to adenovirally express parvalbumin, view localization and expression levels, and study live cell function through calcium imaging and contractility. In this way one can have a full understanding of the functional implications of parvalbumin in the cell.

4 Notes

1. Prepare and use KHB on the same day.
2. Weigh collagenase and keep in Erlenmeyer flask covered with thin foil until needed.
3. Only add warm 37 °C media to cells. M199 media can be stored at 4 °C for weeks.
4. During retrograde perfusion visualize the heart to verify that it is not turning white. This is a sign of ischemic tissue most likely due to a small bubble blocking perfusion.
5. It is important to prevent the heart from drying out while being cut down and minced.
6. The supernatant can be recovered in a 50 mL conical tube to be reused if one runs out of the collagenase solution in the Erlenmeyer flask.
7. The amount and velocity of trituration are critical as too forceful and fast will result in dead myocytes, while too slow and weak will result in not many myocytes.
8. The cell pellet color is also a good indicator of cell viability; a pink pellet likely contains a large number of viable cells, whereas in a white or slightly gray pellet, most cells are

probably dead. Typically, cell pellets get pink (>70% viable cells) around fraction 5.

9. The surface tension should maintain the droplet on the coverslip. Make sure to not disturb the droplet when moving the plate to the incubator to avoid cells dispersion and to obtain a good yield of live cells.
10. If using polyvinylidene fluoride (PVDF), pre-wet in methanol. This membrane is very hydrophobic, and water will not gain access to the filter unless it is pre-treated with an alcohol. In contrast, nitrocellulose is very soluble in high concentrations of alcohol and will dissolve if placed in these solutions.
11. If the cells are not stimulating, make sure all grease is removed from the edges. If they still are not stimulating, check the pulse duration (should be 2–4 ms) and increase the voltage.
12. Make sure the myocyte is rod shaped and is pacing correctly with no random waves of contraction.
13. Collecting pairwise increases the statistical power of the experiment therefore using a motorized stage, and recalling cells is optimal.
14. If one is not getting good fits with the baseline or other portions of the transient, alter the duration and offset to match the trace.
15. Common parameters to graph include baseline (BL), normalized peak height (bl% peak h), peak amplitude (peak h), times to peak, and times to baseline (usually 25%, 50%, and 75% for to peak and to bl). Be careful if you want to plot the exponential fits and tau. You need to make sure that those parameters fit well on the trace. This is not always the case.
16. Caution should be used for sarcomere length data in Fura-2 loaded myocytes as it can buffer Ca^{+2} and affect contractility. The above loading procedure limits this, but every cell is different. That said, monitoring sarcomere length changes relative to Ca^{+2} changes is very informative.
17. If perfusion is for an entire well, then only one cell is used per coverslip. Otherwise using micropipettes with a three-barrel end and a micromanipulator one can switch solutions at one cell, and use multiple cells on one coverslip.

Acknowledgments

This work was supported by funds from NIH.

References

1. Davis J, Westfall MV, Townsend D, Blankinship M, Herron TJ, Guerrero-Serna G, Wang W, Devaney E, Metzger JM (2008) Designing heart performance by gene transfer. Physiol Rev 88:1567–1651
2. Wang W, Barnabei MS, Asp ML, Heinis FI, Arden E, Davis J, Braunlin E, Li Q, Davis JP, Potter JD et al (2013) Noncanonical EF-hand motif strategically delays Ca2+ buffering to enhance cardiac performance. Nat Med 19:305–312
3. Schwaller B (2009) The continuing disappearance of "pure" Ca2+ buffers. Cell Mol Life Sci 66:275–300
4. Pauls TL, Cox JA, Berchtold MW (1996) The Ca2+–binding proteins parvalbumin and oncomodulin and their genes: new structural and functional findings. Biochim Biophys Acta 1306:39–54
5. Coutu P, Bennett CN, Favre EG, Day SM, Metzger JM (2004) Parvalbumin corrects slowed relaxation in adult cardiac myocytes expressing hypertrophic cardiomyopathy-linked alpha-tropomyosin mutations. Circ Res 94:1235–1241
6. Rodenbaugh DW, Wang W, Davis J, Edwards T, Potter JD, Metzger JM (2007) Parvalbumin isoforms differentially accelerate cardiac myocyte relaxation kinetics in an animal model of diastolic dysfunction. Am J Physiol Heart Circ Physiol 293:H1705–H1713
7. Szatkowski ML, Westfall MV, Gomez CA, Wahr PA, Michele DE, DelloRusso C, Turner II, Hong KE, Albayya FP, Metzger JM (2001) In vivo acceleration of heart relaxation performance by parvalbumin gene delivery. J Clin Invest 107:191–198
8. Coutu P, Metzger JM (2002) Optimal range for parvalbumin as relaxing agent in adult cardiac myocytes: gene transfer and mathematical modeling. Biophys J 82:2565–2579
9. Wang W, Metzger JM (2008) Parvalbumin isoforms for enhancing cardiac diastolic function. Cell Biochem Biophys 51:1–8
10. Asp ML, Sjaastad FV, Siddiqui JK, Davis JP, Metzger JM (2016) Effects of modified parvalbumin EF-hand motifs on cardiac myocyte contractile function. Biophys J 110:2094–2105

Chapter 13

Expression and Purification of Calmodulin for NMR and Other Biophysical Applications

Benjamin M. M. Grant, Christopher B. Marshall, and Mitsuhiko Ikura

Abstract

Calmodulin (CaM) is a ubiquitous calcium-sensing protein that has one of the most highly conserved sequences among eukaryotes. CaM has been a useful tool for biologists studying calcium signaling for decades. In recent years, CaM has also been implicated in numerous cancer-associated pathways, and rare CaM mutations have been identified as a cause of human cardiac arrhythmias. Here, we present a collection of our most recent and effective protocols for the expression and purification of recombinant CaM from *Escherichia coli*, including various isotopic labeling schemes, primarily for nuclear magnetic resonance (NMR) spectroscopy and other biophysical applications.

Key words Calmodulin, CaM, Calcium, Expression, Purification, NMR, Isotopic labeling

1 Introduction

Calmodulin (CaM) was first discovered as an activator of cyclic AMP phosphodiesterase in the brain by Cheng [1] and Kakiuchi [2] independently. The Ca^{2+} dependency of CaM activity demonstrated by Kakiuchi opened a totally new field of calcium-dependent signaling regulators and their cellular pathways. CaM is a small, 16.7kD protein that acts as an important master regulator of Ca^{2+} signaling in cells [3, 4]. The CaM sequence is conserved among all vertebrates and is highly conserved throughout kingdoms Animalia and Plantae. CaM is highly acidic, soluble, and stable, both in Ca^{2+}-free and Ca^{2+}-bound states, and interacts with binding targets under both conditions [5]. CaM binds Ca^{2+} through motifs called "EF-hands," first discovered in carp parvalbumin by Kresinger et al. [6]. There are four EF-hand motifs in CaM, each of which forms a 29-amino acid helix-loop-helix structure that coordinates Ca^{2+} through side chains and backbones of five residues in a twelve-residue loop; one pair of EF-hands forms an N-terminal lobe, and the other forms a C-terminal lobe. Binding of Ca^{2+} induces a

Claus W. Heizmann (ed.), *Calcium-Binding Proteins of the EF-Hand Superfamily: From Basics to Medical Applications*, Methods in Molecular Biology, vol. 1929, https://doi.org/10.1007/978-1-4939-9030-6_13,

structural change that opens a hydrophobic pocket in each lobe, which mediates the interaction with Ca^{2+}-dependent binding partners [5]. EF-hands have been identified in 1613 protein sequences [7], including 221 proteins encoded in the human genome.

Due in part to its size and stability, CaM was an early target for solution protein NMR and was used to test and develop many of the core NMR experiments employed today [8–10]. There are currently ~200 structures of CaM in the Protein Data Bank (www.rcsb.org). In early years, CaM was purified from various tissues (e.g., the heart, brain, and red blood cells) using combinations of anion exchange, gel filtration chromatography, and ammonium sulfate precipitation, with many schemes taking advantage of the high stability of CaM through precipitation of undesired proteins by heating [11] and trichloroacetic acid (TCA), followed by hydrophobic column chromatography [12].

With the development of recombinant DNA technology, researchers moved to overexpression of CaM in bacterial cells in order to obtain large quantities of protein, as well as to produce mutants to probe structure/function relationships. In the 1990s, *Drosophila* CaM was cloned into a heat-inducible pAS system for expression in *E. coli* to allow incorporation of isotopic labels [13]. This construct did not contain any affinity tags and was purified through a multi-step process involving TCA precipitation of unwanted proteins (3% TCA), followed by TCA precipitation of CaM (6% TCA), Ca^{2+}-dependent binding of resuspended CaM to Phenyl Sepharose, and elution with ethylenediaminetetraacetic acid (EDTA) after extensive washing. The *Drosophila melanogaster* CaM sequence has three amino acid substitutions relative to vertebrate CaM; thus, many studies adopted recombinant expression of *Xenopus laevis* CaM, which has an amino acid sequence that is identical to that of all known vertebrates. However, it should be noted that recombinant CaM differs from endogenous CaM in that the initiating Met residue is not removed, and it lacks acetylation of the N-terminus and trimethylation of Lys115.

Recent implications of CaM in cancer-associated pathways have renewed and expanded interest in this protein for in vitro biophysical and biochemical characterization. Here we present our laboratory's most recent and effective protocols for the expression and purification of recombinant CaM from *E. coli* and include techniques for various isotopic amino acid labeling schemes for NMR. We use a T7-inducible promoter system to express a hexa-histidine-tagged CaM protein, which we purify through a two-step process: Nickel-nitrilotriacetic acid (Ni-NTA) affinity chromatography followed by size-exclusion liquid chromatography (Fig. 1). The entire protocol can be performed within a week and yields approximately 30 mg/L of highly pure, functional protein.

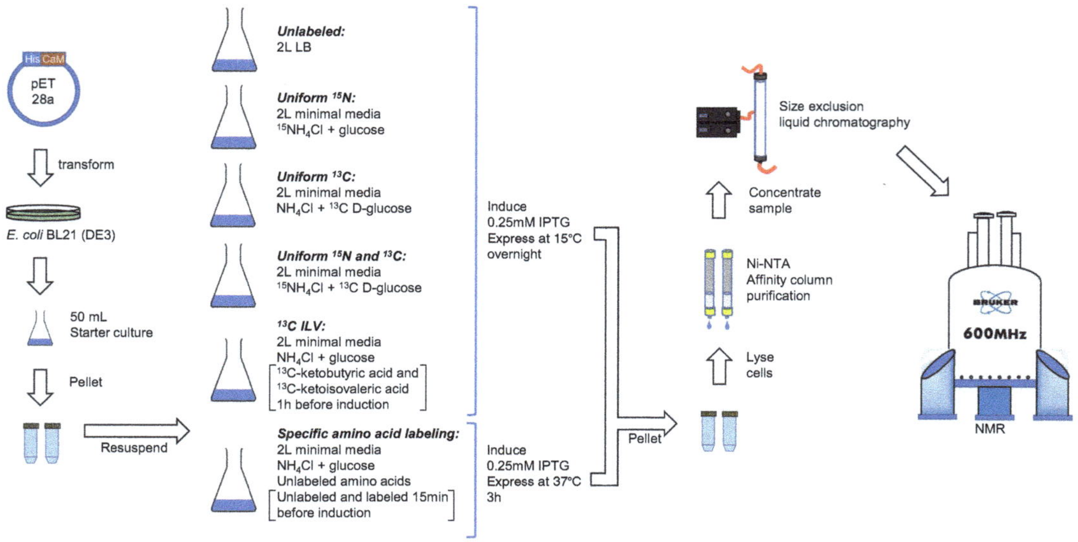

Fig. 1 Schematic flow of calmodulin expression and purification protocol

2 Materials

2.1 CaM in pET28 Vector

CaM construct: CaM has been engineered with an N-terminal hexa-histidine (His) tag and linker containing a specific cleavage motif that is recognized by thrombin. This construct was cloned into the pET28a vector with its stop codon using the NdeI and BamHI restriction endonuclease sites. This encodes a reversibly His-tagged CaM protein under the control of a T7 promoter such that expression can be induced by isopropyl β-D-1-thiogalactopyranoside (IPTG) in DE3$^+$ *E. coli* strains, with selection through kanamycin resistance. Following cleavage of the His-tag by thrombin, three extra residues (Gly-Ser-His) remain at the N-terminus.

2.2 Competent Cells

The construct is expressed in *E. coli* BL21 (DE3$^+$) which is compatible with the T7 expression system employed by pET vectors. Strains with additional copies of specific tRNA genes that are rare in the *E. coli* genome (e.g., codon + or Rosetta) can improve yields but are not necessary. These strains carry an additional antibiotic resistance marker (e.g., chloramphenicol).

2.3 Stock Solutions

1. LB broth: Dissolve in water 10 g/L tryptone, 5 g/L yeast extract, and 10 g/L NaCl, or purchase premixed broth powder and follow instructions.
2. M9 Media (10× stock): Dissolve in water Na_2HPO_4, 60 g/L, KH_2PO_4, 30 g/L, and NaCl, 5 g/L, and adjust pH to 7.4.
3. Trace elements (1000×): Add 5 mL of 0.5 M EDTA, pH 8.0, to 80 mL of water. Add trace elements one at a time in the

following order, with stirring, waiting at least 10 min between the addition of each compound: Ferric chloride ($FeCl_3{\cdot}6H_2O$) 833 mg, zinc chloride ($ZnCl_2$) 50 mg, cupric chloride ($CuCl_2{\cdot}2H_2O$) 12.6 mg, cobalt chloride ($CoCl_3{\cdot}6H_2O$) 10 mg, boric acid (H_3BO_3) 10 mg, and manganese chloride ($MnCl_2{\cdot}4H_2O$) 2500 mg. Do not readjust pH or there will be precipitation. Top up to 100 mL and filter (0.22 μm).

4. Kanamycin (1000×): Prepare 50 mg/mL stock in water; store in aliquots at −20 °C.
5. Chloramphenicol (1000×): Prepare 30 mg/mL solution in ethanol; store at −20 °C.
6. Biotin (1000×): Prepare a 1 mg/mL solution in 50% ethanol, ideally stored at 4 °C.
7. Thiamine (1000×): 1 mg/mL in water.
8. Magnesium sulfate ($MgSO_4$) (1000×): 1 M solution in water.
9. Ammonium chloride (NH_4Cl) and/or ^{15}N ammonium chloride (for uniform ^{15}N labeling).
10. 20% (w/v) glucose in water or ^{13}C D-glucose (for uniform ^{13}C labeling).
11. ^{13}C-ketobutyric acid (for methyl labeling of Ile) and ^{13}C-ketoisovaleric acid (for methyl labeling of Leu and Val).
12. ^{15}N and/or ^{13}C amino acids (for labeling of specific amino acids).
13. Minimal media: Add 200 mL of 10× M9 to 1800 mL of water in a 6 L Erlenmeyer flask and autoclave. After media has cooled, add 1 mL/L of kanamycin, [chloramphenicol for codon +], biotin, thiamine, magnesium sulfate, and trace elements (1000× stocks), as well as 0.3 mL/L of 1 M $CaCl_2$. Calcium and metals in the trace elements solution will form a visible precipitate with the phosphate in M9, which is not a cause for concern. The remaining components of the minimal media are dependent on isotopic labeling schemes which are covered in *Methods*.
14. Lysis buffer: Tris pH 8.0, 50 mM, NaCl, 150 mM, NP-40 (detergent), 0.1% v/v, glycerol, 10% v/v, imidazole, 10mM. Make 1 L and store. Before use, take 50 mL in a Falcon tube and add β-mercaptoethanol, 10 mM (pure liquid β-mercaptoethanol is 14.3 M), lysozyme (egg white), DNase I (both enzymes can be added by scooping a small amount of powder with a pipette tip), and phenylmethylsulfonyl fluoride (PMSF), 1 mM (toxic) (*see* **Note 1**).
15. His-wash buffer 1: Tris pH 8.0, 50 mM, NaCl, 500 mM, glycerol, 10% v/v, imidazole, 10 mM. Make 1 L and store.

Before use, aliquot 100 mL and add β-mercaptoethanol, 10 mM.

16. His-wash buffer 2: Tris pH 8.0, 50 mM, NaCl, 150 mM, glycerol, 10% v/v, imidazole, 10 mM. Make 1 L and store. Before use, aliquot 100 mL and add β-mercaptoethanol, 10 mM.
17. His-elution buffer: Tris pH 8.0, 50 mM, NaCl, 150 mM, glycerol, 10% v/v, imidazole, 250 mM. Make 1 L and store. Before use, aliquot 50 mL and add β-mercaptoethanol, 10 mM.
18. General dialysis buffer: Tris pH 8.0, 50 mM, NaCl, 150 mM, glycerol, 20% v/v, DTT, 1 mM. Make 4 L in a beaker and store short term at 4 °C.
19. Gel filtration buffer: 4-(2-hydroxyethyl)-1-piperazineethane-sulfonic acid (HEPES) pH 7.4, 20 mM, NaCl, 100 mM, tris (2-carboxyethyl)phosphine (TCEP), 2 mM. TCEP is a reducing agent available as a highly acidic hydrochloride salt (TCEP-HCl); thus, the pH of TCEP-containing buffers must be adjusted before topping up to final volume. Filter (0.22 μm) and degas 20 min. Store at 4 °C.
20. Chelation buffer: HEPES pH 7.4, 20 mM, NaCl, 100 mM, EDTA, 20 mM, TCEP, 2 mM. Prepare in smaller batches as needed.

3 Methods

3.1 Transformation

1. Thaw 50 μL *E. coli* BL21 (DE3$^+$) [codon + optional] competent cells on ice.
2. Add 50–200 ng of plasmid to cells, mixing gently. Store on ice for 30 min.
3. Heat shock at 42 °C for 1 min. Return to ice for 5 min.
4. Add 450 μL of LB to tube and incubate at 37 °C for 1 h.
5. Plate 100 μL on kanamycin LB plate [add chloramphenicol for codon +], and incubate at 37 °C overnight (*see* **Note 2**).

3.2 Expression

1. Prepare a starter culture: pick one colony of transformed cells with a sterile loop and inoculate 50 mL of LB containing 1× kanamycin (50 μg/L) [add 30 μg/L chloramphenicol for codon +]. Shake at 37 °C overnight (*see* **Note 3**).
2. Prepare media based on desired isotopic labeling scheme.
   (a) *Unlabeled CaM*: LB broth is used for preparation of unlabeled protein. Autoclave 2 L of LB in a 6 L Erlenmeyer flask prior to inoculation. The volume and number of flasks can be adjusted for the desired protein yield;

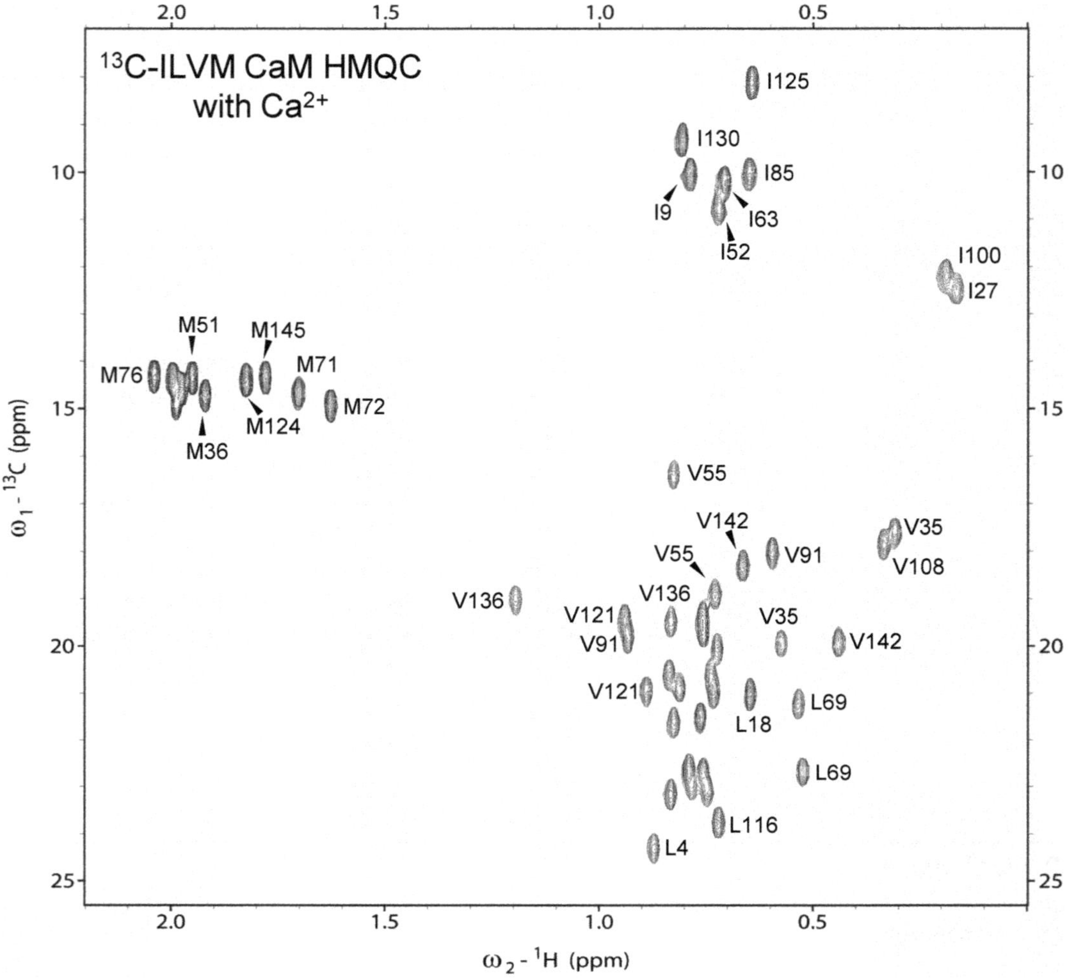

Fig. 2 ^{1}H-^{13}C HMQC of specifically labeled ^{13}C-methyls of Ile, Leu, Val, and Met in calmodulin bound to Ca^{2+}. Sample contains 0.75 mM calmodulin in gel filtration buffer (20 mM HEPES 7.4, 100 mM NaCl, 1 mM TCEP) with 10 mM $CaCl_2$. HMQC collected at 25 °C

however, for efficient aeration, the flasks should not be filled beyond ~1/3.

(b) *Uniform ^{15}N-labeling*: For preparation of isotopically labeled CaM, minimal M9 media is used to control available isotopes. In this case, $^{15}NH_4Cl$ is the only source of nitrogen present in the media and will therefore be incorporated in newly synthesized amino acids and proteins. Prepare up to 2 L of minimal media, as described in *Materials*. For uniform ^{15}N labeling, add 10 mL/L glucose (0.2% w/v) and 1 g/L $^{15}NH_4Cl$.

(c) *Uniform ^{13}C-labeling*: Prepare minimal media as described, and add 1 g/L unlabeled NH_4Cl and 2 g/L

^{13}C D-glucose. As glucose is the principle carbon source in the media, ^{13}C will be incorporated in all newly synthesized amino acids and proteins.

(d) *Uniform ^{15}N- and ^{13}C-labeling*: Prepare minimal media as outlined above using both isotopically labeled $^{15}NH_4Cl$, 1 g/L, and ^{13}C D-glucose, 2 g/L.

(e) *Specific methyl labeling—ILV*: For specific methyl labeling, prepare minimal media with unlabeled NH_4Cl (1 g/L) and glucose (0.2% w/v). Isotopic labeling of Ile, Leu, and Val (ILV) methyl groups is achieved by adding isotopically labeled metabolic precursors of these amino acids. ^{13}C-ketobutyric acid (for Ile—50 mg/L) and ^{13}C-ketoisovaleric acid (for Leu and Val—100 mg/L) are added 1 h prior to the anticipated induction of CaM expression, i.e., when the OD_{600} is approximately 0.5.

(f) *Specific amino acid labeling*: For specific labeling of amino acids, prepare minimal media with unlabeled NH_4Cl (1 g/L) and glucose (0.2% w/v) and 100 mg/L of each amino acid not to be isotopically labeled. Isotopic labeling of specific amino acids is achieved by supplying a source of the labeled amino acid(s) at a crucial time to ensure incorporation in proteins and minimize isotopic scramble to other amino acids due to metabolic processes. Immediately prior to the anticipated induction of CaM expression, i.e., when the OD_{600} is approximately 0.6–0.8, add 500 mg/L addition unlabeled amino acids and 100 mg/L of desired specific labeled amino acid(s). Wait 15 min and induce. Gly, Ala, and Ser labeling is inefficient due to rapid interconversion with metabolites, and Gln, Glu, Asn, and Asp will readily scramble.
An example of a specific isotopic labeling scheme for CaM (methyl labeling of ILVM residues) using these techniques is shown below (Fig. 2).

3. Set aside 1 mL of media for use as a spectrophotometer blank later on. Transfer the starter culture to two 50 mL Falcon tubes, and centrifuge at 2300 × *g* for 10 min. Resuspend pellets in 5–10 mL of new media and inoculate the large flask.
4. Shake flask at 37 °C and 180 rpm (*see* **Note 4**). Measure OD_{600} of cultures roughly every hour, using the blank set aside in **step 3**. Cell cultures grow much more slowly in M9 media than in LB. When OD_{600} reaches roughly 0.4, lower the temperature to 15 °C. It takes some time to cool the shaker and significantly longer for the culture to equilibrate to this temperature, during which the cells will continue to grow. The goal is to cool the culture to 15 °C when the OD_{600} reaches between 0.6 and 0.8, to optimize the yield of soluble protein. At this time, induce

protein expression by adding 0.25 mL/L 1 M IPTG, and continue shaking at 15 °C overnight (*see* **Notes 5** and **6**). For specific amino acid labeling, overnight expression promotes isotopic scrambling. In this case, induce expression at 37 °C and express protein for 3 h.

5. The following morning, harvest cells by centrifugation in 1 L centrifuge bottles at 8500 × *g* for 20 min. Discard supernatant and transfer cell pellets into 50 mL Falcon tubes. These cell pellets can be used for protein purification immediately or frozen at −80 °C for later use (*see* **Notes 7** and **8**).

3.3 Purification

1. Add DNase I, lysozyme, PMSF (*see* **Note 1**), and β-mercaptoethanol to 50 mL lysis buffer. Add fresh lysis buffer to cell pellets, frozen or fresh, leaving 5–10 mL in the Falcon tube for air. Rock or rotate the tube at 4 °C to resuspend the pellet for at least 30 min but no more than 2 h. Lysis efficiency is improved when the DNase I and lysozyme have time to work. Before continuing, ensure the remaining cell pellet is fully resuspended by pipetting until homogenous.
2. Lyse cells by sonication on ice using intermittent pulses (e.g., 0.5 s on and 2.0 s off) to prevent heating of the lysate (*see* **Notes 9** and **10**). Sonication can be performed twice to ensure complete lysis.
3. Transfer lysate to 35 mL ultracentrifuge bottles, and spin at 45,000 × *g* for 40 min (*see* **Note 11**). Samples can be collected from the pellet and supernatant to check for effective lysis (*see* **Note 12**).
4. Wash 5–10 mL of Ni-NTA resin by pouring the resin slurry (in 20% ethanol storage solution) into a 50 mL Falcon tube and pelleting gently at 200 × *g* for 3 min. Pour off the ethanol, rinse with water, and repeat centrifugation. Resuspend Ni-NTA resin with the lysate supernatant, and stir at 4 °C for 1 h.
5. Pour lysate-resin slurry into a gravity flow column, and collect flow-through as Ni-NTA resin settles. Resin should not be allowed to run dry. Samples can be collected throughout purification for subsequent analysis by sodium dodecyl sulfate–polyacrylamide gel electrophoresis (SDS-PAGE).
6. Wash Ni-NTA resin with His Wash 1 (high salt). Slowly pipette wash buffer down the side of the column to prevent disruption of the resin bed (*see* **Note 13**). Wash with approximately 10–20 column volumes (volume of resin bed) of His Wash 1 until clean (*see* **Note 14**).
7. Wash Ni-NTA resin with His Wash 2 (low salt) for 10 column volumes.

8. Elute CaM from Ni-NTA resin by slowly adding His-elution buffer, 2 column volumes at a time. Monitor protein levels (*see* **Note 14**) until no more protein is liberated. Elution volume is generally around 40 mL (*see* **Notes 15** and **16**).
9. Add 10 U thrombin (bovine thrombin is commercially available at low cost) to elution fraction for cleavage of His-tag *during* dialysis. Collect samples for SDS-PAGE before addition of thrombin, and after dialysis, to confirm complete cleavage of the His-tag. If downstream applications require immobilization through the His-tag, perform dialysis without thrombin to remove imidazole.
10. Prepare 4 L general dialysis buffer (stored at 4 °C). Measure a sufficient length of 10,000 molecular weight cutoff (MWCO) dialysis membrane to accommodate the sample with space for clamps. Wet the dialysis membrane in dialysis buffer, and fill with elution fraction after clamping one end (*see* **Note 17**). Clamp other side, and affix a buoyant object to keep the dialysis membrane clear of the magnetic stir rod. Dialyze overnight.
11. Carefully empty dialysis bag into a beaker on ice. At this point, check pre/post cleavage samples by SDS-PAGE before continuing. Concentrate sample to less than 10 mL using 50 mL 10,000 MWCO centrifugal concentrators (*see* **Note 18**). Centrifuge at 2300 × *g* for 10 min at a time, pausing between spins to thoroughly mix the sample to prevent aggregation and to add more protein. Before discarding the flow-through, check for protein by testing with Bradford dye, to ensure the membrane has not ruptured. As the protein becomes more concentrated, the flow rate will decrease. The protein solution may also become visibly viscous or appear faint yellow, which are not causes of concern; however, it should not become cloudy.
12. Prepare, filter, and degas gel filtration buffer (this can be performed in advance), and equilibrate size-exclusion (also known as gel filtration) chromatography column such as Superdex 75 26/60 run on an AKTA (GE Healthcare Life Sciences), or similar fast protein liquid chromatography (FPLC) (*see* **Note 19**). Filter the sample (0.22 μm), load into a 10 mL loop, and run 1.2 column volumes (~350 mL) of buffer at <2 mL/min. CaM should elute at a volume of ~200 mL.
13. Collect elution fractions, and concentrate once more to a working concentration of CaM for the given purpose. CaM is stable for NMR at concentrations above 1 mM, though 100–200 μM is sufficient for most experiments (*see* **Note 20**). Samples can be aliquoted, flash frozen in liquid nitrogen or ethanol/dry ice bath, and stored at −80 °C indefinitely.

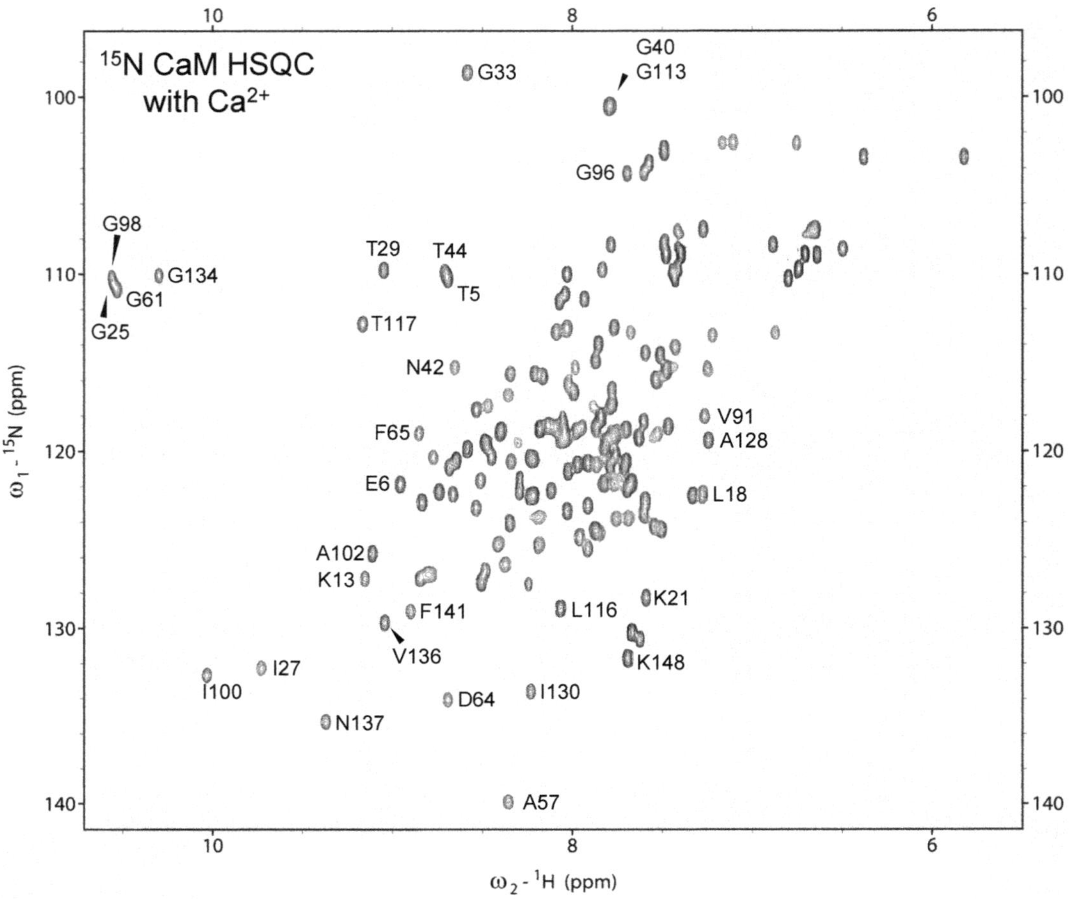

Fig. 3 ^1H-^{15}N HSQC of uniformly ^{15}N-labeled calmodulin bound to Ca^{2+}. Sample contains 0.75 mM calmodulin in gel filtration buffer (20 mM HEPES 7.4, 100 mM NaCl, 1 mM TCEP) with 10 mM $CaCl_2$. HSQC collected at 25 °C in 10 min

3.4 Preparation Methods

1. CaM from *E. coli* will be purified bound partially to Mg^{2+}. If Ca^{2+}-bound CaM is required, add 10 mM $CaCl_2$ to sample before use. Binding is fast and will displace other cations that may be present in the sample. An example of a ^1H-^{15}N heteronuclear single-quantum correlation (HSQC) spectrum of Ca^{2+} loaded, uniformly ^{15}N-labeled CaM is included (Fig. 3). If apo-CaM is required, a buffer exchange must first be performed, either on a single sample or with a batch, into chelation buffer. Use 10,000 MWCO centrifugal concentrators for buffer exchange (available in 15 mL, 4 mL, or 500 μL sizes). Add chelation buffer until it makes up at least 50% of the volume before switching back to gel filtration buffer. Thoroughly exchange buffer back into gel filtration buffer to remove the EDTA. An example of a ^1H-^{15}N HSQC of uniformly ^{15}N-labeled apo-CaM is shown below (Fig. 4) (*see* **Note 21**).

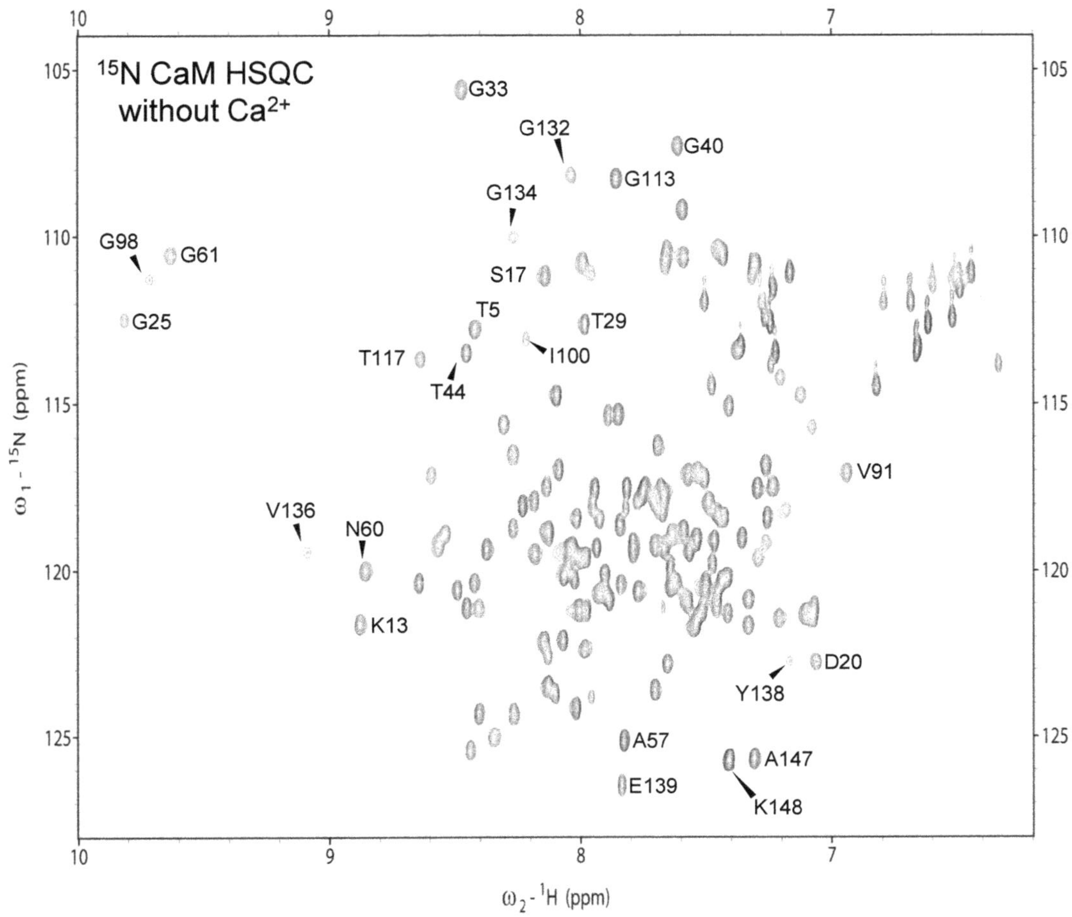

Fig. 4 ^{1}H-^{15}N HSQC of uniformly ^{15}N-labeled Ca^{2+}-free calmodulin. Sample contains 0.5 mM calmodulin in gel filtration buffer (20 mM HEPES 7.4, 100 mM NaCl, 1 mM TCEP). HSQC collected at 25 °C in 20 min

2. For formation of a CaM-peptide complex, mix Ca^{2+}-bound or apo-CaM and peptide at low concentrations, and then co-concentrate the mixture in a 3000 MWCO centrifugal concentrator to the working concentration. Mixing at low concentrations reduces the risk of precipitation.

3. Lyophilization allows long-term storage of CaM preparations. First, exchange apo-CaM (Ca^{2+}-containing buffers are not amenable to this technique) into a buffer of 25 mM ammonium bicarbonate. Ammonium bicarbonate provides sufficient buffering capacity but is volatile and thus dissipates during lyophilization. Following freeze-drying, pure apo-CaM powder can be stored at −20 °C for very long periods. To resuspend, place a drop of buffer beside the CaM powder, and allow

the powder to be slowly "drawn in." Adding buffer too quickly produces clumps that are difficult to dissolve.

4 Notes

1. PMSF is a protease inhibitor and is extremely toxic. Make a stock solution in ethanol to reduce routine exposure to the powder, and always wear personal protective equipment (PPE). It is rapidly degraded in water and thus must be added to lysis buffer immediately before use. There are also commercially available protease inhibitor cocktails that contain PMSF and are easier to handle.
2. This is a very flexible protocol that can be adjusted according to time constraints. Initial incubation on ice should be at least 15 min. The heat shock is the most critical step and should be performed at 42 °C for 1 min before returning the tube to ice, but the post-heat shock incubation on ice can be reduced to as little as 2 min. After addition of LB, shaking at 37 °C should be longer than 15 min but less than 1.5 h. Resuspension and plating volumes can be adjusted to optimize colony density. The plate can be incubated at room temperature over a weekend in place of overnight incubation at 37 °C.
3. Expression can be expedited by resuspending the entire plate, rather than a single colony, with 1–2 mL of media and inoculating the 50 mL LB starter culture. Shake at 37 °C for approximately 3 h (rather than overnight) to reach an appropriate cell density to pellet and proceed.
4. Alternatively, aeration and mixing of the *E. coli* culture can be achieved by bubbling compressed air in a system such as a LEX bioreactor (Epiphyte3).
5. Due to slow growth in M9 media, the one-day expression method described (*see* **Note 3**) can take as long as 12 h to achieve the OD required for induction.
6. The expression system used is highly efficient and can produce very large quantities of recombinant protein, which can form insoluble inclusion bodies. Often the overall yield of soluble protein can be improved by reducing the IPTG concentration and lowering the induction temperature. If a refrigerated incubator is not available or time is an issue, the induction of expression can also be performed at 37 °C for 3–4 h. Because CaM is highly soluble, this produces acceptable protein yields.
7. Freezing the cell pellets at −80 °C has no adverse impact on protein yield or quality and promotes cell lysis.
8. It is important, especially when unfamiliar with the protocol, to monitor every stage of the expression and purification by SDS-

PAGE. Several samples should be collected during expression and purification. The first is to test that induction of protein expression occurred. A useful trick is to save a cuvette used to measure OD_{600} preinduction until the following morning, so that samples of the culture both pre- and post-induction have similar cell densities. Take 100 μL of each sample, and centrifuge in Eppendorf tubes at max speed for 1 min. Discard supernatant and resuspend pellets in 50 μL of 1× SDS-PAGE sample buffer. When run on a gel, these samples should have similar background protein levels and a single strong band in the post-induction sample around 17 kD corresponding to CaM.

9. Heat generated by the sonication process will melt the ice around the sample beaker. Periodically check that the beaker is supported and that the tip of the sonicator remains ~1 cm from the bottom. A plastic beaker is preferable to glass, which may break.
10. Alternative lysis methods such as a French press are suitable.
11. Lipids can be loosely pelleted and contaminate the supernatant. To avoid clogging columns and filters, decant supernatant carefully and immediately following centrifugation.
12. Similar to monitoring expression as outlined above (*see* **Note 8**), it is important to check the proportion of protein found in the supernatant. For this to be reliable, the pellet and supernatant fractions need to be normalized: before centrifugation, take 50 μL of lysate, and centrifuge separately in an Eppendorf tube, max speed for 1 min. Separate supernatant from pellet, and solubilize both in SDS-PAGE sample buffer, to the same volume. As a highly soluble protein, CaM should only be present in the supernatant. If a large amount of CaM is found in the pellet, next time try slower expression (lower IPTG induction concentration) and/or more thorough lysis.
13. The resin bed is easily disrupted by the flow of buffer, which can create voids in the resin bed, allowing the buffer to channel rather than flowing evenly through the resin. If this happens, it is best to let the resin settle before continuing.
14. Bradford dye can be used to monitor the washing or elution progress. Make 10 mL of fresh Bradford dye before beginning column purification, and aliquot 400 μL into Eppendorf tubes. Collect one drop from the gravity flow column in a tube to qualitatively assess protein levels and decide when washing or elution is complete. When protein is present, Bradford dye turns blue.
15. The most efficient method to elute protein is to let one column volume elute at a time, with brief incubations in between to allow for protein dissociation from the resin. As the protein

becomes more dilute, extend the wait period. Elution is complete when these flow samples do not react with Bradford dye.

16. The Ni-NTA purification step can be automated using a prepacked Ni-NTA column on an FPLC system, such as AKTA (GE Healthcare Life Sciences).
17. Wet the dialysis membrane in dialysis buffer, and tease one side between fingers until it opens. Pipette buffer through the open membrane, making sure that it can pass freely through the entire length before clamping and adding sample. To monitor thrombin cleavage, run samples taken before and after dialysis on SDS-PAGE. Cleaved CaM runs with an approximate molecular weight just below 15 kD on SDS-PAGE due to its high negative charge.
18. Samples may be concentrated by alternative means to centrifugal concentrators.
19. Size-exclusion chromatography can be performed without an FPLC system, using a pump (e.g., peristaltic) and a column manually packed with size exclusion resin. The UV absorbance of each fraction can be measured manually to detect protein.
20. We use two methods to determine the concentrations of CaM, which are generally consistent: UV absorbance at 280 nm and Bradford. CaM lacks tryptophan; thus, its UV absorbance is relatively low, with predicted extinction coefficient at 280 nm of 2980 $M^{-1}cm^{-1}$. For Bradford reagent, the concentration can be determined using a standard curve (e.g., Abs = 0.0447 [concentration (mg/mL)] – 0.0597).
21. For sensitive analysis of apo CaM, it is important to decalcify all buffers. Ca^{2+} contamination can be present even in highly purified water and high-grade reagents. Buffers can be passed through a Chelex® 100 Resin (Bio-Rad) column to remove residual cations. This is not required for all purposes; however, the affinity of CaM for Ca^{2+} is enhanced in the presence of high-affinity Ca^{2+}-dependent target peptides; thus, traces of Ca^{2+} contamination may be sufficient to support Ca^{2+}-dependent interactions.

Acknowledgments

This work was supported by grants from Canadian Institutes of Health Research, Natural Sciences and Engineering Research Council of Canada, and Heart and Stroke Foundation of Canada. M.I. holds the Canada Research Chair in Cancer Structural Biology. The NMR facility at UHN is supported by Canada Foundation for Innovation and the Princess Margaret Foundation.
This article is in memory of Dr. Claude B. Klee for her encouragement and support of MI.

References

1. Cheung WY (1970) Cyclic 3′ -5′-nucleotide phosphodiesterase. Demonstration of an activator. Biochem Biophys Res Commun 38 (3):533–538
2. Kakiuchi S, Yamazaki R (1970) Calcium dependent phosphodiesterase activity and its activating factor (PAF) from brain studies on cyclic 3′,5′-nucleotide phosphodiesterase. Biochem Biophys Res Commun 41 (5):1104–1110
3. Cohen P, Klee CB (eds) (1988) Calmodulin. Molecular aspects of cellular regulation, vol Vol. 5. Elsevier, Amsterdam
4. Marshall CB et al (2015) Calmodulin and STIM proteins: two major calcium sensors in the cytoplasm and endoplasmic reticulum. Biochem Biophys Res Commun 460(1):5–21
5. O'Connell DBM, Marshall CB, Ikura M, Linse S (2013) Calmodulin. In: Kretsinger RH, Uversky VN, Permyakov EA (eds) Encyclopedia of metalloproteins. Springer, Berlin
6. Moews PC, Kretsinger RH (1975) Terbium replacement of calcium in carp muscle calcium-binding parvalbumin: an x-ray crystallographic study. J Mol Biol 91(2):229–232
7. Kawasaki H, Kretsinger RH (2017) Structural and functional diversity of EF-hand proteins: Evolutionary perspectives. Protein Sci 26:1898–1920
8. Ikura M, Kay LE, Bax A (1990) Novel approach for sequential assignment of 1H, 13C, 15N spectra of proteins: heteronuclear triple-resonance three-dimensional NMR spectroscopy. Application to calmodulin. Biochemistry 29(19):4659–4667
9. Ikura M, Kay LE, Bax A (1991) Improved three-dimensional 1H-13C-1H correlation spectroscopy of a 13C-labeled protein using constant-time evolution. J Biomol NMR 1 (3):299–304
10. Ikura M, Spera S, Barbato G, Kay LE, Krinks M, Bax A (1991) Secondary structure and side-chain 1H and 13C resonance assignments of calmodulin in solution by heteronuclear multidimensional NMR spectroscopy. Biochemistry 30(38):9216–9228
11. Muallem S, Karlish SJ (1979) A simple, rapid and efficient procedure for purification of calmodulin from human red blood cells. FEBS Lett 107(1):209–212
12. Yoshida M, Minowa O, Yagi K (1983) Divalent cation binding to wheat germ calmodulin. J Biochem 94(6):1925–1933
13. Ikura M, Marion D, Kay LE, Shih H, Krinks M, Klee CB, Bax A (1990) Heteronuclear 3D NMR and isotopic labeling of calmodulin. Towards the complete assignment of the 1H NMR spectrum. Biochem Pharmacol 40 (1):153–160

Chapter 14

The Use of Cre/loxP Inducible Mouse Models to Dissect the Specific Roles of Calcineurin Signaling in Myeloid Cells

Teresa Zelante and Jan Frič

Abstract

The calcineurin/nuclear factor of activated T-cell (NFAT) signaling pathway is important for the development and function of different leukocyte subsets. Calcineurin is activated through increased intracellular levels of calcium that is triggered by cell receptors. The phosphatase activity of calcineurin dephosporylates the members of NFAT transcription factor family and activates their translocation to the nucleus and further starts the transcription of NFAT-regulated genes. Although calcineurin is expressed ubiquitously, its functions are cell- and tissue-specific. The genetically engineered murine models allow to dissect the calcineurin roles in tissue-specific manner and as such have been essential to pursue the understanding of important role of calcineurin/NFAT pathway in orchestrating the developmental and immune response processes. This protocol describes several examples of the use of Cre/loxP inducible mouse models deficient in calcineurin in different myeloid subsets to dissect the cell- and tissue-specific roles of calcineurin signaling.

Key words Cre/lox breeding, Dendritic cells, Calcineurin, Calcium, Macrophages, Granulocytes, NFAT, Myeloid cells

1 Introduction

Calcineurin is ubiquitously expressed in most of the cell types; therefore, dissecting its specific roles in different tissues is challenging. Use of cre/lox system is one of the strategies used genetic engineering to enable tissue-specific or inducible knockouts allowing better control of calcineurin expression. This also provides a useful experimental approach, as the full calcineurin knockout mouse (CNA$^{-/-}$) is lethal in 3–4 weeks of age.

In the immune system, the most of the calcineurin functions have been associated to T lymphocytes. Several findings how calcineurin functions in other immune cell types were achieved using the genetically engineered models based on cre/loxP deletion models. This can be demonstrated on example of progress in the field of calcineurin signaling in myeloid cells where the findings in

Claus W. Heizmann (ed.), *Calcium-Binding Proteins of the EF-Hand Superfamily: From Basics to Medical Applications*, Methods in Molecular Biology, vol. 1929, https://doi.org/10.1007/978-1-4939-9030-6_14,

specific calcineurin knockout showed essential role of calcineurin in antifungal protection [1–4]. The success of immunosuppressive drugs targeting calcineurin such as cyclosporine A and tacrolimus has increased the need for detailed understanding of calcineurin role in different subsets of immune cells.

This protocol aims to describe the strategy on how to obtain tissue-specific calcineurin knockout mice including the breeding strategy and the genotyping protocol for each strain and further provide guidelines on how to isolate and analyze the calcineurin knockout cells.

1.1 Cre/lox Strategy for Genetic Engineering of Tissue-Specific Knockout Genetic Engineering

The principle of cre/lox system breeding requires generating mice that are heterozygous for a loxP-flanked allele and hemizygous/heterozygous for the cre transgene. This leads to the need to generate double-mutant/transgenic mice. The most efficient strategy is to mate a homozygous loxP-flanked mouse to a cre transgenic mouse strain. This results in approximately 50% of the offspring being heterozygous for the loxP allele and heterozygous for the cre transgene. In the subsequent crossing, these mice are mated back to the homozygous loxP-flanked mice. Approximately 25% of the progeny from this mating will be experimental mice, homozygous for the loxP-flanked allele and heterozygous for the cre transgene. Another 25% mice among the offspring can be used as control because they are homozygous for loxP allele but lack any cre recombinase expression. This approach indeed requires routine genotyping of offspring animals.

1.2 Calcineurin Conditional and Calcineurin Knockout Mice

Several breeding strategies and mice strains allow obtaining the tissue-specific knockout of calcineurin. Here we explain some approaches on how to deplete calcineurin specifically in some myeloid cell types, mainly dendritic cells, macrophages, and neutrophils. Calcineurin molecule is a heterodimer composed by a catalytic subunit (A) that binds to calcium sensor calmodulin and a regulatory subunit (B) that contains four calcium ion-binding domains. Genetic engineering—knockout of regulatory subunit (*cnb1*)—is a broadly accepted strategy to impair the function of calcineurin molecule. To knock out the *cnb1*, we used mouse strain with loxP sites flanking exons 2–4 of the endogenous cnb gene ($Cnb1^{loxP}$) [5]. Crossbreeding of the strain with transgenic mice expressing cre recombinase under specific promoter will result in tissue-specific knockout. Some researchers used also alternative $Cnb1^{loxP}$ strains to obtain conditional calcineurin knockout [6].

In general, the myeloid cell subsets cannot be distinguished by a single marker; therefore, the conditional knockout introduced by cre recombinase under specific marker promoter has always a limited cell specificity. Here we provide as an example a strategy to knock down the calcineurin under CD11c and LysM promoter.

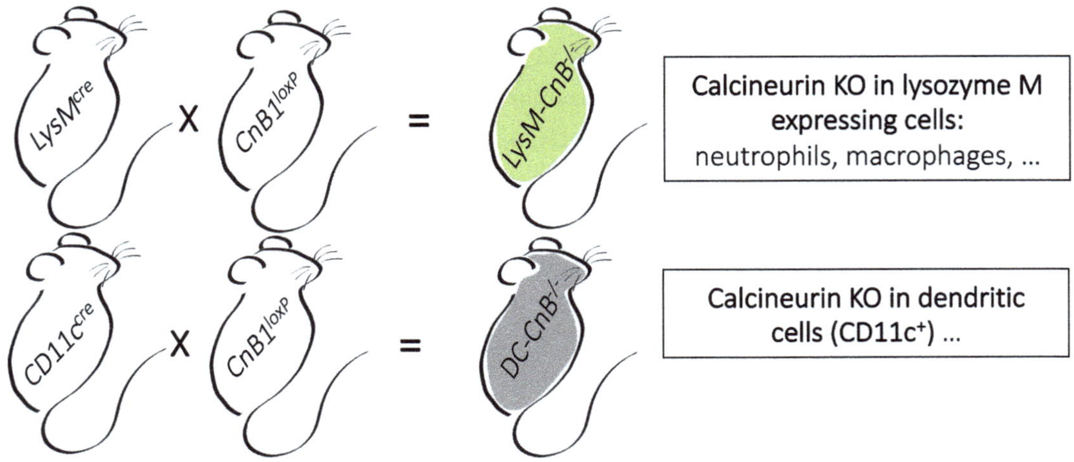

Fig. 1 Graphical overview of cre/lox calcineurin knockout in $CD11c^+$ and $LysM^+$ myeloid cells

In order to achieve a specific deletion of calcineurin in $CD11c^+$ cells (dendritic cells and some tissue macrophages), the mice expressing cre recombinase under control of the CD11c promoter ($CD11c^{cre}$) [7] can be crossed with the previously described mice carrying a floxed allele of the calcineurin regulatory subunit B ($cnb1^{loxP}$). This breeding generates $CD11c^{cre}$ $cnb1^{loxP}$ mice. As $CD11c^+$ is not a unique marker of dendritic cells but shows expression also in other myeloid cell subsets such as alveolar macrophages, an alternative strategy of depleting calcineurin in different pool of myeloid cells can be adopted using cre expression controlled by lysozyme M promoter ($LysM^{CRE}$) [8]. The final breeding leads to mice with conditional calcineurin knockout in populations of granulocytes and macrophages, schematically shown at Fig. 1.

2 Materials

2.1 Mouse Strains

1. C57BL/6-$Ppp3r1^{tm1Stl}$/J ($cnb1^{loxP}$) The Jackson Laboratory (#006581) [5].
2. B6.Cg-$Tg^{(Itgax-cre)1-1Reiz}$/J ($CD11c^{CRE}$), The Jackson Laboratory (#008068) [7].
3. B6.129P2-$^{Lyz2tm1(cre)Ifo}$/J) ($LysM^{CRE}$) The Jackson Laboratory (#004781) [8].

2.2 Genotyping

1. dNTP mix—U1515, Promega,
2. GoTaq(R) Flexi DNA Polymerase, M8295, Promega.
3. $MgCl_2$ A3511, Promega.
4. DNA Loading Dye, R0611, Fermentas.

2.3 Tissue Digestion for Cell Sorting

1. Collagenase P (1.6 mg/mL), Roche.
2. Hanks' balanced salt solution (HBSS), Roche.
3. 70-μm cell strainers
4. Medium RPMI 1640.
5. Fetal bovine serum (10%), Gibco.
6. L-glutamine (1%), Gibco.
7. Penicillin-streptomycin (1%), Gibco.

3 Methods

3.1 Genotyping Protocol—fCNB—C57BL/6-Ppp3r1^{tm1Stl}/J

1. Cnb1^{loxP} genotyping—PCR mix (Table 1).
2. Cnb1^{loxP} genotyping—PCR cycle setting (Table 2).

3.1.1 Primers

1. oIMR6604 (Forward): CAA CAC CAG AAA CTC TGT CAT G,
2. oIMR6605 (Reverse): GCA CCA CTA TCA TAC TGG TCT.

3.1.2 Expected Results

1. Mutant, 520 bp; heterozygote, 350 bp and 520 bp; wild type, 350 bp.
2. Separated by gel electrophoresis on a 1.5% agarose gel.

3.2 Genotyping Protocol—CD11c^{CRE}

1. CD11c^{CRE} genotyping—PCR mix (Table 3).
2. CD11c^{CRE} genotyping—PCR cycle setting (Table 4).

Table 1
Cnb1^{loxP} genotyping—PCR mix

Component	Stock Conc.	Final Conc.	Volume (μL)
DNA			2
ddH_2O			3.96
dNTP (mM)	2.5	0.2	0.96
$MgCl_2$ (mM)	25	2	0.96
DNA Loading Dye (mM)	5	0.69	1.66
10× PCR Buffer	10×	1	1.2
oIMR6604 (μM)	20	1	0.6
oIMR6605 (μM)	20	1	0.6
Taq Pol (U/μL)	5	0.03	0.06
Total			12

Table 2
Cnb1^{loxP} genotyping—PCR cycle setting

Temp (°C)	Time	
94	3 min	
94	30 s	
60	1 min	35 cycles
72	1 min	
72	2 min	
10	—	

Table 3
CD11c^{CRE} genotyping—PCR mix

Component	Stock Conc.	Final Conc.	Volume (μL)
DNA			2
H_2O			15.45
dNTP (mM)	2.5	0.2	2
$MgCl_2$ (mM)	25	2	2
10× PCR Buffer	10	1	2.5
Cre #1_1084 (μM)	100	1	0.25
Cre #2_1085 (μM)	100	1	0.25
Internal ctrl F′_7338 (μM)	100	1	0.25
Internal ctrl R′_7339 (μM)	100	1	0.25
Taq Pol (U/μL)	5	0.01	0.05
Total			25

Table 4
CD11c^{CRE} genotyping—PCR cycle setting

Temp (°C)	Time	
94	3 min	
94	30 s	
60	1 min	35 cycles
72	1 min	
72	2 min	
4	—	

3.2.1 Primers

1. Cre #1_1084 (Forward): 5′-GCG GTC TGG CAG TAA AAA CTA TC-3′.
2. Cre #2_1085 (Reverse): 5′-GTG AAA CAG CAT TGC TGT CAC TT-3′.
3. Internal ctrl F′_7338 (Forward): 5′-CTA GGC CAC AGA ATT GAA AGA TCT-3′.
4. Internal ctrl R′_7339 (Reverse): 5′-GTA GGT GGA AAT TCT AGC ATC ATC C-3′.

3.2.2 Expected Results

1. Wild type, 324 bp band; transgenic, 100 bp band.
2. Separated by gel electrophoresis on a 1.5% agarose gel (*see* **Note 1**).

3.3 Genotyping Protocol—LysM^{CRE}

1. LysM^{CRE} genotyping—PCR mix (Table 5).
2. LysM^{CRE} genotyping—PCR cycle setting (Table 6).

3.3.1 Primers

1. LC MUT_3066: 5′-CCC AGA AAT GCC AGA TTA CG-3′.
2. LC Common_3067: 5′-CTT GGG CTG CCA GAA TTT CTC-3′.
3. LC WT_3068: 5′-TTA CAG TCG GCC AGG CTG AC-3′.

3.3.2 Expected Results

1. Wild type, 350 bp band; transgenic, 700 bp band.
2. Separated by gel electrophoresis on a 1.5% agarose gel.

3.4 Isolation of Specific Myeloid Cell Subsets

The myeloid cells form the conditional knockout mice show strong downregulation of calcineurin in the cells with expression of the selected promoter (CD11c or LysM) as demonstrated in Fig. 2. Here we provide a protocol to sort the tissue-specific myeloid cells. It is optimized for dendritic cell isolation [1, 2, 4] and shows the KO validation (*see* **Note 2**).

1. Cut tissue (lungs) into small pieces.
2. Digest tissue for 30 min at 37 °C with collagenase P (1.6 mg/mL in HBSS).
3. Pass through 70-μm cell strainers.
4. Wash with RPMI 1640 (supplemented with FBS, L-glutamine, and penicillin-streptomycin).
5. Lyse the red blood cells using ammonium-chloride-potassium lysing buffer.
6. Label single-cell suspensions for flow cytometry or eventually magnetic beads enrichment.

Table 5
LysM^{CRE} genotyping—PCR mix

Component	Stock Conc.	Final Conc.	Volume (μL)
Mutant mix			
DNA			2
H_2O			16.03
dNTP (mM)	2.5	0.2	2
$MgCl_2$ (mM)	25	2	2
10× PCR Buffer	10	1	2.5
LC MUT_3066 (μM)	100	0.5	0.13
LC Common_3067 (μM)	100	1	0.25
Taq Pol (U/μL)	5	0.02	0.1
Total			25
Wild-type mix			
DNA			2
H_2O			11.44
dNTP (mM)	2.5	0.2	2
$MgCl_2$ (mM)	25	2	2
10× PCR Buffer	10	1	2.5
LC Common_3067 (μM)	100	1	0.25
LC WT_3068 (μM)	100	0.04	0.01
Taq Pol (U/μL)	5	0.02	0.1
Total			25

Table 6
LysM^{CRE} genotyping—PCR cycle setting

Temp (°C)	Time	
94	3 min	
94	20 s	
60	30 s	35 cycles
72	50 s	
72	2 min	
4	—	

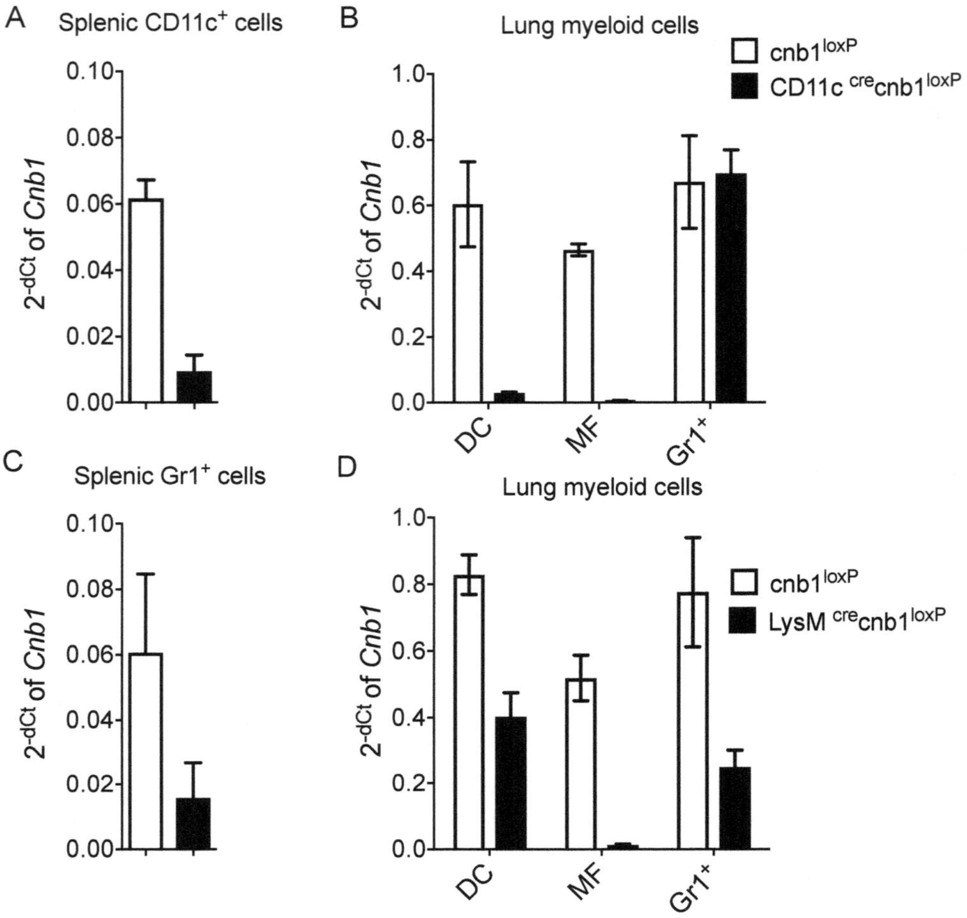

Fig. 2 Validation of calcineurin downregulation in different tissue-specific myeloid cells subsets. Cre/lox calcineurin knockout in CD11c$^+$ and LysM$^+$ (**b**) myeloid cells. Mice lacking calcineurin in CD11c-expressing cells: (**a**, **b**) efficiency of calcineurin deletion determined as the relative change in cnb1 mRNA expression in: (**a**) splenic dendritic cells (DCs) (CD11c$^+$) and (**b**) lung myeloid cells, all sorted from CD11c^{cre}Cnb1^{loxP} mouse. Subsets were sorted as following CD11c$^+$ MHCII$^+$ as DC-like cells, CD11b$^+$CD206$^+$ or SiglecF$^+$ cells as macrophages and CD11b$^+$ Gr1$^+$ myeloid cells. Mice lacking calcineurin in lysozyme M (LysM)-expressing cells: (**c**, **d**) efficiency of calcineurin deletion determined as the relative change in cnb1 mRNA expression in (**c**) bone marrow (BM)-isolated neutrophils (GR-1$^+$) from LysM^{cre}Cnb1^{loxP} mouse and (**d**) sorted lung myeloid cells from LysM^{cre}Cnb1^{loxP} mice

4 Notes

1. This assay will *not* distinguish hemizygous from homozygous transgenic animals.
2. The level of target protein downregulation (calcineurin knockout) always depends on the level of the promoter (CD11c or LysM) in specific cell type; therefore the deficiency of the knock out should by carefully assessed for each cell type.

Acknowledgments

Supported by European Social Fund and European Regional Development Fund, Project MAGNET (No. CZ.02.1.01/0.0/0.0/15_003/0000492), and by the Italian Grant "Programma per Giovani Ricercatori - Rita Levi Montalcini 2013".

References

1. Fric J, Lim CX, Mertes A, Lee BT, Vigano E, Chen J, Zolezzi F, Poidinger M, Larbi A, Strobl H, Zelante T, Ricciardi-Castagnoli P (2014) Calcium and calcineurin-NFAT signaling regulate granulocyte-monocyte progenitor cell cycle via Flt3-L. Stem Cells 32:3232–3244
2. Zelante T, Wong AY, Ping TJ, Chen J, Sumatoh HR, Vigano E, Hong Bing Y, Lee B, Zolezzi F, Fric J, Newell EW, Mortellaro A, Poidinger M, Puccetti P, Ricciardi-Castagnoli P (2015) CD103(+) dendritic cells control Th17 cell function in the lung. Cell Rep 12:1789–1801
3. Herbst S, Shah A, Mazon Moya M, Marzola V, Jensen B, Reed A, Birrell MA, Saijo S, Mostowy S, Shaunak S, Armstrong-James D (2015) Phagocytosis-dependent activation of a TLR9-BTK-calcineurin-NFAT pathway co-ordinates innate immunity to Aspergillus fumigatus. EMBO Mol Med 7:240–258
4. Zelante T, Wong AY, Mencarelli A, Foo S, Zolezzi F, Lee B, Poidinger M, Ricciardi-Castagnoli P, Fric J (2017) Impaired calcineurin signaling in myeloid cells results in downregulation of pentraxin-3 and increased susceptibility to aspergillosis. Mucosal Immunol 10:470–480
5. Zeng H, Chattarji S, Barbarosie M, Rondi-Reig L, Philpot BD, Miyakawa T, Bear MF, Tonegawa S (2001) Forebrain-specific calcineurin knockout selectively impairs bidirectional synaptic plasticity and working/episodic-like memory. Cell 107:617–629
6. Neilson JR, Winslow MM, Hur EM, Crabtree GR (2004) Calcineurin B1 is essential for positive but not negative selection during thymocyte development. Immunity 20:255–266
7. Caton ML, Smith-Raska MR, Reizis B (2007) Notch-RBP-J signaling controls the homeostasis of CD8- dendritic cells in the spleen. J Exp Med 204:1653–1664
8. Clausen BE, Burkhardt C, Reith W, Renkawitz R, Forster I (1999) Conditional gene targeting in macrophages and granulocytes using LysMcre mice. Transgenic Res 8:265–277

Chapter 15

Calpain Purification Through Calpastatin and Calcium: Strategy and Procedures

Hung Huy Nguyen, Peter Tompa, and Kris Pauwels

Abstract

This chapter describes the strategy and procedures for the calcium-mediated affinity purification of calpain. The affinity capture method exploits the reversible binding properties of calpain's intrinsically disordered protein (IDP) inhibitor, calpastatin. IDPs are easily produced in heterologous expression hosts and purified to homogeneity. Combining these properties with in vivo biotinylation leads to a simplified purification strategy whereby biotinylated human calpastatin domain 1 (hCSD1) can capture calpain efficiently from a complex biological mixture with only a single chromatographic step and in a considerably reduced time. Our approach is generally applicable through the in vivo biotinylation of any IDP of interest in order to capture its binding partner in a calcium- and chelator-based protocol.

Key words Intrinsically disordered proteins, IDPs, Affinity chromatography, Calpain-2, Human m-calpain, In vivo biotinylation, AviTag, Chemical biotinylation, BirA co-expression, Affinity capture

1 Introduction

Human m-calpain is an intracellular cysteine-protease that irreversibly modulates the activity of many proteins [1]. Its activity is tightly controlled by calcium and by an endogenous intrinsically disordered protein (IDP) inhibitor called calpastatin. The role of calpain activity in muscular dystrophies, wound healing, cancer, neurodegeneration, and inflammation makes it an appealing protein for structural/functional studies and drug development [2]. Despite active research for several decades, an in-depth in vitro biophysical characterization of calpain (in particular the human isoform) has been hampered by complicated, multistep, and labor-intensive purification protocols. Sample purity is crucial since heterogeneity due to poor sample preparation is often catastrophic for profound molecular and biophysical characterization.

In this chapter, we describe a strategy that exploits the reversible calcium-dependent binding properties of the IDP human

Claus W. Heizmann (ed.), *Calcium-Binding Proteins of the EF-Hand Superfamily: From Basics to Medical Applications*, Methods in Molecular Biology, vol. 1929, https://doi.org/10.1007/978-1-4939-9030-6_15,

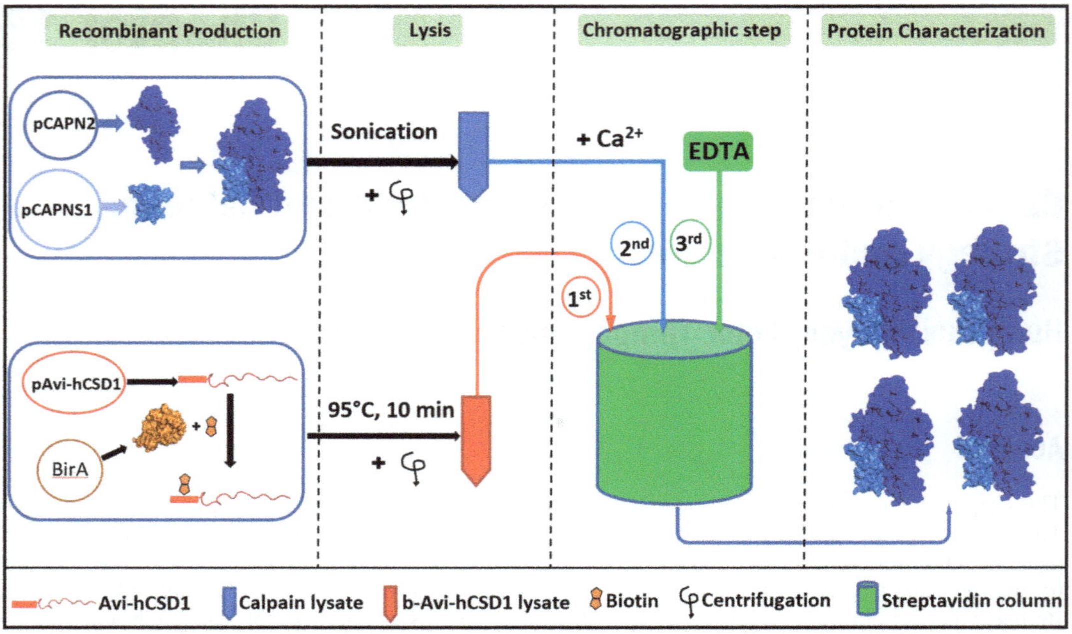

Fig. 1 Schematic representation of the purification workflow of the calcium-mediated affinity purification of calpain based on the recombinant co-expression of Avi-tagged hCSD1 and the biotin-ligase BirA that covalently attaches biotin to Avi-hCSD1. The b-Avi-hCSD1 is compatible with heat treatment typically employed for IDP purification and can be immobilized directly onto the streptavidin-based resin without any further manipulation. This immobilized b-Avi-hCSD1 allows to capture recombinantly produced human m-calpain from the bacterial lysate (sonicated, centrifuged, and supplemented with calcium). The complex between calpain and b-Avi-hCSD1 is dissociated by applying EDTA as a third step onto the column, leading to the elution of calpain

calpastatin domain 1 (hCSD1), to isolate human m-calpain and selected variants thereof [3, 4] (Fig. 1). This affinity capture methodology markedly improved the production yield and simplified the traditional multistep purification procedure [5–7] to a single chromatographic step. The method is based upon the ease of calpastatin purification that exploits thermal denaturation treatment as a very efficient cell lysis step after the bacterial fermentation. In addition, we are the first to present the expression and purification of an in vivo biotinylated IDP [4]. Therefore, this chapter showcases a unique example of the effective “affinity capture” of an enzyme of interest, and it provides detailed and technically sound information to improve purification strategies whereby the reversible protein-protein interaction is calcium-mediated. Altogether, we have established an elegant framework for the efficient purification of calpain. Beyond that, this approach is widely applicable for other IDP-based and calcium-mediated affinity purifications.

2 Materials

Chemical substances were of highest purity, unless otherwise indicated.

1. The expression plasmids of the wild-type (WT) 80 kDa human m-calpain large subunit (CAPN2 in pET24b(+)) and the truncated 23 kDa small subunit (CAPNS1ΔG2 in pACpET) are generously provided by Dr. Hiroyuki Sorimachi (Tokyo Metropolitan Institute of Medical Science, Japan) [8].
2. The pT7pol26 is an IPTG-inducible expression vector to produce T7 RNA polymerase [9] and is a generous gift of Dr. Henri De Greve (Vrije Universiteit Brussel, Belgium) (*see* **Note 1**).
3. The Avi-tagged version of hCSD1 (corresponding to region A137-K277 of the human calpastatin amino acid sequence) (*see* **Note 2**) is designed in silico and obtained through a commercial gene synthesis service (GenScript). The genetic construct design is done using CLC workbench v7 for reverse translation and codon optimization by frequency distribution for *Escherichia coli*. This gene is subcloned into pET22b vector using the NdeI and EcoRI restriction sites.
4. *Escherichia coli* BL21(DE3) (Invitrogen).
5. The *E. coli* AVB100 strain (Avidity LLC) is an *E. coli* K12-derived strain [*MC1061 araD139 δ(ara-leu)7696 δ(lac) l74 galU galK hsdR2(rK-mK+) mcrB1 rpsL(StrR)*], whereby the *birA* gene (encoding for the biotin ligase) is integrated into the bacterial chromosome with an arabinose-inducible promoter (*see* **Note 3**).
6. *E. coli* AVB100 containing pT7pol23.
7. *E. coli* AVB100 containing pT7pol23 and pAvi-hCSD1.
8. *E. coli* BL21(DE3) containing both the CAPN2 gene in pET24b(+) and the CAPNS1ΔG2 gene in pACpET.
9. Lysogeny broth (LB): add 25 g of commercially available powder (Duchefa Biochemie) to 1 L water and autoclave.
10. Terrific broth (TB): add 100 mL of a sterilized solution of 0.17 M KH_2PO_4 and 0.72 M K_2HPO_4 to 900 mL sterilized solution containing 12 g tryptone, 24 g yeast extract, and 4 mL glycerol.
11. Petri plates.
12. Ampicillin: dissolve 1.0 g in 10 mL water, filter sterilize, aliquot, and store at −20 °C.
13. Kanamycin: dissolve 0.5 g in 10 mL water, filter sterilize, aliquot, and store at −20 °C.

14. LB agar plates: add 35 g LB broth with agar (Sigma) into 1 L water, sterilize by autoclaving, cool down sufficiently prior to the addition of the appropriate antibiotics, and pour into petri plates to solidify.
15. LB agar plates supplemented with 100 μg/mL ampicillin.
16. LB agar plates supplemented with 50 μg/mL kanamycin.
17. LB agar plates supplemented with 100 μg/mL ampicillin and 50 μg/mL kanamycin.
18. 0.1 M $CaCl_2$: add 11.1 g $CaCl_2$ to 100 mL water, filter sterilize, and cool on ice or store at 4 °C.
19. 0.1 M $CaCl_2$, 15% glycerol: dissolve 11.1 g in 85 mL water, add 15 mL glycerol, mix well, filter sterilize, and cool on ice or store at 4 °C.
20. Isopropyl β-D-1-thiogalactopyranoside (IPTG; Duchefa Biochemie): dissolve 4.77 g in 20 mL water and filter-sterilize with 0.22 μm sterile filters. Aliquot this 1 M IPTG stock solution into microtubes and store at −20 °C until use.
21. L-(+)-arabinose: dissolve 20 g in 100 mL water, filter-sterilize, and store at −20 °C until use.
22. Protease inhibitor cocktail (Roche Diagnostics; one tablet per 50 mL lysis buffer solution) (*see* **Note 4**).
23. Tris(hydroxymethyl)aminomethane (Tris; trizma base), pH-adjusted with HCl (*see* **Note 5**).
24. 3-(N-morpholino)propanesulfonic acid (MOPS), pH-adjusted with NaOH (*see* **Note 5**).
25. Tris(2-carboxyethyl)phosphine (TCEP).
26. Dithiothreitol (DTT).
27. Ethylenediaminetetraacetic acid (EDTA).
28. NaCl.
29. $CaCl_2$ is used to mediate the capture of calpain via calpain-calpastatin interaction.
30. Calpastatin lysis buffer (buffer A): 50 mM Tris–HCl, pH 7.5, 0.1 mM EDTA, 2 mM DTT, cocktail inhibitor tablets.
31. Calpain lysis buffer (buffer B): 20 mM MOPS, pH 7.5, 1 mM TCEP, 150 mM NaCl, cocktail inhibitor tablets (*see* **Notes 4** and **6**).
32. Calpain binding buffer (buffer C): 20 mM MOPS, pH 7.5, 1 mM TCEP, 150 mM NaCl, 5 mM $CaCl_2$ (*see* **Note 7**).
33. Calpain elution buffer (buffer D): 20 mM MOPS, pH 7.5, 1 mM TCEP, 10 mM EDTA.
34. Calpain desalting buffer (buffer E): 20 mM MOPS, pH 7.5, 1 mM TCEP.

35. Streptavidin HP 1 mL column (GE Healthcare).
36. HiTrap Desalting 26/10 column (GE Healthcare).
37. Minitron orbital shaker-incubator (InFors HT).
38. Centrifuge.
39. Sterile 1.5 mL microtubes (Eppendorf).
40. Sonicator: 130Watt Ultrasonic Processor (Sonic & Materials, Inc.).
41. Thermostated water bath (Haake SC100 water bath, Thermo Scientific).
42. 0.22 μm cutoff polyethersulfone membrane filter (Sarstedt).
43. 50 mL syringes.
44. High-performance liquid chromatography system (e.g., Akta Avant, GE Healthcare).
45. PAGE-Blue staining (Thermo Scientific).

3 Methods

3.1 Generation of Chemically Competent Cells

1. Grow the bacterial cells in 50 mL of LB supplemented with appropriate antibiotics at 37 °C until the optical density at 600 nm reaches 0.6.
2. Centrifuge the bacterial culture at 5000 × *g* and 4 °C for 10 min and discard the supernatant.
3. Resuspend the bacterial pellet in 5 mL of 0.1 M ice-cold $CaCl_2$, vortex to dissolve the pellets completely, and incubate on ice for 1 h.
4. Centrifuge the resuspended cells at 5000 × *g* for 10 min, and discard the supernatant.
5. Resuspend the pellets in 1 mL of ice-cold 0.1 M $CaCl_2$ containing 15% glycerol without vortexing, and incubate on ice for 20 min.
6. Make aliquots of 50 μL of the resuspended cells into sterile microtubes and store at −80 °C until use.

3.2 Plasmid Transformation

1. Thaw 50 μL of chemically competent cells for 10 min by placing the tube on ice.
2. Add 20 ng of plasmid DNA to the competent cells and incubate for 30 min on ice.
3. Perform a heat shock during 45 s by incubating the tube in a thermostated water bath at 42 °C.
4. After the heat shock, put the transformation vial immediately on ice for 2 min.

5. Add 500 μL of LB into the vial, and incubate the cells for 1 h at 37 °C with vigorous shaking to allow for phenotypic expression.
6. Plate out the transformants onto a LB agar plate supplemented with the appropriate antibiotics, and grow for 16–24 h at 37 °C in an incubator.
7. Upon successful transformation, use a single colony for the next steps of recombinant expression.

3.3 Transformation and Expression of In Vivo Biotinylated avi-hCSD1

1. Transform the pT7pol26 into chemically competent *E. coli* AVB100 cells (*see* Subheadings 3.1 and 3.2) (*see* **Notes 1** and **3**).
2. Select the successful transformants on a LB agar plate supplemented with 50 μg/mL kanamycin.
3. Transform the pAvi-hCSD1 into chemically competent *E. coli* AVB100 cells that already contain the pT7pol26 (chemically competent cells are generated using the protocol in Subheading 3.1 and the transformation procedure is described in Subheading 3.2) (*see* **Note 8**).
4. Select the successful transformants on the LB-agar plate supplemented with 50 μg/mL kanamycin and 100 μg/mL ampicillin.
5. Inoculate 20 mL of LB that contains 50 μg/mL kanamycin and 100 μg/mL ampicillin with a single colony of those *E. coli* AVB100 containing pT7pol23 and pAvi-hCSD1 and grow overnight at 37 °C while shaking: this will function as a pre-culture for the large-scale expression of the in vivo biotinylated Avi-hCSD1 (b-Avi-hCSD1).
6. Inoculate 20 mL of pre-culture into 1 L of LB medium supplemented with D-biotin at a final concentration of 50 μM (*see* **Note 9**).
7. Grow the cells at 37 °C and 180 rpm.
8. Add 0.4% L-(+)-arabinose (final concentration) to the bacterial culture when the OD_{600} reaches 0.4 to induce expression of the BirA biotin ligase, and continue the bacterial fermentation at 37 °C.
9. When the OD_{600} reaches 0.6, add 1 mM IPTG (final concentration) to induce the expression of Avi-hCSD1 and continue the bacterial fermentation at 37 °C for 5 h.
10. Collect the cell culture after 5 h of induced protein expression by centrifugation at 5000 × *g* for 15 min, discard the supernatant, and store the cell pellet at −20 °C (*see* **Notes 10** and **11**).

3.4 Expression of Human m-Calpain

1. Use *E. coli* BL21(DE3), which harbors both the CAPN2 gene in pET24b(+) plasmid and the CAPNS1ΔG2 gene in pACpET to co-express the large and small subunit of WT human m-calpain [3, 4].
2. Prepare a preculture of 20 mL LB supplemented with 50 μg/mL kanamycin and 100 μg/mL ampicillin.
3. Add the 20 mL preculture to 1 L TB supplemented with 50 μg/mL kanamycin and 100 μg/mL ampicillin, and grow the bacterial cells at 27 °C while shaking at 180 rpm in an orbital shaker-incubator.
4. Induce the protein expression by the addition of 0.2 mM IPTG when the OD_{600} reaches 0.6, and lower the temperature to 16 °C. Continue the bacterial growth for 16–20 h at 16 °C.
5. Collect the cells by centrifuging the cell culture at 5000 × *g* for 15 min, discard the supernatant, and store the cell pellet at −80 °C.

3.5 Preparation of b-Avi-hCSD1 and Human m-Calpain Cleared Lysates

1. Thaw and resuspend the biotinylated Avi-hCSD1 (b-Avi-hCSD1) containing cell pellet (originating from 1 L growth culture) in 30 mL of buffer A by vortexing and heat that sample at 90 °C for 15 min in a water bath (*see* **Note 12**).
2. Thaw and resuspend the cell pellet containing human m-calpain (originating from 1 L of growth culture) in 60 mL of buffer B, followed by a pulse sonication (2 s on, 5 s off) for 1 min (60% amplitude) using a tip sonicator while keeping the sample on ice (*see* **Note 6**).
3. Centrifuge the above described cell lysates at 20,000 × *g* and at 4 °C for 45 min to separate the soluble and insoluble fraction. Discard the pellets and retain the supernatants, which constitute the cleared lysates that contain the proteins of interest.
4. Filter the cleared lysates through a 0.22 μm cutoff polyethersulfone membrane using a 50 mL syringe.

3.6 Affinity Purification of Human m-Calpain

1. It is recommended to use a high-performance liquid chromatography system for the purification to allow you to control multiple parameters (i.e., UV absorbance at different wavelengths, conductivity, etc.) (*see* **Note 6**).
2. Apply the b-Avi-hCSD1-containing lysate onto a HiTrap Streptavidin HP 1 mL column that was equilibrated with buffer C (*see* **Notes 12** and **13**).
3. After loading the b-Avi-hCSD1 onto the resin, wash the column with five column volumes of buffer C.
4. Add 5 mM $CaCl_2$ (final concentration) to the calpain-containing lysate, mix gently but thoroughly by pipetting up

and down, and load this sample immediately onto the column where b-Avi-hCSD1 is immobilized (*see* **Notes 7**, **14**, and **15**).

5. Upon loading the calpain-containing lysate, wash the resin with buffer C for five column volumes.
6. Elute calpain with five column volumes of buffer D and collect the elution peak.
7. Collect the peak fractions of the eluting proteins, pool them, and subject the purified calpain sample to a desalting step using a HiTrap Desalting 26/10 column equilibrated with buffer E. Evaluate the purity of the protein using sodium dodecyl sulfate polyacrylamide gel electrophoresis (SDS-PAGE) with PAGEBlue staining (*see* **Note 16**).
8. Flash-freeze the pure enzyme with liquid nitrogen and stored at −80 °C (*see* **Note 17**).

4 Notes

1. We do not recommend to use the pT7pol23, which is a heat-inducible version of pT7pol26, because of leaky expression and the insolubility of the BirA at elevated temperatures [4, 9, 10].
2. The choice of the affinity tag for the subsequent immobilization is important. The high-affinity biotin-streptavidin interaction is excellent to immobilize IDPs to probe biomolecular interaction kinetics (e.g., by bio-layer interferometry (BLI)). We found that the AviTag in combination with in vivo biotinylation is a great choice, based on our observations with BLI when analyzing the interaction between calpain and hCSD1 [4]. In this context, it is worth to mention that in our own experience, immobilizing calpain as analyte with the intrinsically disordered hCSD1 as the free ligand in solution does not yield a binding signal when performing BLI experiments. This observation is generally true for IDPs used in our lab (unpublished results), so immobilizing the IDP as the analyte and detecting binding of a well-folded protein as the ligand is the best strategy to apply. Other polypeptide tags can also be used, such as glutathione-S-transferase (GST), chitin-binding protein (CBP), or maltose-binding protein (MBP), although their oligomeric state will complicate the BLI data analysis due to avidity effects. From our previous experience [3], we have learned that the GST-tag for hCSD1 immobilization (as fusion protein) is a bad choice, since GST elutes from the GSTrap column (GE Healthcare) with EDTA (resulting in the co-elution of GST-hCSD1 and calpain rather than eluting the enzyme alone). Unexpectedly, we also observed that GST alone (without fusion with hCSD1) can interact with Ni^{2+}-Sepharose

(HisTrap column). We could rationalize this observation based on the crystal structure of GST (1UA5.pdb) where a histidine-rich patch on the GST surface can be identified. GST did not adsorb simply in a non-specific manner, since it eluted with EDTA. To prevent this type of binding, more specific or stringent binding conditions should be identified (inclusion of EDTA, imidazole, altered salt concentrations, etc.).

3. There is a commercially available AVB101 strain, a variant of AVB100, whereby a BirA expression plasmid is present instead of integrating the *birA* gene into the bacterial chromosome. The pBirA is derived from pECYC184, which has a p15A ori and contains chloramphenicol-resistant gene. The compatibility of those expression plasmids is not a problem since pET22b (phCSD1 and pavi-hCSD1) has a F1 ori and ampicillin resistance gene, where pT7pol26 has a pSC101 ori and kanamycin resistance gene. This strategy requires the parallel use of multiple antibiotics.
4. The addition of protease inhibitor cocktail does not influence the calpain protease activity.
5. Our preferred buffer for calpain-containing samples is MOPS as it has low ionization enthalpy, and it is compatible with calcium [11]. Tris buffer can be used, yet it may not be suitable for biophysical studies because of its high ionization enthalpy [11]. Phosphate buffer should be avoided because it is incompatible with the use of calcium (i.e., formation of the insoluble calcium phosphate).
6. All steps that involve calpain manipulation should be performed at 4 °C, and the protein sample should be kept on ice whenever possible. We have performed all the chromatographic steps at 4 °C using an Akta Avant system that is operating in a cold cabinet.
7. Upon addition of calcium, it is recommended to process the calcium-activated calpain sample as soon as possible due to its autolysis activity. The active calpain purified by the method described in this protocol is approximately 20% less active than the one purified by traditional method, where calcium is not employed during the purification processes [3].
8. We have tried to co-transform both plasmids simultaneously into chemically competent *E. coli* AVB100 cells: these attempts were unsuccessful since no bacterial colonies were observed on the LB agar plates that are supplemented with kanamycin and ampicillin. Therefore, we have applied the strategy that consists of two subsequent transformations (*see* Subhaeding 3.2).
9. Bacterial expression of hCSD1 (or tagged versions thereof) can be performed in minimal medium for, e.g., ^{15}N-isotope labeling. The subsequent purification (after recombinant expression

in minimal medium) did not reveal any additional experimental obstacles, e.g., degradation or presence of contaminants [4].

10. It is desirable to perform an expression test by taking an aliquot of the bacterial fermentation broth prior to the addition of the inducing agent: by comparing equal amounts of cells via an SDS-PAGE, the appearance of a band of the protein-of-interest can be visualized when the expression is induced successfully (*see* reference [4]).
11. hCSD1 has an aberrant electrophoretic mobility because of the bias in its amino acid composition [12, 13]. Consequently, it runs at a higher apparent molecular weight (~23 kDa) compared to its actual mass of 14,764 Da [4]. Multi-biotinylated hCSD1 (upon chemical biotinylation of hCSD1) results in a more pronounced and visible shift of electrophoretic mobility in SDS-PAGE, whereas the shift of a mono-biotinylated Avi-hCSD1 is hardly noticeable on a 12% SDS-PAGE.
12. We identified an endogenous bacterial protein, biotin carboxyl carrier protein (BCCP), that was also biotinylated along with b-Avi-hCSD1. Hence, this protein might occupy streptavidin binding sites and reduce the binding capacity of the column for the biotinylated IDP. We found that this BCCP can simply be removed by heating the b-Avi-hCSD1-containing lysate (which induces the aggregation of BCCP). Another contaminant consistently detected in the hCSD1-containing lysate was flagellin. This protein is heat resistant, and hence it remains in the bacterial lysate. However, it does not have a negative effect on the purification as it does not bind to the Streptavidin column during b-Avi-hCSD1 immobilization. This contaminant is only observed when using the *E. coli* K12 strain, while it is not observed in the case of the BL21(DE3) strain.
13. As an alternative biotinylation method, in vitro chemically biotinylated hCSD1 can be immobilized onto the Streptavidin column, instead of b-Avi-hCSD1. The hCSD1 can be easily expressed in the *E. coli* BL21-AI strain and purified by boiling the cell suspension, followed by clearance by centrifugation and a single anion-exchange chromatography step. The chemical biotinylation is routinely done using EZ-link NHS-biotin (*see* reference [3] for experimental details). We found that chemical biotinylation influences the binding properties of hCSD1 due to the heterogeneity of the modification of the primary amines of lysines or the amino-terminus of the polypeptides (which affects the reproducibility of the in vitro biotinylation) [4]. Due to reduced functionality, the chemically biotinylated hCSD1 (which requires additional production steps) can lead to a reduced yield of calpain purification. In contrast, in vivo biotinylation by BirA (*see* Subheading 3.3)

occurs specifically at the primary amine of the lysine side chain in the AviTag sequence, and we could not detect any adverse effect to the b-Avi-hCSD1 binding properties.

14. In case of the enzymatically inactive C105A mutant of calpain, the addition of calcium can lead to aggregation/precipitation and even subunit disassociation [14]. To avoid aggregation, 150–200 mM NaCl (final concentration) should be added to the buffer.
15. This purification protocol has been applied successfully to purify recombinantly expressed human m-calpain both in its active and inactive mutant forms and also for the purification of individual calpain domains [3]. We are confident that it can also be used to isolate other calpain isoforms from any crude biological mixture that contains calpain (recombinantly expressed in cells, as well as from tissue or blood, etc.) [15]. This would enable to address the type and effect of posttranslational modifications of calpain. However, it is plausible that due to the binding promiscuity of IDPs [13], this could lead to the copurification of multiple other proteins that can bind to calpastatin. By using different calpastatin isoforms or domains, such tailor-made purification strategy could be optimized.
16. After elution from the streptavidin column, the protein sample should not be stored as such (i.e., in elution buffer) because we have observed the loss of enzymatic activity. To preserve its activity, it is necessary to perform a buffer exchange to the designated buffer by a desalting step, and store it immediately at −80 °C after flash freezing into liquid nitrogen.
17. Repeated freezing and thawing should be avoided due to a reported loss in enzyme activity [15].

Acknowledgments

The authors thank Ivo Van Overstraeten for excellent technical assistance. This work was supported in part by a pilot grant of Stichting Alzheimer Onderzoek (P#12017), and K.P. is the recipient of a FWO long-term postdoctoral fellowship (#1218713).

References

1. Ono Y, Sorimachi H (2012) Calpains: an elaborate proteolytic system. Biochim Biophys Acta 1824:224–236. https://doi.org/10.1016/j.bbapap.2011.08.005
2. Ono Y, Saido TC, Sorimachi H (2016) Calpain research for drug discovery: challenges and potential. Nat Rev Drug Discov 15:854–876. https://doi.org/10.1038/nrd.2016.212
3. Nguyen HH, Varadi M, Tompa P, Pauwels K (2017) Affinity purification of human m-calpain through an intrinsically disordered inhibitor, calpastatin. PLoS One 12: e0174125. https://doi.org/10.1371/journal.pone.0174125
4. Nguyen HH, Volkov AN, Vandenbussche G et al (2018) In vivo biotinylated calpastatin

improves the affinity purification of human m-calpain. Protein Expr Purif 145:77–84. https://doi.org/10.1016/j.pep.2018.01.002

5. Hata S, Ueno M, Kitamura F, Sorimachi H (2012) Efficient expression and purification of recombinant human m-calpain using an Escherichia coli expression system at low temperature. J Biochem 151:417–422. https://doi.org/10.1093/jb/mvs002
6. Elce JS (2000) Calpain methods and protocols. Humana Press, Totowa, NJ
7. Anagli J, Vilei EM, Molinari M et al (1996) Purification of active calpain by affinity chromatography on an immobilized peptide inhibitor. Eur J Biochem 241:948–954
8. Yang HQ, Ma H, Takano E et al (1994) Analysis of calcium-dependent interaction between amino-terminal conserved region of calpastatin functional domain and calmodulin-like domain of mu-calpain large subunit. J Biol Chem 269:18977–18984
9. Mertens N, Remaut E, Fiers W (1995) Versatile, multi-featured plasmids for high-level expression of heterologous genes in Escherichia coli: overproduction of human and murine cytokines. Gene 164:9–15. https://doi.org/10.1016/0378-1119(95)00505-Z
10. Li Y, Sousa R (2012) Expression and purification of E. coli BirA biotin ligase for in vitro biotinylation. Protein Expr Purif 82:162–167. https://doi.org/10.1016/j.pep.2011.12.008
11. Goldberg RN, Kishore N, Lennen RM (2002) Thermodynamic quantities for the ionization reactions of buffers. J Phys Chem Ref Data 31:231–370. https://doi.org/10.1063/1.1416902
12. Tompa P (2002) Intrinsically unstructured proteins. Trends Biochem Sci 27:527–533
13. Uversky VN (2011) Intrinsically disordered proteins from A to Z. Int J Biochem Cell Biol 43:1090–1103. https://doi.org/10.1016/j.biocel.2011.04.001
14. Thompson VF, Goll DE (2000) Purification of μ-calpain, m-calpain, and calpastatin from animal tissues. In: Methods in molecular biology. Humana Press, Totowa, NJ, pp 3–16
15. Tantos A, Friedrich P, Tompa P (2009) Cold stability of intrinsically disordered proteins. FEBS Lett 583:465–469. https://doi.org/10.1016/j.febslet.2008.12.054

Chapter 16

Characterization of the EF-Hand Calcium-Binding Domains of Human Plastins

Oleg Khassan, Katharine V. Jensen, Andrew G. Woodman, Hans J. Vogel, and Hiroaki Ishida

Abstract

The three human plastins (L-plastin, T-plastin, and I-plastin) are important regulatory Ca^{2+}-binding proteins that belong to the family of actin-binding proteins. Plastins are involved in the regulation of the actin cytoskeleton as well as the cross-linking of actin filaments. In addition to four calponin-homology (CH) domains, all three plastins contain two N-terminal EF-hand Ca^{2+}-binding motifs which together are homologous to a single lobe of the well-known calcium-regulatory protein calmodulin. This part of the protein allows for the regulation of the actin bundling activity in response to elevated calcium levels. In this protocol, we describe the purification of the EF-hand headpiece domains of all three plastins, as well as SPR studies, ITC studies, and NMR interaction studies with different peptides and calcium. In combination, these three experimental techniques provide detailed insights into a novel regulatory mechanism, involving the linker region between the EF-hand domain and the first CH domain of the plastins.

Key words L-plastin, T-plastin, I-plastin, Fimbrin, Calcium-binding domain, Actin bundling, EF-hand motifs, NMR spectroscopy, SPR, ITC

1 Introduction

Plastins are unique multi-domain proteins of ~70 kDa, and all mammals contain three paralogs (L-plastin, T-plastin, and I-plastin) that are expressed in a cell-type-specific manner [1, 2]. In humans, L-plastin (LPL, plastin 2, PLS2) is normally exclusively expressed in hematopoietic cells, T-plastin (TPL, plastin 3, PLS3) is expressed in solid tissues, and I-plastin (IPL, plastin 1, PLS1) is expressed in the small intestine, colon, and kidneys [3]. These proteins share the property of being able to bind Ca^{2+}, regulate actin cytoskeleton, and thereby cross-link actin filaments to give rise to tight bundling of actin [4, 5]. The localized expression of these three isoforms of plastin allows for regulation in a tissue-specific manner due to the distinct properties of each isoform.

Claus W. Heizmann (ed.), *Calcium-Binding Proteins of the EF-Hand Superfamily: From Basics to Medical Applications*, Methods in Molecular Biology, vol. 1929, https://doi.org/10.1007/978-1-4939-9030-6_16,

L-plastin is normally only expressed in hematopoietic cell lineages where it plays an important role in the regulation of white blood cells, such as the T-cells and the B-cells, which play an important role in the immune response [6, 7]. L-plastin is also known as lymphocyte cytosolic protein 1 (LCP1) because of its abundance in white blood cells. Deletion of the L-plastin gene in mouse models or zebra fish leads to an impaired immune response [8–10]. Moreover, neutrophils that are devoid of L-plastin display reduced motility and an inability to kill bacterial pathogens [11]. L-plastin has also been found to be highly expressed in nonhematopoietic cells that have transformed into metastatic cancerous cells. It has been shown by RT-PCR that activation of LPL is increased in most cancer cells in humans [12]. LPL is also expressed in 68% of cancers derived from epithelia and 53% of cancers derived from nonepithelial origin [13]. In these cells, L-plastin expression gives rise to increased cell motility, and it promotes invasiveness [14–17]. Inhibition of L-plastin in PC3 prostate cancer cells through co-expression with camelid antibodies blocks invasion [18], as does inhibition by expression of an antisense gene [19]. Similar effects are seen in other cancers and transformed cell lines [20, 21], suggesting that inhibition of L-plastin could have beneficial anti-metastatic effects, by blocking the invasiveness of circulating cancer cells. LPL has also been recognized as a crucial biomarker in efforts of early detection of various forms of cancer, such as choroid plexus tumors [22], colon cancer [23], and urinary bladder cancer [24]. Moreover, it can be used as a local inflammatory marker for periodontitis [25], highlighting the many important functions for this protein.

T-plastin, when highly expressed in neurons, can act as a protective modifier of spinal muscular atrophy which is the most frequent genetic cause of lethality seen in early childhood. The overexpression of TPL can rescue the length of axons and outgrowth defects in neurons of an SMA mouse [26]. In addition, previous studies have demonstrated that TPL has the ability to lessen the toxicity of polyglutamine proteins in neurons, such as ataxin and huntingtin, which can cause neurodegenerative disorders [27]. T-plastin has also been shown to play a role in a human bone health, as genetic variants of the protein can give rise to osteoporosis and fractures [28, 29]. While the molecular level mechanisms for this process remain unknown, these findings indicate the need for further studies of T-plastin. Interestingly, TPL can be used as an important marker to differentiate mycosis fungoides (MF) and Sézary syndrome (SS) which are both the most common forms of cutaneous T-cell lymphoma. The expression of TPL in SS is significantly higher than in MF without atypical cells. TPL has also been found to be upregulated in UV radiation-resistant cells [30]. These UV radiation-resistant cells have a higher capacity to repair DNA damage [31], and TPL may thus be directly or indirectly involved in DNA repair as its expression is upregulated. T-plastin has also been implicated in other forms of cancer. For example, T-plastin is a marker for circulating cancer cells, and it has been found to play a role in breast cancer and in colorectal cancer [32, 33], where it may predict long-term prognosis. Hence, T-plastin is a vital protein in the regulation of various physiological conditions in the human body.

To date, I-plastin has not been studied as much as the other two paralogs. It was found to be expressed in the small intestine, colon, and kidney in humans and has been studied in part by using mice deficient of the I-plastin gene. These studies have shown that mice lacking I-plastin had decreased transepithelial resistance, increased intestinal epithelium fragility, and elevated sensitivity to dextran sodium sulfate-induced colitis [34]. This is important as it highlights the role of I-plastin as a regulator of the morphology of the intestinal brush border and its stability. It has been reported that I-plastin may also play a role in hearing loss, as the protein is required to form proper stereocilia [35, 36].

The plastin proteins, also known as fimbrins in other organisms [37], have a rather peculiar domain structure with four ~110-residue calponin homology (CH) domains (Fig. 1). Two tandem CH domains together form an actin-binding domain (ABD) [38], and hence plastins can bind to two F-actin filaments simultaneously, thereby causing actin bundling. In white blood cells, this process leads to the formation of filopodia, long protruding thin cell extensions that play a major role in cell migration and motility [39]. In this process, plastin works together with fascin, another protein that promotes actin bundling [40]. In cancer cells, these extended filopodia structures are often referred to as invadopodia. The crystal structures for the ABD regions of plant and yeast fimbrins have been determined, revealing a horseshoe-shaped structure and two separate interfaces for binding to actin [41]. All mammalian plastins include two N-terminal EF-hand Ca^{2+}-binding motifs which are homologous (34.7% identical) to a single lobe of the well-known calcium-regulatory protein calmodulin (CaM) [4]. The EF-hand domains are critical for the calcium-regulated actin bundling

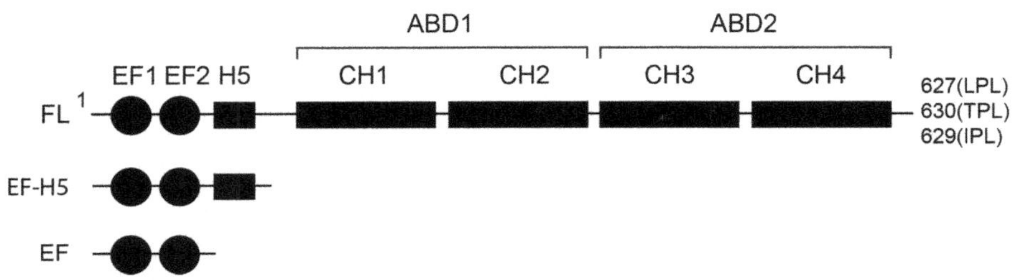

Fig. 1 Schematics of the human plastins and their constructs discussed in here. Note the location of the regulatory H5 region in the linker region between EF2 and CH1. (FL means full length protein)

activity of plastins [42]. Structural information for the EF-hand domain region of L-plastin has recently become available [43, 44]. Unexpectedly, the calcium-saturated EF-hand domain was shown to bind directly to the linker region (helix 5 or switch helix) between the headpiece and the first CH domain. In the apoprotein, this region is released from the EF-hand domain and becomes flexible, likely allowing it to bind to actin. Although the multi-domain structure of plastins is conserved throughout evolution in all eukaryotes [1], it should be noted that the EF-hand motifs in plant and yeast, while still present, can no longer bind calcium due to substitutions in the EF-hand calcium-binding loops [1] (unpublished observations). In contrast, all mammalian plastin EF-hand motifs, as well as the protein from the amoeba *Dictyostelium discoideum* [45], bind calcium and can regulate actin bundling in this manner.

The aim of studying the calcium-binding EF-hand domains of the three different plastins was to better understand their interactions with their individual binding partners by three biophysical methods: surface plasmon resonance (SPR), isothermal titration calorimetry (ITC), and nuclear magnetic resonance (NMR) spectroscopy. SPR has the unique advantage of analyzing biomolecular interactions without the requirement for a tag or a label, and it produces kinetic information such as the on- and off-rates, which allows for estimation of the dissociation constants (K_d). ITC also looks at ligand binding but has an additional benefit of accurately determining the stoichiometry of the binding as well as providing thermodynamic information. Solution NMR spectroscopy has unique advantages as it can reveal structural and dynamics information at the level of individual protein residues, an approach that can also be used to study protein-ligand or protein-protein interactions. The latter is usually done by recording multiple heteronuclear single-quantum coherence (HSQC) $^1H,^{15}N$ NMR spectra. Moreover, the binding of Ca^{2+} to the EF-hands of the protein can be studied by NMR and ITC as well. The application of these experimental strategies in combination provides a valuable starting point for understanding the interactions of the plastin proteins and the structural/functional relationships of the entire plastin family.

2 Materials

2.1 Bacterial Strain and Expression Vector

Synthetic genes for the EF-hand domains of L-plastin (LPL), T-plastin (TPL), and I-plastin (IPL) were purchased from GeneArt (Regensburg, Germany). The synthetic plastin genes contained codons optimized for protein expression in *Escherichia coli* and were cloned into a pET15 vector (Invitrogen) using the NdeI and XhoI sites. The EF-hand domain construct contains an N-terminal 6His-tag followed by a TEV-protease cleavage site. The His-tag can

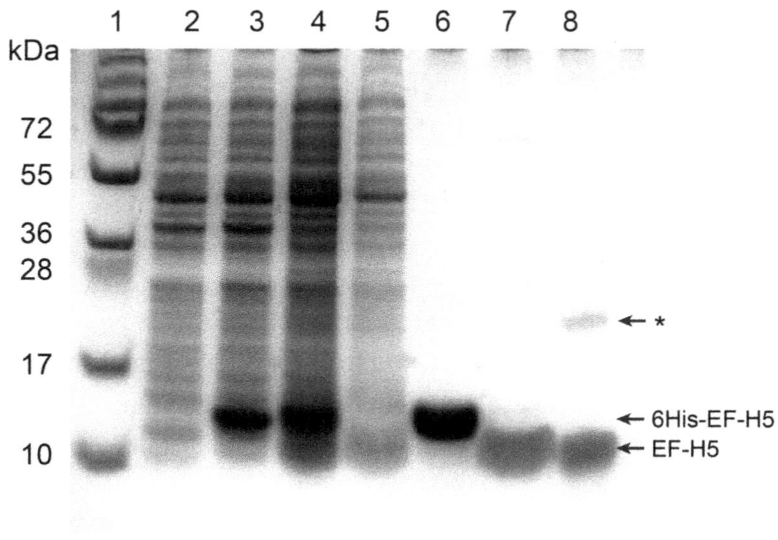

Fig. 2 SDS-PAGE for the purification of the LPL-EF-H5 using a Ni-column: (1) Protein molecular markers. (2 and 3) *E.coli* cell lysate before and after IPTG induction, respectively. (4) Supernatant after French press and centrifugation. (5) Flow through during loading of the supernatant onto the Ni-column. (6) Elution with the elution buffer. (7) After TEV protease digestion. (8) Purified EF-H5 after concentration. * A weak band for the dimeric protein may appear

be proteolytically removed after the first step of the two-step purification method using a Ni-column. Each plastin plasmid was transformed into competent *E. coli* BL21 (DE3) cells for protein expression. Each purification step is assessed by sodium dodecyl sulfate-polyacrylamide gel electrophoresis (SDS-PAGE, Fig. 2).

2.2 Protein Expression, Labeling, and Purification

1. Luria-Bertani (LB) broth.
2. Sterile syringe filters with a 0.22 μm pore size.
3. 100 mM ampicillin solution (filtered).
4. 10× salts (1 L): 34 g Na_2HPO_4, 15 g KH_2PO_4 and 2.5 g NaCl.
5. M9 minimal medium (1 L): Mix 900 mL distilled H_2O, 0.5 g/L of $^{15}NH_4Cl$, and 100 mL of 10× salts. This mix should be sterilized by autoclaving prior to addition of other ingredients. Add 500 μL of 10% thiamine (filtered), 4 g of glucose (dissolved in 10 mL of dH_2O and filtered), 1 mL of 1 M $MgSO_4$ (autoclaved), 100 μL of 1 M $CaCl_2$ (filtered), 50 μL of 1 M $MnCl_2$ (filtered), 30 μL of 1 M $FeCl_2$ (filtered), and 1 mL of vitamin mix (optional, *see* **Note 1**).
6. Isopropyl β-d-1-thiogalactopyranoside (IPTG).
7. Binding buffer: 20 M Tris–Cl, 0.1 M NaCl, and 0.05 M imidazole, pH 8.0.

8. Elution buffer: 20 mM Tris–Cl, 0.1 M NaCl, and 0.3 M imidazole, pH 8.0.
9. TEV protease (*see* **Note 2**).
10. Dialysis buffer (4 L): 20 mM Tris, 100 mM NaCl, 1 mM β-mercaptoethanol and dH_2O, pH 7.5.
11. Laboratory incubator shaker for bacterial growth at 37 °C.
12. Laboratory large-capacity refrigerated centrifuge (*see* **Note 3**).
13. Chelating Sepharose fast-flow resin (IMAC-Sepharose, GE Healthcare) charged with Ni^{2+}.
14. Complete His-Tag nickel column (Roche Diagnostics).
15. Laboratory peristaltic pump.
16. 3.5 kDa molecular weight cutoff (MWCO) dialysis membrane (Spectrum Laboratories).
17. Both plastic and quartz cuvettes to monitor bacterial growth and measure protein concentration, respectively, using a UV/visible spectrophotometer.
18. Bio-Rad protein assay dye reagent (Bio-Rad).

2.3 Surface Plasmon Resonance (SPR) Experiments

1. Immobilization sample buffer: 10 mM sodium formate (pH 4.5).
2. 40 μm vacuum filtration unit.
3. Immobilization running buffer (1 L): 0.01 M HEPES (pH 7.5), 0.15 M NaCl, 0.005% Tween 20. The buffer is filtered using a vacuum filtration unit.
4. Running buffer (1 L): 20 mM HEPES (pH 7.5), 150 mM KCl, 1 mM $CaCl_2$, 0.005% Tween 20. The buffer is filtered using a vacuum filtration unit.
5. Regeneration solution: 350 mM EDTA, 300 mM NaCl.
6. CM5 sensor chip (GE Healthcare).
7. Amicon® 3 kDa Ultra Centrifugal filters (Millipore Sigma).
8. Biacore X100 instrument (GE Healthcare).

2.4 Isothermal Titration Calorimetry (ITC) Experiments

1. Chelex chelating agent (Sigma-Aldrich).
2. ITC buffer: 20 mM HEPES (pH 7.2), 100 mM KCl, 1 mM Tris(2-carboxyethyl) phosphine (TCEP, *see* **Note 4**). The buffer is incubated with Chelex for over 3 days in a plastic container to obtain Ca^{2+} buffer (*see* **Note 5**).
3. MicroCal VP-ITC microcalorimeter or a MicroCal PEAK-ITC microcalorimeter.

2.5 NMR Spectroscopy Experiments

1. NMR spectrometer with a proton frequency of 500 MHz or higher.
2. ^{1}H, ^{15}N, ^{13}C triple resonance probe or a similar cryoprobe.

3 Methods

3.1 Expression of Plastin EF-Hand Domains in 1 L in LB Medium

1. Prepare 50 mL LB broth with 100 μg/mL ampicillin in a 125 mL flask. Inoculate the flask with a single colony or glycerol stock, and grow overnight at 37 °C in a laboratory incubator shaker.
2. Transfer 20 mL of overnight culture into a 2.8 L baffled culture flask containing 1 L LB broth with 100 μg/mL ampicillin. For isotope labeling for NMR experiments, transfer 25 mL of overnight culture into 2.8 L baffled culture flask containing 1 L M9 minimal medium with 100 μg/mL ampicillin. Incubate at 37 °C in the laboratory incubator shaker.
3. Once an optical density at 600 nm (OD_{600}) of ~0.6 is reached (typically takes 2–3 h or 4–5 h in LB or M9 medium, respectively), the protein was induced with 1.0 mM IPTG for 4 h at 37 °C. The OD_{600} absorbance is monitored using a UV/visible spectrophotometer and a plastic cuvette.
4. Cells are to be harvested by centrifugation at 12,000 × *g* for 15 min, and the pellet was stored at −20 °C.

3.2 First Step of Ni-Column Purification

The purification of the EF-hand domains of all plastins followed a simple two-step purification procedure using a Ni-column, and we obtained high-purity EF-hand constructs. It should be noted that the protocols for the expression and the purification of the EF and EF-H5 constructs are the same.

1. The pellet is re-suspended in 30 mL of binding buffer and homogenized.
2. Cells of *E. coli* are lysed via French press (French Pressure Cell Press FA-078, SLM Aminco) for three consecutive rounds (*see* **Note 6**).
3. Centrifuge the lysed cell extract at 43,000 × *g* for 45 min.
4. Equilibrate the Ni-column with 3–4 column volumes of binding buffer.
5. Apply the supernatant of cell lysate after the high-speed centrifugation onto the equilibrated Ni-column using a laboratory peristaltic pump.
6. Wash the column with at least 10 column volumes of binding buffer.
7. Elute protein with the elution buffer; collect fractions (fraction volume 10 mL) that display the presence of protein by using the Bio-Rad protein assay or by a positive A280 absorbance. The absorbance is measured using a UV/visible spectrophotometer and a quartz cuvette.

8. Dialyze protein using a 3.5 kDa MWCO dialysis membrane in 4 L of dialysis buffer supplemented with 0.5 mM EDTA overnight at 4 °C.

3.3 Digestion

1. Empty the contents of dialysis membrane into a 50 mL Falcon tube. Measure concentration of protein using the extinction coefficient at A280 (*see* **Note 7**).
2. Add 1 mM dithiothreitol into the tube (*see* **Note 8**).
3. Add sufficient units of TEV protease to cleave the His-tag (*see* **Note 9**). The Falcon tube is to be incubated with rotation at 30 °C for 1 h.
4. Once digestion is complete, dialyze the protein using a 3.5 kDa MWCO dialysis membrane, and dialyze once more in 4 L of dialysis buffer overnight at 4 °C.

3.4 Second Step of Ni-Column Purification

1. Equilibrate the complete column with 2–4 column volumes of the buffer containing 20 mM Tris (pH 7.5) and 100 mM NaCl (*see* **Note 10**).
2. Empty the contents of the dialysis membrane into a 50 mL Falcon tube, and apply onto the complete column; collect fractions (fraction volume 10 mL) that display the presence of protein by using the Bio-Rad protein assay or by a positive A280 absorbance.
3. Pool the fractions containing the protein, and measure the concentration using the extinction coefficient at A280. A UV/visible spectrophotometer and quartz cuvettes can be used to read the absorbance.
4. The pooled factions are to be separated into 10 mL aliquots in 15 mL Falcon tubes, and then flash frozen using liquid nitrogen followed by storage at −80 °C.

3.5 Peptide Interaction Studies by Surface Plasmon Resonance

In order to perform SPR experiments, the EF-hand domain of plastin needs to contain a cysteine for specific thiol coupling onto the CM5 sensor chip. This can be done by using an existing single cysteine residue (e.g. Cys42 in the case of LPL) or by introducing a single Cys at a desired position by site-directed mutagenesis and purification of the mutant protein using the same steps as mentioned above (*see* **Note 11**).

1. The purified EF hand domain of plastin is subjected to the buffer exchanged with immobilization sample buffer using Amicon 3 kDa Ultra Centrifugal filters (*see* **Note 12**).
2. The CM5 sensor chip is docked into the Biacore X100 instrument and equilibrated with the immobilization running buffer.
3. The protein is immobilized onto flow cell 2 of the CM5 chip at a flow rate of 5 μL/min using thiol coupling.

4. The sensor chip surface was activated with NHS/EDC followed by modification of ester groups by PDEA (2-(2pyridinyldithio)ethaneamine) to become reactive disulfide groups and injection of the EF-hand construct of plastin to couple it to the surface through reaction of its free SH group (*see* **Note 13**).
5. The unreacted disulfide groups were then deactivated by injecting 20 μL of 50 mM L-cysteine, 1 M NaCl in 0.1 M formate buffer pH 4.3.
6. Flow cell 1 is used as reference and underwent the same procedure as flow cell 2 except for the injection of the protein.
7. Switch the buffer to the running buffer and equilibrate the instrument.
8. Prepare peptide samples in running buffer in five different concentrations by threefold serial dilutions starting at a 10 times higher concentration than the estimated K_d value (*see* **Note 14**).
9. Peptide samples are injected at a flow rate of 30 μL/min with a contact time of 1 min, and the dissociation is recorded for 3 min at 25 °C.
10. The CM5 chip can be regenerated readily by injecting 120 μL of regeneration solution.
11. The BIAevaluation software 2.0 (GE Healthcare) is used to evaluate the association and dissociation rates of the peptides, followed by processing of the SPR sensorgrams to fit the curves and determine the dissociation constants (K_d) (Fig. 3, *see* **Note 15**).

3.6 Isothermal Titration Calorimetry (ITC) Experiments for Ca^{2+} Binding to EF-H5 and EF

1. Ca^{2+}-free protein samples can be prepared by addition of 1 mM EDTA followed by intensive buffer exchange with the ITC buffer using Amicon 3 kDa Ultra Centrifugal filter. Protein concentration is adjusted to 40 μM using molar extinction coefficients.
2. Prepare 1.2 mM Ca^{2+} solution by dissolving $CaCl_2$ in the ITC buffer. A highly accurate concentration of Ca^{2+} can be determined for the stock solution by using an atomic absorption spectrometer.
3. ITC experiments are carried out at 25 °C. 40 μM protein sample in the sample cell is titrated with 1.2 μM Ca^{2+} solution in the syringe (*see* **Note 16**).
4. The titration consists of 57 sequential injections with 5 μL of Ca^{2+}. Each injection occurs over 10 s and with 300 s of interval between injections to ensure the equilibration of the system before the next injection.

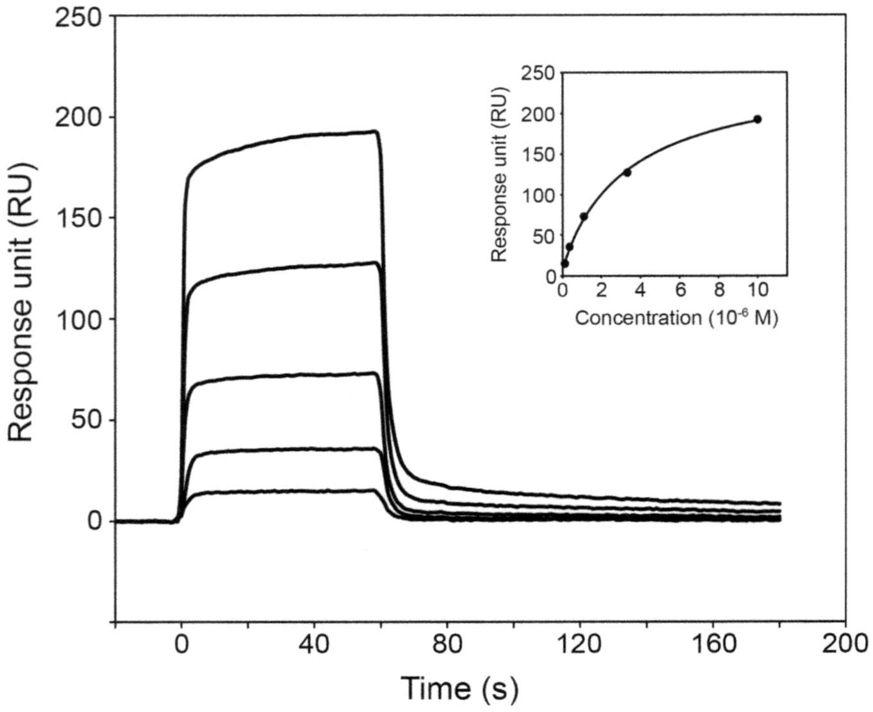

Fig. 3 SPR measurements of the binding of the synthetic peptide corresponding to the helix 5 region of TPL (Ac-SSDIAKTFRKAINRKEGI-NH2) to TPL-EF in the presence of Ca^{2+}. The peptide was injected at concentrations of 10, 3.33, 1.11, 0.37, and 0.12 μM. The K_d value obtained for the TPL-H5 peptide from the affinity fitting (inset) is $3.5 \pm 0.4 \times 10^{-6}$ M. A similar experiment with the bee venom peptide melittin gave a K_d of 2.1×10^{-7} M, showing that melittin binds stronger to TPL-EF than the synthetic TPL peptide. However, melittin can not displace the H5 region in the TPL-EF-H5 construct, unlike what was seen for LPL-EF-H5 (unpublished observations)

5. Obtain the heat of dilution data by injecting Ca^{2+} solution into the ITC buffer (*see* **Note 17**).
6. After subtraction of the heat of dilution data, the experimental data can be analyzed using the MicroCal Origin software. The experimental data can be fit to either one or two sets of site models to obtain the apparent stoichiometry (N), association constants (K), enthalpy changes (dH), and entropy changes (dS) of the binding (Fig. 4).

3.7 NMR Measurements

1. Obtain $^{1}H,^{15}N$ HSQC NMR spectra for samples containing 0.5–1.0 mM ^{15}N-labeled EF-H5 or EF, 100 mM KCl, 20 mM Bis-Tris (pH 6.8), 5 mM $CaCl_2$ or 1 mM EDTA, 0.03% NaN_3, 0.5 mM DSS, and 10 mM ^{2}H-DTT in 10% D_2O/90% H_2O at 25 °C (*see* **Note 18**).
2. Backbone sequential assignments: Uniformly $^{15}N,^{13}C$-labeled EF-H5, and EF were used for the backbone assignment.

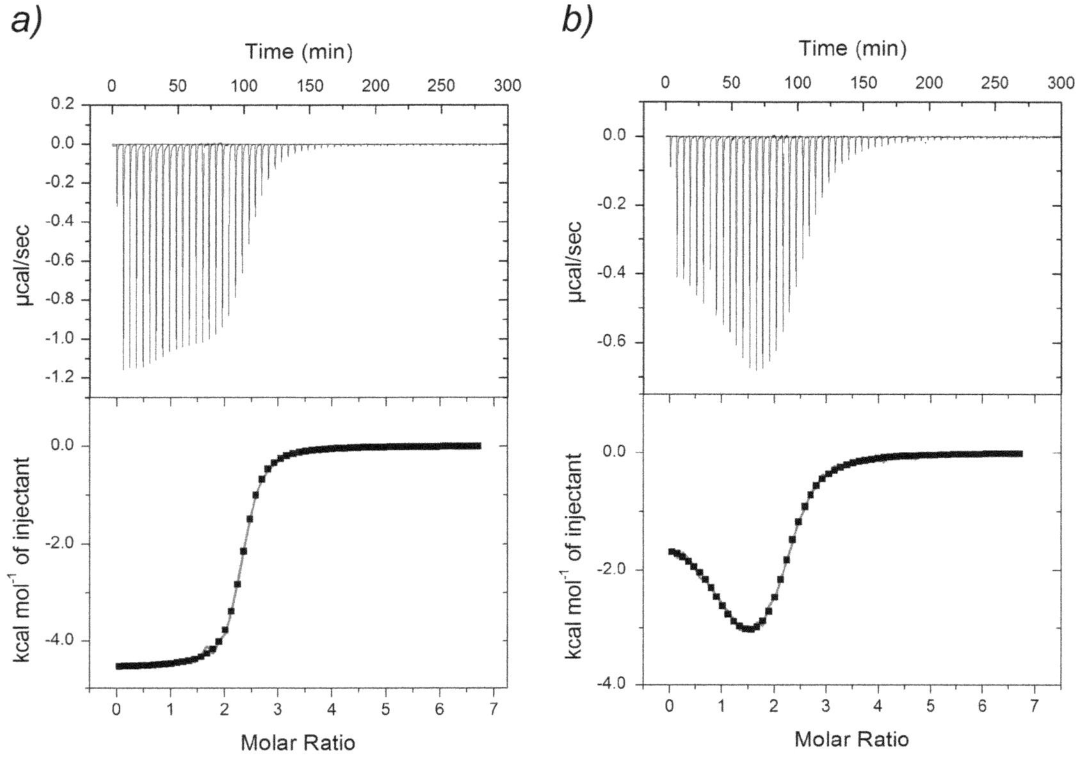

Fig. 4 Ca^{2+} binding to the (**a**) LPL-EF-H5 and (**b**) LPL-EF constructs was monitored by ITC [43]. The baseline-corrected raw calorimetric data are shown in the top panel. The derived binding isotherms were fitted to obtain the thermodynamic parameters (bottom panel). For EF-H5, the isotherm was fitted using a "one set of sites" model which produced a K_d of 0.7 μM. For the EF construct that lacks the H5 region, the isotherm was fitted using a "two sets of sites" model which produced two K_ds: $K_d1 = 0.3$ μM, and $K_d2 = 2.6$ μM

Analyzing a series of three-dimensional NMR experiments (*see* **Note 19**), all backbone amide resonances except for the Pro residues were unambiguously assigned (Fig. 5).

3. Obtain the HSQC NMR spectrum for a sample containing ~0.2 mM ^{15}N-labeled EF-H5, 100 mM KCl, 20 mM Bis-Tris (pH 6.8), 5 mM $CaCl_2$, 0.5 mM DSS, and 10 mM ^{2}H-DTT in 10% D_2O/90% H_2O at 25 °C.
4. Titrate a concentrated melittin peptide solution dissolved in the same buffer as the protein into the NMR sample. The HSQC spectra are acquired at 1:0.5, 1:1, and 1:1.5 protein/peptide ratios.
5. Overlay the spectra to identify the signals with a significant chemical shift change upon binding melittin (Fig. 6). In this manner, the binding site for the peptide on the protein can be mapped.

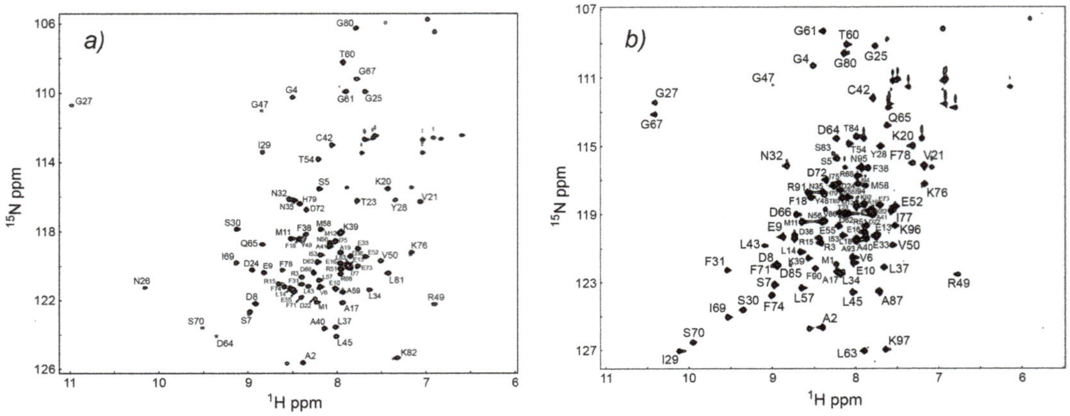

Fig. 5 The completely assigned ^{1}H,^{15}N-HSQC spectra of (**a**) Ca^{2+}-free LPL-EF and (**b**) Ca^{2+}-bound LPL-EF-H5 [43]

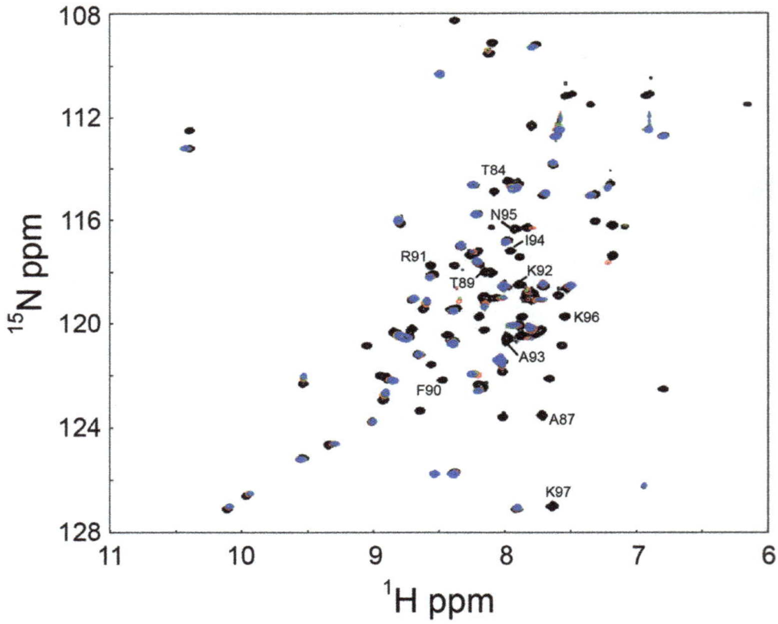

Fig. 6 Ca^{2+}-LPL-EF-H5 is titrated with the bee venom peptide melittin [43]. The ^{1}H,^{15}N HSQC spectrum of LPL-EF-H5 (black) is overlaid with the spectrum of LPL-EF-H5 with 0.5 (red), 1.0 (green), and 1.5 (blue) molar excess of melittin. Only peaks from the H5 region that have moved upon peptide binding are labeled. These data show that binding of melittin can displace the H5 region from the EF-H5 domain of LPL [43]

4 Notes

1. Vitamin mix: pyridoxine hydrochloride, biotin, d-pantothenate, cytosine, choline chloride. 50 mg each is dissolved in 50 mL of ddH_2O and filtered.
2. TEV protease was expressed and purified from the pRK793 plasmid as previously described [46]. The plasmid to express TEV protease mutant S219 V is available from Addgene.
3. Sorvall Lynx 6000 superspeed centrifuge (Thermo Scientific™) using a Fiberlite™ F9–6 × 1000 LEX Fixed Angle Rotor or a Fiberlite™ F21–8 × 50y Fixed-Angle Rotor.
4. An ITC-compatible reducing agent such as TCEP is required to prevent possible intermolecular disulfide bond. More commonly used reducing agents, such as DTT, will cause an erratic baseline drift in ITC experiments.
5. Chelex can be recycled using the following protocol: (1) Wash with two bed volumes of 1 M HCl. (2) Wash with two bed volumes of 1 M NaOH. (3) Wash with Milli-Q water until the pH of the waste becomes neutral.
6. Other cell disruption/lysis methods such as sonication can be used instead of the French press method.
7. The extinction coefficient for LPL-EF is 2980 M^{-1} cm^{-1} at 280 nm, while 4470 M^{-1} cm^{-1} is used for TPL-EF and IPL-EF.
8. The reaction buffer for TEV protease is 50 mM Tris–HCl 7.5, 100 mM NaCl, 1 mM DTT, and 0.5 mM EDTA.
9. 1 μg of TEV can cleave 25–100 μg of substrate.
10. The complete column can be substituted in principle with any Ni-column; however the higher tolerance against residual DTT of the complete column will provide better results.
11. In order to obtain homogeneous attachment of the protein to the SPR chip, when immobilizing the protein using a newly introduced Cys residue, any indigenous Cys residues need to be replaced first with Ser.
12. The pH of the protein sample must be lower than the isoelectric point (pI) of the protein. For example, the pI of LPL-EF is 5.12.
13. More information is available online at SPR pages [47].
14. The peptides used in these experiments are synthetic peptides that were purchased at >95% purity from GenScript (San Diego, CA).
15. Two different concentrations in each experiment were repeated in order to obtain the fitting errors (SEM).

16. Both the ITC sample cell and the syringe should be thoroughly rinsed with the ITC buffer prior to loading of the samples.
17. This can be achieved by either completing a separate control experiment or by using the average heat from several injections.
18. The NMR spectrometers used in our studies were Bruker Avance 600 or 700 MHz instruments equipped with a $^{1}H,^{15}N,^{13}C$-TXI probe.
19. The three-dimensional NMR experiments used for spectral assignment of the protein backbone were HNCACB, CBCA (CO)NH, HN(CA)CO, HNCO, HNCA, and HN(CO)CA.

Acknowledgments

This research is currently supported by a Discovery grant from the Natural Sciences and Engineering Research Council of Canada.

References

1. Delanote V, Vandekerckhove J, Gettemans J (2005) Plastins: versatile modulators of actin organization in (patho)physiological cellular processes. Acta Pharmacol Sin 26:769–779
2. Shinomiya H (2012) Plastin family of actin-bundling proteins: its functions in leukocytes, neurons, intestines, and cancer. Int J Cell Biol 2012:213492
3. Lin CS, Aebersold RH, Kent SB, Varma M, Leavitt J (1988) Molecular cloning and characterization of plastin, a human leukocyte protein expressed in transformed human fibroblasts. Mol Cell Biol 8:4659–4668
4. Miyakawa T, Shinomiya H, Yumoto F, Miyauchi Y, Tanaka H, Ojima T, Kato YS, Tanokura M (2012) Different Ca2+-sensitivities between the EF-hands of T- and L-plastins. Biochem Biophys Res Commun 429:137–141
5. Dos Remedios CG, Chhabra D, Kekic M, Dedova IV, Tsubakihara M, Berry DA, Nosworthy NJ (2003) Actin binding proteins: regulation of cytoskeletal microfilaments. Physiol Rev 83:433–473
6. Morley SC (2012) The actin-bundling protein L-plastin: a critical regulator of immune cell function. Int J Cell Biol 2012:935173
7. Morley SC (2013) The actin-bundling protein L-plastin supports T-cell motility and activation. Immunol Rev 256:48–62
8. Wang C, Morley SC, Donermeyer D, Peng I, Lee WP, Devoss J, Danilenko DM, Lin Z, Zhang J, Zhou J, Allen PM, Brown EJ (2010) Actin-bundling protein L-plastin regulates T cell activation. J Immunol 185:7487–7497
9. Todd EM, Deady LE, Morley SC (2011) The actin-bundling protein L-plastin is essential for marginal zone B cell development. J Immunol 187:3015–3025
10. Kell MJ, Riccio RE, Baumgartner EA, Compton ZJ, Pecorin PJ, Mitchell TA, Topczewski J, LeClair EE (2018) Targeted deletion of the zebrafish actin-bundling protein L-plastin (lcp1). PLoS One 13:e0190353
11. Chen H, Mocsai A, Zhang H, Ding R-X, Morisaki JH, White M, Rothfork JM, Heiser P, Colucci-Guyon E, Lowell CA, Gresham HD, Allen PM, Brown EJ (2003) Role for plastin in host defense distinguishes integrin signaling from cell adhesion and spreading. Immunity 19:95–104
12. Park T, Chen ZP, Leavitt J (1994) Activation of the leukocyte plastin gene occurs in most human cancer cells. Cancer Res 54:1775–1781
13. Lin CS, Park T, Chen ZP, Leavitt J (1993) Human plastin genes. Comparative gene structure, chromosome location, and differential expression in normal and neoplastic cells. J Biol Chem 268:2781–2792
14. Riplinger SM, Wabnitz GH, Kirchgessner H, Jahraus B, Lasitschka F, Schulte B, van der Pluijm G, van der Horst G, Hämmerling GJ, Nakchbandi I, Samstag Y (2014) Metastasis of prostate cancer and melanoma cells in a preclinical in vivo mouse model is enhanced by

L-plastin expression and phosphorylation. Mol Cancer 13:10

15. Koide N, Kasamatsu A, Endo-Sakamoto Y, Ishida S, Shimizu T, Kimura Y, Miyamoto I, Yoshimura S, Shiiba M, Tanzawa H, Uzawa K (2017) Evidence for critical role of lymphocyte cytosolic protein 1 in oral cancer. Sci Rep 7:43379
16. Foran E, McWilliam P, Kelleher D, Croke DT, Long A (2006) The leukocyte protein L-plastin induces proliferation, invasion and loss of E-cadherin expression in colon cancer cells. Int J Cancer 118:2098–2104
17. Klemke M, Rafael MT, Wabnitz GH, Weschenfelder T, Konstandin MH, Garbi N, Autschbach F, Hartschuh W, Samstag Y (2007) Phosphorylation of ectopically expressed L-plastin enhances invasiveness of human melanoma cells. Int J Cancer 120:2590–2599
18. Delanote V, Vanloo B, Catillon M, Friederich E, Vandekerckhove J, Gettemans J (2010) An alpaca single-domain antibody blocks filopodia formation by obstructing L-plastin-mediated F-actin bundling. FASEB J 24:105–118
19. Zheng J, Rudra-Ganguly N, Powell WC, Roy-Burman P (1999) Suppression of prostate carcinoma cell invasion by expression of antisense L-plastin gene. Am J Pathol 155:115–122
20. Samstag Y, Klemke M (2007) Ectopic expression of L-plastin in human tumor cells: diagnostic and therapeutic implications. Adv Enzym Regul 47:118–126
21. Lommel MJ, Trairatphisan P, Gäbler K, Laurini C, Muller A, Kaoma T, Vallar L, Sauter T, Schaffner-Reckinger E (2016) L-plastin Ser5 phosphorylation in breast cancer cells and in vitro is mediated by RSK downstream of the ERK/MAPK pathway. FASEB J 30:1218–1233
22. Hasselblatt M, Böhm C, Tatenhorst L, Dinh V, Newrzella D, Keyvani K, Jeibmann A, Buerger H, Rickert CH, Paulus W (2006) Identification of novel diagnostic markers for choroid plexus tumors: A microarray-based approach. Am J Surg Pathol 30:66–74
23. Li J, Zhao R (2011) Expression and clinical significance of l-plastin in colorectal carcinoma. J Gastrointest Surg 15:1982–1988
24. Harris LD, De La Cerda J, Tuziak T, Rosen D, Xiao L, Shen Y, Sabichi AL, Czerniak B, Grossman HB (2008) Analysis of the expression of biomarkers in urinary bladder cancer using a tissue microarray. Mol Carcinog 47:678–685
25. Öztürk VÖ, Emingil G, Osterwalder V, Bostanci N (2015) The actin-bundling protein L-plastin: a novel local inflammatory marker associated with periodontitis. J Periodontal Res 50:337–346
26. Oprea GE, Kröber S, McWhorter ML, Rossoll W, Müller S, Krawczak M, Bassell GJ, Beattie CE, Wirth B (2008) Plastin 3 is a protective modifier of autosomal recessive spinal muscular atrophy. Science 320:524–527
27. Ralser M, Nonhoff U, Albrecht M, Lengauer T, Wanker EE, Lehrach H, Krobitsch S (2005) Ataxin-2 and huntingtin interact with endophilin-A complexes to function in plastin-associated pathways. Hum Mol Genet 14:2893–2909
28. van Dijk FS, Zillikens MC, Micha D, Riessland M, Marcelis CLM, de Die-Smulders CE, Milbradt J, Franken AA, Harsevoort AJ, Lichtenbelt KD, Pruijs HE, Rubio-Gozalbo ME, Zwertbroek R, Moutaouakil Y, Egthuijsen J, Hammerschmidt M, Bijman R, Semeins CM, Bakker AD, Everts V, Klein-Nulend J, Campos-Obando N, Hofman A, te Meerman GJ, Verkerk AJMH, Uitterlinden AG, Maugeri A, Sistermans EA, Waisfisz Q, Meijers-Heijboer H, Wirth B, Simon MEH, Pals G (2013) *PLS3* mutations in X-linked osteoporosis with fractures. N Engl J Med 369:1529–1536
29. Mäkitie RE, Kämpe AJ, Taylan F, Mäkitie O (2017) Recent discoveries in monogenic disorders of childhood bone fragility. Curr Osteoporos Rep 15:303–310
30. Higuchi Y, Kita K, Nakanishi H, Wang XL, Sugaya S, Tanzawa H, Yamamori H, Sugita K, Yamaura A, Suzuki N (1998) Search for genes involved in UV-resistance in human cells by mRNA differential display: Increased transcriptional expression of nucleophosmin and T-plastin genes in association with the resistance. Biochem Biophys Res Commun 248:597–602
31. Suzuki N (1984) A UV-resistant mutant without an increased repair synthesis activity, established from a UV-sensitive human clonal cell line. Mutat Res 125:55–63
32. Yokobori T, Iinuma H, Shimamura T, Imoto S, Sugimachi K, Ishii H, Iwatsuki M, Ota D, Ohkuma M, Iwaya T, Nishida N, Kogo R, Sudo T, Tanaka F, Shibata K, Toh H, Sato T, Barnard GF, Fukagawa T, Yamamoto S, Nakanishi H, Sasaki S, Miyano S, Watanabe T, Kuwano H, Mimori K, Pantel K, Mori M (2013) Plastin3 is a novel marker for circulating tumor cells undergoing the epithelial-mesenchymal transition and is associated with

colorectal cancer prognosis. Cancer Res 73:2059–2069

33. Ueo H, Sugimachi K, Gorges TM, Bartkowiak K, Yokobori T, Müller V, Shinden Y, Ueda M, Ueo H, Mori M, Kuwano H, Maehara Y, Ohno S, Pantel K, Mimori K (2015) Circulating tumour cell-derived plastin3 is a novel marker for predicting long-term prognosis in patients with breast cancer. Br J Cancer 112:1519–1526
34. Grimm-Gunter E-MS, Revenu C, Ramos S, Hurbain I, Smyth N, Ferrary E, Louvard D, Robine S, Rivero F (2009) Plastin 1 binds to keratin and is required for terminal web assembly in the intestinal epithelium. Mol Biol Cell 20:2549–2562
35. Taylor R, Bullen A, Johnson SL, Grimm-Günter EM, Rivero F, Marcotti W, Forge A, Daudet N (2015) Absence of plastin 1 causes abnormal maintenance of hair cell stereocilia and a moderate form of hearing loss in mice. Hum Mol Genet 24:37–49
36. Krey JF, Krystofiak ES, Dumont RA, Vijayakumar S, Choi D, Rivero F, Kachar B, Jones SM, Barr-Gillespie PG (2016) Plastin 1 widens stereocilia by transforming actin filament packing from hexagonal to liquid. J Cell Biol 215:467–482
37. Bretscher A, Weber K (1980) Fimbrin, a new microfilament-associated protein present in microvilli and other cell surface structures. J Cell Biol 86:335–340
38. Stradal T, Kranewitter W, Winder SJ, Gimona M (1998) CH domains revisited. FEBS Lett 431:134–137
39. Mattila PK, Lappalainen P (2008) Filopodia: molecular architecture and cellular functions. Nat Rev Mol Cell Biol 9:446–454
40. Van Audenhove I, Denert M, Boucherie C, Pieters L, Cornelissen M, Gettemans J (2016) Fascin rigidity and L-plastin flexibility cooperate in cancer cell invadopodia and Filopodia. J Biol Chem 291:9148–9160
41. Klein MG, Shi W, Ramagopal U, Tseng Y, Wirtz D, Kovar DR, Staiger CJ, Almo SC (2004) Structure of the actin crosslinking core of fimbrin. Structure 12:999–1013
42. Namba Y, Ito M, Zu Y, Shigesada K, Maruyama K (1992) Human T cell L-plastin bundles actin filaments in a calcium-dependent manner. J Biochem 112:503–507
43. Ishida H, Jensen KV, Woodman AG, Hyndman ME, Vogel HJ (2017) The calcium-dependent switch helix of L-plastin regulates actin bundling. Sci Rep 7:40662
44. Schwebach CL, Agrawal R, Lindert S, Kudryashova E, Kudryashov DS (2017) The roles of actin-binding domains 1 and 2 in the calcium-dependent regulation of actin filament bundling by human plastins. J Mol Biol 429:2490–2508
45. Prassler J, Stocker S, Marriott G, Heidecker M, Kellermann J, Gerisch G (1997) Interaction of a dictyostelium member of the plastin/fimbrin family with actin filaments and actin-myosin complexes. Mol Biol Cell 8:83–95
46. Kapust RB, Tözsér J, Fox JD, Anderson DE, Cherry S, Copeland TD, Waugh DS (2001) Tobacco etch virus protease: mechanism of autolysis and rational design of stable mutants with wild-type catalytic proficiency. Protein Eng Des Sel 14:993–1000
47. SPR-Pages Thiol coupling. https://www.sprpages.nl/immobilization/immobilization-procedures/thiol. Accessed June 4 2018

Chapter 17

Expression and Characterization of MICU2, a Ca^{2+} Sensor Protein

Wenping Wu, Jimin Zheng, and Zongchao Jia

Abstract

MICU2 is a Ca^{2+} sensor protein of mitochondrial uniporter which is a highly selective Ca^{2+} channel mediating mitochondrial Ca^{2+} uptake to regulate cell death, metabolism, and cytoplasmic Ca^{2+} signaling. Here we describe the procedures for protein preparation of various MICU2 constructs, which have enabled successful in vitro characterizations of MICU2 including interaction with MICU1 using pull-down assays and oligomerization using multi-angle laser light scattering.

Key words MICU2, Ca^{2+} sensor, EF hand, Mitochondrial, SEC, MALLS, Pull-down, Purification

1 Introduction

The mitochondrial uniporter is a highly selective Ca^{2+} channel mediating mitochondrial Ca^{2+} uptake to regulate cell death, metabolism, and cytoplasmic Ca^{2+} signaling [1]. Mitochondrial Ca^{2+} uptake from cytoplasm is driven by the inner membrane (IMM) potential ($\Delta\psi_m$) which is generated by the respiratory chain [2–4]. The uniporter is a macromolecular complex located in the inner membrane which includes the mitochondrial Ca^{2+} uniporter (MCU) [5], MCU regulatory subunit b (MCUb) [6], essential MCU regulator (EMRE) [7], and the mitochondrial Ca^{2+} uniporter protein 1/2 (MICU1/2) [8, 9]. Crystal structure shows that transmembrane MCU is a pore-forming pentamer mediating Ca^{2+} uptake [10]. MCUb acts as an endogenous domain-negative MCU subunit. EMRE has dual functions that activate MCU via helix-helix interactions from both proteins and retains MICU1 via its C-terminal polyaspartate tail [11]. MICU1 and MICU2 sense Ca^{2+} concentration in intermembrane space through their EF hands to regulate the activities of MCU. Whereas the roles of MICU-regulated MCU are controversial, it was convincingly shown that MICU1 activates MCU at higher $[Ca^{2+}]_c$, evidenced by both

Claus W. Heizmann (ed.), *Calcium-Binding Proteins of the EF-Hand Superfamily: From Basics to Medical Applications*, Methods in Molecular Biology, vol. 1929, https://doi.org/10.1007/978-1-4939-9030-6_17,

planar lipid bilayer experiments and intact cell assay [12]. Conversely, MICU2 is a proven negative regulator to shut down MCU activity [12]. Both cooperativity and gatekeeping require EF-hands binding Ca^{2+} to activate MCU or relieve channel inhibition, whereas another study showed that MICU1 and MICU2 do not have overlapping function and together set the $[Ca^{2+}]$ threshold to regulate MCU activity [13]. The heterodimer of MICU1 and MICU2 inhibits Ca^{2+} transport when the cytosolic $[Ca^{2+}]_c$ is below the threshold and activates the channel when the cytosolic $[Ca^{2+}]_c$ exceeds the threshold [13]. Regardless what models are at play, EF hands play crucial roles in regulating the activities of MCU. It was reported that MICU1 and MICU2 form a 95 kDa heterodimer through a disulfide bond by cysteine residues in the C-helix domain [12]. In our pull-down assays, however, we demonstrated that MICU1 and MICU2 can perform strong heterodimer in vitro even lacking the cysteine residues [14]. In addition, the heterodimer of MICU1 and MICU2 formation is independent of Ca^{2+}-binding. This is consistent with a recently published paper [15].

Here, we present experiment procedures including design of a series of recombinant MICU2 truncations, their expression in *Escherichia coli* cells, and purification in the presence or absence of Ca^{2+} by size exclusion chromatography (SEC). Pull-down experiments were used to explore the interaction between MICU2 and MICU1 in vitro, in which GST beads binding protein with GST-tag were used to detect another protein lacking GST-tag. Furthermore, we performed multi-angle laser light scattering (MALLS) experiment to characterize the oligomerization states of MICU2 in both conditions (Ca^{2+}-bound and Ca^{2+}-free). Altogether, those methods provide useful tools for MICU2 characterization.

2 Materials

All tubes, pipette tips, and medium must be autoclaved at 121 °C for 30 min. All buffers need to be prepared using ultrapure water (resistivity reached 18.25 MΩ at 25 °C) and filtered by micropore (0.22 μm) membrane. Particularly, the ÄKTA buffer needs to be extremely fresh and vacuum pumped for 30 min. All reagents are of high purity grade, and all buffers are stored at 4 °C.

2.1 Cloning

1. Expression vector: pET-28a (+), pGEX-6p-1.
2. cDNA: human MICU1 (BC004190.2), human MICU2 (BC031089.1).
3. 1× TAE electrophoresis buffer: 40 mM Tris, 20 mM acetate, 1 mM EDTA, pH 8.3; store at room temperature.
4. Separate plate: Weigh 1.5% (w/v) agarose in LB medium, and autoclave at 121 °C for 30 min. Add appropriate antibiotics

(100 μg/mL) (*see* **Note 1**), and mix well when the medium is not hot to touch (*see* **Note 2**). Pour about 15 mL medium into 90 mm Petri dishes, and wait to congeal. Store upside down at 4 °C, and avoid drying.

5. DNA gel: Weigh 1% (w/v) agarose in 1× TAE buffer, and heat the mixture until melting completely. Add 0.01% 10,000× gel red and mix uniformly. Finally, transfer to the DNA gel mold.
6. Primers of MICU1 and MICU2 truncations and mutants: The detailed information is shown in Table 1 (*see* **Note 3**).

2.2 Expression

1. Lysogeny broth (LB): 10 g tryptone, 5 g yeast extract, 10 g NaCl. Make up to 1 L with ddH_2O.
2. Kanamycin antibiotic stock: 100 mg/mL; store at −40 °C.
3. Ampicillin antibiotic stock: 100 mg/mL; store at −40 °C.

2.3 Purification and Pulldown

1. Lysis buffer: 20 mM Tris, pH 7.0, 300 mM NaCl, 20 mM imidazole, 0.3% Triton X-100, with or without 2 mM $CaCl_2$.
2. Wash buffer: 20 mM Tris, pH 7.0, 300 mM NaCl, 50 mM imidazole, with or without 2 mM $CaCl_2$.
3. Elution buffer: 20 mM Tris, pH 7.0, 300 mM NaCl, 400 mM imidazole, with or without 2 mM $CaCl_2$.
4. ÄKTA buffer: 20 mM MES, pH 6.8, 300 mM NaCl, 2 mM EGTA, or 2 mM $CaCl_2$ (*see* **Note 4**).
5. Affinity column: Ni-NTA agarose.

3 Methods

3.1 Cloning of MICU1 and MICU2

1. Amplify target genes by PCR. PCR reaction system (*see* **Notes 5** and **6**) and condition (*see* **Note 7**) are given in Tables 2 and 3, respectively.
2. Run the DNA gel electrophoresis at 150 V for 20 min to separate the amplified products.
3. DNA purification. The gel purification is carried out using DNA gel extraction kit.
4. Restriction enzyme cutting. Details of the protocol for digestion are given in Table 4. Mix the reagents referred in Table 4 gently, spin down for a few seconds, and then incubate at 37 °C for 1 h (*see* **Note 8**).
5. Homologous recombination. Details of the protocol for recombination reaction are shown in Table 5. Mix the reagents referred in Table 5 gently, spin down for a few seconds, and then incubate at 50 °C for 30 min.

Table 1
Primers of MICU1 and MICU2 truncations and mutants. MICU1 and MICU2 truncation primers are applied to assemble enzyme homologous recombination

Clone	Vector	Restriction sites	Primer
MICU1-NΔ54	pET-28a (+)	NcoI/XhoI	F-AAGAAGGAGATATACCATGGGCCCATGCAGAATCTCCA R-TGGTGGTGGTGGTGCTCGAGTTACTGTTTGGGTAAAGC
MICU1-NΔ96	pET-28a (+)	NcoI/XhoI	F-AAGAAGGAGATATACCATGGAAGAGAAAAAGAAGAAA R-TGGTGGTGGTGGTGCTCGAGTTACTGTTTGGGTAAAGC
MICU1-NΔ96-CΔ32	pET-28a (+)	NcoI/XhoI	F-TGCCGCGCGGCAGCCATATGGAAGAGAAAAAGAAGAAA R-TGGTGGTGGTGGTGCTCGAGTTAGCCTCTCATCAGCCGTTG
GST-MICU1-NΔ54	pGEX-6p-1	BamHI/XhoI	F-TCCAGGGGCCCCTGGGATCCGCCCATGCAGAATCTCCA R-TCACGATGCGGCCGCTCGAGTTACTGTTTGGGTAAAGCGAA
GST-MICU1-NΔ96	pGEX-6p-1	BamHI/XhoI	F-TCCAGGGGCCCCTGGGATCCGAAGAGAAAAAGAAGAAA R-TCACGATGCGGCCGCTCGAGTTACTGTTTGGGTAAAGCGAA
GST-MICU1-NΔ96-CΔ32	pGEX-6p-1	BamHI/XhoI	F-TCCAGGGGCCCCTGGGATCCGAAGAGAAAAAGAAGAAA R-TCACGATGCGGCCGCTCGAGTTAGCCTCTCATCAGCCGTTG
MICU2-NΔ66	pET-28a (+)	NdeI/XhoI	F-TGCCGCGCGGCAGCCATATGGTCTCCGCACAGAAAAAT R-TGGTGGTGGTGGTGCTCGAGTTAAAAAAGACCTTTTCC
MICU2-NΔ84	pET-28a (+)	NdeI/XhoI	F-TGCCGCGCGGCAGCCATATGCTTCGTAAGCAGCGCTTC R-TGGTGGTGGTGGTGCTCGAGTTAAAAAAGACCTTTTCC
MICU2-NΔ66-CΔ32	pET-28a (+)	NdeI/XhoI	F-TGCCGCGCGGCAGCCATATGGTCTCCGCACAGAAAAAT R-TGGTGGTGGTGGTGCTCGAGTTATGGTACCCATAAACCTCG
MICU2-NΔ84-CΔ28	pET-28a (+)	NdeI/XhoI	F-TGCCGCGCGGCAGCCATATGCTTCGTAAGCAGCGCTTC R-TGGTGGTGGTGGTGCTCGAGTTAACTCTGATGTTGTGGTAC
MICU2-NΔ84-CΔ28^{EF1mut} (D185A & E196K)	pET-28a (+)	NdeI/XhoI	F-TTTCATGTTGCTTTTAAAATGCTGGCTACAGATGGTAATGA R- TTCAATCATCTCATTACCATCTGTAGCCAGCATTTTAAAAG F-TGGTAATGAGATGATTGAAAAAAGGAAATTTTTTAAGCTGC R-ATGATCTTCTGCAGCTTAAAAAATTTCCTTTTTTCAATCAT
MICU2-NΔ84-CΔ28^{EF2mut} (D375A & E386K)	pET-28a (+)	NdeI/XhoI	F-TTGGACACTGTCTTTAAGATCTTTGCTTTGGATGGTGATGA R-ACTAAGACATTCATCACCATCCAAAGCAAAGATCTTAAAGA F- TGGTGATGAATGTCTTAGTCATGAAAAGTTTCTTGGGGTGT R-CTGTTTTTTAACACCCCAAGAAACTTTTCATGACTAAGACA

Table 2
Protocol of polymerase chain reaction (PCR). The pfu DNA polymerase includes dNTPs and $MgSO_4$

Reagents	50 μL reaction system	Final concentration
2× pfu DNA polymerase (master mix)	25 μL	1×
Forward primer (10 μM)	1.5 μL	0.3 μM
Reverse primer (10 μM)	1.5 μL	0.3 μM
DNA template	<0.5 μg	<0.5 μg/50 μL
Nuclease-free water	up to 50 μL	

Table 3
Procedures of the polymerase chain reaction (PCR)

Procedure	Temperature	Time	Number of cycles
Initial denaturation	95 °C	2 min	1 cycle
Denaturation Annealing Extension	95 °C 45–65 °C 72 °C	1 min 0.5 min 1–4 min	40 cycles
Final extension	72 °C	10 min	1 cycle
Soak	4 °C	Indefinite	1 cycle

Table 4
Protocol of the restriction enzyme cutting

Reagents	20 μL reaction system
10× buffer	2 μL
NdeI/NcoI	1 μL
XhoI	1 μL
Vector (pET-28a (+))	10 μL ~ 1 μg
Nuclease-free water	up to 20 μL

3.2 Transformation of Recombination Mixture into Chemically Competent Cell

1. Fetch 100 μL high transformation efficiency Top10 or DH5α cells from −80 °C freezer, and immediately place on ice for 5 min.
2. Pipette each recombination mixture directly into competent cells, and mix by tapping gently (*see* **Note 9**).
3. Incubate the competent cells on ice for 30 min.

Table 5
Protocol of the homologous recombination

Reagents	20 μL reaction system
Inserts (PCR products)	50–100 ng
Linear cloning vector	50–100 ng
Molar ratio (inserts/vector)	1:3
2× seamless cloning mix	10 μL
Nucleases-free water	Up to 20 μL

4. Heat shock the competent cells for 55 s in the 42 °C water bath.
5. Remove the cells from the 42 °C water bath and place on ice for 3 min.
6. Add 500 μL LB medium in competent cells, and shake at 37 °C for 45 min at 150 rpm in a shaking incubator.
7. Dry the separate plates containing appropriate (100 μg/mL) antibiotics.
8. Spin down competent cells at 3000 rpm (~860 × *g*) for 2 min in order to enrich cells, and then remove 300 μL supernatant. Mix the enriched cells well, and pipette onto separate plates. Use glass beads to coat uniformly.
9. Incubate plates upside down overnight in a 37 °C incubator.
10. Pick single colony, add to 10 mL LB medium with appropriate antibiotics (100 μg/mL), and shake at 37 °C for 12 h at 150 rpm/min.
11. Send for sequencing to ensure the colonies are correct.
12. Purify plasmids of correct sequence by mini-plasmid isolation kit.

3.3 Expression in E. coli

1. Repeat Subheading 3.2, **steps 1–10** with two changes: Use BL21 DE3 competent cells instead of Top10 or DH5α cells. Add 1 μL plasmids to competent cells described in Subheading 3.2, **step 2**.
2. Take 1 mL bacterial liquid from the Subheading 3.2, **step 10**, add to 100 mL LB medium with appropriate antibiotics (100 μg/mL), and incubate overnight at 37 °C at 150 rpm.
3. The following morning, transfer 30 mL overnight grown cells into 1 L LB medium with appropriate antibiotics (100 μg/mL) to amplify culture.
4. Incubate at 37 °C at 150 rpm until the OD_{600} reaches 0.6–0.8 (*see* **Note 10**).

5. Induce with isopropyl β-D-thiogalactoside (IPTG) at final concentration of 0.5 mM (*see* **Note 11**).
6. Incubate additional 20 h at 16 °C at 150 rpm.
7. Harvest cells by centrifugation using 500 mL centrifuge bottles at 8000 × *g* for 15 min.
8. Remove the supernatant and scoop the bacteria pellet into 50 mL tube; store at −40 °C (*see* **Note 12**).

3.4 Ni-NTA Purification

1. Take the frozen bacteria pellet from the −40 °C freezer, and resuspend the cells in 80 mL lysis buffer (*see* **Note 13**).
2. Vortex the cells until all are thawed, and solubilize into lysis buffer; ensure there are no chunks.
3. Add protease inhibitor phenylmethylsulfonyl fluoride (PMSF) at a final concentration of 1 mM and mix well.
4. Sonicate the pellet on ice with pulse of 10 s on/50 s off at 40% amplitude for a total time of 30 min.
5. Centrifuge the lysate at 18000 × *g* for 30 min at 4 °C (*see* **Note 14**).
6. Transfer the supernatant to a 100 mL beaker for isolation. Remove 30 μL supernatant, and precipitate in a 1.5 mL EP tube for 12% SDS-PAGE analysis. Label "S" and "P," respectively.
7. Load the supernatant onto the affinity column slowly, and allow the solution drip through the stopcock valve. Collect 30 μL flow-through fluid in a 1.5 mL EP tube, and label "Ft."
8. Wash the column with 100 mL wash buffer. Collect 30 μL flow-through from drip through the stopcock valve at three points when wash buffer is loaded at 20 mL, 50 mL, and 100 mL to 1.5 mL EP tubes and label "W_1," "W_2," and "W_3," respectively.
9. Elute the protein with 20 mL elution buffer and collect the solution. Place 30 μL elute solution in the 1.5 mL EP tube, and label "E."
10. Pipette 10 μL 4× loading buffer into each sample and mix well. Boil the samples for 10 min at 100 °C to denature the proteins.
11. Load 10 μL of each sample into electrophoresis gel, and run for 50 min at 220 V (*see* **Note 15**).
12. Analyze the results (Fig. 1).

3.5 Size Exclusion Chromatography (SEC)

1. Equilibrate the 120 mL Hiload Superdex 200 10/300 GL column on the ÄKTA purifier system with 140 mL ÄKTA buffer. Set up parameter of ÄKTA system for equilibration; the details are listed in Table 6.

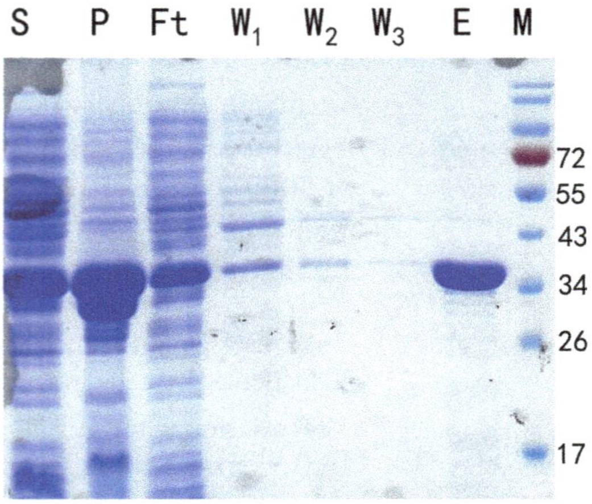

Fig. 1 SDS-PAGE analysis of the purification of MICU2-NΔ84-CΔ28 by Ni-NTA. Lane "S," the supernatant after centrifugation; lane "P," the precipitant after centrifugation; lane "Ft," the sample flow through from stopcock; lanes "W_1," "W_2," and "W_3," the samples from stopcock when 20 mL, 50 mL, and 100 mL wash buffers were loaded; lane "E," the elute solution; lane "M," the molecular weight marker

Table 6
Parameter setting for ÄKTA equilibration

Parameter	Setting
Flow	1 mL/min
Pump wash basic	on
Alarm pressure (MPa)	0.5 MPa
End time (volume)	140 mL

2. Concentrate the protein solution from the Subheading 3.4, **step 9** using 10 kDa ultrafiltration at 3000 × *g* until it reaches 1 mL (*see* **Note 16**).
3. Transfer the protein from ultrafiltration to 1.5 mL EP tube, and centrifuge at 12000 × *g* for 10 min.
4. Wash the sample loop (1 mL) with ÄKTA buffer using 1 mL syringe for five times, and place the sample tubes.
5. Inject the sample, and run the ÄKTA purifier system with 1 mL/min flow rate for 150 mL volume (*see* **Note 17**).
6. Fractionate the sample at appropriate volume with 1 mL fractionation size.
7. Save the running result (Fig. 2).
8. Fetch 30 μL each sample from the main peak, and add 10 μL loading buffer for SDS-PAGE analysis (*see* **Note 18**). Repeat the Subheading 3.4, **steps 10–12** (Fig. 3).

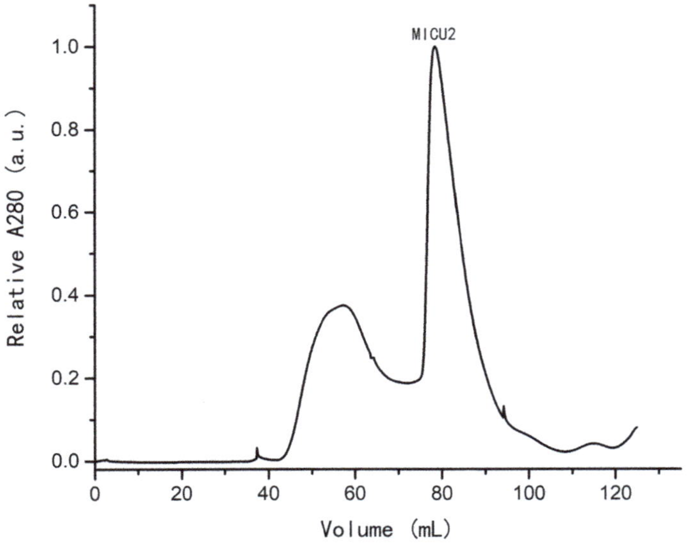

Fig. 2 SEC profile of MICU2-NΔ84-CΔ28 in Ca^{2+}-free condition. The sharp peak corresponds to MICU2 at 280 nm absorption wavelength

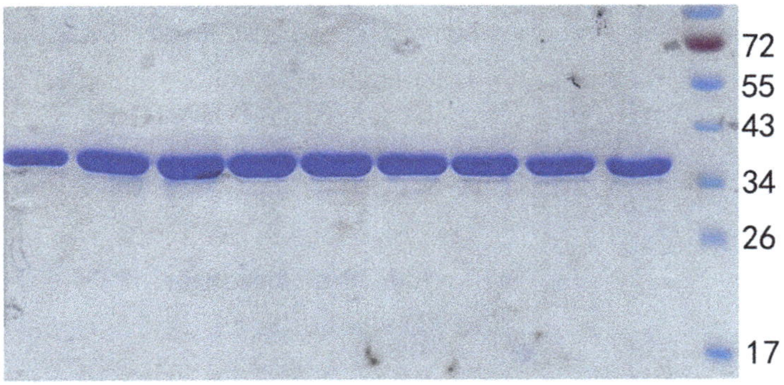

Fig. 3 SDS-PAGE analysis of the fractions of MICU2-NΔ84-CΔ28 in Ca^{2+}-free condition taken from the SEC peak (Fig. 2). The molecular weight of MICU2 truncation is ~40 kDa

3.6 Pull-Down Assay

1. Prepare two 5 mL gravity columns, and pipette 100 μL GST beads (or Ni-NTA) into each column. Equilibrate with lysis buffer (*see* **Note 13**).
2. Take the frozen cell pellets from the −40 °C freezer, and weigh 0.5 g pellet of GST-tagged MICU1 to a 20 mL container.
3. Suspend the pellet with 10 mL lysis buffer, and vortex the cells until all solubilize into lysis buffer without any chunks.
4. Sonicate the pellet on ice with pulse of 10 s on/50 s off at 40% amplitude for the total time of 10 min.
5. Centrifuge the lysate at 18000 × *g* for 30 min at 4 °C.

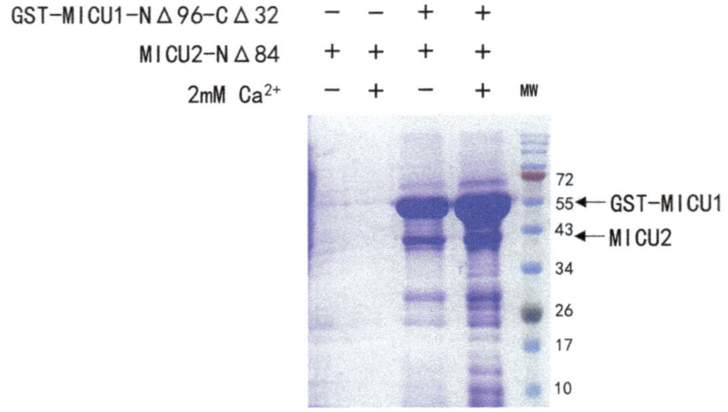

Fig. 4 Pull-down analysis using reducing SDS-PAGE. MICU2 can interact with MICU1 both in Ca^{2+}-free and Ca^{2+}-bound conditions

6. Take 30 μL supernatant as a sample of the input. Load the supernatant onto the GST column slowly as Subheading 3.4, **step** 7. Simultaneously, weigh 1 g MICU2 pellet, and suspend with 20 mL appropriate lysis buffer. Repeat Subheading 3.6, **steps 3–5**.
7. Wash the GST column with 50 mL wash buffer.
8. Fetch 30 μL MICU2 supernatant as a sample of the input. Split the supernatant of MICU2 into two parts with equal volume. Load the supernatant onto both GST columns (with and without MICU1) slowly as Subheading 3.4, **step** 7.
9. Wash both GST columns with 50 mL wash buffer.
10. Add 30 μL beads into 1.5 mL EP tubes, and add 10 μL 4× loading buffer into all samples. Boil the samples for 10 min at 100 °C.
11. Analyze the result using SDS-PAGE, and repeat Subheading 3.4, **steps 10–12** (Fig. 4).

3.7 Multi-angle Laser Light Scattering (MALLS)

1. Purify the protein samples as described above. Equilibrate the column with ÄKTA buffer (*see* **Note 19**).
2. Collect the protein samples corresponding to the main peak, and concentrate using a 10 kDa ultrafiltration at 3000 × *g* until it reaches 2 mg/mL.
3. Filter the protein through a 0.22 μm filter.
4. Wash the sample loop (200 μL) with ÄKTA buffer using a 1 mL syringe for five times. Inject 200 μL of the protein sample, and run the system with 0.5 mL/min flow rate for 40 min (Fig. 5).
5. Calculate the molecular weight by peak integration (ASTRA software).

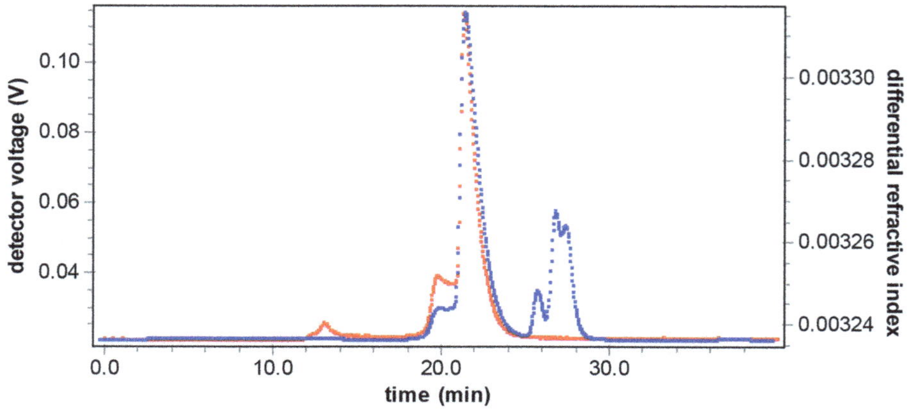

Fig. 5 MALLS results for MICU2-NΔ84-CΔ28 in Ca^{2+}-free condition. The red and blue lines represent the laser light scattering data and differential refractive index data, respectively

4 Notes

1. pET-28a (+) and pGEX-6p-1 contain kanamycin and ampicillin resistance genes, respectively.
2. If the medium congeals, reheat it in microwave.
3. For all pET-28a (+) clones, the reverse primers are designed with a termination codon to ensure a lack of His_6-tag in the C-terminus. In our clones, all of MICU2 constructions are designed with N-terminal His_6-tag due to NdeI restriction site. Conversely, MICU1 with NcoI restriction site lacks His_6-tag.
4. ÄKTA buffers are required to be extremely fresh and vacuum pumped for 30 min to remove the air dissolved in the buffer.
5. If the pfu polymerase does not contain $MgSO_4$ and dNTPs, add 2 μL 2 mM $MgSO_4$ and 5 μL 2 mM dNTPs to the reaction system.
6. If it still fails, add 1–2 μL DMSO, betaine, or 0.1–1% Triton X-100 to the reaction mixture to enhance the efficiency of PCR.
7. The pfu DNA polymerase takes approximately 1 min for every 1 kb to be amplified. Thus, the extension time will depend on the length of the target gene. Also, the annealing temperature for an amplification reaction will rely on the T_m of the two primers.
8. High concentration of the enzyme or prolonged reaction time could contribute to star activity.
9. Do not mix by pipetting up and down.

10. When the OD_{600} reaches 0.6–0.8, change the temperature from 37 °C to 16 °C.
11. Add IPTG when the temperature reaches 16 °C.
12. 1 L culture can be collected in a 50 mL tube.
13. Depending on the condition (Ca^{2+}-bound or Ca^{2+}-free), add the appropriate buffer with or without 2 mM $CaCl_2$.
14. Simultaneously, pipette 2 mL Ni-NTA resin into 20 mL gravity column, and equilibrate with lysis buffer. Prepare the ÄKTA buffer.
15. 2 μL molecular weight marker is sufficient.
16. If the final volume is smaller than 1 mL, add the ÄKTA buffer to reach 1 mL.
17. Injection of the sample must avoid air bubbles or precipitate.
18. The samples can be stored at −80 °C until future use.
19. In this section, equilibration buffer must be identical to the ÄKTA buffer.

References

1. Rizzuto R, De Stefani D, Raffaello A, Mammucari C (2012) Mitochondria as sensors and regulators of calcium signalling. Nat Rev Mol Cell Biol 13(9):566–578. https://doi.org/10.1038/nrm3412
2. O'Rourke B (2007) Mitochondrial ion channels. Annu Rev Physiol 69:19–49. https://doi.org/10.1146/annurev.physiol.69.031905.163804
3. Santo-Domingo J, Demaurex N (2010) Calcium uptake mechanisms of mitochondria. Biochim Biophys Acta 1797(6–7):907–912. https://doi.org/10.1016/j.bbabio.2010.01.005
4. Drago I, Pizzo P, Pozzan T (2011) After half a century mitochondrial calcium in- and efflux machineries reveal themselves. EMBO J 30 (20):4119–4125. https://doi.org/10.1038/emboj.2011.337
5. De Stefani D, Raffaello A, Teardo E, Szabo I, Rizzuto R (2011) A forty-kilodalton protein of the inner membrane is the mitochondrial calcium uniporter. Nature 476(7360):336–340. https://doi.org/10.1038/nature10230
6. Raffaello A, De Stefani D, Sabbadin D, Teardo E, Merli G, Picard A, Checchetto V, Moro S, Szabo I, Rizzuto R (2013) The mitochondrial calcium uniporter is a multimer that can include a dominant-negative pore-forming subunit. EMBO J 32(17):2362–2376. https://doi.org/10.1038/emboj.2013.157
7. Sancak Y, Markhard AL (2013) EMRE is an essential component of the mitochondrial calcium uniporter complex. Science 342 (6164):1379–1382. https://doi.org/10.1126/science.1242993
8. Perocchi F, Gohil VM, Girgis HS, Bao XR, McCombs JE, Palmer AE, Mootha VK (2010) MICU1 encodes a mitochondrial EF hand protein required for Ca(2+) uptake. Nature 467(7313):291–296. https://doi.org/10.1038/nature09358
9. Plovanich M, Bogorad RL, Sancak Y, Kamer KJ, Strittmatter L, Li AA, Girgis HS, Kuchimanchi S, De Groot J, Speciner L, Taneja N, Oshea J, Koteliansky V, Mootha VK (2013) MICU2, a paralog of MICU1, resides within the mitochondrial uniporter complex to regulate calcium handling. PLoS One 8(2):e55785. https://doi.org/10.1371/journal.pone.0055785
10. Oxenoid K, Dong Y, Cao C, Cui T, Sancak Y, Markhard AL, Grabarek Z, Kong L, Liu Z, Ouyang B, Cong Y, Mootha VK, Chou JJ (2016) Architecture of the mitochondrial calcium uniporter. Nature 533(7602):269–273. https://doi.org/10.1038/nature17656
11. Tsai MF, Phillips CB, Ranaghan M, Tsai CW, Wu Y, Willliams C, Miller C (2016) Dual functions of a small regulatory subunit in the mitochondrial calcium uniporter complex. elife 5: e15545. https://doi.org/10.7554/eLife.15545

12. Patron M, Checchetto V, Raffaello A, Teardo E, Vecellio Reane D, Mantoan M, Granatiero V, Szabo I, De Stefani D, Rizzuto R (2014) MICU1 and MICU2 finely tune the mitochondrial Ca2+ uniporter by exerting opposite effects on MCU activity. Mol Cell 53 (5):726–737. https://doi.org/10.1016/j.molcel.2014.01.013
13. Kamer KJ, Mootha VK (2014) MICU1 and MICU2 play nonredundant roles in the regulation of the mitochondrial calcium uniporter. EMBO Rep 15(3):299–307. https://doi.org/10.1002/embr.201337946
14. Li D, Wu W, Pei H, Wei Q, Yang Q, Zheng J, Jia Z (2016) Expression and preliminary characterization of human MICU2. Biol Open 5 (7):962–969. https://doi.org/10.1242/bio.018572
15. Kamer KJ, Grabarek Z, Mootha VK (2017) High-affinity cooperative Ca2+ binding by MICU1-MICU2 serves as an on-off switch for the uniporter. EMBO Rep 18 (8):1397–1411. https://doi.org/10.15252/embr.201643748

Chapter 18

S100 Proteins in the Innate Immune Response to Pathogens

Natalia Kozlyuk, Andrew J. Monteith, Velia Garcia, Steven M. Damo, Eric P. Skaar, and Walter J. Chazin

Abstract

S100 proteins are distinct dimeric EF-hand Ca^{2+}-binding proteins that can bind Zn^{2+}, Mn^{2+}, and other transition metals with high affinity at two sites in the dimer interface. Certain S100 proteins, including S100A7, S100A12, S100A8, and S100A9, play key roles in the innate immune response to pathogens. These proteins function via a "nutritional immunity" mechanism by depleting essential transition metals in the infection that are required for the invading organism to grow and thrive. They also act as damage-associated molecular pattern ligands, which activate pattern recognition receptors (e.g., Toll-like receptor 4, RAGE) that mediate inflammation. Here we present protocols for these S100 proteins for high-level production of recombinant protein, measurement of binding affinities using isothermal titration calorimetry, and an assay of antimicrobial activity.

Key words S100 proteins, S100A7, S100A12, S100A8, S100A9, Calprotectin, Nutritional immunity, Metal binding, Host-pathogen interaction, Inflammatory response, Protein expression, Protein purification, Isothermal titration calorimetry, Antimicrobial growth assay

1 Introduction

1.1 S100 Proteins

S100 proteins are a unique subfamily of EF-hand calcium (Ca^{2+})-binding proteins that have an integrated dimeric structure and can form higher-order oligomers [1]. Although EF-hand proteins are widely known as second messengers that transduce Ca^{2+} signals, S100 proteins also function in the extracellular milieu [2, 3]. Remarkably, several S100 proteins have been shown to play central roles in the innate immune response to infection by pathogenic organisms, with dual functions via "nutritional immunity" and activation of inflammation [4]. Of the 25 different S100 genes, only S100A7, S100A8, S100A9, and S100A12 have been identified as innate immune modulators.

Natalia Kozlyuk, Andrew J. Monteith, and Velia Garcia are contributed equally to the chapter.

Claus W. Heizmann (ed.), *Calcium-Binding Proteins of the EF-Hand Superfamily: From Basics to Medical Applications*, Methods in Molecular Biology, vol. 1929, https://doi.org/10.1007/978-1-4939-9030-6_18,

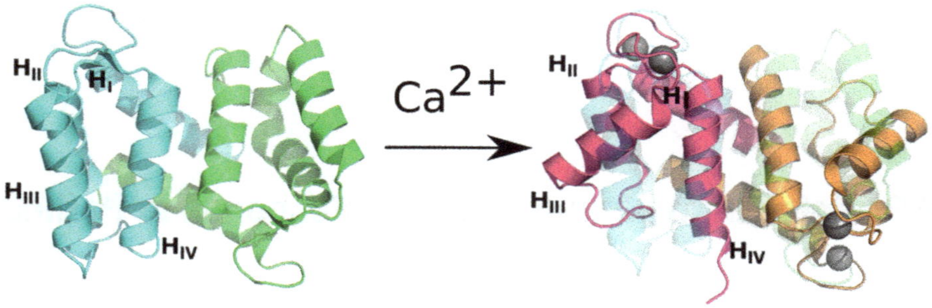

Fig. 1 The effect of Ca^{2+} binding on the structure of S100A12. The Ca^{2+}-free protein (green and cyan) is shown on the left, and the Ca^{2+}-bound state (magenta and orange) is shown on the right, superimposed on the Ca^{2+}-free state. Note the change in position of helix 3 between the Ca^{2+}-free and Ca^{2+}-bound states. PDB ID: 2WCE and 2M9G

1.2 Structure and Function of S100 Proteins

S100 proteins contain N-terminal S100-specific and C-terminal canonical, "EF-hand" helix-loop-helix calcium-binding motifs that come together to form the stable four-helix bundle characteristic of all EF-hand proteins. However, S100 proteins are distinct among the EF-hand proteins in that the basic structural unit is a completely integrated, obligate dimer [1]. All S100 proteins except the ancestral and C-terminally truncated S100G form homodimers, and a select few form heterodimers. However, S100A8 and S100A9 are unique in that they preferentially form heterodimers [5].

Ca^{2+} binding results in a significant shift in the orientation of helix 3 (Fig. 1), exposing a hydrophobic cleft that mediates binding to target proteins [6]. Because the levels of Ca^{2+} present in the extracellular milieu are so high (in the mM range), when outside cells S100 proteins are constitutively bound to Ca^{2+} and activated for target binding. Besides binding Ca^{2+}, many S100 proteins also bind transition metal ions (e.g., Mn^{2+}, Cu^{2+}, Fe^{2+}, Ni^{2+}, and Zn^{2+}) [7]. Each S100 dimer has two transition metal-binding sites at symmetrically disposed locations across the dimer interface (Fig. 2) [7, 8]. Binding of transition metals has been characterized for S100A12 [9], S100A7 [10], and S100A8/S100A9 heterodimer (calprotectin, CP) [8]. S100A7 and S100A12 have canonical sites comprised of three histidine and one aspartic acid residue that coordinate Zn^{2+} or Cu^{2+} in a tetrahedral fashion [7]. CP has two distinct transition metal-binding sites. One is specific to CP and is comprised of six histidine residues arranged in an octahedral coordination sphere, capable of binding Mn^{2+}, Zn^{2+}, Ni^{2+}, Cu^{2+} and Fe^{2+} (S1). The other (S2) is a canonical site that can bind Zn^{2+} and Cu^{2+}. Although binding of transition metal ions minimally perturbs S100 protein structures, high concentrations of these ions (and also Ca^{2+}) induce the formation of tetramers and higher-order oligomers. In some cases, it appears that these oligomeric states have a key role for function [3, 11].

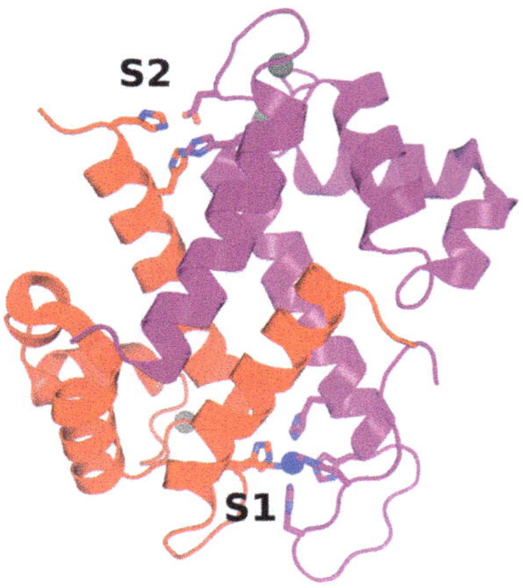

Fig. 2 Structure of Ca^{2+}- and Mn^{2+}-bound CP. S100A8 is colored red, and S100A9 is colored purple. Ca^{2+} ions are gray spheres and the Mn^{2+} ion is a blue sphere. The side chains chelating transition metal ions in S1 and S2 are shown. PDB ID: 4GGF

1.3 S100 Proteins in Nutritional Immunity

During infections, pathogens acquire nutrient metals from the host, while the host in turn limits the availability of nutrient metals [12, 13]. S100 proteins play a central role in restricting the bioavailability of nutrient transition metals, a mechanism termed "nutritional immunity" [14]. In keeping with this, increasing dietary transition metals can overcome this defense and in turn cause detrimental effects for the host during infection. For example, dietary iron supplementation enhances *Mycobacterium avium* [15] and *Mycobacterium tuberculosis* [16] infection, excess dietary zinc heightens *Clostridium difficile* colonization of the gut [17], and elevated dietary manganese increases susceptibility of the heart to *Staphylococcus aureus* infection [18]. Consistent with CP playing a critical role during infection, a deficiency in CP enhances susceptibility of the murine host to numerous bacterial and fungal pathogens [19–24]. S100A12 inhibits *Campylobacter jejuni* [25] and *Helicobacter pylori* [26] through Zn^{2+}-dependent sequestration. Additionally, S100A12 exhibits antimicrobial activity against multiple parasites, possibly through the binding of copper, which enhances the production of superoxide [27, 28]. In 2007, it was reported that S100A7 and S100A15 have antimicrobial activity toward *Escherichia coli* during infections of the skin [29, 30]. In addition to limiting metal nutrients, S100A7 can directly adhere to and reduce the survival of *E. coli* in the skin [31]. These observations suggest that the ability of S100 proteins to bind a variety of nutrient metal ions is critical in establishing nutritional immunity and protection during infections.

1.4 S100 Proteins in the Inflammatory Response

In addition to nutritional immunity, S100 proteins play diverse roles as immunomodulators. S100A7 and S100A15 proteins are expressed in keratinocytes [32, 33], whereas CP and S100A12 are primarily expressed in monocytes, macrophages, and neutrophils [28, 34, 35]. However, during the inflammatory response, CP and S100A12 are also expressed in keratinocyte, endothelial, and epithelial cells [36–39]. Secreted CP can act as a potent chemoattractant for monocytes, macrophages, and neutrophils during inflammation [40, 41]. As a result, CP inhibits immune cell growth or induces apoptosis [17]. In addition, S100 proteins can act as damage-associated molecular patterns by binding pattern recognition receptors including TLR4 [42], RAGE [43, 44], and CD33 [45]. Misregulation of S100 protein expression is implicated in chronic inflammation [46], as overactivation of the pattern recognition receptors TLR4 and RAGE drives NF-κB production and recruitment of neutrophils, monocytes, and macrophages. For example, high levels of calprotectin and S100A12 have been associated with inflammatory bowel disease and rheumatoid arthritis and high levels of S100A7 with inflammatory skin diseases (e.g., psoriasis, atopic dermatitis) [17, 30, 44]. Altered expression of S100 proteins also correlates with multiple autoimmune disorders including systemic lupus erythematosus, Still's, and Sjogren's [47–50]. Given the numerous associations to inflammation and infections, S100 proteins have been targeted for potential therapeutic effects [51].

2 Materials

2.1 Expression and Purification of S100 Proteins

2.1.1 Reagents and Solutions

1. pGEMEX plasmid containing S100A12.
2. BL21 (DE3) competent cells.
3. LB/agar plates supplemented with 100 μg/mL ampicillin.
4. Autoclaved Luria broth (LB) growth media.
5. 100 mg/mL ampicillin solution sterile-filtered prior to use
6. Ammonium sulfate.
7. Lysis buffer: 20 mM Tris–HCl (pH 8.0).
8. Q buffer A: 20 mM Tris–HCl (pH 8.0).
9. Q buffer B: 20 mM Tris–HCl (pH 8.0), 1 M NaCl.
10. SEC buffer: 20 mM Tris–HCl (pH 8.0), 100 mM NaCl.
11. SDS-PAGE buffers and Coomassie stain.

2.1.2 Equipment

1. Autoclave.
2. 0.45 and 0.2 μm syringe filters.
3. Plastic syringes.
4. Appropriate flasks and petri dishes.

5. Shaking incubator.
6. Sonication system.
7. Centrifuges.
8. Dialysis tubing with a 1 kDa MWCO.
9. 10 kDa MWCO concentrators.
10. FPLC system.
11. Q-sepharose column.
12. S75 column.
13. SDS-PAGE system and gels.

2.2 Determination of Transition Metal Affinities of S100 Proteins

2.2.1 Reagents and Solutions

1. CP protein (~2 mL at 10–50 μM) in ITC buffer.
2. ITC buffer: 20 mM HEPES (pH 7.5), 75 mM NaCl.
3. $Zn(OAc)_2$ in ITC buffer.
4. 1 M $CaCl_2$.

2.2.2 Equipment

1. VP-ITC titration calorimeter.
2. Degassing station.

2.3 Antimicrobial Growth Assay

2.3.1 Reagents and Solutions

1. Tryptic Soy Agar (TSA) plate.
2. Tryptic Soy Broth (TSB).
3. CP protein in CP buffer.
4. CP buffer: 20 mM Tris (pH 7.5), 100 mM NaCl, 3 mM $CaCl_2$, 10 mM BME, filtered.

2.3.2 Equipment

1. 96-well microtiter plate.
2. Plastic wrap.
3. UV/Vis spectrophotometer.

3 Methods

3.1 Expression and Purification of S100 Proteins

A lac operon-based system is used for bacterial expression of S100 proteins. Following standard protocols, *Escherichia coli* (*E. coli*) cells are transformed with a plasmid containing the gene of interest, grown to a high optical density, and stimulated to express the protein of interest by the addition of isopropyl β-D-1-thiogalacto-pyranoside (IPTG). After cell lysis, the protein is purified from the soluble lysate. Purification takes advantage of the extraordinary structural stability of S100 proteins. The name S100 is derived from the observation that these proteins remain soluble in 100% ammonium sulfate [52]. Hence, the first step of the purification is an ammonium sulfate precipitation that increases the ionic strength of the solution and results in the precipitation of most of the endogenous *E. coli* proteins, while S100 remains in solution. The next step is anion-exchange chromatography for further separation of impurities by differences in isoelectric point. The last step of the purification involves size-exclusion chromatography to separate the S100 protein from its impurities by size. Finally, mass spectrometry of the purified product confirms the identity of the protein of interest. Protocols for S100A7, S100A8, S100A9, S100A12, and CP have been published [5, 14, 53–55]. Here we provide a specific detailed example, an optimized 1 L expression and purification protocol for human S100A12.

3.1.1 Procedure

Expression

1. Transform pGEMEX-S100A12 into BL21 (DE3) chemically competent cells using a standard transformation protocol [56] (*see* **Note 1**).
2. Plate 100 μL of transformation solution on LB/agar plate supplemented with ampicillin, and grow 12–16 h at 37 °C.
3. Pick a single colony from the freshly transformed plate, and inoculate 25 mL of LB media supplemented with antibiotic.
4. Grow this culture in a shaking incubator at 250 rpm for 12–16 h at 37 °C.
5. In the morning, inoculate 10 mL of the culture to inoculate 1 L of LB supplemented with antibiotic (*see* **Note 2**).
6. Grow this culture at 37 °C in a shaking incubator at 250 rpm.
7. While shaking the culture, monitor cell growth. Once the optical density at 600 nm ($O.D._{600}$) is between 0.6 and 0.8 absorbance units, add IPTG to a concentration of 0.5 mM.
8. Allow the culture to grow at 37 °C in a shaking incubator at 250 rpm for 12–16 h.
9. Harvest the cells by centrifugation at 5000 × *g* for 20 min, and freeze the pellets at −80 °C.

Purification

1. Thaw the frozen cell pellet at room temperature, and resuspend in 30 mL of lysis buffer.
2. Sonicate on ice at 40% amplitude (5 s on, 5 s off) for 5 min (*see* **Note 3**).
3. Clarify the lysate by centrifugation at 20,000 × *g* for 25 min at 4 °C.
4. The S100A12 protein will be in the supernatant (*see* **Note 4**). Decant supernatant into 100 mL glass beaker and place the beaker on ice. Add a stir bar and slowly add ammonium sulfate to 60% saturation (*see* **Note 5**). Allow the solution to stir in the cold room for 1 h.

5. Clarify the solution by centrifugation at 20,000 × *g* for 25 min at 4 °C.
6. Transfer the supernatant into dialysis tubing. Dialyze against 2 L of Q buffer A at 4 °C. Change the dialysis buffer twice, allowing at least 4 h in between changes (*see* **Note 6**).
7. Using the FPLC system, equilibrate a Q-Sepharose column with 4 column volumes (CV) of Q Buffer A.
8. Load the protein sample and collect flow through.
9. Wash the column with four CV of Q buffer A.
10. Elute proteins with a 0–30% NaCl gradient over 15–20 CV using Q buffer B, collecting 5 mL fractions.
11. Take an aliquot of each fraction and run SDS-PAGE with Coomassie staining.
12. Pool the fractions containing S100A12 protein. The protein has a molecular weight of ~10.5 kDa.
13. Concentrate the fractions to <5 mL using a 10 kDa MWCO concentrator.
14. Equilibrate an S75 column with 1 CV of SEC buffer (*see* **Note 7**).
15. Load the concentrated S100A12 protein onto the SEC column.
16. Elute the protein over 1 CV, collecting 5 mL fractions.
17. Take an aliquot of each fraction, and run SDS-PAGE with Coomassie staining.
18. Pool together the S100A12 protein fractions, and validate the protein identity using mass spectrometry.
19. Measure the absorbance of the protein using a spectrophotometer in order to obtain the concentration. The extinction coefficient of S100A12 is 5960 M^{-1} cm^{-1} (*see* **Note 8**).
20. Concentrate protein to ~5–10 mg/mL. Concentrated protein will reduce the risk of degradation. Aliquot into 1.5 mL Eppendorf tubes, and flash freeze using liquid nitrogen. Store the protein at −80 °C (*see* **Note 9**).

3.2 Determination of Transition Metal Affinities of S100 Proteins by Isothermal Titration Calorimetry

Determining transition metal ion-binding affinities of S100 proteins is one key to understanding their role at the host-pathogen interface as this informs on their ability to function via the nutritional immunity mechanism, i.e., outcompete bacterial metal ion transporters for the nutrient metals [53]. Many techniques have been implemented to determine the dissociation constants (K_d) for S100 proteins, which have been reported in the μM-pM range [7]. Isothermal titration calorimetry (ITC), which measures the amount of heat released or absorbed upon binding, has been used

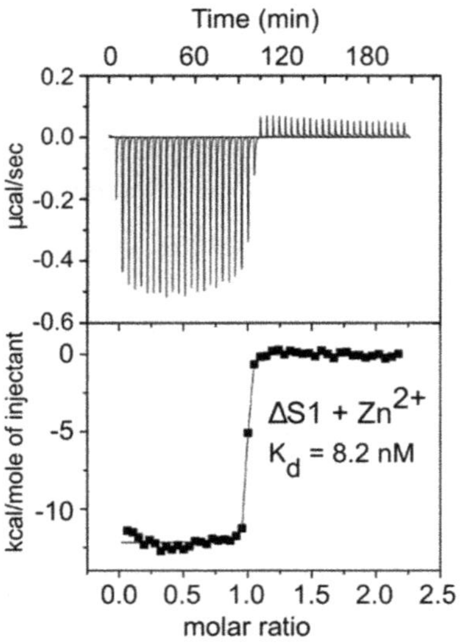

Fig. 3 The binding of Zn^{2+} to CP monitored by ITC. ITC isotherms of CP as Zn^{2+} are titrated from the syringe into the cell with S1 or S2 knockout mutants (ΔS1, ΔS2). Note the modest differences in K_d values between S1 and S2. These panels were taken from Fig. 1 in the original publication in Damo et al. [8]

to measure μM-nM transition metal ion-binding affinities of S100 proteins [8, 57, 58]. ITC estimates of the binding affinities of CP for Zn^{2+} (S1-3 nM, S2-8 nM) and Mn^{2+} (S1-6 nM) have been reported (Fig. 3). Additional methods for determining metal ion affinities include optical absorption spectroscopy [53], electron paramagnetic resonance spectroscopy [53], competition chelator fluorescence experiments [53, 59], and equilibrium gel filtration [60, 61], each method having its own set of advantages and limitations. Competition chelator fluorescence measurements are the most sensitive and are able to measure K_d values in the pM range, although this method requires transition metal-specific fluorescence dyes and can be complicated by interference from Ca^{2+} [62]. Here, we provide a protocol for measuring transition metal binding to CP protein using ITC. This is the only experimental method that allows for the complete thermodynamic characterization of a reaction (ΔH, ΔG, and ΔS) and also provides the stoichiometry of the reaction.

3.2.1 Procedure

1. Dialyze CP protein (10–50 μM) in ITC buffer for 4–6 h. Replace with fresh buffer twice for a total of three dialysis exchanges (*see* **Note 10**).
2. Prepare a 100–750 μM $Zn(OAc)_2$ solution using the dialysis ITC buffer. Save the remaining dialysis buffer for rinsing the syringe and sample cell of the ITC instrument (*see* **Note 11**).

3. Add stoichiometric amount of $CaCl_2$ to both the protein and the metal solutions.
4. Degas the protein and metal ion solutions for 10 min.
5. Rinse the ITC sample cell three times with ITC buffer.
6. Add the CP protein to the sample cell of the ITC instrument. Ensure that no air bubbles are present in the syringe.
7. Load the $Zn(OAc)_2$ solution into the ITC titration syringe.
8. Titrate the metal ion solution into the protein. We recommend starting with 50 injections, each of 6 μL and 210 s between each injection. The number of injections, injection volumes, and the time between injections must be optimized depending on the metal-binding affinities and the heat absorbed or released upon binding of the system. Before performing another injection, make sure that the heat from each injection is fully dissipated (*see* **Note 12**).
9. To correct for the heat of dilution, perform a metal solution to buffer titration after thoroughly rinsing out the sample cell with the ITC buffer. Repeat **steps 2–8**, except add only ITC buffer to the sample cell (*see* **Note 13**).
10. To process the data, subtract the buffer isotherm from the protein isotherm. Fit a single-site binding model to the data using the VP-ITC Origin software (Northampton, MA), and obtain the thermodynamic parameters (*see* **Note 14**). An example of data depicting the titration of CP mutants with $ZnSO_4$ is shown in Fig. 3.

3.3 Antimicrobial Growth Assay

The ability of S100 proteins to sequester nutrient transition metals inhibits growth of pathogens [20, 22, 26]. Here we describe an antimicrobial growth assay that measures the activity of WT and mutant CP, monitoring the ability of CP to inhibit the growth of *Staphylococcus aureus.* This assay verifies the ability of CP to sequester transition metal ions. The results of the assay are reported as $O.D._{600}$ plotted over time, quantifying bacterial growth in the presence of decreasing concentrations of CP (Fig. 4) [20]. Typically, CP at a concentration $\geq$ 75 μg/mL is sufficient to inhibit the growth of *S. aureus* in minimal media; however, the inhibition of bacterial growth is also species dependent with different concentrations of CP required for inhibition. *Salmonella* is greater than 125 μg/mL. *Acinetobacter baumannii* is greater than 37.5 μg/mL. *Helicobacter pylori* is greater than 300 μg/mL. *Pseudomonas aeruginosa* and *Clostridium difficile* require 1 mg/mL. Inhibition of bacterial growth by CP could vary based on transition metal enrichment of the media chosen to facilitate bacterial growth. This same assay can be implemented to probe the antimicrobial activity of other S100 proteins such as S100A12 [26].

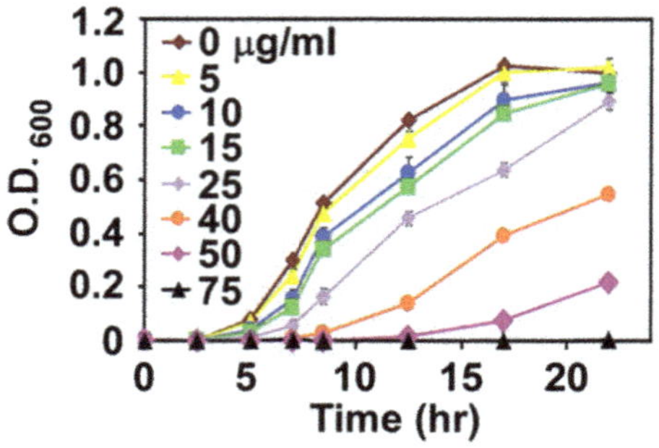

Fig. 4 Representative growth assay of *S. aureus* in the presence of increasing concentrations of recombinant CP. Growth was quantified by O.D.$_{600}$. Error bars represent the standard deviation of at least three replicates. Panel is adapted from the original publication in Corbin et al. [20]

3.3.1 Procedure

1. Streak a TSA plate with *S. aureus*, and incubate at 37 °C for 16–18 h (*see* **Note 15**).
2. Select a single colony of *S. aureus*, start a 5 mL overnight culture in TSB, and incubate at 37 °C on a shaker for 16–18 h at 180 rpm.
3. Back dilute the overnight culture 1:50 in fresh TSB, and incubate at 37 °C on a roller drum for 1 h at 40 rpm.
4. During the incubation, prepare a 96-well microtiter plate as follows:
   (a) Rows B–H will each contain 62 μL CP buffer and 38 μL TSB (*see* **Note 16**).
   (b) Row A will contain the highest concentration of CP with CP buffer to a total volume of 124 μL. In addition, each well in this row will have 76 μL TSB. For example, if you want 200 μg of protein in row A and have a protein stock that is 10 μg/μL, you would add 20 μL protein stock to 104 μL CP buffer.
   (c) The CP stock should be thawed and supplemented with 3 mM $CaCl_2$ prior to use (*see* **Note 17**).
   (d) To create a 1:2 serial dilution of CP for this assay, transfer 100 μL from row A into row B. Pipet to mix, and transfer 100 μL from row B to row C. Repeat until you transfer 100 μL from row F into row G. After mixing row G, remove 100 μL from row G so that there is a final volume of 100 μL. Do not transfer row G to row H. Row H should only contain CP buffer and TSB without protein (*see* **Note 18**).
   (e) After adding all the media components to the wells, briefly centrifuge (800 r.c.f., 2 min) the plate to remove bubbles.

5. After 1 h, add 1 μL of back-diluted culture from **step 3** to each well.
6. Measure initial O.D.$_{600}$ to establish background and 0 h time point (*see* **Note 19**).
7. Incubate the microtiter plate at 37 °C with shaking. Wrap the microtiter plate in plastic wrap to prevent evaporation (*see* **Note 20**).
8. Take time points every 1–2 h. Vortex the microtiter plate prior to reading the O.D.$_{600}$ to resuspend any bacteria that have settled to the bottom. Representative data for antimicrobial activity of WT CP against *S. aureus* are shown in Fig. 4.

4 Notes

1. S100 proteins may also be purified from *E. coli* using a lac operon-based system with a plasmid-containing N-terminal 6× His tag. The protein can then be purified by immobilized metal affinity chromatography (IMAC). This is the case for S100A7 and S100A15 [55]. A plasmid-containing N-terminal 6× His tag has been reported for S100A4 with purification via IMAC followed by anion-exchange chromatography [63].
2. The use of auto-inducing media in place of LB can significantly increase yields for S100A12 [54, 64]. Whether this is true for other S100 proteins must be experimentally verified.
3. S100 proteins are stable enough to perform the purification at room temperature with the exception of the sonication, centrifugation, and ammonium sulfate precipitation steps.
4. Purification of CP requires a protein-refolding step before the anion-exchange column [5, 14, 53], as S100A8–S100A9 are expressed into inclusion bodies rather than in the supernatant.
5. The optimum ammonium sulfate percentage must be optimized for each protein.
6. For S100 proteins containing cysteine residues, it may be necessary to include BME (or other reducing agents) to the buffers right before use to prevent disulfide bond formation.
7. Adding 0.5–5 mM of $CaCl_2$ to the chromatography buffers may increase protein stability and yields. However, this must be removed by the addition of EDTA and dialysis prior to storing the protein, as Ca^{2+} induces S100 protein oligomerization.
8. Typical yields for S100 proteins using this protocol are 35–45 mg per 1 L of culture. However, the optimal induction time must be empirically determined for each protein.

9. To maintain S100 protein activity, avoid repeated freeze-thaw cycles.
10. The protein and metal ion concentrations have to be accurately determined as this will affect the results. Protein concentrations can be experimentally determined with the use of the extinction coefficients and absorbance at 280 nm.
11. It is critical that the protein in the sample cell and the metal ion solutions in the syringe have the *exact* same buffer. The best way to ensure that the buffer is exactly the same is to prepare the syringe solutions in the dialysis buffer that the protein was dialyzed in. If the buffers are not exactly the same, fluctuations in the data will give false binding results.
12. This ITC procedure can be used for all S100 proteins; however, the number of injections and concentrations might need to be adjusted. It should also be noted that the typical sensitivity of ITC places a lower limit on K_d in the nM range. Methods described above might need to be used if the binding affinities fall outside of this range.
13. Reducing agents can adversely affect the linearity of the baseline. Whether your system can tolerate the reducing agent in the buffer must be determined empirically. In general, TCEP is preferable to BME, which is preferable to DTT. Another alternative is to make Cys to Ser mutations in your protein.
14. Repeat the experiments in triplicate.
15. This assay can use a variety of different bacterial strains, other than *S. aureus*. If another strain is used, the appropriate media for the chosen bacteria needs to be implemented.
16. A lower amount of CP buffer relative to growth media can be used in order to facilitate bacterial growth; however, this will usually increase the concentration of CP necessary to inhibit growth.
17. Add BME to the CP buffer right before use.
18. For most assays, a twofold dilution of CP starting at a concentration of 100 μg/100 μL in rows A–G will suffice for *S. aureus*.
19. With *S. aureus*, there will be some initial growth at the highest dose of CP, most likely due to residual metal ions stored within the bacteria.
20. It is possible to use an automated plate reader to quantify the $O.D._{600}$. In this situation, it is recommended for the lid to remain closed to avoid evaporation.

Acknowledgments

This work was supported by the National Institutes of Health R01AI101171 (EPS and WJC) and the National Science Foundation HRD1547757 (SMD).

References

1. Nelson MR, Chazin WJ (1998) Structures of EF-hand Ca(2+)-binding proteins: diversity in the organization, packing and response to Ca2 + binding. Biometals 11(4):297–318
2. Donato R (2003) Intracellular and extracellular roles of S100 proteins. Microsc Res Tech 15:540–551
3. Donato R, Cannon BR, Sorci G, Riuzzi F, Hsu K, Weber DJ, Geczy CL (2013) Functions of S100 proteins. Curr Mol Med 13:24–57
4. Zackular JP, Chazin WJ, Skaar EP (2015) Nutritional immunity: S100 proteins at the host-pathogen interface. J Biol Chem 290:18991–18998
5. Hunter MJ, Chazin WJ (1998) High level expression and dimer characterization of the S100 EF-hand proteins, migration inhibitory factor-related proteins 8 and 14. J Biol Chem 273:12427–12435
6. Bhattacharya S, Bunick CG, Chazin WJ (2004) Target selectivity in EF-hand calcium binding proteins. Biochim Biophys Acta 1742 (1–3):69–79
7. Gilston BA, Skaar EP, Chazin WJ (2016) Binding of transition metals to S100 proteins. Science China 59:792–801
8. Damo SM, Kehl-Fie TE, Sugitani N, Holt ME, Rathi S, Murphy WJ, Zhang Y, Betz C, Hench L, Fritz G, Skaar EP, Chazin WJ (2013) Molecular basis for manganese sequestration by calprotectin and roles in the innate immune response to invading bacterial pathogens. Proc Natl Acad Sci U S A 110 (10):3841–3846
9. Moroz OV, Blagova EV, Wilkinson AJ, Wilson KS, Bronstein IB (2009) The crystal structures of human S100A12 in apo form and in complex with zinc: new insights into S100A12 oligomerisation. J Mol Biol 391:536–551
10. Brodersen DE, Nyborg J, Kjeldgaard M (1999) Zinc-binding site of an S100 protein revealed. Two crystal structures of Ca2 +–bound human psoriasin (S100A7) in the Zn2+–loaded and Zn2+–free states. Biochemistry 38:1695–1704
11. Moroz OV, Burkitt W, Wittkowski H, He W, Ianoul A, Novitskaya V, Xie J, Polyakova O, Lednev IK, Shekhtman A, Derrick PJ, Bjoerk P, Foell D, Bronstein IB (2009) Both Ca2+ and Zn2+ are essential for S100A12 protein oligomerization and function. BMC Biochem 10:1–18
12. Hood MI, Skaar EP (2012) Nutritional immunity: transition metals at the pathogen-host interface. Nat Rev Microbiol 10(8):525–537
13. Palmer LD, Skaar EP (2016) Transition metals and virulence in bacteria. Annu Rev Genet 50:67–91
14. Kehl-Fie TE, Chitayat S, Hood MI, Damo S, Restrepo N, Garcia C, Munro KA, Chazin WJ, Skaar EP (2011) Nutrient metal sequestration by calprotectin inhibits bacterial superoxide defense enhancing neutrophil killing of Staphylococcus aureus. Cell Host Microbe 10:158–164
15. Dhople AM, Ibanez MA, Poirier TC (1996) Role of iron in the pathogenesis of Mycobacterium avium infection in mice. Microbios 87 (351):77–87
16. McDermid JM, Hennig BJ, van der Sande M, Hill AV, Whittle HC, Jaye A, Prentice AM (2013) Host iron redistribution as a risk factor for incident tuberculosis in HIV infection: an 11-year retrospective cohort study. BMC Infect Dis 13:e48
17. Zackular JP, Moore JL, Jordan AT, Juttukonda LJ, Noto MJ, Nicholson MR, Crews JD, Semler MW, Zhang Y, Ware LB, Washington MK, Chazin WJ, Caprioli RM, Skaar EP (2016) Dietary zinc alters the microbiota and decreases resistance to Clostridium difficile infection. Nat Med 22(11):1330–1334
18. Juttukonda LJ, Berends ETM, Zackular JP, Moore JL, Stier MT, Zhang Y, Schmitz JE, Beavers WN, Wijers CD, Gilston BA, Kehl-Fie TE, Atkinson J, Washington MK, Peebles RS, Chazin WJ, Torres VJ, Caprioli RM, Skaar EP (2017) Dietary manganese promotes staphylococcal infection of the heart. Cell Host Microbe 22(4):531–542
19. Besold AN, Gilston BA, Radin JN, Ramsoomair C, Culbertson EM, Li CX, Cormack BP, Chazin WJ, Kehl-Fie TE, Culotta VC (2018) Role of calprotectin in withholding zinc

and copper from Candida albicans. Infect Immun 86:e00779–e00717

20. Corbin BD, Seeley EH, Raab A, Feldmann J, Miller MR, Torres VJ, Anderson KL, Dattilo BM, Dunman PM, Gerads R, Caprioli RM, Nacken W, Chazin WJ, Skaar EP (2008) Metal chelation and inhibition of bacterial growth in tissue abscesses. Science 319 (5865):962–965
21. Gaddy JA, Radin JN, Loh JT, Piazuelo MB, Kehl-Fie TE, Delgado AG, Ilca FT, Peek RM, Cover TL, Chazin WJ, Skaar EP, Scott Algood HM (2014) The host protein calprotectin modulates the Helicobacter pylori cag type IV secretion system via zinc sequestration. PLoS Pathog 10(10):e1004450
22. Hood MI, Mortensen BL, Moore JL, Zhang Y, Kehl-Fie TE, Sugitani N, Chazin WJ, Caprioli RM, Skaar EP (2012) Identification of an Acinetobacter baumannii zinc acquisition system that facilitates resistance to calprotectin-mediated zinc sequestration. PLoS Pathog 8 (12):e1003068
23. Kehl-Fie TE, Zhang Y, Moore JL, Farrand AJ, Hood MI, Rathi S, Chazin WJ, Caprioli RM, Skaar EP (2013) MntABC and MntH contribute to systemic Staphylococcus aureus infection by competing with calprotectin for nutrient manganese. Infect Immun 81(9):3395–3405
24. Urban CF, Ermert D, Schmid M, Abu-Abed U, Goosmann C, Nacken W, Brinkmann V, Jungblut PR, Zychlinsky A (2009) Neutrophil extracellular traps contain calprotectin, a cytosolic protein complex involved in host defense against Candida albicans. PLoS Pathog 5(10):e1000639
25. Shank JM, Kelley BR, Jackson JW, Tweedie JL, Franklin D, Damo SM, Gaddy JA, Murphy CN, Johnson JG (2018) The host antimicrobial protein Calgranulin c participates in the control of campylobacter jejuni growth via zinc sequestration. Infect Immun 86(6): e00234–e00218
26. Haley KP, Delgado AG, Piazuelo MB, Mortensen BL, Correa P, Damo SM, Chazin WJ, Skaar EP, Gaddy JA (2015) The human antimicrobial protein calgranulin C participates in control of Helicobacter pylori growth and regulation of virulence. Infect Immun 83(7):2944–2956
27. Moroz OV, Antson AA, Grist SJ, Maitland NJ, Dodson GG, Wilson KS, Lukanidin E, Bronstein IB (2003) Structure of the human S100A12-copper complex: implications for host-parasite defence. Acta Crystallogr D Biol Crystallogr 59(Pt 5):859–867
28. Ravasi T, Hsu K, Goyette J, Schroder K, Yang Z, Rahimi F, Miranda LP, Alewood PF, Hume DA, Geczy C (2004) Probing the S100 protein family through genomic and functional analysis. Genomics 84(1):10–22
29. Buchau AS, Hassan M, Kukova G, Lewerenz V, Kellermann S, Wurthner JU, Wolf R, Walz R, Walz M, Gallo RL, Ruzicka T (2007) S100A15, an antimicrobial protein of the skin: regulation by E. coli through Toll-like receptor 4. J Invest Dermatol 127:2596–2604
30. Glaser R, Harder J, Lange H, Bartels J, Christophers E, Schroder JM (2005) Antimicrobial psoriasin (S100A7) protects human skin from Escherichia coli infection. Nat Immunol 6:57–64
31. Lee KC, Eckert RL (2007) S100A7 (Psoriasin)- mechanism of antibacterial action in wounds. J Invest Dermatol 127:945–957
32. Boniface K, Bernard F-X, Garcia M, Gurney AL, Lecron J-C, Morel F (2005) IL-22 inhibits epidermal differentiation and induces proinflammatory gene expression and migration of human keratinocytes. J Immunol 174:3695–3702
33. Wolf R, Mirmohammadsadegh A, Walz M, Lysa B, Tartler U, Remus R, Hengge U, Michel G, Ruzicka T (2003) Molecular cloning and characterization of alternatively spliced mRNA isoforms from psoriatic skin encoding a novel member of the S100 family. FASEB J 17:1969–1971
34. Hsu K, Champaiboon C, Guenther BD, Sorenson BS, Khammanivong A, Ross KF, Geczy CL, Herzberg MC (2009) Anti-infective protective properties of S100 calgranulins. Antiinflamm Antiallergy Agents Med Chem 8 (4):290–305
35. Lagasse E, Clerc RG (1988) Cloning and expression of two human genes encoding calcium-binding proteins that are regulated during myeloid differentiation. Mol Cell Biol 8(6):2402–2410
36. Goebeler M, Roth J, van den Bos C, Ader G, Sorg C (1995) Increase of calcium levels in epithelial cells induces translocation of calcium-binding proteins migration inhibitory factor-related protein 8 (MRP8) and MRP14 to keratin intermediate filaments. Biochem J 309(Pt 2):419–424
37. Grimbaldeston MA, Geczy CL, Tedla N, Finlay-Jones JJ, Hart PH (2003) S100A8 induction in keratinocytes by ultraviolet A irradiation is dependent on reactive oxygen intermediates. J Invest Dermatol 121 (5):1168–1174
38. Mork G, Schjerven H, Mangschau L, Soyland E, Brandtzaeg P (2003) Proinflammatory cytokines upregulate expression of calprotectin (L1 protein, MRP-8/MRP-14) in

cultured human keratinocytes. Br J Dermatol 149(3):484–491
39. Stoll SW, Zhao X, Elder JT (1998) EGF stimulates transcription of CaN19 (S100A2) in HaCaT keratinocytes. J Invest Dermatol 111 (6):1092–1097
40. Cornish CJ, Devery JM, Poronnik P, Lackmann M, Cook DI, Geczy CL (1996) S100 protein CP-10 stimulates myeloid cell chemotaxis without activation. J Cell Physiol 166(2):427–437
41. Lackmann M, Cornish CJ, Simpson RJ, Moritz RL, Geczy CL (1992) Purification and structural analysis of a murine chemotactic cytokine (CP-10) with sequence homology to S100 proteins. J Biol Chem 267(11):7499–7504
42. Vogl T, Tenbrock K, Ludwig S, Leukert N, Ehrhardt C, van Zoelen MA, Nacken W, Foell D, van der Poll T, Sorg C, Roth J (2007) Mrp8 and Mrp14 are endogenous activators of Toll-like receptor 4, promoting lethal, endotoxin-induced shock. Nat Med 13 (9):1042–1049
43. Foell D, Wittkowski H, Vogl T, Roth J (2007) S100 proteins expressed in phagocytes: a novel group of damage-associated molecular pattern molecules. J Leukoc Biol 81(1):28–37
44. Leclerc E, Fritz G, Vetter SW, Heizmann CW (2009) Binding of S100 proteins to RAGE: an update. Biochim Biophys Acta 1793 (6):993–1007
45. Chen X, Eksioglu EA, Zhou J, Zhang L, Djeu J, Fortenbery N, Epling-Burnette P, Van Bijnen S, Dolstra H, Cannon J, Youn JI, Donatelli SS, Qin D, De Witte T, Tao J, Wang H, Cheng P, Gabrilovich DI, List A, Wei S (2013) Induction of myelodysplasia by myeloid-derived suppressor cells. J Clin Invest 123 (11):4595–4611
46. Bresnick AR, Weber DJ, Zimmer DB (2015) S100 proteins in cancer. Nat Rev 15:96–109
47. Choi IY, Gerlag DM, Holzinger D, Roth J, Tak PP (2014) From synovial tissue to peripheral blood: myeloid related protein 8/14 is a sensitive biomarker for effective treatment in early drug development in patients with rheumatoid arthritis. PLoS One 9(8):e106253
48. Frosch M, Ahlmann M, Vogl T, Wittkowski H, Wulffraat N, Foell D, Roth J (2009) The myeloid-related proteins 8 and 14 complex, a novel ligand of toll-like receptor 4, and interleukin-1beta form a positive feedback mechanism in systemic-onset juvenile idiopathic arthritis. Arthritis Rheum 60 (3):883–891
49. Haga HJ, Brun JG, Berntzen HB, Cervera R, Khamashta M, Hughes GR (1993) Calprotectin in patients with systemic lupus erythematosus: relation to clinical and laboratory parameters of disease activity. Lupus 2 (1):47–50
50. Nordal HH, Brun JG, Halse AK, Madland TM, Fagerhol MK, Jonsson R (2014) Calprotectin (S100A8/A9), S100A12, and EDTA-resistant S100A12 complexes (ERAC) in primary Sjogren's syndrome. Scand J Rheumatol 43 (1):76–78
51. Pietzsch J (2011) S100 proteins in health and disease. Amino Acids 41:755–760
52. Moore BW (1965) A soluble protein characteristic of the nervous system. Biochem Biophys Res Commun 19:739–744
53. Brophy MB, Hayden JA, Nolan EM (2012) Calcium ion gradients modulate the zinc affinity and antibacterial activity of human calprotectin. J Am Chem Soc 134:18089–18100
54. Jackson E, Little S, Franklin DS, Gaddy JA, Damo SM (2017) Expression, purification, and antimicrobial activity of S100A12. J Vis Exp 123:6–11
55. Murray JI, Tonkin ML, Whiting AL, Peng F, Farnell B, Cullen JT, Hof F, Boulanger MJ (2012) Structural characterization of S100A15 reveals a novel zinc coordination site among S100 proteins and altered surface chemistry with functional implications for receptor binding. BMC Struct Biol 12:16–23
56. Froger A, Hall JE (2007) Transformation of plasmid DNA into E. coli using the heat shock method. J Vis Exp 6:e253
57. Sivaraja V, Kumar TK, Rajalingam D, Graziani I, Prudovsky I, Yu C (2006) Copper binding affinity of S100A13, a key component of the FGF-1 nonclassical copper-dependent release complex. Biophys J 91:1832–1843
58. Wilder PT, Varney KM, Weiss MB, Gitti RK, Weber DJ (2005) Solution structure of zinc- and calcium-bound rat S100B as determined by nuclear magnetic resonance spectroscopy. Biochemistry 44:5690–5702
59. Koch M, Bhattacharya S, Kehl T, Gimona M, Vasak M, Chazin W, Heizmann CW, Kroneck PM, Fritz G (2007) Implications on zinc binding to S100A2. Biochim Biophys Acta 1773:457–470
60. Schäfer BW, Fritschy JM, Murmann P, Troxler H, Durussel I, Heizmann CW, Cox JA (2000) Brain S100A5 is a novel calcium-, zinc-, and copper ion-binding protein of the EF-hand superfamily. J Biol Chem 275:30623–30630
61. Sturchler E, Cox JA, Durussel I, Weibel M, Heizmann CW (2006) S100A16, a novel

calcium-binding protein of the EF-hand superfamily. J Biol Chem 281:38905–38917

62. Cuden LS, Gaillard A, Nolan EM (2016) Calcium ions tune the zinc-sequestering properties and antimicrobial activity of human S100A12. Chem Sci 7:1338–1348

63. Ecsédi P, Kiss B, Gógl G, Radnai L, Buday L, Koprivanacz K, Liliom K, Leveles I, Vértessy B, Jeszenői N, Hetényi C, Schlosser G, Katona G, Nyitray L (2017) Regulation of the equilibrium between closed and open conformations of Annexin A2 by N-terminal phosphorylation and S100A4-binding. Structure 25:1195–1207

64. Studier FW (2014) Chapter 2: Stable expression clones and auto-induction for protein production in E. coli. Methods Mol Biol 1091:17–32

Chapter 19

Targeting S100 Calcium-Binding Proteins with Small Molecule Inhibitors

Paul T. Wilder, Kristen M. Varney, and David J. Weber

Abstract

S100B is a small, dimeric, calcium-binding protein that is implicated in various diseases, most significantly cancer; therefore, there is interest in identifying S100B inhibitors that may have therapeutic value (Bresnick et al. Nat Rev Cancer 15:96–109, 2015; Chong et al. Curr Med Chem 23:1571–1596). Two fluorescence polarization competition assays (FPCA) are described here for S100B and S100A1 that are amenable to high-throughput screening (HTS) campaigns and can be used to determine the binding affinity (K_i) of the inhibitors. One FPCA is used to identify and characterize inhibitors of S100B with the aim of finding new therapeutics, and the other was developed as a counter-screen to avoid inhibitors of S100A1 due to its role in regulating skeletal and cardiac muscle function. Also outlined are methods for expressing and purifying S100B and S100A1 in quantities needed for performing large HTS campaigns.

Key words High-throughput screening, Fluorescence polarization competition assay, S100 protein, Small molecule, Protein-protein interface, Inhibitors, Calcium binding, S100B, S100A1, Cancer, Melanoma

1 Introduction

The calcium-regulated S100B protein is a prognostic marker for poor survival in several cancers including malignant melanoma [1]. In addition to its role as a cancer marker, S100B is involved in promoting tumor growth and metastasis. Specifically, S100B binds and regulates several proteins and enzymes involved in the progression of cancer including RSK, Hdm2, Hdm4, and p53, to name a few, and S100B modulates the tumor microenvironment when elevated in cancer [1, 3–5]. As a result, S100B is considered a therapeutic target for treating cancer, particularly in malignant melanoma where a majority of patients have elevated S100B [1, 2]. With this in mind, it is important to develop S100B inhibitors that are highly specific since there are 21 members of the S100

Claus W. Heizmann (ed.), *Calcium-Binding Proteins of the EF-Hand Superfamily: From Basics to Medical Applications*, Methods in Molecular Biology, vol. 1929, https://doi.org/10.1007/978-1-4939-9030-6_19,

protein family that have a similar global fold and high levels of sequence homology. Like S100B, S100A1 is involved in a wide array of cellular activities, including regulating calcium signaling in heart and skeletal muscle via interaction(s) with the ryanodine receptor [6–8]. Thus, when developing S100B inhibitors, care must be taken to avoid blocking S100A1, which is 60% homologous to S100B, and could potentiate problems with skeletal and cardiac muscle function. To this end, fluorescence polarization competition assays (FPCA) were developed to identify highly specific S100B inhibitors that do not block S100A1.

FPCAs developed to block S100-target interactions are well suited for high-throughput screening (HTS) campaigns due to their sensitivity, speed, homogenous format, and the need for relatively simple instrumentation [9–11]. The FPCA takes advantage of the ability to distinguish the fluorescence polarization of small peptides (<3 kDa) labeled with probes such as fluorescein isothiocyanate (FITC) or 5,6-carboxytetramethylrhodamine (TAMRA) in the free state versus bound to a S100 homodimer (>20 kDa). Peptide probes for both S100B and S100A1 were identified by testing several peptides derived from full-length S100 targets. Those labeled peptides deemed acceptable had a binding affinity (K_D) of 0.1–5 μM, an interpretable stoichiometry ($n = 1$, per monomeric subunit), and the fluorescently labeled peptide was shown to bind in the same site as the unlabeled peptide via direct competition experiments. The concept for the FPCA is straightforward; the fluorescent probe is highly polarized when bound to the S100 protein; however, upon being displaced by the unlabeled peptide, or a small molecule inhibitor, the fluorescent peptide probe is released, and its fluorescence polarization (FP) value decreases significantly (*see* Fig. 1). The final FPCAs developed here are robust due to the stability of the S100 proteins themselves and the relative simplicity of the method. Below is a description of

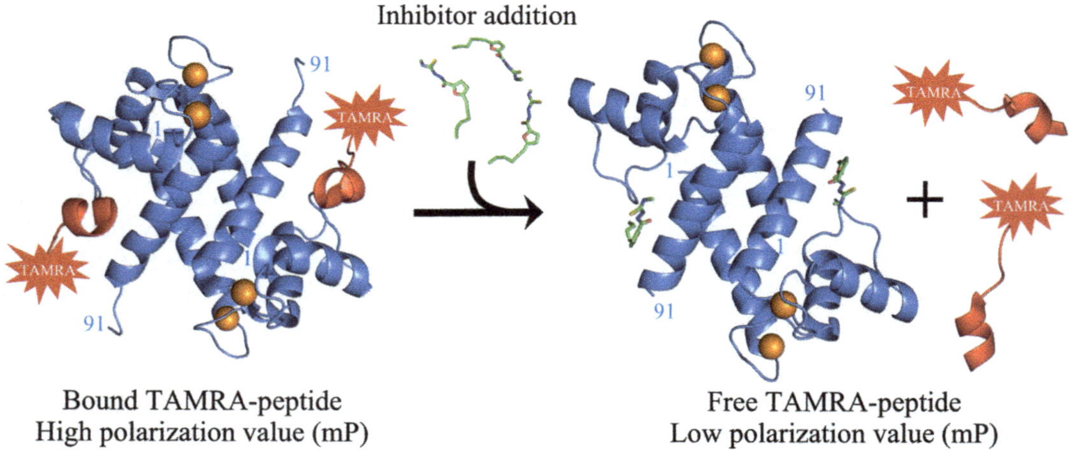

Fig. 1 Schematic illustration of the S100B TAMRA-TRTK FPCA

the FPCA for identifying inhibitors of S100B versus S100A1, calculations to determine the inhibitor affinity (K_i), and for the expression and purification of recombinant S100 proteins needed for the assay.

2 Materials

All chemical reagents are American Chemical Society grade or higher unless otherwise indicated, and the solutions need to be prepared using ultrapure water.

2.1 Recombinant S100B and S100A1 Production and Purification

2.1.1 Equipment

1. Misonix Sonicator 3000 (600 W with ½" horn) or equivalent sonicator.
2. Fraction collector.
3. UV monitor.
4. SDS-PAGE electrophoresis equipment.
5. Sorvall Superspeed refrigerated centrifuge equipped with both a SS-34 and GS-3 fixed angle rotor or equivalent.
6. Oak Ridge high-speed, polypropylene centrifuge tubes, 50 mL.
7. Centrifuge bottles, 500 mL, to fit Sorvall GS-3 rotor.
8. Shaker incubator for 2.5 L culture flasks.

2.1.2 Common Materials

1. 10 kDa MWCO Amicon Ultra-15 centrifugal filter units.
2. 8–10 kDa MWCO dialysis tubing: Dry dialysis tubing is hydrated and washed to remove metals and glycerin ahead of time by (1) boiling the tubing in 0.4 M EDTA, pH 8.0, for 30 min; (2) boiling the tubing in 0.5 M NaCl for 30 min; (3) boiling the tubing in 0.1 M EDTA, pH 8.0, for 30 min, twice; (4) boiling in water at least three times or until water remains clear; and (5) storing at 4 °C in 0.35 mM NaN_3, 0.5 mM EDTA. Rinse tubing with high-quality water inside and out before using.
3. Chelex 100 resin slurry: Clean the resin by (1) hydrating it in water and transferring to a chromatography column or Büchner funnel, (2) washing the resin with two bed volumes of 1 M HCl, (3) washing the resin with five bed volumes of water, (3) washing the resin with two bed volumes of 1 M NaOH, (4) washing the resin with five bed volumes of water, and then (5) washing the resin with five bed volumes of water with 0.35 mM NaN_3. Make a 50% slurry in the final buffer and store at 4 °C.
4. Bradford protein assay.
5. BSA protein standards or S100 standards with known concentrations as determined by amino acid analyses.

2.1.3 Glycerol Stocks, Cell Growth, Protein Expression, and Cell Harvesting

1. LB-Amp: Luria-Bertani broth with 100 μg/mL ampicillin.
2. LB-Amp plates: Luria-Bertani agar plates with 100 μg/mL ampicillin.
3. Ampicillin stock: 100 mg/mL ampicillin water and sterilize by passing thru a 0.2 μm filter.
4. Sterile 80% glycerol: Autoclaved 80% glycerol in water.
5. Isopropyl-1-thio-ß-D-galactopyranoside (IPTG) solution: 0.1 g/mL prepared from powder day of use.
6. pET11b S100 protein expression vectors (*see* **Note 1**).
9. Competent HMS174(DE3) cells (*see* **Note 2**).
10. Tunair full-baffle shake flask, 2.5 L.

2.1.4 Cell Lysis, Ammonium Sulfate Purification, and Heat Precipitation

1. Lysis buffer: 0.5 mM AEBSF added to DEAE buffer A right before use.
2. Lysozyme solution: 10 mg/mL lysozyme in water.
3. DNase solution: 1 mg/mL DNase I, 5 mM NaOAc, 1 mM $CaCl_2$, 50% glycerol. This can be stored at −20 °C for 3 months.
4. 1 g/mL $MgCl_2 \cdot 6H_2O$.
5. 500 mM EDTA.
6. 10% streptomycin sulfate solution in water.
7. Ammonium sulfate.

2.1.5 DEAE Anion Exchange

1. DEAE Sepharose fast-flow resin in a 2.5-cm-diameter column, ~6 cm bed height, ~30 mL bed volume, or equivalent.
2. DEAE buffer A: 50 mM Tris–HCl, pH 7.5, 0.5 mM DTT.
3. DEAE buffer B: 2 M NaCl, 50 mM Tris–HCl, pH 7.5, 0.5 mM DTT.
4. 0.10 M NaCl DEAE buffer: 95.00 mL DEAE buffer A + 5.00 mL DEAE buffer B.
5. 0.17 M NaCl DEAE buffer: 27.45 mL DEAE buffer A + 2.55 mL DEAE buffer B.
6. 0.22 M NaCl DEAE buffer: 26.70 mL DEAE buffer A + 3.30 mL DEAE buffer B.
7. 0.35 M NaCl DEAE buffer: 24.75 mL DEAE buffer A + 5.25 mL DEAE buffer B.
8. 0.50 M NaCl DEAE buffer: 45.00 mL DEAE buffer A + 15.00 mL DEAE buffer B.
9. Resin storage buffer: 20% ethanol, 0.02% NaN_3.

2.1.6 G-75 Size Exclusion

1. Sephadex G-75 medium resin (GE Healthcare 17-0050-01) in a 2.5 cm diameter column, 100 cm length, ~480 mL bed volume, or equivalent.
2. G-75 column buffer: 10 mM Tris, 50 mM NaCl, 0.25 mM DTT.
3. Chelex slurry: 50% Chelex 100 in water.
4. G-75 resin storage buffer: G-75 column buffer, 0.02% NaN_3.

2.1.7 Phenyl-Sepharose

1. Phenyl-Sepharose CL-4B resin in a 2.5-cm diameter column, ~6 cm bed height, ~30 mL bed volume, or equivalent.
2. Phenyl-Sepharose buffer A: 10 mM Tris, pH 7.5, 500 mM NaCl, 10 mM CaCl2, 0.25 mM DTT.
3. Phenyl-Sepharose buffer B: 10 mM Tris, pH 7.5, 500 mM NaCl, 10 mM EDTA, 0.25 mM DTT.

2.1.8 G-25 Size Exclusion

1. Sephadex G-25 medium resin (GE Healthcare 17-0033-01) in a 1.5 cm diameter column, 30–50 cm length, 45–80 mL bed volume, or equivalent.
2. G-25 column buffer: 10 mM HEPES, pH 7.2, 15 mM NaCl, 0.25 mM DTT.
3. G-25 resin storage buffer: G-25 column buffer, 0.02% NaN_3.

2.2 FPCA

2.2.1 Equipment

1. Microplate reader with fluorescence polarization measurement capabilities to read 10 nM TAMRA, 20 μL, in a 384-well black microplate from the top (*see* **Note 3**).
2. Origin software (OriginLab Corp., Northampton, MA) or other software for determining the IC_{50} values from a dose titration plot using nonlinear least squares analysis.
3. Multichannel pipette.

2.2.2 Common Materials

1. 384-well black polypropylene microplate(s) (*see* **Note 4**).
2. 10× FPCA buffer: 0.5 M HEPES, pH 7.2, 1 M KCl, 0.15 M NaCl, 0.1 M $CaCl_2$, 0.1% Triton X-100. Store at 4 °C.

2.2.3 S100A1 Buffers

1. TAMRA-$HDM4^{25-43}$ peptide: Peptide corresponds to HDM4 residues 25–43 with a single N-terminal 5,6-carboxytetramethylrhodamine (TAMRA) and an amidated C-terminus. Make stock in water up to 1 mM, and immediately adjust to a pH of ~7 (*see* **Note 5**).
2. S100A1 full FPCA buffer: 1X FPCA buffer, 10 nM TAMRA-$HDM4^{25-43}$, 6 μM S100A1 protein. Make at least 0.02 mL per well for 384-well microplate (0.1 mL per well for 96-well microplate) fresh.

3. S100A1 no S100 control: 1X FPCA buffer, 10 nM TAMRA-HDM4$^{25–43}$. Make at least 0.02 mL per well for 384-well microplate (0.1 mL per well for 96-well microplate) fresh.

2.2.4 S100B Buffers

1. TAMRA-TRTK: Peptide corresponds to the CapZ protein residues 265–276 [12] with a single N-terminal 5,6-carboxy-tetramethylrhodamine (TAMRA) and an amidated C-terminus. Make stock in water up to 1 mM, and immediately adjust to a pH of ~7 (*see* **Note 5**).
2. S100B full FPCA buffer: 1× FPCA buffer, 10 nM TAMRA-TRTK, 1 μM S100B protein. Make at least 0.02 mL per well for 384-well microplate (0.1 mL per well for 96-well microplate) fresh.
3. S100B no S100 control: 1× FPCA buffer, 10 nM TAMRA-TRTK. Make at least 0.02 mL per well for 384-well microplate (0.1 mL per well for 96-well microplate) fresh.

3 Methods

3.1 Recombinant S100B and S100A1 Production and Purification

3.1.1 Glycerol Stocks

1. Transform competent HMS174(DE3) cells with pET11b S100 protein expression vector construct, and spread plate them on LB-Amp plates overnight at 37 °C (*see* **Note 2**).
2. Inoculate 5 mL LB-Amp medium with a single colony. Repeat this for five to ten unique colonies. Incubate overnight at 37 °C with orbital shaking.
3. Dilute an aliquot of the overnight cultures (1:100) into 10 mL LB-Amp medium, and incubate at 37 °C with orbital shaking until the A_{600} is 0.6–0.8. Save the overnight cultures at 4 °C.
4. Remove 500 μL, sediment the bacteria, suspend the cell pellet in 100 μL SDS-PAGE sample buffer, and store at −20 °C.
5. Add 20 μL IPTG solution to the remaining culture, and incubate for another 2–4 h at 37 °C with orbital shaking, 150 rpm.
6. Remove 500 μL, sediment the bacteria, and suspend the cell pellet in 100 μL SDS-PAGE sample buffer.
7. Run SDS-PAGE analysis on 15 μL of the preinduction (**step 4**) and post-induction (**step 6**) cultures, and visually determine the lane with the highest expression of the S100 protein (*see* **Note 6**).
8. Dilute the overnight culture of the highest expression colony (1:100) in 10 mL LB-Amp medium, and incubate at 37 °C with orbital shaking until the A_{600} is 0.6–0.8.
9. Aliquot 100 μL sterile 80% glycerol into pre-labeled sterile 1.5 mL Eppendorf tubes (three to five tubes).

10. Add 900 μL of the bacterial culture, mix quickly by pipetting or vortex mixing, and rapidly freeze by placing in crushed dry ice powder, a dry ice ethanol bath, or into liquid nitrogen.
11. After 5 min transfer to −80 °C for long-term storage (*see* **Note 7**).

3.1.2 Cell Growth, Protein Expression, and Cell Harvesting for S100B and S100A1

1. Inoculate a 5 mL LB-Amp culture with a scraping from the glycerol stock using a sterile pipet tip or inoculating loop and grow at 37 °C with orbital shaking of 150 rpm overnight (*see* **Note 7**).
2. The next morning, inoculate 200 mL sterile LB-Amp (2 L flask) with 2 mL of the overnight 5 mL LB culture and at 37 °C, 150 rpm, and grow to an A_{600} of ~0.6.
3. Inoculate 5 × 1 L cultures in baffled 2.5 L flask with 30 mL from the 200 mL culture, and grow at 37 °C with 200 rpm orbital shaking.
4. Start monitoring the A_{600} of the cultures every hour, and graph read on a semilog plot. Cultures will be ready for induction when $A_{600} = 0.8–1.0$. To avoid overshooting, take more frequent measurements when the A_{600} reaches ~0.6.
5. During this period prepare 2 mL IPTG solution for each 1 L culture induction by dissolving 0.1 g/mL IPTG in ddH_2O for each 6 L flask in use.
6. Before induction, take a 15 μL aliquot from each flask for SDS-PAGE analysis.
7. Induce cells by adding 2 mL of IPTG solution per 1 L culture.
8. Incubate 3 h at 37 °C with 200 rpm orbital shaking.
9. During this time weigh six empty centrifuge bottles and record their total weight.
10. Record the A_{600} and take a 15 μL aliquot from each flask for SDS-PAGE analysis.
11. Load the centrifuge bottles to their maximum volume, balance, and spin at 6000 × *g* for 20 min at 4 °C in the GS-3 rotor. Keep remaining culture suspension on ice.
12. Carefully remove the media from the cell pellet, and add the remaining cell culture suspension to the same centrifuge bottles containing the pellet, balance, and spin at 6000 × *g* for 20 min at 4 °C in the GS-3 rotor. Repeat until all the cells are pelleted.
13. After final supernatant is decanted, remove as much media as possible with a pipette, and weigh bottles, subtracting the weight of the empty bottles to determine the pellet weight.
14. Place centrifuge bottles in −80 °C freezer (*see* **Note 8**).

3.1.3 Cell Lysis for S100B and S100A1

1. If frozen, thaw pellets in water bath at room temperature, and immediately put on ice.
2. After this point all steps are to be performed at or near 4 °C by using ice bath or cold room unless specified. Use only plastic to avoid introducing metals and losing protein (*see* **Note 9**).
3. Dissolve AEBSF in DEAE buffer A to make lysis buffer.
4. Suspend the pellets using 1 mL lysis buffer per 1 g of pellet. Work quickly but gently to avoid lather.
5. Pour cell suspension into 250 mL plastic beaker with stir bar, and add 80 mL lysozyme solution per 1 g of pellet. Stir gently on ice for 20 min. The viscosity should increase due to DNA release.
6. Add 300 μL DNase solution and 100 μL $MgCl_2$ solution. Continue stirring on ice for 15–20 min. The viscosity should decrease as the DNA is degraded; if not make new DNase solution and repeat.
7. Add sufficient volume of EDTA solution to make suspension 5 mM in EDTA (i.e., add 0.01× cell suspension volume of 500 mM EDTA solution).
8. Remove the stir bar. Set the sonicator to 70% of maximum power. Sonicate 0.5 min/mL solution. Allow 20–30 s in an ice bath after each minute of sonication or as necessary to keep the liquid cool.
9. Centrifuge the solution for 45 min at 27,000 × *g* and 4 °C in the SS-34 rotor.
10. Pour the supernatant into a 100 mL plastic beaker with a stir bar. Slowly add streptomycin solution over the next 30 min in ten aliquots; each aliquot should be 0.01× volume of supernatant.
11. Continue to stir for an additional 30 min.
12. Centrifuge the solution for 45 min at 27,000 × *g* and 4 °C in the SS-34 rotor.
13. For S100B proceed with ammonium sulfate purification (*see* Subheading 3.1, **step 4**), and for S100A1 proceed with heat precipitation (*see* Subheading 3.1, **step 5**).

3.1.4 Ammonium Sulfate Purification for S100B

1. Measure the volume of the supernatant and transfer into a 100 mL plastic beaker.
2. With constant stirring, make the solution 80% $(NH_4)_2SO_4$ by slow addition over a period of 1 h on ice or in cold room at 4 °C.
3. Continue stir for an additional 30 min.

4. Centrifuge the solution for 45 min at 27,000 × *g* and 4 °C in the SS-34 rotor.
5. Transfer the supernatant into 8–10 kDa MW cutoff dialysis tubing allowing for threefold volume expansion during dialysis.
6. Dialyze the supernatant twice against 4 L DEAE buffer A at 4 °C for a minimum of 4 h each with gentle mixing. It can be left overnight.
7. Proceed with DEAE anion exchange (*see* Subheading 3.1, **step 6**).

3.1.5 Heat Precipitation for S100A1

1. Transfer the supernatant to a new centrifuge tube.
2. Suspend the uncapped centrifuge tube in boiling water.
3. Heat until the internal sample temperature reaches 60 °C.
4. Remove from the boiling water, allow to cool, and cap.
5. Centrifuge the solution for 45 min at 27,000 × *g* and 4 °C in the SS-34 rotor, and save the supernatant.
6. Proceed with DEAE anion exchange (*see* Subheading 3.1, **step 6**).

3.1.6 DEAE Anion Exchange for S100B and S100A1

1. Clean DEAE Sepharose fast-flow resin with 100 mL DEAE buffer B.
2. Equilibrate with five column volumes (CV) (~150 mL) DEAE buffer A at ≤5 mL/min (60 cm/h linear flow rate; maximum for this resin is 750 cm/h).
3. Load sample, and run through column at a rate not to exceed that of the equilibration step above. Collect and save the eluent.
4. Follow the sample with 100 mL DEAE buffer A to wash the column. Collect and save the eluent.
5. Set up fraction collector for 60 5 mL (150-drop) fractions.
6. Load the buffer solutions in order of increasing salt concentrations making sure to wait until each buffer has run into the resin bed before loading the next one and recording the fractions collected for each.
7. Add 100 mL of 0.1 M NaCl DEAE buffer.
8. Add 30 mL of 0.17 M NaCl DEAE buffer.
9. Add 30 mL of 0.22 M NaCl DEAE buffer.
10. Add 30 mL of 0.35 M NaCl DEAE buffer.
11. Add 60 mL of 0.5 M NaCl DEAE buffer.
12. Wash column with ~60 mL DEAE buffer B, saving the eluent.

13. Wash column with 60–100 mL of DEAE buffer A, saving the eluent.
14. Measure the UV absorbance at 280 nm and 260 nm of fractions, and run SDS-PAGE analysis on the samples containing protein based on UV.
15. Pool fractions containing S100 protein verified by SDS-PAGE. The bulk is released with 0.17 M NaCl and will be in fractions 29–41 but can range from 16 to 47.
16. After it has been determined that S100B has been eluted, equilibrate column with 150 mL resin storage buffer. Resin should be stored at 4 °C, either in the column or another container.
17. For S100A1 proceed with Phenyl-Sepharose (*see* Subheading 3.1, **step 7**), and for S100B proceed with G-75 size exclusion (*see* Subheading 3.1, **step 8**).

3.1.7 Phenyl-Sepharose for S100A1

1. Dialyze the pooled fractions twice against 4 L Phenyl-Sepharose buffer A at 4 °C for a minimum of 4 h each with gentle mixing. This can be left over overnight.
2. Wash the Phenyl-Sepharose CL-4B column with five column volumes of 1:9 dilution of Phenyl-Sepharose buffer B in Phenyl-Sepharose buffer A.
3. Equilibrate the Phenyl-Sepharose CL-4B column with five CV of Phenyl-Sepharose buffer A.
4. Load S100A1 protein sample and collect the flow through.
5. Set up the fraction collector to collect 3 mL (100-drop) fractions.
6. Wash the column with five CV of Phenyl-Sepharose buffer A, and collect the flow through in 3 mL fraction.
7. Elute with three CV buffer B and save in 3 mL fraction.
8. Run an SDS page gel on all of the flow through samples and fractions.
9. Pool pure S100A1 samples, and proceed with G-25 size exclusion (*see* Subheading 3.1, **step 9**).
10. Pool remaining samples with S100A1, and proceed with G-75 size exclusion to further purify (*see* Subheading 3.1, **step 8**).
11. After it has been determined that S100A1 protein has been eluted, equilibrate column with two CV of water followed by two CV of resin storage buffer. Resin should be stored at 4 °C.

3.1.8 G-75 Size Exclusion for S100B and S100A1

1. Transfer the pooled fractions containing S100 protein into 8–10 kDa MW cutoff dialysis tubing allowing for threefold volume expansion during dialysis.

2. Dialyze the pooled fractions twice against 4 L G-75 column buffer containing 2 mL Chelex slurry at 4 °C for a minimum of 4 h each with gentle mixing. This can be left overnight.
3. Concentrate the sample using Amicon Ultra-15 centrifugal filter units 10 kDa MWCO to 0.5–1.0% of the G-75 column bed volume.
4. Equilibrate G-75 column with at least two CV of G-75 column buffer at 4 °C using no more than 1.5 mL/min (18 cm/h linear flow rate).
5. Set up the fraction collector for 175 3 mL (100-drop) fractions.
6. Load sample, and allow to run through column at a rate not to exceed that of the equilibration step above. Begin the fraction collector as the sample is loaded.
7. Measure the UV absorbance at 260–280 nm of fractions, and run SDS-PAGE analysis on the samples containing protein based on UV.
8. Pool fractions containing S100 protein.
9. After it has been determined that S100 protein (S100B or S100A1) has been eluted, equilibrate column with 500 mL G-75 resin storage buffer. Resin should be stored at 4 °C in the column.
10. Proceed to G-25 size exclusion (*see* Subheading 3.1, **step 9**).

3.1.9 G-25 Size Exclusion S100B and S100A1

1. Transfer the pooled fractions containing S100 protein into 8–10 kDa MW cutoff dialysis tubing allowing for threefold volume expansion during dialysis.
2. Dialyze the pooled fractions twice against 4 L G-25 column buffer containing 2 mL Chelex slurry at 4 °C for a minimum of 4 h each with gentle mixing. This can be left overnight.
3. Concentrate the sample using Amicon Ultra-15 centrifugal filter units 10 kDa MWCO to 0.5–1.0% of the G-25 column bed volume (*see* **Note 10**).
4. Equilibrate with five CV G-25 column buffer at 4 °C using at no more than 2 mL/min.
5. Set up the fraction collector with enough tubes to collect one bed volume in 30-drop (1 mL) fractions.
6. Load sample, and allow to run through column at a rate not to exceed that of the equilibration step above. Begin the fraction collector as the sample is loaded.
7. Measure the UV absorbance at 260–280 nm of fractions, and run SDS-PAGE analysis on the samples containing protein based on UV.

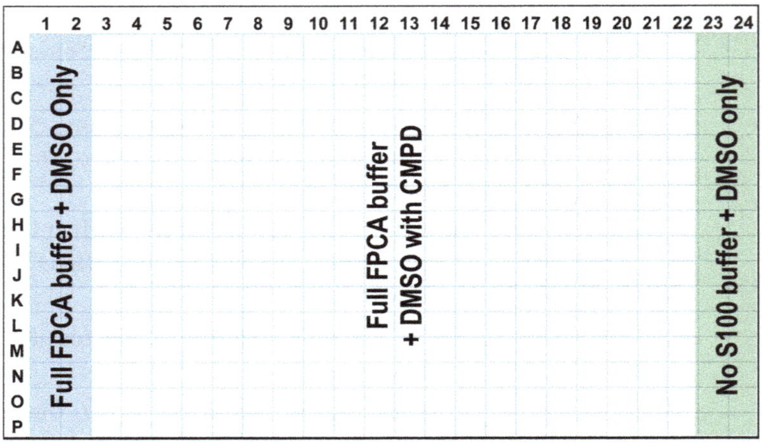

Fig. 2 Layout of typical HTS assay in a 384-well microplate for screening compound libraries that normally have DMSO on the first and last columns (CMPD, test compound)

8. Pool fractions containing S100 protein.
9. After it has been determined that S100 protein has been eluted, equilibrate column with two column volumes of G-25 resin storage buffer. Resin should be stored at 4 °C in the column.
10. Concentrate the sample using Amicon Ultra-15 centrifugal filter units 10 kDa MWCO to 0.5–1.0% of the G-25 column bed volume.
11. Determine an accurate protein concentration of the final concentrated protein using the Bradford protein assay. Freeze protein stock at −20 °C at 0.5–10 mM S100 protein (*see* **Note 10**).

3.2 Fluorescence Polarization Competition Assay (FPCA)

3.2.1 FPCA for Screening

1. *See* Fig. 2 for assay setup in a 384-well microplate.
2. Add 20 μL no S100 control + DMSO to columns 23–24.
3. Add 20 μL full FPCA buffer to every other well, except those in columns 23–24.
4. Add up to 1 μL of the test compounds per well so that the final concentration of DMSO will not exceed 5% (*see* **Note 12**).
5. Add an equivalent amount of DMSO to the control wells so that all of the wells have the same final DMSO concentration.
6. Take care not to introduce bubbles while pipetting, but the plate can be centrifuged ~5 min to remove bubbles.
7. Seal the plates with sealing tape.
8. Incubate the plates for at least 4 h room temperature (25 °C).
9. Remove the sealing tape carefully as not to introduce bubbles.

10. For the first assay plate only, calibrate the G-factor for the fluorescence microplate reader with polarization capabilities on one of the wells containing the no S100 control + DMSO, such as well B23, setting the polarization as 100 mP using 544 nm excitation and 590 nm emission wavelength (*see* **Note 11**).
11. Read the fluorescence polarization for all the plates using those settings.

3.2.2 Z-Factor

1. Determine μ_b: the mean polarization values for all the wells containing full FPCA buffer + DMSO only (bound probe).
2. Determine μ_f: the mean polarization values for all the wells containing no S100 control + DMSO only (free probe).
3. Determine σ_b: the standard deviation of the polarization values for wells containing full FPCA buffer + DMSO only (bound probe).
4. Determine σ_f: the standard deviation of the polarization values for wells containing full FPCA buffer + DMSO only (free probe).
5. Determine the Z-factor: $1-((3 \times \sigma_b + 3 \times \sigma_f)/(\mu_b-\mu_f))$
6. The Z-factor can be any value ≤ 1, with a value of 1 being an ideal assay with a value ≥ 0.5–1.0 being acceptable for screening [9]; however, below 0.5 the assay should be redone.

3.2.3 Hit Identification

1. Determine the mean (μ) and standard deviation (σ) of the polarization values for all the wells except the no S100 Control + DMSO only.
2. A "hit" (i.e., perspective inhibitor) is defined to result in a polarization value $\leq \mu-(3 \times \sigma)$.

3.2.4 FPCA Titration for Determining IC_{50}

1. *See* Fig. 3 for assay setup in a 384-well microplate.
2. Add 40 μL of the highest concentration of the test compound in full FPCA buffer into wells in the first row (row A) of the 384-well microplate plate with each test compound done in triplicate (*see* **Note 13**).
3. Add 40 μL of full FPCA buffer with the equivalent amount of DMSO as the test compound(s) samples in the first row of the first three columns (row A, columns 1–3).
4. Add 20 μL of the full FPCA buffer in the wells below the compound samples and DMSO control. The same pipette tips can be used repeatedly taking care not to create bubbles.
5. Add 40 μL of no S100 FPCA buffer with the equivalent amount of DMSO as the test compound(s) samples in the first row of the first three columns (row A, columns 22–24).

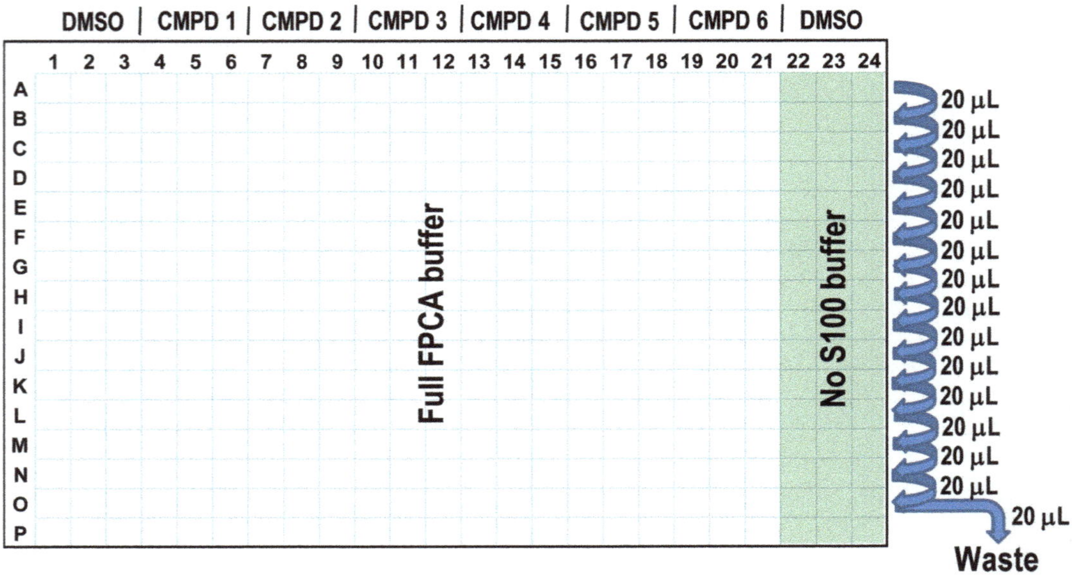

Fig. 3 Layout of typical compound titrations performed in a 384-well microplate where the sample is serial diluted down the plate with row A having the highest, row O the lowest, and row P containing no compound (CMPD, test compound)

6. Add 20 μL of the no S100 FPCA buffer in the remaining wells in columns 22–24 below DMSO control. The same pipette tips can be used repeatedly taking care not to create bubbles.
7. Serial dilute the sample by pipetting 20 μL from the first wells (row A) into the next wells (row B) mixing slowly and carefully pipetting up and down two to four times with the existing 20 μL buffer in the well. Be careful not to blow out the sample and introduce bubbles.
8. Using the same pipette tips, transfer 20 μL from these current wells (row B) to the next wells (row C) mixing slowly and carefully pipetting up and down two to four times with the existing 20 μL buffer in the well. Be careful not to blow out the sample and introduce bubbles.
9. Repeat **step 8** until you are in the next-to-final wells (row O).
10. Remove 20 μL from the next-to-final well (row O) and dispose it off. The final well should not contain any compound.
11. Seal the plates with a sealing tape.
12. Incubate the plates for at least 4 h room temperature (25 °C).
13. Remove the sealing tape carefully as not to introduce bubbles.
14. Calibrate the G-factor for the fluorescence microplate reader with polarization capabilities on one of the wells containing the no S100 control + DMSO, such as well B23, setting the

polarization as 100 mP using 544 nm excitation and 590 nm emission wavelength (*see* **Note 11**).

15. Read the fluorescence polarization for all the plates using those settings.
16. Convert the raw polarization values, typically given as mP, into % inhibition:

$$\%Inhibition = \left(1 - \left({}^{SAMPLE}mP - {}^{NO\ S100}mP\right) \div \left({}^{S100}mP - {}^{NO\ S100}mP\right)\right) \times 100\%$$

where:

${}^{SAMPLE}mP =$ the polarization value of experimental sample.

${}^{NO\ S100}mP =$ mean polarization value of the no S100 Control + DMSO wells.

${}^{S100}mP =$ mean polarization value of the S100 Control + DMSO wells.

17. Determine the IC50 by fitting the data to a single-site binding model using nonlinear least-squares analysis.

3.2.5 K_i Calculation

Determine the K_i from the experimentally determined IC_{50}, the variables in Table 1, using the Nikolovska-Coleska et al. equation as follows (*see* **Note 14**):

$$K_i = [I_{50}]/([L_{50}]/K_D + [P_0]/K_D) + 1)$$

where:

$P =$ S100 protein.

$L =$ fluorescence-labeled peptide probe.

$I =$ inhibitor.

$[P_0] = (-(K_D + [L_T]-[R_T]) + ((K_D + [L_T]-[R_T])^2 + 4[R_T]K_D)^{0.5})/2$

$[PL_0] = [P_T]-[P_0]$

$[PL_{50}] = [PL_0]/2$

$[L_{50}] = [L_T]-[PL_{50}]$

$[I_{50}] = IC_{50}-[P_T] + [PL_{50}] \times (1 + K_D/[L_{50}])$

$[P_T] = [P] + [PL] + [PI]$ (*see* Table 1)

$[L_T] = [L] + [PL]$ (*see* Table 1)

$K_D =$ dissociation constant of the PL complex (*see* Table 1)

Table 1
Values for K_i calculation

	$[P_T]$ (μM)	$[L_T]$ (μM)	K_D (μM)
S100A1	6.0	0.01	3.67 ± 0.07
S100B	1.0	0.01	0.13 ± 0.01

3.2.6 Hit Confirmation

It is necessary to have an independent secondary assay to confirm that the interaction of S100 occurs directly with the compound (i.e., one not depending on the same technique). This would rule out the possibility that the compound is a false positive due to fluorescence overlap or quenching of the fluorescent probe, interaction with the peptide, or other indirect effects.

4 Notes

1. The methods described here were performed successfully with human, bovine, and rat S100B and S100A1. The S100A1 and S100B pET11b expression vectors were synthesized using codon optimization for expression in *E. coli*.
2. BL21(DE3) or BL21(DE3) Star cells are also suitable *E. coli* for expression.
3. The assay can be run on any instrument capable of measuring TAMRA fluorescence polarization accurately at 10 nM with a 544 nm excitation and 590 nm emission wavelength. The assay could be adapted easily to run in a 96-well microplate or in a cuvette; however the throughput is highly limited using a single cuvette system. Also, since the final sample volume would have to be increased in these lower-throughput methods, more reagents including S100 protein, compound, and fluorescently-labeled peptide would be needed. For example, just going from the 384-well protocol to a typical 96-well microplate would necessitate increasing the sample from 20 μL to 100 μL and requires fivefold more sample. The use of a BMG PHERAstar FS multimode microplate reader is recommended, but Tecan, BioTek, Molecular Devices, PerkinElmer, and other manufacturers produce instruments with similar capabilities.
4. Microplates are manufactured with different materials to either minimize or maximize the binding of proteins or other biomolecules. For the FPCA, minimizing binding of the fluorescently labeled peptide, S100 proteins, and compounds of interest to the microplate is desired. Untreated polystyrene should not be used. While there are polystyrene microplates with surfaces engineered to have very low binding that can be purchased, the polypropylene microplates are sufficient for this application at a fraction of the cost.
5. Both peptides were synthesized using solid-state peptide synthesis, and their purity was determined to be >95% by high-pressure liquid chromatography (HPLC) and mass spectrometry. Since the peptides used are not terminal in the full-length proteins, the C- and N-terminus residues are modified to

neutralize the charge. A TAMRA label was chosen as the fluorescent probe since it is readily available, most fluorimeters are capable of measuring its FP, it is relatively inexpensive compared to other fluorophores, and its fluorescence properties are not as susceptible to pH and oxidation as is the case for fluorescein isothiocyanate (FITC). The concentration of the peptide solutions are determined using the extinction coefficient for TAMRA, $\varepsilon_{547} = 65{,}000\ \text{cm}^{-1}\ \text{M}^{-1}$ [11]. Most synthetic peptides have TFA counter ions, so the pH must be adjusted to ~pH 7 immediately upon suspending the powder in water, or else peptide degradation can occur. Titrations of the respective S100 proteins were done to determine the respective TAMRA-peptide binding affinity (K_D; *see* Table 1), and the fluorescently labeled peptide was competed off with the addition of the unlabeled version of the same peptide, as needed to demonstrate that the label does not cause the peptide to bind to an alternative site(s).

6. A 4–20% gradient gel for SDS-PAGE is recommended, since it readily separates S100A1 and S100B from the buffer front; however, even with a high percent gel, it is necessary to run the buffer front completely to the bottom of the gel to obtain good separation. Be aware that S100B and S100A1 will run at ~6 kDa by the SDS-PAGE rather than the expected ~10.7 kDa (expected monomer size) because this family of proteins are typically highly negatively charged.
7. Do not thaw the glycerol stocks, and return them to −80 °C immediately after using. If stored and handled correctly, the glycerol stocks can be used for over 10 years.
8. The cells can be stored at −80 °C for up to 3 months. Even though freeze-thawing the cells helps lyse the cells and release the overexpressed protein of interest, this step can be skipped if the plan includes immediately performing the cell lysis step (*see* Subheading 3.1, **step 3**).
9. S100 proteins bind calcium and in some cases zinc and copper. This protocol will produce mostly S100 proteins in the metal-free state (<1%), which is sufficient since the FPCA assay will typically be run with saturating calcium concentrations to observe peptide binding and the competition experiments. Additional steps can be taken to further lower contaminating calcium for "Ca^{2+}-free" control experiments. Such steps include using only plastic containers when possible, acid washing glass that contacts the sample or buffers, starting with highest-purity metal-free water, Chelex treating all buffers and reagents, and adding 5 mM EGTA to the G-75 column buffer for the G-75 size exclusion step. For full removal of the 5 mM EGTA, it will be necessary to dialyze twice against 4 L of

G-75 column buffer +100 mM NaCl before the G-25 size exclusion, **step 2**. If this is not done, the EGTA can contaminate the final sample. A longer G-25 column, ~80 mL column volume, will more efficiently remove EGTA.

10. The purified S100 protein stock can be stored in 5–50 mM HEPES, Tris, or other nonfluorescent buffer at pH 6.8–7.5. An accurate protein concentration is critical for the FPCA particularly if you wish to use the assay to determine the *Ki* of the compound. A Bradford protein assay can be used to determine the stock concentration. The S100 concentration is calculated for a single subunit since S100 proteins exists typically as a symmetric dimer with a single fluorescent peptide probe bound per S100 subunit.
11. Most microplate readers do not determine the absolute polarization, instead the manufacturer instructions say to calibrate the instrument to a fluorophore sample with a known polarization value such as FITC (27 mP) or TAMRA (50 mP) to determine a grating factor (G-factor). It is recommended to use the "no S100 control" sample set for 100 mP to calibrate the G-factor and to avoid having to make a separate sample. (The absolute values of the free peptides were determined in these assay conditions, 72.9 ± 5.7 mP for TAMRA-HDM4$^{25-43}$ and 134.2 ± 5.1 for TAMRA-TRTK, but since the only concern is with the change in polarization rather than the absolute value, for the FPCA, the 100 mP value is fine.)
12. There are several sources of small molecule compound collections available commercially and from public sources (NCI) for screening. Most will be supplied in 96- or 384-well plates and will be suspended at 10–20 mM in 100% DMSO. Since both the S100 FPCAs can be performed with ≤5% DMSO, the compounds concentration for the inhibition studies can approach the high μM range. The test compounds can be added via a multichannel pipette; however, the use of automated liquid handlers using 96 or 384 channel pipetting head or pin tools will greatly increase the throughput.
13. Start with a maximum concentration of compound ~20-fold higher than the expected IC_{50} as long as the total DMSO does not exceed 5% and the compound is soluble (i.e., does not precipitate). For example, if the primary screen was as 10 μM compounds and the hit decreased the mP by 50% inhibition, then your IC_{50} is about 10 μM, and then start with a maximum compound concentration of at least 200 μM. It may require multiple trials to get the best concentration range, resulting in a good curve fit to the data.
14. Once a putative S100 inhibitor is identified in a single-point screening assay, a titration of the compound can be carried out

to determine the IC_{50}. From that, the K_i of the inhibitor for the S100 can be calculated. While the IC_{50} values can be used to compare the potency of inhibitors within the same assay, the K_i is more useful for comparing binding affinities determined in different assay conditions or types. An equation developed by Nikolovska-Coleska et al. [13] is used here to calculate the K_i (i.e., versus the Cheng-Prusoff equation) [14] as FPCA fails to meet the basic assumption of the latter equation. While both equations give estimates of the K_i, the Nikolovska-Coleska et al. equation was shown to be closer to the true K_i. The authors maintain a publicly accessible web-based computer program for determining the K_i using the Nikolovska-Coleska et al. equation at http://sw16.im.med.umich.edu/software/calc_ki/.

References

1. Bresnick AR, Weber DJ, Zimmer DB (2015) S100 proteins in cancer. Nat Rev Cancer 15 (2):96–109. https://doi.org/10.1038/nrc3893
2. Chong ZZ, Changyaleket B, Xu H, Dull RO, Schwartz DE (2016) Identifying S100B as a biomarker and a therapeutic target for brain injury and multiple diseases. Curr Med Chem 23(15):1571–1596
3. Wilder PT, Lin J, Bair CL, Charpentier TH, Yang D, Liriano M, Varney KM, Lee A, Oppenheim AB, Adhya S, Carrier F, Weber DJ (2006) Recognition of the tumor suppressor protein p53 and other protein targets by the calcium-binding protein S100B. Biochim Biophys Acta 1763(11):1284–1297. https://doi.org/10.1016/j.bbamcr.2006.08.024
4. Hartman KG, Vitolo MI, Pierce AD, Fox JM, Shapiro P, Martin SS, Wilder PT, Weber DJ (2014) Complex formation between S100B protein and the p90 ribosomal S6 kinase (RSK) in malignant melanoma is calcium-dependent and inhibits extracellular signal-regulated kinase (ERK)-mediated phosphorylation of RSK. J Biol Chem 289 (18):12886–12895. https://doi.org/10.1074/jbc.M114.561613
5. Wang H, Zhang L, Zhang IY, Chen X, Da Fonseca A, Wu S, Ren H, Badie S, Sadeghi S, Ouyang M, Warden CD, Badie B (2013) S100B promotes glioma growth through chemoattraction of myeloid-derived macrophages. Clin Cancer Res 19(14):3764–3775. https://doi.org/10.1158/1078-0432.CCR-12-3725
6. Prosser BL, Wright NT, Hernandez-Ochoa EO, Varney KM, Liu Y, Olojo RO, Zimmer DB, Weber DJ, Schneider MF (2008) S100A1 binds to the calmodulin-binding site of ryanodine receptor and modulates skeletal muscle excitation-contraction coupling. J Biol Chem 283(8):5046–5057. https://doi.org/10.1074/jbc.M709231200
7. Wright NT, Prosser BL, Varney KM, Zimmer DB, Schneider MF, Weber DJ (2008) S100A1 and calmodulin compete for the same binding site on ryanodine receptor. J Biol Chem 283 (39):26676–26683. https://doi.org/10.1074/jbc.M804432200
8. Cannon BR, Zimmer DB, Weber DJ (2011) S100A1 (S100 calcium binding protein A1). Atlas Genet Cytogenet Oncol Haematol 15 (10):873–876. https://doi.org/10.4267/2042/46035
9. Hall MD, Yasgar A, Peryea T, Braisted JC, Jadhav A, Simeonov A, Coussens NP (2016) Fluorescence polarization assays in high-throughput screening and drug discovery: a review. Methods Appl Fluoresc 4(2):022001. https://doi.org/10.1088/2050-6120/4/2/022001
10. Huang X, Aulabaugh A (2016) Application of fluorescence polarization in HTS assays. Methods Mol Biol 1439:115–130. https://doi.org/10.1007/978-1-4939-3673-1_7
11. Wilder PT, Charpentier TH, Liriano MA, Gianni K, Varney KM, Pozharski E, Coop A, Toth EA, Mackerell AD, Weber DJ (2010) In vitro screening and structural characterization of inhibitors of the S100B-p53 interaction. Int J High Throughput Screen 2010(1):109–126. https://doi.org/10.2147/IJHTS.S8210
12. Ivanenkov VV, Jamieson GA Jr, Gruenstein E, Dimlich RV (1995) Characterization of S-100b binding epitopes. Identification of a

novel target, the actin capping protein, CapZ. J Biol Chem 270(24):14651–14658

13. Nikolovska-Coleska Z, Wang R, Fang X, Pan H, Tomita Y, Li P, Roller PP, Krajewski K, Saito NG, Stuckey JA, Wang S (2004) Development and optimization of a binding assay for the XIAP BIR3 domain using fluorescence polarization. Anal Biochem 332(2):261–273. https://doi.org/10.1016/j.ab.2004.05.055
14. Cheng Y, Prusoff WH (1973) Relationship between the inhibition constant (K1) and the concentration of inhibitor which causes 50 per cent inhibition (I50) of an enzymatic reaction. Biochem Pharmacol 22(23):3099–3108

Chapter 20

Monitoring Interactions Between S100B and the Dopamine D2 Receptor Using NMR Spectroscopy

Yuning Wang, Roya Tadayon, and Gary S. Shaw

Abstract

S100B is a dimeric EF-hand protein that undergoes a calcium-induced conformational change and interacts with a wide range of proteins to modulate their functions. The dopamine D2 receptor is one potential S100B binding partner that may play a key role in neurological processing. In this chapter, we describe the use of NMR spectroscopy to examine the interaction between calcium-bound S100B and the third intracellular loop (IC3) from the dopamine D2 receptor. We provide details that allow the strength of the interaction (K_d) between the two proteins to be determined and the IC3 site of interaction on the structure of S100B to be identified. Both these characteristics can be identified from a single series of nondestructive experiments.

Key words S100B, Dopamine receptor, Protein interactions, Binding site, Protein structure

1 Introduction

Protein-protein interactions are necessary for nearly all cellular processes including cell growth, cell differentiation and apoptosis, protein and nucleic acid metabolism, cellular architecture, and signal transduction [1, 2]. In many cases these events are controlled through calcium binding to a modulatory protein to regulate its specific target interactions. S100B is a member of the EF-hand calcium-binding S100 protein family shown to have both extracellular and intracellular regulatory activities [3]. It is expressed by astrocytes in the brain and has been associated with neurological diseases such as Alzheimer's disease, Parkinson's disease, and schizophrenia [4, 5]. Three-dimensional structures of S100B and other S100 proteins show these proteins are composed of two EF-hand motifs, which each bind calcium and form dimeric structures [6–9]. Each S100 protomer is comprised of an N-terminal

---

Yuning Wang and Roya Tadayon contributed equally to this work.

Claus W. Heizmann (ed.), *Calcium-Binding Proteins of the EF-Hand Superfamily: From Basics to Medical Applications*, Methods in Molecular Biology, vol. 1929, https://doi.org/10.1007/978-1-4939-9030-6_20,

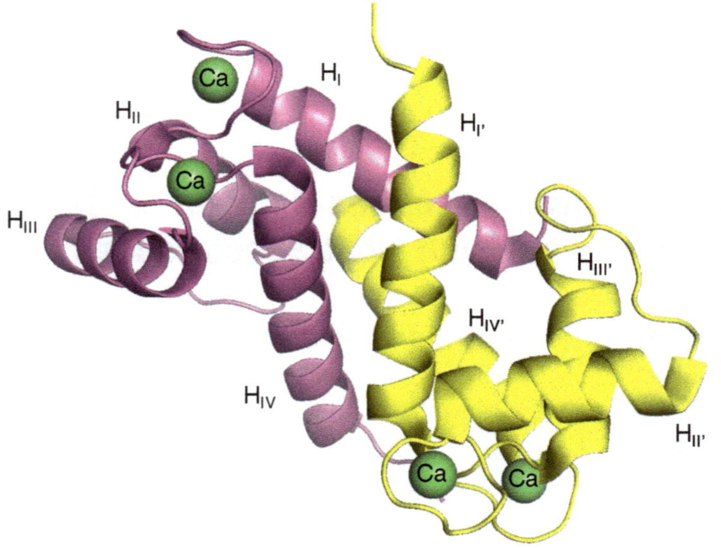

Fig. 1 Ribbon diagram showing the homodimeric structure of Ca^{2+}-S100B using PDB coordinates 2H61 [10]. The structure shows two S100B protomers (magenta, yellow) that each bind two Ca^{2+} ions (green spheres). Helices in each protomer are labeled H_I–H_{IV} and $H_{I'}$–$H_{IV'}$. This figure was made using PyMol (pymol.org)

EF-hand (helix-loop-helix) with a 14-residue Ca^{2+} binding loop (site I) and a C-terminal EF-hand with a canonical 12-residue Ca^{2+} binding loop (site II) that are separated by a hinge region and flanked by amino- and carboxy-terminal helices (Fig. 1).

In many S100 proteins, including S100B, calcium binding causes a conformational change that alters the arrangement between helices III and IV in site II, uncovering a hydrophobic protein-interaction surface. This is the target binding region for a variety of proteins and small molecules such as TRTK-12 [11, 12], ribosomal S6 kinase 1 [13], p53 [14, 15], and many others in a Ca^{2+}-dependent manner [6, 7, 16]. One potential target for S100B is the dopamine D2 receptor, identified through two-hybrid and co-immunoprecipitation experiments [17]. The dopamine D2 receptor belongs to the G protein-coupled receptor family with seven transmembrane domains and is widely distributed in neurons of the central nervous system [18]. Downstream activation of multiple kinases (glycogen synthase kinase-3, mitogen-activated protein kinases) and recruitment of the scaffolding protein β-arrestin2 are important functions of the dopamine D2 receptor. Dysregulation of the D2 receptor has been associated with a number of neuropsychiatric disorders including Parkinson's disease and schizophrenia [19]. The activity of the dopamine D2 receptor is also linked to stimulation of K^+ channels and inhibition of Ca^{2+} channels (Fig. 2a). The D2 receptor has three isoforms which are distinguishable by differential splicing of the pre-mRNA in the

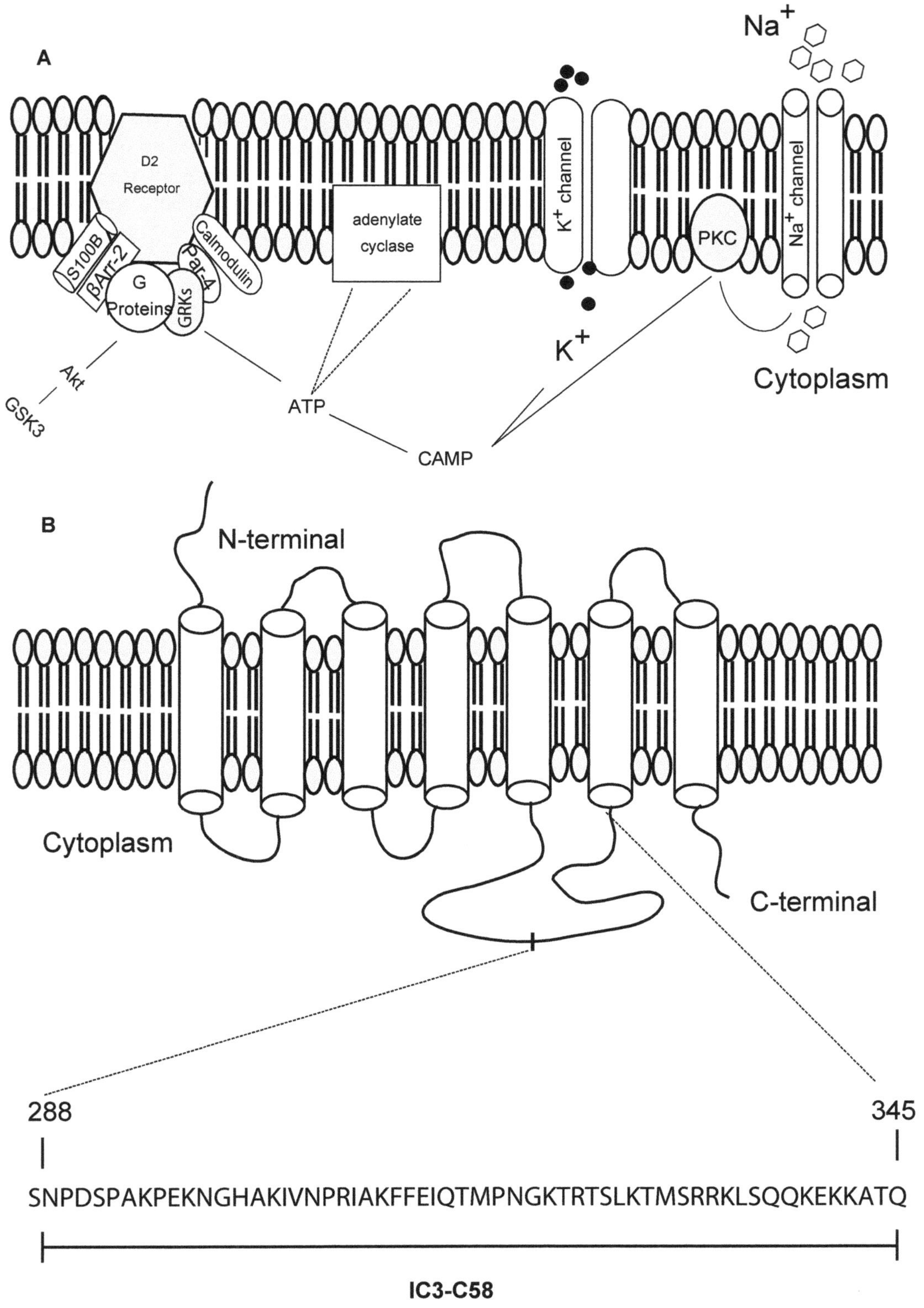

Fig. 2 A schematic diagram of the generalized dopamine D2 receptor protein in a cell membrane. (**a**) Interactions between potential binding partners of the dopamine D2 receptor and links to small molecule

region coding for their third intracellular loop (IC3) (R217-Q345) [22]. The IC3 loop connects the fifth and sixth transmembrane helices (Fig. 2b) and displays the largest diversity in the dopamine receptor family. A variety of methods have shown that the IC3 loop of the dopamine D2 receptor interacts with S100B and that this interaction is nearly 300-fold stronger upon calcium binding to the S100 protein [17, 23]. The extreme C-terminal 58-residue section (IC3-C58) of the IC3 loop displays the major interaction site for calcium-bound S100B (Fig. 2b).

Providing detailed information of protein-protein interactions requires an assessment of the strength of the interaction and its location(s). Many methods to identify and possibly quantify an interaction both in vitro and in vivo are available including co-immunoprecipitation, far-Western blot analysis, isothermal titration calorimetry, yeast two-hybrid, protein pull-down experiments, Förster resonance energy transfer (FRET) experiments, and chemical cross-linking [24]. Nuclear magnetic resonance (NMR) spectroscopy is unique because it allows the site of the interaction and the strength of the interaction to be identified [25] from a single experiment, and the method is nondestructive. This chapter will describe the practical use of NMR spectroscopy to quantify the interactions between the calcium-bound form of S100B (Ca^{2+}-S100B) and the IC3 loop region (IC3-58) of the dopamine D2 receptor (Fig. 2b).

2 Materials

2.1 S100B and IC-58 Proteins

1. Solution of recombinant ^{15}N-labeled human S100B expressed and purified from *Escherichia coli* N99 cells in M9 media. The plasmid (pSS2; plasmid #26774) containing this construct is available from Addgene (Addgene.org). The solution is collected from final purification on an anion exchange column at 4 °C and stored at −80 °C. For complete expression and purification of recombinant human S100B, see Smith et al. [26].
2. Solution of the recombinant IC3-C58 (S288-Q345) region of the dopamine D2 receptor expressed and purified from *Escherichia coli* BL21(DE3) cells in Luria broth media. The solution is collected from final purification on a cation exchange column

Fig. 2 (continued) effectors are shown. Not all known interacting proteins are depicted for simplicity (Adapted from [20, 21]). (**b**) Two-dimensional representation of the dopamine D2 receptor showing the location and amino acid sequence of the IC3-C58 binding loop, the site identified for Ca^{2+}-S100B interaction at 4 °C and stored at −80 °C. For complete expression and purification of recombinant IC3-C58, *see* Dempsey and Shaw [23].

2.2 NMR Samples

1. NMR buffer: 50 mM MOPS, 50 mM KCl, 1 mM TCEP, 1 mM DTT, 5 mM NaN_3, 6 mM $CaCl_2$ at pH 7.0 in deionized H_2O.
2. D_2O (99.9%, available from commercial sources).
3. NMR internal reference solution: 10 mM DSS (4,4-dimethyl-4-silapentane-1-sulfonic acid) dissolved in deionized H_2O.
4. NMR internal pH indicator solution: 2 M imidazole (pH 7.0) dissolved in deionized H_2O.
5. Thin-walled, precision 5 mm NMR sample tubes (i.e., Wilmad 535-PP-8).
6. Calibrated 10 μL Hamilton syringe capable of reproducibly dispensing 1 μL aliquots.

2.3 Instrumentation and Software

1. High-resolution solution NMR spectrometer capable of ^1H direct observation and ^{15}N indirect detection. The instrument should be capable of acquiring ^1H-^{15}N HSQC spectra (*see* **Note 1**).
2. Assigned ^1H-^{15}N HSQC spectra of human Ca^{2+}-S100B [27]. This data is available from the BioMagResBank (BMRB) (bmrb.wisc.edu) under the accession number for Ca^{2+}-S100B (BMRB ID 5206).
3. Personal computer and software to process and analyze NMR spectra using NMRPipe [28] and NMRViewJ [29]. These programs are available online for multiple platforms (NMRPipe; NMRViewJ).
4. Graphing and data analysis software capable of nonlinear regression analysis and fitting (Origin, GraphPad Prism).
5. Three-dimensional structure of Ca^{2+}-S100B using coordinates (PDB 2H61) available from the RCSB Protein Data Bank (PDB) [10].
6. PyMOL or other program for visualization and analysis of molecular structures.

3 Methods

Here we describe a step-by-step protocol for the titration of Ca^{2+}-S100B with the IC3-C58 peptide from the dopamine D2 receptor using NMR spectroscopy. Data processing for the determination of the S100B/IC3-C58 dissociation constant (K_d) is described. Along with the three-dimensional structure of Ca^{2+}-S100B, the data can be used to map and identify the IC3-C58 binding interface.

3.1 NMR Sample Preparation

1. Prepare a Ca^{2+}-S100B solution by dialyzing purified uniformly ^{15}N-labeled S100B into the NMR buffer at 4 °C overnight (with gentle stirring of the buffer). Prepare an IC3-C58 peptide solution in the identical buffer.
2. Concentrate the Ca^{2+}-S100B solution to a higher concentration than required for NMR titration experiments (100–200 μM). Filter the protein with a 0.22 μm filter. When complete determine an accurate protein concentration of the S100B protein (*see* **Note 2**). In this example the concentration of this solution was 931 μM Ca^{2+}-S100B.
3. Make a stock solution of IC3-C58 containing a concentration of the peptide that is approximately 50–100 times higher than the target Ca^{2+}-S100B concentration that will be used in the titration. The actual peptide stock solution is dependent on the Ca^{2+}-S100B concentration used in the experiment and the predicted binding affinity with the peptide. In these experiments a stock solution of 6 mM IC3-C58 was prepared in NMR buffer (Table 1). When complete determine an accurate protein concentration of the IC3-C58 protein (*see* **Note 2**).
4. Prepare the NMR sample of Ca^{2+}-S100B in a plastic 1.5 mL Eppendorf tube. Add 86 μL of Ca^{2+}-S100B solution, 613 of NMR buffer, 80 μL 99.9% D_2O, 1 μL of 2 M imidazole solution, and 20 μL of 10 mM DSS to the tube. The final sample volume should be ≥600 μL. Gently mix the sample by hand. The concentration of this solution used for NMR experiments was 100 μM Ca^{2+}-S100B.
5. Use a long syringe or Pasteur pipette to transfer 600 μL of the mixed Ca^{2+}-S100B sample into a clean 5 mm NMR tube. Cover with the cap. Gently tap the tube to make sure all the samples reach the bottom of the tube.

3.2 Titration Experiments

1. Insert the NMR tube into the NMR spectrometer following the NMR manufacturer's protocol. Adjust the sample temperature if necessary to 35 °C, and let the sample equilibrate in the spectrometer for at least 15 min. Make appropriate adjustments to the sample to optimize the ^1H signal and linewidth of the sample.
2. Collect a one-dimensional ^1H spectrum, and reference the ^1H chemical shift using the DSS signal set at 0.00 ppm. Measure the position (chemical shift) of the imidazole H2 signal that should be located between 7.77 and 8.68 ppm. Use this signal position to determine the pH of the sample [30] (*see* **Note 3**).
3. Acquire a ^1H-^{15}N HSQC spectrum of the Ca^{2+}-S100B sample according to the spectrometer manufacturer instructions. For the ^{15}N-labeled Ca^{2+}-S100B described here, typical parameters were 16 transients (repetitions), ^1H spectral width of 7000 Hz

Table 1
Titration data for the dopamine D2 IC3-C58 peptide to Ca^{2+}-S100B (100 μM)

Titration point	[IC3-C58] stock (mM)	IC3-C58 added (μL)	Final NMR volume (μL)	Final [IC3-C58] (μM)	[IC3-C58]/[Ca^{2+}-S100B]
0	0	0	600	0	0
1	6	1	601	10	0.1
2	6	1	602	20	0.2
3	6	1	603	30	0.3
4	6	1	604	40	0.4
5	6	2	606	59	0.6
6	6	2	608	79	0.8
7	6	3	611	108	1.1
8	6	5	616	156	1.6

for 1024 complex points, and 1944 Hz with 64 increments in the ^{15}N dimension (*see* **Note 4**). When complete, inspect the spectrum and compare to the assigned NMR spectrum for Ca^{2+}-S100B. Save the raw data.

4. For each titration point, eject the sample from the spectrometer, and carefully add 1–5 μL of the prepared IC3-C58 stock solution (Table 1) (*see* **Note 5**) into the tube by using a 10 μL Hamilton syringe (*see* **Note 6**). Inject the peptide solution along the wall of the NMR tube, and carefully tip the tube to distribute the solution without introducing air bubbles. When the titration is complete, the total volume of the sample should have changed by less than 10% of the starting sample volume.
5. Place the sample back in the spectrometer, wait for at least 15 min for equilibration, and acquire ^{1}H-^{15}N HSQC spectra after each IC3-C58 addition. The parameters used to collect each spectrum should be identical to the original data set. Ideally, eight to ten experiments are carried out for each titration experiment.

3.3 Data Processing and Dissociation Constant Determinations

1. Process all the data identically using the NMRPipe software [28]. This software allows the spectra to be correctly referenced in ^{1}H and ^{15}N dimensions. Details on processing can be found on the NMRPipe website and online tutorial as there are some differences depending on NMR spectrometer manufacturer and the specific ^{1}H-^{15}N HSQC experiment used to collect the data.

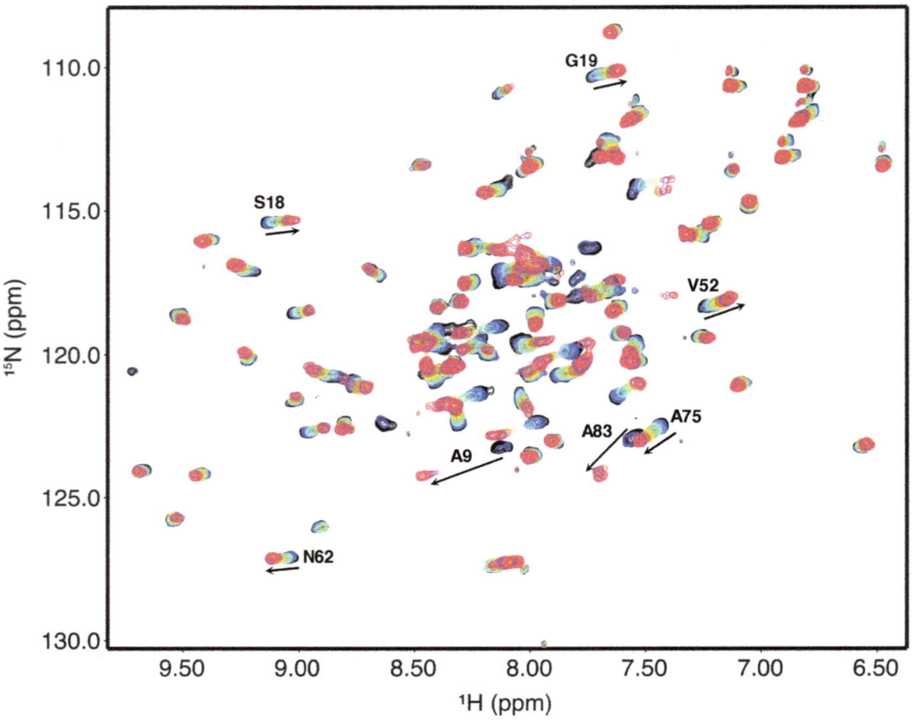

Fig. 3 Superposition of several ^{1}H-^{15}N HSQC spectra for Ca^{2+}-S100B during a titration with the IC3-C58 peptide. The spectra show 100 μM Ca^{2+}-S100B (black) and additions of 10 μM (blue), 20 μM (dark blue), 30 μM (cyan), 40 μM (green), 60 μM (yellow), 80 μM (orange), 110 μM (pink), and 160 μM (red) IC3-C58. Residue assignments indicated beside some signals are from the Ca^{2+}-S100B data in the BMRB (BMRB ID 5206)

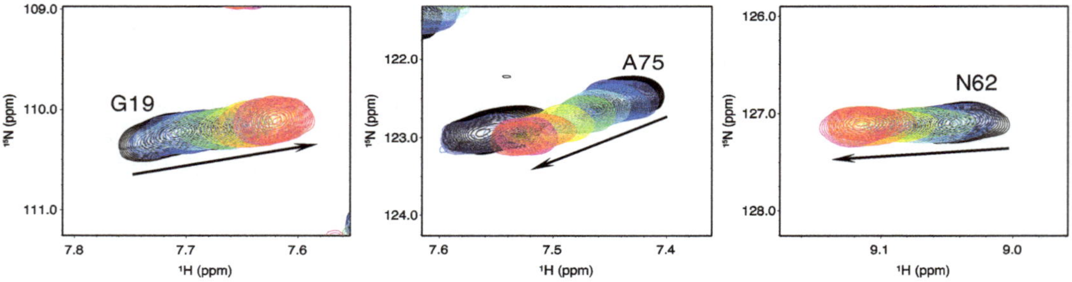

Fig. 4 Closeup regions of superimposed ^{1}H-^{15}N HSQC spectra for Ca^{2+}-S100B showing chemical shift changes of residues G19, N62, and A75 upon addition of IC3-C58 peptide. Colors are as indicated in Fig. 3

2. Display and superimpose all the spectra in NMRViewJ. Details can be found on the NMRViewJ website and accompanying tutorials and Johnson et al. [29].
3. Use NMRViewJ and the BMRB data for Ca^{2+}-S100B to identify all the signals in the initial ^{1}H-^{15}N HSQC spectrum of Ca^{2+}-S100B prior to addition of the IC3-C58 peptide.

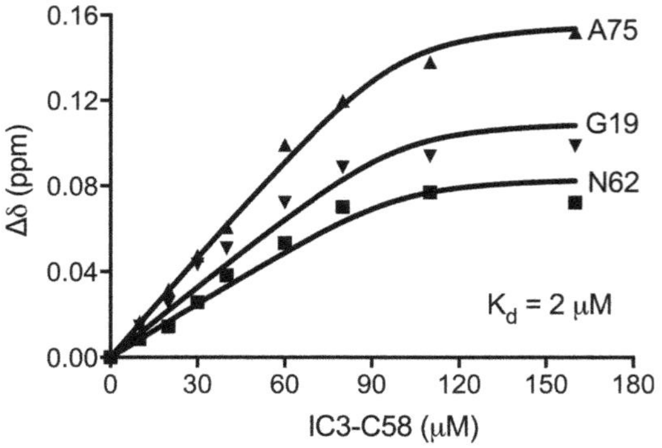

Fig. 5 Plot of the chemical shift changes ($\Delta\delta$) of A75, G19, and N62 at each titration point as a function of concentration of IC3-C58 peptide added to the Ca^{2+}-S100B sample. Global fitting was performed using Eq. (2) and yielded dissociation constant Kd = 2 μM

4. Examine the series of IC3-C58 titration spectra. The interaction with IC3-C58 peptide causes some signals to shift from their original position (Fig. 3).
5. Examine the spectra, and select signals that are visible in all spectra, and gradually move from their original location (Fig. 4). These signals are characterized as having fast exchange on the chemical shift time scale (*see* **Note** 7). Check that these signals move in random directions (*see* **Note 8**).
6. Measure the positions (chemical shift) of the selected peaks in both ^{1}H and ^{15}N dimensions at each titration point. In the example shown, resonances corresponding to N62, A75, and G19 were chosen due to their obvious chemical shift changes (Fig. 4). Determine the change in chemical shift ($\Delta\delta$) at each addition of IC3-C58 according to the Eq. (1).

$$\Delta\delta = \sqrt{\left(\Delta^2 H\right)^2 + \left(\frac{\Delta^{15} N}{5}\right)^2} \qquad (1)$$

In this equation, $\Delta^1 H$ and $\Delta^{15} N$ refer to differences in signal position in the absence and presence of a known amount of IC3–C58 peptide.

7. Plot the chemical shift changes ($\Delta\delta$) of selected peaks as a function of concentration of total IC3–C58 peptide added to the Ca^{2+}-S100B sample. The data should form a hyperbolic curve similar to that shown in Fig. 5. A global K_d value can be calculated by simultaneously fitting the chemical shift changes for the selected residues using Eq. (2) (*see* **Note 9**):

$$\Delta\delta = N\frac{(P_t + L_t + K_d) - \sqrt{(P_t + L_t + K_d)^2 - 4P_tL_t}}{2P_t} \qquad (2)$$

where P_t is the total Ca^{2+}-S100B concentration, L_t is the total IC3-C58 peptide concentration at a given titration point, K_d is the dissociation constant, and N is the maximum chemical shift change that would occur upon saturation of Ca^{2+}-S100B by IC3–C58. In our example, simultaneous fitting of $\Delta\delta$ of N62, A75, and G19 using GraphPad Prism provided a $K_d = 2$ μM (Fig. 5).

3.4 Determination of IC3–C58 Interaction Site

1. Measure the chemical shift change for every signal in Ca^{2+}-S100B between the initial and final addition of IC3-C58 peptide according to Eq. (1).
2. Identify the residues that have the largest chemical shift changes. Calculate the average chemical shift change over all residues. Identify residues that have a chemical shift change >0.25 standard deviations above this average.
3. Display the three-dimensional structure of Ca^{2+}-S100B in PyMOL. Select the residues identified in the last step, and highlight these on the structure. Examine the location and distribution of the residues. The interaction of IC3–C58 with Ca^{2+}-S100B shows a contiguous surface near the interface of helices I, III, and IV of S100B (Fig. 6).

4 Notes

1. In this example we used a 14.1 T (600 MHz) NMR spectrometer with gradient probe capable of ^{1}H direct observation and ^{15}N indirect detection. For proteins of similar size to S100B, the magnetic field strength of the spectrometer should be 11.75 T (500 MHz) or higher to achieve sufficient resolution between signals.
2. Accurate protein concentrations are critical for dissociation constant determination. In our experience absorbance spectra at 280 nm provides poor quantitation unless the protein contains several tryptophan residues. S100B lacks tryptophan and has a low extinction coefficient that can be quite variable depending on calcium, pH, and ionic strength. We typically use quantitative amino acid analysis for protein concentration determination in our experiments due to its high sensitivity, accuracy, and precision. Analyses are done in triplicate using alanine and leucine residues to measure protein concentration against an internal standard (norleucine).

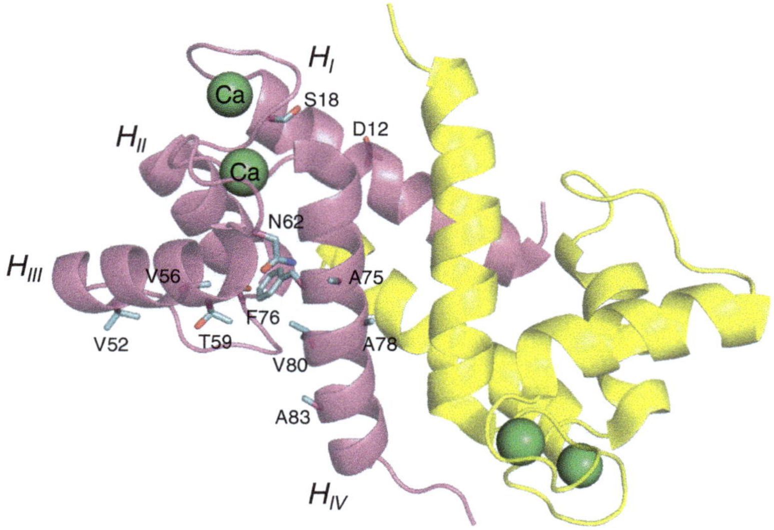

Fig. 6 Ribbon diagram structure of Ca^{2+}-S100B as shown in Fig. 1 with the binding surface for the IC3-C58 peptide mapped to the surface of Ca^{2+}-S100B. Residues that showed the largest changes in chemical shift upon IC3-C58 addition are labeled and shown as sticks

3. Imidazole has a pK_a near 7.1. The H2, H4, and H5 signals from the imidazole ring are sensitive to the pH of a solution between pH 5.8 and 8.6. These signals shift according to the Henderson-Hasselbalch equation and can be used to calculate the pH internally without the use of a pH meter [30]. Typically the H2 protein is more useful because it exhibits a larger chemical shift change. In NMR titration experiments, this approach allows the pH to be checked at each titration point and prevents sample loss or contamination from an external pH meter probe. During the experiment the pH should not change by more than 0.2 units. If this occurs it is suggested to choose a buffer with better capacity or increase the buffer concentration.
4. A preliminary ^{1}H-^{15}N HSQC spectrum in the absence of IC3-C58 should be recorded to determine the optimal conditions for the titration. This will typically involve increasing or decreasing the number of transients (repetitions) depending on the actual concentration and behavior of the sample. In the initial spectrum, signals should be clearly discernable from background noise, and the spectrum should have a good baseline. In our experiments, the acquisition time of each spectrum was about 30 min so that the titration could be completed within 2 days.
5. Prior to performing the titration experiment, an estimate of the binding affinity is useful so that during the titration Ca^{2+}-S100B saturation or near saturation can be obtained. This is

needed to obtain acceptable fitting during the data analysis stage. For NMR titration experiments, we estimated the binding affinity of IC3-C58 peptide by performing a course experiment where we acquire ^{1}H-^{15}N HSQC spectra using 0, 1, and 2 equivalents of IC3-C58 peptide and examine the data for signal changes. If there is little difference between spectra with 1 and 2 equivalents of IC3-C58 peptide, this would suggest binding is tight relative to the K_d, and titration points should be planned between 0 and 2 equivalents IC3-C58. In our experiments this was observed. If there are measurable differences between the spectra at 1 and 2 equivalents of IC3-C58 peptide, this would suggest binding is weak relative to the K_d, and more IC3-C58 will be needed to saturate the Ca^{2+}-S100B.

6. Adding reproducible amounts of peptide solution at each titration point is crucial for the experiment. We have found that using a Hamilton syringe is more reliable than a micropipette. The Hamilton syringe should be calibrated beforehand using deionized water and a balance. It is good practice to weigh 1 μL aliquots of deionized water to determine dispensing volume and reproducibility. During the titration clean and dry the Hamilton syringe before each use. Air bubbles in the NMR tube can be removed by centrifuging the tube at low speed using a benchtop centrifuge for a few minutes.
7. In the peptide titration to Ca^{2+}-S100B, the majority of the peaks exhibit fast exchange behavior as shown in Fig. 4. The analysis of these signals is described in this chapter. Some signals show different exchange processes ("slow intermediate") due to larger chemical shift differences between the free and IC3-C58 bound signals (A9, A83) that make analysis more complicated. Methods to analyze these signals have been previously described for S100 proteins [31]. These signals however should be considered when identifying the IC3-C58 binding interface.
8. During the titration, signals should shift randomly to indicate the environments of the ^{1}H and ^{15}N nuclei are affected differently by the IC3-C58 peptide added. If signals appear to shift in the same direction (especially in the ^{1}H dimension), this may indicate a systematic change in pH, salt concentration, or temperature rather than interaction with the peptide. In our experience pH can be problematic and can typically be addressed by increasing buffer concentration or choosing a different buffer. *See* also **Note 2**.

9. Equation (2) is derived from the following two equations. In Eq. (3), PL is the concentration of the complex formed between Ca^{2+}-S100B and the IC3-C58 peptide.

$$\mathrm{PL} = \frac{(P_t + L_t + K_d) - \sqrt{(\mathrm{P}_t + \mathrm{L}_t + K_d)^2 - 4P_tL_t}}{2} \quad (3)$$

$$\Delta\delta = N\frac{\mathrm{PL}}{\mathrm{P}_t} \quad (4)$$

Acknowledgment

This research was supported by a research grant (MOP 93520) from the Canadian Institutes of Health Research (GSS). We thank Brian Dempsey for the helpful discussions. Yuning Wang and Roya Tadayon contributed equally to this work.

References

1. Luker KE, Piwnica-Worms D (2004) Optimizing luciferase protein fragment complementation for bioluminescent imaging of protein-protein interactions in live cells and animals. Methods Enzymol 385:349–360. https://doi.org/10.1016/S0076-6879(04)85019-5
2. Nohe A, Petersen NO (2004) Analyzing protein-protein interactions in cell membranes. BioEssays 26:196–203. https://doi.org/10.1002/bies.10380
3. Donato R, Sorci G, Riuzzi F et al (2009) S100B's double life: intracellular regulator and extracellular signal. Biochim Biophys Acta 1793:1008–1022. https://doi.org/10.1016/j.bbamcr.2008.11.009
4. Liu J, Wang H, Zhang L et al (2011) S100B transgenic mice develop features of Parkinson's disease. Arch Med Res 42:1–7. https://doi.org/10.1016/j.arcmed.2011.01.005
5. Leclerc E, Sturchler E, Vetter SW (2010) The S100B/RAGE axis in Alzheimer's disease. Cardiovasc Psychiatry Neurol 2010:539581. https://doi.org/10.1155/2010/539581
6. Rezvanpour A, Shaw GS (2009) Unique S100 target protein interactions. Gen Physiol Biophys 28:F39–F46
7. Santamaria-Kisiel L, Rintala-Dempsey AC, Shaw GS (2006) Calcium-dependent and -independent interactions of the S100 protein family. Biochem J 396:201–214. https://doi.org/10.1042/BJ20060195
8. Heizmann CW, Fritz G, Schäfer BW (2002) S100 proteins: structure, functions and pathology. Front Biosci 7:d1356–d1368. https://doi.org/10.2741/heizmann
9. Fritz G, Heizmann CW (2006) 3D Structures of the calcium and zinc binding S100 proteins. In: Handbook of metalloproteins. John Wiley & Sons, Ltd, Hoboken, NJ. https://doi.org/10.1002/0470028637
10. Ostendorp T, Leclerc E, Galichet A et al (2007) Structural and functional insights into RAGE activation by multimeric S100B. EMBO J 26:3868–3878. https://doi.org/10.1038/sj.emboj.7601805
11. McClintock KA, Shaw GS (2003) A novel S100 target conformation is revealed by the solution structure of the Ca^{2+}-S100B-TRTK-12 complex. J Biol Chem 278:6251–6257. https://doi.org/10.1074/jbc.M210622200
12. Inman KG, Yang R, Rustandi RR et al (2002) Solution NMR structure of S100B bound to the high-affinity target peptide TRTK-12. J Mol Biol 324:1003–1014. https://doi.org/10.1016/S0022-2836(02)01152-X
13. Gógl G, Alexa A, Kiss B et al (2016) Structural Basis of Ribosomal S6 Kinase 1 (RSK1) Inhibition by S100B Protein. J Biol Chem 291:11–27. https://doi.org/10.1074/jbc.M115.684928
14. Rustandi RR, Drohat AC, Baldisseri DM et al (1998) The Ca^{2+}-dependent interaction of S100B(ββ) with a peptide derived from p53.

Biochemistry 37:1951–1960. https://doi.org/10.1021/bi972701n

15. Rustandi RR, Baldisseri DM, Weber DJ (2000) Structure of the negative regulatory domain of p53 bound to S100B(ββ). Nat Struct Biol 7:570–574. https://doi.org/10.1038/76797
16. Donato R, Cannon BR, Sorci G et al (2013) Functions of S100 proteins. Curr Mol Med 13:24–57. https://doi.org/10.2174/1566524011307010024
17. Liu Y, Buck DC, Neve KA (2008) Novel interaction of the dopamine D2 receptor and the Ca^{2+} binding protein S100B: role in D2 receptor function. Mol Pharmacol 74:371–378. https://doi.org/10.1124/mol.108.044925
18. Moritz AE, Benjamin Free R, Sibley DR (2018) Advances and challenges in the search for D2 and D3 dopamine receptor-selective compounds. Cell Signal 41:75–81. https://doi.org/10.1016/j.cellsig.2017.07.003
19. Urs NM, Peterson SM, Caron MG (2017) New concepts in dopamine D2 receptor biased signaling and implications for Schizophrenia therapy. Biol Psychiatry 81:78–85. https://doi.org/10.1016/j.biopsych.2016.10.011
20. Katzung BG (2001) Introduction to autonomic pharmacology. In: Basic and clinical pharmacology, 8th edn. The McGraw Hill Companies, Inc, New York, NY
21. Stanwood GD (2008) Protein-protein interactions and dopamine D2 receptor signaling: a calcium connection. Mol Pharmacol 74:317–319. https://doi.org/10.1124/mol.108.049098
22. Senogles SE, Heimert TL, Odife ER, Quasney MW (2004) A region of the third intracellular loop of the short form of the D2 dopamine receptor dictates Gi coupling specificity. J Biol Chem 279:1601–1606. https://doi.org/10.1074/jbc.M309792200
23. Dempsey BR, Shaw GS (2011) Identification of calcium-independent and calcium-enhanced binding between S100B and the dopamine D2 receptor. Biochemistry 50:9056–9065. https://doi.org/10.1021/bi201054x
24. Xing S, Wallmeroth N, Berendzen KW, Grefen C (2016) Techniques for the analysis of protein-protein interactions in vivo. Plant Physiol 171:727–758. https://doi.org/10.1104/pp.16.00470
25. O'Connell MR, Gamsjaeger R, Mackay JP (2009) The structural analysis of protein-protein interactions by NMR spectroscopy. Proteomics 9:5224–5232. https://doi.org/10.1002/pmic.200900303
26. Smith SP, Barber KR, Dunn SD, Shaw GS (1996) Structural influence of cation binding to recombinant human brain S100b: Evidence for calcium-induced exposure of a hydrophobic surface. Biochemistry 35:8805–8814. https://doi.org/10.1021/bi952698c
27. Smith SP, Shaw GS (1997) Assignment and secondary structure of calcium-bound human S100B. J Biomol NMR 10:77–88
28. Delaglio F, Grzesiek S, Vuister GW et al (1995) NMRPipe: a multidimensional spectral processing system based on UNIX pipes. J Biomol NMR 6:277–293. https://doi.org/10.1007/BF00197809
29. Johnson BA, Blevins RA (1994) NMR View: A computer program for the visualization and analysis of NMR data. J Biomol NMR 4:603–614. https://doi.org/10.1007/BF00404272
30. Baryshnikova OK, Williams TC, Sykes BD (2008) Internal pH indicators for biomolecular NMR. J Biomol NMR 41:5–7. https://doi.org/10.1007/s10858-008-9234-6
31. Rintala-Dempsey AC, Santamaria-Kisiel L, Liao Y et al (2006) Insights into S100 target specificity examined by a new interaction between S100A11 and annexin A2. Biochemistry 45:14695–14705. https://doi.org/10.1021/bi061754e

Chapter 21

Isolation and Characterization of S100 Protein-Protein Complexes

Bence Kiss, Péter Ecsédi, Márton Simon, and László Nyitray

Abstract

S100 proteins are small, mostly dimeric, EF-hand Ca^{2+}-binding proteins. Upon Ca^{2+} binding, a conformational change occurs resulting in the exposure of a shallow hydrophobic binding groove in each subunit. Interestingly, S100 proteins can interact with their partners in two ways: symmetrically, when the two partners identically bind into each groove, or asymmetrically, when only one partner binds to the S100 dimer occupying both binding pockets. Here we present a heterologous expression and purification protocol for all known human S100 proteins as well as for their partner peptides. Moreover, we provide a detailed description of three in vitro methods to determine the affinity, stoichiometry, and kinetics of S100 protein-protein interactions.

Key words EF-hand, Ca^{2+}-binding, Protein expression, Protein purification, Isothermal titration calorimetry (ITC), Fluorescence polarization (FP), Surface plasmon resonance (SPR), Fluorescence labelling, S100 proteins

1 Introduction

Members of the S100 protein family are small, mostly homodimer EF-hand Ca^{2+}-binding proteins. They appear exclusively in vertebrates and are represented by 20 closely related paralogues in human. They are involved in many pathologies: carcinogenesis and metastasis formation, chronic inflammations such as rheumatoid arthritis and various forms of fibrosis, and heart and neurodegenerative diseases. However, the exact role of S100 proteins in these diseases is still unclear [1–3]. The conformational change caused by Ca^{2+}-binding results in the exposure of two identical hydrophobic binding surfaces [4]. S100A10 is an exception in the family since it remains in the open conformation in the apo form making its activity Ca^{2+}-independent. Note that there are S100

Claus W. Heizmann (ed.), *Calcium-Binding Proteins of the EF-Hand Superfamily: From Basics to Medical Applications*, Methods in Molecular Biology, vol. 1929, https://doi.org/10.1007/978-1-4939-9030-6_21,

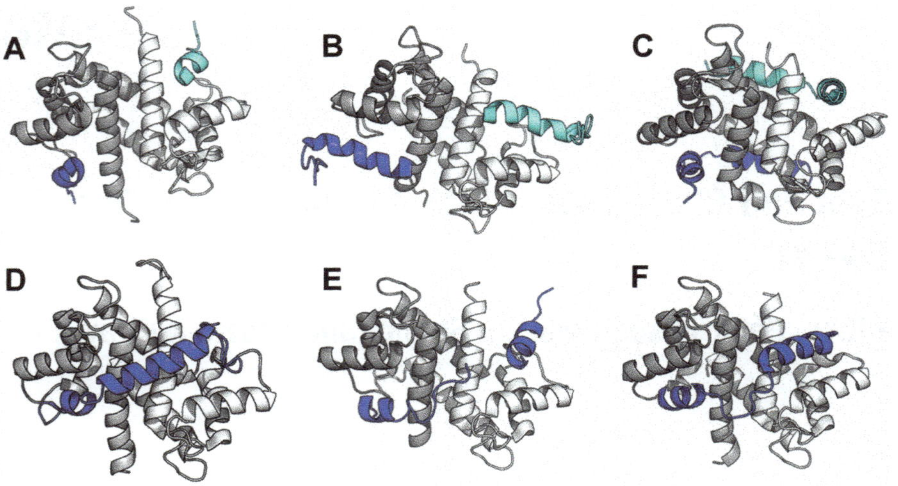

Fig. 1 Crystal structures of different S100 complexes. S100 homodimers (subunits are white and light gray) can form symmetric and asymmetric complexes with their partners (blue and cyan). In the structures of S100B—p53 (**a**) (PDB: 1DT7), S100A10—ANXA2 (**b**) (PDB: 4HRH) and S100A6—SIP (**c**) (PDB: 2JTT), the two partner peptides bind into the two identical binding pockets of the dimer. In the case of S100A4—NMIIA (**d**) (PDB: 3ZWH), S100B—RSK1 (**e**) (PDB: 5CSJ) and S100A4—ANXA2 (**f**) (PDB: 5LPU) complexes, a long fragment of the interaction partner wraps around the S100 homodimer and binds into both binding pockets more or less occupying a binding surface formed at the dimer interface

proteins (such as S100A16 or S100A7) that lack one of their two Ca^{2+}-binding EF-hands [5]. S100 proteins have a single function; they interact with partner proteins in a Ca^{2+}-dependent manner. In order to understand the physiological and pathological roles of S100 proteins, structures of their complexes should be characterized. Interestingly, although they are symmetric homodimers (known exceptions are the S100A8/A9 and S100A1/B heterodimers, as well as the monomeric S100 G), in their complexes they are either symmetric or asymmetric (Fig. 1). Such asymmetric complexes have been recently recognized [6–8], and it would be interesting to know how widespread these complexes are, what their binding partner specificity is, and how the structural diversity in the family is related to their various intra- and extracellular functions.

For in vitro characterization, proteins and protein complexes have to be produced in a highly pure, homogenous form before biochemical, biophysical, or structural biological studies. Because of this, there is a high need for developing rapid and cost-effective isolation protocols. In this section we describe the detailed expression and purification methods of human S100 proteins and their binding partners, usually representing small segments/domains of the full-length partner proteins, which are routinely used in our laboratory and meet the above criteria. Moreover, we describe in detail, with ample notes, three methods to characterize S100 complexes, to determine their stoichiometry, as well as thermodynamic and kinetic parameters of the interactions. Typically, these experiments are performed prior to high-resolution structural biology studies.

2 Materials

2.1 Heterologous Expression in Escherichia coli

1. Human S100 paralogues (UniProt accession codes: **S100A1**: P23297, **S100A2**:P29034, **S100A3**:P33764, **S100A4**: P26447, **S100A5**:P33763, **S100A6**:P06703, **S100A7**: P31151, **S100A8**:P05109, **S100A9**:P06702, **S100A10**: P60903, **S100A11**:P31949, **S100A12**:P80511, **S100A13**: Q99584, **S100A14**:Q9HCY8, **S100A15**:Q86SG5, **S100A16**:Q96FQ6, **S100B**:P04271, **S100G**:P29377, **S100P**:P25815, and **S100Z**:Q8WXG8) were either cloned from isolated RNA, obtained from Addgene, or as gifts from colleagues.
2. S100 proteins were expressed from recombinant constructs. S100 coding sequences were subcloned into the pET15b-derived vector pEV to produce TEV-protease cleavable His_6-tagged proteins (not detailed here).
3. S100 partner peptides were subcloned into the pGEX-derived vector pETARA to produce TEV-protease cleavable GST fusion peptides (not detailed here).
4. Luria-Bertani (LB) medium: dissolve 12 g tryptone, 5 g yeast extract, and 10 g NaCl in 1 L deionized water. Pour 990 and 10 mL to a 2.8 L Fernbach flask and to a 100 mL Erlenmeyer flask, respectively. Sterilize them using an autoclave.
5. Antibiotic solution: to prepare a 1000× stock solution, dissolve 1 g ampicillin in 10 mL Milli-Q water, and then filter-sterilize. Store at −20 °C in 500 μL aliquots.
6. Isopropyl β-D-1-thiogalactopyranoside (IPTG) solution: to prepare a 2000× stock solution, dissolve 2.38 g IPTG in 10 mL Milli-Q water, and then filter-sterilize. Store at −20 °C in 500 μL aliquots.
7. Competent *Escherichia coli* cells used are BL21(DE3) and Rosetta(DE3) (New England Biolabs).

2.2 Purification of Proteins and Peptides

1. Tris(2-carboxyethyl)phosphine (TCEP) solution pH = 7.5: Add 500 μL TCEP-HCl to a plastic vial and adjust its pH to 7.5 with 10 M NaOH. Store at 4 °C.
2. Phenylmethanesulfonyl fluoride (PMSF): to prepare a 100× stock solution, dissolve 1 g PMSF in 50 mL isopropanol and store at room temperature.

3. Ethylenediaminetetraacetic acid (EDTA): to prepare a 0.5 M stock solution, dissolve 7.31 g EDTA in 40 mL Milli-Q water. Upon adjusting the pH to 7.5 with 10 M NaOH, it will be completely dissolved. Complete the volume to 50 mL, and then store at 4 °C.
4. $CaCl_2$: to prepare a 1 M stock solution, dissolve 11.1 g $CaCl_2$ in 100 mL Milli-Q water. Store at 4 °C.
5. Bradford reagent: dissolve 100 mg Coomassie Brilliant Blue G-250 (100 mg) in 50 mL 96% ethanol and add 100 mL 85% (w/v) phosphoric acid. Adjust the volume to 1 L when the dye has completely dissolved, and filter through filter paper.
6. Ni^{2+} affinity chromatography:
   (a) Lysis and wash buffer (Buffer A): dissolve 2.42 g Tris–HCl and 29.22 g NaCl in 950 mL deionized water. Add 50 μL 0.5 M TCEP and adjust the pH to 8. Complete the volume to 1 L, and then store at 4 °C.
   (b) Elution buffer (Buffer B): dissolve 8.51 g imidazole, 1.21 g Tris–HCl, and 8.77 g NaCl in 450 mL deionized water. Add 25 μL 0.5 M TCEP and adjust the pH to 8. Complete the volume to 0.5 L, and then store at 4 °C.
   (c) Use Profinity IMAC resin (Bio-Rad) in glass chromatography column.
7. Phenyl Sepharose affinity chromatography.
   (a) Wash buffer (Buffer C): dissolve 4.76 g Hepes and 2.92 NaCl in 950 mL deionized water. Add 500 μL 1 M $CaCl_2$ and 50 μL 0.5 M TCEP, adjust the pH to 7.5, and then complete the volume to 1 L. Store at 4 °C.
   (b) Elution buffer (Buffer D): dissolve 2.38 g Hepes and 1.46 NaCl in 450 mL deionized water. Add 5 mL 0.5 M EDTA and 25 μL 0.5 M TCEP. Adjust the pH to 7.5, and then complete the volume to 0.5 L. Store at 4 °C.
   (c) Use Phenyl Sepharose 6 Fast Flow resin (GE Healthcare) in glass chromatography column.
8. Anion exchange chromatography.
   (a) Low salt buffer (Buffer E): dissolve 2.42 g Tris–HCl and 1.17 g NaCl in 950 mL Milli-Q water. Adjust the pH to 8.5, complete the volume to 1 L, and then sterile-filter.
   (b) High salt buffer (Buffer F): dissolve 1.21 g Tris–HCl and 14.61 g NaCl in 450 mL Milli-Q water. Adjust the pH to 8.5, complete the volume to 1 L, and then sterile-filter.
   (c) Use HiTrap Q HP anion exchange column (GE Healthcare).

9. Size exclusion chromatography.
   (a) Running buffer (Buffer G): dissolve 4.76 g Hepes and 8.77 g NaCl in 950 mL deionized water. Add 0.1 mL 0.5 M TCEP solution, adjust the pH to 7.5, complete the volume to 1 L with Milli-Q water, and then sterile-filter.
   (b) Use Superdex-75 10/300 chromatography column (GE Healthcare).
10. GST affinity chromatography.
   (a) Lysis and wash buffer (Buffer H): dissolve 2.42 g Tris–HCl and 8.77 g NaCl in 950 mL deionized water. Add 50 μL 0.5 M TCEP and adjust the pH to 7.5. Complete the volume to 1 L, and then store at 4 °C.
   (b) Elution buffer (Buffer I): dissolve 150 mg reduced glutathione in 100 mL Buffer H, and then adjust the pH to 7.5. Always prepare prior to use!
   (c) Use Protino Glutathione Agarose 4B resin (Macherey-Nagel) in glass chromatography columns.
11. RP-HPLC.
   (a) Eluent A: add 1 L Milli-Q water into a graduated cylinder, and then supplement it with 1 mL trifluoroacetic acid.
   (b) Eluent B: add 500 mL HPLC grade acetonitrile into a graduated cylinder, and then supplement it with 450 μL trifluoroacetic acid.
   (c) Use Jupiter 300 C5 RP-HPLC column (Phenomenex), 0.45 μm pore size syringe filters.

2.3 Labelling of Peptides

1. 5-iodoacetamidofluorescein (5-IAF) solution: to prepare a 20 mM stock solution, dissolve 1 mg 5-IAF in 97 μL DMSO. Protect from light, and store at −20 °C.
2. Labelling buffer (Buffer J): dissolve 2.38 g Hepes and 0.88 g NaCl in 90 mL Milli-Q water. Adjust the pH to 7.0, and then complete the volume to 100 mL. Store at −20 °C in 10 mL aliquots.
3. Use HiTrap Desalting column 5 mL (GE Healthcare) or RP-HPLC column and eluents (*see* above).

2.4 Fluorescence Polarization (FP) Measurements

1. FP buffer (Buffer K): dissolve 0.95 g Hepes and 0.88 g NaCl in 95 mL Milli-Q water. Add 20 μL 0.5 M TCEP solution, 100 μL 1 M $CaCl_2$, and 100 μL 10% Tween-20. Adjust the pH to 7.5, and then complete the volume to 100 mL with Milli-Q water. Store at −20 °C in 10 mL aliquots.
2. 384-well non-binding microplates, black (Greiner), fluorescently labelled peptide (*see* above).

2.5 Isothermal Titration Calorimetry (ITC) Measurements

1. ITC buffer (Buffer L): dissolve 4.76 g Hepes and 8.77 g NaCl in 950 mL deionized water. Add 100 μL 0.5 M TCEP and 1 mL 1 M $CaCl_2$, adjust the pH to 7.5, and then complete the volume to 1 L with Milli-Q water.

2.6 Surface Plasmon Resonance (SPR) Spectroscopy

1. Running buffer (Buffer M): dissolve freshly 4.76 g Hepes and 17.54 g NaCl in 990 mL Milli-Q water. Add 1 mL 1 M $CaCl_2$, 100 μL 0.5 M TCEP, and 50 μL 10% Tween-20, adjust the pH to 7.5, and then complete the volume to 1 L with Milli-Q water, and then sterile-filter. Separate 50 mL buffer for ligand and analyte dilution.
2. Activation solution: dissolve 0.24 g $NiCl_2 \times 6H_2O$ in 100 mL Milli-Q water, and then sterile-filter. Store at 4 °C.
3. Regeneration solution: dissolve freshly 0.48 g Hepes and 1.76 g NaCl in 90 mL Milli-Q water. Add 100 μL 0.5 M EDTA, 10 μL 0.5 M TCEP, and 50 μL 10% Tween-20, adjust the pH to 7.5, complete the volume to 100 mL with Milli-Q water, and then sterile-filter.
4. HTG sensor chip (Bio-Rad) that allows the immobilization of His-tagged proteins.

3 Methods

3.1 Expression and Purification of S100 Proteins

1. Transform 100 μL *Escherichia coli* BL21(DE3) cells (*see* **Note 1**) with S100 constructs. Inoculate 10 mL LB broth supplemented with 100 $\mu g{\cdot}mL^{-1}$ ampicillin with the transformants, and incubate the culture in a shaking incubator at 37 °C overnight.
2. Transfer the overnight culture into 1 L sterile LB supplemented with 100 $\mu g{\cdot}mL^{-1}$ ampicillin, and grow at 37 °C until the cell density reaches $OD_{600} = 0.6$ (or McFarland standard value of 3). Depending on *E.coli* strain and status, it takes approximately 3–4 h.
3. Induce the recombinant protein expression by adding 0.5 mL 1 M IPTG. Incubate at 37 °C for 3 h.
4. Pellet the cells with 5000 × *g*, and resuspend them in Buffer A supplemented with 20 $\mu g{\cdot}mL^{-1}$ PMSF (*see* **Note 2**). Place the resuspended cells on ice.
5. Disintegrate the cells by ultrasonication for 3 × 1 min (*see* **Note 3**).
6. Clarify the cell lysate by centrifugation at 50,000 × *g* for 20 min at 4 °C.
7. Pour the Ni^{2+} affinity chromatography resin slurry in a glass chromatography column. 10 mL resin per 1 L culture is

recommended. After washing out the ethanol using deionized water, pre-equilibrate the resin with Buffer A.

8. Apply the clarified cell lysate onto the resin, and wash it thoroughly with at least 10 column volumes of Buffer A.
9. Elute the His_6-tagged proteins with Buffer B (collect several fractions, and check their protein content with Bradford reagent).
10. Unite the S100-containing fractions, supplement the solution with 0.5 mM TCEP, and then add TEV protease at 1:100 molar ratio to remove the His_6-tag. Incubate the sample for 4 h at room temperature or at 4 °C overnight.
11. Using SDS-PAGE check if the cleavage has been completely occurred.
12. Perform the following purification steps:
    (a) The fully cleaved **S100A1**, **S100A2**, **S100A4**, **S100A5**, **S100A6**, **S100A11**, **S100A12**, **S100B**, and **S100P** should be applied onto Phenyl Sepharose resin pre-equilibrated with Buffer C (*see* **Note 4**). Supplement the TEV protease digested S100 protein solution with 1 mM $CaCl_2$ before applying onto the Phenyl Sepharose resin. Thoroughly wash the column with at least 10 column volumes Buffer C. Elute the samples with Buffer D.
    (b) Dialyze the fully cleaved **S100A3**, **S100A8**, **S100A9**, **S100A10**, **S100A13**, **S100A14**, **S100A16**, and **S100G** against Buffer E. Pre-equilibrate the anion exchange chromatography column with Buffer E, and apply the samples onto the column. Wash the column with 10 column values of Buffer E, and then elute the proteins by a linear gradient of Buffer F. The S100 proteins elute at a conductance of 80–350 $mS{\cdot}cm^{-1}$. **S100G** is to be separated by isocratic elution (*see* **Note 5**).
    (c) Dialyze **S100A7**, **S100A15**, and **S100Z** in Buffer A after complete cleavage. To remove the His_6-tag, the His_6-tagged TEV protease and the uncleaved S100 protein, apply the dialysed sample onto the Ni^{2+} affinity resin pre-equilibrated by Buffer A and collect the flow-through. Concentrate the samples using Amicon Ultra (Merck) centrifugation filter units, and then separate them by size exclusion chromatography on Superdex 75 column using Buffer G.
13. Concentrate the purified protein to a few millimolar concentration with Amicon Ultra centrifugation filter units, add equimolar amount of TCEP and determine the concentration based on the absorbance at 280 nm (*see* **Note 6**). Store the purified proteins at −80 °C (Fig. 2).

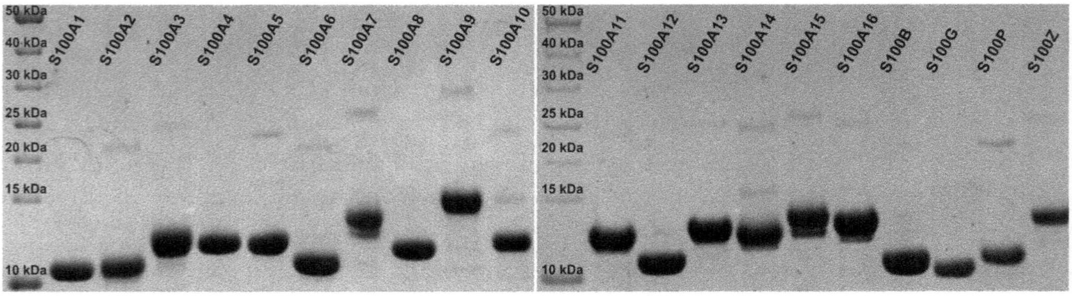

Fig. 2 SDS-PAGE analysis of human recombinant S100 proteins purified using the two-step protocol. The first columns on the left and right gels contain protein molecular weight markers. Note that the small bands above of many S100 proteins correspond to a small amount of dimeric chains (>2%)

3.2 Expression and Purification of Binding Peptides for In Vitro Analysis

1. Transform 100 μL *Escherichia coli* BL21(DE3) cells with the chosen peptide construct. Inoculate 10 mL LB broth supplemented with 100 $\mu g{\cdot}mL^{-1}$ ampicillin with the transformants, and incubate the culture in a shaking incubator at 37 °C overnight.
2. Transfer the overnight culture into 1 L sterile LB broth supplemented with 100 $\mu g{\cdot}mL^{-1}$ ampicillin, and grow at 37 °C until the cell density reaches $OD_{600} = 0.6$ (or McFarland standard value of 3). Depending on *E.coli* strain and status, it takes approximately 3–4 h.
3. Induce the recombinant protein expression by adding 0.5 mL 1 M IPTG. Incubate at 28 °C for 4 h or at 18 °C overnight.
4. Pellet the cells with 5000 × *g*, and resuspend them in Buffer H supplemented with 20 $\mu g{\cdot}mL^{-1}$ PMSF. Place the resuspended cells on ice.
5. Disintegrate the cells by ultrasonication for 3 × 1 min.
6. Clarify the cell lysate by centrifugation at 50,000 × *g* for 20 min at 4 °C.
7. Pour the GST affinity resin slurry in a glass chromatography column. 10 mL resin per 1 L culture is recommended. After washing out the ethanol using deionized water, pre-equilibrate the resin with Buffer H.
8. Apply the clarified cell lysate onto the resin, and wash it thoroughly with at least 10 column volumes of Buffer H.
9. Elute the GST-tagged peptides with Buffer I (collect several fractions, and check their protein content with Bradford reagent).
10. Unite the fractions containing the GST-peptide, supplement the solution with 0.5 mM TCEP, and then add TEV protease at 1:100 molar ratio to remove the GST-tag. Incubate the sample for 4 h at room temperature or at 4 °C overnight.

11. Using SDS-PAGE check if the cleavage has been completely occurred.
12. Precipitate GST by heat denaturation at 90 °C for 10 min. Pellet the precipitation by 10,000 × *g*.
13. Equilibrate the RP-HPLC column with Eluent A, and then inject the clarified and filtered peptide solution onto the column. Wash thoroughly the column with Eluent A until the baseline stabilizes, and then elute the peptide with the linear gradient of Eluent B. Peptides typically elute between 30 and 45% Eluent B.
14. Freeze the fractions at −80 °C, and then lyophilize them overnight. Store at −80 °C.
15. Optional: check the molecular mass of the fractions by mass spectrometry.

3.3 Fluorescent Labelling of Peptides for In Vitro Analysis

If naturally occurring Cys is not present, the introduction of an N-terminal Cys residue is necessary for the selective labelling of S100 partner peptides.

1. Dissolve the lyophilized peptide in Buffer J at a concentration of 100 μM, and then add fivefold excess of 5-IAF.
2. Incubate the reaction mixture in dark for 2 h.
3. Apply the reaction mixture onto a desalting column pre-equilibrated with Buffer J. Collect the peak eluting at approximately 1.5–2 mL.
4. Separate the fraction eluted at 1.5–2 mL with RP-HPLC (*see* **Note 7**).

3.4 Investigation of S100 Protein Interactions by Fluorescence Polarization

FP method is a rapid, cost-effective, and high-throughput protein-protein interaction measurement tool that requires a fluorescently labelled interaction partner to be small in size. FP measurements can be performed in a direct and in a competitive way. In a direct assay, the formation of the labelled peptide—S100 protein complex—is monitored as a function of S100 protein concentration. In a competitive assay, the dissociation of the labelled peptide—S100 protein complex—is investigated as a function of unlabelled S100 partner peptide concentration. The advantage of the former is its simplicity; however, the fluorescent moiety can affect the affinity of the peptide. The competitive assay benefits from the need of only one tracer peptide per S100 paralogue and the binding of the competitors are not biased by the fluorescent moiety. However, data analysis is sometimes challenging due to the more complicated setup (e.g., uncertainties in the concentrations).

1. Dissolve the lyophilized labelled peptide (tracer) in DMSO to ten to hundred micromolar concentration (stock solution).

2. Prepare 500 μL tracer solution (50 nM final concentration) by diluting the stock solution in Buffer K. Pipette 100 μL into a vial and then 50 μL into other seven vials.
3. Add the S100 protein (use at least millimolar concentration of stock solution) into the first vial to get 50 μM dimer concentration, and then perform twofold serial dilution. Do not dilute the last well (*see* **Note 8**).
4. Pipette 10–10 μL from each well of the serial dilution into the 384-well non-binding plate, and then measure the fluorescence polarization.
5. Analyze the data using Origin software (or other data analyzer software package) using a quadratic binding equation.
6. To carry out competitive assay, prepare 500 μL tracer solution (50 nM final concentration) that is saturated 70–80% by the given S100 protein. Pipette 100 μL into a vial and then 50 μL into other seven vials.
7. Add the unlabelled competitor peptide (use at least millimolar concentration of stock solution) into the first vial to a final concentration of 50 μM, and then perform twofold serial dilution. Do not dilute the last well.
8. Pipette 10–10 μL from each well of the serial dilution into the 384-well non-binding plate, and then measure the fluorescence polarization.
9. Analyze the data using Origin software (or other data analyzer software package) using a competitive binding model [9].

For further information and examples, *see* Gogl et al. [7].

3.5 Investigation of S100 Protein Interactions by Isothermal Titration Calorimetry

Due to the relatively high protein concentration in the ITC sample cell (typically 10–20-fold higher than the dissociation constant of the investigated interaction) that is required to detect high enough heat changes, ITC provides not only the thermodynamic parameters of the interaction but also the stoichiometry.

1. Dialyze the S100 protein (50–100 μM dimer, 2 mL) and the partner peptide (0.6–1.2 mM, 500 μL) against Buffer K at 4 °C overnight, and then determine the protein concentration (*see* **Note 9**). Alternatively, the lyophilized peptide can be dissolved in the dialysis buffer.
2. Fill the VP-ITC cell with S100 protein solution (*see* **Note 10**).
3. Fill the syringe with the partner peptide solution (*see* **Note 11**).
4. Start equilibration, and then run the ITC program which typically consists of 30–60 injections of 5–10 μL and takes 4 h.
5. Analyze the data using the built-in model "One set of sites" in Origin for ITC software.

For further information and examples, *see* Kiss et al. [8].

3.6 Investigation of S100 Protein Interactions by Surface Plasmon Resonance

SPR, like ITC, is a label-free technique, but unlike ITC it allows the determination of the kinetics of a protein-protein interaction while requires a relatively low amount of interaction partners. However, the interaction takes place on a surface, which implies the immobilization of one of the binding partners called ligand. Ligand immobilization is challenging when a covalent immobilization method is used. A significant portion of the ligand generally loses its binding capability, or a heterogenic population is formed on the surface of the sensor chip where ligand molecules with different binding capabilities reside. The other challenge is to avoid the non-specific binding (NSB) of the other binding partner called analyte, which phenomenon is concentration dependent. Working with ProteOn XPR36 (Bio-Rad) SPR system, our experience is that interactions in the micromolar affinity range (requiring analyte injections at concentrations in the ten micromolar range) can be hardly measured accurately due to NSB. The protocol below demonstrates an effective non-covalent immobilization mode of S100A4 protein through His_6-tag, the measurement of its interaction with a tight-binding NMIIA fragment, and a selective regeneration method that does not affect the interaction of His_6-S100A4 with the HTG sensor chip (Fig. 3).

1. Using a modified purification protocol (does not add TEV protease to the S100 solution after **Step 9**) produce His_6-tagged S100 proteins.
2. Dilute the ligand to 0.25–0.5 μM in Buffer M.

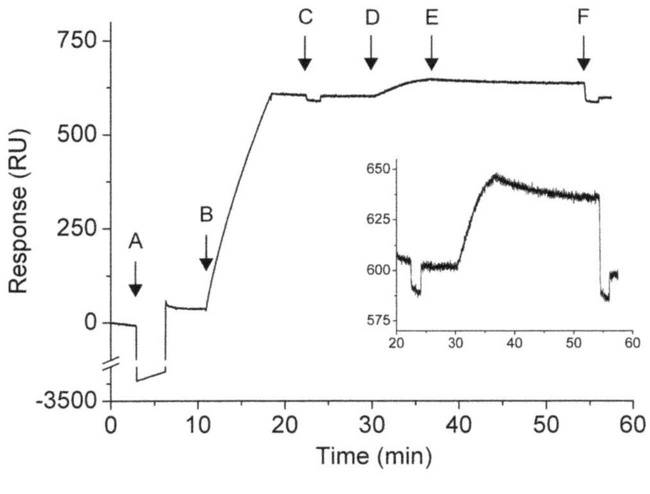

Fig. 3 Measuring protein-protein interactions of S100 proteins with SPR. **A**: injection of the activation solution. **B**: His_6-S100A4 injection. **C**: Injection of the regeneration solution. **D**: Analyte injection (NMIIA peptide, 1 nM), association phase. E: Analyte injection, dissociation phase. F: Injection of the regeneration solution. The magnified view of **steps C–F** is shown in the inset

3. Perform the conditioning of HTG sensor chip according to the manufacturer's protocol, and then inject the activation solution for 200 s at a flow rate of 30 μL·s^{-1}.
4. Inject the diluted ligand onto the activated sensor chip at a flow rate of 30 μL·s^{-1} until the ligand density reaches 500–1000 RU (*see* **Note 12**).
5. Inject regeneration solution for 100 s at a flow rate of 60 μL·s^{-1} for surface stabilization.
6. Inject the analyte at a flow rate of 60 μL·s^{-1} (*see* **Note 13**).
7. Inject regeneration solution for 100 s at a flow rate of 60 μL·s^{-1} to dissociate selectively the analyte from the immobilized S100 protein.
8. Perform kinetic analysis using the built-in model "Langmuir" in XPR36 software. Alternatively, equilibrium analysis can be also performed if the sensorgrams have reached a steady state.

For further information and examples, *see* [10–13].

4 Notes

1. Expression of S100A7 and S100A15 in Rosetta(DE3) cells produces much higher yields. If Rosetta(DE3) cells are used, incubate the cells for 30 min in 0.5 mL LB broth before inoculating the transformants into 10 mL LB broth supplemented with 25 μg·mL^{-1} chloramphenicol (prepare the 1000× stock solution in ethanol) and 100 μg·mL^{-1} ampicillin.
2. Avoid using phosphate buffers. On the one hand, it forms precipitate with Ca^{2+} that will be applied at subsequent steps. On the other hand, TCEP, which is proved to be more effective reducing agent than DTT or β-mercaptoethanol, is not stable in the presence of phosphates. PMSF and other protease inhibitors can be omitted.
3. Freezing the resuspended cells at −20 °C before ultrasonication further enhances the disintegration of cells leading to increased yields.
4. The amount of resin is to be used depends on the quantity of S100 protein in the elution fraction of Ni^{2+} affinity chromatography.
5. Do not dialyze S100A16 as it tends to precipitate during dialysis. Instead, apply low salt Ni^{2+} affinity chromatography elution buffer (e.g., with 50 mM NaCl).
6. Calculating S100 protein concentration based on the theoretical extinction coefficients of aromatic residues at 280 nm is

often heavily biased due to fact that the vast majority of S100 proteins lacks Trp residues.

7. Based on our experience, the coupled fluorescein moiety significantly affects the retention of the peptides enabling the separation of the labelled and unlabelled forms. The color of the labelled peptide should be yellowish after lyophilization, while 5-IAF and the unlabelled peptides are brown orange and white, respectively.
8. The 50 μM dimer concentration should be decreased in case of strong interactions.
9. During dialysis samples are diluted to some extent. Never use DTT as a reducing agent during ITC measurement.
10. The VP-ITC sample cell has a volume of 1.4 mL; however, 2 mL sample is required to fill the cell securely. Be aware to avoid the presence of air bubbles in the sample cell.
11. For the VP-ITC, syringe has a volume of 300 μL; however, 500 μL sample is required to fill the syringe securely.
12. Because of an S100 dimer comprises two His_6-tags, despite the non-covalent immobilization technique, a highly stable ligand coupling is obtained.
13. Analyte injection parameters such as analyte concentrations and injection times should be optimized for the given interaction.

Acknowledgment

This work was supported by the National Research, Development and Innovation Office (NKFIH) grants K119359 (to LN). MS was supported through the New National Excellence Program of the Hungarian Ministry of Human Capacities. We also thank the support of the MedInProt program of the Hungarian Academy of Sciences. Project no. FIEK_16-1-2016-0005 has been implemented with the support provided from the National Research, Development and Innovation Fund of Hungary, financed under the FIEK_16 funding scheme. This work was also supported by the National Development Agency Grant KMOP4.2.1/B-10-2011.

References

1. Boye K, Maelandsmo GM (2010) S100A4 and metastasis: a small actor playing many roles. Am J Pathol 176(2):528–535. https://doi.org/10.2353/ajpath.2010.090526
2. Bresnick AR, Weber DJ, Zimmer DB (2015) S100 proteins in cancer. Nat Rev Cancer 15 (2):96–109. https://doi.org/10.1038/nrc3893
3. Yammani RR (2012) S100 proteins in cartilage: role in arthritis. Biochim Biophys Acta 1822(4):600–606. https://doi.org/10.1016/j.bbadis.2012.01.006
4. Donato R (2001) S100: a multigenic family of calcium-modulated proteins of the EF-hand type with intracellular and extracellular

functional roles. Int J Biochem Cell Biol 33 (7):637–668

5. Marenholz I, Heizmann CW, Fritz G (2004) S100 proteins in mouse and man: from evolution to function and pathology (including an update of the nomenclature). Biochem Biophys Res Commun 322(4):1111–1122. https://doi.org/10.1016/j.bbrc.2004.07.096
6. Ecsedi P, Kiss B, Gogl G, Radnai L, Buday L, Koprivanacz K, Liliom K, Leveles I, Vertessy B, Jeszenoi N, Hetenyi C, Schlosser G, Katona G, Nyitray L (2017) Regulation of the equilibrium between closed and open conformations of annexin A2 by N-terminal phosphorylation and S100A4-binding. Structure 25 (8):1195–1207 e1195. https://doi.org/10.1016/j.str.2017.06.001
7. Gogl G, Alexa A, Kiss B, Katona G, Kovacs M, Bodor A, Remenyi A, Nyitray L (2016) Structural basis of ribosomal S6 kinase 1 (RSK1) inhibition by S100B protein: modulation of the extracellular signal-regulated kinase (ERK) signaling cascade in a calcium-dependent way. J Biol Chem 291(1):11–27. https://doi.org/10.1074/jbc.M115.684928
8. Kiss B, Duelli A, Radnai L, Kekesi KA, Katona G, Nyitray L (2012) Crystal structure of the S100A4-nonmuscle myosin IIA tail fragment complex reveals an asymmetric target binding mechanism. Proc Natl Acad Sci U S A 109(16):6048–6053. https://doi.org/10.1073/pnas.1114732109
9. Wang ZX (1995) An exact mathematical expression for describing competitive binding of two different ligands to a protein molecule. FEBS Lett 360(2):111–114. https://doi.org/10.1016/0014-5793(95)00062-E
10. Kiss B, Kalmar L, Nyitray L, Pal G (2016) Structural determinants governing S100A4-induced isoform-selective disassembly of nonmuscle myosin II filaments. FEBS J 283 (11):2164–2180. https://doi.org/10.1111/febs.13728
11. Leclerc E, Fritz G, Weibel M, Heizmann CW, Galichet A (2007) S100B and S100A6 differentially modulate cell survival by interacting with distinct RAGE (receptor for advanced glycation end products) immunoglobulin domains. J Biol Chem 282 (43):31317–31331. https://doi.org/10.1074/jbc.M703951200
12. Dattilo BM, Fritz G, Leclerc E, Kooi CW, Heizmann CW, Chazin WJ (2007) The extracellular region of the receptor for advanced glycation end products is composed of two independent structural units. Biochemistry 46 (23):6957–6970. https://doi.org/10.1021/bi7003735
13. Ostendorp T, Leclerc E, Galichet A, Koch M, Demling N, Weigle B, Heizmann CW, Kroneck PM, Fritz G (2007) Structural and functional insights into RAGE activation by multimeric S100B. EMBO J 26(16):3868–3878. https://doi.org/10.1038/sj.emboj.7601805

Chapter 22

The Multifaceted S100A4 Protein in Cancer and Inflammation

Noona Ambartsumian, Jörg Klingelhöfer, and Mariam Grigorian

Abstract

The metastasis-promoting S100A4 protein, a member of the S100 family, has recently been discovered as a potent factor implicated in various inflammation-associated diseases. S100A4 is involved in a range of biological functions such as angiogenesis, cell differentiation, apoptosis, motility, and invasion. Moreover, S100A4 is also a potent trigger of inflammatory processes and induces the release of cytokines and growth factors under different pathological conditions.

Indeed, the release of S100A4 upon stress and mainly its pro-inflammatory role emerges as the most decisive activity in disease development, such as rheumatoid arthritis (RA), systemic sclerosis (SSc) allergy, psoriasis, and cancer. In the scope of this review, we will focus on the role of S100A4 as a mediator of pro-inflammatory pathways and its associated biological processes involved in the pathogenesis of various human noncommunicable diseases (NCDs) including cancer.

Key words S100A4 protein, Calcium, EF-hand, DAMP proteins, Inflammation, Cancer progression, Noncommunicable diseases

1 Introduction

The S100A4 protein belongs to the S100 superfamily of small Ca^{2+}-binding proteins. The most well-established function of S100A4 is associated with the induction and promotion of tumor metastasis [1]. However, this protein is also involved in the pathogenesis of major human NCDs, such as autoimmune, fibrotic, and inflammatory disorders [2, 3].

As many other S100 proteins, S100A4 exhibits both intra- and extracellular activities. Inside the cell, the S100A4 expression is associated with cell migration, apoptosis, and maintaining of stemness [2, 4–9]. Extracellularly, S100A4 activates various processes mainly by stimulating pro-inflammatory pathways and the expression of various molecules, like cytokines, growth factors, ECM proteins, and matrix metalloproteinases (MMPs) [10].

Claus W. Heizmann (ed.), *Calcium-Binding Proteins of the EF-Hand Superfamily: From Basics to Medical Applications*, Methods in Molecular Biology, vol. 1929, https://doi.org/10.1007/978-1-4939-9030-6_22,

Thus S100A4's extracellular function mainly accounts for its pro-inflammatory and metastasis-promoting activity [3, 11–13]. S100 proteins [14, 15], including S100A4, can be categorized as a group of molecules known as alarmins or damage-associated molecular patterns (DAMPs), which have critical functions in the body's inflammatory response [16]. These molecules are typically released to the extracellular space from cells in response to stress. They bind to specific receptors on immune cells to promote the activation of innate immunity, cell differentiation, cell death, or secretion of inflammatory mediators, thus amplifying the inflammatory response [17].

Inflammation is a part of the natural defense system involving immune cells that fight against pathogens and injuries caused by various agents and stresses. However, sustained inflammation is associated with a wide range of pathological conditions (NCDs) including cancer and metastasis [18–21]. There are many similarities in molecular processes that underlie chronic inflammation and cancer development. However, many of the molecular and cellular mechanisms mediating this relationship remain unresolved [1].

In the present review, we mostly concentrate on S100A4 and its role as DAMP protein enhancing inflammation in cancer and a wide range of pathological conditions.

2 The Role of S100A4 in Stimulation of Metastasis

The importance of S100A4 in the process of cancer metastasis has been validated by several experimental approaches. S100A4 was originally cloned as a gene differentially expressed in metastatic vs. nonmetastatic mammary cancer cell lines [22].

Studies of metastatic cell lines revealed comprehensive data on the involvement of S100A4 in the metastatic process. Overexpression of S100A4 in a benign rat mammary epithelial cell and in nonmetastatic human breast cancer cells promotes metastasis to the lungs and lymph nodes. Suppression of S100A4 in metastatic human osteosarcoma cells and Lewis lung carcinoma cells is associated with a decrease in experimental metastases [23–26].

Further evidence that highlights a central role of S100A4 in metastasis came from genetically engineered mice. Studies of two transgenic mouse models convincingly demonstrate a causal role of S100A4 in stimulation of metastasis. When transgenic mice were crossed into a tumorigenic background, its offspring displayed a markedly increased frequency of lung metastases [27, 28].

Likewise, mice carrying null alleles for S100A4 demonstrate a significant suppression of metastasis formation. Indeed, S100A4-deficient mice, grafted with metastatic mouse mammary carcinoma cells, demonstrate severe impairments in tumor development and cessation of metastasis formation [29]. In TRAMP/S100A4(+/−)

mice, a transgenic model of prostate cancer with heterozygous deletion of S100A4, the tumor latency, and metastasis formation is impaired. Yet, spontaneous mammary tumors in polyoma middle T oncogene (PyMT) transgenic mice deficient in S100A4 (PyMT-S100A4 −/− mice) result in reduced metastasis formation, while the tumor latency and development are not affected [30, 31].

Finally, a causal role of S100A4 in tumor progression is confirmed by utilizing an anti-S100A4 function-blocking antibody. Antibody treatment significantly reduces the metastatic burden in the lungs of experimental animals [32, 33].

In human, the S100A4 expression is associated with the progressed stages of different cancers [34, 35]. The S100A4 expression is strongly correlated with poor prognosis in breast cancer and colorectal cancer patients [36, 37]. Further investigations have demonstrated an association between S100A4 expression and patient survival for several other tumor types, including ovarian carcinoma, pancreatic cancer, malignant melanoma, bladder cancer, non-small cell lung cancer, gallbladder cancer, esophageal squamous cell carcinoma, and gastric cancer, as reviewed in [1, 11, 34].

3 The Role of S100A4 in the Tumor Microenvironment

In recent years it became evident that the tumor microenvironment is deeply engaged in determining the metastatic fate of the tumor. Different cells and components of the stroma influence the spread of tumor cells by modulating the molecular network in the tumor milieu [38]. Likewise, the primary tumor sends specific signals to distant organs that initiate the formation of the pre-metastatic niche in the microenvironment of the secondary organ [39].

The S100A4 protein is suggested to modulate the microenvironment, both at the site of the primary tumor and the pre-metastatic niche [29, 30, 32, 40].

S100A4 is released into the tumor microenvironment by both cancer and nonmalignant cells. Thus, it has been shown that macrophages, fibroblasts, and tumor cells in culture release S100A4 [40–44]. Tumor cells also stimulate the release of S100A4 from stromal cells [34]. In vivo, an increased level of S100A4 is detected in human breast cancer interstitial fluid and in the serum of aging S100A4 transgenic mice [41, 42]. Evidence for the essential role of stromal cells produced S100A4 is obtained by S100A4 knockout mice. The metastatic potential of orthotopic mammary tumors is significantly reduced in *S100A4*-deficient mice [29, 45], and it is rescued by co-injection of tumor cells with S100A4-positive fibroblasts. The administrated fibroblasts substitute the metastasis-promoting effects of S100A4-positive cells in the tumor microenvironment [30]. The effect of S100A4 on metastasis is associated with T-cell attraction to a premalignant tumor and to the pre-metastatic lung [30]. Moreover, the S100A4-dependent deposition of fibronectin (FN) and accumulation of T cells in pre-metastatic lungs can be neutralized by an S100A4-specific antibody [32]. These data suggest that increased concentrations of S100A4 in the tumor microenvironment mediate the creation of a pro-metastatic milieu. Co-injection of S100A4-negative tumor cells with the recombinant S100A4 protein increases the metastatic capacity of tumor cells, confirming the data obtained on S100A4-deficient mice [40].

4 The Role of S100A4 in Non-cancer Disorders (NCDs)

Mounting evidence gained from the experimental and clinical studies revealed significant implication of S100A4 in various NCDs, such as autoimmune disorders; cardiovascular, nervous, pulmonary, and gastroenteric pathologies; and diabetes [3, 13]. Here we will review the pathologies with the established role of S100A4 in the disease pathogenesis.

These include rheumatoid arthritis (RA), psoriasis, systemic sclerosis (SS), and allergy.

4.1 Rheumatoid Arthritis

Differential expression of S100A4 in synovial tissues from patients with RA compared to normal individuals served as the first indication of its implication in RA [46]. Fibroblasts and immune and vascular cells contribute to the production of S100A4 in synovial tissue consistent with the high concentrations of the protein in RA patient plasma and synovial fluid. Importantly, abundant extracellular oligomeric S100A4 was detected in RA, whereas, in osteoarthritis, the less severe form of joint destruction, the majority of extracellular S100A4 was detected in the less-active dimeric form [47, 48]. S100A4 levels are also elevated with the increase of radiographic damage and disease progression. High S100A4 levels are also associated with a poor clinical response to anti-TNF-α therapy [49]. Moreover, persistently high S100A4 levels predict poor treatment outcome and are suggested as the promising biomarker for assessing treatment response in RA patients [50].

4.2 Psoriasis

Psoriasis is a chronic inflammatory skin disease with epidermal hyperplasia and vascular and joint alterations. In psoriatic skin, the S100A4 expression is significantly upregulated in various cell types of the upper dermal compartment as compared to healthy tissue. High expression of S100A4 correlates with a massive release of oligomeric S100A4 into the interstitial fluid of psoriatic skin. Neutralization of S100A4 by means of specific anti-S100A4 antibodies in a human psoriasis xenograft SCID mouse model results in diminishing of the main features of the disease. These data directly implicate S100A4 in the pathogenesis of psoriasis [51].

4.3 Systemic Sclerosis

Systemic sclerosis (SS) or scleroderma is an autoimmune disease characterized by fibrotic scar formation with excessive deposition of ECM by fibroblasts.

S100A4 protein is upregulated in the skin of SSc patients. High expression of S100A4 is attributed to fibroblasts but also to cells composing vessel, perivascular inflammatory infiltrated cells, and cells of the basal layer of the epidermis. The S100A4 expression is also increased in different mouse models of SSc. TGF-β was shown as a major activator of S100A4 via the Smad3-dependent pathway. Furthermore, S100A4 promotes the production and release of collagen that is abrogated in vitro by knocking down S100A4 by specific siRNA. Moreover, deficiency of S100A4 protects fibrosis development in two experimental models of SSc, the murine bleomycin-induced dermal fibrosis model, and SSc activation in the tight skin (Tsk-1) mice. The data collectively point to S100A4 as an essential factor in SSc disease progression and potential therapeutic target [52]. Moreover, inhibition of the pro-fibrotic activity in primary cultures of SSc fibroblasts/myofibroblasts significantly reduces protein synthesis and gene expression of S100A4 and collagen-1 [53].

4.4 Allergy

Seasonal allergic rhinitis (SAR) is a common IgE-mediated disorder of extremely high prevalence. For the adult population in Europe, the overall prevalence of allergic rhinitis is about 21% [54].

S100A4 is also an important protein involved in allergy [55]. *S100A4* is highly expressed in cells of relevance for allergy, including CD8+ T cells, monocytes, and eosinophils. S100A4 is increased in nasal fluids from allergic patients and in the skin of patients with allergic dermatitis. Functional studies using animal model and PMBCs of allergy patients link S100A4 to allergic reactions and Th2 cell activation.

Thus, S100A4 deficiency or S100A4 neutralization by a specific antibody diminished local and systemic signs of allergic inflammation that implicates the protein directly in the pathogenesis of allergic reaction.

In summary, accumulating data on the implication and correlation of S100A4 in numerous human NCDs allow one to speculate on the key role of S100A4 in the inflammatory pathways which is, most likely, shared by various serious life-debilitating and life-threatening conditions.

5 Functional Roles of the S100A4 Protein

Although the exact mechanism how S100A4 acts under pathological conditions is not yet clear [10], it has been shown that S100A4 affects various cellular processes, like motility, EMT, angiogenesis, cell differentiation, apoptosis, and autophagy. The effects of

S100A4 could be exhibited both by extracellular and intracellular S100A4.

5.1 S100A4 as an Intracellular and Extracellular Protein

As many other members of the S100 family, the S100A4 protein can be found both intra- and extracellular. Inside the cell, the vast majority of the protein is localized in the cytoplasm but is also detected in the nucleus. The variation of S100A4's functionality is dependent on the protein's topography and is preconditioned by the interaction and modulation of diverse intracellular protein targets located in distinct cellular compartments.

The interaction with non-muscle myosin heavy chain (NMMHC) IIA leads to increased cell motility and invasiveness [56–58]. Mechanistically, S100A4 binds in a Ca^{2+}-dependent manner to the C-terminal coiled-coil domain of the myosin-IIA heavy chain. By this interaction S100A4 promotes filament depolymerization and increases the amount of soluble myosin-IIA, resulting in the formation of stable protrusions facilitating chemotaxis [59].

S100A4 also binds rhotekin, and together they form a complex with RhoA. This complex modulates stress fiber formation and stimulates membrane ruffling and consequently cancer cell invasion [54].

The direct interaction of S100A4 with tumor suppressor p53 protein preconditions its participation in fundamental pro-tumorigenic events, like apoptosis, proliferation, and differentiation [4, 60–62]. S100A4 modulates the pro-apoptotic function of p53 by binding to its C-terminal transactivation domain [63].

Another intracellular molecular target of the S100A4 protein is liprin beta1 a transmembrane tyrosine-phosphatase LAR-interacting protein [64]. As a component of the LAR-liprin α1-liprin β1 network, S100A4 modulates the liprin-α1/liprin β1 complex formation in vivo and affects LAR activity in regulating LAR function involved in cell adhesion and metastasis [65].

Yet, numerous reports attribute a crucial function of S100A4 to its extracellular species.

Externalization of S100A4 is associated with various types of stress, like neoplasia, inflammation, and injury, and considered as a part of the pathological process. Indeed, under non-pathological conditions, S100A4 is expressed scarcely in certain cell types, such as neutrophils, monocytes, T lymphocytes (CD4+ and CD8+), fibroblasts, and stem cells [47, 66–69]. Importantly, under the non-pathological conditions, the protein is localized mostly inside the cell, and only low level of the extracellular protein is detected in body fluids [47].

Mounting evidence strongly associate elevated levels of S100A4 in the extracellular space not only with the metastatic progression of cancer [29, 70, 71] but also with various other human inflammatory indications [47, 72–74].

Functionally the extracellular S100A4 induces pro-inflammatory factors, such as cytokines and Serum Amyloid A (SAA), other

S100 proteins (S100A8/S100A9), and MMPs. It stimulates angiogenesis, cell invasion, and chemotaxis and promotes neurite outgrowth and neuron survival and differentiation [10, 75].

However, in spite of the growing evidence of externalization of S100A4 in vitro and in vivo, there are scarce publications on the mechanism of the S100A4 secretion [40, 41, 47].

The S100A4 protein lacks a secretion signaling peptide and is not secreted by the classical ER/Golgi route [76, 77]. The exploration of alternative mechanisms excluded also the involvement of lysosomes, which are involved in the secretory process of different cell types.

Cytokine-mediated pathways have been reported in stimulation of S100A4 from various normal and tumor cell types. Thus, IL-7 mediates the secretion of S100A4 without the involvement of classical Golgi route in chondrocytes [77]. Another cytokine, RANTES/CCL5 stimulates the externalization of S100A4 via microvesicle (MV) shedding from the plasma membrane of a tumor and normal cells. The importance of the RANTES-S100A4 axis has been shown in in vivo experiments, using wild-type and S100A4-deficient mouse models, inoculated with tumor cells overproducing RANTES. A substantial influence of a tumor cell-derived RANTES on S100A4 release into blood circulation has been documented and correlated with an ultimate increase of the metastatic burden in mice [76]. The transcription factors Oct-1A, in particular, its isoform Oct-1X, are also significant for the secretion of S100A4 [78].

Most likely, stress-induced and externalized S100A4 molecules form high-molecular-weight aggregates, so-called oligomers/multimers, as they have been detected in the plasma of rheumatoid arthritis, psoriasis, and cancer patients [41, 47, 48, 51].

Numerous data indicate that the biological activity of the protein is attributed to the oligomeric forms of S100A4 rather than the dimeric. Thus, oligomeric forms of S100A4 exhibit neuritogenic effects on neurons as well as stimulate motility of endothelial cells [41, 79]. Oligomeric S100A4 also affects the survival of cerebellar and hippocampal neurons in contrast to the dimeric form of the protein [80]. Furthermore, oligomeric recombinant S100A4 strongly activates the transcription of pro-inflammatory SAA, cytokines, other members of S100 proteins family, and MMPs, whereas genetically modified protein unable to form oligomers revealed no stimulatory activity [21].

5.2 Receptors and Signal Transduction

S100A4, like several other members of the S100 family, is a multireceptor ligand. The receptors reported to be involved in extracellular S100A4 signaling are associated with the activation of pro-inflammatory pathways and factors.

The most studied putative receptor for S100A4 is the receptor for advanced glycation end-products (RAGE). RAGE is a member of the immunoglobulin protein family of cell surface molecules and nowadays considered as a pattern-recognition receptor (PRR). PRRs bind PAMPs (pathogen-associated molecular patterns) and DAMP/alarmins resulting in the activation of innate immune responses [81]. RAGE plays an important role in tumorigenesis and in the pathogenesis of different acute and chronic inflammatory diseases [82, 83].

The extracellular S100A4 protein interacts with RAGE in an autocrine and paracrine manner in tumor cells, chondrocytes, and endothelial cells [84–87]. The interaction of S100A4 with RAGE leads to physiological cell responses. Thus, it induces phenotypic changes in coronary artery smooth muscle cells and mediates the pro-migratory response of extracellular S100A4 [88, 89]. Although both dimeric and oligomeric S100A4 forms bind to RAGE in vitro [90] and the S100A4-RAGE axis induces cell response in vivo, it seems that the S100A4 signaling is more complex as several tumor and non-tumor systems suggest RAGE-independent S100A4 response [79, 91, 92].

Another receptor which is attributed to S100A4 is the Toll-like receptor 4 (TLR4). The transmembrane Toll-like receptor family belongs to the multi-ligand pathogen recognition receptors (PRR). TLRs usually are expressed in sentinel cells of the innate immune system such as macrophages and dendritic cells, as well as in endothelial and tumor cells, and also serve as receptors to several members of S100 family proteins [93, 94].

TLR4 was also identified as an S100A4-mediated signal transducer in tumor and immune cells [95, 96]. Triggering of S100A4-TLR4-NFκB axis leads to activation of cytokines, SAA, MMPs, and S100A8/S100A9 activation [21]. The activation of IL-1β, IL-6, and TNF-α peripheral blood monocytes (PMBC) by recombinant S100A4 is TLR-4-dependent as well [91].

These observations once more stress the complexity of S100A4 signaling, which most likely involves the cooperation of different receptors and depends on the particular cellular context.

Several other receptors are also suggested to engage S100A4. Among them is the interleukin 10 receptor (IL10R). Direct interaction of S100A4 with IL10R was shown by surface plasmon resonance (SPR) analysis and cellular assays with neurons. IL10R, most likely, in cooperation with other receptors, mediates S100A4 signaling via STAT pathway leading to neuronal protection and survival [97].

Extracellular S100A4 appears to be also involved in epidermal growth factor receptor (EGFR) signaling. EGFR is a transmembrane protein and the receptor for members of the epidermal growth factor family (EGF family) ligands [98]. Screening of a phage display peptide library using the biologically active oligomeric form of S100A4 as bait identified the mature growth factors/ligands of the EGFR family and links S100A4 with EGFR signaling. S100A4 binds directly to several EGFR ligands, including amphiregulin (AR), epidermal growth factor (EGF), basal cell EGF (BC-EGF), and heparin-binding EGF (HB-EGF), with the highest affinity for AR. Moreover, the interaction between S100A4 and AR has functional consequences through the stimulation of cell proliferation [99]. Furthermore, the inhibition of S100A4-EGF-mediated induction of EGFR signaling by amlexanox, an anti-allergic immunomodulator, suppresses cell proliferation [100]. A systematic sensitivity analysis of the S100A4 signaling network integrating the available knowledge showed that the interaction between EGFR and S100A4 modulates the sensitivity of angiogenesis development [101].

S100A4 also directly binds to Annexin A2, an endothelial plasminogen co-receptor. Extracellular S100A4 stimulates the translocation of Annexin A2 to the cell surface of endothelial cells that facilitate the formation of an S100A4/Annexin A2 complex. The complex formation enhances the tPA (tissue plasminogen activator)-mediated plasmin production and influences MMP activation. This chain leads to enhanced angiogenesis and tumor invasion [5]. In agreement with the abovementioned, protein levels of S100A4 and Annexin A2 are upregulated in tumor compared with adjacent non-tumor tissues and associated with invasion depth, lymph node metastasis, and distant metastasis [102].

Although the interaction of S100A4 with its putative receptors is documented for different cell types, it leads to activation of similar pro-inflammatory signal transduction pathways. Studies of the effect of extracellular S100A4 on various malignant and non-malignant cells reveal activation of mitogen-activated protein kinase pathway (MAP-kinase pathway), the major pro-inflammatory pathway in the cell [43, 92, 97, 103]. The MAP-kinase pathway leads to activation of NF-κB, which is one of the important regulators of the cellular inflammatory responses. RAGE stimulation by various DAMPs, including proteins of S100 family, leads to the activation of the MAPK/NF-κB signaling pathway [104]. Additionally, the activation of STAT3, triggered by various stimuli including RAGE, amplifies extended initiation of NF-κB, pointing on NF-κB as a central effector of RAGE signaling [105, 106].

Interaction of both recombinant and cell-derived extracellular S100A4 with RAGE results in ERK1/ERK2 phosphorylation and activates the pro-inflammatory NF-κB axis [84, 107–109].

Induction of phosphorylation and subsequent degradation of the NF-κB inhibitor IκBα, as well as nuclear translocation of NF-κB in response to S100A4 treatment, were demonstrated for both cancer and normal cells including cells of the immune system [110]. S100A4-TLR4 signaling in PBMC results in activation of the transcription factor NF-κB and the MAP kinases p38 and ERK1/ERK2 [91].

Interestingly, in immune cells, the interaction with S100A4 also leads to activation of JAK/STAT signaling pathway [32, 97]. This pathway is recognized as a major regulator of immune cell differentiation, being in good agreement with S100A4's role in modulating the differentiation of immune cells [111].

5.3 Motility

Cellular motility and invasion are processes closely linked to epithelial-mesenchymal transition (EMT) and play a determinative role in the metastatic spread of cancer cells as well as in inflammation [112]. Intracellular S100A4 interacts with non-muscle myosin heavy chain IIA, rhotekin, tropomyosin, and actin that are integral parts of the cellular machinery regulating cell migration and invasion [75]. Interaction with the C-terminal end of non-muscle myosin affects cell motility and further on invasion and metastasis of cancer cells [57]. Disruption of this interaction with sulfiredoxin results in microfilament remodeling and altered cellular motility and adhesion [113]. Disruption of S100A4-rhotekin interaction affects the ability of breast cancer cells to invade in 3D matrix [114]. Accordingly, deletion of S100A4 in macrophages and fibroblasts results in the disruption of cellular mobility [29, 59].

The interaction of S100A4 with RAGE has a significant effect on cellular motility and leads to the modulation of motility of vascular smooth muscle cells, colon and thyroid cancer cells through MAPK/ERK signaling, and hypoxia [84, 89, 115]. Treatment of endothelial cells with recombinant S100A4 stimulates their motility and invasion [41].

5.4 Epithelial-Mesenchymal Transition and Stem Cell Differentiation

The involvement of S100A4 in EMT is well documented [1]. EMT is characterized by changes in cell morphology, and it is associated with inflammation, embryogenesis, tissue fibrosis, and cancer progression [112]. Several pathways regulate the process of EMT, and upon those the most well-known are TGF-β, Wnt, Notch, and Hedgehog pathways [116].

S100A4 is known as a transcriptional target of β-catenin, one of the major components of the WNT pathway, suggesting the mechanism by which S100A4 participates in the EMT [6].

In recent years, much attention has been paid to the association of EMT with the acquisition of stem cell characteristics. This concept was specifically developed for cancer stem cells (CSC) that acquire mesenchymal phenotype to progress to more aggressive tumors [117].

It has been shown that S100A4 expression is increased in certain classes of stem cells including CSCs. So, in glioblastoma S100A4 is a critical regulator of glioblastoma stem cells. Moreover, it appears to be a novel upstream regulator of SNAIL2 and ZEB1 and mesenchymal signature genes in glioblastoma [8, 118]. Similarly, in triple-negative breast cancer, CSCs acquire the ability to colonize brain by activating CAMKII/S100A4 pathway [119]. The S100A4 expression is also required to maintain stemness in head and neck cancer CSCs [120, 121].

Furthermore, the involvement of S100A4 in stem cell differentiation is also documented in non-cancer NCDs. Thus, in idiopathic pulmonary sclerosis, an increased expression of S100A4 is detected in mesenchymal progenitor cells that serve as a cell of origin for disease-mediating myofibroblasts [122].

Stroma cell-produced S100A4 also plays a role in supporting stem cell-like phenotype of hepatocellular carcinoma. This study also suggests a cross talk between inflammation and stemness [123]. An extracellular function of S100A4 in supporting stem cells is also suggested by its high expression in bulge cells of regenerating hair follicles [7].

5.5 S100A4 in Autophagy and Apoptosis

Among important advances in cancer research in recent years is the realization that circumvention of apoptosis has a profound effect on tumor progression and metastasis. p53 is a tumor suppressor protein that induces apoptosis when cells are exposed to various forms of stress. Loss of functional p53 blocks the cell's ability to undergo apoptosis after exposure to cellular stress and by this contributes to tumor formation. P53 is also regarded as a potential master regulator of metastasis, which controls the transcription of genes that are involved in canonical metastasis pathways, including cell adhesion, motility, invasion, EMT, stemness, ECM interactions, and anoikis [124].

S100A4 cooperates with the wild-type p53 in apoptosis induction suggesting that it can contribute to the loss of p53 function in tumors [4]. It also interacts with mutant p53 stimulating its accumulation in the cells [125]. The tumor suppressor function of PRAME, reported in breast cancer and leukemia, is largely dependent on its interference with the regulating of S100A4/p53 signaling [126, 127]. Knockout of S100A4 in mice leads to increased cancer predisposition which is associated with functional destabilization of p53 in cells [128].

Interaction of S100A4 with p53 takes place in the nucleus. It is suggested that this interaction promotes, similar to MDM2, degradation of wild-type p53 [61].

Apart from its multiple regulatory functions in cells, p53 has been shown as a main regulator of autophagy. Autophagy is an indispensable mechanism of the eukaryotic cell, facilitating the removal and renewal of cellular components and thereby balancing the cell's energy consumption and homeostasis. Deregulation of autophagy is now regarded as a characteristic feature contributing to the development of tumors [129].

In this connection, it has been shown that S100A4 affects gastric cancer cell autophagy and differentiation via interaction with mutant p53 [125]. Moreover, a large-scale RNA interference screen in K562 human chronic myeloid leukemia cells reveals S100A4 in an autophagy-regulating gene pattern [130].

Recently presented data shows that both extracellular treatment and intracellular overexpression of S100A4 inhibit autophagy via the β-catenin pathway activation in a RAGE-dependent manner [107].

5.6 Angiogenesis

Angiogenesis, the growth of blood vessels, is for years regarded as a key player in tumor progression that supplies growing tumor with oxygen, nutrition, and escape routes for cancer cells. High vascular index of tumors is associated with poor prognosis. The de novo growth of blood vessels is also an integral part of inflammatory processes [131–133].

Studies of transgenic mice revealed a specific angiostimulatory function of S100A4, which could impact tumor development and dissemination [41]. The S100A4 expression is associated with an increased vessel density in mammary tumors of S100A4 transgenic mice and with S100A4 expression in melanoma and prostate cancer xenograft models [41, 87]. Furthermore, vessel density and S100A4 expression in primary tumors are positively correlated in patients with breast cancer [134].

One suggested mechanism could be that extracellular S100A4 stimulates angiogenesis via interaction with annexin A2 on the surface of endothelial cells [5]. Another suggestion is that S100A4 executes its angiogenic function via interaction with RAGE [87]. Extracellular S100A4 stimulates the production of MMPs from endothelial cells, suggesting that ECM remodeling underlies the stimulation of motility and invasion of endothelial cells by S100A4 [43, 87].

Similar to other processes, the angiostimulatory function of S100A4 could also be attributed to its intracellular activity. So, it has been shown that blocking of S100A4 interaction with methionine aminopeptidase 2 in endothelial cells leads to the inhibition of angiogenesis in vivo and in vitro [135].

5.7 S100A4 and Inflammation

Perhaps the most important functions ascribed to S100A4 are its profound effects on the body's immune system as a response to pathological stimuli. These activities are characteristic both for the intracellular and extracellular S100A4 and influence the adaptive and innate immunity in several ways. First, the protein is expressed in multiple cell types of lymphoid and myeloid lineages—macrophages, neutrophils, mast cells, and memory T cells—suggesting a role of S100A4 in the regulation of immune cells [42, 66, 68].

The endogenous S100A4 expression is indispensable for macrophage chemotaxis. Studies of S100A4-deficient mice establish that S100A4 is a regulator of physiological macrophage motility and demonstrate that S100A4 mediates macrophage recruitment and chemotaxis in vivo [59]. In contrast, mouse T-cell motility is not dependent on the endogenous S100A4 expression. The S100A4 protein is specifically expressed by memory T cells, but their migration toward chemokine gradients and the capacity to generate T-cell response and differentiation of T helper subsets are not affected by endogenous S100A4 deficiency [68]. S100A4 deficiency also has a strong impact on the development of both humoral and cellular immunity at least after immunization with cholera toxin [136].

In contrast, silencing of S100A4 in human T cells, leads to decreased mRNA expression of IL-5 and IL-13, in human TH2-polarized cells [55], indicating that there could be species-specific variations in the immune cell's response to the S100A4 expression. S100A4 is also differentially expressed in the human, but not in the mouse regulatory T cells [137, 138].

A second way that S100A4 interferes with body's immune response is its secretion from cells of the immune system in response to stress. Accordingly, S100A4 was found to be secreted by a subpopulation of inflammatory macrophages and consequently promotes the development of liver fibrosis [139, 140]. In vitro cultivated human macrophages release S100A4 into extracellular space. Since increased levels of S100A4 are observed in a tumor interstitial fluid of breast cancer, one can speculate that immune cells, at least of myeloid origin, are the source of extracellular S100A4 in the damaged tissues [42].

The most pronounced effect of S100A4 on immune regulation that attracts particular attention is associated with its extracellular activity, especially due to the fact that the extracellular S100A4 is a well-known activator of pro-inflammatory pathways and stimulator of cytokine production from various cells. Extracellular S100A4 exhibits its effects on immune cells in two major ways—it affects differentiation of cells of innate and adaptive immune systems, and it activates these cells to produce numerous cytokines which many of them are linked to pro-tumorigenic functions.

Thus, prolonged exposure to S100A4 potentiates differentiation of monocytes into the pro-tumorigenic M2 phenotype and stimulates chemotactic migration of monocytes [141]. Monocytes, activated by S100A4, produce inflammatory cytokines which contribute to aggravated disease activity of rheumatoid arthritis [48, 91]. Similarly, S100A4 indirectly influence the differentiation of myeloid cells by triggering breast cancer cells to produce cytokines that convert monocytes into the tumor-associated macrophages [142]. S100A4 stimulates generation of osteoclasts, cells of monocyte/macrophage lineage, in bones. Moreover, S100A4 treatment stimulated osteoclastogenesis [143].

Extracellular S100A4 exhibits analogous effects on T-cell population. Thus, the presence of extracellular S100A4 in tumor microenvironment facilitates T-cell infiltration at an early premalignant stage of tumor development [30]. In vitro studies support a function of S100A4 in the attraction and activation of T cells [29]. Moreover, inhibition of S100A4 by neutralizing antibody in vivo significantly reduces metastasis and T lymphocyte infiltration at the site of a primary tumor and pre-metastatic niche [32]. In vitro cultivation of T cells in the presence of S100A4 alters the T-cell polarization balance toward the Th2 phenotype, indicating that not only it activates T cells but also affects their differentiation [32].

In the study of allergy, S100A4 activates T cells and thereby influence allergic inflammation [55]. Additionally, human $CD4^+$ T cells show increased production of IL-13 protein after treatment with recombinant S100A4 [55]. Extracellular S100A4 also serves as a target molecule for TCR γδ-T-cell-mediated lysis [144]. S100A4 also affects indirectly regulatory T cells (Tregs). S100A4 pulsed dendritic cells induce the proliferation and increase the metabolic activity of Tregs [145]. Tregs play a fundamental role in immune suppression by downregulating the function of effector T cells.

In recent years attention was paid to the role of S100A4 as a survival factor for myeloid-derived suppressor cells (MDSC). MDSCs represent a pathologic state of activation of monocytes and immature neutrophils. These cells are generated in response to a large array of pathologic conditions ranging from cancer to obesity. The prominent feature of these cells is their ability to inhibit T-cell function and thus contribute to the pathogenesis of various diseases [146]. Extracellular S100A4 activates TLR-4/Erkl/Erk2 pathway to abrogate intrinsic caspase-9-dependent apoptosis induction in MDSCs, thus supporting their survival [95]. Furthermore, S100A4-dependent accumulation of MDSCs to the tumor site correlates with tumor growth of fibrosarcoma xenografts [95].

In summary, extracellular S100A4 has a profound effect on cells of the lymphoid and myeloid lineages, stimulating their activation and aiming their differentiation toward pro-tumorigenic phenotype.

S100A4-mediated activation of cells from the innate and adaptive immune system is manifested by the production of cytokines. This network of cytokines consequently modulates body's immune response. The most noticeable cytokines, whose production is stimulated by S100A4, are eotaxin-2, G-SCF, IL-4, IL-6, and RANTES. Interestingly, these cytokines are released in response to S100A4 from various cell types including cancer cells, melanoma, T cells, and monocytes (*see* Table 1).

In addition to cytokines, S100A4 stimulates the production of DAMP proteins such as the acute-phase response SAAs and S100A8/S100A9 [21, 141].

Table 1
Cytokines produced by different cells in response to extracellular S100A4

Cytokine	Source	Reference
Eotaxin-2	Pre-metastatic lungs, T cells, tumor in mice	[30, 32]
G-CSF	Cancer cells, pre-metastatic lungs, T cells, tumor in mice, melanoma	[30, 32, 76, 141]
IL-6	Pre-metastatic lungs, T cells, melanoma, monocytes	[32, 91, 141]
IL-4	Pre-metastatic lungs, T cells, tumor in mice	[30, 32]
SAA	melanoma, cancer cells, T cells, macrophages	[21, 141]
CCL-2	Melanoma, monocytes	[141]
IL-10	T cells, monocytes	[32, 141]
IL-13	T cells, $CD4^+$ T cells	[32, 55]
IL-9	Pre-metastatic lungs, T cells [32]	[62]
RANTES (CCL-5)	Cancer cells, T cells	[30, 76]
IL-8	melanoma	[141]
IL-1β	Melanoma, monocytes	[91, 141]
S100A8/S100A9	Cancer cells	[21]
MIP	Cancer cells, T cells	[30, 76]
VEGF	T cells	[30]
NF-α	Monocytes	[91]
IL-5	Human T cells	[55]
Fractalkine	Pre-metastatic lungs	[32]
SDF-1	Pre-metastatic lungs	[32]
Lymphotactin	Pre-metastatic lungs	[32]
CXCL-5	Cancer cells	[76]
MCP-1	Pre-metastatic lungs	[32]
CXCL-1	Cancer cells	[21]
IL-17A	T cells	[32]
CXCL16	Tumor cells KO mice	[30]

Altogether, it is well documented that S100A4 plays a substantial role in modulation of the immune system in many different ways indicating that this protein could be a key player in a multitude of NCDs where the inflammatory pathway(s) play a largely contributing role.

5.8 Remodeling of the Extracellular Matrix

Apart from changes in cytokine repertoire during inflammation and tumor development, the ECM plays a substantial role in supporting immune responses at the affected sites [147, 148]. Numerous molecules, including cytokines, are docked in the structure of the ECM and presented to cells at pathological sites. The S100A4 protein activity strongly contributes to the process of ECM remodeling by stimulating the production of ECM structure molecules as well as ECM remodeling proteases [75].

Thus, treatment of fibroblasts with S100A4 induces the secretion of FN and collagen [32, 52], major components of the ECM. S100A4 stimulates the deposition of FN in lungs, thus the formation of the pre-metastatic niche [32]. In osteosarcoma cells, intracellular S100A4 induces expression and secretion of osteopontin [149]. As an ECM protein and pro-inflammatory cytokine, osteopontin facilitates the recruitment of monocytes/macrophages and mediates cytokine secretion from leukocytes [150].

Another mode of action of S100A4 in ECM remodeling is the stimulation of MMP production. MMPs play an essential role in the proteolytic degradation of ECM, a process that is necessary for cancer cell invasion and metastatic spread. MMPs also substantially contribute to the inflammatory milieu [151].

S100A4 modulates the MMP production both as intra- and extracellular molecule. However, the major effect seems to be mediated by extracellular S100A4 [40, 43, 86].

S100A4-dependent MMP activation is observed both in cancer and nonmalignant cells.

Downregulation of the S100A4 expression in osteosarcoma cells leads to a reduction in the expression of MMP2 and MT1-MMP with a subsequent reduction in the MMP2 activity and a reduced ability to migrate through Matrigel-coated filters [152]. Endogenous S100A4 expression contributes to hepatocellular carcinoma metastasis, squamous cell carcinoma, and human prostate cancer cell invasion by activation of MMP9 expression [153–155]. In clinical specimens of human breast cancer, elevated levels of the MMP2 and MMP13 are associated with a decrease in patient survival and closely correlated with S100A4 expression levels [156]. S100A4 participates in EMT in breast cancer via targeting MMP2 [157]. Endogenous S100A4 also regulates the expression of MMP2 and MMP9 in the thoracic aortic aneurysm [158].

As an extracellular protein, S100A4 stimulates the production of MMP13 from osteosarcoma cells, mouse mammary carcinoma cells, and human breast cancer cells [31, 91, 98]. Extracellular S100A4 stimulates expression of MMPs and subsequent differentiation of smooth muscle cells [88]. Mouse microvascular endothelial cells stimulated by S100A4 induce a spike in expression and secretion of MMP13 [43]. Human endothelial cells produce MMP9 in response to the S100A4-dependent activation, whereas articular

Table 2
Comparison extracellular and intracellular functions of the S100A4 protein

Function	Extracellular	Intracellular
Motility	Yes	Yes
EMT	Yes	Yes
Stem cell differentiation	Yes	Yes
Autophagy	Yes	Yes
Apoptosis	No	Yes
Angiogenesis	Yes	Yes
Regulation of immune system	Yes	Yes
ECM remodeling	Yes	Yes

chondrocytes produce MMP13 [77, 87]. Synovial fibroblasts treated with the S100A4 multimer upregulate the expression and secretion of MMP2, MMP9, and MMP13 [159]. S100A4 also activates MMP3 in mouse mammary epithelial cells, contributing to the mammary gland branching morphogenesis [160]. Interestingly, the computer simulation of the S100A4 network revealed that S100A4 enhances the activity of MMPs, causing higher cell dissociation [101].

The fact that the S100A4 protein is internalized by cells [161] raises the question which function of S100A4 is executed by the intracellular, the secreted, or the released and internalized form of S100A4. A comprehensive overview of extracellular and intracellular functions of S100A4 is presented in Table 2.

6 Concluding Remarks

Integrating the knowledge on the S100A4 protein, primarily on its extracellular oligomeric form, specific cell surface receptors (RAGE, TLR4, EGFR), and strong propensity to activate pro-inflammatory pathways, is categorizing it as an effective DAMP molecule [16].

Being released into extracellular space in response to stress, S100A4 triggers a sequence of events that activates inflammatory pathways and leading to sustained inflammation. The fact that S100A4 is increased and plays a pivotal role in various pathological conditions including cancer makes it an attractive therapeutic target.

Indeed, in recent studies, neutralization of S100A4 activity by anti-S100A4 antibodies or by specific small molecules has proven S100A4 as an excellent target to combat metastatic disease [33, 87,

162, 163]. Two small molecule S100A4 inhibitors, sulindac and niclosamide, are currently in clinical trials [164]. Available data also suggests that anti-S100A4 therapy could be instrumental in combating non-cancer NCDs.

References

1. Fei F, Qu J, Zhang M, Li Y, Zhang S (2017) S100A4 in cancer progression and metastasis: a systematic review. Oncotarget 8 (42):73219–73239. https://doi.org/10.18632/oncotarget.18016
2. Kim EJ, Helfman DM (2003) Characterization of the metastasis-associated protein, S100A4. Roles of calcium binding and dimerization in cellular localization and interaction with myosin. J Biol Chem 278 (32):30063–30073. https://doi.org/10.1074/jbc.M304909200
3. Fei F, Qu J, Li C, Wang X, Li Y, Zhang S (2017) Role of metastasis-induced protein S100A4 in human non-tumor pathophysiologies. Cell Biosci 7:64. https://doi.org/10.1186/s13578-017-0191-1
4. Grigorian M, Andresen S, Tulchinsky E, Kriajevska M, Carlberg C, Kruse C, Cohn M, Ambartsumian N, Christensen A, Selivanova G, Lukanidin E (2001) Tumor suppressor p53 protein is a new target for the metastasis-associated Mts1/S100A4 protein: functional consequences of their interaction. J Biol Chem 276(25):22699–22708. https://doi.org/10.1074/jbc.M010231200
5. Semov A, Moreno MJ, Onichtchenko A, Abulrob A, Ball M, Ekiel I, Pietrzynski G, Stanimirovic D, Alakhov V (2005) Metastasis-associated protein S100A4 induces angiogenesis through interaction with Annexin II and accelerated plasmin formation. J Biol Chem 280(21):20833–20841. https://doi.org/10.1074/jbc.M412653200
6. Dahlmann M, Kobelt D, Walther W, Mudduluru G, Stein U (2016) S100A4 in cancer metastasis: Wnt signaling-driven interventions for metastasis restriction. Cancers (Basel) 8(6). https://doi.org/10.3390/cancers8060059
7. Ito M, Kizawa K (2001) Expression of calcium-binding S100 proteins A4 and A6 in regions of the epithelial sac associated with the onset of hair follicle regeneration. J Invest Dermatol 116(6):956–963. https://doi.org/10.1046/j.0022-202x.2001.01369.x
8. Chow KH, Park HJ, George J, Yamamoto K, Gallup AD, Graber JH, Chen Y, Jiang W, Steindler DA, Neilson EG, Kim BYS, Yun K (2017) S100A4 is a biomarker and regulator of glioma stem cells that is critical for mesenchymal transition in glioblastoma. Cancer Res 77(19):5360–5373. https://doi.org/10.1158/0008-5472.CAN-17-1294
9. Tumbar T, Guasch G, Greco V, Blanpain C, Lowry WE, Rendl M, Fuchs E (2004) Defining the epithelial stem cell niche in skin. Science 303(5656):359–363. https://doi.org/10.1126/science.1092436
10. Bresnick AR, Weber DJ, Zimmer DB (2015) S100 proteins in cancer. Nat Rev Cancer 15 (2):96–109. https://doi.org/10.1038/nrc3893
11. Boye K, Maelandsmo GM (2010) S100A4 and metastasis: a small actor playing many roles. Am J Pathol 176(2):528–535. https://doi.org/10.2353/ajpath.2010.090526
12. Nasser MW, Elbaz M, Ahirwar DK, Ganju RK (2015) Conditioning solid tumor microenvironment through inflammatory chemokines and S100 family proteins. Cancer Lett 365 (1):11–22. https://doi.org/10.1016/j.canlet.2015.05.002
13. Grigorian M, Ambartsumian N, Lukanidin E (2008) Metastasis-inducing S100A4 protein: implication in non-malignant human pathologies. Curr Mol Med 8(6):492–496
14. Bertheloot D, Latz E (2017) HMGB1, IL-1alpha, IL-33 and S100 proteins: dual-function alarmins. Cell Mol Immunol 14 (1):43–64. https://doi.org/10.1038/cmi.2016.34
15. Austermann J, Zenker S, Roth J (2017) S100-alarmins: potential therapeutic targets for arthritis. Expert Opin Ther Targets 21 (7):739–751. https://doi.org/10.1080/14728222.2017.1330411
16. Chan JK, Roth J, Oppenheim JJ, Tracey KJ, Vogl T, Feldmann M, Horwood N, Nanchahal J (2012) Alarmins: awaiting a clinical response. J Clin Invest 122(8):2711–2719. https://doi.org/10.1172/JCI62423
17. Boteanu RM, Suica VI, Uyy E, Ivan L, Dima SO, Popescu I, Simionescu M, Antohe F (2017) Alarmins in chronic noncommunicable diseases: atherosclerosis, diabetes and cancer. J Proteome 153:21–29. https://doi.org/10.1016/j.jprot.2016.11.006

18. Raposo TP, Beirao BC, Pang LY, Queiroga FL, Argyle DJ (2015) Inflammation and cancer: till death tears them apart. Vet J 205 (2):161–174. https://doi.org/10.1016/j.tvjl.2015.04.015
19. Hanahan D, Weinberg RA (2011) Hallmarks of cancer: the next generation. Cell 144 (5):646–674. https://doi.org/10.1016/j.cell.2011.02.013
20. Wu Y, Zhou BP (2009) Inflammation: a driving force speeds cancer metastasis. Cell Cycle 8(20):3267–3273. https://doi.org/10.4161/cc.8.20.9699
21. Hansen MT, Forst B, Cremers N, Quagliata L, Ambartsumian N, Grum-Schwensen B, Klingelhofer J, Abdul-Al A, Herrmann P, Osterland M, Stein U, Nielsen GH, Scherer PE, Lukanidin E, Sleeman JP, Grigorian M (2015) A link between inflammation and metastasis: serum amyloid A1 and A3 induce metastasis, and are targets of metastasis-inducing S100A4. Oncogene 34 (4):424–435. https://doi.org/10.1038/onc.2013.568
22. Ebralidze A, Tulchinsky E, Grigorian M, Afanasyeva A, Senin V, Revazova E, Lukanidin E (1989) Isolation and characterization of a gene specifically expressed in different metastatic cells and whose deduced gene product has a high degree of homology to a Ca2+-binding protein family. Genes Dev 3 (7):1086–1093
23. Takenaga K, Nakamura Y, Endo H, Sakiyama S (1994) Involvement of S100-related calcium-binding protein pEL98 (or mts1) in cell motility and tumor cell invasion. Jpn J Cancer Res 85(8):831–839
24. Grigorian M, Ambartsumian N, Lykkesfeldt AE, Bastholm L, Elling F, Georgiev G, Lukanidin E (1996) Effect of mts1 (S100A4) expression on the progression of human breast cancer cells. Int J Cancer 67 (6):831–841. https://doi.org/10.1002/(SICI)1097-0215(19960917)67:6<831::AID-IJC13>3.0.CO;2-4
25. Davies BR, Davies MP, Gibbs FE, Barraclough R, Rudland PS (1993) Induction of the metastatic phenotype by transfection of a benign rat mammary epithelial cell line with the gene for p9Ka, a rat calcium-binding protein, but not with the oncogene EJ-ras-1. Oncogene 8(4):999–1008
26. Maelandsmo GM, Hovig E, Skrede M, Engebraaten O, Florenes VA, Myklebost O, Grigorian M, Lukanidin E, Scanlon KJ, Fodstad O (1996) Reversal of the in vivo metastatic phenotype of human tumor cells by an anti-CAPL (mts1) ribozyme. Cancer Res 56 (23):5490–5498
27. Ambartsumian NS, Grigorian MS, Larsen IF, Karlstrom O, Sidenius N, Rygaard J, Georgiev G, Lukanidin E (1996) Metastasis of mammary carcinomas in GRS/A hybrid mice transgenic for the mts1 gene. Oncogene 13(8):1621–1630
28. Davies MP, Rudland PS, Robertson L, Parry EW, Jolicoeur P, Barraclough R (1996) Expression of the calcium-binding protein S100A4 (p9Ka) in MMTV-neu transgenic mice induces metastasis of mammary tumours. Oncogene 13(8):1631–1637
29. Grum-Schwensen B, Klingelhofer J, Berg CH, El-Naaman C, Grigorian M, Lukanidin E, Ambartsumian N (2005) Suppression of tumor development and metastasis formation in mice lacking the S100A4(mts1) gene. Cancer Res 65(9):3772–3780. https://doi.org/10.1158/0008-5472.CAN-04-4510
30. Grum-Schwensen B, Klingelhofer J, Grigorian M, Almholt K, Nielsen BS, Lukanidin E, Ambartsumian N (2010) Lung metastasis fails in MMTV-PyMT oncomice lacking S100A4 due to a T-cell deficiency in primary tumors. Cancer Res 70(3):936–947. https://doi.org/10.1158/0008-5472.CAN-09-3220
31. Xue C, Plieth D, Venkov C, Xu C, Neilson EG (2003) The gatekeeper effect of epithelial-mesenchymal transition regulates the frequency of breast cancer metastasis. Cancer Res 63(12):3386–3394
32. Grum-Schwensen B, Klingelhofer J, Beck M, Bonefeld CM, Hamerlik P, Guldberg P, Grigorian M, Lukanidin E, Ambartsumian N (2015) S100A4-neutralizing antibody suppresses spontaneous tumor progression, pre-metastatic niche formation and alters T-cell polarization balance. BMC Cancer 15:44. https://doi.org/10.1186/s12885-015-1034-2
33. Klingelhofer J, Grum-Schwensen B, Beck MK, Knudsen RS, Grigorian M, Lukanidin E, Ambartsumian N (2012) Anti-S100A4 antibody suppresses metastasis formation by blocking stroma cell invasion. Neoplasia 14(12):1260–1268
34. Chen H, Xu C, Jin Q, Liu Z (2014) S100 protein family in human cancer. Am J Cancer Res 4(2):89–115
35. Sherbet GV (2009) Metastasis promoter S100A4 is a potentially valuable molecular target for cancer therapy. Cancer Lett 280 (1):15–30. https://doi.org/10.1016/j.canlet.2008.10.037

36. Rudland PS, Platt-Higgins A, Renshaw C, West CR, Winstanley JH, Robertson L, Barraclough R (2000) Prognostic significance of the metastasis-inducing protein S100A4 (p9Ka) in human breast cancer. Cancer Res 60(6):1595–1603
37. Gongoll S, Peters G, Mengel M, Piso P, Klempnauer J, Kreipe H, von Wasielewski R (2002) Prognostic significance of calcium-binding protein S100A4 in colorectal cancer. Gastroenterology 123(5):1478–1484
38. Joyce JA, Pollard JW (2009) Microenvironmental regulation of metastasis. Nat Rev Cancer 9(4):239–252. https://doi.org/10.1038/nrc2618
39. Peinado H, Zhang H, Matei IR, Costa-Silva B, Hoshino A, Rodrigues G, Psaila B, Kaplan RN, Bromberg JF, Kang Y, Bissell MJ, Cox TR, Giaccia AJ, Erler JT, Hiratsuka S, Ghajar CM, Lyden D (2017) Pre-metastatic niches: organ-specific homes for metastases. Nat Rev Cancer 17(5):302–317. https://doi.org/10.1038/nrc.2017.6
40. Schmidt-Hansen B, Klingelhofer J, Grum-Schwensen B, Christensen A, Andresen S, Kruse C, Hansen T, Ambartsumian N, Lukanidin E, Grigorian M (2004) Functional significance of metastasis-inducing S100A4 (Mts1) in tumor-stroma interplay. J Biol Chem 279(23):24498–24504. https://doi.org/10.1074/jbc.M400441200
41. Ambartsumian N, Klingelhofer J, Grigorian M, Christensen C, Kriajevska M, Tulchinsky E, Georgiev G, Berezin V, Bock E, Rygaard J, Cao R, Cao Y, Lukanidin E (2001) The metastasis-associated Mts1 (S100A4) protein could act as an angiogenic factor. Oncogene 20(34):4685–4695. https://doi.org/10.1038/sj.onc.1204636
42. Cabezon T, Celis JE, Skibshoj I, Klingelhofer J, Grigorian M, Gromov P, Rank F, Myklebust JH, Maelandsmo GM, Lukanidin E, Ambartsumian N (2007) Expression of S100A4 by a variety of cell types present in the tumor microenvironment of human breast cancer. Int J Cancer 121 (7):1433–1444. https://doi.org/10.1002/ijc.22850
43. Schmidt-Hansen B, Ornas D, Grigorian M, Klingelhofer J, Tulchinsky E, Lukanidin E, Ambartsumian N (2004) Extracellular S100A4(mts1) stimulates invasive growth of mouse endothelial cells and modulates MMP-13 matrix metalloproteinase activity. Oncogene 23(32):5487–5495. https://doi.org/10.1038/sj.onc.1207720
44. Pedersen KB, Andersen K, Fodstad O, Maelandsmo GM (2004) Sensitization of interferon-gamma induced apoptosis in human osteosarcoma cells by extracellular S100A4. BMC Cancer 4:52. https://doi.org/10.1186/1471-2407-4-52
45. O'Connell JT, Sugimoto H, Cooke VG, MacDonald BA, Mehta AI, LeBleu VS, Dewar R, Rocha RM, Brentani RR, Resnick MB, Neilson EG, Zeisberg M, Kalluri R (2011) VEGF-A and Tenascin-C produced by S100A4+ stromal cells are important for metastatic colonization. Proc Natl Acad Sci U S A 108 (38):16002–16007. https://doi.org/10.1073/pnas.1109493108
46. Masuda K, Masuda R, Neidhart M, Simmen BR, Michel BA, Muller-Ladner U, Gay RE, Gay S (2002) Molecular profile of synovial fibroblasts in rheumatoid arthritis depends on the stage of proliferation. Arthritis Res 4 (5):R8. https://doi.org/10.1186/ar427
47. Klingelhofer J, Senolt L, Baslund B, Nielsen GH, Skibshoj I, Pavelka K, Neidhart M, Gay S, Ambartsumian N, Hansen BS, Petersen J, Lukanidin E, Grigorian M (2007) Up-regulation of metastasis-promoting S100A4 (Mts-1) in rheumatoid arthritis: putative involvement in the pathogenesis of rheumatoid arthritis. Arthritis Rheum 56(3):779–789. https://doi.org/10.1002/art.22398
48. Oslejskova L, Grigorian M, Hulejova H, Vencovsky J, Pavelka K, Klingelhofer J, Gay S, Neidhart M, Brabcova H, Suchy D, Senolt L (2009) Metastasis-inducing S100A4 protein is associated with the disease activity of rheumatoid arthritis. Rheumatology (Oxford) 48(12):1590–1594. https://doi.org/10.1093/rheumatology/kep316
49. Erlandsson MC, Forslind K, Andersson SE, Lund A, Bokarewa MI (2012) Metastasin S100A4 is increased in proportion to radiographic damage in patients with RA. Rheumatology (Oxford) 51 (5):932–940. https://doi.org/10.1093/rheumatology/ker362
50. Senolt L, Cerezo LA, Sumova B, Pecha O, Plestilova L, Forejtova S, Ruzickova O, Husakova M, Zavada J, Pavelka K, Vencovsky J, Mann H (2015) High levels of metastasis-inducing S100A4 protein and treatment outcome in early rheumatoid arthritis: data from the PERAC cohort. Biomarkers 20(1):47–51. https://doi.org/10.3109/1354750X.2014.989544
51. Zibert JR, Skov L, Thyssen JP, Jacobsen GK, Grigorian M (2010) Significance of the S100A4 protein in psoriasis. J Invest Dermatol 130(1):150–160. https://doi.org/10.1038/jid.2009.206

52. Tomcik M, Palumbo-Zerr K, Zerr P, Avouac J, Dees C, Sumova B, Distler A, Beyer C, Cerezo LA, Becvar R, Distler O, Grigorian M, Schett G, Senolt L, Distler JH (2015) S100A4 amplifies TGF-beta-induced fibroblast activation in systemic sclerosis. Ann Rheum Dis 74(9):1748–1755. https://doi.org/10.1136/annrheumdis-2013-204516
53. Cutolo M, Ruaro B, Montagna P, Brizzolara R, Stratta E, Trombetta AC, Scabini S, Tavilla PP, Parodi A, Corallo C, Giordano N, Paolino S, Pizzorni C, Sulli A, Smith V, Soldano S (2018) Effects of selexipag and its active metabolite in contrasting the profibrotic myofibroblast activity in cultured scleroderma skin fibroblasts. Arthritis Res Ther 20(1):77. https://doi.org/10.1186/s13075-018-1577-0
54. Janson C, Anto J, Burney P, Chinn S, de Marco R, Heinrich J, Jarvis D, Kuenzli N, Leynaert B, Luczynska C, Neukirch F, Svanes C, Sunyer J, Wjst M, European Community Respiratory Health Survey II (2001) The European Community Respiratory Health Survey: what are the main results so far? European Community Respiratory Health Survey II. Eur Respir J 18(3):598–611
55. Bruhn S, Fang Y, Barrenas F, Gustafsson M, Zhang H, Konstantinell A, Kronke A, Sonnichsen B, Bresnick A, Dulyaninova N, Wang H, Zhao Y, Klingelhofer J, Ambartsumian N, Beck MK, Nestor C, Bona E, Xiang Z, Benson M (2014) A generally applicable translational strategy identifies S100A4 as a candidate gene in allergy. Sci Transl Med 6(218):218ra214. https://doi.org/10.1126/scitranslmed.3007410
56. Kriajevska MV, Cardenas MN, Grigorian MS, Ambartsumian NS, Georgiev GP, Lukanidin EM (1994) Non-muscle myosin heavy chain as a possible target for protein encoded by metastasis-related mts-1 gene. J Biol Chem 269(31):19679–19682
57. Ismail TM, Fernig DG, Rudland PS, Terry CJ, Wang G, Barraclough R (2008) The basic C-terminal amino acids of calcium-binding protein S100A4 promote metastasis. Carcinogenesis 29(12):2259–2266. https://doi.org/10.1093/carcin/bgn217
58. Naiya G, Kaypee S, Kundu TK, Roy S (2015) A constrained helical peptide against S100A4 inhibits cell motility in tumor cells. Chem Biol Drug Des 86(4):945–950. https://doi.org/10.1111/cbdd.12553
59. Li ZH, Dulyaninova NG, House RP, Almo SC, Bresnick AR (2010) S100A4 regulates macrophage chemotaxis. Mol Biol Cell 21(15):2598–2610. https://doi.org/10.1091/mbc.E09-07-0609
60. Fernandez-Fernandez MR, Veprintsev DB, Fersht AR (2005) Proteins of the S100 family regulate the oligomerization of p53 tumor suppressor. Proc Natl Acad Sci U S A 102(13):4735–4740. https://doi.org/10.1073/pnas.0501459102
61. Orre LM, Panizza E, Kaminskyy VO, Vernet E, Graslund T, Zhivotovsky B, Lehtio J (2013) S100A4 interacts with p53 in the nucleus and promotes p53 degradation. Oncogene 32(49):5531–5540. https://doi.org/10.1038/onc.2013.213
62. Tamaki Y, Iwanaga Y, Niizuma S, Kawashima T, Kato T, Inuzuka Y, Horie T, Morooka H, Takase T, Akahashi Y, Kobuke K, Ono K, Shioi T, Sheikh SP, Ambartsumian N, Lukanidin E, Koshimizu TA, Miyazaki S, Kimura T (2013) Metastasis-associated protein, S100A4 mediates cardiac fibrosis potentially through the modulation of p53 in cardiac fibroblasts. J Mol Cell Cardiol 57:72–81. https://doi.org/10.1016/j.yjmcc.2013.01.007
63. Xu Y, Rong LJ, Meng SL, Hou FL, Zhang JH, Pan G (2016) PRAME promotes in vitro leukemia cells death by regulating S100A4/p53 signaling. Eur Rev Med Pharmacol Sci 20(6):1057–1063
64. Kriajevska M, Fischer-Larsen M, Moertz E, Vorm O, Tulchinsky E, Grigorian M, Ambartsumian N, Lukanidin E (2002) Liprin beta 1, a member of the family of LAR transmembrane tyrosine phosphatase-interacting proteins, is a new target for the metastasis-associated protein S100A4 (Mts1). J Biol Chem 277(7):5229–5235. https://doi.org/10.1074/jbc.M110976200
65. Serra-Pages C, Medley QG, Tang M, Hart A, Streuli M (1998) Liprins, a family of LAR transmembrane protein-tyrosine phosphatase-interacting proteins. J Biol Chem 273(25):15611–15620
66. Grigorian M, Tulchinsky E, Burrone O, Tarabykina S, Georgiev G, Lukanidin E (1994) Modulation of mts1 expression in mouse and human normal and tumor cells. Electrophoresis 15(3-4):463–468
67. Kalluri R, Zeisberg M (2006) Fibroblasts in cancer. Nat Rev Cancer 6(5):392–401. https://doi.org/10.1038/nrc1877
68. Weatherly K, Bettonville M, Torres D, Kohler A, Goriely S, Braun MY (2015) Functional profile of S100A4-deficient T cells. Immun Inflamm Dis 3(4):431–444. https://doi.org/10.1002/iid3.85

69. Greene WA, Burke TA, Por ED, Kaini RR, Wang HC (2016) Secretion profile of induced pluripotent stem cell-derived retinal pigment epithelium during wound healing. Invest Ophthalmol Vis Sci 57(10):4428–4441. https://doi.org/10.1167/iovs.16-19192
70. Herwig N, Belter B, Pietzsch J (2016) Extracellular S100A4 affects endothelial cell integrity and stimulates transmigration of A375 melanoma cells. Biochem Biophys Res Commun 477(4):963–969. https://doi.org/10.1016/j.bbrc.2016.07.009
71. Rasanen K, Sriswasdi S, Valiga A, Tang HY, Zhang G, Perego M, Somasundaram R, Li L, Speicher K, Klein-Szanto AJ, Basu D, Rustgi AK, Speicher DW, Herlyn M (2013) Comparative secretome analysis of epithelial and mesenchymal subpopulations of head and neck squamous cell carcinoma identifies S100A4 as a potential therapeutic target. Mol Cell Proteomics 12(12):3778–3792. https://doi.org/10.1074/mcp.M113.029587
72. Gong XJ, Song XY, Wei H, Wang J, Niu M (2015) Serum S100A4 levels as a novel biomarker for detection of acute myocardial infarction. Eur Rev Med Pharmacol Sci 19 (12):2221–2225
73. Abu El-Asrar AM, Nawaz MI, De Hertogh G, Alam K, Siddiquei MM, Van den Eynde K, Mousa A, Mohammad G, Geboes K, Opdenakker G (2014) S100A4 is upregulated in proliferative diabetic retinopathy and correlates with markers of angiogenesis and fibrogenesis. Mol Vis 20:1209–1224
74. Christensen MH, Fenne IS, Nordbo Y, Varhaug JE, Nygard KO, Lien EA, Mellgren G (2015) Novel inflammatory biomarkers in primary hyperparathyroidism. Eur J Endocrinol 173(1):9–17. https://doi.org/10.1530/EJE-14-1038
75. Donato R, Cannon BR, Sorci G, Riuzzi F, Hsu K, Weber DJ, Geczy CL (2013) Functions of S100 proteins. Curr Mol Med 13 (1):24–57
76. Forst B, Hansen MT, Klingelhofer J, Moller HD, Nielsen GH, Grum-Schwensen B, Ambartsumian N, Lukanidin E, Grigorian M (2010) Metastasis-inducing S100A4 and RANTES cooperate in promoting tumor progression in mice. PLoS One 5(4):e10374. https://doi.org/10.1371/journal.pone.0010374
77. Yammani RR, Long D, Loeser RF (2009) Interleukin-7 stimulates secretion of S100A4 by activating the JAK/STAT signaling pathway in human articular chondrocytes. Arthritis Rheum 60(3):792–800. https://doi.org/10.1002/art.24295
78. Dukhanina EA, Portseva TN, Pankratova EV, Soshnikova NV, Stepchenko AG, Dukhanin AS, Georgieva SG (2016) Oct-1 modifies S100A4 exchange between intra- and extracellular compartments in Namalwa cells and increases their sensitivity to glucocorticoids. Cell Cycle 15(11):1471–1478. https://doi.org/10.1080/15384101.2016.1175260
79. Novitskaya V, Grigorian M, Kriajevska M, Tarabykina S, Bronstein I, Berezin V, Bock E, Lukanidin E (2000) Oligomeric forms of the metastasis-related Mts1 (S100A4) protein stimulate neuronal differentiation in cultures of rat hippocampal neurons. J Biol Chem 275(52):41278–41286. https://doi.org/10.1074/jbc.M007058200
80. Pedersen MV, Kohler LB, Grigorian M, Novitskaya V, Bock E, Lukanidin E, Berezin V (2004) The Mts1/S100A4 protein is a neuroprotectant. J Neurosci Res 77(6):777–786. https://doi.org/10.1002/jnr.20221
81. Beutler B, Jiang Z, Georgel P, Crozat K, Croker B, Rutschmann S, Du X, Hoebe K (2006) Genetic analysis of host resistance: Toll-like receptor signaling and immunity at large. Annu Rev Immunol 24:353–389. https://doi.org/10.1146/annurev.immunol.24.021605.090552
82. Pusterla T, Nemeth J, Stein I, Wiechert L, Knigin D, Marhenke S, Longerich T, Kumar V, Arnold B, Vogel A, Bierhaus A, Pikarsky E, Hess J, Angel P (2013) Receptor for advanced glycation endproducts (RAGE) is a key regulator of oval cell activation and inflammation-associated liver carcinogenesis in mice. Hepatology 58(1):363–373. https://doi.org/10.1002/hep.26395
83. Riehl A, Nemeth J, Angel P, Hess J (2009) The receptor RAGE: bridging inflammation and cancer. Cell Commun Signal 7:12. https://doi.org/10.1186/1478-811X-7-12
84. Dahlmann M, Okhrimenko A, Marcinkowski P, Osterland M, Herrmann P, Smith J, Heizmann CW, Schlag PM, Stein U (2014) RAGE mediates S100A4-induced cell motility via MAPK/ERK and hypoxia signaling and is a prognostic biomarker for human colorectal cancer metastasis. Oncotarget 5 (10):3220–3233. https://doi.org/10.18632/oncotarget.1908
85. Siddique HR, Adhami VM, Parray A, Johnson JJ, Siddiqui IA, Shekhani MT, Murtaza I, Ambartsumian N, Konety BR, Mukhtar H, Saleem M (2013) The S100A4 oncoprotein promotes prostate tumorigenesis in a transgenic mouse model: regulating NFkappaB through the RAGE receptor. Genes Cancer 4

(5–6):224–234. https://doi.org/10.1177/1947601913492420
86. Yammani RR, Carlson CS, Bresnick AR, Loeser RF (2006) Increase in production of matrix metalloproteinase 13 by human articular chondrocytes due to stimulation with S100A4: role of the receptor for advanced glycation end products. Arthritis Rheum 54(9):2901–2911. https://doi.org/10.1002/art.22042
87. Hernandez JL, Padilla L, Dakhel S, Coll T, Hervas R, Adan J, Masa M, Mitjans F, Martinez JM, Coma S, Rodriguez L, Noe V, Ciudad CJ, Blasco F, Messeguer R (2013) Therapeutic targeting of tumor growth and angiogenesis with a novel anti-S100A4 monoclonal antibody. PLoS One 8(9):e72480. https://doi.org/10.1371/journal.pone.0072480
88. Chaabane C, Heizmann CW, Bochaton-Piallat ML (2015) Extracellular S100A4 induces smooth muscle cell phenotypic transition mediated by RAGE. Biochim Biophys Acta 1853(9):2144–2157. https://doi.org/10.1016/j.bbamcr.2014.07.022
89. Medapati MR, Dahlmann M, Ghavami S, Pathak KA, Lucman L, Klonisch T, Hoang-Vu C, Stein U, Hombach-Klonisch S (2015) RAGE mediates the pro-migratory response of extracellular S100A4 in human thyroid cancer cells. Thyroid 25(5):514–527. https://doi.org/10.1089/thy.2014.0257
90. Kiryushko D, Novitskaya V, Soroka V, Klingelhofer J, Lukanidin E, Berezin V, Bock E (2006) Molecular mechanisms of Ca(2+) signaling in neurons induced by the S100A4 protein. Mol Cell Biol 26(9):3625–3638. https://doi.org/10.1128/MCB.26.9.3625-3638.2006
91. Cerezo LA, Remakova M, Tomcik M, Gay S, Neidhart M, Lukanidin E, Pavelka K, Grigorian M, Vencovsky J, Senolt L (2014) The metastasis-associated protein S100A4 promotes the inflammatory response of mononuclear cells via the TLR4 signalling pathway in rheumatoid arthritis. Rheumatology (Oxford) 53(8):1520–1526. https://doi.org/10.1093/rheumatology/keu031
92. Grotterod I, Maelandsmo GM, Boye K (2010) Signal transduction mechanisms involved in S100A4-induced activation of the transcription factor NF-kappaB. BMC Cancer 10:241. https://doi.org/10.1186/1471-2407-10-241
93. Hiratsuka S, Watanabe A, Sakurai Y, Akashi-Takamura S, Ishibashi S, Miyake K, Shibuya M, Akira S, Aburatani H, Maru Y (2008) The S100A8-serum amyloid A3-TLR4 paracrine cascade establishes a pre-metastatic phase. Nat Cell Biol 10(11):1349–1355. https://doi.org/10.1038/ncb1794
94. Vogl T, Tenbrock K, Ludwig S, Leukert N, Ehrhardt C, van Zoelen MA, Nacken W, Foell D, van der Poll T, Sorg C, Roth J (2007) Mrp8 and Mrp14 are endogenous activators of Toll-like receptor 4, promoting lethal, endotoxin-induced shock. Nat Med 13(9):1042–1049. https://doi.org/10.1038/nm1638
95. Li Q, Dai C, Xue R, Wang P, Chen L, Han Y, Erben U, Qin Z (2018) S100A4 protects myeloid-derived suppressor cells from intrinsic apoptosis via TLR4-ERK1/2 signaling. Front Immunol 9:388. https://doi.org/10.3389/fimmu.2018.00388
96. Li J, Yang F, Wei F, Ren X (2017) The role of toll-like receptor 4 in tumor microenvironment. Oncotarget 8(39):66656–66667. https://doi.org/10.18632/oncotarget.19105
97. Dmytriyeva O, Pankratova S, Owczarek S, Sonn K, Soroka V, Ridley CM, Marsolais A, Lopez-Hoyos M, Ambartsumian N, Lukanidin E, Bock E, Berezin V, Kiryushko D (2012) The metastasis-promoting S100A4 protein confers neuroprotection in brain injury. Nat Commun 3:1197. https://doi.org/10.1038/ncomms2202
98. Herbst RS (2004) Review of epidermal growth factor receptor biology. Int J Radiat Oncol Biol Phys 59(2 Suppl):21–26. https://doi.org/10.1016/j.ijrobp.2003.11.041
99. Klingelhofer J, Moller HD, Sumer EU, Berg CH, Poulsen M, Kiryushko D, Soroka V, Ambartsumian N, Grigorian M, Lukanidin EM (2009) Epidermal growth factor receptor ligands as new extracellular targets for the metastasis-promoting S100A4 protein. FEBS J 276(20):5936–5948. https://doi.org/10.1111/j.1742-4658.2009.07274.x
100. Cho CC, Chou RH, Yu C (2016) Amlexanox blocks the interaction between S100A4 and epidermal growth factor and inhibits cell proliferation. PLoS One 11(8):e0161663. https://doi.org/10.1371/journal.pone.0161663
101. Buetti-Dinh A, Pivkin IV, Friedman R (2015) S100A4 and its role in metastasis - computational integration of data on biological networks. Mol BioSyst 11(8):2238–2246. https://doi.org/10.1039/c5mb00110b
102. Zhang Q, Zhao Z, Ma Y, Wang H, Ma J, He X, Zhang D (2014) Combined expression of S100A4 and Annexin A2 predicts disease progression and overall survival in patients

with urothelial carcinoma. Urol Oncol 32 (6):798–805. https://doi.org/10.1016/j.urolonc.2014.02.009

103. Boye K, Grotterod I, Aasheim HC, Hovig E, Maelandsmo GM (2008) Activation of NF-kappaB by extracellular S100A4: analysis of signal transduction mechanisms and identification of target genes. Int J Cancer 123 (6):1301–1310. https://doi.org/10.1002/ijc.23617
104. Rojas A, Figueroa H, Morales E (2010) Fueling inflammation at tumor microenvironment: the role of multiligand/RAGE axis. Carcinogenesis 31(3):334–341. https://doi.org/10.1093/carcin/bgp322
105. Terzic J, Grivennikov S, Karin E, Karin M (2010) Inflammation and colon cancer. Gastroenterology 138(6):2101–2114 e2105. https://doi.org/10.1053/j.gastro.2010.01.058
106. Ahmad S, Khan H, Siddiqui Z, Khan MY, Rehman S, Shahab U, Godovikova T, Silnikov V, Moinuddin (2018) AGEs, RAGEs and s-RAGE; friend or foe for cancer. Semin Cancer Biol 49:44–55. https://doi.org/10.1016/j.semcancer.2017.07.001
107. Hou S, Tian T, Qi D, Sun K, Yuan Q, Wang Z, Qin Z, Wu Z, Chen Z, Zhang J (2018) S100A4 promotes lung tumor development through beta-catenin pathway-mediated autophagy inhibition. Cell Death Dis 9(3):277. https://doi.org/10.1038/s41419-018-0319-1
108. Wolf S, Haase-Kohn C, Lenk J, Hoppmann S, Bergmann R, Steinbach J, Pietzsch J (2011) Expression, purification and fluorine-18 radiolabeling of recombinant S100A4: a potential probe for molecular imaging of receptor for advanced glycation endproducts in vivo? Amino Acids 41(4):809–820. https://doi.org/10.1007/s00726-010-0822-x
109. Haase-Kohn C, Wolf S, Herwig N, Mosch B, Pietzsch J (2014) Metastatic potential of B16-F10 melanoma cells is enhanced by extracellular S100A4 derived from RAW264.7 macrophages. Biochem Biophys Res Commun 446(1):143–148. https://doi.org/10.1016/j.bbrc.2014.02.126
110. Ambartsumian N, Grigorian M (2016) [S100A4, a link between metastasis and inflammation]. Mol Biol (Mosk) 50 (4):577–588. https://doi.org/10.7868/S0026898416040029
111. Seif F, Khoshmirsafa M, Aazami H, Mohsenzadegan M, Sedighi G, Bahar M (2017) The role of JAK-STAT signaling pathway and its regulators in the fate of T helper cells. Cell Commun Signal 15(1):23. https://doi.org/10.1186/s12964-017-0177-y
112. Smith BN, Bhowmick NA (2016) Role of EMT in metastasis and therapy resistance. J Clin Med 5(2). https://doi.org/10.3390/jcm5020017
113. Bowers RR, Manevich Y, Townsend DM, Tew KD (2012) Sulfiredoxin redox-sensitive interaction with S100A4 and non-muscle myosin IIA regulates cancer cell motility. Biochemistry 51(39):7740–7754. https://doi.org/10.1021/bi301006w
114. Chen M, Bresnick AR, O'Connor KL (2013) Coupling S100A4 to Rhotekin alters Rho signaling output in breast cancer cells. Oncogene 32(32):3754–3764. https://doi.org/10.1038/onc.2012.383
115. Spiekerkoetter E, Guignabert C, de Jesus Perez V, Alastalo TP, Powers JM, Wang L, Lawrie A, Ambartsumian N, Schmidt AM, Berryman M, Ashley RH, Rabinovitch M (2009) S100A4 and bone morphogenetic protein-2 codependently induce vascular smooth muscle cell migration via phospho-extracellular signal-regulated kinase and chloride intracellular channel 4. Circ Res 105 (7):639–647, 613 p following 647. https://doi.org/10.1161/CIRCRESAHA.109.205120
116. Talbot LJ, Bhattacharya SD, Kuo PC (2012) Epithelial-mesenchymal transition, the tumor microenvironment, and metastatic behavior of epithelial malignancies. Int J Biochem Mol Biol 3(2):117–136
117. Pradella D, Naro C, Sette C, Ghigna C (2017) EMT and stemness: flexible processes tuned by alternative splicing in development and cancer progression. Mol Cancer 16(1):8. https://doi.org/10.1186/s12943-016-0579-2
118. Harris MA, Yang H, Low BE, Mukherjee J, Guha A, Bronson RT, Shultz LD, Israel MA, Yun K (2008) Cancer stem cells are enriched in the side population cells in a mouse model of glioma. Cancer Res 68(24):10051–10059. https://doi.org/10.1158/0008-5472.CAN-08-0786
119. Ren D, Zhu X, Kong R, Zhao Z, Sheng J, Wang J, Xu X, Liu J, Cui K, Zhang XH, Zhao H, Wong STC (2018) Targeting brain-adaptive cancer stem cells prohibits brain metastatic colonization of triple-negative breast cancer. Cancer Res 78 (8):2052–2064. https://doi.org/10.1158/0008-5472.CAN-17-2994
120. Cheng LH, Hung KF, Huang TF, Hsieh HP, Wang SY, Huang CY, Lo JF (2016) Attenuation of cancer-initiating cells stemness

properties by abrogating S100A4 calcium binding ability in head and neck cancers. Oncotarget 7(48):78946–78957. https://doi.org/10.18632/oncotarget.12935

121. Lo JF, Yu CC, Chiou SH, Huang CY, Jan CI, Lin SC, Liu CJ, Hu WY, Yu YH (2011) The epithelial-mesenchymal transition mediator S100A4 maintains cancer-initiating cells in head and neck cancers. Cancer Res 71 (5):1912–1923. https://doi.org/10.1158/0008-5472.CAN-10-2350

122. Xia H, Gilbertsen A, Herrera J, Racila E, Smith K, Peterson M, Griffin T, Benyumov A, Yang L, Bitterman PB, Henke CA (2017) Calcium-binding protein S100A4 confers mesenchymal progenitor cell fibrogenicity in idiopathic pulmonary fibrosis. J Clin Invest 127(7):2586–2597. https://doi.org/10.1172/JCI90832

123. Jiao J, Gonzalez A, Stevenson HL, Gagea M, Sugimoto H, Kalluri R, Beretta L (2018) Depletion of S100A4(+) stromal cells does not prevent HCC development but reduces the stem cell-like phenotype of the tumors. Exp Mol Med 50(1):e422. https://doi.org/10.1038/emm.2017.175

124. Powell E, Piwnica-Worms D, Piwnica-Worms H (2014) Contribution of p53 to metastasis. Cancer Discov 4(4):405–414. https://doi.org/10.1158/2159-8290.CD-13-0136

125. Shen W, Chen D, Liu S, Chen L, Yu A, Fu H, Sun X (2015) S100A4 interacts with mutant p53 and affects gastric cancer MKN1 cell autophagy and differentiation. Int J Oncol 47(6):2123–2130. https://doi.org/10.3892/ijo.2015.3209

126. van Dieck J, Teufel DP, Jaulent AM, Fernandez-Fernandez MR, Rutherford TJ, Wyslouch-Cieszynska A, Fersht AR (2009) Posttranslational modifications affect the interaction of S100 proteins with tumor suppressor p53. J Mol Biol 394(5):922–930. https://doi.org/10.1016/j.jmb.2009.10.002

127. Sun Z, Wu Z, Zhang F, Guo Q, Li L, Li K, Chen H, Zhao J, Song D, Huang Q, Li L, Xiao J (2016) PRAME is critical for breast cancer growth and metastasis. Gene 594 (1):160–164. https://doi.org/10.1016/j.gene.2016.09.016

128. EL naaman C, Grum-Schwensen B, Mansouri A, Grigorian M, Santoni-Rugiu E, Hansen T, Kriajevska M, Schafer BW, Heizmann CW, Lukanidin E, Ambartsumian N (2004) Cancer predisposition in mice deficient for the metastasis-associated Mts1 (S100A4) gene. Oncogene 23 (20):3670–3680. https://doi.org/10.1038/sj.onc.1207420

129. Mrakovcic M, Frohlich LF (2018) p53-mediated molecular control of autophagy in tumor cells. Biomol Ther 8(2). https://doi.org/10.3390/biom8020014

130. Guo S, Pridham KJ, Virbasius CM, He B, Zhang L, Varmark H, Green MR, Sheng Z (2018) A large-scale RNA interference screen identifies genes that regulate autophagy at different stages. Sci Rep 8(1):2822. https://doi.org/10.1038/s41598-018-21106-5

131. Bian XW, Chen JH, Jiang XF, Bai JS, Wang QL, Zhang X (2004) Angiogenesis as an immunopharmacologic target in inflammation and cancer. Int Immunopharmacol 4 (12):1537–1547. https://doi.org/10.1016/j.intimp.2004.07.017

132. Jackson JR, Seed MP, Kircher CH, Willoughby DA, Winkler JD (1997) The codependence of angiogenesis and chronic inflammation. FASEB J 11(6):457–465

133. Ono M (2008) Molecular links between tumor angiogenesis and inflammation: inflammatory stimuli of macrophages and cancer cells as targets for therapeutic strategy. Cancer Sci 99(8):1501–1506. https://doi.org/10.1111/j.1349-7006.2008.00853.x

134. de Silva Rudland S, Martin L, Roshanlall C, Winstanley J, Leinster S, Platt-Higgins A, Carroll J, West C, Barraclough R, Rudland P (2006) Association of S100A4 and osteopontin with specific prognostic factors and survival of patients with minimally invasive breast cancer. Clin Cancer Res 12 (4):1192–1200. https://doi.org/10.1158/1078-0432.CCR-05-1580

135. Ochiya T, Takenaga K, Asagiri M, Nakano K, Satoh H, Watanabe T, Imajoh-Ohmi S, Endo H (2015) Efficient inhibition of tumor angiogenesis and growth by a synthetic peptide blocking S100A4-methionine aminopeptidase 2 interaction. Mol Ther Methods Clin Dev 2:15008. https://doi.org/10.1038/mtm.2015.8

136. Sun JB, Holmgren J, Larena M, Terrinoni M, Fang Y, Bresnick AR, Xiang Z (2017) Deficiency in calcium-binding protein S100A4 impairs the adjuvant action of cholera toxin. Front Immunol 8:1119. https://doi.org/10.3389/fimmu.2017.01119

137. Duguet F, Locard-Paulet M, Marcellin M, Chaoui K, Bernard I, Andreoletti O, Lesourne R, Burlet-Schiltz O, Gonzalez de Peredo A, Saoudi A (2017) Proteomic analysis of regulatory T cells reveals the importance of Themis1 in the control of their suppressive function. Mol Cell Proteomics 16

(8):1416–1432. https://doi.org/10.1074/mcp.M116.062745

138. Pfoertner S, Jeron A, Probst-Kepper M, Guzman CA, Hansen W, Westendorf AM, Toepfer T, Schrader AJ, Franzke A, Buer J, Geffers R (2006) Signatures of human regulatory T cells: an encounter with old friends and new players. Genome Biol 7(7):R54. https://doi.org/10.1186/gb-2006-7-7-r54
139. Chen L, Li J, Zhang J, Dai C, Liu X, Wang J, Gao Z, Guo H, Wang R, Lu S, Wang F, Zhang H, Chen H, Fan X, Wang S, Qin Z (2015) S100A4 promotes liver fibrosis via activation of hepatic stellate cells. J Hepatol 62(1):156–164. https://doi.org/10.1016/j.jhep.2014.07.035
140. Osterreicher CH, Penz-Osterreicher M, Grivennikov SI, Guma M, Koltsova EK, Datz C, Sasik R, Hardiman G, Karin M, Brenner DA (2011) Fibroblast-specific protein 1 identifies an inflammatory subpopulation of macrophages in the liver. Proc Natl Acad Sci U S A 108(1):308–313. https://doi.org/10.1073/pnas.1017547108
141. Bettum IJ, Vasiliauskaite K, Nygaard V, Clancy T, Pettersen SJ, Tenstad E, Maelandsmo GM, Prasmickaite L (2014) Metastasis-associated protein S100A4 induces a network of inflammatory cytokines that activate stromal cells to acquire pro-tumorigenic properties. Cancer Lett 344(1):28–39. https://doi.org/10.1016/j.canlet.2013.10.036
142. Prasmickaite L, Tenstad EM, Pettersen S, Jabeen S, Egeland EV, Nord S, Pandya A, Haugen MH, Kristensen VN, Borresen-Dale AL, Oslo Breast Cancer Research Consortium, Engebraten O, Maelandsmo GM (2018) Basal-like breast cancer engages tumor-supportive macrophages via secreted factors induced by extracellular S100A4. Mol Oncol. https://doi.org/10.1002/1878-0261.12319
143. Mah SJ, Lee J, Kim H, Kang YG, Baek SH, Kim HH, Lim WH (2015) Induction of S100A4 in periodontal ligament cells enhances osteoclast formation. Arch Oral Biol 60(9):1215–1221. https://doi.org/10.1016/j.archoralbio.2015.05.014
144. Kondo N, Ichimiya S, Tamura Y, Tonooka A, Koshiba S, Torigoe T, Kamiguchi K, Takenaga K, Sato N (2005) A calcium binding protein, S100A4, mediates T cell dependent cytotoxicity as a transformation-associated antigen. Microbiol Immunol 49(1):49–56
145. Lotfi R, Wiegmann D, Asseck L, Nienhaus C, Erle A, Schrezenmeier H (2012) Necrosis-Associated factors (DAMPs) not only promote tumor proliferation per se but also impact adaptive immune response. Impact of S100A4 within tumor microenvironment. Onkologie 35:81–81
146. Veglia F, Perego M, Gabrilovich D (2018) Myeloid-derived suppressor cells coming of age. Nat Immunol 19(2):108–119. https://doi.org/10.1038/s41590-017-0022-x
147. Sorokin L (2010) The impact of the extracellular matrix on inflammation. Nat Rev Immunol 10(10):712–723. https://doi.org/10.1038/nri2852
148. Wight TN, Frevert CW, Debley JS, Reeves SR, Parks WC, Ziegler SF (2017) Interplay of extracellular matrix and leukocytes in lung inflammation. Cell Immunol 312:1–14. https://doi.org/10.1016/j.cellimm.2016.12.003
149. Berge G, Pettersen S, Grotterod I, Bettum IJ, Boye K, Maelandsmo GM (2011) Osteopontin – an important downstream effector of S100A4-mediated invasion and metastasis. Int J Cancer 129(4):780–790. https://doi.org/10.1002/ijc.25735
150. Kahles F, Findeisen HM, Bruemmer D (2014) Osteopontin: a novel regulator at the cross roads of inflammation, obesity and diabetes. Mol Metab 3(4):384–393. https://doi.org/10.1016/j.molmet.2014.03.004
151. Nissinen L, Kahari VM (2014) Matrix metalloproteinases in inflammation. Biochim Biophys Acta 1840(8):2571–2580. https://doi.org/10.1016/j.bbagen.2014.03.007
152. Bjornland K, Winberg JO, Odegaard OT, Hovig E, Loennechen T, Aasen AO, Fodstad O, Maelandsmo GM (1999) S100A4 involvement in metastasis: deregulation of matrix metalloproteinases and tissue inhibitors of matrix metalloproteinases in osteosarcoma cells transfected with an anti-S100A4 ribozyme. Cancer Res 59(18):4702–4708
153. Saleem M, Kweon MH, Johnson JJ, Adhami VM, Elcheva I, Khan N, Bin Hafeez B, Bhat KM, Sarfaraz S, Reagan-Shaw S, Spiegelman VS, Setaluri V, Mukhtar H (2006) S100A4 accelerates tumorigenesis and invasion of human prostate cancer through the transcriptional regulation of matrix metalloproteinase 9. Proc Natl Acad Sci U S A 103(40):14825–14830. https://doi.org/10.1073/pnas.0606747103
154. Zhang J, Zhang DL, Jiao XL, Dong Q (2013) S100A4 regulates migration and invasion in hepatocellular carcinoma HepG2 cells via NF-kappaB-dependent MMP-9 signal. Eur Rev Med Pharmacol Sci 17(17):2372–2382

155. Zhang W, Liu Y, Wang CW (2014) S100A4 promotes squamous cell laryngeal cancer Hep-2 cell invasion via NF-kB/MMP-9 signal. Eur Rev Med Pharmacol Sci 18 (9):1361–1367

156. Ismail TM, Bennett D, Platt-Higgins AM, Al-Medhity M, Barraclough R, Rudland PS (2017) S100A4 elevation empowers expression of metastasis effector molecules in human breast cancer. Cancer Res 77 (3):780–789. https://doi.org/10.1158/0008-5472.CAN-16-1802

157. Xu H, Li M, Zhou Y, Wang F, Li X, Wang L, Fan Q (2016) S100A4 participates in epithelial-mesenchymal transition in breast cancer via targeting MMP2. Tumour Biol 37 (3):2925–2932. https://doi.org/10.1007/s13277-015-3709-3

158. Cao J, Geng L, Wu Q, Wang W, Chen Q, Lu L, Shen W, Chen Y (2013) Spatiotemporal expression of matrix metalloproteinases (MMPs) is regulated by the Ca2+-signal transducer S100A4 in the pathogenesis of thoracic aortic aneurysm. PLoS One 8(7): e70057. https://doi.org/10.1371/journal.pone.0070057

159. Senolt L, Grigorian M, Lukanidin E, Simmen B, Michel BA, Pavelka K, Gay RE, Gay S, Neidhart M (2006) S100A4 is expressed at site of invasion in rheumatoid arthritis synovium and modulates production of matrix metalloproteinases. Ann Rheum Dis 65(12):1645–1648. https://doi.org/10.1136/ard.2005.047704

160. Andersen K, Mori H, Fata J, Bascom J, Oyjord T, Maelandsmo GM, Bissell M (2011) The metastasis-promoting protein S100A4 regulates mammary branching morphogenesis. Dev Biol 352(2):181–190. https://doi.org/10.1016/j.ydbio.2010.12.033

161. Schneider M, Kostin S, Strom CC, Aplin M, Lyngbaek S, Theilade J, Grigorian M, Andersen CB, Lukanidin E, Lerche Hansen J, Sheikh SP (2007) S100A4 is upregulated in injured myocardium and promotes growth and survival of cardiac myocytes. Cardiovasc Res 75(1):40–50. https://doi.org/10.1016/j.cardiores.2007.03.027

162. Sack U, Walther W, Scudiero D, Selby M, Kobelt D, Lemm M, Fichtner I, Schlag PM, Shoemaker RH, Stein U (2011) Novel effect of antihelminthic Niclosamide on S100A4-mediated metastatic progression in colon cancer. J Natl Cancer Inst 103(13):1018–1036. https://doi.org/10.1093/jnci/djr190

163. Stein U, Arlt F, Smith J, Sack U, Herrmann P, Walther W, Lemm M, Fichtner I, Shoemaker RH, Schlag PM (2011) Intervening in beta-catenin signaling by sulindac inhibits S100A4-dependent colon cancer metastasis. Neoplasia 13(2):131–144

164. Burock S, Daum S, Keilholz U, Neumann K, Walther W, Stein U (2018) Phase II trial to investigate the safety and efficacy of orally applied niclosamide in patients with metachronous or synchronous metastases of a colorectal cancer progressing after therapy: the NIKOLO trial. BMC Cancer 18(1):297. https://doi.org/10.1186/s12885-018-4197-9

Chapter 23

Interaction of S100A6 with Target Proteins In Vitro and in Living Cells

Kyohei Sakane, Fuminori Yamaguchi, Mitsumasa Tsuchiya, Rina Kondo, Naoki Kanayama, Masaki Magari, Naoya Hatano, Ryoji Kobayashi, and Hiroshi Tokumitsu

Abstract

S100A6 is a member of the EF-hand Ca^{2+}-binding protein family, which plays important roles in a wide variety of Ca^{2+} signaling in the cells, as well as in pathophysiological conditions. Herein, we describe analytical protocols for evaluating the interaction of S100A6 with multiple target proteins in vitro, including biotinylated S100A6 overlay, glutathione-*S*-transferase (GST)-precipitation, surface plasmon resonance, and a GST-precipitation assay in living cells. These methods will elucidate the detailed molecular mechanisms of S100A6/target interactions and further improve our understanding of the physiological significance of S100A6-mediated Ca^{2+} signaling. Moreover, they may be used to evaluate other physical S100/target interactions.

Key words S100A6, Ca^{2+}-binding, EF-hand, Protein–protein interaction, FOR20, Protein phosphatase 5, GST-precipitation assay, S100A6-overlay, Transfection, SPR analysis

1 Introduction

S100A6 (also known as calcyclin) is a member of EF-hand calcium-binding protein family [1]. First discovered as the growth-regulated gene (2A9), the mRNA levels of S100A6 increase in quiescent fibroblasts following treatment with serum, PDGF, and EGF [2]. S100A6 is thought to be involved in various cellular responses as a Ca^{2+}-signal transducer through direct interaction with its target proteins. Over ten target proteins for S100A6 are known today, including annexin XI [3], CacyBP/SIP [4], the receptor for advanced glycation-end products (RAGE) [5], the recently identified protein phosphatase 5 (PP5) [6], and a centrosomal protein, FOR20 [7]. The growing number of its targets underlines the physiological significance of S100A6 as a multifunctional signal transducer. Therefore, evaluation of S100A6/target

Claus W. Heizmann (ed.), *Calcium-Binding Proteins of the EF-Hand Superfamily: From Basics to Medical Applications*, Methods in Molecular Biology, vol. 1929, https://doi.org/10.1007/978-1-4939-9030-6_23,

interactions in intact cells, as well as in vitro, is critical for identification and characterization of novel Ca^{2+}-dependent signaling pathways. Various methods for analysis of protein–protein interaction are available using recombinant proteins expressed in *E. coli* or cultured mammalian cells. We have used recombinant S100A6 and glutathione-*S*-transferase (GST)-fused target proteins to evaluate their interaction using a biotinylated S100A6 overlay under denatured conditions and surface plasmon resonance (SPR) analysis and GST-precipitation assays under native conditions [6, 7]. We also describe an analytical method for S100A6/target interactions in living cells using exogenous expression of GST-target protein in cultured mammalian cells, followed by GST-precipitation assay. Detection of endogenous S100A6 that interacts with its targets in particular cell lines, such as COS-7 cells, is readily detected by immunoblotting using anti-S100A6 antibodies.

2 Materials

2.1 Expression and Purification of Recombinant S100A6

1. *Escherichia coli* BL-21 (DE3) competent cells.
2. Luria–Bertani (LB) broth + ampicillin: 10 g tryptone, 5 g dried yeast extract, and 10 g NaCl are added to 1 L H_2O and autoclaved at 121 °C for 15 min, and ampicillin is then added (100 μg/mL final concentration) when LB broth is cooled to room temperature.
3. 100 mM isopropyl β-D-1-thiogalactopyranoside (IPTG).
4. *E. coli* lysis buffer: 500 mM NaCl, 20 mM Tris–HCl (pH 7.5), 5 mM $CaCl_2$.
5. Elution buffer: 500 mM NaCl, 20 mM Tris–HCl (pH 7.5), 5 mM EGTA.
6. Dialysis buffer: PBS (1.83 g $Na_2HPO_4{\cdot}12H_2O$, 0.20 g $NaH_2PO_4{\cdot}2H_2O$, and 9.0 g NaCl in 1 L H_2O).
7. Phenyl Sepharose 6 Fast Flow (low substitution) resin (GE Healthcare).
8. Poly-Prep® chromatography columns (Bio-Rad Laboratories).

2.2 S100A6-Overlay Method

1. 20 mM biotinoyl-ε-aminocaproic acid *N*-hydroxysuccinimide (NHS) ester in *N*,*N*-dimethylformamide
2. PD-10 Desalting Column (GE Healthcare).
3. Buffer A:150 mM NaCl, 20 mM Tris–HCl (pH 7.5).
4. Blocking buffer: Buffer A + 5% skimmed milk.
5. Binding buffer: Blocking buffer in the presence of either 2 mM $CaCl_2$ or 2 mM EGTA.

6. Streptavidin-HRP reaction buffer: Buffer A + streptavidin-HRP conjugate (1:2000 dilution, GE Healthcare) in the presence of either 2 mM $CaCl_2$ or 2 mM EGTA.
7. Enhanced chemiluminescence (ECL) reaction solution: 1 mL enhanced luminol reagent plus +1 mL oxidizing reagent plus (Western Lightning Plus-ECL, PerkinElmer, Inc.) in the presence of either 2 mM $CaCl_2$ or 2 mM EGTA.
8. CCD camera-based imager (ChemiDoc XRS, Bio-Rad Laboratories).

2.3 GST-Precipitation Assay

1. Binding buffer: 150 mM NaCl, 20 mM Tris–HCl (pH 7.5), 0.05% Tween 20 in the presence of either 2 mM $CaCl_2$ or 2 mM EGTA.
2. Glutathione-Sepharose 4B (GE Healthcare), 50% slurry in H_2O.
3. 1× SDS-polyacrylamide gel electrophoresis sample buffer: 50 mM Tris–HCl (pH 6.8), 1% SDS, 2.5% mercaptoethanol, 12.5% glycerol, and 0.005% bromophenol blue.

2.4 Analysis of S100A6/Target Interaction by SPR

2.4.1 Purification of Recombinant His-PP5 Protein

1. *E. coli* lysis buffer: 150 mM NaCl, 20 mM NaH_2PO_4 (pH 8.0), 0.5% Triton X-100.
2. Wash buffer: 25 mM imidazole, 150 mM NaCl, 20 mM NaH_2PO_4 (pH 8.0), 0.5% Triton X-100.
3. Elution buffer: 250 mM imidazole, 150 mM NaCl, 20 mM NaH_2PO_4 (pH 8.0).
4. Ni-NTA agarose (Qiagen).
5. Protease inhibitor: 4-(2-aminoethyl) benzenesulfonyl fluoride hydrochloride (AEBSF).
6. Poly-Prep® chromatography columns.

2.4.2 SPR Analysis

1. Biacore 2000 analyzer (GE Healthcare).
2. A CM5 sensor chip (GE Healthcare) carrying a matrix of carboxymethylated dextran.
3. Amine coupling buffer (GE Healthcare): 750 mg *N*-ethyl-*N*-′-(3-dimethylaminopropyl) carbodiimide hydrochloride (EDC), 115 mg *N*-hydroxysuccinimide (NHS), and 10.5 mL 1 M ethanolamine hydrochloride (pH 8.5). For preparation, dissolve the NHS and EDC by adding 10.0 mL of filtered, deionized water to each vial, and completely dissolve. Dispense the NHS and EDC solutions separately in aliquots. Store the NHS and EDC at −18 °C or below. Ethanolamine hydrochloride may be stored at +2 to −8 °C.
4. Ligand dilution buffer: 20 mM ammonium acetate (pH 4.2).

5. Running buffer: 20 mM HEPES (pH 7.4), 150 mM NaCl, and 0.005% Tween 20 with 1 mM $CaCl_2$ or 1 mM EGTA.
6. Chip regeneration buffer: 50 mM NaOH.
7. Analysis software: BIAevaluation 3.0 (GE Healthcare).

2.5 S100A6/Target (GST-FOR20) Interaction in Living Cells

1. COS-7 cells maintained in Dulbecco's modified Eagle's medium (DMEM) supplemented with 10% fetal bovine serum (FBS).
2. Mammalian expression plasmids for GST alone or the GST-S100A6 target (GST-FOR20) are constructed using pME-18s vectors.
3. "Polyethylenimine Max" (Polysciences, Inc.).
4. Lysis buffer: 150 mM NaCl, 20 mM Tris–HCl (pH 7.5), 0.05% Tween 20, 0.2 mM phenylmethylsulfonyl fluoride (PMSF), and either 2 mM EGTA or 2 mM $CaCl_2$.
5. Anti-S100A6 rabbit monoclonal antibody (Epitomics).
6. Anti-GST goat polyclonal antibody (GE Healthcare).

3 Methods

***3.1 Expression and Purification of S100A6 in* E. coli**

1. Inoculate *E. coli* BL-21 (DE3) harboring pET11a-S100A6 into 5 mL of LB (+100 μg/mL ampicillin), and shake at 37 °C overnight to grow *E.coli*. Inoculate 1 mL of this *E. coli* culture into 100 mL of LB (+100 μg/mL ampicillin) in a flask, and grow further until the OD_{600} reaches 0.6. Then add IPTG to a final concentration of 1 mM, and shake it at 37 °C for another 5 h.
2. Collect *E. coli* pellet by centrifugation at 7000 × *g* for 15 min, and lyse the bacterial pellet by sonication (40% amplitude, 5 times with 15 second bursts, and 1 min on ice between intervals) in 20 mL of ice-cold *E. coli* lysis buffer in the presence of 0.2 mM PMSF, and centrifuge at 28,420 × *g* for 20 min to obtain the supernatant (cell lysates).
3. Apply the supernatant containing recombinant S100A6 to a phenyl-Sepharose column (0.5 mL bed volume in Poly-Prep® chromatography column) that has been pre-equilibrated with *E. coli* lysis buffer, and then wash the column extensively with *E. coli* lysis buffer containing 5 mM $CaCl_2$. Elution is carried out by adding elution buffer containing 5 mM EGTA, and collect eluted fractions (0.5 mL each fraction). After this, check the purified S100A6 in eluted fractions by Tris-Tricine SDS-10% PAGE, followed by Coomassie Brilliant Blue staining, and dialyze the fractions containing S100A6 against dialysis buffer (PBS). Typically, ~1.5 mg of S100A6 with >95% homogeneity can be purified from 100 mL *E. coli* culture.

3.2 Biotinylation of Recombinant S100A6 and S100A6-Overlay Method

3.2.1 Preparation of Biotinylated S100A6

1. Incubate 189 μL of 20 mM biotinoyl-ε-aminocaproic acid NHS ester in *N,N*-dimethylformamide with purified recombinant S100A6 protein (2.4 mg) in 2 mL PBS at molar ratio of 16:1 at 4 °C for 2 h with constant stirring [8].
2. Apply the reaction mixture containing biotinylated S100A6 protein to a PD-10 Desalting Column, and collect eluted fractions (0.5 mL each fraction) to remove unreacted biotin using PBS as an elution buffer, followed by measuring the protein concentration of purified biotinylated S100A6 using the Bradford method.
3. Biotinylated S100A6 is subjected to Tris-Tricine SDS-10% PAGE and transferred to a nitrocellulose membrane.
4. Incubate the membrane with 10 mL of blocking buffer for 1 h at room temperature, followed by incubation with streptavidin-HRP reaction buffer for 1 h at room temperature.
5. After washing at least five times with Buffer A, allowing 5 min per wash, incubate the membrane with ECL reaction solution with constant stirring for 1 min at room temperature, followed by detection of biotinylated S100A6 using a CCD camera-based imager (Fig. 1b). This step is critical to examine the quality of biotinylated protein using overlay method (*see* **Note 1**).

3.2.2 S100A6 Overlay

1. Samples (recombinant proteins or cell lysates) are subjected to SDS-PAGE (7.5%–10%), followed by transfer to a nitrocellulose membrane via a semidry transfer system (*see* **Note 1**).
2. Incubate the blotted membrane with 10 mL of blocking buffer for 1 h at room temperature (Fig. 1a, blocking), followed by incubation with 0.36 μg/mL biotinylated S100A6 in binding buffer in the presence of 2 mM $CaCl_2$ or 2 mM EGTA for 1 h at room temperature or overnight at 4 °C (Fig. 1a, biotinylated S100A6).
3. After washing the membrane with Buffer A at least five times in the presence of 2 mM $CaCl_2$ or 2 mM EGTA, allowing 5 min per wash, incubate the membrane with streptavidin-HRP reaction buffer for 1 h at room temperature (Fig. 1a, avidin-HRP).
4. After extensive washing with Buffer A in the presence of 2 mM $CaCl_2$ or 2 mM EGTA, incubate the membrane with ECL reaction solution for 1 min at room temperature with constant stirring, followed by detection of biotinylated S100A6-binding signal using a CCD camera-based imager (Fig. 1a, ECL detection) (*see* **Note 2**).

3.3 GST-Precipitation Assay

1. Incubate purified recombinant S100A6 (13 μg) with its target protein (GST-FOR20-His_6, 22 μg) in binding buffer (500 μL) containing either 2 mM $CaCl_2$ or 2 mM EGTA in a 1.5 mL tube at room temperature for 1 h.

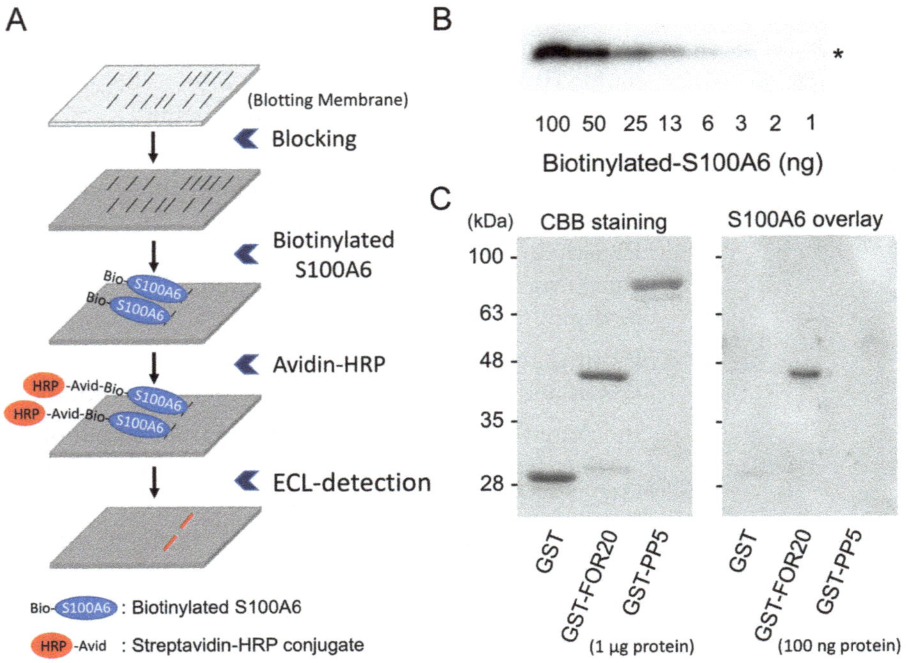

Fig. 1 (**a**) Schematic representation of the S100A6-overlay method. (**b**) Biotinylated S100A6 proteins (1–100 ng) on a nitrocellulose membrane were detected by HRP-conjugated streptavidin-mediated ECL detection. An asterisk indicates biotinylated S100A6. (**c**) GST, GST-FOR20, and GST-PP5 were separated by SDS-10% PAGE, followed by Coomassie Brilliant Blue staining (*left panel*, 1 μg protein) and an S100A6-overlay analysis (*right panel*, 0.1 μg protein) in the presence of 2 mM $CaCl_2$ (reproduced from Ref. [7] with permission from Elsevier)

2. Add 50 μL of glutathione-Sepharose (50% slurry) to the sample containing S100A6 and its targets (recombinant proteins or cell lysates), and incubate the reaction tube at room temperature for 1 h with gentle end-over-end mixing.
3. Precipitate the glutathione-Sepharose beads by centrifugation at 880 × *g* for 5 min, discard the supernatant carefully, and add 1 mL of binding buffer containing either 2 mM $CaCl_2$ or 2 mM EGTA to wash the resin with gentle end-over-end mixing for 10 min.
4. This washing step should be repeated at least five times.
5. Add 50 μL of 1× SDS-polyacrylamide gel electrophoresis sample buffer to glutathione-Sepharose beads, and then heat the samples at 95 °C for 10 min.
6. Samples (10 μL) are subjected to Tris-Tricine SDS-10% PAGE, followed by Coomassie Brilliant Blue staining (Fig. 2a).

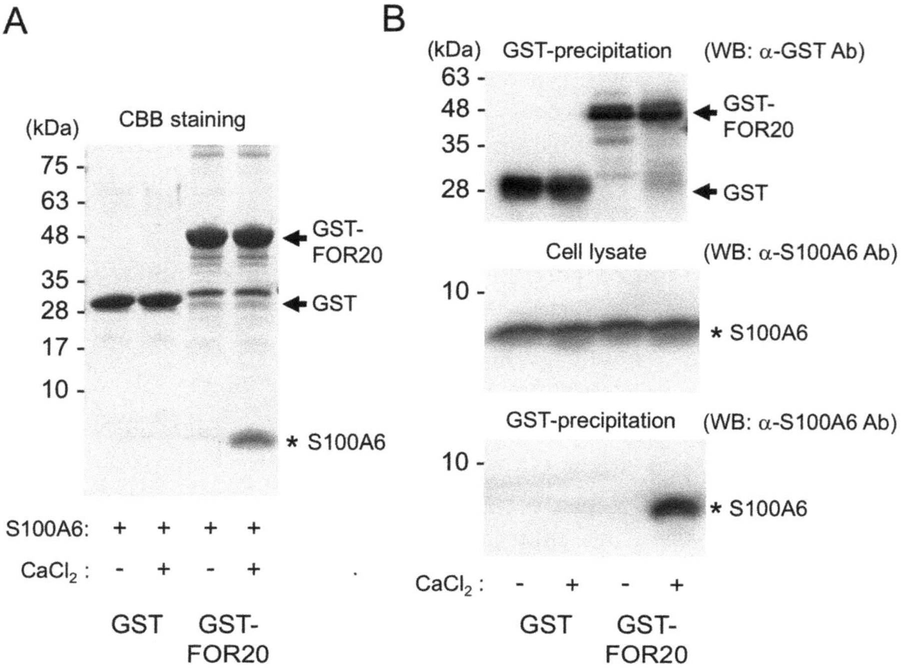

Fig. 2 Evaluation of Ca^{2+}-dependent interaction of S100A6 with FOR20 in vitro (**a**) and in transfected COS-7 cells (**b**) using the GST-precipitation method. (**a**) GST and GST-FOR20 proteins (22 μg) were incubated with S100A6 (13 μg) and precipitated with glutathione-Sepharose beads in the presence (+) or absence (–) of $CaCl_2$. The precipitated samples were analyzed using Tris-Tricine SDS-10% PAGE, followed by Coomassie Brilliant Blue staining. An asterisk indicates S100A6. (**b**) COS-7 cells were transfected with expression vectors encoding GST or wild-type GST-FOR20. After a 48 h incubation, GST-fusion proteins were precipitated with glutathione-Sepharose in the presence (+) or absence (–) of $CaCl_2$, and the precipitates were subjected to immunoblot analysis using either anti-GST antibody (*upper panel*) or anti-S100A6 antibody (*bottom panel*). The cell lysates (10 μL) from the transfected COS-7 cells were analyzed by immunoblotting with anti-S100A6 antibodies (*middle panel*) (reproduced from Ref. [7] with permission from Elsevier)

3.4 Analysis of S100A6/Target Interaction by SPR

3.4.1 Expression and Purification of Recombinant His-PP5 Protein

1. Inoculate *E. coli* BL-21 (DE3) containing pET16a-PP5 into 100 mL of LB (+ 100 μg/mL ampicillin) in a flask, and shake it at 37 °C overnight. Then, inoculate 10 mL of this *E. coli* culture into 400 mL of LB (+100 μg/mL ampicillin), and grow further until the OD_{600} reaches 0.8. Then add IPTG to a final concentration of 1 mM, and shake it overnight at room temperature.
2. Lyse collected *E. coli* pellet by sonication in 10 mL of ice-cold *E. coli* lysis buffer in the presence of 0.5 mM AEBSF, and centrifuge to obtain the supernatant as described in Subheading 3.1.
3. Apply the supernatant onto a Ni-NTA agarose column (1 mL bed volume in a Poly-Prep® chromatography column), which has been pre-equilibrated with *E. coli* lysis buffer, and wash the column with 10 mL of *E. coli* lysis buffer and 10 mL of wash

buffer. Elution is carried out by adding elution buffer, and collect eluted fractions (1 mL each fraction). After this, check the purified His-PP5 in eluted fractions by SDS-PAGE, followed by Coomassie Brilliant Blue staining, and dialyze the fractions containing His-PP5 against the dialysis buffer (10 mM Tris–HCl, pH 7.5) at 4 °C. Aliquots of protein were kept frozen at −80 °C until use.

3.4.2 Coupling of His-PP5 to the Surface of the CM5 Chip

Purified His-PP5 protein is used for the amine coupling to the dextran surface of the CM5 chip on Biacore 2000 analyzer.

1. Dilute purified His-PP5 (100 μg/mL, 140 μL) in 20 mM ammonium acetate (pH 4.2).
2. Activate the surface of the chip by injecting NHS/EDC solution at a flow rate of 5 μL/min for 7 min.
3. Inject the ligand (His-PP5) solution until 1000–3000 resonance units (RU), and a stable baseline is obtained.
4. Deactivate the surface of the chip by injecting the ethanolamine solution at a flow rate of 5 μL/min for 7 min.

3.4.3 Binding Analysis of S100A6 and PP5 by SPR

1. The analyte (S100A6) is prepared at various concentrations (156 nM, 313 nM, 625 nM, and 1.25 μM) by dilution in running buffer containing 1 mM $CaCl_2$ (Fig. 3a, b) or 1 mM EGTA (Fig. 3a).
2. Inject the analyte on Biacore 2000 analyzer at a flow rate of 20 μL/min.
3. Following each injection, the sensor chip is regenerated with 50 mM NaOH until the response unit baseline returns to the preinjection level.
4. The signal generated simultaneously on the control flow cell is used to fit the response curves (*see* **Note 3**).

3.4.4 Kinetic Evaluation of S100A6 Binding Data

The experimental sensorgram data (Fig. 3b) is used for kinetic analysis using the BIAevaluation software (BIAevaluation 3.0) as briefly described below. The data are analyzed by the nonlinear fitting method, and the signal generated simultaneously on the control flow cell is used to fit the response curves.

1. Open the experimental sensorgrams.
2. Click "Fit Kinetics Simultaneous K_a/K_d."
3. Choose the model (1:1 [Langmuir] binding), input the concentration of analytes, and click "Fit."
4. Calculated parameters are shown in Fig. 3c (*see* **Note 3**). k_a (Ms^{-1}), association rate constant; k_d (s^{-1}), dissociation rate constant; K_D (M), dissociation constant; R_{max}, maximum binding amount; Chi2, statistical measure of the closeness of fit.

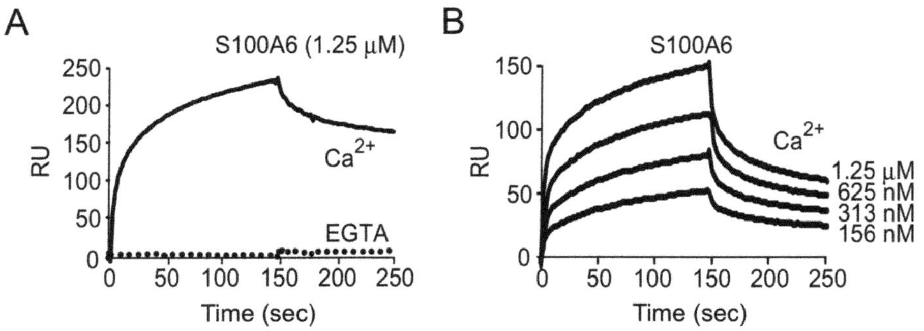

C

	k_a (Ms^{-1})	k_d (s^{-1})	K_D (M)	Chi2	R_{max} (RU)
S100A6	6.89×10^4	1.32×10^{-2}	1.92×10^{-7}	13.7	165

Fig. 3 Surface plasmon resonance (SPR) analysis of the interaction between His-PP5 and S100A6. (**a**) Purified His-PP5 was immobilized on the CM5 chip in 20 mM ammonium acetate (pH 4.2) until 1377 (0.16 pmoL) resonance units (RU) were bound and a stable baseline was obtained. Recombinant S100A6 (1.25 μM) was injected in the presence of 1 mM $CaCl_2$ (*solid line*) or EGTA (*broken line*). Response curves were prepared for fitting by subtracting the signal generated simultaneously on the control flow cell. (**b**) Recombinant S100A6 was injected at various concentrations (1.25 μM, 625 nM, 313 nM, and 156 nM), and each of the response curves is overlaid. The sensorgram curves were evaluated in BIAevaluation 3.0 using a 1:1 Langmuir model. *RU*, resonance unit. (**c**) The kinetic parameters of S100A6 for His-PP5 binding were calculated from the SPR sensorgrams shown in panel B (reproduced from Ref. [6] with permission from the American Society for Biochemistry and Molecular Biology)

3.5 S100A6/Target (GST-FOR20) Interaction in Living Cells

1. COS-7 cells cultured in DMEM supplemented with 10% FBS on 10 cm petri dishes are transfected with 3 μg of either GST expression plasmid (pME-GST) or GST-FOR20 expression plasmid (pME-GST-FOR20), using "polyethylenimine Max," according to the manufacturer's protocol.
2. After 24 h culture, lyse the cells with 1 mL of ice-cold lysis buffer with or without $CaCl_2$, followed by brief sonication, and centrifuge at 19,060 × *g* for 10 min to obtain the cell lysate. The GST-precipitation assay is performed by incubation with 50 μL of glutathione-Sepharose beads (50% slurry) as described in Subheading 3.3.
3. GST-precipitation samples in 1× SDS-polyacrylamide gel electrophoresis sample buffer (10 μL) and cell lysates are subjected to Tris-Tricine SDS-10% PAGE, followed by immunoblotting using anti-S100A6 antibody (Fig. 2b, *lower and middle panel*), and anti-GST antibody (Fig. 2b, *upper panel*) to confirm the amount of GST or GST-fused S100A6 target protein precipitated (*see* **Note 4**).

4 Notes

1. Sub-nanogram levels of biotinylated S100A6 are readily detectable using a streptavidin-HRP-mediated ECL method (Fig. 1b). Rather than using biotinylated S100A6, ^{125}I-labeled A100A6 may be used for the conventional gel overlay method to detect S100A6/target interaction although it is time-consuming [9].
2. The S100A6-overlay method using an SDS-PAGE gel or blotted membrane is a rapid and sensitive method for analysis of S100A6/target interaction, including FOR20 and annexin XI (calcyclin-associated protein 50 kDa, CAP-50). However, some target proteins (such as PP5) cannot be detected using this method (Fig. 1c) [3, 7]. This is likely because of distinct binding modes of S100A6 target proteins, that is, some contain a short stretch of binding sequence (including FOR20, residues 1–30), which can interact with biotinylated S100A6 in a Ca^{2+}-dependent manner under denaturation conditions, and others (including PP5) that require a tetratricopeptide repeat motif for S100A6 binding [6] cannot be detected by the overlay method. Therefore, the GST-precipitation assay under native conditions should be performed in parallel.
3. To examine the real-time binding kinetics of S100A6 to its targets, such as PP5, SPR analysis using recombinant 6× histidine-tagged target protein is suitable. The sensorgram of SPR analysis (Fig. 3a) clearly shows the Ca^{2+}-dependent interaction of S100A6 with His-PP5. The binding curves of S100A6/His-PP5 are fitted to the 1:1 ligand-binding model. When we used other members of the S100 protein family, the kinetic parameters for each S100 protein, including association rate constants (k_a) and dissociation rate constants (k_d), were easily compared [6]. It has been shown that the dissociation constant of S100 protein family members for their targets is approximately 1 μM [10], consistent with that of S100A6 for His-PP5, as calculated from SPR analysis (Fig. 3c).
4. COS-7 cells are suitable for S100A6/target interaction experiments using the GST-precipitation method in intact cells, since S100A6 is overexpressed in these cells and is readily detected by immunoblot analysis in the cell extract (Fig. 2b, *middle panel*).

Acknowledgment

This work was supported in part by a grant (to H.T.) from the Ryobi Teien Memory Foundation, Japan.

References

1. Marenholz I, Heizmann CW, Fritz G (2004) S100 proteins in mouse and man: from evolution to function and pathology (including an update of the nomenclature). Biochem Biophys Res Commun 322:1111–1122
2. Calabretta B, Battini R, Kaczmarek L, de Riel JK, Baserga R (1986) Molecular cloning of the cDNA for a growth factor-inducible gene with strong homology to S-100, a calcium-binding protein. J Biol Chem 261:12628–12632
3. Tokumitsu H, Mizutani A, Minami H, Kobayashi R, Hidaka H (1992) A calcyclin-associated protein is a newly identified member of the Ca^{2+}/phospholipid-binding proteins, annexin family. J Biol Chem 267:8919–8924
4. Filipek A, Jastrzebska B, Nowotny M, Kuznicki J (2002) CacyBP/SIP, a calcyclin and Siah-1-interacting protein, binds EF-hand proteins of the S100 family. J Biol Chem 277:28848–28852
5. Leclerc E, Fritz G, Weibel M, Heizmann CW, Galichet A (2007) S100B and S100A6 differentially modulate cell survival by interacting with distinct RAGE (receptor for advanced glycation end products) immunoglobulin domains. J Biol Chem 282:31317–31331
6. Yamaguchi F, Umeda Y, Shimamoto S, Tsuchiya M, Tokumitsu H, Tokuda M, Kobayashi R (2012) S100 proteins modulate protein phosphatase 5 function: a link between Ca^{2+} signal transduction and protein dephosphorylation. J Biol Chem 287:13787–13798
7. Sakane K, Nishiguchi M, Denda M, Yamagchi F, Magari M, Kanayama N, Morishita R, Tokumitsu H (2017) Identification and characterization of a centrosomal protein, FOR20 as a novel S100A6 target. Biochem Biophys Res Commun 491:980–985
8. Billingsley ML, Pennypacker KR, Hoover CG, Brigati DJ, Kincaid RL (1985) A rapid and sensitive method for detection and quantification of calcineurin and calmodulin-binding proteins using biotinylated calmodulin. Proc Natl Acad Sci U S A 82:7585–7589
9. Tokumitsu H, Kobayashi R, Hidaka H (1991) A calcium-binding protein from rabbit lung cytosol identified as the product of growth-regulated gene (2A9) and its binding-proteins. Arch Biochem Biophys 288:202–207
10. Santamaria-Kisiel L, Rintala-Dempsey AC, Shaw GS (2006) Calcium-dependent and -independent interactions of the S100 protein family. Biochem J 396:201–214

Chapter 24

Preparation of the Oxidized and Reduced Forms of Psoriasin (S100A7)

Lisa S. Cunden and Elizabeth M. Nolan

Abstract

Human S100A7 (psoriasin) is a metal-chelating host-defense protein expressed by epithelial cells. S100A7 possesses two Cys residues that generate two redox isoforms of the protein. In the oxidized form ($S100A7_{ox}$), Cys47 and Cys96 form an intramolecular disulfide bond, whereas these residues exist as free thiols in the reduced form ($S100A7_{red}$). In this chapter, we provide a step-by-step protocol for the purification of $S100A7_{ox}$ and $S100A7_{red}$ that affords each protein in high yield and purity. In this procedure, S100A7 is expressed in *Escherichia coli* BL21(DE3), and the homodimer is obtained following ammonium sulfate precipitation, folding, and column chromatography. Treatment of S100A7 with 1,4-dithiothreitol (DTT) affords $S100A7_{red}$. A Cu(II)-catalyzed oxidation reaction is employed to obtain $S100A7_{ox}$. A RP-HPLC method that allows for baseline separation of $S100A7_{ox}$ and $S100A7_{red}$ is provided, as well as a biochemical Zn(II)-binding assay that can be employed to evaluate the functional integrity of S100A7.

Key words S100 protein, EF-hand, Bacterial expression, Zn(II)-binding protein, Disulfide bond

1 Introduction

Human S100A7 (psoriasin) is a metal-chelating host-defense protein that is a member of the S100 family of Ca(II)-binding proteins [1–4]. Its antimicrobial activity has been the subject of several investigations. S100A7 has been reported to be an *E. coli*-cidal factor on the skin, human tongue, and female genital tract [5–7] and to have the capacity to permeabilize bacterial membranes [8]. Furthermore, the growth inhibitory activity of S100A7 against select bacterial and fungal species has been attributed to Zn(II) withholding [9, 10]. S100A7 may also act as a solid-phase barrier at the surface of epithelial cells, preventing microbial invasion during wound healing [11]. These examples indicate that the contributions of S100A7 to host defense are complex and multifaceted, and we reason that further biochemical and biophysical

Claus W. Heizmann (ed.), *Calcium-Binding Proteins of the EF-Hand Superfamily: From Basics to Medical Applications*, Methods in Molecular Biology, vol. 1929, https://doi.org/10.1007/978-1-4939-9030-6_24,

studies of the protein will inform the molecular basis for how it performs known and as-yet unidentified biological functions.

In its native state, apo S100A7 exists as a 22-kDa homodimer composed of two non-covalently linked antiparallel monomer subunits (101 residues, 11 kDa) [12, 13]. Each S100A7 monomer contains two Ca(II)-binding EF-hand domains. The C-terminal EF-hand domain is referred to as "canonical" and it binds Ca(II) in a 7-coordinate geometry. The "noncanonical" N-terminal EF-hand coordinates Ca(II) in a 6-coordinate geometry [1, 14]. In addition to the Ca(II)-binding sites, S100A7 houses two His_3Asp motifs at the dimer interface, composed of residues His87 and His91 of one monomer and residues His18 and Asp25 of the other monomer. The His_3Asp sites each coordinate Zn(II) in a distorted tetrahedral geometry to afford a 2:1 Zn(II):S100A7 homodimer complex [13].

Each S100A7 polypeptide contains two Cys residues, Cys47 and Cys96, that can exist as free thiols ($S100A7_{red}$) or can form an intramolecular disulfide bond ($S100A7_{ox}$). Cys96, hence the disulfide bond, is in close proximity to His91 of the metal-binding motif [9, 12, 13], and disulfide bond formation and reduction are reported to affect the transition-metal-binding properties and antimicrobial activity of S100A7 [9, 10]. Furthermore, both $S100A7_{ox}$ and $S100A7_{red}$ have been detected in human specimens [5–7, 10], and biochemical studies support the notion that both redox forms of S100A7 are physiologically relevant [9, 10].

Herein, we present a procedure for the purification of $S100A7_{ox}$ and $S100A7_{red}$ and provide an overview of methods that can be employed to evaluate the redox state and integrity of the purified proteins. These methods are based on published biochemical studies of S100A7 [9]. Briefly, S100A7 is obtained in high yield and purity following recombinant expression in *E. coli* BL21 (DE3), two chromatography steps, and dialysis. The chromatography steps involve anion-exchange chromatography (AEC), followed by size-exclusion chromatography (SEC). $S100A7_{red}$ is obtained by dialyzing and storing the protein in the presence of DTT, and a procedure that involves the Cu(II)-catalyzed oxidation of thiols to disulfides is employed to obtain $S100A7_{ox}$. While unusual, this oxidation method has been utilized to generate intra- and intermolecular disulfide bonds in other proteins, including S100B and oxyhemoglobin [15–17]. The assays highlighted in this chapter include analytical RP-HPLC, mass spectrometry, thiol quantification, and a competition titration for determining the stoichiometry of Zn(II) binding by S100A7.

2 Materials

1. All solvents and chemicals are obtained from commercial suppliers and used as received.

2. All aqueous solutions are prepared using Milli-Q water (18.2 MΩ·cm, 0.22-μm filter).
3. Protein concentrations are routinely quantified by using the calculated extinction coefficients for the S100A7 homodimer (ProtParam, $\varepsilon_{280} = 9190\ M^{-1}\ cm^{-1}$ for S100A7$_{ox}$; $\varepsilon_{280} = 8940\ M^{-1}\ cm^{-1}$ for S100A7$_{red}$). All protein concentrations are for the S100A7 homodimer, and reported stoichiometries are for the S100A7 homodimer.

2.1 Commercial Material and Reagent Preparations

1. The S100A7 synthetic gene is obtained from ATUM (formerly DNA2.0).
2. The pET41a expression vector is obtained from Invitrogen.
3. To reduce metal-ion contamination (*see* **Note 1**), plastic spatulas are used to transfer reagents.
4. All metal stock solutions are prepared in acid-washed volumetric glassware and transferred to sterile polypropylene tubes for long-term storage.
5. Stock solutions of Zn(II) (100 mM, 100 mL) and Cu(II) (100 mM, 100 mL) are prepared from 99.999% anhydrous $ZnCl_2$ and 99.999% $CuCl_2$, respectively, and Milli-Q water (*see* **Note 2**). Working solutions are prepared by diluting the stock solutions in Milli-Q water.
6. Stock solutions of Ca(II) (1 M, 100 mL) are prepared from 99.999% $CaCl_2$ and Milli-Q water. Working solutions are prepared by diluting the stock solutions in Milli-Q water.
7. Reagents employed for the indicated purposes were purchased from the following suppliers:

Purpose	Reagent	Supplier
Protein overexpression and purification	Luria-Bertani (LB) 99.5% HEPES TraceSELECT NaCl Guanidinium hydrochloride (GuHCl) Dithiothreitol (DTT) EDTA disodium salt dihydrate Chelex resin	Becton Dickinson MilliporeSigma MilliporeSigma MilliporeSigma VWR Life Science VWR Life Science Bio-Rad
Solution studies	Ultrol grade HEPES (free acid) Tris hydrochloride DMSO Zincon monosodium salt (*see* **Note 3**)	Calbiochem J. T. Baker MilliporeSigma MilliporeSigma
pH adjustment	TraceSELECT aqueous NaOH	Fisher Scientific

2.2 Zincon Preparation

For the Zn(II) competition titration experiment, the competitor Zincon is employed [18–20].

1. Stock solutions (≈10 mM) of Zincon are prepared in methanol, aliquoted into 200-μL portions, and lyophilized to dryness (*see* **Note 3**).
2. Each aliquot is thawed only once and dissolved in 200 μL of Milli-Q water.
3. The Zincon concentration is determined (≈10 mM) by preparing a sample of Zn(II)-bound Zincon (≈20 μM) and using the reported extinction coefficient of this complex at 621 nm [18–20].

2.3 Buffer for Protein Purification

Buffers employed for the purification of S100A7 are prepared with the compositions described below and sterile-filtered through a 0.22-μm bottle-top filter prior to use.

Buffer (1 L)	Composition
Lysis buffer A	50 mM Tris, 100 mM NaCl, 1 mM EDTA, 0.5% Triton X-100, pH 8.0, 5 mM DTT, 1 mM PMSF[a]
Lysis buffer B	50 mM Tris, 100 mM NaCl, 4 M GuHCl, pH 8.0
Dialysis buffer	20 mM HEPES, pH 8.0
Mono Q buffer A	20 mM HEPES, pH 8.0
Mono Q buffer B	20 mM HEPES, 1 M NaCl, pH 8.0
S75 buffer	20 mM Tris, 100 mM NaCl, pH 7.5
Metal-binding buffer	75 mM HEPES, 100 mM NaCl, pH 7.0

[a]PMSF stock solutions (100 mM) are prepared in ethanol, stored at −20 °C, and added to lysis buffer A prior to lysis

2.4 Equipment for Protein Purification

1. An ÄKTA purifier (GE Lifesciences) housed at 4 °C and outfitted with a 150-mL Superloop (GE Lifesciences) is employed for chromatographic purification.
2. S100A7 is purified in two chromatographic steps: anion-exchange chromatography (AEC) is first performed using a Mono Q 10/100 GL column (GE Lifesciences), followed by size-exclusion chromatography (SEC) using a Superdex 75 26/600 GL column (GE Lifesciences).

2.5 Equipment for Optical Absorption Spectroscopy

1. Optical absorption spectra are collected on a Beckman Coulter DU 800 spectrophotometer thermostatted at 25 °C with a Peltier temperature controller.

2. Quartz cuvettes (1-cm path length, Starna) are employed for all optical absorption measurements of samples of purified proteins.
3. Plastic cuvettes (1-cm path length, VWR) are employed for monitoring bacterial culture growth by the culture optical density at 600 nm (OD_{600}).
4. Observed OD_{600} values are reported throughout this protocol as benchmarks; these values will vary from instrument to instrument.

3 Methods

The general methods described in this section can also be used to purify S100A7 variants obtained by site-directed mutagenesis and S100A15 (*see* **Note 4**). S100A15 shares high sequence homology with S100A7, and the S100A15 homodimer has two His_3 sites that coordinate Zn(II) [21–24].

3.1 Design of the Synthetic Gene for S100A7

A synthetic gene for human S100A7 optimized for *E. coli* codon usage includes an N-terminal *Nde*I restriction site (N-terminal Met residue encoded by the *Nde*I site) and a C-terminal stop codon followed by an *Xho*I restriction site.

E. coli optimized nucleotide sequence **NdeI**-S100A7-Stop-**XhoI:** **CATATG**AGCAACACCCAGGCAGAACGTAGCATTA TTGGTATGATTGACATGTTTCACAAATACACGCGCCGTG ATGATAAGATCGACAAACCGTCGCTGCTGACGATGATG AAAGAGAACTTCCCGAATTTTCTGTCTGCCTGCGATAA GAAAGGCACCAATTATCTGGCGGACGTGTTCGAAAAG AAAGACAAAAACGAGGACAAGAAGATCGACTTTAGCG AGTTCTTGTCCCTGCTGGGTGATATCGCGACCGATTA CCATAAGCAAAGCCACGGCGCTGCGCCGTGTAGCGG TGGTAGCCAGTAA**CTCGAG**

Translated sequence for **NdeI**-S100A7-Stop-**XhoI**: **H M** S N T Q A E R S I I G M I D M F H K Y T R R D D K I D K P S L L T M M K E N F P N F L S A C D K K G T N Y L A D V F E K K D K N E D K K I D F S E F L S L L G D I A T D Y H K Q S H G A A P C S G G S Q Stop **L E**

3.2 Subcloning of the Human S100A7 Gene

1. The codon-optimized gene is obtained in the pJ201 vector from DNA 2.0.
2. The gene is amplified by PCR, digested with *Nde*I/*Xho*I, and ligated into the pET41a vector with T4 DNA ligase to afford pET41a-*S100A7*, and the resulting plasmids are transformed into chemically competent *E. coli* TOP10 cells. This plasmid affords full-length S100A7 with no tags attached.

3. The pET41a-*S100A7* expression plasmids (verified by DNA sequencing) are transformed into chemically competent *E. coli* BL21(DE3) cells for protein overexpression.
4. Cell stocks of *E. coli* BL21(DE3) containing pET41a-*S100A7* are prepared by growing *E. coli* in LB containing 50 μg/mL kanamycin to saturation. Aliquots are frozen in 25% glycerol and stored at −80 °C.

3.3 Purification of Human S100A7

The protein purification is described for an overexpression carried out in 2 L of culture, and the procedure affords protein yields that range from ≈30 to 50 mg/L.

Day 1: Plating of E. coli BL21(DE3)

1. Using an inoculating loop, streak *E. coli* BL21(DE3) that contain pET41a-*S100A7* onto a LB/agar plate containing 50 μg/mL kanamycin, and incubate the plate for ≈20 h at 37 °C.

Day 2: Overnight culture

2. To a sterile 250-mL baffled flask, add 40 mL of LB medium supplemented with 50 μg/mL kanamycin, inoculate the medium with a single colony from the LB/agar plate, and incubate the culture overnight (14–16 h, 37 °C, 150 rpm).
3. Prepare and autoclave 2 L of LB medium in a 4-L flask, and incubate the medium overnight at 37 °C.

Day 3: Overexpression

4. Measure the optical density of the overnight culture at 600 nm (OD_{600}), which should be ≈1.5–2.5.
5. Add 50 μg/mL kanamycin (final concentration, 1:100 dilution from a 50-mg/mL stock solution in water) into the 4-L baffled flask containing 2 L of LB medium, and dilute 0 mL (1:100 dilution) of the overnight culture into the medium. Transfer the resulting culture to an incubator shaker (37 °C, 150 rpm).
6. Record the OD_{600} of the bacterial culture every 30 min, and induce overexpression at $OD_{600} \approx 0.4$ by adding IPTG to a final concentration of 125 μM. The cultures should reach an $OD_{600} \approx 0.4$ after ≈1.5 h if the LB medium was pre-warmed overnight.
7. Incubate the culture (37 °C, 150 rpm) for an additional 4–4.5 h after induction, and measure the OD_{600} of the cell culture, which should be ≈3.
8. Harvest the cells by centrifugation (3210 × *g*, 15 min, 4 °C), and discard the supernatant.

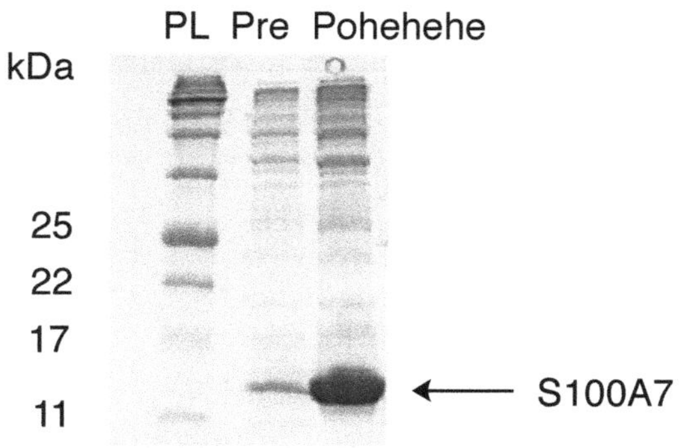

Fig. 1 Expression of S100A7. SDS-PAGE gel (15% Tris– tricine gel) of pre- and post-induction samples of S100A7. Pre-induction sample (Pre); post-induction sample (Post); protein ladder (PL) is P7712s color prestained protein standard broad range (New England Biolabs)

9. Transfer the cell pellet to a sterile 50-mL centrifuge tube. Flash freeze the pellet in liquid nitrogen, and store at −80 °C until future use.

   This procedure typically yields ca. 4 g of cells / 2 L of culture (wet weight), which can be stored at −80 °C for up to 4 months. The overexpression can be evaluated by SDS-PAGE analysis (Coomassie stain) of the pre- and post-induction cell samples (whole cell lysate) using a 15% Tris-tricine gel. Successful overexpression should show an intense band at ≈11 kDa in the post-induction sample, corresponding to the S100A7 monomer (Fig. 1).

Days 4–5: Cell lysis and protein refolding.

The following steps should be carried out at 4 °C. Buffers should be equilibrated and maintained at 4 °C or on ice).

10. Thaw a cell pellet from a 2-L culture on ice, and resuspend the pellet in 100 mL of lysis buffer A.
11. Transfer the resuspension to an ice-cold stainless-steel beaker, and sonicate the resuspension on ice at 40% amplitude for 5 min (30 s on, 10 s off).
12. Centrifuge the resulting cell lysate (20,200 × *g*, 10 min, 4 °C).
13. Transfer the supernatant, which contains soluble aggregated S100A7, to a 150-mL glass beaker housing a magnetic stir bar that is maintained at 4 °C.
14. Weigh out 38 g of ammonium sulfate, and slowly add the entire portion to the supernatant over 15–20 min with constant stirring. This step yields a 60% saturated ammonium sulfate solution and causes contaminating proteins to precipitate (*see* **Note 5**).

S100A7 is soluble in up to 80% saturated ammonium sulfate and remains in the soluble portion of the mixture.

15. Once the entire portion of ammonium sulfate has been added, allow the solution to stir gently at 4 °C for 60 min.
16. Centrifuge the mixture (20,200 × *g*, 10 min, 4 °C), and quickly transfer the supernatant to an ice-cold clean glass beaker. If necessary, use a plastic spatula to remove any greasy clumps.
17. Weigh out 25 g of ammonium sulfate, and slowly add the entire portion to the supernatant over 10 min while stirring to bring the final saturation level of ammonium sulfate to 90%.

    S100A7 precipitates at an ammonium sulfate saturation level of 90%.
18. Once the entire portion of ammonium sulfate has been added, allow the solution to stir gently at 4 °C for 60 min.
19. Centrifuge the mixture (20,200 × *g*, 10 min, 4 °C). Quickly decant the supernatant to prevent the pellet, which contains S100A7, from redissolving.
20. Keep the pellet at 4 °C, and add a 100-mL portion of lysis buffer B to resolubilize the pellet. Incubate the mixture on ice or in a cold room with occasional and gentle swirling to dissolve the S100A7 pellet.

    This step typically requires ca. 30 min and affords S100A7 in a soluble and denatured state. The following step allows for the refolding of S100A7.
21. Transfer the resulting supernatant, which contains denatured S100A7, to a dialysis bag (Spectrapor3 10 KDa MWCO) for 3 rounds of dialysis against 4 L of dialysis buffer for >12 h each. Include 1 mM DTT in the dialysis buffer for the first round of dialysis, but not for the latter rounds (*see* **Note 6**).

    After dialysis, the refolded protein is purified by AEC and SEC—*see* days 6–7.
22. The day before the chromatography steps, pre-equilibrate the S75 column with 1 column volume (CV) of S75 buffer (*see* **Note 7**).

Days 6–7: Protein purification using an FPLC ÄKTA purifier (AEC and SEC).

Over the course of the refolding process described above, a small portion of S100A7 forms an insoluble aggregate and accumulates at the bottom of the dialysis bag as a white precipitate.

23. Pellet the precipitated protein from the dialysate by centrifugation (20,200 × *g*, 10 min, 4 °C). Decant and vacuum filter the supernatant using a 0.22-μm bottle-top filter (*see* **Note 8**).
24. Load the filtrate (≈120 mL) into the Superloop.

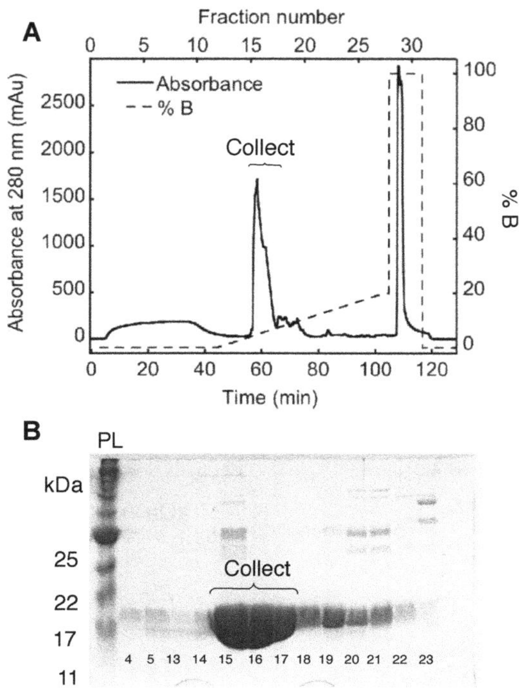

Fig. 2 Elution profile of S100A7 from the Mono Q column. (**a**) S100A7 was eluted over 12 CV with a gradient of 0–10% Mono Q buffer B at a flow rate of 2 mL/min. Fractions 15–17 were collected and pooled for the next chromatography step. (**b**) SDS-PAGE gel (15% Tris–tricine gel) of select fractions from the Mono Q purification step. Protein Ladder (PL): P7712s color prestained protein standard broad range (New England Biolabs). As a precaution, the flow through is routinely collected, which corresponds to fractions 1–10

25. Wash the Mono Q column with 2 CV of Milli-Q water, followed by equilibration with 2 CV of Mono Q buffer A, 2 CV of Mono Q buffer B, and 2 CV of Mono Q buffer A (2 mL/min flow rate).
26. Load ≈60 mL of the protein solution from the Superloop onto the Mono Q column, and elute the proteins with a gradient of 0–10% Mono Q buffer B over 12 CV at a flow rate of 2 mL/min (Fig. 2). This step should take between 110 and 130 min depending on the injection volume. The fractions corresponding to the folded S100A7 homodimer elute at ≈5% B (Fig. 2).
27. Repeat **steps 25** and **26**, loading the rest of the protein sample from the Superloop.
28. Collect and pool the fractions containing dimeric S100A7 (Fig. 2), and concentrate the protein by centrifugation to a final volume of ≈10 mL (15-mL Amicon 10-kDa MWCO spin

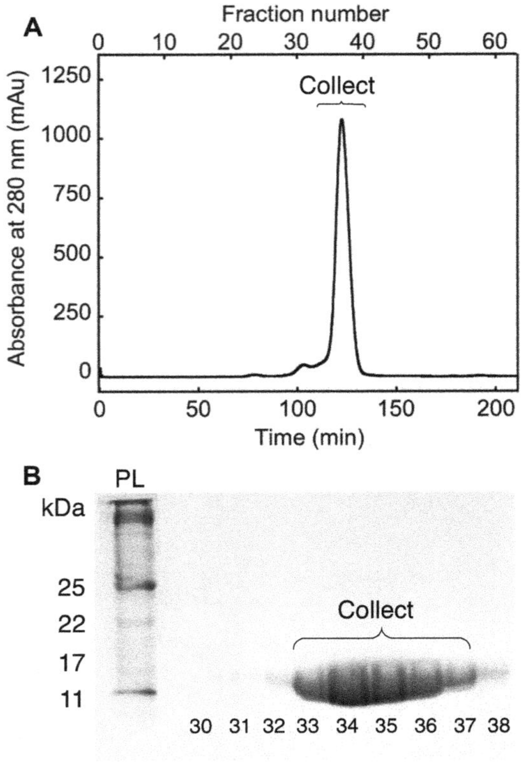

Fig. 3 Elution profile of S100A7 from the S75 column. (**a**) S100A7 was eluted over 1 CV in S75 buffer at a flow rate of 1 mL/min. Fractions 33–37 were collected and pooled for the subsequent step. (**b**) SDS-PAGE gel (15% Tris–tricine gel) of select fractions from the S75 purification step. Protein Ladder (PL): P7712s color prestained protein standard broad range (New England Biolabs)

filter). This step typically requires 2–3 rounds of centrifugation (3210 × *g*, 20 min per round, 4 °C).

29. Load the concentrated S100A7 sample onto the pre-equilibrated S75 column (*see* **Note 7**), and elute the protein over 1 CV at a flow rate of 1 mL/min.

30. Collect and pool the fractions containing S100A7 (Fig. 3), and measure the protein concentration using the calculated extinction coefficient for S100A7$_{red}$ (Subheading 2). At this stage, the protocol should yield ≈20–25 mL of ≈90 μM protein that is predominantly S100A7$_{red}$ (Fig. 4).

    Steps 31–35 describe the step to obtain homogeneous S100A7red and S100A7ox (*see* **Note 9**).

31. For S100A7$_{red}$, dialyze the purified protein against 1 L of metal-binding buffer, containing 1 mM DTT and ≈10 g Chelex resin (Biorad), at 4 °C for ≈12 h (Spectrapor3 3500 Da MWCO dialysis bag). Analytical HPLC (Table 1, Fig. 4) may

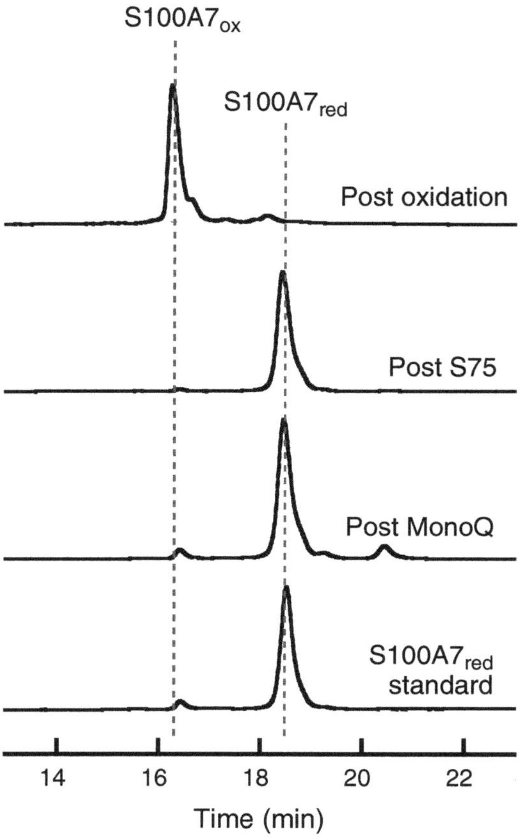

Fig. 4 Analytical HPLC traces showing S100A7 over the course of the purification and Cu(II)-catalyzed oxidation. A 45-μL aliquot of 30 μM protein was mixed with 100 μL of 6 M GuHCl and 5 μL of 6% aqueous TFA. The sample was centrifuged, and a 120-μL portion was transferred to an HPLC vial. A 100-μL aliquot was injected onto the HPLC outfitted with a C4 column. A gradient of 35–50% B over 30 min at 0.5 mL/min was employed. Solvent A, water +0.1% TFA; solvent B, MeCN +0.1% TFA

be employed to ascertain the speciation of S100A7 and confirm it is fully reduced prior to the next steps.

32. The dialysate containing $S100A7_{red}$ is then passed through a 0.22-μm syringe filter to remove any contaminating Chelex resin.
33. Go to **step 38** for the storage of $S100A7_{red}$.
34. For the Cu(II)-catalyzed oxidation to obtain $S100A7_{ox}$, dilute the purified protein from **step 30** ≈4-fold to yield ≈100 mL of ≈20 μM protein in S75 buffer (equilibrated at room temperature), and transfer the solution to a 150-mL glass beaker.
35. Add three equivalents of Cu(II) (60 μL from a 100-mM Cu(II) stock solution for 100 mL of 20 μM S100A7) dropwise to the

Table 1
Characterization of S100A7

Protein	HPLC retention time (min)[a]	Free thiol[b]	Calculated mass (g/mol)	Observed mass (g/mol)[c]
$S100A7_{ox}$	16.4	0.11 ± 0.03	11454.9	11455.4 11324.0 (−Met1)
$S100A7_{red}$	18.4	4.16 ± 0.45	11456.9	11457.4 11326.0 (−Met1)

[a]Retention times determined by analytical RP-HPLC. A 45-μL aliquot of 30 μM protein was mixed with 100 μL of 6 M GuHCl and 5 μL of 6% aqueous TFA. The sample was centrifuged (5 min, 20,200 × *g*, 4 °C), and a 120-μL portion of supernatant was transferred to an HPLC vial. Analytical RP-HPLC was performed using a Proto C4 column (5-μm pore, 4.6 × 250 mm, Higgins Analytical Inc.), a 100-μL injection volume, a gradient of 35–50% B over 30 min, and a flow rate of 1 mL/min (solvent A, water +0.1% TFA; solvent B, MeCN +0.1% TFA)

[b]Free thiol content determined by using the DTDP assay [9]. The reported errors are the SDM ($n = 2$)

[c]High-resolution mass spectrometry was performed on an Agilent 1290 series LC system equipped with an Agilent 6230 TOF system housing Agilent Jetstream ESI source. An Agilent Poroshell 300SB-C18 column (5-μm pore) was employed with a flow rate of 0.2 mL/min (solvent A, water +0.1% formic acid; solvent B, MeCN +0.1% formic acid). Protein samples (10 μM) were prepared in Milli-Q water, and 1 μL was injected for each analysis

protein solution, and incubate the mixture at room temperature for 2 h with stirring (*see* **Note 10**). Some precipitate may form over the course of the oxidation and can be removed by centrifugation (3210 × *g*, 10 min, 4 °C). After 2 h, analyze a reaction aliquot by analytical RP-HPLC to confirm that the reaction reached completion (Fig. 4).

36. Transfer the resulting solution of $S100A7_{ox}$ to a Spectrapor3 3500 MWCO dialysis bag, and dialyze overnight against 1 L of metal-binding buffer containing ≈10 g Chelex resin and 1 mM EDTA at 4 °C.
37. Remove the EDTA by two rounds of dialysis against the metal-binding buffer (2 × 4 L, ≈12 h per dialysis, 4 °C).
38. Following dialysis, concentrate $S100A7_{red}$ and $S100A7_{ox}$ using a 15-mL Amicon 10-kDa MWCO spin filter (3210 × *g*, 4 °C) to >500 μM, partitioned into sterile microcentrifuge tubes as 50-μL aliquots, flashed frozen in liquid nitrogen, and stored at −80 °C until further use. Prior to any experiments, the proteins are analyzed by SDS-PAGE and analytical RP-HPLC (Fig. 5) (*see* **Note 11**).

3.4 Zinc Competition Titration Between S100A7 and Zincon

This procedure allows for the Zn(II):S100A7 homodimer stoichiometry to be ascertained. The titration should provide a 2:1 stoichiometry. This titration can be performed at 25 °C or room temperature. Experiments with Zincon are performed in the dark.

1. In an acid-washed quartz cuvette, prepare a 2-mL solution containing Zincon (20 μM) from the Zincon stock solution in the metal-binding buffer, ±2 mM Ca(II), and record the

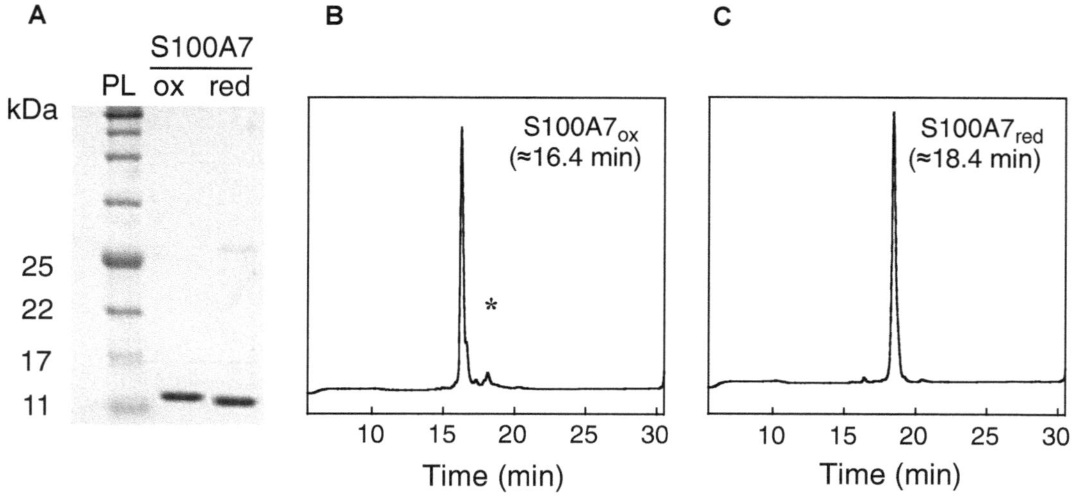

Fig. 5 SDS-PAGE purity gel and analytical HPLC traces of S100A7$_{ox}$ and S100A7$_{red}$. (**a**) SDS-PAGE purity gel (15% Tris–tricine gel) of S100A7$_{ox}$ and S100A7$_{red}$. Protein Ladder (PL): P7712s color prestained protein standard broad range (New England Biolabs). (**b** and **c**) HPLC chromatograms (220 nm absorption) of S100A7$_{ox}$ and S100A7$_{red}$, respectively. A 45-μL aliquot of 30 μM protein was mixed with 100 μL of 6 M GuHCl and 5 μL of 6% aqueous TFA. The sample was centrifuged, and a 120-μL portion was transferred to an HPLC vial. A 100-μL aliquot was injected onto the HPLC outfitted with a C4 column. A gradient of 35–50% B over 30 min at 0.5 mL/min was employed. Solvent A, water +0.1% TFA; solvent B, MeCN +0.1% TFA. An asterisk (*) indicates the position of the isoform of S100A7 missing the N-terminal methionine

optical absorption spectrum. Check that the spectrum agrees with the literature spectrum of the metal-free Zincon [20].

2. Add an aliquot of the S100A7 stock solution (10 μM final concentration, reduced or oxidized form) to the Zincon solution, gently swirl the cuvette to mix the sample, and record the optical absorption spectrum of this mixture. Provided a negligible dilution factor, the Zincon spectrum should be unchanged in the absence and presence of apo S100A7.
3. Titrate the mixture containing Zincon and S100A7 with Zn(II) by adding Zn(II) in increments of 2.5 μM until the Zn(II) concentration is 40 μM (1 μL from a 5-mM Zn(II) stock solution per addition). After each Zn(II) addition, gently swirl the cuvette, and allow the sample to equilibrate for 2 min before the optical absorption spectrum is collected.
4. Plot the absorbance at 621 nm versus the [Zn(II)]/[S100A7] ratio (Fig. 6).

For the competition titration with S100A7$_{red}$, the thawed protein sample must be immediately buffer-exchanged to remove DTT (3 × 5 min, 20,200 × *g*, 4 °C), mixed with Zincon, and then titrated with Zn(II). We observed no air-oxidation of S100A7$_{red}$ over the course of the titration (≈30 min) as ascertained by analytical RP-HPLC.

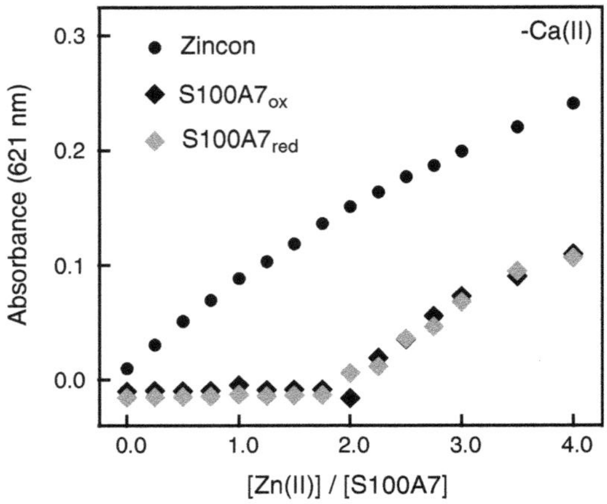

Fig. 6 S100A7 outcompetes Zincon for two equivalents of Zn(II). Representative plot showing the response of 20 μM Zincon to Zn(II) in the presence of 10 μM S100A7 proteins in metal-binding buffer at 25 °C. S100A7 also binds Ca(II) ions at EF-hand domains and these titrations were performed without Ca(II) added to the metal-binding buffer

3.5 Biochemical Characterization of Human S100A7

Following the purification of S100A7, we recommend employing standard biochemical techniques to evaluate the integrity of the purified protein [9]. These methods include SDS-PAGE, mass spectrometry, metal analysis by inductively coupled plasma mass spectrometry (ICP-MS), analytical RP-HPLC, and thiol quantification assays using DTDP (Table 1, Fig. 5). The experimental conditions for these procedures, as well as circular dichroism spectroscopy and inductively coupled plasma mass spectrometry, are found elsewhere [9].

A competition titration between S100A7 and the colorimetric Zn(II) indicator Zincon ($K_{d,Zn} \approx 10$ μM) [18–20] should be carried out to confirm that the purified protein has the correct Zn(II)-binding stoichiometry (Subheading 3.4). Zincon is a colorimetric probe that displays an increase in absorbance at 621 nm upon binding Zn(II) (Fig. 6). When Zn(II) is titrated to a solution mixture containing Zincon and S100A7, both S100A7$_{ox}$ and S100A7$_{red}$ outcompete Zincon for two equivalents of Zn(II). This behavior is expected because the protein has two His_3Asp sites that coordinate Zn(II) with sub-nanomolar affinity [9]. If the competition shows that S100A7 outcompetes Zincon for less than two equivalents of Zn(II), then the purified protein may have a metal contamination.

4 Notes

1. Because S100A7 is a metal-binding protein, we employ high-purity buffer reagents and plastic or Teflon-coated spatulas to minimize the risk of metal contamination from buffers. These

precautions, as well as extensive dialysis against Chelex resin (and EDTA for $S100A7_{ox}$ obtained by Cu(II)-catalyzed oxidation), afford apo S100A7 in high yield and purity.

2. $ZnCl_2$ is hygroscopic and is therefore stored in a desiccator.
3. We have found that the quality of the Zincon reagent depends on the supplier. We recommend purchasing Zincon monosodium salt from MilliporeSigma. Furthermore, we have observed decomposition of the compound when stored at −20 °C in DMSO. If stored as a stock solution at −20 or −80 °C, we recommend checking the integrity of the Zincon stock solution by optical absorption spectroscopy prior to use. We also recommend storing Zincon as a powder at −80 °C if the reagent will be used infrequently, and making a new stock solution (10 mM) immediately prior to each experiment.
4. The expression and purification procedure described herein can be adapted for other S100A7 variants as follows:
   (a) For non-Cys containing variants of S100A7, including S100A7-Ala and S100A7-Ser that contain Cys47→Ser/Ala and Cys96→Ser/Ala mutations [9], the protocol for purification of $S100A7_{red}$ can be followed, with the addition of DTT in the first dialysis step after SEC (**step 31**) omitted.
   (b) We have designed and evaluated S100A7Δ, a variant of S100A7 that lacks the His_3Asp metal-binding sites because the four residues that comprise the His_3Asp motif of S100A7 (His18, Asp25, His87, and His91) have been replaced by noncoordinating Ala residues [9]. $S100A7\Delta_{ox}$ can be obtained by modifying the Cu(II)-catalyzed oxidation protocol for S100A7 such that two equivalents of Cu(II) are employed rather than three equivalents. During the oxidation reaction, a precipitate will slowly appear. Following a 2-h incubation at room temperature, the precipitate is removed by centrifugation, and the resulting soluble protein-Cu mixture is allowed to stir overnight at room temperature. This procedure results in the complete oxidation of S100A7Δ to $S100A7\Delta_{ox}$.
   (c) The protocols described for preparing $S100A7_{ox}$ and $S100A7_{red}$ can be used to obtain $S100A15_{ox}$ and $S100A15_{red}$ (unpublished work). The Cys residues of S100A7 and S100A15 are conserved.
5. This step should be carried out with caution by slow addition of ammonium sulfate in a scoop-by-scoop manner and gentle stirring, in order to minimize S100A7 precipitation.

6. For the dialysis procedure described in **step 21**, DTT is preferred as a reducing agent over TCEP because it is less expensive and because more protein precipitation occurs during the refolding process when TCEP is employed. Furthermore, we find that including 1-mM DTT only in the first round of dialysis minimizes protein precipitation during refolding, as well as formation of an S100A7 adduct (observed by SDS-PAGE) that appears to form and persist when DTT is included in two or more rounds of dialysis (data not shown). We suspect that the identity of the protein adduct is S100A7-DTT.
7. We chose to purify S100A7 in Tris buffer for the SEC step (S75 buffer; 20 mM Tris, 100 mM NaCl, pH 7.0) because we observed significant precipitation during the subsequent Cu(II)-catalyzed oxidation step when using HEPES buffer.
8. We recommend using vacuum filtration over gravity filtration for **step 23** because the sample tends to clog the filter membrane, making gravity filtration more time-consuming.
9. The protein collected after SEC can be split into two batches for subsequent purification of $S100A7_{ox}$ and $S100A7_{red}$. The volume afforded in **step 34** (≈100 mL of 20 μM S100A7) assumes that the protein batch is not split at **step 30**.
10. Upon addition of Cu(II), the clear and colorless solution turns yellowish.
11. The metal content of purified $S100A7_{ox}$ can be analyzed by inductively coupled plasma mass spectrometry (ICP-MS). We routinely find that $S100A7_{ox}$ contains ≈0.03 and ≈0.04 equivalents of Cu and Zn, respectively. These results indicate that the dialysis step removed the Cu(II) introduced during the oxidation step.

Acknowledgment

Our current studies of psoriasin are supported by the National Science Foundation (CHE-1352132).

References

1. Donato R, Canon BR, Sorci G et al (2013) Functions of S100 proteins. Curr Mol Med 13(1):24–57
2. Zackular JP, Chazin WJ, Skaar EP (2015) Nutritional immunity: S100 proteins at the host-pathogen interface. J Biol Chem 290(31):18991–18998
3. Cunden LS, Nolan EM (2018) Bioinorganic explorations of Zn(II) sequestration by human S100 host-defense proteins. Biochemistry 57(11):1673–1680
4. Jinquan T, Vorum H, Larsen CG et al (1996) Psoriasin: a novel chemotactic protein. J Invest Dermatol 107(1):5–10

5. Meyer JE, Harder J, Sipos B et al (2008) Psoriasin (S100A7) is a principal antimicrobial peptide of the human tongue. Mucosal Immunol 1 (3):239–243
6. Mildner M, Stichenwirth M, Abtin A et al (2010) Psoriasin (S100A7) is a major *Escherichia coli*-cidal factor of the female genital tract. Mucosal Immunol 3(6):602–609
7. Gläser R, Harder J, Lange H et al (2005) Antimicrobial psoriasin (S100A7) protects human skin from *Escherichia coli* infection. Nat Immunol 6(1):57–64
8. Michalek M, Gelhaus C, Hecht O et al (2009) The human antimicrobial protein psoriasin acts by permeabilization of bacterial membranes. Dev Comp Immunol 33(6):740–746
9. Cunden LS, Brophy MB, Rodriguez GE et al (2017) Biochemical and functional evaluation of the intramolecular disulfide bonds in the zinc-chelating antimicrobial protein human S100A7 (Psoriasin). Biochemistry 56 (43):5726–5738
10. Hein KZ, Takahashi H, Tsumori T et al (2015) Disulphide-reduced psoriasin is a human apoptosis-inducing broad-spectrum fungicide. Proc Natl Acad Sci U S A 112 (42):13039–13044
11. Lee KC, Eckert RL (2007) S100A7 (Psoriasin)—mechanism of antibacterial action in wounds. J Invest Dermatol 127(4):945–957
12. Brodersen DE, Etzerodt M, Madsen P et al (1998) EF-hands at atomic resolution: the structure of human psoriasin (S100A7) solved by MAD phasing. Structure 6(4):477–489
13. Brodersen DE, Nyborg J, Kjeldgaard M (1999) Zinc-binding site of an S100 protein revealed. Two crystal structures of Ca^{2+}-bound human psoriasin (S100A7) in the Zn^{2+}-loaded and Zn^{2+}-free states. Biochemistry 38 (6):1695–1704
14. Gifford JL, Walsh MP, Vogel HJ (2007) Structures and metal-ion-binding properties of the Ca^{2+}-binding helix-loop-helix EF-hand motifs. Biochem J 405(2):199–221
15. Lee ISM, Suzuki M, Hayashi N et al (2000) Copper-dependent formation of disulfide-linked dimer of S100B protein. Arch Biochem Biophys 374(2):137–141
16. Harrison CA, Raftery MJ, Walsh J et al (1999) Oxidation regulates the inflammatory properties of the murine S100 protein S100A8. J Biol Chem 274(13):8561–8569
17. Smith RC, Reed VD (1992) Inhibition by thiols of copper(II)-induced oxidation of oxihemoglobin. Chem Biol Interact 82 (2):209–217
18. Talmard C, Bouzan A, Faller P (2007) Zinc binding to amyloid-β: Isothermal titration calorimetry and Zn competition experiments with Zn sensors. Biochemistry 46(47): 13658–13666
19. Maret W, Vallee BL (1998) Thiolate ligands in metallothionein confer redox activity on zinc clusters. Proc Natl Acad Sci U S A 95 (7):3478–3482
20. Kocyla A, Pomorski A, Krężel A (2017) Molar absorption coefficients and stability constants of Zincon metal complexes for determination of metal ions and bioinorganic applications. J Inorg Biochem 176:53–65
21. Wolf R, Mirmohammadsadegh A, Walz M et al (2003) Molecular cloning and characterization of alternatively spliced mRNA isoforms from psoriatic skin encoding a novel member of the S100 family. FASEB J 17(13):1969–1971
22. Büchau AS, Hassan M, Kukova G et al (2007) S100A15, an antimicrobial protein of the skin: regulation by *E. coli* through Toll-like Receptor 4. J Invest Dermatol 127(11):2596–2604
23. Wolf R, Ruzicka T, Yuspa SH (2011) Novel S100A7 (psoriasin)/S100A15 (koebnerisin) subfamily: highly homologous but distinct in regulation and function. Amino Acids 41 (4):789–796
24. Murray J, Tonkin ML, Whiting AL et al (2012) Structural characterization of S100A15 reveals a novel zinc coordination site among S100 proteins and altered surface chemistry with functional implications for receptor binding. BMC Struc Biol 12:16

Chapter 25

Preparation and Iron Redox Speciation Study of the Fe(II)-Binding Antimicrobial Protein Calprotectin

Rose C. Hadley and Elizabeth M. Nolan

Abstract

Calprotectin (CP, S100A8/S100A9 heterooligomer) is an abundant metal-sequestering host-defense protein expressed by neutrophils, other white blood cells, and epithelial cells. The apoprotein is a S100A8/S100A9 heterodimer that contains two sites for transition metal binding at the S100A8/S100A9 interface: a His_3Asp motif (site 1) and a His_6 motif (site 2). In this chapter, we provide a step-by-step protocol for the overexpression and purification of the human and murine orthologues of CP that affords each apo heterodimer in high yield and purity. In these procedures, the S100A8 and S100A9 subunits are overexpressed in *Escherichia coli* BL21(DE3), and each apo heterodimer is obtained following cell lysis, folding, column chromatography, and dialysis against Chelex resin to reduce metal contamination. Recent studies demonstrated that human CP coordinates Fe(II) and that the protein affects the redox speciation of Fe in solution. An Fe redox speciation assay employing ferrozine is described that demonstrates the ability of both the human and murine orthologues of CP to shift the redox speciation of Fe from the ferric to the ferrous oxidation state over time.

Key words S100 protein, EF-hand, Bacterial expression, Fe(II)-binding protein, Iron redox speciation

1 Introduction

Calprotectin (CP, S100A8/S100A9 heterooligomer) is a member of the S100 family of Ca(II)-binding proteins and a topic of interest in the metal homeostasis community because of its contributions to the metal-withholding innate immune response, which is often termed nutritional immunity [1, 2]. CP is a cytoplasmic protein produced by neutrophils, macrophages, monocytes, and epithelial cells. In response to infection, CP is released into the extracellular space where it sequesters essential nutrient metal ions to limit microbial growth. Our current understanding of its metal-sequestering function largely comes from mouse models of infectious disease [3–8], including studies that involve a $S100A9^{-/-}$ or

Claus W. Heizmann (ed.), *Calcium-Binding Proteins of the EF-Hand Superfamily: From Basics to Medical Applications*, Methods in Molecular Biology, vol. 1929, https://doi.org/10.1007/978-1-4939-9030-6_25,

CP knockout mouse [9], and molecular characterization of the recombinant human protein [10–20]. A recombinant expression and purification of murine CP was recently reported, which provides a foundation for future evaluation of this orthologue [21].

Biochemical, biophysical, and structural studies of human CP (hCP) demonstrated that the protein is a heterooligomer of S100A8 and S100A9 [10, 11, 22, 23]. Each subunit possesses two EF-hand Ca(II)-binding sites: a canonical C-terminal EF-hand and a noncanonical N-terminal EF-hand [24]. In the absence of Ca (II) and transition metal ions, hCP is a heterodimer [10]. Two transition-metal-binding sites form at the dimer interface [11–18, 20, 25]. Site 1 is a His_3Asp motif composed of (A8)His83, (A8) His87, (A9)His20, and (A9)Asp30 [11]. Site 2 is a His_6 motif composed of (A8)His17, (A8)His27, (A9)His91, (A9)His95, (A9)His103, and (A9)His105 [13–15]. The His_6 site of hCP has gained significant attention in recent years because it can sequester a range of divalent first-row transition metal ions including Mn(II), Fe(II), Zn(II), and Ni(II) [1, 13–20]. Like the human protein, murine CP (mCP) is also a heterooligomer of S100A8 and S100A9, and each subunit contains a canonical C-terminal EF-hand and a noncanonical N-terminal EF-hand domain. Amino acid sequence alignment of the human and murine S100A8 and S100A9 polypeptides indicates that the His_3Asp and His_6 sites are conserved [21].

Early studies of hCP revealed that Ca(II) binding causes two heterodimers to self-associate and form a heterotetramer [22, 23]. In addition to this change in quaternary structure, Ca (II) binding enhances the transition metal affinities, antimicrobial activity, and proteolytic stability of hCP [1, 12, 26]. Initial biochemical evaluation of mCP revealed that Ca(II) binding also causes this orthologue to form heterotetramers [21], and further investigations are required to decipher whether Ca(II) ions modulate other structural and functional properties of the protein. The studies of hCP provide the basis for a working model where CP responds to the high extracellular Ca(II) concentration ($\approx$2 mM) at an infection site and becomes a tetramer with high transition metal affinities [1, 12].

In this chapter, we first present the overexpression and purification protocols for mCP and hCP-Ser, a cysteine-null variant [(hS100A8(C42S)/hS100A9(C3S)], which has been used extensively in biochemical, biophysical, and structural studies of the human orthologue. These protocols are based on published work [12, 21], provide each protein as the apo heterodimer, and can also be used for preparing hCP, as well as variants of either orthologue with single or multiple point mutations. Because the His_6 site of hCP was recently shown to sequester Fe(II) and the protein was found to affect the redox speciation of Fe in aerobic solution [18, 19], we also provide a protocol for an Fe speciation assay that we first designed and utilized to study hCP-Ser [19]. Herein, we extend this assay to mCP and report that this protein also shifts the redox speciation of Fe from Fe(III) to Fe(II) under aerobic conditions.

2 Materials

1. All chemicals are purchased from commercial suppliers and used as received.
2. All solutions are prepared using Milli-Q water (18.2 MΩ·cm, 22-μm filter).
3. Protein concentrations are determined using the calculated extinction coefficients of the S100A8/S100A9 (calprotectin) homodimer (ProtParam: $\varepsilon_{280} = 5{,}960\ M^{-1}\ cm^{-1}$ for mCP (mS100A8/mS100A9), 18,450 M^{-1} cm^{-1} for hCP (hS100A8/hS100A9) and hCP-Ser (hS100A8(C42S)/ hS100A9(C3S)). All concentrations reported are for the heterodimer.

2.1 Preparation of Expression Plasmids

1. The synthetic genes for each protein subunit (mS100A8, mS100A9, hS100A8, hS100A9) are optimized for *Escherichia coli* codon usage and obtained from ATUM (formerly DNA2.0). These genes, as well as procedures for site-directed mutagenesis to obtain genes encoding hS100A8(C42S) and hS100A9(C3S), are described in the literature [12, 21]. Protein nomenclature and expression plasmids are listed in Table 1.
2. The pET41a expression vector is obtained from Invitrogen.
3. The genes are inserted into the *Nde*I and *Xho*I restriction sites of pET41a, which affords the untagged, full-length proteins

Table 1
Protein nomenclature and expression plasmids

Protein	Abbreviation	Subunits	Expression plasmid	Reference[a]
Human calprotectin	hCP	hS100A8 hS100A9	pET41a-*hS100A8* pET41a-*hS100A9*	[12] [12]
Human calprotectin Cys → Ser variant	hCP-Ser	hS100A8(C42S) hS100A9(C3S)	pET41a-*hS100A8(C42S)* pET41a-*hS100A9(C3S)*	[12] [12]
Murine calprotectin	mCP	mS100A8 mS100A9	pET41a-*mS100A8* pET41a-*mS100A9*	[21] [21]

[a]The original names for the plasmids containing the human S100A8 and S100A9 genes did not include "*h*" to designate "human"

with no additional amino acids after IPTG-induced overexpression [12, 21].

4. The pET41a plasmids containing mS100A8, mS100A9, hS100A8(C42S), and hS100A9(C3S) are transformed into chemically competent *E. coli* BL21(DE3) cells for overexpression.
5. To prepare cell stocks of *E. coli* BL21(DE3), the cells are grown in LB to saturation, diluted 1:1 into a 50% glycerol solution, frozen in liquid nitrogen, and stored at −80 °C.

2.2 Commercial Materials and Preparation of Reagents

1. To reduce metal ion contamination, plastic spatulas are used to transfer reagents (*see* **Note 1**).
2. Stock solutions of metal ions and sodium citrate (100 mM) are prepared in nitric acid-washed volumetric glassware and transferred to polypropylene tubes for storage (*see* **Note 1**).
3. Stock solutions of Fe(III) (100 mM) are prepared from ≥99.99% trace metal basis anhydrous $FeCl_3$, Milli-Q water, and trace metal basis 37% HCl (*see* **Note 2**).
4. Stock solutions of Ca(II) (1 M) are prepared from 99.999% $CaCl_2$ and Milli-Q water.
5. Other reagents utilized in this protocol are listed in Table 2.

2.3 Ferrozine Solution Preparation

For the Fe speciation assay, ferrozine is used to determine the Fe(II) content of solutions via detection of the $[Fe(ferrozine)_3]^{4-}$ complex by optical absorption spectroscopy [19, 27, 28].

1. Stock solutions (≈100 mM, 20 mL) of ferrozine are prepared in Milli-Q water, aliquoted into 800-μL portions, and stored at −80 °C.
2. Each aliquot is thawed only once and subsequently diluted to the working concentration of 6.17 mM (10 mL) in 0.1 M HCl immediately before use.

2.4 Buffers for Protein Purification and the Iron Speciation Assay

Buffers employed for the purification of mCP and hCP-Ser are prepared as indicated below with Milli-Q water and are subsequently filtered (0.2 μm) and stored at 4 °C. Buffer compositions are given in Table 3. For buffers used in the preparation of mCP or hCP, DTT (5 mM final concentration) is added to all buffers immediately before use because these proteins contain cysteine residues (*see* **Note 3**). The Fe speciation assay buffer is prepared from high-purity reagents (*see* Subheading 2.2) in acid-washed volumetric glassware and subsequently transferred to a polypropylene container (*see* **Note 1**).

Table 2
Reagents and supplier information

Purpose	Reagent	Supplier
Protein overexpression and purification	Luria-Bertani (LB)	Becton Dickinson
	Kanamycin sulfate	VWR
	Isopropyl β-D-1-thiogalactopyranoside (IPTG)	BACHEM
	99.5% 4-(2-hydroxyethyl)-1-piperazineethanesulfonic acid (HEPES)	MilliporeSigma
	Sodium chloride (NaCl)	MilliporeSigma
	Sodium hydroxide (NaOH)	Macron
	Triton X-100	EMD
	Ethylenediaminetetraacetic acid disodium salt dihydrate (EDTA)	VWR
	Phenylmethylsulfonyl fluoride (PMSF)	VWR
	Dithiothreitol (DTT)	VWR
	Guanidinium hydrochloride (GuHCl)	MilliporeSigma
	Ammonium sulfate	MilliporeSigma
	Chelex resin	Bio-rad
Fe speciation assay	Ultrol grade HEPES (free acid)	Calbiochem
	TraceSELECT sodium chloride (NaCl)	MilliporeSigma
	37% hydrochloric acid (HCl), trace metal basis	MilliporeSigma
	≥99.5% sodium citrate tribasic dihydrate	MilliporeSigma
	99.99% anhydrous ferric chloride	MilliporeSigma
	≥99.0% sodium ascorbate	MilliporeSigma
	≥99.5% trichloroacetic acid (TCA)	MilliporeSigma
	≥99.99% ammonium acetate	MilliporeSigma
	97% 3-(2-Pyridyl)-5,6-diphenyl-1,2,4-triazine-*p,p′*-disulfonic acid monosodium salt hydrate (ferrozine)	MilliporeSigma

2.5 Equipment for Protein Purification

1. An ÄKTA purifier (GE life sciences) housed in a 4 °C room and equipped with a 150-mL Superloop (GE life sciences) is employed for chromatographic purification.
2. Both human and murine CP orthologues are purified in two chromatographic steps: Anion exchange chromatography using a MonoQ 10/100 GL column (GE life sciences) followed by size exclusion chromatography using a Superdex 75 26/600 GL column (GE life sciences).

2.6 Equipment for Optical Absorption Spectroscopy

1. A Beckman coulter DU 800 spectrophotometer thermostatted at 25 °C with a Peltier temperature controller is used for optical absorption spectroscopy.
2. Plastic cuvettes (1-cm pathlength, polystyrene, VWR) are employed for optical absorption measurements during the Fe speciation assay. New cuvettes are used for each measurement.

Table 3
Buffer composition

Buffer (1 L)	Composition
Lysis buffer A	50 mM Tris, 100 mM NaCl, 1 mM EDTA, 0.5% Triton X-100, 1 mM PMSF[a], pH 8.0
Lysis buffer B	50 mM Tris, 100 mM NaCl, 4 M GuHCl, pH 8.0
Dialysis buffer	20 mM HEPES, pH 8.0
MonoQ buffer A	20 mM HEPES, pH 8.0
MonoQ buffer B	20 mM HEPES, 1 M NaCl, pH 8.0
S75 buffer	20 mM HEPES, 100 mM NaCl, pH 8.0
Fe speciation assay buffer	75 mM HEPES, 100 mM NaCl, 2 mM Ca(II)[b], pH 7.0

[a]PMSF is added immediately before use. It is dissolved in anhydrous ethanol before use or can be stored at −20 °C
[b]Ca(II) (80 μL of a 1-M stock solution) is added to 40 mL of the buffer before use

Acid-washed quartz cuvettes can also be used for these measurements.

3. Plastic cuvettes (1-cm pathlength, polystyrene, VWR) are used for monitoring the bacterial culture optical density at 600 nm (OD_{600}, absorption at 600 nm).
4. Observed OD_{600} values of bacterial cultures are given throughout this protocol as benchmarks; however, these values will vary from instrument to instrument.

3 Methods

The methods for protein overexpression and purification described in this section are based on published procedures [12, 21] and can be employed to produce human and murine CP variants obtained by site-directed mutagenesis.

3.1 Protein Preparation

The protein purification procedures are described for overexpressions carried out using a 1-L culture for each subunit. Typical yields are ≈60 mg/2 L culture for hCP-Ser and ≈45 mg/2 L culture for mCP.

3.1.1 Day 1: Preparation for Protein Overexpression

1. Prepare and autoclave Luria-Bertani (LB) medium for protein overexpression according to the manufacturer's instructions. Prepare 1 L of LB medium in a 2-L baffled flask for each S100 subunit to be overexpressed. Place the flasks in a 37 °C incubator shaker or warm room overnight.

2. Prepare 50 mg/mL kanamycin by dissolving kanamycin sulfate in Milli-Q water. Filter the solution (0.2-μm syringe filter), collecting the filtrate in a polypropylene tube, and store the solution at −20 °C. The pET41a vector contains a kanamycin resistance cassette.
3. Prepare 0.5 M IPTG by dissolving IPTG in Milli-Q water. Filter the solution (0.2-μm syringe filter), collect the filtrate in a polypropylene tube, and store at −20 °C. Protein expression will be induced by IPTG addition on day 2.
4. Prepare starter cultures of *E. coli* BL21(DE3) containing the desired pET41a expression plasmids. To a sterile 250-mL baffled flask, add 30 mL of LB and 50 μg/mL kanamycin (final concentration). Inoculate the medium from a freezer stock (Subheading 2.1) of the *E. coli* overexpression strain or from a single colony that was grown on an agar plate. Incubate the cultures (37 °C, 150 rpm) for ≈16 h.

3.1.2 Day 2: Overexpression of S100A8 and S100A9 Subunits

5. Measure the OD_{600} of the starter culture, which should be >1.5.
6. Add 50 μg/mL kanamycin (final concentration, 1 mL of a 50 mg/mL stock solution in Milli-Q water) into the 2 L baffled flask containing 1 L of LB medium, and dilute a 10 mL volume (1:100 dilution) of the overnight culture into the medium. Place the resulting culture in an incubator shaker (37 °C, 150 rpm).
7. Monitor the OD_{600}.
8. When OD_{600} ≈ 0.6–0.7, add 125 μM IPTG (250 μL of 0.5 M IPTG stock solution) to the culture to induce protein expression.
9. Incubate the culture (37 °C, 150 rpm) for an additional 3.5–4 h after induction. At this time, measure the OD_{600} of the culture, which should be >1.5.
10. Harvest the cells by centrifugation (2246 × *g*, 15 min, 4 °C), discard the supernatant, and transfer the cell pellet to a sterile 50-mL polypropylene tube. Flash freeze the pellet in liquid nitrogen and store at −80 °C until use.

This procedure typically yields ≈2 g cells/1 L culture (wet weight) for each subunit, which can be stored at −80 °C for over 4 months without an impact on protein yield. The overexpression can be evaluated by SDS-PAGE analysis (Coomassie stain) of the pre- and post-induction cell samples (whole cell lysate) using a 15% Tris-glycine gel (Fig. 1). Successful overexpression of the S100A8 subunit should show an intense band at ≈11 kDa in the post-induction sample, corresponding to the S100A8 monomer. Successful overexpression of the S100A9 subunit should show an intense band at ≈13 kDa, corresponding to the S100A9 monomer.

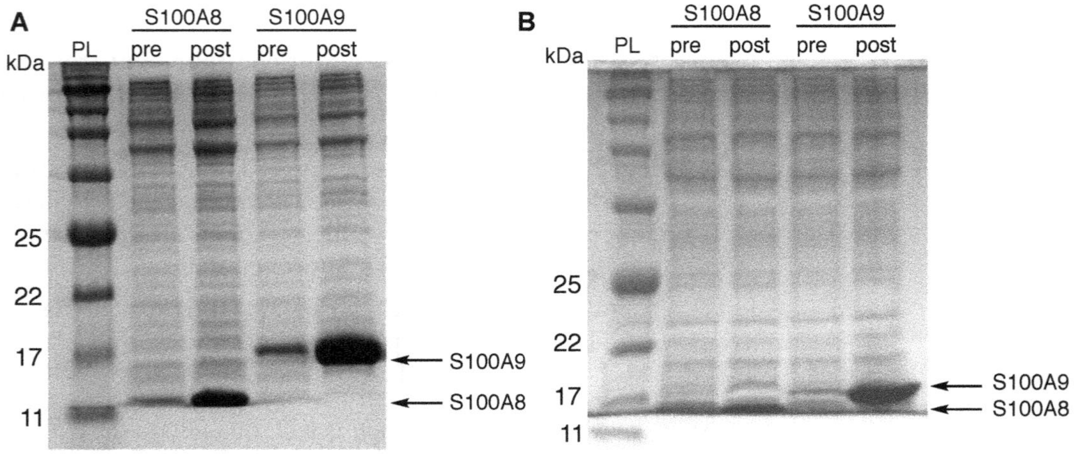

Fig. 1 Representative 15% Tris-glycine SDS-PAGE gel of whole cell lysate of *E. coli* BL21(DE3) obtained from the overexpression of the S100A8 and S100A9 subunits of mCP (**a**) and hCP-Ser (**b**). Samples pre- and post-induction with IPTG are shown. The samples were prepared using the B-PER reagent (Thermo Fisher Scientific, manufacturer protocol). The protein ladder (PL) is p7712S (New England Biolabs)

These overexpression conditions result in insoluble hS100A8, hS100A8(C42S), hS100A9, hS100A9(C3S), and mS100A8 and soluble mS100A9.

3.1.3 Day 3: Lysis and Refolding of mCP

The following steps should be carried out at 4 °C or on ice, and buffers should be equilibrated and stored at 4 °C or on ice.

11. Thaw one cell pellet of mS100A8 and one cell pellet of mS100A9, each from a 1 L overexpression culture, on ice.
12. Add 116 mg DTT to 150 mL of lysis buffer A, and stir to dissolve, affording a final concentration of 5 mM (*see* **Note 4**). Keep lysis buffer A at 4 °C or on ice.
13. Suspend each cell pellet in 30 mL of lysis buffer A containing 5 mM DTT. Once the cell pellets are fully suspended, combine the suspensions to yield a 60-mL mixture containing both mS100A8 and mS100A9.
14. Transfer the combined suspension to a stainless steel beaker on ice. Lyse the cells by sonication on ice (40% amplitude, 2.5 min, 30 s on, 10 s off; Branson digital sonifier).
15. Centrifuge the mixture (22,000 × *g*, 4 °C, 10 min). Transfer the supernatant, which contains soluble mS100A9, to a glass beaker on ice.
16. Repeat **steps 13–15** for a total of two times. Keep the final pellets (containing insoluble mS100A8) on ice, and combine the supernatant from each round (containing soluble mS100A9) in a glass beaker on ice.

17. To precipitate contaminating proteins from the combined supernatant containing soluble mS100A9, add ≈45 g of ammonium sulfate to the solution (≈120 mL) in one portion, and stir rapidly for ≈1 h at 4 °C. This quantity of ammonium sulfate affords a final concentration of 60% ammonium sulfate.
18. Centrifuge the mixture obtained from **step 17** (22,000 × *g*, 4 °C, 20 min). This procedure affords a greasy pellet that contains contaminating proteins and mS100A9 in the supernatant. Either (i) quickly decant and save the supernatant on ice or (ii) vacuum filter (Büchner funnel) the centrifuged mixture and collect the filtrate, which contains soluble mS100A9, on ice.
19. Add ≈38 g of ammonium sulfate to the combined filtrate in one portion to increase the ammonium sulfate concentration to 100%. Stir the mixture at 4 °C for ≈1 h to precipitate mS100A9.
20. Centrifuge the mixture obtained in **step 19** (22,000 × *g*, 4 °C, 20 min). This procedure affords a greasy pellet that contains insoluble mS100A9. Either (i) quickly decant the superntatnt and collect the pellet or (ii) vacuum filter (Büchner funnel) the centrifuged mixture and collect the precipitate, which contains mS100A9.
21. Add 77 mg DTT to 100 mL lysis buffer B (*see* **Note 4**), stir to dissolve, and keep the buffer on ice. Resuspend the mS100A8 pellets (**step 16**) and the mS100A9 precipitate (**step 20**) together in this solution by using a tissue homogenizer or by gently stirring the mixture with a stir bar at 4 °C.
22. Transfer the suspension to a stainless steel beaker, and sonicate for 5 min (40% amplitude, 30 s on, 10 s off) on ice.
23. Centrifuge the mixture (22,000 × *g*, 4 °C, 10 min). Transfer the supernatant to a dialysis bag (3500 MWCO), and dialyze against 3 × 4 L of dialysis buffer containing 5 mM DTT at 4 °C for at least 12 h each (*see* **Note 4**).

3.1.4 Day 3: Lysis and Refolding of hCP-Ser

The following steps should be carried out at 4 °C, and buffers should be equilibrated and stored at 4 °C or on ice.

24. Thaw one cell pellet of hS100A8(C42S) and one hS100A9 (C3S), each from a 1-L overexpression culture, on ice.
25. Keep 200 mL of lysis buffer A at 4 °C or on ice. Suspend each cell pellet in 30 mL (60 mL total) lysis buffer A. Once the cell pellets are fully suspended, combine the suspensions to yield a 60-mL mixture containing both hS100A8(C42S) and hS100A9(C3S).
26. Transfer the mixture to a stainless steel beaker on ice. Lyse the cells by sonication (40% amplitude, 2.5 min, 30 s on, 10 s off; Branson digital sonifier).

27. Centrifuge the crude lysate (22,000 × *g*, 4 °C, 10 min). Discard the supernatant. The pellet contains the hS100A8(C42S) and hS100A9(C3S) subunits.
28. Repeat **steps 25–27** for a total of three times.
29. Suspend the resulting cell pellets together in 100 mL lysis buffer B using a tissue homogenizer or by gently stirring the mixture with a stir bar at 4 °C.
30. Transfer the solution to a steel beaker on ice, and sonicate for 5 min (40% amplitude, 30 s on, 10 s off).
31. Centrifuge the mixture (22,000 × *g*, 4 °C, 10 min). Transfer the supernatant to a dialysis bag (3500 MWCO), and dialyze against 3 × 4 L of dialysis buffer at 4 °C for at least 12 h each.

3.1.5 Days 4 and 5: Change Dialysis Buffer

32. Change the dialysis buffer. If it is an mCP or hCP dialysis, add 5 mM DTT to the new dialysis buffer (3.1 g to 4 L dialysis buffer).

    After dialysis, the refolded protein is purified by AEC and SEC—*see* Day 6.
33. The day before the chromatography steps, pre-equilibrate the S75 column with 1 column volume (CV) of S75 buffer.

3.1.6 Day 6: Protein Purification of mCP and hCP-Ser

Over the course of the refolding process described above, some insoluble aggregates form and accumulate at the bottom of the dialysis bag as a white precipitate.

34. Centrifuge the dialysate to pellet the precipitate (22,000 × *g*, 4 °C, 10 min). Decant and vacuum filter the supernatant using a 0.2-μm bottle-top filter, and collect the filtrate in a polypropylene bottle stored on ice.
35. Load the filtrate (≈110 mL; dialysis can cause some increase in the volume) solution into the Superloop.
36. Equilibrate the MonoQ column with 2 CV Milli-Q water, 2 CV MonoQ buffer A, 2 CV MonoQ buffer B, and 2 CV MonoQ buffer A. Use a flow rate of 2 mL/min. Include 5 mM DTT in the MonoQ A and B buffers when purifying a cysteine-containing protein (*see* **Note 4**).
37. Load the column with the protein solution, and elute at 2 mL/min using MonoQ buffer A and a gradient of:
    (a) 0–15% MonoQ buffer B over 15 CV for mCP (Fig. 2a, c). or
    (b) 0–30% MonoQ buffer B over 30 CV for hCP-Ser (Fig. 2b, d).

    Monitor protein elution at 280 nm and collect 5 mL fractions. SDS-PAGE analysis of the fractions is used to determine which

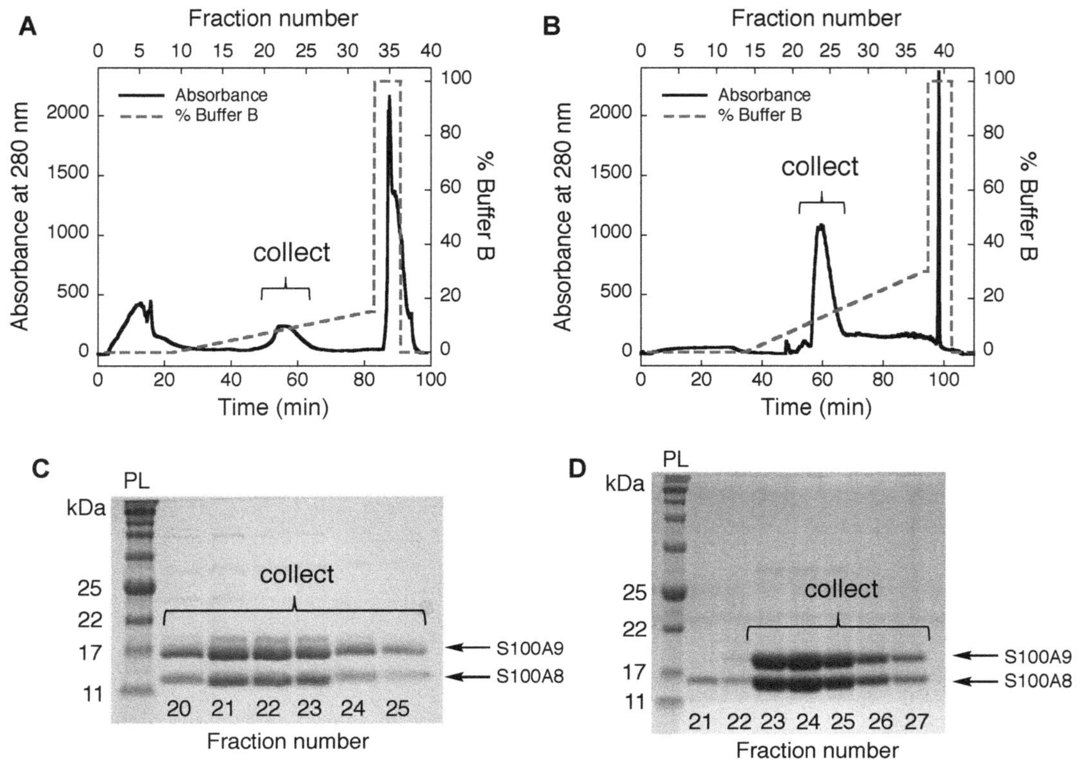

Fig. 2 Purification of mCP (**a**, **c**) and hCP-Ser (**b**, **d**) by anion exchange chromatography using a MonoQ column. Elution profiles (**a**, **b**) and corresponding 15% Tris-glycine SDS-PAGE gels of the fractions (**c**, **d**), indicating the fractions collected for subsequent purification. The protein ladder (PL) is p7712S (New England Biolabs)

fractions contain the CP heterodimer, which is evidenced by the presence of both the S100A8 and S100A9 subunits in about equal abundance on the gel (Fig. 2). We typically perform two or three MonoQ runs for a protein purification on this scale. The column is equilibrated (**step 36**) between runs.

38. Combine fractions containing both S100A8 and S100A9 (Fig. 2), and concentrate the protein by centrifugation to ≈10 mL using an Amicon spin filter (10 kDa MWCO) and centrifuging (3210 × *g*, ≈25 min, 4 °C).
39. Load all of the concentrated protein onto the Superloop.
40. Elute the heterodimer over 1 CV on the pre-equilibrated S75 column at a flow rate of 1 mL/min using S75 buffer (Fig. 3). Include 5 mM DTT in the S75 buffer when purifying a cysteine-containing protein (*see* **Note 4**).
41. Collect the fractions containing CP, and dialyze for at least 12 h in 1 L of S75 buffer containing ≈10 g Chelex resin at 4 °C. Include 5 mM DTT in this buffer when purifying a cysteine-containing protein (*see* **Note 4**).

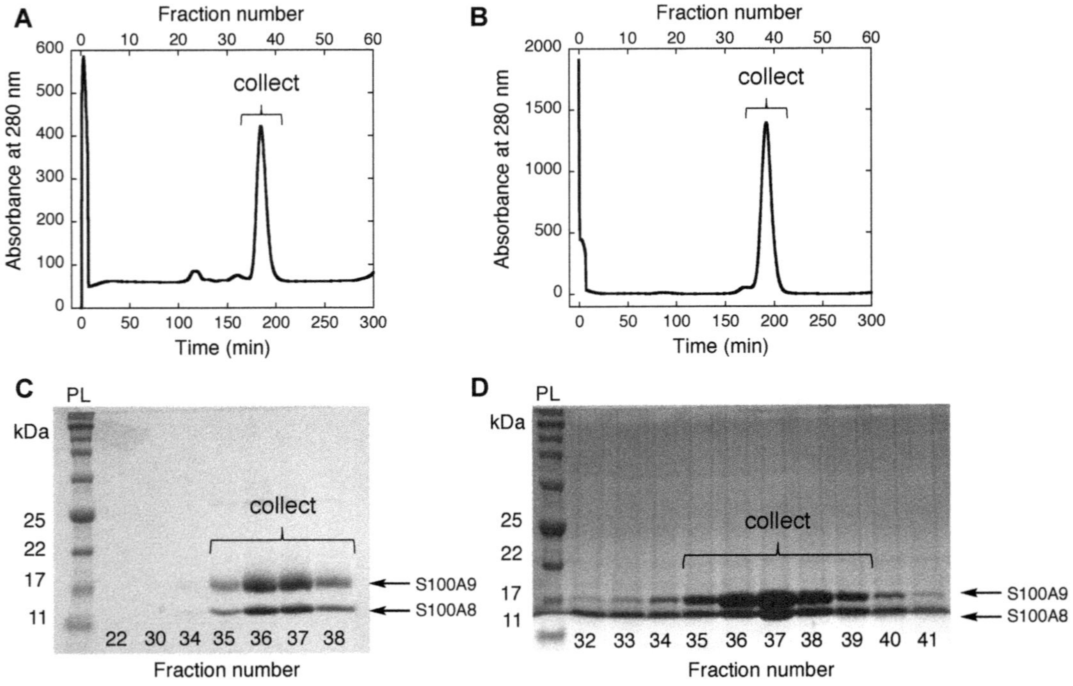

Fig. 3 Purification of mCP (**a**, **c**) and CP-Ser (**b**, **d**) by size exclusion chromatography using a S75 column. Elution profiles (**a**, **b**) and corresponding 15% Tris-glycine SDS-PAGE gels of the fractions (**c**, **d**), indicating fractions collected for subsequent dialysis. The protein ladder (PL) is p7712s (New England Biolabs)

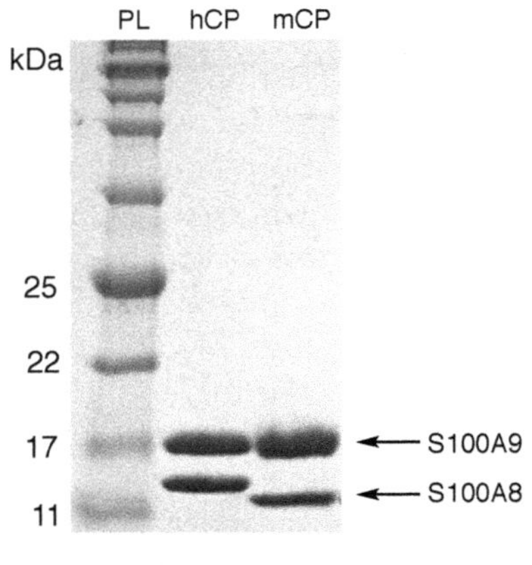

Fig. 4 Representative 15% Tris-glycine SDS-PAGE gel of purified hCP-Ser (hCP) and mCP. The protein ladder (PL) is p7712S (New England Biolabs)

3.1.7 Day 7: Protein Storage

42. Concentrate the protein solution to >500 μM (≈3 mL). Aliquot the protein into microcentrifuge tubes, flash freeze the protein aliquots in liquid nitrogen, and store at −80 °C.

3.2 Biochemical Characterization of Calprotectin

Following the purification of human or murine CP, we recommend employing standard biochemical techniques to evaluate the integrity of the purified protein [12]. These methods include SDS-PAGE (Fig. 4) to ascertain purity, analytical size exclusion chromatography to monitor oligomeric state, and metal analysis by inductively coupled plasma mass spectrometry (ICP-MS). The experimental conditions for these procedures, as well as circular dichroism spectroscopy, are described elsewhere [12, 21].

3.3 Fe Speciation Assay Using Ferrozine

The Fe speciation assay allows quantification of Fe(II) as well as total Fe content in solutions containing this metal ion. The protocol described below is modified from prior reports [19, 27, 28].

Day 1: Preparation of Samples

1. Combine 1 mL of the 100 mM $FeCl_3$ stock solution (Subheading 2.2) with 1 mL of the 100 mM sodium citrate stock solution (Subheading 2.2) to afford a 50 mM Fe(III)-citrate solution (2 mL).
2. Dilute 1 mL of the 50 mM Fe(III)-citrate solution with 9 mL Milli-Q water to afford a 5 mM Fe(III)-citrate solution (10 mL).
3. Buffer-exchange mCP and hCP-Ser into the Fe speciation assay buffer (4×, 500-μL spin filter) (*see* **Note 5**).
4. Prepare 5-mL samples that contain both 30 μM Fe(III) citrate and 20 μM protein in 15-mL polypropylene tubes by diluting the 5 mM Fe(III)-citrate solution (**step 2**) and the buffer-exchanged protein (**step 3**) into the Fe speciation assay buffer. Cap the tubes and incubate in an incubator shaker (30 °C, 150 rpm).
5. Prepare Fe standards for a standard curve. Dilute the 50 mM Fe(III)-citrate stock solution (**step 1**) with Fe speciation assay buffer to afford 10 mL aliquots of 32 μM and 25 μM Fe(III) citrate. Serially dilute the 32 μM Fe(III)-citrate solution (1:1 dilution series) with Fe speciation assay buffer to prepare standards containing 16, 8, 4, and 2 μM Fe(III) citrate. The standard curve is generated using standards containing 32, 25, 16, 8, 4, and 2 μM Fe as well as a buffer-only sample for 0 μM Fe.

Day 1: Fe Speciation Assay of Calibration Standards

6. Before initiating the assay, set a heating block to 95 °C, and turn on the optical absorption spectrophotometer.
7. Prepare 0.4 M TCA and 1.3 M ammonium acetate solutions in Milli-Q water (or thaw a solution prepared the same day and stored at −20 °C) (*see* **Note 6**).

8. Thaw an aliquot of 100 mM ferrozine.
9. Prepare 0.2 M HCl (10 mL) in a 15 mL polypropylene tube by diluting 166.6 μL of 37% HCl into 10 mL of Milli-Q water (*see* **Note 7**).
10. Prepare 40 mL of 1.5 mM sodium ascorbate in 0.2 M HCl. Dilute 667 μL 37% HCl into 40 mL of Milli-Q water, and add 12 mg of sodium ascorbate. This solution must be prepared immediately before use (*see* **Notes 6** and 7).
11. Prepare 6.17 mM ferrozine in 0.1 M HCl. Dilute 83.3 μL of 37% HCl and 617 μL of 100 mM ferrozine into 10 mL of Milli-Q water in a 15-mL polypropylene tube (*see* **Notes 6** and 7).
12. Transfer 200 μL of each standard solution to a separate 1.7 mL microcentrifuge tube. The total Fe content of these standards will be quantified. Ascorbate will be added to each sample to reduce all Fe(III) to Fe(II) before absorption of the Fe(II)-ferrozine complex is measured by optical absorption spectroscopy. Label each these tubes with an **a**.
13. Add 200 μL of 1.5 mM ascorbate solution to each tube labeled **a**.
14. Add 200 μL of 0.4 M TCA to each tube labeled **a**.
15. Vortex each tube labeled **a** for 10 s.
16. Heat all tubes labeled **a** for 5 min at 95 °C.
17. Centrifuge all tubes labeled **a** (15,000 × *g*, 5 min, 4 °C).
18. Transfer 300 μL of each supernatant to a new microcentrifuge tube.

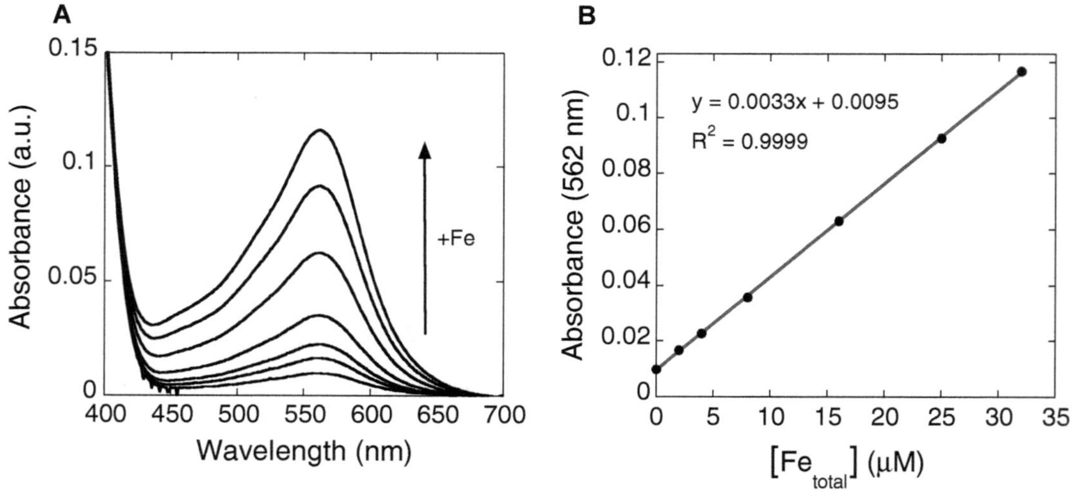

Fig. 5 Absorbance profile (**a**) of Fe standards in Fe speciation assay buffer (0, 2, 4, 8, 16, 25, and 32 μM total Fe). A calibration curve is generated by plotting A_{562} vs. $[Fe_{total}]$ and applying a linear regression fit (**b**)

19. Add 400 μL of 1.3 M ammonium acetate to each tube.
20. Add 100 μL of 6.17 mM ferrozine in 0.1 M HCl to each tube. Vortex to mix.
21. Transfer 500 μL of each solution to disposable semi-micro polystyrene cuvettes.
22. Blank the optical absorption spectrometer using Milli-Q water, and collect the optical absorption spectrum of each solution (Fig. 5a).
23. Plot A_{562} vs [Fe_{total}] of the calibration standards, and apply a linear regression to generate a calibration curve (Fig. 5b).

Days 1–4: Fe Speciation Assay of Protein Samples

This assay is conducted at 0, 24, 48, and 72 h after the protein assay setup.

24. On each day, complete **steps 6–11** from *Fe speciation assay of calibration standards*, above.
25. Immediately after preparing the samples that contain both Fe(III) citrate and protein (*Day 1: Preparation of samples*, **step 4**), and on each day after, add 200 μL of each sample to *each* of two microcentrifuge tubes, labeled **a** and **b**, respectively. Samples labeled **a** will be treated with an ascorbate solution to reduce all Fe to Fe(II) and quantify total Fe. Samples labeled **b** will not be treated with ascorbate to quantify Fe(II).
26. Add 200 μL of 0.2 M HCl to all tubes labeled **b**.
27. Add 200 μL of 1.5 mM ascorbate solution to all tubes labeled **a**.
28. Add 200 μL of 0.4 M TCA to all tubes labeled **a** and **b**.
29. Vortex each tube for 10 s.
30. Heat all tubes for 5 min at 95 °C.
31. Centrifuge all tubes (15,000 × *g*, 5 min, 4 °C). A small, white protein pellet is observed in the bottom of the tubes containing protein.
32. Transfer 300 μL of the supernatant to new microcentrifuge tubes.
33. Add 400 μL of 1.3 M ammonium acetate to each tube.
34. Add 100 μL of 6.17 mM ferrozine in 0.1 M HCl to each tube. Vortex to mix.
35. Transfer 500 μL of each solution to disposable semi-micro polystyrene cuvettes.
36. Blank the optical absorption spectrometer using Milli-Q water, and collect the optical absorption spectrum of each solution 400–700 nm (Fig. 6a, b). Record the absorbance of the Fe(II)-ferrozine complex in each solution at 562 nm.

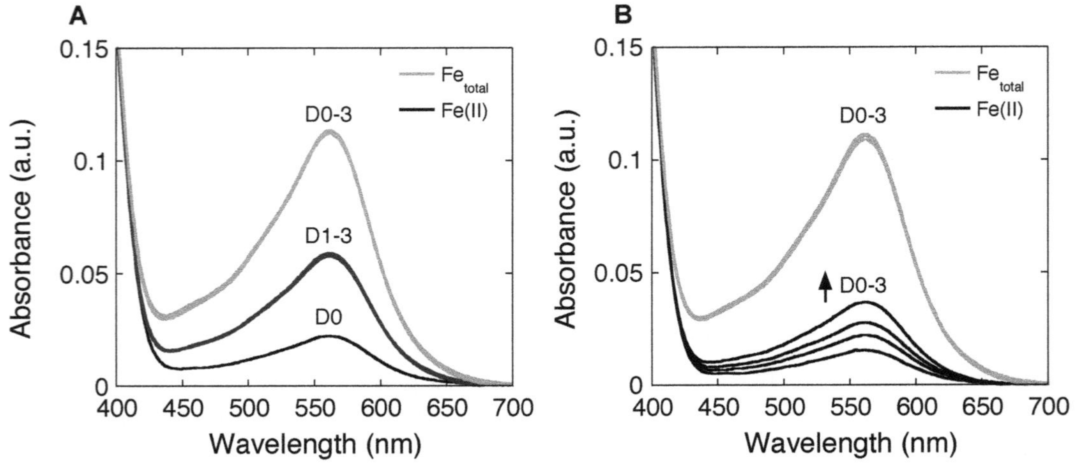

Fig. 6 Absorbance profiles of mCP (**a**) and hCP-Ser (**b**) samples analyzed by the Fe speciation assay. Spectra obtained for the Fe_{total} quantification samples (tubes labeled **a**, gray traces) and the Fe(II) quantification samples (tubes labeled **b**, black traces) over days 0–3 (D0–3) are shown

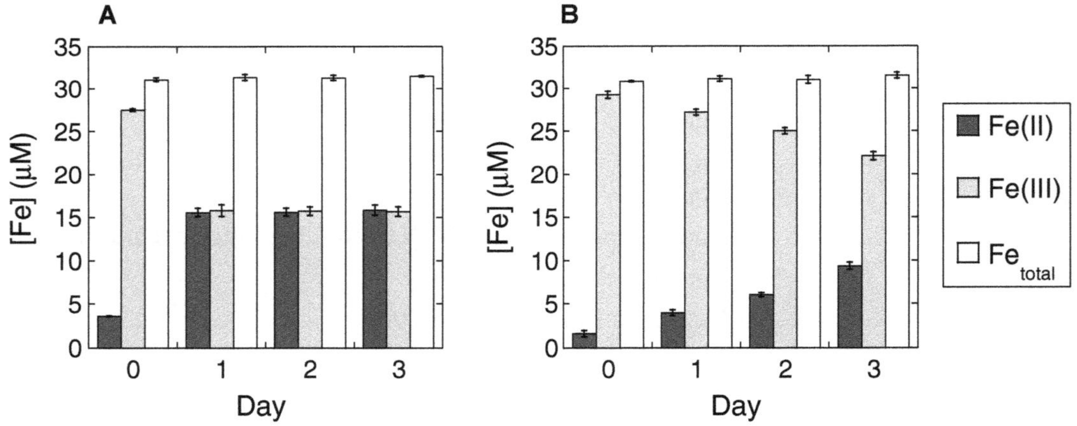

Fig. 7 Bar plots showing the concentrations of Fe(II), Fe(III), and total Fe in solutions containing mCP (**a**) and hCP-Ser (**b**) quantified on days 0–3 using the Fe speciation assay (average ± SDM, $n = 3$)

37. Use the calibration curve to determine the [Fe(II)] and [Fe_{total}] in the mCP and CP-Ser samples. Subtract the measured [Fe(II)] from the [Fe_{total}] to determine the [Fe(III)]. Plot the concentrations of Fe(II), Fe(III), and total Fe for the samples over days 0–3 (Fig. 7) (*see* **Note 8**).

4 Notes

1. Both human and murine CP coordinate transition metal ions with high affinity. Thus, precautions must be taken to reduce metal contamination when working with these proteins and to ensure the apo heterodimers are obtained from the purification. Because calcium ions affect the quaternary structure and transition metal affinities of CP, care must be taken to avoid calcium contaminations during the preparation of the apo heterodimer. After column purification, the protein is dialyzed in buffer containing Chelex resin to limit metal ion contamination. Additionally, for solution studies such as the Fe speciation assay, high-purity reagents are employed, and precautions are taken to avoid metal ion contamination. These precautions include the use of plastic spatulas to transfer reagents, the exclusive use of Milli-Q water, and the use of acid-washed glassware for the preparation of buffer and metal ion solutions. Because metal ions can leach from glassware, solutions are transferred to polypropylene containers for long-term storage.
2. Anhydrous $FeCl_3$ is purchased in an ampule and is hygroscopic. Upon opening the ampule, the powder should be weighed and the stock solution prepared immediately.
3. If the protein contains one or more cysteine residues, 5 mM DTT is included in all buffers used for protein purification [12, 21]. This modification is employed for purifying hCP, which contains two cysteine residues, as well as variants of mCP, and should be applicable for variants that have cysteine residues incorporated at non-native positions by site-directed mutagenesis. It is important to determine whether DTT in the storage buffer must be removed prior to an experiment. Note that DTT can affect the redox speciation of metal ions like Fe and Cu.
4. DTT is not infinitely stable, and thus buffer containing DTT should not be stored for extended periods of time. We recommend storing buffers without DTT and adding DTT powder to buffers immediately before use.
5. To set up the Fe speciation assay, proteins are buffer-exchanged into the Fe speciation assay buffer. Remember that mCP is stored in buffer that contains 5 mM DTT, and this reducing agent must be removed before setting up the Fe speciation assay.
6. We strongly recommend that the TCA, ammonium acetate, and sodium ascorbate reagents used in the ferrozine assay are prepared immediately before use. Ferrozine solutions can be stored at −80 °C and thawed daily for use in the assay.

7. We strongly recommend using high-purity reagents for the iron speciation assay, including trace metals basis HCl solution, to avoid metal contamination. In particular, we have found that use of lower purity HCl compromises the Fe speciation assay.
8. We recommend using at least 30 μM total Fe concentration in the assay. This concentration will provide absorbance values >0.1 a.u. when measuring the total iron content of the solution with a 1-cm pathlength cuvette.

Acknowledgments

Our current studies of calprotectin are supported by the National Science Foundation (CHE-1352132) and the National Institutes of Health (R01GM118695 and R01GM126376).

References

1. Zygiel EM, Nolan EM (2018) Transition metal sequestration by the host-defense protein calprotectin. Annu Rev Biochem 87:621–643
2. Hood MI, Skaar EP (2012) Nutritional immunity: transition metals at the pathogen-host interface. Nat Rev Microbiol 10(8):525–537
3. Corbin BD, Seeley EH, Raab A et al (2008) Metal chelation and inhibition of bacterial growth in tissue abscesses. Science 319 (5865):962–965
4. Hood MI, Mortensen BL, Moore JL et al (2012) Identification of an *Acinetobacter baumannii* zinc acquisition system that facilitates resistance to calprotectin-mediated zinc sequestration. PLoS Pathog 8(12):e1003068
5. Liu JZ, Jellbauer S, Poe AJ et al (2012) Zinc sequestration by the neutrophil protein calprotectin enhances *Salmonella* growth in the inflamed gut. Cell Host Microbe 11 (3):227–239
6. De Jong HK, Achouiti A, Koh GC et al (2015) Expression and function of S100A8/A9 (calprotectin) in human typhoid fever and the murine *Salmonella* model. PLoS Negl Trop Dis 9(4):e0003663
7. Wakeman CA, Moore JL, Noto MJ et al (2016) The innate immune protein calprotectin promotes *Pseudomonas aeruginosa* and *Staphylococcus aureus* interaction. Nat Commun 7:11951
8. Clark HL, Jhingran A, Sun Y et al (2016) Zinc and manganese chelation by neutrophil S100A8/A9 (calprotectin) limits extracellular *Aspergillus fumigatus* hyphal growth and corneal infection. J Immunol 196(1):336–344
9. Hobbs JA, May R, Tanousis K et al (2003) Myeloid cell function in MRP-14 (S100A9) null mice. Mol Cell Biol 23(7):2564–2576
10. Hunter MJ, Chazin WJ (1998) High level expression and dimer characterization of the S100 EF-hand proteins, migration inhibitory factor-related proteins 8 and 14. J Biol Chem 273(20):12427–12435
11. Korndörfer IP, Brueckner F, Skerra A (2007) The crystal structure of the human $(S100A8/S100A9)_2$ heterotetramer, calprotectin, illustrates how conformational changes of interacting α-helices can determine specific association of two EF-hand proteins. J Mol Biol 370 (5):887–898
12. Brophy MB, Hayden JA, Nolan EM (2012) Calcium ion gradients modulate the zinc affinity and antibacterial activity of human calprotectin. J Am Chem Soc 134(43):18089–18100
13. Hayden JA, Brophy MB, Cunden LS et al (2013) High-affinity manganese coordination by human calprotectin is calcium-dependent and requires the histidine-rich site formed at the dimer interface. J Am Chem Soc 135 (2):775–787
14. Brophy MB, Nakashige TG, Gaillard A et al (2013) Contributions of the S100A9 C-terminal tail to high-affinity Mn (II) chelation by the host-defense protein human calprotectin. J Am Chem Soc 135 (47):17804–17817
15. Damo SM, Kehl-Fie TE, Sugitani N et al (2013) Molecular basis for manganese sequestration by calprotectin and roles in the innate

immune response to invading bacterial pathogens. Proc Natl Acad Sci 110(10):3841–3846
16. Gagnon DM, Brophy MB, Bowman SEJ et al (2015) Manganese binding properties of human calprotectin under conditions of high and low calcium: X-ray crystallographic and advanced electron paramagnetic resonance spectroscopic analysis. J Am Chem Soc 137 (8):3004–3016
17. Nakashige TG, Stephan JR, Cunden LS et al (2016) The hexahistidine motif of host-defense protein human calprotectin contributes to zinc withholding and its functional versatility. J Am Chem Soc 138(37):12243–12251
18. Nakashige TG, Zhang B, Krebs C et al (2015) Human calprotectin is an iron-sequestering host-defense protein. Nat Chem Biol 11 (10):765–771
19. Nakashige TG, Nolan EM (2017) Human calprotectin affects the redox speciation of iron. Metallomics 9(8):1086–1095
20. Nakashige TG, Zygiel EM, Drennan CL et al (2017) Nickel sequestration by the host-defense protein human calprotectin. J Am Chem Soc 139(26):8828–8836
21. Hadley RC, Gu Y, Nolan EM (2018) Initial biochemical and functional evaluation of murine calprotectin reveals Ca(II)-dependence and its ability to chelate multiple nutrient transition metal ions. Biochemistry 57 (19):2846–2856
22. Vogl T, Roth J, Sorg C et al (1999) Calcium-induced noncovalently linked tetramers of MRP8 and MRP14 detected by ultraviolet matrix-assisted laser desorption/ionization mass spectrometry. J Am Soc Mass Spectrom 10(11):1124–1130
23. Strupat K, Rogniaux H, Van Dorsselaer A et al (2000) Calcium-induced noncovalently linked tetramers of MRP8 and MRP14 are confirmed by electrospray ionization-mass analysis. J Am Soc Mass Spectrom 11(9):780–788
24. Gifford JL, Walsh MP, Vogel HJ (2007) Structures and metal-ion-binding properties of the Ca^{2+}-binding helix-loop-helix EF-hand motifs. Biochem J 405(2):199–221
25. Kehl-Fie TE, Chitayat S, Hood MI et al (2011) Nutrient metal sequestration by calprotectin inhibits bacterial superoxide defense, enhancing neutrophil killing of *Staphylococcus aureus*. Cell Host Microbe 10(2):158–164
26. Stephan JR, Nolan EM (2016) Calcium-induced tetramerization and zinc chelation shield human calprotectin from degradation by host and bacterial extracellular proteases. Chem Sci 7(3):1962–1975
27. Stookey LL (1970) Ferrozine- a new spectrophotometric reagent for iron. Anal Chem 42 (7):779–781
28. Carter P (1971) Spectrophotometric determination of serum iron at the submicrogram level with a new reagent (ferrozine). Anal Biochem 40:450–458

Chapter 26

Structural Analysis of S100A8 Complex with Zinc and Calcium: A General Protocol for the Study of S100 Proteins in the Presence of Divalent Cations by X-Ray Crystallography

Laure Yatime

Abstract

Ions are important regulators for the cellular function of many proteins. This holds particularly true for S100 proteins whose function is not only calcium-dependent but also appears to be modulated by other divalent cations such as zinc, manganese, or copper. One way ions are thought to influence the function of S100 proteins (and any protein in general) is by changing their three-dimensional organization, through modifications in either their monomeric shape, their oligomeric state, or both. X-ray crystallography is a very powerful technique to study the effect of ions on the 3D architecture of macromolecules since it gives a direct visualization of where ions bind and how the protein structure is affected upon ion binding. Taking the example of human S100A8, I describe here the complete procedure to obtain a highly pure and homogenous S100 protein sample, crystallize it in the presence of divalent cations, and derive a 3D structural model from diffraction images. I further detail computational methods used to determine precisely the nature and position of the divalent cations within S100A8 structure. This methodology can easily be applied to any ion-binding protein, provided that the ion anomalous scattering properties allow to identify it unambiguously.

Key words S100 proteins, Divalent cations, Oligomerization, Chromatography, X-ray crystallography, Protein 3D structure, Anomalous scattering

1 Introduction

Divalent cations are important cofactors for numerous proteins achieving essential cellular functions, including the large S100 family of EF-hand Ca^{2+}-binding proteins found exclusively in vertebrates [1, 2]. It has long been acknowledged that the function of these proteins is highly dependent on calcium [3], which allows them to adopt an active conformation for ligand binding [4]. Over the years, the idea that other divalent cations can modulate specific functions of these proteins has also emerged [5–7]. Zn^{2+} and Mn^{2+}

Claus W. Heizmann (ed.), *Calcium-Binding Proteins of the EF-Hand Superfamily: From Basics to Medical Applications*, Methods in Molecular Biology, vol. 1929, https://doi.org/10.1007/978-1-4939-9030-6_26,

are, for example, essential substrates for the antimicrobial activity of several S100 proteins [8, 9]. Zn^{2+} and Cu^{2+} have also been proposed to modify the oligomeric state of various S100 proteins, thereby affecting their interactions with target proteins [10–14]. The presence of additional divalent cations is also known to alter S100 protein affinity for calcium through small conformational rearrangements of the two EF-hand motifs, thereby modulating target recognition [10, 15–17]. Finally, divalent cations may be important for the interaction of S100 proteins with their cognate receptor RAGE [18–20]. Analyzing the structure–function relationships for divalent cation binding to S100 proteins is therefore critical to understand how these ions affect S100 cellular functions. X-ray crystallography is a method of choice to evaluate these structure–function relationships, and it has been widely used for the study of S100 proteins as illustrated by the availability of more than 100 crystallographic structures of S100 proteins in the Protein Data Bank (PDB).

X-ray crystallography relies on the specific properties of ordered matter with respect to light. When a ray of light hits a solid object, it interacts with the encountered atoms. Two phenomena occur as a result of this interaction: first absorption and then diffusion of the incoming beam in all directions of space. For electromagnetic waves in the X-ray range, the diffusion of the incoming beam is essentially due to its interaction with the electronic cloud of the atoms. When the object encountered displays a periodical arrangement (such as in a protein crystal), a special phenomenon called diffraction occurs. The diffused beams coming from each scattering element interfere with each other. As a result, they cancel out each other (i.e., their intensity is reduced to zero) except in very specific spatial directions (constructive interferences). The direction of the diffracted beams is given by Bragg's law and only depends on (1) the wavelength of the incoming beam (λ), (2) the angle between the incoming beam and the plans in the crystal bearing the scattering elements (θ), and (3) the distance separating the parallel plans containing the scattering elements (d) (Fig. 1).

In a protein crystal, the periodical lattice formed by the electrons present in each molecule of protein is characterized by different family of plans. The spatial position of these plans is defined by a set of indices (*hkl*), called the Miller indices. When subjected to X-rays, each (*hkl*) family of plans will give rise to a diffracted beam which can be mathematically described by the function F_{hkl}, called structure factor (Fig. 1). F_{hkl} is a complex variable characterized by an amplitude, $|F_{hkl}|$, and a phase, Φ_{hkl}. Because the diffracted beams for each encountered family of plans (*hkl*) strictly depend on the position of these electrons within the crystal, there is a direct relationship between the electronic density $\rho_{(x,y,z)}$ within the crystal

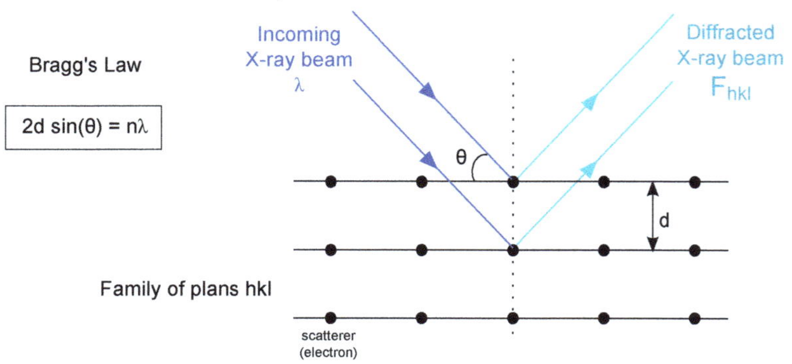

Fig. 1 Principle of X-ray diffraction by a protein crystal. Due to the periodical organization of atoms within the protein crystal, electrons are arranged in a lattice that defines multiple families of plans noted with the (*hkl*) Miller indices. Each (*hkl*) family of plans will diffract the incident X-ray beam (dark blue) in only one particular direction of space (light blue). This direction is given by Bragg's law indicated on the left side of the figure where λ is the wavelength of the incident beam, θ is the angle between the incident beam and the (*hkl*) plans, d is the distance between two consecutive plans of the (*hkl*) family, and n is a positive integer. The diffracted X-ray beam is mathematically described by the structure factor $F_{hkl} = |F_{hkl}| \times e^{ihkl}$

and the F_{hkl} function, given by the following Fourier transform equation:

$$F_{hkl} = \int\int\int \rho_{(x,y,z)} e^{2\Pi i(hx+ky+lz)} \quad (1)$$

Therefore, once all the structure factors F_{hkl} for the considered crystal are known, i.e., once all the diffracted X-ray beams have been measured and characterized mathematically, it is possible to determine the electronic density within the crystal by applying the inverse Fourier transform of Eq. (1) and thereby derive a 3D atomic model of the protein present in the crystal. The amplitude of the structure factors, $|F_{hkl}|$, is directly calculated from a diffraction experiment. It is extrapolated from the intensity of the diffracted X-ray beams which are recorded as spots on a plate detector. However, because intensities are always positive integers, the diffraction pattern is centrosymmetric, and only the absolute value of the amplitude can be obtained. This translates into Friedel's law which states that reflections of opposite (*hkl*) indices have equal amplitudes and opposite phases (i.e., $|F_{hkl}| = |F_{-h-k-l}|$ and $\Phi_{hkl} = -\Phi_{-h-k-l}$). The phase of the structure factors, Φ_{hkl}, is obtained either by using the structure of a known structural homolog of the protein present in the crystal (molecular replacement method MR [21]) or by derivatizing the crystallized protein with heavy atoms (multiple isomorphous replacement, MIR [22, 23]) and/or anomalous scatterers (single/multiple wavelength anomalous dispersion SAD/MAD [24]; see below).

Carbon, nitrogen, and oxygen, the main components of polypeptidic chains, have similar scattering properties and scatter X-rays only weakly due to the small number of electrons present in their electronic shells. In contrast, atoms containing a large number of electrons, such as heavy atoms and/or transition metals, interfere more strongly with X-rays. In particular, if the wavelength of the incoming X-ray beam is close to the absorption edge of the atom, X-rays will be absorbed more strongly by the specific atom, leading to stronger scattering as well as shifts in the phase of the corresponding diffracted beams. This phenomenon is known as anomalous dispersion. Due to this anomalous contribution in the diffracted beams, Friedel's law is broken, and the centrosymmetry of the diffraction pattern is lost. This property can be used advantageously to identify the nature and position of the anomalous scatterers in the crystal. Indeed, when an anomalous scatterer is present in a family of (*hkl*) plans, it will contribute anomalously to the resulting structure factor F_{hkl}, thereby breaking the relationship $|F_{hkl}| = |F_{-h-k-l}|$. The difference between Friedel's pair, $\Delta F_{hkl} = |F_{hkl}| - |F_{-h-k-l}|$, is called D_{ano} and directly measures this anomalous contribution to the scattering. Similarly to the electron density map that can be extrapolated by inverse Fourier transform of the ensemble of amplitudes $|F_{hkl}|$ and phases Φ_{hkl} characterizing the space of structure factors F_{hkl}, a difference electron density map can be extrapolated by inverse Fourier transform of an ensemble of difference amplitudes ΔF_{hkl} and phases Φ_{hkl}. Intuitively, one can easily comprehend that a difference map extrapolated from the difference D_{ano} and the corresponding phases will generate a difference electron density map only showing the contribution, to the electronic cloud, of the anomalous scatterers. As the anomalous scattering will occur at a different wavelength for each anomalous atom, this difference map will allow not only to identify the position of the anomalous scatterers in the crystal but also to precisely determine their nature.

Ca^{2+}, Zn^{2+}, Mn^{2+}, Cu^{2+}, and other divalent cations known to bind to S100 proteins display absorption edges for X-rays at different wavelengths. The generation of an anomalous difference electron density map from the diffraction data obtained for a crystal of S100 protein in complex with any of these ions can therefore be used to identify unambiguously where the ions are located within the S100 protein fold. In this chapter, I will present general methodologies for the X-ray structural analysis of S100 proteins in the presence of divalent cations, taking the example of human S100A8 (hS100A8) in complex with zinc and calcium [14]. In particular, I will describe the computational procedure allowing to calculate an anomalous difference electron density map and the subsequent analysis to identify where Ca^{2+} and Zn^{2+} ions are bound to the S100A8 protein.

2 Materials

Unless otherwise stated, all solutions and buffers are prepared with ultrapure water (with a resistivity of 18.2 MΩ cm at 25 °C) and filtered with 0.2 μm filters before use. All manipulations with bacteria and sterile medium are performed under a hood with laminar flow.

2.1 Protein Expression and Purification

1. LB agar plates containing the appropriate antibiotics.
2. 2YT medium: 10 g yeast extract, 16 g Bacto peptone, 5 g NaCl, dissolved in 1 L of distilled water buffered at pH 7.5 by adding 10 mM Tris–HCl pH 7.5. Medium is then autoclaved 20 min at 120 °C for sterilization.
3. 50 mg/mL sterile kanamycin.
4. *Escherichia coli* BL21 (DE3) competent cells.
5. 1 M Isopropyl β-D-1-thiogalactopyranoside (IPTG).
6. Binding buffer: 50 mM HEPES pH 7.5, 200 mM NaCl, 30 mM imidazole, 1 mM phenylmethylsulfonyl fluoride (PMSF).
7. High-salt wash buffer: 50 mM HEPES pH 7.5, 1 M NaCl, 30 mM imidazole, 1 mM PMSF.
8. Elution buffer: 50 mM HEPES pH 7.5, 200 mM NaCl, 250 mM imidazole, 1 mM PMSF.
9. Dialysis buffer: 50 mM Tris–HCl pH 8.8, 50 mM NaCl.
10. Size exclusion chromatography (SEC) buffer: 20 mM HEPES pH 7.5, 100 mM NaCl, 5 mM $CaCl_2$.
11. Apparatus for sonication.
12. Peristaltic pump.
13. Fast Protein Liquid Chromatography (FPLC) system in a refrigerated cabinet.
14. Chromatographic columns: 5 mL HisTrap FF Ni-column, 9 mL Source 15Q column, 24 mL Superdex 75 Increase columns (all columns are from GE Healthcare Life Sciences).

2.2 Protein Crystallization and Crystal Handling

1. 500 μL centrifugal protein concentrator.
2. Commercial crystallization screens of 96 conditions each (from Molecular Dimensions Ltd., Hampton Research, or equivalent companies) aliquoted in 96-well trays with deep wells (2 mL).
3. 96-Well crystallization plates (e.g., 96-well 2-drop Swissci plates).
4. Crystal clear tape with high sealing performance.

5. Cryoprotectant solution: 0.2 M ammonium acetate, 0.1 M Na acetate pH 4.0, 35% (w/v) PEG 3350, 0.4% (v/v) 2,2,2-trifluoroethanol.
6. Multichannel pipette.
7. Glass cover slides.
8. Magnetic cryogenic vials and caps with 18 mm SPINE standard pins.
9. Cryogenic loops mounted on 18 mm pins and fitting the size of your protein crystals.
10. Tools for crystal handling and freezing: cryogenic tong, magnetic wand, cryocane, dewars for manipulation and storage in liquid nitrogen.
11. Stereomicroscope with sufficient magnification to monitor crystal growth.

2.3 Data Collection, Structure Determination, and Analysis

Collection of diffraction images from the frozen protein crystals requires a powerful X-ray source and the proper detection apparatus. These experiments are generally performed on dedicated beamlines at a synchrotron, which implies to be affiliated to a structural biology group which has access to such facilities. Processing of the collected datasets can then be performed in your home laboratory providing that you are equipped with a computer with sufficient computing power. Generally, crystallographers are using graphic workstations combining high-computing power and high-quality display of three-dimensional objects. Software dedicated to protein crystallography need to be installed on the computer. As a start, the following programs/packages should be installed: XDS [25] or MOSFLM [26] (for data reduction), PHENIX [27] (for structure determination and model refinement), CCP4 [28] (for structural analysis), and Coot [29] (for structure rebuilding). Most programs have now versions that can run on any type of operating system (Windows, Mac, or Unix) although they were originally designed for Unix systems.

3 Methods

A prerequisite for the crystallographic analysis of a protein is to obtain a highly pure and homogenous protein sample in reasonable amount (generally at least a few milligrams of protein). For this purpose, the purification protocol usually encompasses several chromatographic steps to reach a purity of at least 95–97%.

3.1 Protein Expression and Purification

3.1.1 Protein Expression

1. Transform the plasmid encoding the studied S100 protein into *E. coli* BL21 (DE3) competent cells. In our case, the gene coding for hS100A8 was cloned into pETM13 (EMBL vector collection; *see* **Note 1**). Plate an aliquot of the transformation mix on LB agar plates supplemented with the appropriate antibiotic (50 μg/mL kanamycin for the hS100A8-pETM13 construct), and incubate overnight at 37 °C.
2. Select a few positive colonies to inoculate a 50 mL culture of 2YT medium supplemented with 50 μg/mL kanamycin, and shake overnight at 37 °C and 200 rpm.
3. Transfer 10–15 mL of the overnight culture per 2 L of 2YT culture medium supplemented with 50 μg/mL kanamycin, and allow the cultures to grow at 37 °C with shaking (200 rpm) until ^{600}OD reaches a 0.6–0.8 value.
4. Induce protein expression by adding 1 mM IPTG in each culture flask, and allow the cells to incubate at 18 °C overnight with more gentle shaking (140–160 rpm).
5. Harvest the cells by centrifugation at 6000 × *g* for 15 min, and transfer each 2 L culture pellet in a 50 mL tube. At this stage, cell pellets can be stored at −20 °C until further use or one can directly proceed with protein purification.

3.1.2 Protein Purification: Affinity Chromatography

1. Resuspend each 2 L culture pellet in 50 mL of binding buffer (*see* **Note 2**) by vortexing, and keep the bacterial suspensions on ice.
2. Open the cells by sonication (three times for 5 min for each tube of bacterial suspension kept on ice).
3. Centrifuge the sonicated samples at 30,000 × *g* and 4 °C for 30 min.
4. Recover the supernatant, and apply it via a peristaltic pump onto a 5 mL HisTrap column equilibrated in binding buffer. This step should be performed in a cold room or in a refrigerated cabinet.
5. Remove nonspecifically bound proteins by washing the column with 30 mL of high-salt wash buffer.
6. Elute the protein of interest with 30 mL of elution buffer.
7. Analyze the different fractions on a 16% acrylamide SDS-PAGE gel, and collect fractions containing the S100 protein.

3.1.3 Protein Purification: Anion Exchange Chromatography

1. The pooled fractions are then dialyzed overnight at 4 °C against 1 L of dialysis buffer.
2. In the meantime, connect the 9 mL Source 15Q column to an FPLC system in a refrigerated cabinet, and equilibrate in dialysis buffer.

3. Recover the dialyzed sample and apply it onto the Source 15Q column.
4. Elute the hS100A8 protein with a 100 mL linear gradient from 50 to 300 mM NaCl in 50 mM Tris–HCl pH 8.8.
5. Analyze selected fractions on a 16% acrylamide SDS-PAGE gel, and collect the fractions containing the hS100A8 protein.

3.1.4 Protein Purification: SEC Chromatography

1. Concentrate the pooled fractions to a volume lower than 5 mL using a Vivaspin 20 centrifugal protein concentrator (Sartorius) of 5 kDa molecular weight cutoff.
2. In the meantime, connect the 24 mL Superdex 75 Increase column to the FPLC system, and equilibrate in SEC buffer.
3. Apply the concentrated sample onto the Superdex 75 column, and elute with one column volume of SEC buffer. Only 500 μL of sample can be injected at a time on the SEC column. Several runs must therefore be performed to pass the whole sample through the column. Alternatively, a column with larger bed volume (typically above 100 mL) can be used to allow for injection of the entire sample at once and thereby minimize protein loss due to multiple injections.
4. Analyze selected fractions on a 16% acrylamide SDS-PAGE gel, and collect the fractions containing the hS100A8 protein (*see* **Note 3**). The different oligomeric forms of the protein must be pooled separately (*see* **Note 4**). Figure 2 shows a typical chromatogram obtained for the last step of purification of hS100A8 and the corresponding analysis by SDS-PAGE of the peak fractions, in the presence or in the absence of reducing agent.
5. For each pool, measure the protein concentration by recording the absorbance of the solution at 280 nm on a NanoDrop spectrophotometer (Thermo Scientific). For hS100A8, the molar extinction coefficient at 280 nm was 11,460, and the molecular weight was 10,834 Da (*see* **Note 5**).
6. Aliquot the purified protein by 1 mL fractions, and flash freeze in liquid nitrogen for storage at −80 °C (*see* **Note 6**).

3.2 Protein Crystallization and Crystal Handling

3.2.1 Protein Crystallization

1. Thaw an aliquot of purified protein and keep on ice.
2. Concentrate the protein to the desired concentration using a 500 μL centrifugal protein concentrator. A concentration of 10 mg/mL is generally a good starting point for most proteins (*see* **Note 7**).
3. Add 3 mM $ZnCl_2$ in the concentrated protein sample, and leave it to incubate at room temperature for 1 h (*see* **Note 8**).
4. In the meantime, aliquot the crystallization conditions in the crystallization plates using a multichannel pipette. Commercial kits encompassing a broad variety of precipitants are available

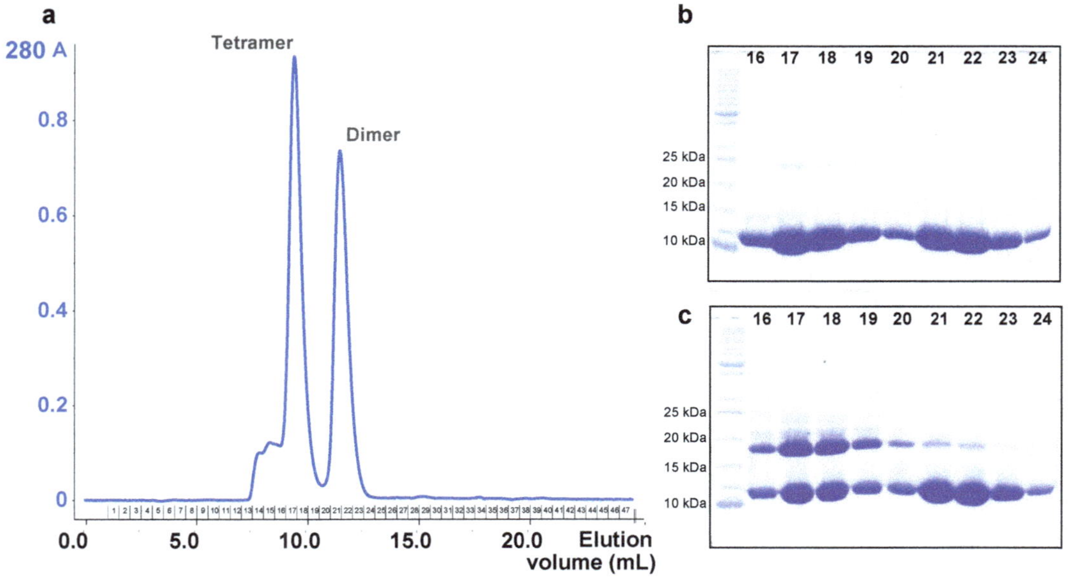

Fig. 2 Final step of purification of hS100A8 on size exclusion chromatography. (**a**) Elution profile of hS100A8 on a 24 mL Superdex 75 Increase column equilibrated in SEC buffer. The protein elutes as two major oligomeric species corresponding, respectively, to a homotetramer and a homodimer. (**b**) SDS-PAGE analysis of the peak fractions for the elution of hS100A8 under reducing conditions. The fraction number is indicated above each lane. Each fraction displays a single band at around 10 kDa on the gel corresponding to the size of the hS100A8 monomer. (**c**) SDS-PAGE analysis of the peak fractions for the elution of hS100A8 under nonreducing conditions. Higher molecular weight species appear on the gel corresponding mainly to disulfide cross-linking between two hS100A8 monomers

from various distributors. For initial screening, one prefers to use 96-well crystallization trays to minimize the amount of protein consumed. In that case, a volume of 50 μL of crystallization solution is added in each well of the plate.

5. Before setting up the crystallization experiment, centrifuge the protein sample for 5 min at 11,000 × *g* to remove possible aggregates. The supernatant is then transferred into a clean tube.
6. Set up the crystallization experiment by mixing the protein solution and the crystallization solution in each crystallization well of the 96-well plate. In general, the protein and crystallization solutions are mixed in a 1:1 volume ratio (*see* **Note 9**).
7. Once all crystallization drops have been generated, seal the plate with crystal clear tape.
8. Incubate the crystallization plate at a constant temperature (generally around 18–20 °C or at 4 °C), and monitor regularly under a stereomicroscope until crystals appear.
9. When crystals appear in a particular condition, reproducibility of the crystal formation and optimization of their shape and

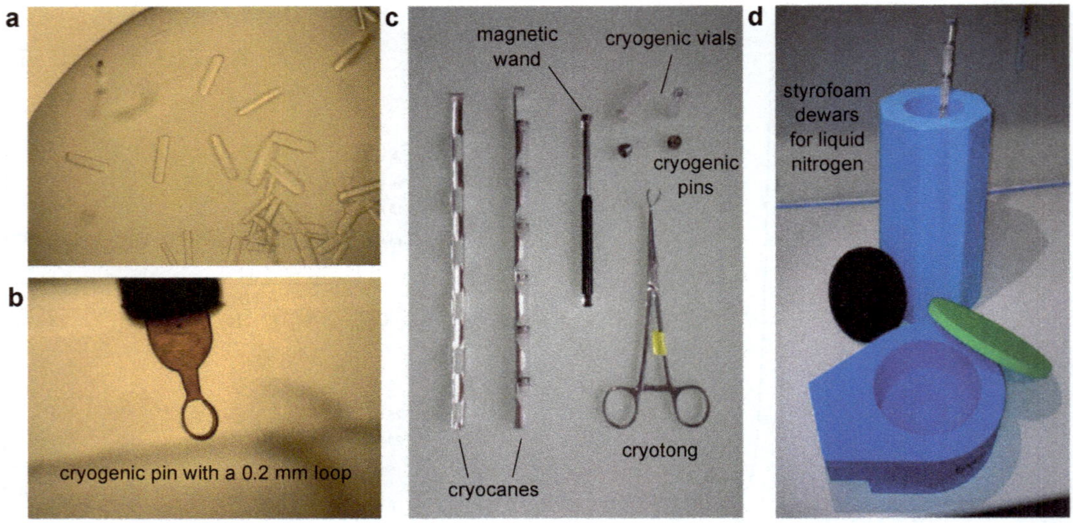

Fig. 3 Equipment required for crystal freezing. (**a**) Crystallization drop displaying typical crystals obtained for hS100A8 in complex with zinc and calcium. (**b**) Cryogenic pin holding a 0.2 mm loop as visualized under the stereomicroscope. (**c**) Tools for the manipulation and storage of protein crystals in liquid nitrogen. (**d**) Liquid nitrogen containers used during crystal freezing. The taller container is used to store cryocanes bearing the cryogenic vials with the crystals. The smaller container is used to fill the cryogenic vials with liquid nitrogen and perform flash-freezing once the crystal has been fished from the crystallization drop

size are then tested. For this purpose, prepare a new set of crystallization conditions derived from the initial hit by varying the pH, salt, precipitant, and any other component concentration, and grow new crystals over these conditions. The size of the crystals, the sharpness of their edges, and the absence of visible defects are generally good indicators that the crystal properties have been improved. Figure 3a shows crystals obtained for hS100A8 in the presence of calcium and zinc.

3.2.2 Crystal Fishing and Flash Cooling

Once crystals have grown to their maximal size, they must be frozen in liquid nitrogen in order to subject them to X-ray beams without too much damage.

1. Prepare a cryoprotectant solution that will allow to flash freeze the protein crystal without ice formation and crystal damage. This solution corresponds to the crystallization condition in which the crystal appeared supplemented with a cryoprotectant agent (*see* **Note 10**).
2. Prepare containers with liquid nitrogen to manipulate the crystal, and store the cryocanes bearing the cryogenic vials (Fig. 3).
3. Select a cryogenic pin with a loop fitting the size of the crystal you want to fish (Fig. 3b), and attach it to a magnetic wand. Attach the corresponding cryovial to a cryotong, and plunge it into liquid nitrogen so that the vial is filled with liquid nitrogen.

4. Place a 3 μL drop of the cryoprotectant solution on a glass cover slide, and place the slide on top of the crystallization plate under the microscope.
5. Remove the sealing tape around the crystallization reservoir containing the crystal using a scalpel.
6. Plunge the cryogenic pin in the crystallization drop, place the crystal in the loop using capillarity, and transfer it into the drop of cryoprotectant solution.
7. After a few seconds in the cryoprotectant solution, fish the crystal again in the loop, and plunge rapidly the cryogenic pin in liquid nitrogen.
8. Once the crystal is frozen, transfer the cryogenic pin into the cryogenic vial. Make sure to keep the crystal inside liquid nitrogen during the whole procedure.
9. Using the cryogenic tong, transfer the vial containing the crystal onto a cryocane. Each cryocane can usually hold five or six vials.
10. Frozen crystals on cryocanes can be stored in a large dewar filled with liquid nitrogen until further use.

3.3 Collection of Diffraction Data

The diffraction properties of the crystals are generally analyzed on a dedicated X-ray crystallography beamline at a synchrotron. Figure 4a represents the classical experimental setup allowing to record the diffraction pattern of the crystal of interest at the beamline. The cryogenic pin holding the crystal is mounted on a goniometer head which allows the crystal to rotate in the monochromatic X-ray beam. The mounting is either performed manually or with the help of a robotized sample changer. During the entire mounting procedure and later on, once placed on the goniometer head, the cryogenic pin is maintained in liquid nitrogen to keep the crystal frozen. The crystal is then aligned in the path of the incoming X-ray beam using computer-assisted procedures. Finally, the crystal is subjected to X-rays while rotating on the goniometer head, and the diffracted X-ray beams are recorded on a plate detector (Fig. 4b). To obtain a complete dataset, one needs to rotate the crystal of an angle sufficient to collect the minimal set of diffracted X-ray beams that will allow to generate the entire set of diffracted X-rays by applying the symmetry elements of the crystal. For example, if the crystal presents a fourfold symmetry element, one needs in principle to only rotate the crystal by 90° in total to generate the entire reciprocal space of diffracted X-ray beams. The wavelength of the incoming X-ray is generally set close to 1.0 Å. If anomalous scatterers are present in the crystal, this value can be modified so that it is in the vicinity of the absorption edge of the anomalous elements (at the absorption edge, the

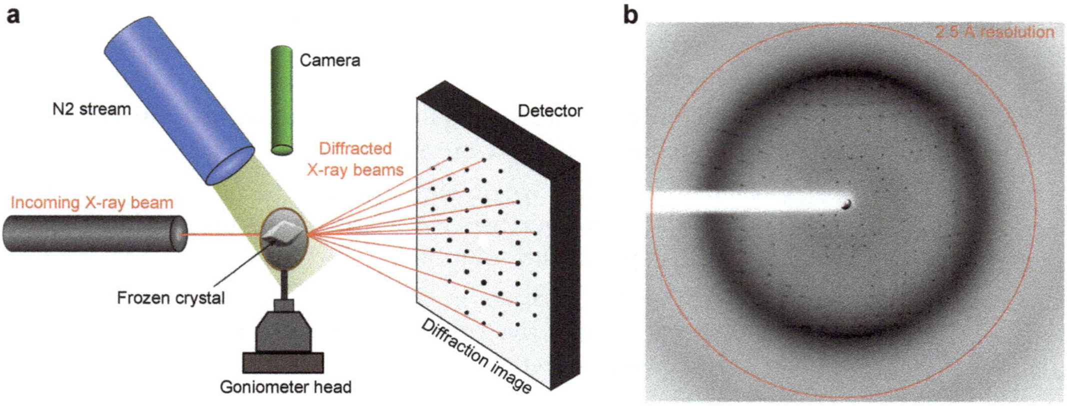

Fig. 4 Measurement of the X-ray diffraction pattern of a protein crystal. (**a**) Experimental setup for the recording of X-ray diffraction images at a synchrotron beamline. The cryogenic pin holding the frozen crystal is mounted on a goniometer head which allows the crystal to rotate in the X-ray beam. The crystal is maintained under a permanent flux of liquid nitrogen in order to remain frozen during the entire measurement. A camera placed above the crystal allows to align the crystal in the path of the X-ray beam using dedicated beamline software. A monochromatic X-ray beam with wavelength generally close to 1 Å is then sent onto the crystal, on an axis perpendicular to the axis of the goniometer head. On the opposite side of the crystal, the diffracted X-ray beams are recorded as diffraction spots on the surface of a plate detector. (**b**) Typical diffraction pattern obtained for a crystal of Ca^{2+}/Zn^{2+}-bound hS100A8 during a 1.0° oscillation on the goniometer head. The diffraction image was collected on beamline I911-3 at MAX-lab in Lund (Sweden)

anomalous signal will be maximum; at a wavelength higher than the absorption edge, no more anomalous signal will be measured).

3.4 Data Reduction, Structure Determination, and Refinement

3.4.1 Data Reduction

This step consists in analyzing all the diffraction images of a single dataset in order to derive the amplitude $|F_{hkl}|$ of each diffracted X-ray beam and to attribute it to the corresponding family of plans (*hkl*) in the crystal. Various dedicated programs are available to perform this analysis, including XDS [25], iMOSFLM [26], or the HKL package [30]. They all function similarly following a four-step procedure: indexation, integration, scaling, and truncation. Input files to run these programs are generated automatically during data acquisition, or input parameters can be extracted by the programs from the diffraction images. During indexation, the program scans the diffraction images to (1) find the diffraction spots, (2) derive the cell parameters and orientation of the crystal based on the respective positions of the spots, and (3) propose the most likely space group. Based on this prediction, each diffraction spot is indexed with the corresponding set of *hkl* indices. The second step allows to integrate the intensity of each diffraction spot on the entire dataset, keeping in mind that a diffraction spot can appear on several diffraction images. The measured intensities are then scaled to take into account variations that can happen intrinsically or during the measurement (variations in crystal exposure to the X-ray beam, changes in X-ray intensity, radiation damage, etc.). The scaling step should in particular ensure that symmetry equivalent reflections have identical intensities. Finally, the amplitude $|F_{hkl}|$ of each diffracted X-ray beam is derived from the measured intensities with an associated standard deviation $\sigma(|F_{hkl}|)$, and a so-called mtz file listing the measured $|F_{hkl}|$ and $\sigma(|F_{hkl}|)$ for each hkl index is generated.

If the crystal contains anomalous scatterers (such as certain divalent cations), the centrosymmetry is broken, i.e., Friedel's pairs of reflection (F_{hkl} and F_{-h-k-l}) do not have equal intensities anymore. In that case, the difference between these pairs represents the anomalous contribution to the scattering of the particular atoms. This anomalous signal can be used later on to identify the position of the anomalous scatterers in the crystal. When anomalous signal is expected to be present, one should therefore specify in the data reduction program that Friedel's law is broken (default assumption is that Friedel's law is followed). Data processing with nonequivalent Friedel reflections generates two additional columns in the final mtz file, one encompassing the measured anomalous difference D_{ano} and the other giving the standard deviation of this measurement $\sigma(D_{\text{ano}})$.

3.4.2 Structure Determination

Structural models for almost all S100 proteins are available in the Protein Data Bank. Thus, new structures are generally solved with the technique of molecular replacement (MR). As the name implies, this technique consists in using the structure of a molecule (molecule A) closely resembling the molecule present in our protein crystal (molecule B). MR programs such as Phaser [31] will then place molecule A in a virtual crystal having the same cell parameters, orientation, and space group as the one obtained experimentally for molecule B. The program therefore tries to find the most likely rotational and translational operations that will allow such placement. Once this is achieved, a theoretical set of structure factors is generated for this virtual crystal of molecule A. The resulting phases are used in combination with the amplitudes measured experimentally to solve the structure of molecule B and calculate an initial electron density map of the crystal by inverse Fourier transformation. The more closely molecule A resembles molecule B, the more accurate the initial set of phases derived from MR will be. To solve the structure of hS100A8 in complex with Ca^{2+} and Zn^{2+}, we used the published structure of human S100A8 in complex with Ca^{2+} [32] as a template for MR search.

3.4.3 Structure Refinement

Once an initial electron density map has been calculated, structure refinement allows to rebuild and optimize the corresponding atomic model. When the structure has been solved by MR, the MR search model (molecule A in the example) is already positioned in the initial electron density map to serve as a template for the rebuilding of the real structure (molecule B). Refinement is then performed by using iterative cycles of manual rebuilding in a 3D-visualization software such as Coot [29] and energy minimization of the resulting atomic model using the PHENIX program [27], until convergence is reached, i.e., until the rebuilt structural model fits as best as possible the diffraction data measured experimentally.

3.5 Characterization of the Bound Ions Through Analysis of the Anomalous Signal

This analysis should be performed once a partially refined atomic model of the protein structure has been obtained. For each ion suspected to be present in the crystal, two complete X-ray diffraction datasets must be collected: one at a wavelength close (but inferior) to the absorption edge of the atom and the other one at a wavelength above the absorption edge (where there won't be any remaining anomalous signal for the considered atom). Both datasets are processed as described in Subheading 3.4.1, specifying that Friedel's law is not followed (i.e., Friedel's pair will not have equal amplitudes and should therefore not be merged). This generates a set of D_{ano} values and the corresponding standard deviations $\sigma(D_{ano})$ for each hkl index, which reflect the overall anomalous contribution in the X-ray scattered beams. These D_{ano} values are then combined to the set of phases Φ_{hkl} calculated from the current atomic model of the protein structure (best refined model). First, the phases calculated from the refined structural model are weighted using the program SIGMAA [33] from the CCP4 package [28]. This step allows to remove bias on the phases that can arise from the fact that the structural model is incomplete and/or partially inaccurate. The phases and their associated weights are then combined in a single file with the anomalous differences D_{ano} and their associated standard deviations $\sigma(D_{ano})$, with the help of the CAD program from the CCP4 package [28]. Finally, an anomalous Fourier difference map is calculated from this set of D_{ano}/Φ_{hkl} values with the program FFT [34], resulting in an electron density map which only corresponds to the anomalous scatterers present in the protein crystal, at the considered wavelength.

Figure 5 summarizes the analysis of the anomalous signal from crystals of hS100A8 in complex with zinc and calcium. As shown in panel a, zinc presents an absorption edge of X-rays at 1.2837 Å, whereas calcium has no specific absorption edge but rather displays an increasing X-ray absorption with increasing X-ray wavelengths. Accordingly, a complete dataset for these crystals was first recorded at a wavelength just before the absorption edge of zinc (1.27 Å) at which the anomalous scattering of Zn^{2+} ions is maximal whereas Ca^{2+} ions have only minor contribution. A second dataset was collected at a wavelength just after the absorption edge of zinc (1.30 Å) at which the anomalous contribution of Zn^{2+} ions is lost, whereas Ca^{2+} ions have a slightly enhanced contribution to the anomalous signal. Figure 5b shows the stage of refinement (atomic model and corresponding electron density map) used to

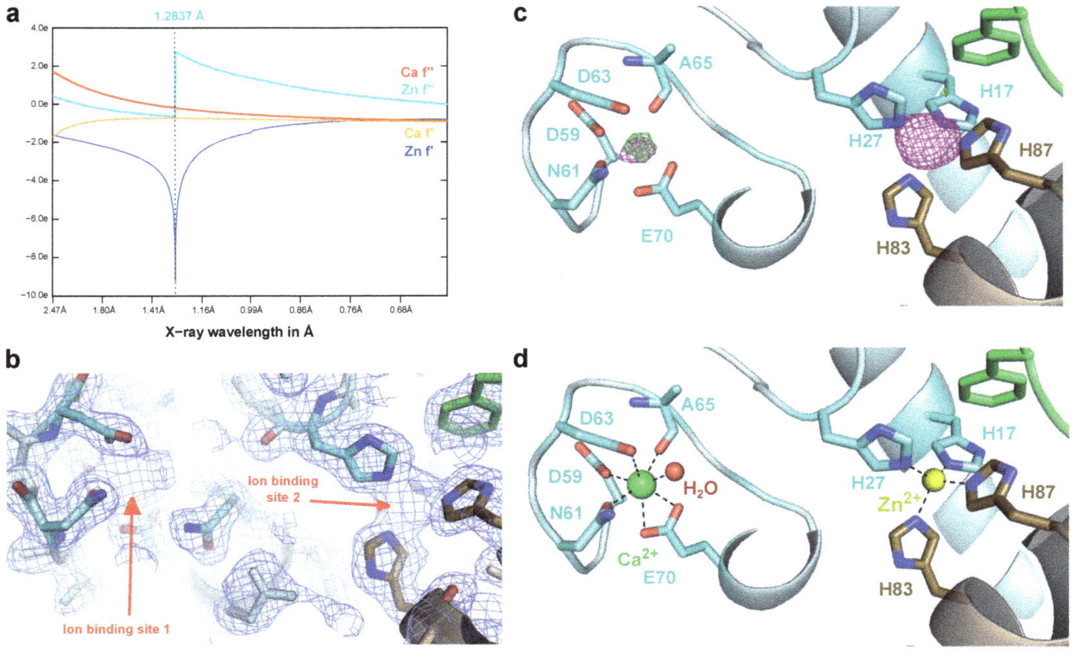

Fig. 5 Analysis of the anomalous signal to locate divalent cations in the structure of hS100A8. (**a**) Plot of the anomalous scattering coefficients f′ and f″ for Zn and Ca against the wavelength of the incoming X-ray beam in Å (adapted from http:/skuld.bmsc.washington.edu/cgi-bin/edgeplots). (**b**) Partially refined structural model for hS100A8 in complex with zinc and calcium and the corresponding 2mFo-DFc electron density map at the stage where anomalous signal analysis was performed. On the particular region displayed on the figure, two ion-binding sites are clearly visible, but the corresponding ions have not yet been placed in the model. (**c**) Anomalous difference Fourier maps calculated using phases and weight from the best refined atomic model (without ions) obtained with the native dataset collected at $\lambda = 1.0$ Å and anomalous differences from the dataset collected at $\lambda = 1.27$ Å (purple mesh, contour at 4 σ) or at $\lambda = 1.30$ Å (green mesh, contour at 4 σ). For the ion-binding site on the right, the anomalous signal clearly disappears at a wavelength above the Zn absorption edge ($\lambda = 1.30$ Å), indicating that the ion present is Zn^{2+}. In contrast, the anomalous signal is weak but still present at both wavelengths for the ion-binding site on the left, suggesting that the ion bound is Ca^{2+}. (**d**) Final refined model where ions and water molecules have been placed

extrapolate weighted Φ_{hkl} phases. The resulting anomalous Fourier difference maps calculated at 1.27 and 1.30 Å are then displayed in Fig. 5c. Anomalous signal for the ion present in the electron density surrounded by four histidines, on the right side of the panel, is quite strong at $\lambda = 1.27$ Å and disappears at $\lambda = 1.30$ Å, which allows to identify the ion as Zn^{2+}. In contrast, the ion at the center of the loop, on the left side of the panel, displays weak anomalous signal at both wavelengths and is therefore a Ca^{2+} ion. This analysis allows to unambiguously identify the nature of the two ions present in the observed ion-binding sites and to complete the refinement of the structural model of hS100A8 in complex with zinc and calcium (Fig. 5d).

4 Notes

1. A classical purification protocol for recombinant proteins expressed in bacteria is to first perform an affinity chromatography, generally on a nickel column due to the presence of a polyhistidine (His_6) tag on the engineered protein, followed by the cleavage of the tag with the appropriate protease and the removal of the tag through a second run on the affinity chromatography column. In the case of hS100A8, this procedure was however not successful for us since hS100A8 contains several histidines on its C-terminus. Our first attempts to purify it with an N-terminal cleavable polyhistidine tag therefore did not succeed as the protein remained bound to the nickel column even after proteolytic removal of the His_6 tag. Instead, we chose to take advantage of the presence of the histidine patch on hS100A8 C-terminus. The protein was thus expressed without any additional tag and purified first on a nickel column followed by ion exchange and size exclusion chromatography.
2. Many S100 proteins contain free cysteines that are highly reactive and can associate from two adjacent molecules to form disulfide cross-linked dimers. So far, it has been a common trend to purify these proteins in the presence of a reducing agent or to even mutate away these cysteine residues in order to avoid S–S cross-linking. However, several reports now indicate that these disulfide-linked forms may have specific pro-inflammatory activity [35, 36] and we also uncovered recently the importance of free cysteine for the binding of S100A6 to the receptor for advanced glycation end products (RAGE) [20, 37]. We therefore prefer to purify S100 proteins in their native state, without any reducing agent, in order to mimic more closely the conditions S100 proteins would meet in the extracellular compartment.
3. To visualize the disulfide-cross-linked species obtained during purification, one can prepare the samples to load on the SDS-PAGE gel with a loading buffer containing no reducing agent (*see* Fig. 2c).
4. S100 proteins commonly form both homodimers and higher-order oligomers such as homotetramers. For crystallographic studies, it is crucial to obtain a monodisperse sample corresponding to a single oligomeric state of the protein. The final purification step through size exclusion chromatography is therefore essential to identify and separate the various oligomeric forms of the studied protein.
5. A routine purification from 4 L cultures generally yields between 50 and 100 mg of highly pure S100 protein.

6. Glycerol is often added to protein samples to facilitate their storage at low temperatures. However, addition of any chemical agent may interfere later on with protein crystallization. Instant freezing in liquid nitrogen and subsequent storage at −80 °C without addition of any cryoprotectant compound allows to avoid changing the buffer composition of the sample. One should however check that the protein remains stable after one cycle of flash-freezing and thawing. In any case, aliquots should only be thawed once and then discarded to avoid protein degradation.
7. The optimal protein concentration used to set up crystallization experiments has to be adjusted for each protein. Ideally, one should try several concentrations during the first screening to determine the most suitable range. This will be determined by monitoring the crystallization drops after equilibration has been reached. If too many drops display protein precipitates, one should try to decrease protein concentration. Inversely, if most drops remain clear after a week or two, protein concentration should be increased. For S100 proteins, we generally perform screening at both 20 and 10 mg/mL. However, some S100 proteins are highly soluble, and the concentration may have to be increased up to 50 mg/mL.
8. Many proteins tend to precipitate in the presence of zinc when the protein concentration is elevated. The concentration of zinc chosen for the crystallization experiment must therefore reach a compromise between having enough zinc to saturate most S100 molecules in the protein solution and avoiding massive precipitation which would render the protein sample unsuitable for crystallization. In any case, all traces of precipitate must be removed by centrifugation before setting up the crystallization experiment.
9. If crystallization trays are set up by hands, one usually mixes 0.3–0.5 μL of protein solution and 0.3–0.5 μL of crystallization solution, lower volumes being difficult to pipet reproducibly by hand. Alternatively, a crystallization robot can be used, allowing to pipet much smaller volumes down to 20–50 nL. Advantages are the rapidity of the procedure (a few minutes to fill an entire crystallization plate), the robustness of the method (lower rate of pipetting errors), and the low amount of protein required for crystallization. Very small crystallization drops may however dehydrate quite rapidly and are more difficult to monitor under the microscope. In contrast, larger drops offer the possibility to directly fish and analyze crystals from initial screening, which may be a considerable advantage if crystals are difficult to reproduce.

10. Glycerol, polyethylene glycols (PEGs), and sucrose are commonly used as cryoprotectant agents due to their high viscosity, but several other chemical compounds used for crystallization have also cryopreservative properties. There is no rule as to which cryoprotectant will be more efficient on a particular crystal, so several compounds should be tested. The concentration of cryoprotectant required depends on the composition of the crystallization buffer (e.g., buffers containing high amounts of salt or viscous compounds will require lower concentrations of cryoprotectant). Concentrations of the four most commonly used cryoprotectants (glycerol, PEG 400, ethylene glycol, and 1,2-propanediol) to be used in the 100 most frequent crystallization buffers have been determined and can serve as a basis to design a cryoprotectant solution specific to your own crystals [38].

References

1. Schafer BW, Heizmann CW (1996) The S100 family of EF-hand calcium-binding proteins: functions and pathology. Trends Biochem Sci 21:134–140
2. Donato R, Cannon BR, Sorci G, Riuzzi F, Hsu K, Weber DJ et al (2013) Functions of S100 proteins. Curr Mol Med 13:24–57
3. Calissano P, Moore BW, Friesen A (1969) Effect of calcium ion on S-100, a protein of the nervous system. Biochemistry 8:4318–4326
4. Santamaria-Kisiel L, Rintala-Dempsey AC, Shaw GS (2006) Calcium-dependent and -independent interactions of the S100 protein family. Biochem J 396:201–214
5. Baudier J, Gerard D (1983) Ions binding to S100 proteins: structural changes induced by calcium and zinc on S100a and S100b proteins. Biochemistry 22:3360–3369
6. Gilston BA, Skaar EP, Chazin WJ (2016) Binding of transition metals to S100 proteins. Sci China Life Sci 59:792–801
7. Heizmann CW, Cox JA (1998) New perspectives on S100 proteins: a multi-functional Ca(2+)-, Zn(2+)- and Cu(2+)-binding protein family. Biometals 11:383–397
8. Zackular JP, Chazin WJ, Skaar EP (2015) Nutritional immunity: S100 proteins at the host-pathogen interface. J Biol Chem 290:18991–18998
9. Damo SM, Kehl-Fie TE, Sugitani N, Holt ME, Rathi S, Murphy WJ et al (2013) Molecular basis for manganese sequestration by calprotectin and roles in the innate immune response to invading bacterial pathogens. Proc Natl Acad Sci U S A 110:3841–3846
10. Koch M, Bhattacharya S, Kehl T, Gimona M, Vasak M, Chazin W et al (2007) Implications on zinc binding to S100A2. Biochim Biophys Acta 1773:457–470
11. Yanamandra K, Alexeyev O, Zamotin V, Srivastava V, Shchukarev A, Brorsson AC et al (2009) Amyloid formation by the pro-inflammatory S100A8/A9 proteins in the ageing prostate. PLoS One 4:e5562
12. Fritz G, Botelho HM, Morozova-Roche LA, Gomes CM (2010) Natural and amyloid self-assembly of S100 proteins: structural basis of functional diversity. FEBS J 277:4578–4590
13. Moroz OV, Blagova EV, Wilkinson AJ, Wilson KS, Bronstein IB (2009) The crystal structures of human S100A12 in apo form and in complex with zinc: new insights into S100A12 oligomerisation. J Mol Biol 391:536–551
14. Lin H, Andersen GR, Yatime L (2016) Crystal structure of human S100A8 in complex with zinc and calcium. BMC Struct Biol 16:8
15. Baudier J, Glasser N, Gerard D (1986) Ions binding to S100 proteins. 1. Calcium-binding and zinc-binding properties of bovine brain S100-alpha-alpha, S100a-(alpha,beta), and S100b-(beta-beta), protein – Zn-2+ regulates Ca-2+ binding on S100b protein. J Biol Chem 261:8192–8203
16. Kerkhoff C, Vogl T, Nacken W, Sopalla C, Sorg C (1999) Zinc binding reverses the calcium-induced arachidonic acid-binding capacity of the S100A8/A9 protein complex. FEBS Lett 460:134–138

17. Moroz OV, Burkitt W, Wittkowski H, He W, Ianoul A, Novitskaya V et al (2009) Both Ca2+ and Zn2+ are essential for S100A12 protein oligomerization and function. BMC Biochem 10:11
18. Leclerc E, Fritz G, Vetter SW, Heizmann CW (2009) Binding of S100 proteins to RAGE: an update. Biochim Biophys Acta 1793:993–1007
19. Bjork P, Bjork A, Vogl T, Stenstrom M, Liberg D, Olsson A et al (2009) Identification of human S100A9 as a novel target for treatment of autoimmune disease via binding to quinoline-3-carboxamides. PLoS Biol 7:e97
20. Yatime L, Betzer C, Jensen RK, Mortensen S, Jensen PH, Andersen GR (2016) The structure of the RAGE:S100A6 complex reveals a unique mode of homodimerization for S100 proteins. Structure 24:2043–2052
21. Rossmann MG, Blow DM (1962) Detection of sub-units within crystallographic asymmetric unit. Acta Crystallogr D Biol Crystallogr 15:24–31
22. Perutz MF (1956) Isomorphous replacement and phase determination in non-centrosymmetric space groups. Acta Crystallogr D Biol Crystallogr 9:867–873
23. Kendrew JC, Bodo G, Dintzis HM, Parrish RG, Wyckoff H, Phillips DC (1958) 3-Dimensional model of the myoglobin molecule obtained by X-ray analysis. Nature 181:662–666
24. Taylor GL (2010) Introduction to phasing. Acta Crystallogr D Biol Crystallogr 66:325–338
25. Kabsch W (2010) Xds. Acta Crystallogr D Biol Crystallogr 66:125–132
26. Battye TG, Kontogiannis L, Johnson O, Powell HR, Leslie AG (2011) iMOSFLM: a new graphical interface for diffraction-image processing with MOSFLM. Acta Crystallogr D Biol Crystallogr 67:271–281
27. Adams PD, Afonine PV, Bunkoczi G, Chen VB, Davis IW, Echols N et al (2010) PHENIX: a comprehensive Python-based system for macromolecular structure solution. Acta Crystallogr D Biol Crystallogr 66:213–221
28. Winn MD, Ballard CC, Cowtan KD, Dodson EJ, Emsley P, Evans PR et al (2011) Overview of the CCP4 suite and current developments. Acta Crystallogr D Biol Crystallogr 67:235–242
29. Emsley P, Cowtan K (2004) Coot: model-building tools for molecular graphics. Acta Crystallogr D Biol Crystallogr 60:2126–2132
30. Minor W, Cymborowski M, Otwinowski Z, Chruszcz M (2006) HKL-3000: the integration of data reduction and structure solution – from diffraction images to an initial model in minutes. Acta Crystallogr D Biol Crystallogr 62:859–866
31. McCoy AJ, Grosse-Kunstleve RW, Adams PD, Winn MD, Storoni LC, Read RJ (2007) Phaser crystallographic software. J Appl Crystallogr 40:658–674
32. Ishikawa K, Nakagawa A, Tanaka I, Suzuki M, Nishihira J (2000) The structure of human MRP8, a member of the S100 calcium-binding protein family, by MAD phasing at 1.9 A resolution. Acta Crystallogr D Biol Crystallogr 56:559–566
33. Read RJ (1986) Improved Fourier coefficients for maps using phases from partial structures with errors. Acta Crystallogr A 42:140–149
34. Read RJ, Schierbeek AJ (1988) A phased translation function. J Appl Crystallogr 21:490–495
35. Haase-Kohn C, Wolf S, Lenk J, Pietzsch J (2011) Copper-mediated cross-linking of S100A4, but not of S100A2, results in proinflammatory effects in melanoma cells. Biochem Biophys Res Commun 413:494–498
36. Winningham-Major F, Staecker JL, Barger SW, Coats S, Van Eldik LJ (1989) Neurite extension and neuronal survival activities of recombinant S100 beta proteins that differ in the content and position of cysteine residues. J Cell Biol 109:3063–3071
37. Yatime L (2017) A cystein-dependent activation mechanism for the pro-inflammatory ligands of RAGE? Med Sci (Paris) 33:351–354
38. McFerrin MB, Snell EH (2002) The development and application of a method to quantify the quality of cryoprotectant solutions using standard area-detector X-ray images. J Appl Crystallogr 35:538–545

Chapter 27

Analysis of Ca^{2+}-Dependent Weibel–Palade Body Tethering by Live Cell TIRF Microscopy: Involvement of a Munc13-4/S100A10/Annexin A2 Complex

Nina Criado Santos, Tarek Chehab, Anna Holthenrich, and Volker Gerke

Abstract

Endothelial cells respond to blood vessel injury by the acute release of the procoagulant von Willebrand factor, which is stored in unique secretory granules called Weibel–Palade bodies (WPBs). Stimulated, Ca^{2+}-dependent exocytosis of WPBs critically depends on their proper targeting to the plasma membrane, but the mechanism of WPB-plasma membrane tethering prior to fusion is not well characterized. Here we describe a method to visualize and analyze WPB tethering and fusion in living human umbilical vein endothelial cells (HUVEC) by total internal reflection fluorescence (TIRF) microscopy. This method is based on automated object detection and allowed us to identify components of the tethering complex of WPBs and to monitor their dynamics in space and time. An important tethering factor identified by this means was Munc13-4 that was shown to interact with S100A10 residing in a complex with plasma membrane-bound annexin A2.

Key words Annexin, Calcium, Endothelial cell, Exocytosis, Secretion, S100 protein

1 Introduction

Weibel–Palade bodies (WPBs) are unique secretory organelles of endothelial cells [1]. They serve as storage granules for important factors regulating vascular homeostasis. Factors that are stored in WPBs for acute release on demand include the coagulant glycoprotein von Willebrand factor (VWF) and the leukocyte receptor P-selectin [2, 3]. WPBs form at the trans-Golgi network and are then transported along microtubules toward the cell periphery where they are anchored at the actin cytoskeleton for final maturation. Exocytosis of the fully matured WPBs is evoked by secretagogues such as histamine and thrombin that raise the intracellular Ca^{2+} concentration. Typically, acute Ca^{2+}-dependent WPB secretion is initiated after blood vessel injury or local inflammatory activation. It results in a marked change of endothelial surface properties, rapidly transforming the antiadhesive surface of resting endothelial

Claus W. Heizmann (ed.), *Calcium-Binding Proteins of the EF-Hand Superfamily: From Basics to Medical Applications*, Methods in Molecular Biology, vol. 1929, https://doi.org/10.1007/978-1-4939-9030-6_27,

cells, which supports the unrestricted circulation of blood cells, into an adhesive one capable of capturing platelets and leukocytes. Thus, controlled WPB exocytosis serves an important role in regulating blood vessel homeostasis [4, 5].

Several factors involved in WPB maturation and exocytosis have been described; however, cell components participating in plasma membrane tethering of WPBs have not been identified. Once WPBs are committed to exocytosis, the actual fusion reaction is mediated by one or several different soluble *N*-methylmaleimide-sensitive factor attachment protein receptor (SNARE) complexes. WPB-associated VAMP3 and plasma membrane-localized syntaxin-4 and SNAP-23 form a trans-SNARE complex that mediates WPB exocytosis and possibly functions in conjunction with accessory proteins [6–9]. It is unknown how the cortically anchored WPBs are transferred to the fusion-mediating SNARE machinery after secretagogue stimulation and the experimental analysis of the transient plasma membrane-tethered intermediate that is formed in the course of acute WPB exocytosis is challenging. Here we describe a total internal reflection fluorescence (TIRF) microscopy-based method to visualize and analyze WPB tethering and fusion in living primary human endothelial cells (HUVEC). TIRF microscopy allows a high spatial and temporal resolution and is ideally suited to visualize tethering and fusion due to its property that it only records signals from fluorophores excited within 100–200 nm of the plasma membrane [10]. We use an automated object detection, originally developed to detect Golgi structures, customized to identify WPBs [11]. Using this technique, we characterized the dynamics of Munc13-4 and could show that it serves as tethering factor in histamine-evoked WPB exocytosis: In resting cells, Munc13-4 is homogenously distributed at the plasma membrane as well as present on WPBs. Secretagogue stimulation triggers an increased recruitment of Munc13-4 to WPBs and a clustering of Munc13-4 at sites of WPB-plasma membrane contact. We could also show that this clustering is mediated by an interaction of Munc13-4 with the S100 protein S100A10 [12]. S100A10 itself resides in a tight complex with annexin A2 that Ca^{2+}-dependently translocates to the plasma membrane, most likely serving as a docking site for Munc13-4. Thus, the valuable technique described here was instrumental to build a mechanistic model describing the regulation of WPB tethering. Furthermore, this technique can be adapted and customized to study tethering of other exocytic organelles of interest.

2 Materials

2.1 HUVEC Culture and Transfection

1. Endothelial cell growth medium (ECGM1 [Promocell]) supplemented with 30 μg/mL gentamicin (CytoGen), 15 ng/mL amphotericin B (Biochrom), and 100 I.E. heparin (Ratio-Pharm) and mixed 1:1 with M199 (Biochrom) containing 10% fetal calf serum (Sigma-Aldrich), 30 μg/mL gentamicin, 15 ng/mL amphotericin B, and 100 I.E. heparin.
2. Imaging medium: Mixed endothelial growth medium supplemented with 20 mM HEPES (Biochrom).
3. 0.05% trypsin/0.02% EDTA (Biochrom).
4. Dulbecco's phosphate-buffered saline (DPBS).
5. 100 mM histamine (Sigma-Aldrich).
6. Transfection buffer: Amaxa nucleofection (Lonza) or self-made buffer: 4 mM KCl, 10 mM $MgCl_2$, 10 mM sodium succinate, 100 mM NaH_2PO_4, pH 7.4.
7. Coating reagent: 50 μg/mL rat tail collagen I (Advanced BioMatrix) in 0.02 M acetic acid (Sigma-Aldrich).
8. μ-slide eight-well glass bottom dishes (ibidi).
9. Electroporation cuvettes, 2 mm (Aurora Borealis Control BV).
10. Nucleofector™ AAD 1001 Device (Lonza).
11. 60 mm cell culture dishes, CellBind (Corning).
12. Centrifuge.
13. 15 mL Falcon tubes.
14. Eppendorf tubes.
15. Immersion oil for 37 °C (Zeiss).

2.2 Plasmids

We routinely use VWF as a WPB marker; however, other WPB markers such as P-selectin can also be selected. In the described TIRF assay we investigate the role of Munc13-4 in WPB tethering using plasmids encoding different Munc13-4 versions such as full length and truncation or point mutants. This assay can be used to investigate other potential tethering factors of interest such as other Munc13 proteins or Munc18.

1. YFP-Munc13-4 and Munc13-4-mKate were described previously [13, 14].
2. VWF-GFP and VWF-RFP were described in Babich et al. [15] and were kindly provided by Tom Carter (St. George's University, London, UK).

2.3 Confocal Microscopy

Confocal imaging employed a Zeiss LSM780 confocal microscope and a Plan-Apochromat 63×/1.4 oil immersion objective + heated chamber.

2.4 TIRF Microscopy

TIRF microscopy is ideally suited to investigate exocytosis or other processes occurring at or close to the plasma membrane: When excitation light is totally internally reflected at the coverslip at its liquid interface, an evanescent wave is generated at the coverslip-liquid interface. The intensity of the evanescent wave exponentially decays with distance from the glass coverslip, so that only fluorophores within 100–200 nm of the glass coverslip are efficiently excited.

1. Olympus IX71 microscope with an objective-type TIRFM setup (Till Photonics) and a customized heated incubation chamber.
2. Objective 60×.
3. Lasers 488 and 561.
4. TILL Image QE charge-coupled device camera (Till Photonics).
5. MetaMorph software (Molecular Devices).
6. Manually adjustable TIRF angle.

2.5 MATLAB-Based Automated Object Detection

To detect Munc13-4-positive WPBs over time in stimulated cells from TIRF live recordings, we used the MATLAB-based image-processing algorithm MorphoQuant originally developed to detect Golgi structures [11]. MorphoQuant is freely available at https://github.com/tischi/MorphoQuant. The algorithm detects locally bright large objects of arbitrary shape.

The algorithm used a top-hat filter with a circular structural element whose radius r1 was defined to include a typical WPB ($r1 = 15$ pixel with 1 pixel $\approx$ 133 nm). This filter segmented images into a foreground image containing objects of the size of Munc13-4-positive WPBs and a background image representing the Munc13-4 at the plasma membrane. Subsequent filtering steps improved the distinction between foreground objects and background signal based on their relative fluorescence intensity, defined by an adjustable threshold t1 ($t1 = 8$ a.u.). This procedure identified Munc13-4-positive locally bright (adjusted by $t1$) objects of a defined size (adjusted by $r1$) but otherwise arbitrary shape; we visually confirmed that object detection was accurate and that objects showed the typical elongated shape of WPBs. The algorithm was pre-configured to detect a second class of locally bright objects of smaller size and dot-like shape (defined by $r2 = 2$, $t2 = 2$). These smaller objects were probably also Munc13-4-positive WPBs, but we excluded them from our measurements because they were defined as dot-like and we could not define them as WPBs by their elongated shape.

2.6 Software and Macros for Image Analysis

1. MATLAB.
2. MorphoQuant.
3. Image J.

3 Methods

3.1 HUVEC Culture

For our experiments, we use primary endothelial cells isolated from umbilical cord veins (HUVEC) as described before [16]. HUVECs can also be purchased as cryoconserved pools (Promocell). For attachment of HUVECs to the surface of the culture dishes, coating with reagents such as collagen or fibronectin is necessary. Alternatively, commercial pre-coated dishes such as CellBind dishes (Corning) are available and are routinely utilized by us.

1. Culture cells in mixed endothelial growth medium at 37 °C and 5% CO_2.
2. Use nearly confluent cells at passages 2–4.
3. Passage cells at a maximum dilution of 1:5.

3.2 HUVEC Transfection

To visualize WPBs in a live cell setup, cells are transfected to express fluorophore tagged proteins localizing to WPBs. We routinely use VWF either conjugated to the green fluorophore EGFP or to red mRFP [15]. Alternatively, the luminal part of P-Selectin also localizes nearly exclusively to WPBs [17]. As opposed to VWF, P-Selectin luminal diffuses away from the fusion site rather quickly after exocytosis of WPBs.

Cells are co-transfected with a second protein of interest, in our case YFP-Munc13-4 or Munc13-4-mKate [13, 14]. Munc13-4 localizes to both the plasma membrane and WPBs. A plasma membrane staining facilitates the adjustment of the TIRF angle for the image acquisition. One 60 mm cell culture dish of HUVECs (approximately 21 cm^2) with 70–80% confluency is sufficient for expansion into three eight-well-μ-slide glass bottom dishes (approximately 1 cm^2 per well) when imaging 24 h after transfection.

1. Coat μ-slide glass bottom dishes by covering each well with collagen coating reagent and incubate for 20 min at 37 °C.
2. Wash the μ-slide dishes three times with DPBS.
3. Harvest HUVECs by washing with DPBS, trypsinizing and pelleting cells.
4. Prepare the cuvettes by adding 2–10 μg of each plasmid DNA into the cuvette. Per cuvette, 10–30 cm^2 of HUVECs can be transfected. We routinely use one 60 mm culture dish of

70–80% confluency per cuvette and add the following amount of plasmid DNA: 5 μg for VWF constructs, 3 μg for the P-Selectin luminal construct, and 5 μg for Munc13-4 constructs.

5. Aspirate the medium, and carefully resuspend the cell pellet in transfection buffer. Use 90 μL for each transfection.
6. Transfer 100 μL cell suspension into each cuvette, and mix with DNA by gently vortexing.
7. Electroporate the cells with the nucleofector program U-001.
8. Resuspend the cells in medium to a final volume of 4.8 mL per 60 mm dish. Distribute the cells evenly to the μ-slide wells by adding 200 μL of cell suspension to each well.
9. Culture the cells at 37 °C for 24 h. Cells should be 80% confluent for optimal imaging (*see* **Notes 1** and **2**). Some constructs may express better after a different time frame. Adjust the number of cells seeded per well accordingly. For longer incubation times, change the medium 1 day after transfection to remove dead cells and cellular debris.

3.3 Microscope and Sample Preparation

1. Prewarm the microscope heating chamber to ensure a stable temperature during acquisition.
2. Aspirate the medium, and add 200 μL prewarmed imaging medium to each well. Avoid touching the cells with the aspirating pipette to prevent mechanical prestimulation of the cells (*see* **Note 3**).
3. Prepare a 20 mM histamine solution by diluting the 100 mM stock solution 1:50 in mixed endothelial growth medium as stimulation reagent.
4. Set up the acquisition protocol: Record dual-color movies for 3 min with two frames per second (TIRF) or eight frames per minute (Confocal).
5. Add a drop of immersion oil on the objective and mount the eight-well glass bottom dish on the microscope stage.

3.4 TIRF/Confocal Image Acquisition and Stimulation

We routinely use histamine as stimulation reagent which functions by eliciting intracellular Ca^{2+} signals. Other stimulation reagents that activate the same signaling pathway, such as thrombin, can also be used. Furthermore, other secretagogues, such as epinephrine and vasopressin, which elevate intracellular cAMP can also be applied to study WPB tethering (*see* **Note 4**).

1. Select the first well, adjust the focus, and search for co-transfected cells with appropriate expression level and representative expression pattern.

2. Adjust the TIRF angle. Ideally, a large area of the plasma membrane of the selected cells is visible in the TIRF field.
3. Start the acquisition.
4. Stimulate the cells after 20–30 frames by carefully pipetting 10 μL of the stimulation reagent into the well. Make sure not to touch the well with the pipette tip (*see* **Note 1**). The final histamine concentration is 100 μM. Open and close the heating chamber carefully to avoid focus shifts. Write down the time point of stimulation.
5. Adjust the focus during the acquisition if necessary.
6. Repeat **steps 1–5** for the remaining wells.
7. Save acquired images for analysis.

3.5 Stimulation-Induced Recruitment of Munc13-4 to WPBs

Stimulation-induced recruitment of Munc13-4 to WPBs is measured in TIRF or confocal recordings of cells expressing VWF-RFP or VWF-GFP and YFP-Munc13-4 or Munc13-4-mKate, respectively. We investigate here the recruitment of Munc13-4 to WPBs in response to stimulation; however, other proteins of interest can be analyzed using this method.

1. Create multiple regions of interest (ROIs) for all Munc13-4-positive WPBs in the cell by thresholding image stills.
2. Compare mean fluorescence intensities of all ROIs in the stills immediately before and after stimulation.
3. Ensure that the same WPB is visible in both frames. If necessary delete ROIs that do not include the same WPB in both frames due to its movement out of the ROI or loss of focus.
4. Set mean fluorescence before stimulation as 1, and measure the increase after stimulation as *n*-fold change in mean fluorescence intensities for every individual ROI.

3.6 Automated Object Detection of Munc13-4-Positive Objects

Munc13-4-positive objects are detected in TIRF live recordings before and after stimulation using the MATLAB-based image processing algorithm MorphoQuant (see above). We use this image processing algorithm to detect WPBs, but it can be customized to investigate other objects of interest (*see* **Note 5**).

1. Prepare every recording using ImageJ: Crop image stacks to include cells of interest. Threshold image stacks cautiously if at all. Remove single frames that are out of focus or in other way resulting in unreliable object detection. Save as grayscale tiff stack with lossless compression to reduce file size.
2. Create a *Z*-axis projection (in ImageJ: Image–Stacks–Z project; projection type: average intensity) to generate a single slice image showing each pixel with its average intensity over time.
3. Duplicate the single slice projected image. In one duplicate, create one ROI for every individual cell; you can create multiple

ROIs in this one image. In the other duplicate, create the single inverse ROI including only the background without the cells marked in the first duplicate. Convert both duplicates to binary images so that the respective ROIs have a pixel value of larger than 0.

4. You now have three tiff images: An image stack of the live cell recording and two single slice images containing ROIs for the cells and the background, respectively. The image stack should be named, for example, "MyData.tif," the two single slice images "MyData-mask.tif" and "MyData-mask-bg.tif," respectively.
5. Repeat **steps 1–4** for any number of recordings as they will be batch-analyzed.
6. In MorphoQuant, adjust the parameters with $rl = 15$ and $tl = 8$ as described above, and start the object detection algorithm. Adjust the parameters as you see fit, but these values have proved reliable to detect WPB-sized elongated objects.
7. Use the resulting text file containing all measurements for further analysis to, e.g., track the number of Munc13-4-positive objects over time before and after stimulation.

3.7 Statistical Analysis

Statistical analyses were performed with GraphPad Prism 6. Significance threshold was $P < 0.05$. We tested for normal distribution with the D'Agostino–Pearson omnibus test. When single datasets were not normally distributed, we confirmed the validity of one-way ANOVA or *t*-test by nonparametric Kruskal–Wallis or Mann–Whitney *U* test.

4 Notes

1. Avoid over-confluent cells for TIRF microscopy, thereby allowing each cell sufficient space to tightly attach to the glass coverslip to ensure that the maximum area of the plasma membrane is visible in the TIRF plane.
2. Avoid low confluency to ensure a maximum stimulation efficiency.
3. Handle the cells with care especially on the day of the image acquisition to avoid mechanical stimulation.
4. Fusion of WPBs can be detected by the collapse of the elongated structure into a bright round fusion spot.
5. For details on data preparation and all available parameters and measurements, see the MorphoQuant readme at https://github.com/tischi/MorphoQuant/blob/master/README.md.

References

1. Weibel ER, Palade GE (1964) New cytoplasmic components in arterial endothelia. J Cell Biol 23:101–112
2. Sadler JE (1998) Biochemistry and genetics of von Willebrand factor. Annu Rev Biochem 67:395–424. https://doi.org/10.1146/annurev.biochem.67.1.395
3. Wagner DD, Frenette PS (2008) The vessel wall and its interactions. Blood 111:5271–5281. https://doi.org/10.1182/blood-2008-01-078204
4. Nightingale T, Cutler D (2013) The secretion of von Willebrand factor from endothelial cells; an increasingly complicated story. J Thromb Haemost 11(Suppl. 1):192–201. https://doi.org/10.1111/jth.12225
5. Valentijn KM, Eikenboom J (2013) Weibel-Palade bodies: a window to von Willebrand disease. J Thromb Haemost 11:581–592. https://doi.org/10.1111/jth.12160
6. Matsushita K, Morrell CN, Cambien B et al (2003) Nitric oxide regulates exocytosis by S-nitrosylation of N-ethylmaleimide-sensitive factor. Cell 115:139–150
7. Pulido IR, Jahn R, Gerke V (2011) VAMP3 is associated with endothelial Weibel-Palade bodies and participates in their Ca(2+)-dependent exocytosis. Biochim Biophys Acta 1813:1038–1044. https://doi.org/10.1016/j.bbamcr.2010.11.007
8. van Breevoort D, van Agtmaal EL, Dragt BS et al (2012) Proteomic screen identifies IGFBP7 as a novel component of endothelial cell-specific Weibel-Palade bodies. J Proteome Res 11:2925–2936. https://doi.org/10.1021/pr300010r
9. van Breevoort D, Snijders AP, Hellen N et al (2014) STXBP1 promotes Weibel-Palade body exocytosis through its interaction with the Rab27A effector Slp4-a. Blood 123:3185–3194. https://doi.org/10.1182/blood-2013-10-535831
10. Fish KN (2009) Total internal reflection fluorescence (TIRF) microscopy. Curr Protoc Cytom 12:Unit12.18. doi: https://doi.org/10.1002/0471142956.cy1218s50
11. Schuberth CE, Tängemo C, Coneva C et al (2015) Self-organization of core Golgi material is independent of COPII-mediated endoplasmic reticulum export. J Cell Sci 128:1279–1293. https://doi.org/10.1242/jcs.154443
12. Chehab T, Santos NC, Holthenrich A et al (2017) A novel Munc13-4/S100A10/annexin A2 complex promotes Weibel-Palade body exocytosis in endothelial cells. Mol Biol Cell 28:1688–1700. https://doi.org/10.1091/mbc.E17-02-0128
13. Neeft M, Wieffer M, de Jong AS et al (2005) Munc13-4 is an effector of Rab27a and controls secretion of lysosomes in hematopoietic cells. Mol Biol Cell 16:731–741. https://doi.org/10.1091/mbc.E04-10-0923
14. Boswell KL, James DJ, Esquibel JM et al (2012) Munc13-4 reconstitutes calcium-dependent SNARE-mediated membrane fusion. J Cell Biol 197:301–312. https://doi.org/10.1083/jcb.201109132
15. Babich V, Meli A, Knipe L et al (2008) Selective release of molecules from Weibel-Palade bodies during a lingering kiss. Blood 111:5282–5290. https://doi.org/10.1182/blood-2007-09-113746
16. Jaffe EA, Nachman RL, Becker CG, Minick CR (1973) Culture of human endothelial cells derived from umbilical veins. Identification by morphologic and immunologic criteria. J Clin Invest 52:2745–2756. https://doi.org/10.1172/JCI107470
17. Harrison-Lavoie KJ, Michaux G, Hewlett L et al (2006) P-selectin and CD63 use different mechanisms for delivery to Weibel-Palade bodies. Traffic 7:647–662. https://doi.org/10.1111/j.1600-0854.2006.00415.x

Chapter 28

Analysis of S100A11 in DNA Damage Repair

Franziska Foertsch and Christian Melle

Abstract

DNA damage possesses the capacity to threaten the genomic integrity of an organism. A multitude of proteins are involved in the detection and repair of DNA double-strand breaks (DSBs), a severe kind of DNA damage. The function of DNA repair proteins can be examined by biochemical assays in vitro as well as in cell-based studies. The Ca^{2+}-binding protein S100A11 shows functional interactions with factors involved in the repair of DSBs by homologous recombination (HR), a high-fidelity DNA repair pathway, such as RAD51 and RAD54B. The key enzyme of the homologous recombination repair is RAD51 that catalyzes the invasion of single-stranded DNA (ssDNA) into double-stranded DNA (dsDNA) containing homologous regions and the exchange of these DNA molecules generating heteroduplex DNA (hDNA). In this chapter, we describe a protocol for the purification of S100A11 to near homogeneity. Using purified proteins, we show the ability of S100A11 to stimulate RAD51 in a DNA strand exchange assay. Additionally, we describe a protocol how S100A11 can be localized in sites of DNA repair by immunofluorescence staining. Furthermore, we present a protocol for assessment of chromosomal aberrations after depletion of S100A11 that illustrate the apparent involvement of S100A11 in genome integrity.

Key words DNA damage, Double-strand breaks, Chromosomal aberrations, Repair foci, S100A11

1 Introduction

Eukaryotic genomes are always challenged by specific conditions that impair its integrity. Exogenous sources such as carcinogenic compounds or ionizing radiation and endogenous sources, like radical oxygen species (ROS) or collapsed DNA replication forks, can induce double-strand breaks (DSBs), a severe kind of DNA damage. DSBs can lead to illegitimate genomic rearrangements that can contribute to cellular dysfunction and various disorders. Hence, cells have developed strategies to detect and repair DSBs to avoid these deleterious consequences [1, 2]. The recognition of a DSB leads to the formation of a multiprotein complex comprising MRN and ATM at the site of the DNA damage [3]. Here, a central event in signaling of DNA damage is the phosphorylation of histone H2AX by ATM [4, 5]. The generation of γH2AX results in the

Claus W. Heizmann (ed.), *Calcium-Binding Proteins of the EF-Hand Superfamily: From Basics to Medical Applications*, Methods in Molecular Biology, vol. 1929, https://doi.org/10.1007/978-1-4939-9030-6_28,

recruitment of proteins involved in the processing of the broken double-stranded DNA (dsDNA) ends triggering the composition of repair foci [2, 6].

A main repair pathway to delete DSBs is homologous recombination (HR), a high-fidelity DNA repair pathway. HR is largely error-free as it engages an identical sister chromatid as a template for repair in S or G2 phase cells. In initial HR process, several nucleases and helicases process the broken dsDNA ends to generate resected 3′-single-stranded DNA (ssDNA) overhangs. These ssDNA tails are stabilized by replication protein A (RPA), an ssDNA binding protein. RPA is subsequently replaced by RAD51 assisted by mediator proteins such as BRCA1, BRCA2, RAD51 paralogs, and RAD52 to form a helical nucleoprotein complex termed the presynaptic filament [7, 8]. RAD51 is an evolutionarily conserved key factor of the HR which possesses activity to exchange DNA strands between homologous partners. It has been shown that human RAD51 protein is stimulated by Ca^{2+} by modulation of its ATPase activity [9]. The RAD51/ssDNA nucleoprotein filament searches for a homologous template, followed by invasion into the homologous dsDNA to form a displacement loop (D-loop) structure which performs DNA strand exchange generating heteroduplex DNA (hDNA). The RAD51-mediated formation of the D-loop structure is stimulated by several regulators such as RAD54 and/or its paralog RAD54B. The invading strand conduces to prime DNA repair synthesis on the homologous DNA template. Thereafter, the newly synthesized strand dissociates from the homologous hDNA and reanneals with the other end of the DSB to complete the DSB repair restoring the structure of the damaged DNA [7]. In this chapter, we describe a protocol for a DNA strand exchange assay (SEA), developed for human RAD51 protein that is stimulated by S100A11 (Fig. 1). This assay allows testing of the function of an additional protein factor or conditions presumably influencing the RAD51-dependent recombination [9, 10].

Among distinct cellular processes such as cell growth and motility, and cell differentiation, S100A11 seems to be implicated in the repair of DNA damage. S100A11 responds to DNA damaging agent such as bleomycin, γ-radiation, and ROS by translocation from cytoplasm into the nucleus and is localized here in repair foci (Fig. 2) [10–12]. S100A11 is functionally associated with factors involved in the repair of DSBs. It was shown that S100A11 targets RAD54B to sites of DSB repair [13]. Additionally, S100A11 stimulates RAD51's strand exchange activity and influences the persistence of RAD51 in DNA repair foci. Furthermore, the depletion of S100A11 results in reduced sister chromatid exchange confirming the restriction of cellular recombination capacity as well as in an increase of chromosomal aberrations reflecting the functional requirement of S100A11 in maintenance of the genome integrity

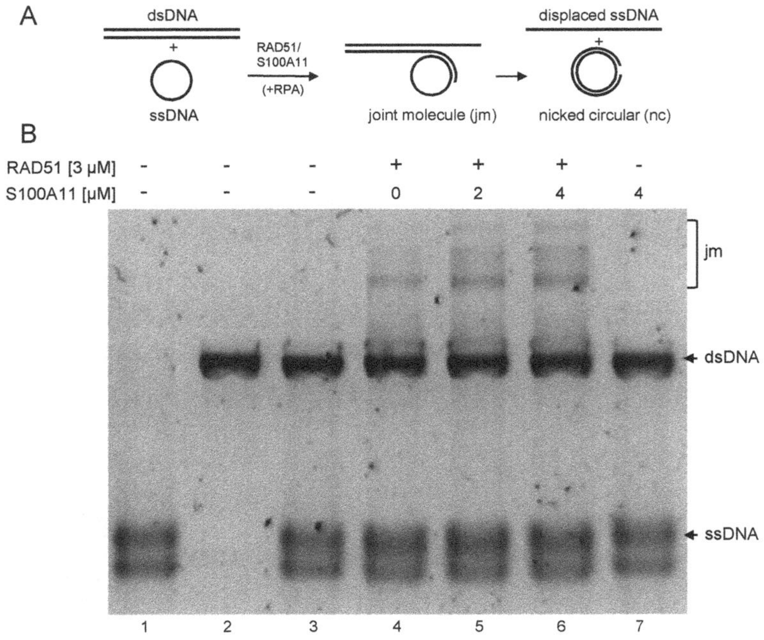

Fig. 1 DNA strand exchange assay. (**a**) Scheme of the strand exchange reaction. (**b**) RAD51 (lanes 4–6, 3 μM) alone or with S100A11 (lane 5, 2 μM; lane 6, 4 μM) was incubated with circular ΦX174 ssDNA (24 μM) in strand exchange buffer for 15 min at 37 °C followed by incubation with 2.4 μM RPA for 5 min and addition of ΦX174 dsDNA (24 μM) to initiate strand exchange reaction for 2 h at 37 °C. As negative control, S100A11 (lane 7, 4 μM) was incubated alone. The joint molecule (JM) products were separated by gel electrophoresis at 25 V for 15 h at 4 °C and visualized by GelStar staining

[10]. Hence, we depict in this chapter how to carry out immunofluorescence (IF) staining to detect S100A11 together with relevant DNA damage repair proteins in DNA repair foci as well as a workflow to analyze chromosomal aberrations after depletion of S100A11.

2 Materials

2.1 Protein Preparation

1. *E. coli* strain BL21 (DE).
2. Liquid LB media: 25 g complete powder into 1 L ddH_2O and autoclave it at 121 °C for 15 min; after autoclaving, LB media is supplemented with 100 μg/mL ampicillin.
3. pGEX-2T-S100A11 protein expression plasmid containing a glutathione *S*-transferase (GST)-tag linked to wild-type human *S100A11*.
4. Heat shaking incubator.

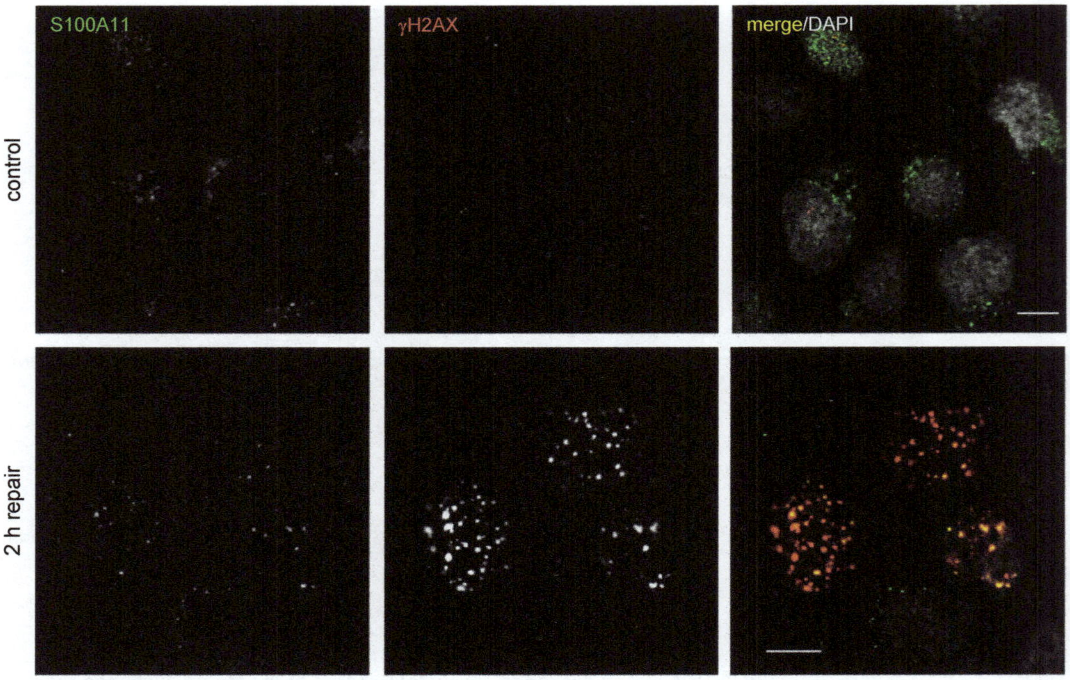

Fig. 2 S100A11 co-localizes with γH2AX at sides of DNA repair. HaCaT cells were synchronized in S phase by double-thymidine block. Afterward, cells were treated with 12.5 μg/mL bleomycin (2 h repair) for 30 min and released in fresh medium for 90 min or were untreated (control) followed by immunostaining using specific antibodies against S100A11 (green) and γH2AX (red). Nuclear DNA was detected using DAPI (white). Bar, 10 μm

5. UV spectrometer (Perkin Elmer, Lambda 650).
6. 0.23 mM IPTG (isopropyl-1-thio-β-D-galactopyranoside).
7. Ultracentrifuge.
8. Sonifier (Branson Digital Sonifier, W-250D).
9. 10× phosphate-buffered saline (PBS): Dissolve 80 g of NaCl, 2 g of KCl, 17.8 g of $Na_2HPO_4 \cdot 2H_2O$, and 2.4 g of KH_2PO_4 in 1 L H_2O. Adjust to pH 7.4 with 1 N HCl.
10. Lysis buffer: 20 mM Tris–HCl, pH 7.5, 200 mM NaCl, 20% sucrose, 10 mM $MgCl_2$, 0.1 mM PMSF, 1 μg/mL leupeptin.
11. Polyethyleneimine (PEI 600; 10% PEI-HCl, pH 7.8) (*see* **Note 1**).
12. Magnetic stirrer (Heidolph, MR3001).
13. Glutathione agarose beads.
14. Thrombin (120 units diluted in 2.5 mL PBS).
15. End-over-end mixer (Scientific Industries, Roto-Shake Genie).
16. Vivaspin 20 MWCO3000 concentrator.

17. Gel filtration buffer: 50 mM HEPES, pH 7.5, 100 mM NaCl, 10% glycerol.
18. Superdex 75 HighLoad 16/60 column (GE Healthcare).
19. Äkta Avant 25 device (GE Healthcare).
20. NanoDrop or UV spectrophotometer to measure protein concentrations.

2.2 DNA Strand Exchange Assay (SEA)

Published protocols were used for the expression and purification of human S100A11 [10] (*see* Subheading 3.1), human RAD51 [10, 14], and human RPA [15].

1. ΦX174 ssDNA (virion) and ΦX174 dsDNA (replicative form I, RFI) are purchased from New England Biolabs (NEB).
2. Restriction enzyme *Xho*I and NEB CutSmart buffer (both NEB).
3. QIAquick PCR Purification Kit (Qiagen).
4. DNA strand exchange buffer: 25 mM Tris–HCl, pH 7.5, 2.5 mM ATP, 1 mM DTT, 50 μg/mL bovine serum albumin (BSA), 2 mM magnesium acetate, 250 mM sodium chloride, 2 mM calcium acetate, 20 mM phosphocreatine, 0.1 μg/mL creatine kinase; 1 μg/μL creatine kinase stock solution is stored at −20 °C.
5. Stop buffer: 0.714% (v/v) sodium dodecyl sulfate (SDS), 4.3 mg/mL proteinase K; prepare 84 μL solution freshly by mixing 6 μL of 10% SDS and 18 μL of 20 mg/mL proteinase K.
6. 4× DNA gel loading buffer: 0.1% bromophenol blue (w/v), 30% glycerol; store at −20 °C.
7. Agarose running apparatus and power supply.
8. TBE buffer: 89 mM Tris base, 89 mM H_3BO_3, 2 mM EDTA, pH 8.0.
9. DNA/oligonucleotide staining: 2000× GelStar (Lonza) (10,000× concentrate in DMSO diluted in TBE buffer).
10. Digital gel imaging system with UV illumination box.
11. Software to quantify images.

2.3 Immuno-fluorescence (IF) and Confocal Microscopy

1. Human keratinocyte cell line HaCaT or normal human epidermal keratinocytes (NHEK).
2. Dulbecco's Modified Eagle's Medium (DMEM), Keratinocyte Growth Medium 2 (ready-to-use, PromoCell); each supplemented with 10% fetal calf serum (FCS).
3. 10× phosphate-buffered saline (PBS): Dissolve 80 g of NaCl, 2 g of KCl, 17.8 g of $Na_2HPO_4 \cdot 2H_2O$, and 2.4 g of KH_2PO_4 in 1 L H_2O. Adjust to pH 7.4 with 1 N HCl.
4. S phase synchronization: 0.1 M thymidine.

5. 10 μg/μL bleomycin (Cayman Chemicals).
6. Parafilm M.
7. Microscope cover slips (13 mm diameter).
8. 6-Well dish.
9. Humidity chamber.
10. Fixation buffer: 4% *para*formaldehyde (PFA).
11. Antibodies: chicken anti-S100A11 (1:50; Acris Antibodies), mouse anti-γH2AX (Ser139) (1:750, clone JBW301; Millipore).
12. Secondary antibodies: species-specific secondary antibodies linked to fluorescein (FITC), Cy3, or Cy5 (all at 1:200).
13. Mounting medium: ProLong Gold antifade reagent with 4,6-diamidino-2-phenylindole (DAPI).
14. Superfrost microscope slide glass.
15. Zeiss LSM 510 laser scanning confocal device attached to an Axioplan 2 microscope using 63× Plan-Apochromat oil objective (Carl Zeiss Jena).

2.4 Chromosomal Aberration Analysis

1. Human keratinocyte cell line HaCaT.
2. Dulbecco's Modified Eagle's Medium (DMEM), supplemented with 10% fetal calf serum (FCS).
3. Trypsin/EDTA solution.
4. S100A11 siRNA, non-silencing control (nsc) siRNA (Qiagen).
5. Amphiphilic delivery system SAINT-RED (Synvolux Therapeutics).
6. 10 μg/μL bleomycin (Cayman Chemicals).
7. 6-Well dish.
8. Colcemid (Seromed).
9. Hypotonic solution: 75 mM KCl.
10. Fixation solution: absolute methanol/glacial acetic acid, 3:1 (v/v).
11. Humidity chamber.
12. Heating plate.
13. Coplin jar.
14. Giemsa staining solution.
15. PBS Gurr buffer, pH 6.88.
16. Metafer platform and IKAROS software (MetaSystems Hard & Software GmbH).

3 Methods

To elucidate the function of proteins involved in the repair of DNA double-strand breaks by homologous recombination, several in vitro biochemical assays such as strand exchange assay or D-loop formation assay using purified proteins have been developed [9, 10, 16]. For analysis of the DNA repair processes in a cellular environment, microscopy-based immunofluorescence studies using specific antibodies recognizing repair proteins are commonly used [10, 17–19]. The impact of the depletion or mutation of a repair factor on the genome maintenance can be assessed by analysis of chromosomal aberrations [10, 20]. Furthermore, photonic approaches able to allow insight in the DNA repair processes in living cells become more important to examine the dynamics of repair factors [21–24].

3.1 S100A11 Purification

1. All the steps involved with purification of S100A11 are to be performed at 4 °C unless stated otherwise.
2. Inoculate 200 mL LB medium (supplemented with 100 μg/mL ampicillin) with one colony of freshly transformed *E.coli* BL21 harboring pGEX-2T-S100A11 expression plasmid, and grow overnight with shaking (235 rpm) at 37 °C.
3. Filled up the cell culture with 1200 mL LB medium (supplemented with 100 μg/mL ampicillin), and continue incubation at 37 °C to OD_{600} 0.6–0.8.
4. Induce protein expression by addition of 0.23 mM IPTG for 18 h at 21 °C.
5. Harvest cells by centrifugation (4000 × *g* for 15 min), wash the cell pellet two times with pre-chilled PBS, and pelletize in each case by centrifugation; process the cell pellet immediately, or store the cell pellet at −80 °C.
6. Resuspend the cell pellet in 40 mL of pre-chilled lysis buffer, and disrupt the cells by two freezing/thawing cycles using liquid nitrogen followed by sonication (three times for 30 s at 20% amplitude).
7. Add slowly 3 mL polyethyleneimine to the bacterial cell lysate under stirring over 20 min (*see* **Note 2**).
8. Separate cell debris from the protein crude extract by centrifugation (10,000 × *g* for 15 min).
9. Mix the protein extract (supernatant) with glutathione agarose beads (pre-equilibrated with lysis buffer without protease inhibitors) in a batch-wise manner in an end-over-end mixer for 3 h.
10. Pelletize the beads binding GST-tagged S100A11 by centrifugation (1000 × *g* for 5 min), discard the supernatant, and wash

the beads four times with ten batch volumes pre-chilled lysis buffer.

11. Separate S100A11 from the GST-tag bound to the glutathione beads by adding 120 units thrombin (diluted in 2.5 mL PBS) for 7 h at room temperature.
12. After centrifugation (4000 × *g* for 5 min), incubate again the supernatant with pre-equilibrated glutathione agarose beads for 8 h to eliminate possible GST contaminations from the protein extract.
13. Clear the solution by centrifugation (4000 × *g* for 5 min)
14. Concentrate the protein solution to 1 mL volume using a Vivaspin 20 concentrator at 4000 × *g*.
15. Load the protein extract containing S100A11 onto a Superdex 75 HiLoad 16/60 column (pre-equilibrated with gel filtration buffer), and perform a gel filtration chromatography at 6 °C using an Äkta Avant device.
16. Pool S100A11 containing fractions, and incubate again with glutathione agarose beads (pre-equilibrated with gel filtration buffer) for 1 h.
17. Clear the solution by centrifugation (4000 × *g* for 5 min), and concentrate samples containing S100A11 using a Vivaspin 20 concentrator at 4000 × *g*.
18. Examine the purity of the S100A11 sample by sodium dodecyl sulfate-polyacrylamide gel electrophoresis (SDS-PAGE) (Fig. 3), and confirm identity by Western blot analysis.
19. Aliquot the S100A11 samples in 10 μL portions, quick-frozen by liquid nitrogen, and store at −80 °C (*see* **Notes 3** and **4**).

3.2 DNA Strand Exchange Assay (SEA)

1. Linearize circular ΦX174 dsDNA by *Xho*I (*see* **Note 5**); incubate 25 μg ΦX174 dsDNA (RFI) with 100 units of *Xho*I in NEB CutSmart buffer for 1 h at 37 °C.
2. Remove *Xho*I using the QIAquick PCR Purification Kit as described in the manufacturer's instructions.
3. Measure the nucleic acid absorbance at A_{260}, and determine the sample concentration using molar extinction coefficient $\varepsilon_M = 6500/M/cm$ at 260 nm.
4. Incubate 24 μM (nucleotides) ΦX174 circular ssDNA with RAD51 protein and S100A11 protein in 10 μL of DNA strand exchange buffer for 15 min at 37 °C.
5. Add 2.4 μM RPA to the reaction; incubate for 5 min followed by the addition of 24 μM (base pairs) linearized ΦX174 dsDNA to initiate strand exchange.
6. Incubate the reaction for 2 h at 37 °C.

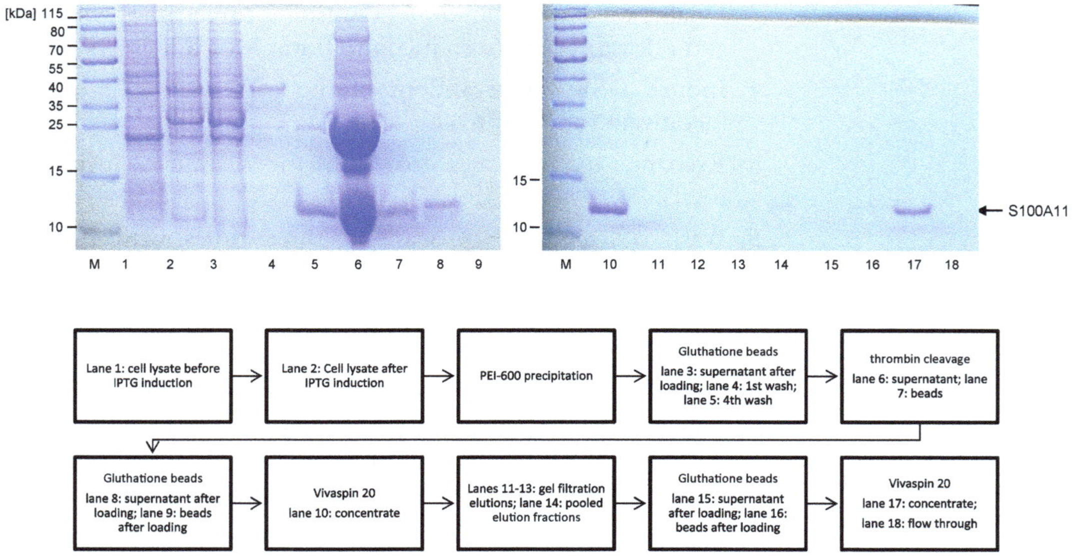

Fig. 3 Protein purification of S100A11. Single steps of the purification of S100A11 are shown. Purification intermediates and purified proteins were subjected to 12% SDS-PAGE and stained by Coomassie brilliant blue staining solution

7. Stop the reaction by addition of 0.7% SDS and 4.3 mg/mL proteinase K for 20 min at 37 °C.
8. Separate strand exchange products by 1% agarose gel electrophoresis in 1× TBE buffer at 25 V for 15 h at 4 °C.
9. Stain gel with 2000× GelStar diluted in TBE (1:1000 v/v) for 1 h at 4 °C followed by one-time rinsing with TBE.
10. Visualize the strand exchange products in the gel, and record the image using a digital gel imaging system.
11. Quantify the intensities of product bands by LabImage-1D software. As shown in Fig. 1, S100A11 stimulates the formation of joint molecule (JM) intermediates by human RAD51 (*see* **Note 6**).

3.3 Immunofluorescence Staining and Confocal Microscopy

1. Prepare cover slips by consecutively washing with 70% ethanol and 96% ethanol for 1 h each. Dry the cleaned cover slips in a heater at 120 °C followed by autoclaving.
2. Plate 1 × 10^5 HaCaT cells or NHEK cells in DMEM or Keratinocyte Growth Medium 2, respectively, on cover slips placed in a 6-well dish. Let the cells grow on the cover slips for 18 h.
3. Synchronize the cells in S phase by double-thymidine block. Hence, treat the cells with 2.5 mM thymidine for 16 h. Release the cells from the block for 9 h by washing three times with PBS. Block again with 2.5 mM thymidine for another 16 h.

Finally, wash off thymidine by washing three times with PBS, and release cells in fresh medium for at least 2 h (*see* **Note 7**).

4. Induce DSBs by treatment of the cells with 12.5 μg/mL bleomycin for 30 min.
5. Exchange medium and grow cells for 90 min (2 h repair in total).
6. Wash off medium by washing three times with PBS (*see* **Note 8**).
7. Fix the cells with 4% paraformaldehyde for 10 min at room temperature.
8. Wash the coverslips containing fixed cells three times with PBS.
9. Prepare the chicken anti-S100A11 antibody at 1:50 (25 μL per cover slip), and drip the antibody solution on a piece of parafilm (1 × 1 cm) placed in a humidity chamber.
10. Put the cover slip upside down on the drop of the antibody solution, and incubate for 45 min (*see* **Note 9**).
11. Put the cover slip upside up outside the humidity chamber on a piece of parafilm (1 × 1 cm), and wash five times with 0.5 mL PBS each. Remove the PBS by aspiration using a suction pump.
12. Prepare the mouse anti-γH2AX antibody at 1:750 (25 μL per cover slip), and drip the antibody solution on a piece of parafilm (1 × 1 cm) placed in a humidity chamber.
13. Put the cover slip upside down on the drop of the antibody solution, and incubate again for 45 min.
14. Wash the cover slip outside the humidity chamber five times with 0.5 mL PBS each.
15. Prepare the anti-chicken secondary antibody and the anti-mouse secondary antibody together in a working solution at 1:200 (*see* **Note 10**).
16. Incubate the cover slip in the secondary antibodies solution in the humidity chamber in a dark environment for 45 min (*see* **Note 11**).
17. Wash the cover slips again five times with PBS. Aspirate completely all liquid from the cover slip. Put the cover slip upside down on a drop of mounting medium that was placed previously on a microscope slide glass. Place the microscope slide in the dark to completely polymerize the mounting medium overnight at room temperature. Afterward, store the slides at −20 °C.
18. Analyze the samples using a confocal laser scanning microscope (Fig. 2) (*see* **Note 12**).

3.4 Chromosomal Aberration Analysis

1. Grow 7.5 × 10^4 HaCaT cells in a 6-well dish for 24 h.
2. Transfect cells with specific S100A11 siRNA using SAINT-RED delivery system as described in the manufacturer's instructions to delete S100A11 for 48 h. Use nsc siRNA in a control approach.
3. Induce DSBs by treatment of the cells with 12.5 μg/mL bleomycin for 30 min.
4. Wash twice with PBS, and culture the cells in fresh medium for another 20 h.
5. Add 0.1 μg/mL colcemid to the cells for 6 h.
6. Harvest the cells using trypsin/EDTA solution to collect metaphases (*see* **Note 13**).
7. Smoothly resuspend the cells in 10 mL of freshly prepared pre-warmed hypotonic solution at 37 °C for 35 min.
8. For prefixation, add 1 mL of freshly prepared pre-chilled fixation solution to the re-suspended cells, swing shortly, and centrifuge (450 × *g*, 5 min).
9. Resuspend the pellet in 10 mL of freshly prepared pre-chilled fixation solution, and centrifuge; repeat twice this procedure.
10. Incubate the metaphases in 10 mL of fixation solution at −20 °C for at least 3 h.
11. Drip the metaphase suspension (40 μL) on pre-chilled glass slides localized in a humidity chamber preheated at 40 °C (*see* **Note 14**).
12. Re-drip with fixation solution (40 μL).
13. Dry shortly (30 s) the glass slides at 45 °C on a heating plate (*see* **Note 15**).
14. Continue drying at room temperature for 24 h.
15. Stain chromosomes by Giemsa-GTG (G-bands by trypsin using Giemsa). Hence, treat the glass slides containing dried chromosomes with trypsin solution for 12 s in a Coplin jar followed by twice wash steps with PBS Gurr buffer. Stain the chromosomes with Giemsa solution for 10 min. Finally, rinse the glass slide with Milli-Q water (three times), and dry at room temperature.
16. Record and assess the chromosomes by Metafer platform and IKAROS software, respectively (Fig. 4).

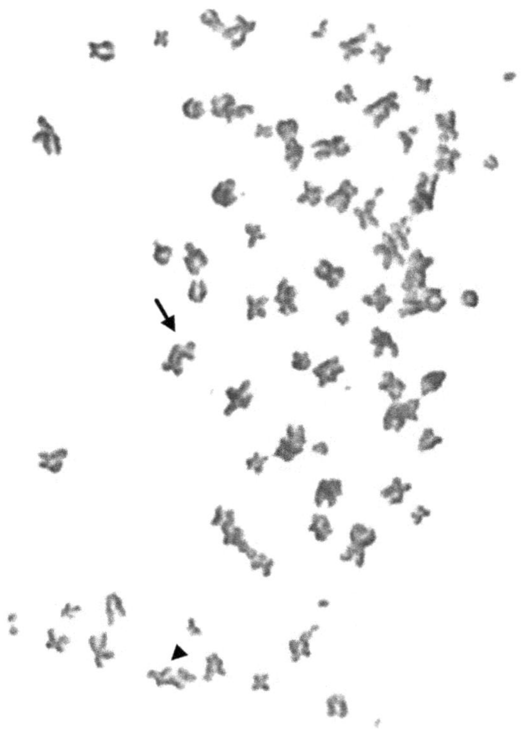

Fig. 4 Depletion of S100A11 by RNAi induces chromosomal aberrations. S100A11 was knocked-down in HaCaT cells by specific S100A11 siRNA. Afterward, cells were treated with bleomycin (12.5 μg/mL) to induce DSBs for 30 min. After further culture in fresh medium for 20 h, 0.1 μg/mL colcemid was added for 6 h to enrich metaphases. An example for chromatid break (arrow) or radial figure (arrowhead), respectively, in a metaphase is shown

4 Notes

1. PEI 600 (3.06 mL polyethyleneimine; 10% PEI-HCl, pH 7.8) as a positively charged substance is used for removal of DNA from the bacterial cell lysate.
2. A beaker glass containing the cell lysate should be placed in a bowl filled with ice that is placed on a magnetic stirrer during addition of PEI to ensure chilling of the lysate all the times.
3. It is very important that proteins used in biochemical assays like strand exchange assay (SEA) are free from nuclease contaminations. The chromatography step using glutathione beads for S100A11 purification and the anion exchange chromatography step in the purification of RAD51 ensure the removal of all nuclease activities that may co-purify with proteins of interest.

4. Dialysis of purified S100A11 impaired its capacity to stimulate RAD51's recombinase activity.
5. Human RAD51 prefers linearized ΦX174 dsDNA with 5′-overhangs as *Xho*I is able to generate.
6. The major end products in DNA strand exchange for human RAD51 are joint molecule (JM) as nicked circle (NC) DNA formation is very low. The yield of DNA strand exchange can be calculated through the equation: product $\% = (JM/1.5 + NC)/(JM/1.5 + NC + \text{dsDNA substrate})$ [16].
7. Cells are enriched in S phase (~75%). Confirm the correct distribution of the cells in distinct cell cycle phases by flow cytometric analysis.
8. It is important to wash properly as medium containing FCS can stick inappropriately on cellular structures in the following fixation step.
9. Keep in mind the side of the cover slip where the cells are grown! Avoid a mixture of the sides of the cover slip! The cover slip should be swim on the drop of antibody solution without drying.
10. By preparing the secondary antibodies solution, use antibodies that are labeled with different dyes regarding their spectral properties (e.g., fluorescein and Cy3 or Cy5, respectively).
11. As the dyes are light sensitive, protect them from light by placing the humidity chamber in the dark (e.g., in a drawer of your lab bench).
12. If you are not familiar with confocal microscope procedure, ask an expert of an imaging facility for support.
13. Collect and pelletize all washing steps using PBS to avoid loss of metaphases as M phase cells are easily detached from the bottom of the 6-well dish.
14. Pre-chill the glass slides in Milli-Q water at 4 °C for 24 h. To drip the metaphase solution, glass slides should be positioned diagonally in the humidity chamber.
15. The heat plate should be covered with wet tissues. During drying of the dripped metaphase solution, methanol of the fixation solution will volatilize resulting in a highly hygroscopic impact of the remaining acetic acid that leads to a swelling of the chromosomes. Thus, the chromosomes can become better visualized [25].

Acknowledgments

We are grateful to Drs Anna Szambowska (FLI Jena) for purified RPA and Anja Weise (UKJ) for microscopic assessment of chromosomal aberrations.

References

1. Jackson SP, Bartek J (2009) The DNA-damage response in human biology and diseases. Nature 461:1071–1078
2. Chapmann JR, Taylor MRG, Boulton SJ (2012) Playing the end game: DNA double-strand repair pathway choice. Mol Cell 47:497–510
3. Ciccia A, Elledge SJ (2008) The DNA damage response: making it safe to play with knives. Mol Cell 40:179–204
4. Rogakou EP, Pilch DR, Orr OH et al (1998) DNA double-strand breaks induce histone H2AX phosphorylation on serine 139. J Biol Chem 273:5858–5868
5. Paull TT (2015) Mechanisms of ATM activation. Annu Rev Biochem 84:711–738
6. Cejka P (2015) DNA end resection: nucleases team up with the right partners to initiate homologous recombination. J Biol Chem 290:22931–22938
7. Heyer WD (2015) Regulation of recombination and genomic maintenance. Cold Spring Harb Perspect Biol 7:a016501
8. Zhao W, Steinfeld JB, Liang F et al (2017) BRCA1-BARD1 promotes RAD51-mediated homologous DNA pairing. Nature 550:360–366
9. Bugreev DV, Mazin AV (2004) Ca^{2+} activates human homologous recombination protein Rad51 by modulating its ATPase activity. Proc Natl Acad Sci U S A 101:9988–9993
10. Foertsch F, Szambowska A, Weise A et al (2016) S100A11 plays a role in homologous recombination and genome maintenance by influencing the persistence of RAD51 in DNA repair foci. Cell Cycle 15:2766–2779
11. Gorsler T, Murzik U, Ulbricht T et al (2010) DNA damage-induced translocation of S100A11 into the nucleus regulates cell proliferation. BMC Cell Biol 11:100
12. Kozlov SV, Waardenberg AJ, Engholm-Keller K et al (2016) Reactive oxygen species (ROS)-activated ATM-dependent phosphorylation of cytoplasmic substrates identified by large scale phosphoproteomics screen. Mol Cell Proteomics 15:1032–1047
13. Murzik U, Hemmerich P, Weidtkamp-Peters S et al (2008) Rad54B targeting to DNA double-strand repair sites requires complex formation with S100A11. Mol Biol Cell 19:2926–2935
14. Cloud V, Chan YL, Grubb J et al (2012) Rad51 is an accessory factor for Dmc1-mediated joint molecule formation during meiosis. Science 337:1222–1225
15. Henricksen LA, Umbrich CB, Wold MS (1994) Recombinant replication protein A: expression, complex formation, and functional characterization. J Biol Chem 269:11121–11132
16. Liu J, Sneeden J, Heyer WD (2011) In vitro assays for DNA pairing and recombination-associated DNA synthesis. Methods Mol Biol 745:363–383
17. Rogako EP, Boon C, Redon C et al (1999) Megabase chromatin domains involved in DNA double-strand breaks in vivo. J Cell Biol 146:905–916
18. Paull TT, Rogakou EP, Yamazaki V et al (2000) A critical role for histone H2AX in recruitment of repair factors to nuclear foci after DNA damage. Curr Biol 10:886–895
19. Bekker-Jensen S, Lukas C, Kitagawa R et al (2008) Spatial organization of the mammalian genome surveillance machinery in response to DNA strand breaks. J Cell Biol 173:195–206
20. Nikkilä J, Parplys AC, Pylkäs K et al (2013) Heterozygous mutations in PALB2 cause DNA replication and damage response defects. Nat Commun 4:2578
21. Essers J, Houtsmuller AB, van Veelen L et al (2002) Nuclear dynamics of RAD52 group homologous recombination proteins in response to DNA damage. EMBO J 21:2030–2037
22. Reuter M, Zelensky A, Smal I et al (2014) BRCA2 diffuses as oligomeric clusters with RAD51 and changes mobility after DNA damage in live cells. J Cell Biol 207:599–613
23. Ito K, Murayama Y, Takahashi M et al (2018) Two three-strand intermediates are processed during Rad51-driven DNA strand exchange. Nat Struct Mol Biol 25:29–36
24. Haas KT, Lee M, Esposito A, Venkitaraman AR (2018) Single-molecule localization microscopy reveals molecular transactions during RAD51 filament assembly at cellular DNA damage sites. Nucleic Acids Res 46:2398–2416
25. Claussen U, Michel S, Muhlig P et al (2002) Demystifying chromosome preparation and the implications for the concepts of chromosome condensation during mitosis. Cytogenet Genome Res 98:136–146

Chapter 29

Fluorine-18 Labeling of S100 Proteins for Small Animal Positron Emission Tomography

Markus Laube, Torsten Kniess, Christin Neuber, Cathleen Haase-Kohn, and Jens Pietzsch

Abstract

The interaction of S100 proteins (S100s), a multigenic family of Ca^{2+}-binding and Ca^{2+}-modulated proteins, with pattern recognition receptors, e.g., Toll-like receptors (TLRs), the receptor for advanced glycation end products (RAGE), or scavenger receptors (SR), is hypothesized to be of high relevance in the pathogenesis of various diseases. This includes chronic inflammatory conditions, atherosclerosis, cardiomyopathies, neurodegeneration, and progression of cancers. However, data concerning the role of circulating S100s in these pathologies are scarce. One reason for this is the shortage of suitable radiolabeling methods for direct assessment of the metabolic fate of circulating S100s in vivo. We report a radiotracer approach using radiolabeling of recombinant human S100s with the positron emitter fluorine-18 (^{18}F) by conjugation with *N*-succinimidyl 4-[^{18}F]fluorobenzoate ([^{18}F]SFB). The methodological radiochemical part focuses on an optimized and automated synthesis of [^{18}F]SFB comprising HPLC purification to achieve higher chemical purity. The respective radioligands, [^{18}F]fluorobenzoylated S100s ([^{18}F]FB-S100s), were obtained with appropriate radiochemical purities, yields, and effective molar activities. Biological applications comprise cell and tissue binding experiments in vitro, biodistribution and metabolite studies in rodents in vivo/ex vivo, and dynamic positron emission tomography studies using dedicated small animal PET systems. Radiolabeling of S100s with ^{18}F and, particularly, the use of small animal PET provide novel probes to delineate both their metabolic fate and the functional expression of their specific receptors under normal and pathophysiological conditions in rodent models of disease.

Key words Bolton-Hunter-type reagent, In vivo imaging, Radiopharmacological characterization, ^{18}F-building block, Module-assisted radiosynthesis, S100 proteins, Calcium, EF hand

1 Introduction

The multigenic S100 protein family comprises to date at least 25 distinct members of Ca^{2+}-binding and Ca^{2+}-modulated proteins of the EF-hand type, differentially expressed in a wide variety of, exclusively, vertebrate cells [1, 2]. In human and in rodents, S100 genes and proteins are strongly conserved [3]. S100 proteins (S100s) have been implicated in Ca^{2+}-dependent regulation of

Claus W. Heizmann (ed.), *Calcium-Binding Proteins of the EF-Hand Superfamily: From Basics to Medical Applications*, Methods in Molecular Biology, vol. 1929, https://doi.org/10.1007/978-1-4939-9030-6_29,

intracellular activities such as protein phosphorylation, enzyme activity, cell growth and differentiation, cell motility, dynamics of cytoskeletal constituents, structural organization of membranes, Ca^{2+} homeostasis, as well as inflammatory stimulation and differentiation. Recent experimental and clinical evidence indicates that S100s not only play important regulatory roles within cells but also exert effects in a cytokine-like manner on definite target cells once released into the extracellular space or the circulating blood [1, 2, 4]. After cellular activation, some S100s are actively secreted following various pathways [4–9]. This potentially qualifies various S100s to be used as biomarkers [4, 10–12] or as potential pharmacological/therapeutic targets [11 *and references therein*]. Functional complexity and diversity of S100s are determined at several levels, including (a) expression in a cell-, tissue-, and organ-specific manner; (b) intracellular localization to both cytoplasm and nucleus; (c) action in both an autocrine and paracrine manner; (d) different affinity to calcium followed by various degrees of conformational changes; (e) binding of other divalent metal ions, such as copper and zinc; and (f) interaction with various, particularly, pattern recognition (multi-ligand) receptors. Target cells of extracellular S100s comprise granulocytes, monocytes, macrophages, mast cells, lymphocytes, dendritic cells, endothelial cells, neurons, astrocytes, microglia, cardiomyocytes, alveolar epithelial cells, and a plethora of tumor cells. Interaction of S100s with these cells both in receptor-dependent and receptor-independent manner results, e.g., in pro- and anti-inflammatory modulation of monocytes, stimulation of angiogenesis, induction of endothelial dysfunction, trophic and pro-survival effects on neurons, stimulation of astrocyte proliferation, stimulation of NO secretion by astrocytes and microglia, as well as pro- and anti-tumorigenic modulation of tumor cells or cells of the tumor microenvironment. Accordingly, increased levels of S100 proteins in the circulating blood have been associated with a number of disease states, e.g., ischemic, infectious, or traumatic injury of cerebral tissue, neurodegenerative disorders, ischemic myocardial injury, cardiomyopathies, diabetes, atherosclerosis, chronic inflammatory disorders, cancer, and metastasis [2 *and references therein*]. An emerging concept of pattern recognition involves certain multi-ligand receptors, particularly, Toll-like receptors (TLRs), the receptor for advanced glycation end products (RAGE), and scavenger receptors (SR), as those transducing the biological effects of most S100s [2]. This all results in a tremendous spectrum of pleiotropic intra- and extracellular functions and determines the modes of interaction with numerous protein and peptide partners. Finally, the fact that multi-ligand receptors and multi-receptor ligands hit each other forms an intricate network of S100-associated pathways. This fact further supports the potential pathophysiological role of S100s and disclosed new routes of investigation in the overall S100 field.

However, the diagnostic significance of different circulating S100s in blood- and tissue-specific S100-receptor interaction during initiation, progression, and manifestation of several human pathologies is still poorly understood. This also applies to a deeper understanding of the effects of pharmacological intervention using compounds targeting S100s. One reason for this is the shortage of suitable radiolabeling methods for direct assessment of the metabolic fate of circulating S100 proteins and, on the other hand, the functional expression of known and putative S100 receptors in vivo.

In order to address this question, the methodical development focused on a novel sensitive and specific radiotracer approach using *no-carrier-added* (*n.c.a.*) radiolabeling of recombinant human S100s. This is exemplified by radiolabeling S100A12 (common aliases calgranulin C and ENRAGE) and S100A4 (Mts1, metastasin) with the positron-emitting nuclide fluorine-18 (^{18}F) and the application of radiolabeled ^{18}F-S100s in dynamic small animal positron emission tomography (PET) studies in rodent models [13–15]. On the one hand, this approach provides novel probes to delineate the metabolic fate in terms of S100s' blood retention, elimination, and interaction with membrane and soluble binding molecules. On the other hand, the functional expression of specific receptors under normal and pathophysiological conditions can be investigated in vivo [13–15]. The selected examples require consideration of specific marginal conditions for the choice of the most appropriate radiolabeling method. It has to be taken into account that physiological formation of S100 dimers or oligomers and interaction with binding molecules should not be affected. Therefore, space-consuming prosthetic groups, e.g., chelating units for metal radionuclides, were excluded. Furthermore, especially in the case of S100A4, adverse oxidation of cysteine residues should be avoided [16]. This essentially precludes classical radiolabeling approaches like oxidative iodination, transition metal-catalyzed reactions, or cysteine-targeted labeling via maleimide-functionalized prosthetic groups [17 *and references therein*, 18, 19]. S100A12, on the other hand, does not contain cysteine residues, a priori favoring a NH_2-targeted labeling approach. Consequently, radiolabeling of recombinant S100s with the Bolton-Hunter-type reagent *N*-succinimidyl 4-[^{18}F]fluorobenzoate ([^{18}F]SFB) [20] was preferred.

N-succinimidyl 4-[^{18}F]fluorobenzoate [^{18}F]SFB is an active ester suitable for the labeling of peptides and proteins which has been used in PET chemistry for years [21–24]. The radiosynthesis of [^{18}F]SFB was at first described by Vaidyanathan and Zalutsky [20] comprising a three-step synthesis starting with the [^{18}F]fluorination of *N,N,N*-trimethylammoniumbenzaldehyde triflate, followed by oxidation with MnO_2 to 4-[^{18}F]fluorobenzoic acid and final conversion to [^{18}F]SFB using *N*-hydroxysuccinimide and dicyclohexylcarbodiimide. Besides this synthetic route and

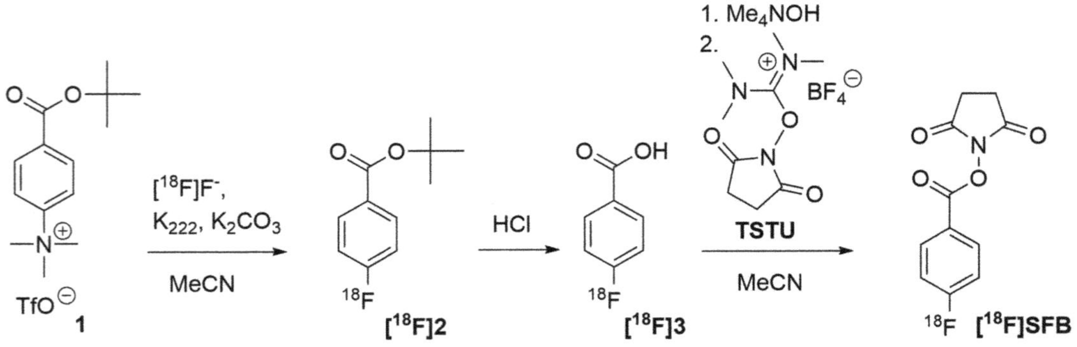

Fig. 1 Synthesis of $[^{18}F]$SFB

optimization thereof [25, 26], $[^{18}F]$SFB was also synthesized by three-step procedures comprising the ^{18}F-labeling of an *O*-protected *N,N,N*-trimethylammoniumbenzoic acid ester triflate, followed by a deprotection step and the conversion into the final *N*-succinimidyl derivative with TSTU or HSTU. For example, Wester et al. introduced the radiosynthesis starting from the respective ethyl ester [27] which was adopted for automatization [28–33] as well as optimized for microfluidic [34, 35] or microwave-based [36] methods. In modifications of this route, the pentamethylbenzyl [37, 38] and, as reported herein, the *tert*-butyl [39] ester was utilized. Recently, the copper-mediated nucleophilic $[^{18}F]$fluorination of aryl boronic esters was reported for a one-step synthesis of $[^{18}F]$SFB [40]. Within this report, the methodological part focuses on the radiosynthesis of $[^{18}F]$SFB starting from the *tert*-butyl protected *N,N,N*-trimethylammoniumbenzoic acid ester triflate **1** (Fig. 1). The radiosynthesis follows the synthetic route [39] and automated procedure [22, 41, 42] described by our group but was optimized by application of a final HPLC purification step to increase the chemical purity of the product in order to raise the labeling yields of S100 proteins with $[^{18}F]$SFB. The latter can then be used for the ^{18}F-labeling of a multitude of different peptides, polypeptides, and proteins but in the context of this article especially for the radiolabeling of different S100s.

2 Materials

2.1 Chemicals Provided or Synthesized

All commercially available chemicals and solvents were purchased in high quality and used without further purification. The precursor, *tert*-butyl protected *N,N,N*-trimethylammoniumbenzoic acid ester triflate **1**, was synthesized as previously reported [39]. The nonradioactive reference compound SFB was prepared as described by Wester et al. [27]. No-carrier-added aqueous $[^{18}F]$fluoride was produced in a CYCLONE 18/9® cyclotron (IBA, Belgium) by irradiation of $[^{18}O]H_2O$ via the $^{18}O(p,n)^{18}F$ nuclear reaction.

2.2 Preparation of Reagents and Eluents

1. 1 M $NaHCO_3$ solution: Weigh 84.01 g $NaHCO_3$, and dissolve in ca. 300 mL water. Dilute solution to 1 L in a volumetric flask.
2. K_2CO_3/K_{222} solution: Weigh 500 mg 4,7,13,16,21,24-Hexaoxa-1,10-diazabicyclohexacosane (K_{222}), and dissolve in 50 mL MeCN to prepare K_{222} solution. Weigh 92.3 mg K_2CO_3 and dissolve in 7 mL water. Transfer K_2CO_3 solution into a 50 mL volumetric flask, and add K222 solution to the 50 mL graduation mark. Store the stock solution in a closed bottle in the fridge.
3. 1 M HCl solution: Add 1 mol HCl (concentrated volumetric solution in an ampoule, PanReac Applichem) to water in a volumetric flask (1 L) and dilute with water to the graduation mark.
4. Precursor solution: Store *tert*-butyl-(4-*N,N,N*-trimethylammonium)benzoate triflate (precursor for radiolabeling) in the fridge until use. Let the precursor warm up to RT, weigh 7 mg *tert*-butyl-(4-*N,N,N*-trimethylammonium)benzoate triflate in an HPLC vial, and dissolve in 1 mL MeCN. Close the vial until the preparation of the synthesis module.
5. Me_4NOH solution: Store 25% Me_4NOH in MeOH (Sigma) in the fridge until use. Let the solution warm up to RT, and add 20 μL of this solution to 0.5 mL MeCN in an HPLC vial to prepare the final Me_4NOH solution. Close the vial until the preparation of the synthesizer module.
6. TSTU solution: Store *O*-(*N*-succinimidyl)-*N,N,N′,N′*-tetramethyluronium tetrafluoroborate (TSTU) in the fridge until use. Let TSTU warm up to RT, weigh 15 mg into a glass vial, and dissolve it in 0.5 mL MeCN. Close the vial until the preparation of the synthesizer module.
7. HPLC eluent: Add 1 mL glacial acetic acid to approx. 200 mL of ultrapure water, and dilute to 1 L in a volumetric flask to prepare 0.1% acetic acid. Mix 650 mL 0.1% acetic acid in ultrapure water with 350 mL MeCN, and degas by ultrasonification for 15 min to prepare MeCN/0.1% acetic acid in water 35/65 (v/v) as HPLC eluent.

2.3 [^{18}F]SFB Synthesis

2.3.1 Equipment for [^{18}F]SFB Synthesis

1. [^{18}F]SFB synthesis is performed in a lead-shielded hot cell with adequate underpressure regulation, inlet and exhaust filters, and radiation monitoring systems by using the commercially available synthesizer TRACERlab Fx_{FDG} for nucleophilic fluorination (GE Medical Systems, Münster, Germany). The system setup and initial position of all valves at the beginning of the synthesis are shown in Fig. 2. Nitrogen 5.0 was used as auxiliary gas (AUX GAS).

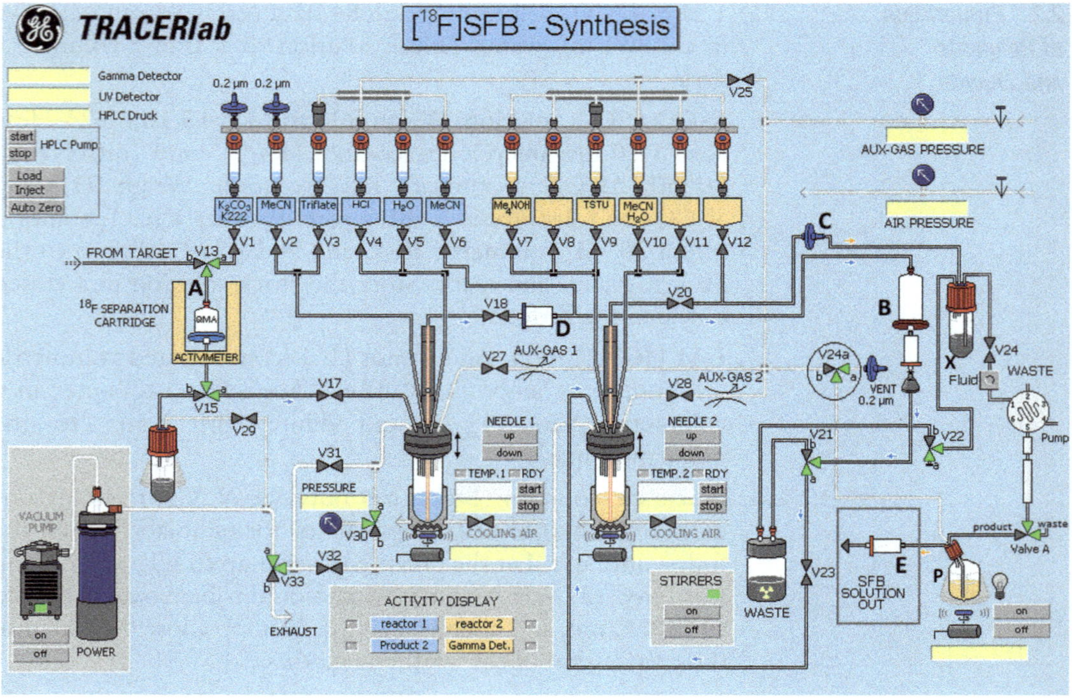

Fig. 2 Schematic representation of the modified TRACERlab Fx_{FDG} synthesis module used for [^{18}F]SFB synthesis. Initial positions of remote-controlled valves are colored gray if closed and green if opened

2. For reasons of clarity and since program structure of commercially available synthesizers differs, explicit time information and program codes have been omitted. Valve positions are highlighted in bold and italic, e.g., ***V1c***, in the text indicating valve number and valve position with the following codes: o = open, c = closed, a = position a (Fig. 2), and b = position b (Fig. 2). Unless otherwise stated, the time between each command for switching valves to the given state should be 0.1–1 s. Longer time intervals between two commands, e.g., X seconds, have been indicated by ***(X s)*** or *wait for* commands.

3. For solid-phase extraction, the following cartridges are used: Sep-Pak light Accell Plus QMA® (130 mg, Part Nr. WAT023525), Lichrolut® RP-18E (500 mg, 3 mL PP, Part Nr. 1.19849.0001), Chromafix® C18ec(S) (270 mg, Part Nr. 731804) , Lichrolut® SCX (500 mg, 3 mL PP, Part Nr. 1.02022.0001), Oasis HLB Plus Short (225 mg, Part Nr. 186000132).

4. Semi-preparative HPLC is integrated in the synthesizer module and performed with a JASCO® system: pump PU-980, degasser DG-1580-53, gradient unit LG-980-02, and UV-detector Knaur K2001; semi-preparative C18 precolumn;

column Discovery HS F5-5 (5 μm, 250 × 10 mm, Supelco®), isocratic mode, HPLC eluent (*see* Subheading 2.2, **item** 7), 4 mL/min flow rate. The products are monitored at $\lambda = 220$ nm and with a radioactivity detector integrated in the synthesizer module.

5. ^{18}F activity of [^{18}F]fluoride, [^{18}F]SFB, and [^{18}F]FB-S100s is measured with an activimeter (Isomed 2000 dose calibrator, raytest, Germany).
6. Manual handling of ^{18}F activity like labeling of S100s with [^{18}F]SFB is performed in a lead-shielded and well-ventilated hood.
7. Analytical HPLC is performed on a Kinetex PFP 100 Å column (5 μm 150 × 4.6 mm, phenomenex) equipped with an analytical C18 precolumn using Agilent 1100 HPLC: binary pump G1312A, autosampler G1313A, column oven G1316A, degasser G1322A, UV detector G1314A, γ detector Gabi Star® (raytest, Straubenhardt, Germany); column temperature, 25 °C; isocratic eluent, MeCN/H_2O 35/65 (v/v); flow rate, 0.5 mL/min; run time, 20 min. The products are monitored at $\lambda = 235$ nm.

2.3.2 Preparation of the Synthesis Module for [^{18}F]SFB Synthesis

1. Fill dewar with liquid nitrogen and immerse cooling trap.
2. Open pressurized air and nitrogen gas supply.
3. Activate QMA cartridge by rinsing with 1 M $NaHCO_3$ solution (10 mL) and then water (10 mL), and install in the activimeter at **position A** (for positions A–E, *see* Fig. 2).
4. Activate both RP18e (500 mg) and Chromafix C18 by rinsing each with ethanol (10 mL) and then water (10 mL). Connect RP18e and Chromafix C18 in a row, and install cartridges at **position B**.
5. Rinse PTFE filter (0.2 μm, diameter 2 cm) with HPLC eluent, and install at **position C**.
6. Activate Lichrolut SCX cartridge with 20 mL water and install at **position D**.
7. Activate HLB cartridge with 10 mL ethanol and then 10 mL water, and install at the product line (**position E**). For radiation safety reasons, it is recommended to place the HLB cartridge within an activimeter in a neighboring hot cell.
8. Install cleaned injection **vial X**.
9. Fill 30 mL water into the cleaned **HPLC product vial P**.
10. Equilibrate analytical and semi-preparative HPLC column. Turn UV detector on, and make sure that gamma and UV detector work properly and the fluidic system has no leaks.

11. Heat an aluminum block placed on a magnetic stirrer with heating plate to 75 °C. The heating block should be equipped with an oil-filled glass vial for temperature measurement and cavities suitable for placing a V-shaped 7 mL glass vial in it.
12. Fill the following reagents and solvents into the respective vials: 1.5 mL K_2CO_3/K_{222} solution (vial 1), 3 mL MeCN *HPLC grade* (vial 2), 1 mL precursor solution (vial 3), 0.5 mL 1 M HCl (vial 4), 12 mL water (vial 5), 3 mL MeCN *HPLC grade* (vial 6), 0.5 mL Me_4NOH solution (vial 7), 0.5 mL TSTU solution (vial 9), 1.5 mL HPLC eluent (vial 10).
13. Perform in cooperation with the cyclotron staff and under respective safety measures a pressure test before radiosynthesis to ensure that tubings are not clogged and reaction vessel is tightly closed. ***V13b, V17o***—Switch valves toward the ^{18}F transportation system. Pressurize the transportation system with nitrogen starting from a suitable point of the transport system. Check that pressure rises to a stable end point. ***V13a, (4s), V17c***—Close connection of reaction vessel toward transportation system. Observe if pressure is stable or only slowly falling to ensure that reaction vessel is tightly closed. ***V33b, V31o***—Open reaction vessel toward exhaust, and wait for pressure release. ***V31c, V33a***—Close reaction vessel 1.

3 Methods

3.1 Synthesis of [^{18}F]SFB

Caution—Handling of ^{18}F activity should be performed considering institutional, local, and governmental regulations. Safety procedures should include but not be restricted to, e.g., using adequate lead shielding (lead walls or hot cells), wearing gloves during manual operations, minimizing time for handling radioactive substances, and increasing distance to the radioactive probe by distance tools. If unfamiliar with handling ^{18}F activity contact your institutional radiation safety officer to define appropriate training and safety measures as well as personal and/or local radiation monitoring procedures.

Start automated procedure comprising the following steps:

1. ***V13b, V15b, V29o***—Pass [^{18}F]fluoride through the QMA cartridge (**A**). Monitor adsorbed activity, and wait until activity has reached a maximum.
2. Heat reaction vessel to 65 °C during **steps 3–6**.
3. ***V15a, V29c, V13a***—Close connection of QMA cartridge toward target.
4. Note [^{18}F]fluoride activity (A_{start}) and the time.

5. ***V25o, V1o***—Open valves toward reaction vessel 1. ***V33b, V17o, V31o***—Elute [^{18}F]fluoride from the QMA cartridge. Wait 40 s for [^{18}F]fluoride transfer to the reaction vessel 1.
6. Note [^{18}F]fluoride activity (A_{rest}) and the time.
7. Start automated azeotropic drying procedure. ***V33a, (20 s), V17c, V1c, V27o***—Close vial 1, and wait for reaction vessel 1 to reach a temperature of 65 °C while purging the reaction vessel under constant nitrogen stream. Evaporate solvent for 2 min.
8. Increase temperature to 95 °C, and wait for temperature. Evaporate solvent for 2 min. ***V27c***—Close nitrogen.
9. ***V33b, V2o***—Add acetonitrile from vial 2 at ambient pressure. Wait 15 s for liquid transfer. ***V2c, V25c***—Close connection to vial 2. ***V27o, V33a***—Evaporate solvent under constant nitrogen flow and reduced pressure at 95 °C for 2 min. ***V27c***—Close nitrogen, and remove residual acetonitrile for 6 min under reduced pressure. ***V31c***—Close reaction vial. Cool reaction vessel to 50 °C. ***V27o, (4 s), V27c, V33b, V31o, (5 s), V31c***—Pressurize reaction vessel with nitrogen, and equalize to ambient pressure.
10. ***V25o, V31o, V3o***—Add precursor solution from vial 3 to the reaction vessel. Wait 10 s for liquid transfer. ***V3c, V31c***—Close vial and reaction vessel 1. Heat to 90 °C and wait for temperature. Hold temperature for 10 min for [^{18}F]fluorination. Cool reactor to 45 °C and wait for temperature.
11. ***V31o, (10s), V31c***—Equalize pressure in reaction vessel 1. ***V4o***—Add HCl from vial 4, and wait 10 s for liquid transfer. ***V4c***—Close vial.
12. Heat reaction vessel to 100 °C, and wait for temperature. Hold temperature for 5 min for saponification. Cool reaction vessel to 40 °C.
13. ***V31o, V5o***—Add water from vial 5 and wait 15 s for liquid transfer. ***V31c, V5c***—Close vial 5 and reaction vessel 1.
14. Stir reaction mixture for 1 min.
15. ***Move needle 1 down. V21b, V27o, V18o***—Pressurize the reaction vessel 1 with nitrogen, and transfer reaction mixture via SCX cartridge (**D**) to RP18E/C18 cartridge (**B**), and route liquid to waste equipped with ventilation needle. Wait 3 min for liquid transfer. ***V18c, V21a, V27c***—Close nitrogen gas and reaction vessel 1.
16. ***V6o, V23o, V32o***—Elute adsorbed 4-[^{18}F]fluorobenzoic acid from the SPE cartridges with acetonitrile from vial 6 into reaction vessel 2. Wait 50 s for liquid transfer. ***V6c, V23c, V21b, V32c***–Close vial 6, connection between SPE and reaction vessel 2, and ventilation valve.

17. ***V30b, V32o, V7o***—Add Me_4NOH solution. Wait 7 s for liquid transfer. ***V7c***—Stop addition from vial 7. ***V33a, V28o***—Evaporate solvent under constant nitrogen stream and reduced pressure while heating reaction vessel 2 to 90 °C. Wait for temperature. Hold temperature for 2 min. ***V28c***—Close nitrogen, and evaporate residual solvent for 2 min under reduced pressure.
18. ***V32c***—Close reaction vessel 2. Cool reaction vessel 2 to 70 °C and wait for temperature. ***V33b, V32o, (4 s), V32c***—Equalize pressure in the reaction vessel 2. Set temperature to 70 °C in reaction vessel 2, and wait for temperature.
19. ***V32o, V9o***—Add TSTU solution from vial 9. Wait 7 s for liquid transfer. ***V9c, (3 s), V32c***—Close reaction vessel. Heat reaction vessel 2 to 90 °C, and wait for temperature. Hold temperature for 2 min to convert 4-[^{18}F]fluorobenzoate into [^{18}F]SFB. Cool reaction vessel 2 to 50 °C, and wait for temperature.
20. ***V32o, V10o***—Add eluent from vial 10. Wait 17 s for liquid transfer. ***V10c, V32c***—Close reaction vessel 2. ***V22b, V20o, V24c***—Stir for 10 s while switching valves for liquid transfer.
21. ***Move needle 2 down, V28o***—Transfer reaction mixture to injection vial by pressurizing reaction vessel 2 with nitrogen. Wait 1 min for liquid transfer. ***V28c, V20c, move needle 2 up***—Stop transfer. ***V30a, V21a, V22a, V33a***—Reset module to initial values.
22. Note the activity level in the injection vial (A_{inj}) and the time.
23. Check if HPLC pump is turned on at a flow rate of 4 mL/min, pressure is stable and within the normal range, and HPLC load/inject valve is in ***load*** position.
24. ***V12o, V24o***—Transfer reaction mixture to the sample loop of the semi-preparative HPLC system via fluid detector. Wait for liquid ON signal. Wait for liquid detector OFF signal.
25. Turn valve immediately to ***inject*** position. Perform auto-zero for UV detector signal, and observe radioactivity signal.
26. ***V12c, V24c, V25c***—Close valves between vial 12 and HPLC load/inject valve.
27. [^{18}F]SFB elutes with the given chromatographic system at approximately t_R = 21–23 min. ***$VA_{product}$***—Start collection of [^{18}F]SFB fraction into HPLC-product vial when radioactivity signal exceeds significantly background levels. Consider delay volume between radioactivity detector and valve A. ***VA_{waste}***—Stop collection of [^{18}F]SFB fraction when radioactivity signal has decreased approximately to background levels. Turn HPLC load/inject valve to ***load*** position.

28. ***V24a–b***—Transfer product solution to HLB cartridge to adsorb [^{18}F]SFB on an HLB cartridge, and route liquid to waste vial equipped with a ventilation needle. Wait for liquid transfer, and observe activity level on HLB cartridge. ***V24a–a***—Stop liquid transfer. Wait 30 s for pressure equalization. For radiation safety reasons, it is recommended to place HLB cartridge and activimeter in a neighbored hot cell and transfer product solution through a connection between these two hot cells.
29. Measure radioactivity on HLB cartridge (A_{HLB}), and note the time.

The following steps are performed manually.

30. Take the HLB cartridge out of the activimeter, and disconnect luer lock connections at both sides of the cartridge. Connect a disposable 5 mL syringe filled with 2 mL MeCN to the female side of the cartridge, and place the cartridge upon a V-shaped glass vial in a lead-shielded container. Pass MeCN through the HLB cartridge to elute [^{18}F]SFB into the glass vial (*Caution! Push gently to elute [^{18}F]SFB, but avoid too high pressure to prevent loosening of the connection*).
31. Measure activity in the eluate (A_{SFB}) and on the HLB cartridge ($\mathrm{HLB}_{\mathrm{rest}}$), and note the time $t_{\mathrm{SFB}/\mathrm{HLB}}$.
32. Take a sample of 50–100 μL with a 200 μL (V_{HPLC}) Eppendorf pipette, and transfer liquid into an HPLC vial. Measure activity of the sample (A_{HPLC}), and note the time (t_{HPLC}). Inject 5 μL (V_{inject}) of the product solution into analytical HPLC. Perform quality control as described in Subheading 3.2, **step 2**.
33. Close the glass vial with a screw cap having a septum punctured with argon inlet and ventilation exhaust tubing. Route exhaust gas to a well-ventilated waste vial.
34. Place the product vial in the heating block preheated to 75 °C. Close hot cell. *Caution! For radiation safety reasons, it is recommended to perform the evaporation while the hot cell is closed because [^{18}F]SFB can evaporate off if product solution is evaporated to dryness. Observe radiation levels in the hood, and stop evaporation immediately if activity levels rise.*
35. Open argon gas stream with a pressure set at 0.5 bar so that a gentle stream of argon is applied. Make sure that argon tubing is placed about 1 cm above the initial liquid level and does not touch the surface. Gently evaporate acetonitrile with visual inspection of the product solution in 30–60 s intervals. Slowly increase argon pressure to 1 bar with advancing evaporation. Concentrate product solution to a volume of 100–200 μL.
36. Stop argon gas flow. Place product vial in a lead-shielded container, and wait for 3 min to cool down product vessel.

Replace screw cap septum with a closed cap. Final [^{18}F]SFB solution is now ready and suitable for further processing, e.g., labeling of S100 proteins.

3.2 Quality Control for [^{18}F]SFB

1. Determine decay-corrected radiochemical yield by using the formula (1):

$$\mathrm{RCY}\ [\%] = \frac{A_{\mathrm{SFB}}}{A_{\mathrm{start}}} \times e^{\frac{(t_{\mathrm{SFB}} - t_{\mathrm{start}})\ \mathrm{min}}{109{,}7\ \mathrm{min}} \times \ln(2)} \times 100. \quad (1)$$

2. Determine radiochemical purity and molar activity by injecting an analytical sample as described in Subheading 3.1 and 3.2 three times into the analytical HPLC (*see* Subheading 2.3.1).
   (a) For radiochemical purity, determine the area under the curve from the radioactivity signal of each peak and the respective percentage for the [^{18}F]SFB peak. Under the specified HPLC conditions, SFB elutes with a retention time t_R of 11.2 min. If it is unclear which peak corresponds to [^{18}F]SFB, an HPLC run either with the nonradioactive reference SFB or with a SFB spiked product solution has to be performed.
   (b) Calculate the molar activity at the time of sample taking by the formula (2):

$$A_m(t_{\mathrm{HPLC}}) = \frac{A_{[^{18}\mathrm{F}]\mathrm{SFB}}}{n_{\mathrm{SFB}}} = \frac{A_{\mathrm{HPLC}} \cdot \frac{V_{\mathrm{inject}}}{V_{\mathrm{HPLC}}}}{n_{\mathrm{SFB(HPLC,inject)}}}. \quad (2)$$

For this, the amount of substance in the injected volume $n_{\mathrm{SFB(HPLC,inject)}}$ is calculated from the integrated UV signal corresponding to SFB and in comparison with a calibration curve, which is determined beforehand (*see* Fig. 3). In brief, SFB standards are prepared by dissolving a definite amount of SFB in MeCN in a volumetric flask followed by serial 1:1 dilution with a 1 mL Eppendorf pipette. A calibration curve is determined by injection of standard solutions in duplicate into the analytical HPLC system under the HPLC specifications given in Subheading 2.3.1, integration of the UV signal corresponding to SFB, and application of a linear fit to the data set.

3.3 Cleaning Procedure for [^{18}F]SFB Synthesizer Module

An automated cleaning procedure is performed twice after each synthesis, the first time by application of the procedure with solvent and second as a dry clean without liquids to remove any solvent residues. *Caution*—depending on the time between synthesis and clean, the synthesizer module may still be contaminated with ^{18}F. We usually performed the clean 24–48 h after the radiosynthesis. In brief, SPE cartridges **A–E** and the **PTFE filter** are removed, and the respective tubings are short-circuited. The [^{18}O]water vial is emptied. Both reaction vessels **R1** and **R2**, the **injection vial X**,

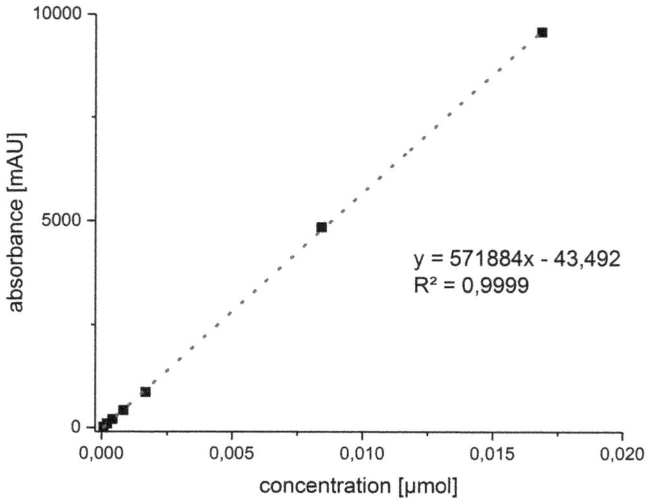

Fig. 3 Representative calibration curve and linear fit obtained from a dilution series of SFB and HPLC analyses

and the **HPLC product vial P** are dissembled, carefully cleaned, dried, and reinstalled into the synthesis module. For the first clean, 20 mL water is filled into **HPLC product vial P**, and the given amount of acetone is filled into each given vial: 2 mL in vials 1, 4, and 7; 1 mL in vials 2, 3, 8, and 9; 5 mL in vials 5 and 10; and 3 mL in vial 6. The cleaning procedure comprises the following steps:

(a) Consecutive addition of solvent from vial 1 and vial 2 to **R1** and vial 7 and vial 8 to **R2** while exhaust is opened via V33, stirring for 90 s, and emptying of **R1** via V18 and V21 to the waste and **R2** via V20, **injection vial X**, and HPLC loop to the waste.

(b) Similar rinsing procedure like (*a*) for V3/V4 and V9/V10 and after heating of both reaction vessels to 85 °C for V5 and V11.

(c) Rinsing lines between **R1** and **R2** via V18, V21, and V23 at 40 °C with solvent from V6 followed by routing this and solvent from V12 via **injection vial X** and V24 to the waste.

(d) Consecutive purging each pair of lines (V1/V7, V2/V8, etc.) via the reaction vessels with nitrogen for 3 min at 120 °C and vacuum suction.

(e) Drying of lines between **R1**, **R2**, **HPLC injection vial X**, V22, and the waste in a stream of nitrogen.

(f) Cooling down of both reaction vessels under vacuum for 5 min and meanwhile emptying of **HPLC product vial P** to clean tubings toward HLB cartridge and waste.

(g) Pressure equalization in vials and reaction vessels.

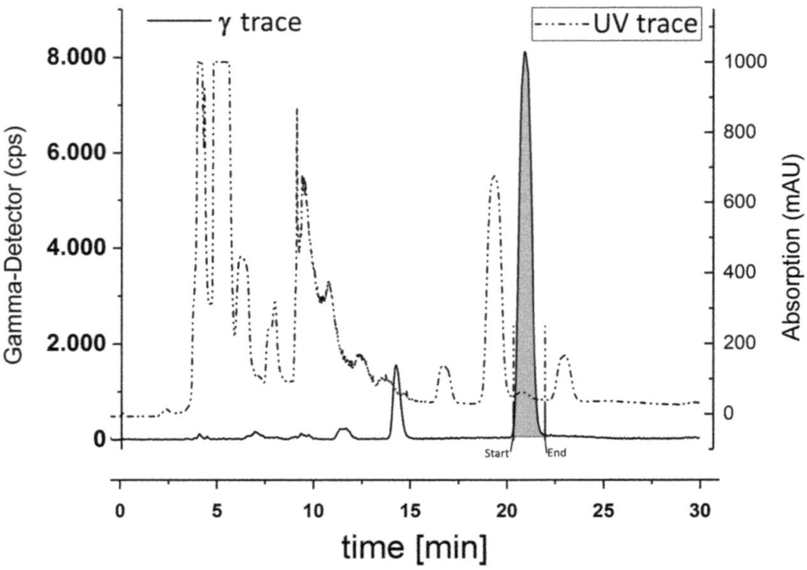

Fig. 4 Representative semi-preparative HPLC chromatogram of [^{18}F]SFB purification. Time points for starting the collection of [^{18}F]SFB by switching valve $VA_{product}$ and ending collection of [^{18}F]SFB by switching valve VA_{waste} are labeled with start and end, respectively, and the collected [^{18}F]SFB fraction is colored in pale gray

3.4 Anticipated Results

A typical experiment started from 15 GBq [^{18}F]fluoride and yielded about 2.9 GBq [^{18}F]SFB after 89 min of synthesis time. The described synthetic procedure was reproducible and gave [^{18}F]SFB in isolated radiochemical yields of 28 ± 3%, molar activity of 52 ± 3 GBq/μmol, and chemical purity of 77 ± 12% ($n = 12$). The radiochemical purity exceeded 99% (*see* Figs. 4 and 5).

3.5 Radiolabeling of S100 Proteins with [^{18}F]SFB

1. In principle, ^{18}F-radiolabeling of either mature or tagged S100s can be performed through acylation of both the chemically accessible *N*-terminal amino acid residues and lysine side-chain residues with [^{18}F]SFB. In the present approach, [^{18}F]fluorobenzoylation of lysine side-chain residues (e.g., 10 lysines in S100A12 and 11 lysines in S100A4) should be avoided.
2. Therefore, labeling reactions are carried out at pH 7.4. Below this pH value, nearly all lysine side-chain residues are supposed to be protonated, hence being not available for [^{18}F]fluorobenzoylation with [^{18}F]SFB. As a consequence, the chemically accessible *N*-terminal residues of S100s are the most privileged site for conjugation. Lysine residues become the preferred site of conjugation with [^{18}F]SFB only under basic conditions (e.g., pH 8.4). For peptide labeling with [^{18}F]SFB, the pH is usually adjusted to 8.2–8.4.

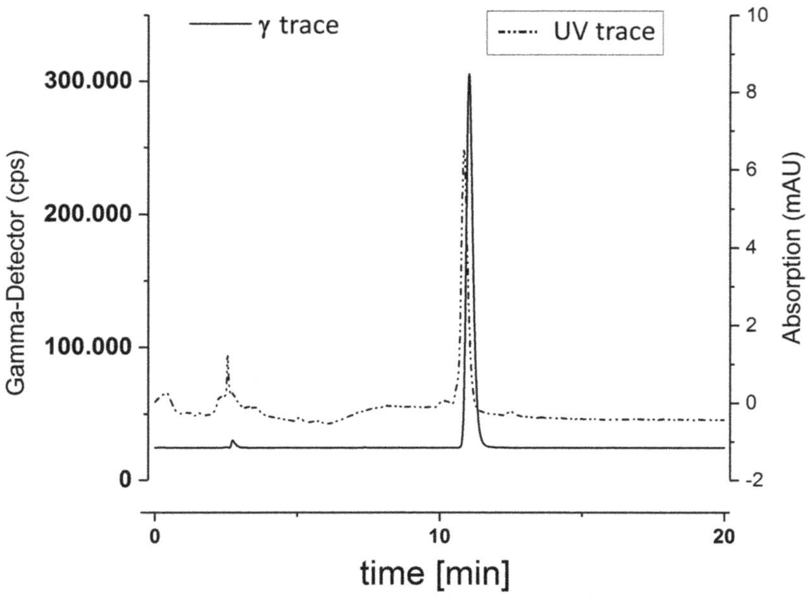

Fig. 5 Representative analytical HPLC chromatogram of 5 μL [^{18}F]SFB solution obtained after isolation of [^{18}F]SFB from the HLB cartridge. The major nonradioactive component in the mixture can be assigned to nonradioactive SFB; the time delay between UV and radioactivity signals is based on the consecutive arrangement of both detectors

3. The obtained [^{18}F]SFB (500–1000 MBq) diluted in acetonitrile then is directly added to 100–400 μg S100 protein dissolved in PBS and incubated at pH 7.4 for 30 min at 37 °C with continuous and gentle shaking in a thermal mixer.
4. The radiolabeled S100 proteins are purified using a HiTrap Desalting column (GE Healthcare; equilibrated with 20 mM HEPES, 150 mM NaCl, 1.2 mM $MgCl_2$, and 1.3 mM $CaCl_2$, pH 7.4) with an ÄKTAprime® plus (GE Healthcare) chromatography system. Gel filtration is carried out at a flow rate of 0.5 mL/min and fractionated in 0.5 mL steps. The radioactivity is determined with an activimeter. Fractions containing the purified ^{18}F-labeled S100 proteins, separated from unreacted [^{18}F]SFB and the by-product [^{18}F]fluorobenzoic acid, will be combined and used for further experiments.
5. The activity of the product is measured using the ISOMED 2000 calibrator (Nuklear-Medizintechnik Dresden GmbH). Because of the short half-life of ^{18}F ($t_{1/2} = 109.77$ min), it is necessary to adjust radioactivity data in order to compare all the time points. Accordingly, all radioactivity measurements in vitro and in vivo should be adjusted and expressed as decay-corrected values [13, 15].

4 Notes

Purified fluorine-18-labeled S100s ([^{18}F]FB-S100s) can be used as probes, e.g., in receptor-binding assays and cell uptake/association experiments in vitro as well as in biodistribution, metabolism, and dynamic PET (imaging) experiments in vivo. The following is an excerpt of essential aspects of the basic methodical approaches employed so far. Special aspects, e.g., for cell-based investigations, were excluded here, since, for example, cell passages, cell numbers, cell density, and protein concentrations depend on the individual cell type. For this and other detailed descriptions, e.g., regarding blocking experiments in vitro and in vivo aiming at binding specificity and differentiation of several binding partners, we refer to the work published elsewhere [13–15, 43].

4.1 In Vitro Experiments

1. Binding of [^{18}F]FB-S100s to certain receptors can be studied in cell-free binding assays. Exemplarily, suitable 96-well plates (Maxisorb TM 96-well plates; Nunc, Langenselbold, Germany) are coated with recombinant sRAGE or albumin (each 2.5 μg/well) in bicarbonate/carbonate buffer (50 mM, pH 9.6) and are incubated overnight at 4 °C. After washing three times with 250 μL phosphate-buffered saline (PBS) containing 0.05% Tween 20, wells are blocked with PBS containing 2% bovine serum albumin for 2 h. Then, wells with adsorbed sRAGE, albumin, or without protein are incubated with 100 μL of radiolabeled [^{18}F]FB-S100s (approximately 300–400 kBq; in 20 mM HEPES buffer with 150 mM NaCl, 1.2 mM $MgCl_2$, 1.3 mM $CaCl_2$, pH 7.5 supplemented with 2% bovine serum albumin) for 1 h at 37 °C with gentle agitation. For this and the following approaches, molar activity and radiochemical yield (>95% minimum) of [^{18}F]FB-S100s must be specified in each case. Subsequently, wells are washed three times with ice-cold PBS on ice. Bound protein is dissolved in 100 μL of 0.1 M NaOH containing 1% SDS at 37 °C for 30 min under shaking conditions. Fluorine-18 activity (in terms of radioactivity concentration) is measured in a gamma counter (Cobra II; Canberra-Packard, Meriden, USA).
2. For cellular association experiments, cells are seeded in 96-well plates at definite time points (24–72 h) and cell numbers (10^3–10^5 cells/well) prior to the experiment. The assay is performed in the presence or absence of nonradioactive S100s. [^{18}F]FB-S100s probes (approximately 100–200 kBq; in PBS) are added, and cells are incubated for 1 h at 37 °C. Subsequently, cells are washed three times with ice-cold PBS and dissolved in 0.1 M NaOH containing 1% SDS. Fluorine-18

activity is measured in a gamma counter, and afterward protein concentration is determined using an appropriate protein assay (Pierce® BCA Protein Assay; Thermo Fisher Scientific, Waltham, USA).

4.2 In Vivo Experiments

1. All animal experiments are carried out according to local guidelines of the national regulations for animal welfare. The protocols must be approved by the local ethical committees for animal experiments. In vivo stability, biodistribution, and dynamic small animal PET studies are assessed, exemplarily, in rats (e.g., Kyoto-Wistar strain; aged 9–12 weeks; 200–230 g).
2. For the investigation of metabolic stability, animals are anesthetized with desflurane (10% and 30% oxygen/air; suprane®; Baxter, Unterschleißheim, Germany), and a catheter is placed into the right common carotid artery or right femoral artery. A volume of 0.5 mL [^{18}F]FB-S100s probes containing 5–20 MBq (PBS, pH 7.4) is injected into the tail vein. Arterial blood samples (0.5 mL) are taken at definite time points (e.g., 5, 10, 20, 40, 60, 120 min post injection). The depleted blood volume is compensated by the injection of saline. Blood samples were immediately centrifuged for 5 min at 10,000 × *g*. Arterial plasma samples were heated using SDS-sample buffer containing β-mercaptoethanol at 95 °C for 5 min, analyzed by SDS-PAGE (radioluminography using, e.g., a BioImaging Analyzer BAS-5000, Fuji Photo Film, Düsseldorf, Germany) and, thereafter, are deproteinated with the twofold volume of water/methanol/TFA (50/45/5; v/v/v). After centrifugation for 2 min at 10,000 × *g*, supernatants and subnatants were measured using an activimeter for fluorine-18 (dose calibrator Isomed 2000 dose calibrator; Nuklear-Medizintechnik Dresden GmbH, Dresden, Germany). Radioluminographic intensities of the electrophoretic bands from SDS-PAGE are quantified using an appropriate software version (AIDA; Raytest, Straubenhardt, Deutschland) and expressed as percentage of total activity amount.
3. For biodistribution experiments, groups of rats ($n \geq 4$) are administered with 500 μL [^{18}F]FB-S100s (1.2–1.5 MBq, PBS, pH 7.4) into the tail vein and sacrificed at 5 and 60 min postinjection (p.i.) by heart puncture under desflurane anesthesia. Organs and tissues of interest are rapidly excised and weighed, and the fluorine-18 activity is determined using a gamma counter cross calibrated with an activimeter for fluorine-18 by aliquots of the injected activity. The radioactivity concentration in organs and selected tissues is calculated as the percentage of the injected dose per gram tissue (%ID/g) or standard uptake value (SUV, g/g) of the tissue. SUV is defined as the tracer concentration at a certain time point normalized

to injected dose per unit body weight and is used for better comparison within animals of different size and weight and with other species.

4. For dynamic small animal PET studies, rats under desflurane anesthesia are positioned prone with thorax and abdominal region (organs of interest: the heart, lung, liver, kidneys, large vessels) in the center of field of view of a dedicated PET scanner for small animals (microPET P4; CTI Concorde Microsystems, Knoxville, USA or NanoScan PET/CT scanner, Mediso, Budapest, Hungary). A bolus injection of 500 μL of [^{18}F]FB-S100s (10–20 MBq, PBS, pH 7.4) is administered via the tail vein within 30 s. Simultaneous with tracer injection, dynamic PET scanning was started for 120 min. Sinogram generation and image reconstruction followed the protocol given by us elsewhere [44]. Images were analyzed by assigning three-dimensional regions of interest (ROI) over the heart region, the liver, and the kidneys using ROVER software (ABX GmbH, Radeberg, Germany). From these ROIs, time-activity curves (TACs) representing the total (decay-corrected) fluorine-18 activity in a defined volume and expressed as radioactivity concentration and percent of maximum (or SUV) were obtained in each rat.

5. TACs from ROIs over the heart region, majorly representing the cardiac blood pool, are used to estimate, e.g., the pharmacokinetic parameter mean residence time. Therefore, the numerical module of SAAM II program (University of Washington, Seattle, USA) using a derivative-free nonlinear regression analysis is used to fit fluorine-18 activity data from ROIs over the heart region from each animal to multiexponential equations. The individual responses are quite consistent, showing a three-exponential equation $A(t) = A1e^{-k1t} + A2e^{-k2t} + A3e^{-k3t}$, where $A(t)$ = % of injected dose per cm^3, to provide the best fit. The distribution/tissue association ($t_{1/2}{}^{d/a}$; $t_{1/2}k1 + t_{1/2}k2$) and elimination ($t_{1/2}{}^{e}$, $t_{1/2}k3$) half-lives were calculated as 0.693/k. For estimation of mean residence time, the area under the heart region TAC and the area under the momentum curve are calculated using standard equations according to Wolfe [45].

5 Concluding Remarks

5.1 Radiopharmacological Studies on [^{18}F]S100A12

Both in vivo and in vitro experiments using [^{18}F]FB-S100A12 have indicated that S100A12 is a physiologically highly affine ligand for the receptor for advanced glycation end products (RAGE). In this regard, the use of fluorine-18 is of particular interest since ^{18}F-labeled radiotracers, including protein-based radioligands, can usually be obtained in high molar activities. By using dedicated small animal PET systems, images of ^{18}F-radiolabeled probes showing their time-dependent organ/tissue distribution and elimination exhibit the highest resolution [44, 46, 47]. In a first study, [^{18}F]FB-S100A12 served as probe for the functional characterization of RAGE [13]. The study showed in a rat model the strong interaction of [^{18}F]FB-S100A12 with organs and tissues showing high expression of RAGE, in particular, the lung and large blood vessels. The in vivo kinetics of [^{18}F]FB-S100A12 was characterized by its high metabolic stability and a relatively long residence time (mean plasma residence time was approximately 5 h) in the circulating blood as well as a sustained tissue-associated temporary retention in the lung and the whole vascular system. Considering an overall faster metabolism in rodents, it can be expected that terminal half-life of S100A12 in human circulation is much more prolonged. In situations where homeostasis of S100A12 secretion and elimination is disturbed, this should have pathophysiological consequences. Of importance, high plasma levels of S100A12 have been described in various inflammatory disorders [4, 48–50]. Interaction of S100A12 with membrane-anchored RAGE in a cytokine-like manner has been proposed to be the key process in initiation or perpetuation of a concerted inflammatory action in various cells and tissues. Continuing secretion of S100A12 by activated granulocytes and macrophages into the blood and, on the other hand, this prolonged terminal half-life of S100A12 would consecutively trigger this process. The PET data in terms of tissue-associated temporary retention correlated well with the observed organ-specific expression of the non-internalizing RAGE in the rat. Highest RAGE expression both on mRNA and protein level was found in the lung. The retention in the vascular system can be explained by the presence of RAGE in endothelial cells and medial smooth cells. Systemic clearance of ^{18}F-radioactivity after intravenous injection of [^{18}F]S100A12 was almost completely explained by uptake in the kidneys and subsequent renal elimination and, to a minor part, by uptake in the liver and hepatobiliary excretion. This can be explained majorly by glomerular filtration of the small acidic S100s and their degradation in the proximal tubule in the kidney. At this point, a hypothesis on the formation of complexes between S100A12 and circulating soluble isoforms of RAGE (sRAGE), which are also present in rats, could be deduced. Potential uptake of these complexes by the liver or the kidneys should further contribute to the systemic elimination of the highly affine S100A12 [13]. In a more detailed study, including cell-binding experiments in human aortic endothelial cells and macrophages exhibiting, besides membrane RAGE, also various SR, we demonstrated substantial SR binding of [^{18}F]FB-S100A12. This finding was also confirmed in subsequent dynamic PET experiments in rats in vivo [15]. At this point, we already hypothesized that the possible interaction of S100A12 with SR-like CD36, which has been reported to mediate pro-inflammatory signaling, is likely to further contribute to the scenarios discussed above. Very recently, by using both fluorescent and radiolabeled derivatives of S100A12, the class B scavenger receptor CD36 actually was identified as a highly affine binding partner of S100A12. Here, cell-binding experiments showed a specific S100A12-CD36 interaction that was blocked by anti-CD36 antibody. The experiments also revealed that S100A12 shares the same binding site (between amino acids 93 and 120) on CD36 as do other CD36 ligands like thrombospondin and collagen. A further finding of this study was that S100A12 upregulates CD36 expression and recruits CD36 to the cell membrane, indicating a potential pathophysiological link between high circulating S100A12 levels and the onset of atherosclerosis, particularly in chronic inflammatory disorders [43, 51]. Furthermore, there are initial considerations hypothesizing possible interaction of S100A12, RAGE, and CD36 in chronic kidney disease [52–54].

5.2 Radiopharmacological Studies on [^{18}F]S100A4

In an in vivo study using rats and a comparable experimental setting, the interaction of [^{18}F]FB-S100A4 with organs and tissues showing high RAGE expression also could be demonstrated. However, from the data, we argued that under physiological conditions in the rat, interaction of extracellular S100A4 with RAGE is weak to moderate when compared to [^{18}F]S100A12 and other S100s investigated by radiotracer approaches ([^{18}F]FB-S100A1, [^{18}F]FB-S100B [13]; [^{18}F]FB-S100B-oligomers, unpublished data). Besides its lower affinity to the receptor, as shown in surface plasmon resonance analyses, a substantially faster metabolic degradation of S100A4 contributes to this observation. Of importance, like for [^{18}F]FB-S100A12, the recognition of [^{18}F]FB-S100A4 by other receptors, e.g., SR, cannot be excluded. Indicative for this assumption was the high uptake of [^{18}F]FB-S100A4 observed in the liver, spleen, and adrenals, organs comprising cells that overexpress SR, e.g., CD36 and SRBI. Furthermore, the cellular binding studies in both an endothelial cell (human aortic endothelial cells) and a RAGE-expressing human melanoma cell (A375) model revealed a substantially higher [^{18}F]FB-S100A4 binding to putative binding sites in endothelial cells than in melanoma cells. This makes it likely that scavenger receptors and other endothelial surface structures, e.g., heparin sulfate moieties on the cell surface of endothelial cells or annexins, contribute to the higher [^{18}F]FB-S100A4 association [55, 56]. Nevertheless, the significant higher association of [^{18}F]FB-S100A4 to A375 cells overexpressing RAGE compared with wild-type A375 cells and mock-transfected cells demonstrated the direct interaction of [^{18}F]FB-S100A4 and RAGE. Taking into account physiological situations with various competing RAGE ligands available, which exhibit higher concentrations and higher affinity to the receptor, the systemic interaction between extracellular S100A4 and RAGE thus is likely to be of minor importance. On the other hand, a typical attribute of RAGE is that it is expressed at relatively low levels in homeostasis, but in situations characterized by enhanced cellular activation or stress, the expression of RAGE is strikingly enhanced, particularly in certain tissue compartments. Under these circumstances, S100A4 could substantially contribute to the overall complex interaction with RAGE. We were later able to elucidate some possible physiological consequences, especially for the tumor microenvironment. First, we could demonstrate that macrophage-derived S100A4 interacted with RAGE in melanoma cells in a paracrine manner potentiating tumor invasion and metastatic lung colonization [57]. Subsequent experiments indicated that interaction of S100A4 with RAGE driving a pro-metastatic phenotype is not limited to paracrine signaling. Involvement of an autocrine S100A4-RAGE signaling has been recognized when it was observed that A375 melanoma cells actively secrete S100A4 in the extracellular space via an endoplasmic reticulum-Golgi-dependent pathway. We were able to demonstrate that pro-metastatic properties of melanoma cells like cell motility were diminished by inhibition of this secretory pathway [8]. We also could demonstrate a role for high levels of extracellular S100A4 in decreasing inter-endothelial tight junction integrity by using the A375 melanoma cell model and an in vitro blood–brain barrier model. Paracrine-mediated signaling of S100A4 secreted from melanoma cells with RAGE expressed on endothelial cells enabled melanoma cell transmigration. The study also revealed that brain metastasis formation by A375 cells injected intracardially into athymic mice was increased with overexpression of S100A4 and RAGE [58, 59]. It has to be clarified whether S100A4-induced endothelial dysfunction is specific for melanoma brain metastasis or rather a general mechanism in S100A4 expressing cancers [60].

5.3 The Radiopharmacologist's Perspective

S100-based radioligand approaches have high potential to contribute to a better understanding of the role of extracellular S100s in specific pathophysiological situations. Gentle radiolabeling under physiological conditions is an essential prerequisite for this. The presented protocol gives a detailed description for the radiosynthesis of [^{18}F]SFB using an automated synthesis module. The synthesis is based on the ^{18}F-labeling of the *tert*-butyl protected *N,N,N*-trimethylammoniumbenzoic acid ester triflate **1**, followed by a hydrolysis step forming 4-[^{18}F]fluorobenzoate and the final conversion to [^{18}F]SFB. High chemical and radiochemical purity have been achieved by applying a semi-preparative HPLC purification of [^{18}F]SFB presented in this protocol as a further optimization of the previously reported method. Ongoing work aims at investigating in more detail the role of S100-RAGE and S100-SR axes and their therapeutic targeting in animal models of inflammatory or neoplastic disease. This also includes research to clarify the role of soluble RAGE or soluble SR as decoy receptors and modulators of circulating S100s levels in certain pathophysiological situations. Furthermore, the importance of S100 proteins as a therapeutic target will be better underlined.

Acknowledgments

We apologize to those researchers whose works have not been mentioned due to space restrictions. We are especially grateful to our former colleagues Susan Hoppmann, Ph.D.; Susann Wolf, Ph.D.; and Nadine Herwig (née Tandler), Ph.D., who all received their doctorate in the field of S100 protein research at the Technische Universität Dresden, Faculty of Chemistry and Food Chemistry, Germany, and also to Christoph Tondera, Ph.D., for the dedicated work and many stimulating and fruitful discussions. The authors thank the staff of the cyclotron and GMP radiopharmaceuticals production units for providing [^{18}F]fluoride. The expert technical assistance of Mareike Barth, Catharina Heinig, Regina Herrlich, Uta Lenkeit, Sebastian Meister, Aline Morgenegg, and Andrea Suhr also is greatly acknowledged. Jens Pietzsch is thankful to the Deutsche Forschungsgemeinschaft (DFG) for supporting this work by research grant PI 304/1-1 "Bildgebende In-vivo-Charakterisierung von Rezeptoren für Advanced Glycation End products mittels Kleintier-Positronen-Emissions-Tomographie" and within the Collaborative Research Center Transregio 67 "Functional Biomaterials for Controlling Healing Processes in Bone und Skin—From Material Science to Clinical Application" (CRC/TRR 67/3). This work also is part of the intramural research initiative "Radiation-Induced Vascular Dysfunction (RIVAD)."

References

1. Hermann A, Donato R, Weiger TM, Chazin WJ (2012) S100 calcium binding proteins and ion channels. Front Pharmacol 3:67
2. Donato R, Cannon BR, Sorci G, Riuzzi F, Hsu K, Weber DJ, Geczy CL (2013) Functions of S100 proteins. Curr Mol Med 13(1):24–57
3. Marenholz I, Heizmann CW, Fritz G (2004) S100 proteins in mouse and man: from evolution to function and pathology (including an update of the nomenclature). Biochem Biophys Res Commun 322(4):1111–1122
4. Pietzsch J, Hoppmann S (2009) Human S100A12: a novel key player in inflammation? Amino Acids 36(3):381–389
5. Rammes A, Roth J, Goebeler M, Klempt M, Hartmann M, Sorg C (1997) Myeloid-related protein (MRP) 8 and MRP14, calcium-binding proteins of the S100 family, are secreted by activated monocytes via a novel, tubulin-dependent pathway. J Biol Chem 272(14):9496–9502
6. Davey GE, Murmann P, Heizmann CW (2001) Intracellular Ca^{2+} and Zn^{2+} levels regulate the alternative cell density-dependent secretion of S100B in human glioblastoma cells. J Biol Chem 276(33):30819–30826
7. Matsunaga H, Ueda H (2006) Evidence for serum-deprivation-induced co-release of

FGF-1 and S100A13 from astrocytes. Neurochem Int 49(3):294–303

8. Herwig N, Belter B, Wolf S, Haase-Kohn C, Pietzsch J (2016) Interaction of extracellular S100A4 with RAGE prompts prometastatic activation of A375 melanoma cells. J Cell Mol Med 20(5):825–835
9. Saho S, Satoh H, Kondo E, Inoue Y, Yamauchi A, Murata H, Kinoshita R, Yamamoto KI, Futami J, Putranto EW, Ruma IM, Sumardika IW, Youyi C, Suzawa K, Yamamoto H, Soh J, Tomida S, Sakaguchi Y, Saito K, Iioka H, Huh NH, Toyooka S, Sakaguchi M (2016) Active secretion of dimerized S100A11 induced by the peroxisome in mesothelioma cells. Cancer Microenviron 9 (2–3):93–105
10. Tandler N, Mosch B, Pietzsch J (2012) Protein and non-protein biomarkers in melanoma: a critical update. Amino Acids 43(6):2203–2230
11. Oesterle A, Bowman MA (2015) S100A12 and the S100/calgranulins: emerging biomarkers for atherosclerosis and possibly therapeutic targets. Arterioscler Thromb Vasc Biol 35 (12):2496–2507
12. Chong ZZ, Changyaleket B, Xu H, Dull RO, Schwartz DE (2016) Identifying S100B as a biomarker and a therapeutic target for brain injury and multiple diseases. Curr Med Chem 23(15):1571–1596
13. Hoppmann S, Haase C, Richter S, Pietzsch J (2008) Expression, purification and fluorine-18 radiolabeling of recombinant S100 proteins – potential probes for molecular imaging of receptor for advanced glycation endproducts (RAGE) in vivo. Protein Expr Purif 57 (2):143–152
14. Wolf S, Haase-Kohn C, Lenk J, Hoppmann S, Bergmann R, Steinbach J, Pietzsch J (2011) Expression, purification and fluorine-18 radiolabeling of recombinant S100A4: a potential probe for molecular imaging of receptor for advanced glycation endproducts in vivo? Amino Acids 41(4):809–820
15. Hoppmann S, Steinbach J, Pietzsch J (2010) Scavenger receptors are associated with cellular interactions of S100A12 in vitro and in vivo. Int J Biochem Cell Biol 42(5):651–661
16. Haase-Kohn C, Wolf S, Lenk J, Pietzsch J (2011) Copper-mediated cross-linking of S100A4, but not of S100A2, results in proinflammatory effects in melanoma cells. Biochem Biophys Res Commun 413(3):494–498
17. Pietzsch J, Bergmann R, Rode K, Hultsch C, Pawelke B, Wuest F, van den Hoff J (2004) Fluorine-18 radiolabeling of low-density lipoproteins: a potential approach for characterization and differentiation of metabolism of native and oxidized low-density lipoproteins in vivo. Nucl Med Biol 31 (8):1043–1050
18. Berndt M, Pietzsch J, Wuest F (2007) Labeling of low-density lipoproteins using the ^{18}F-labeled thiol-reactive reagent N- 6-(4-[^{18}F] fluorobenzylidene)aminooxyhexyl maleimide. Nucl Med Biol 34(1):5
19. Wuest F, Kohler L, Berndt M, Pietzsch J (2009) Systematic comparison of two novel, thiol-reactive prosthetic groups for ^{18}F labeling of peptides and proteins with the acylation agent succinimidyl-4-[^{18}F]fluorobenzoate ([^{18}F]SFB). Amino Acids 36(2):283–295
20. Vaidyanathan G, Zalutsky MR (1992) Labeling proteins with fluorine-18 using *N*-succinimidyl 4-[^{18}F]fluorobenzoate. Int J Rad Appl Instrum B 19(3):275–281
21. Abad S, Nolis P, Gispert JD, Spengler J, Albericio F, Rojas S, Herance JR (2012) Rapid and high-yielding cysteine labelling of peptides with *N*-succinimidyl 4-[^{18}F]fluorobenzoate. Chem Commun 48(49):6118–6120
22. Kuchar M, Pretze M, Kniess T, Jr S, Pietzsch J, Löser R (2012) Site-selective radiolabeling of peptides by ^{18}F-fluorobenzoylation with [^{18}F] SFB in solution and on solid phase: A comparative study. Amino Acids 43(4):1431–1443
23. Rojas S, Nolis P, Gispert JD, Spengler J, Albericio F, Herance JR, Abad S (2013) Efficient cysteine labelling of peptides with N-succinimidyl 4-[^{18}F]fluorobenzoate: stability study and in vivo biodistribution in rats by positron emission tomography (PET). RSC Adv 3(21):8028–8036
24. Matusiak N, Castelli R, Tuin AW, Overkleeft HS, Wisastra R, Dekker FJ, Prly LM, Bischoff RPM, Van Waarde A, Dierckx RAJO, Elsinga PH (2015) A dual inhibitor of matrix metalloproteinases and a disintegrin and metalloproteinases, [^{18}F]FB-ML5, as a molecular probe for non-invasive MMP/ADAM-targeted imaging. Bioorg Med Chem 23(1):192–202
25. Glaser M, Arstad E, Luthra SK, Robins EG (2009) Two-step radiosynthesis of [^{18}F] *N*-succinimidyl-4-fluorobenzoate ([^{18}F]SFB). J Label Compd Radiopharm 52(8):327–330
26. Vaidyanathan G, Zalutsky MR (2006) Synthesis of *N*-succinimidyl 4-[^{18}F]fluorobenzoate, an agent for labeling proteins and peptides with ^{18}F. Nat Protoc 1(4):1655–1661
27. Wester HJ, Hamacher K, Stocklin G (1996) A comparative study of N.C.A. fluorine-18 labeling of proteins via acylation and photochemical conjugation. Nucl Med Biol 23(3):365–372

28. Scott PJH, Shao X (2010) Fully automated, high yielding production of N-succinimidyl 4-[^{18}F]fluorobenzoate ([^{18}F]SFB), and its use in microwave-enhanced radiochemical coupling reactions. J Label Compd Radiopharm 53(9):586–591
29. Tang G, Tang X, Wang X (2010) A facile automated synthesis of N-succinimidyl 4-[^{18}F]fluorobenzoate ([^{18}F]SFB) for ^{18}F-labeled cell-penetrating peptide as PET tracer. J Label Compd Radiopharm 53(8):543–547
30. Thonon D, Goblet D, Goukens E, Kaisin G, Paris J, Aerts J, Lignon S, Franci X, Hustinx R, Luxen A (2011) Fully automated preparation and conjugation of N-succinimidyl 4-[^{18}F]fluorobenzoate ([^{18}F]SFB) with RGD peptide using a GE FASTlab? synthesizer. Mol Imaging Biol 13(6):1088–1095
31. Ackermann U, Yeoh SD, Sachinidis JI, Poniger SS, Scott AM, Tochon-Danguy HJ (2011) A simplified protocol for the automated production of succinimidyl 4-[^{18}F]fluorobenzoate on an IBA Synthera module. J Label Compd Radiopharm 54(10):671–673
32. Tang G, Zeng W, Yu M, Kabalka G (2008) Facile synthesis of N-succinimidyl 4-[^{18}F]fluorobenzoate ([^{18}F]SFB) for protein labeling. J Label Compd Radiopharm 51(1):68–71
33. Johnstrom P, Clark JC, Pickard JD, Davenport AP (2008) Automated synthesis of the generic peptide labelling agent N-succinimidyl 4-[(18)F]fluorobenzoate and application to (18)F-label the vasoactive transmitter urotensin-II as a ligand for positron emission tomography. Nucl Med Biol 35(6):725–731
34. Bejot R, Elizarov AM, Ball E, Zhang J, Miraghaie R, Kolb HC, Gouverneur V (2011) Batch-mode microfluidic radiosynthesis of N-succinimidyl-4-[^{18}F]fluorobenzoate for protein labelling. J Label Compd Radiopharm 54 (3):117–122
35. Nakanishi H, Saiki H, Saji H, Kimura H, Kawashima H, Tomatsu K, Kuge Y (2012) Method of synthesizing [^{18}F]SFB using microsynthesis technique. EP2404903
36. Kimura H, Yagi Y, Ohneda N, Odajima H, Ono M, Saji H (2014) Development of a resonant-type microwave reactor and its application to the synthesis of positron emission tomography radiopharmaceuticals. J Label Compd Radiopharm 57(12):680–686
37. Azarian V, Gangloff A, Seimbille Y, Delaloye S, Czernin J, Phelps ME, Silverman DHS (2006) Synthesis and lipsome encapsulation of a novel F-18-conjugate of omega-conotoxin GVIA for the potential imaging of N-type Ca^{2+} channels in the brain by positron emission tomography. J Label Compd Radiopharm 49(3):269–283
38. Guenther KJ, Yoganathan S, Garofalo R, Kawabata T, Strack T, Labiris R, Dolovich M, Chirakal R, Valliant JF (2006) Synthesis and in vitro evaluation of ^{18}F- and ^{19}F-labeled insulin: a new radiotracer for PET-based molecular imaging studies. J Med Chem 49 (4):1466–1474
39. Wüst F, Hultsch C, Bergmann R, Johannsen B, Henle T (2003) Radiolabelling of isopeptide Nε-(γ-glutamyl)-l-lysine by conjugation with N-succinimidyl-4-[^{18}F]fluorobenzoate. Appl Radiat Isot 59(1):43–48
40. Taylor NJ, Emer E, Preshlock S, Schedler M, Tredwell M, Verhoog S, Mercier J, Genicot C, Gouverneur V (2017) Derisking the Cu-mediated ^{18}F-fluorination of heterocyclic positron emission tomography radioligands. J Am Chem Soc 139(24):8267–8276
41. Mäding P, Füchtner F, Wüst F (2005) Module-assisted synthesis of the bifunctional labelling agent N-succinimidyl 4-[^{18}F]fluorobenzoate ([^{18}F]SFB). Appl Radiat Isot 63(3):329–332
42. Kapty J, Kniess T, Wuest F, Mercer JR (2011) Radiolabeling of phosphatidylserine-binding peptides with prosthetic groups N-[6-(4-[^{18}F]fluorobenzylidene)aminooxyhexyl]maleimide ([^{18}F]FBAM) and N-succinimidyl-4-[^{18}F]fluorobenzoate ([^{18}F]SFB). Appl Radiat Isot 69(9):1218–1225
43. Tondera C, Laube M, Pietzsch J (2017) Insights into binding of S100 proteins to scavenger receptors: class B scavenger receptor CD36 binds S100A12 with high affinity. Amino Acids 49:183–191
44. Pietzsch J, Bergmann R, Wuest F, Pawelke B, Hultsch C, van den Hoff J (2005) Catabolism of native and oxidized low density lipoproteins: in vivo insights from small animal positron emission tomography studies. Amino Acids 29(4):389–404
45. Wolfe RR (1992) Radioactive and stable isotope tracers in biomedicine: principles and practice of kinetic analysis. Wiley-Liss, New York, Wiley-Liss, 471 p
46. Bergmann R, Pietzsch J (2005) Small animal positron emission tomography in food sciences. Amino Acids 29(4):355–376
47. van den Hoff J (2005) Principles of quantitative positron emission tomography. Amino Acids 29(4):341–353
48. Perera C, McNeil HP, Geczy CL (2010) S100 Calgranulins in inflammatory arthritis. Immunol Cell Biol 88(1):41–49
49. van de Logt F, Day AS (2013) S100A12: a noninvasive marker of inflammation in inflammatory bowel disease. J Dig Dis 14(2):62–67

50. Nazari A, Khorramdelazad H, Hassanshahi G, Day AS, Sardoo AM, Fard ET, Abedinzadeh M, Nadimi AE (2017) S100A12 in renal and cardiovascular diseases. Life Sci 191:253–258
51. Farokhzadian J, Mangolian Shahrbabaki P, Bagheri V (2017) S100A12-CD36 axis: a novel player in the pathogenesis of atherosclerosis? Cytokine 6(17):3021–3029
52. Lindholm B (2015) Serum S100A12: a risk marker or risk factor of vascular calcification in chronic kidney disease. Am J Nephrol 42 (1):1–3
53. Isoyama N, Machowska A, Qureshi AR, Yamamoto T, Anderstam B, Heimburger O, Barany P, Stenvinkel P, Lindholm B (2016) Elevated circulating s100a12 associates with vascular disease and worse clinical outcome in peritoneal dialysis patients. Perit Dial Int 36 (3):269–276
54. Yang X, Okamura DM, Lu X, Chen Y, Moorhead J, Varghese Z, Ruan XZ (2017) CD36 in chronic kidney disease: novel insights and therapeutic opportunities. Nat Rev Nephrol 13(12):769–781
55. Kiryushko D, Novitskaya V, Soroka V, Klingelhofer J, Lukanidin E, Berezin V, Bock E (2006) Molecular mechanisms of Ca(2+) signaling in neurons induced by the S100A4 protein. Mol Cell Biol 26(9):3625–3638
56. Semov A, Moreno MJ, Onichtchenko A, Abulrob A, Ball M, Ekiel I, Pietrzynski G, Stanimirovic D, Alakhov V (2005) Metastasis-associated protein S100A4 induces angiogenesis through interaction with annexin II and accelerated plasmin formation. J Biol Chem 280(21):20833–20841
57. Haase-Kohn C, Wolf S, Herwig N, Mosch B, Pietzsch J (2014) Metastatic potential of B16-F10 melanoma cells is enhanced by extracellular S100A4 derived from RAW264.7 macrophages. Biochem Biophys Res Commun 446(1):143–148
58. Herwig N, Belter B, Pietzsch J (2016) Extracellular S100A4 affects endothelial cell integrity and stimulates transmigration of A375 melanoma cells. Biochem Biophys Res Commun 477(4):963–969
59. Syed DN, Aljohani A, Waseem D, Mukhtar H (2018) Ousting RAGE in melanoma: a viable therapeutic target? Semin Cancer Biol 49:20–28
60. Kircher DA, Silvis MR, Cho JH, Holmen SL (2016) Melanoma brain metastasis: mechanisms, models, and medicine. Int J Mol Sci 17 (9):1468

Chapter 30

Reviewing the Crystal Structure of S100Z and Other Members of the S100 Family: Implications in Calcium-Regulated Quaternary Structure

Vito Calderone, Marco Fragai, and Claudio Luchinat

Abstract

This paper takes the cue from the previously solved crystal structure of human apo-S100Z and compares it with that of the calcium-bound S100Z from zebrafish in order to stress, for this particular S100, the significant role of the presence of calcium in promoting supramolecular assemblies with likely biological meaning. This consideration is then expanded through a wider review on analogous situations concerning all other S100s for which there is crystallographic o biochemical evidence of how the presence of calcium promotes the formation of quaternary complexes.

The paper also deals with some considerations on the quality of the crystals obtained for the solved members of this family and on the need for experimental phasing for solving some of the structures where the good general sequence homology among the members of the family would have suggested molecular replacement (MR) as the easiest way to solve them.

These considerations, along with the PCA analysis carried out on all the known S100s, further demonstrate that calcium plays a fundamental role in triggering quaternary structure formation for several members of this family of proteins.

Key words S100Z, EF hand, Oligomerization, Calcium, PCA

1 Introduction

S100 proteins are a family of 21 small acidic proteins bearing 2 - EF-hand motifs, 1 in the N-terminal domain and 1 in the C-terminal, with a large variability in their sequences [1–4]. A functional EF-hand motif is characterized by two helices connected by a calcium(II)-binding loop of 12–14 amino acids, according to a helix-loop-helix scheme [5]. All S100 proteins, but calbindin D9k (S100G) which is monomeric, are organized to form homo- and heterodimers stabilized by van der Waals interactions between conserved hydrophobic residues at the interface of the two monomers [6–8]. All the homo-/heterodimers show a very similar folding but

Claus W. Heizmann (ed.), *Calcium-Binding Proteins of the EF-Hand Superfamily: From Basics to Medical Applications*, Methods in Molecular Biology, vol. 1929, https://doi.org/10.1007/978-1-4939-9030-6_30,

a different shape, charge, and binding surfaces. These low molecular weight proteins (~10–12 kDa) are present only in vertebrates and are codified by four major subgroups of genes conserved also in mammalians [9]. S100 proteins have a specific cellular distribution, but in pathological conditions the expression of a specific member of the family can occur also in cells where it is not physiologically expressed [10, 11]. These proteins have been found to be involved in the regulation of a variety of intracellular functions including cell growth and differentiation, cell mobility and migration, energy metabolism, calcium(II) homeostasis, apoptosis, and inflammation [12–14]. In addition, S100 proteins have also several extracellular activities and functions [12, 14]. In the extracellular space, the different S100 proteins interact with several receptors on the cell surface such as the G protein-coupled receptors, receptor for advanced glycosylation end products (RAGE), fibroblast growth factor receptor, glycoprotein EMMPRIN, activated leukocyte cell adhesion molecule, interleukin-10 receptor, Toll-like receptor-4, and many others [3, 15, 16]. In several members of the family, the biological activities are modulated by the binding to calcium (II) ions and, less frequently, by the interaction with zinc(II) and copper(II) ions [17–19]. The binding of calcium(II) to the EF-hand motifs induces in several S100 proteins a conformational change and a modulation of their binding capability toward their biological targets [20]. The EF-hand motif in the C-terminal domain contains a canonical interconnecting loop of 12 amino acids which usually exhibits a high affinity for calcium(II) ion. Conversely, the longer interconnecting loop present in the N-terminal domain alters the canonical metal coordination and reduces, up two orders of magnitude, the binding affinity for calcium(II) ion [2]. The two calcium-binding sites do not act independently, since a positive cooperativity is often observed. The structural rearrangement observed in some S100s upon calcium(II) binding is due to a reorientation of helix III with respect to helix IV, which allows the protein to adopt an "open" conformation with the exposure of a hydrophobic cleft [21]. Most of the apo S100 proteins are in the closed form, with the exception of S100A10 [22] where the open conformation is already adopted by the apo protein. Conversely, for S100A16 [23] both the apo and the holo forms display a closed conformation. Interesting is the case of S100A14, which exhibits a negligible affinity for calcium(II) ion and adopts a "semi-open" conformation [24]. For some S100 proteins, the equilibrium between "closed" and "open" conformations, controlled by the concentration of calcium(II) ions, has a relevant biological importance since the hydrophobic cleft exposed in the "open" conformational state of the protein is the binding site for proteins, peptides, and drug candidates [25–27]. Moreover, some S100 proteins have been observed to form in vitro, in the presence of millimolar concentrations of calcium, non-covalent oligomers with potential functions in vivo [15, 28–30]. Therefore, the biological relevance of the equilibrium between the two conformational states has prompted the search of more quantitative parameters to describe the open and closed conformations of this family of proteins. To analyze the conformational space sampled by the EF-hand domains of S100 proteins in the apo and holo forms, a principal component analysis (PCA) of the six interhelical angles has been carried out starting from the X-ray and NMR structures available on the Protein Data Bank [23, 24, 31, 32]. The human protein S100Z in its apo form was the last member of the family to be characterized by X-ray crystallography [32]. Although the biological functions of human S100Z are still largely unknown, the analysis of the mRNA codifying for this protein has revealed a different expression level in the various human cells. Interestingly, the protein is downregulated in oral squamous cell carcinoma, whereas high expression levels of S100Z mRNA were found in leukocytes and spleen [33]. The availability of the crystal structure of the homologous calcium-bound S100Z from zebrafish has allowed us to investigate the conformational space spanned by S100Z and the role of calcium(II) ion. The analysis has provided new clues on the role of calcium(II) ion in promoting the supramolecular assemblies of S100 proteins.

2 Results and Discussion

2.1 S100 and S100Z Crystal Facts

A quick browsing of the PDB shows that there are slightly more than 100 entries corresponding to crystal structures of S100s, most of which have been collected at synchrotron sites. Crystals of human apo-S100Z grew at 20 °C from a 0.2 M sodium acetate, 0.1 M sodium cacodylate pH 6.5, and 30% PEG 8000 solution using the sitting drop vapor diffusion technique. The large majority of S100s has been crystallized from solutions containing organic solvents (mainly PEG or some PEG derivatives). About half of them were solved to resolutions from 1.0 to 1.9 Å and the rest to resolutions of 2 Å or worse. This crude data highlights how it is not easy to obtain very good quality crystals for this family of proteins.

Two S100Z crystal structures are known at present: that of zebrafish obtained in the presence of calcium [34] and the human one obtained in the absence of calcium (Fig. 1) [35, 36].

Both structures have been solved by MR but, in the case of the zebrafish protein, structure solution has been achieved in a straightforward fashion by simple application of MR using the untouched calcium-S100A4 model. On the contrary, in the case of human apo-S100Z, structure solution has required a slightly more elaborated approach; the model used was that of apo-S100A12, but in this case, it needed extensive editing in order to be successful since

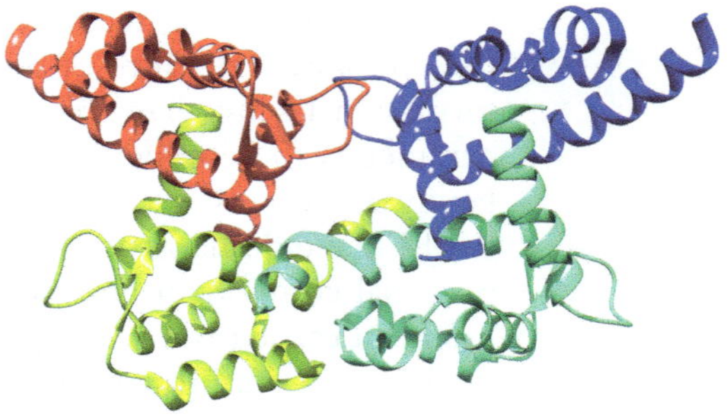

Fig. 1 Overall arrangement of the two homodimers (cyan–green and purple–yellow) of human S100Z in the asymmetric unit

the model as such never lead to structure solution with any of the most popular relevant software.

Despite the good structural homology among the members of the S100 family, structure solution for some of its members has not been easy at all, and it required experimental phasing as in the case of apo-S100A6 [37], calcium-S100A8 [38], calcium-S100A2 [39], calcium-S100A4 [40], apo-S100A16 [23], and of course apo-S100Z [35, 36].

Calcium(II) in fact causes a significant change not only in the tertiary structure of a single monomer but also on the quaternary structure of the physiological homodimer; for this reason, the molecular replacement carried out using calcium-S100Z model was not successful.

As a confirmation of the above, it should be pointed out that the successful model for solving the structure was that of human apo-S100A12, which shows only 43% sequence identity to human S100Z, as compared to almost 70% for zebrafish holo-S100Z. Still, other apo-S100s with analogous sequence homology were not successful in solving the structure.

Figure 2 shows the RMSD of the superposition of C-alphas of the solved human S100Z versus human apo-S100A12 (2WCE), human calcium-S100A12 (1E8A), and zebrafish S100Z (2Y5I). It appears clearly that the zebrafish protein has significantly higher RMSD despite much higher sequence identity with respect to the apo-S100A12 structure; so, as expected, this did not allow for success in MR.

This is also confirmed by the superposition of the backbone human apo-S100Z with that of human apo-S100A12 (2WCE), zebrafish holo-S100Z (2Y5I), and human holo-S100A12 (1E8A), respectively, as shown in Fig. 3.

Interestingly, if the dimer of the successful model (2WCE) is taken without any modification and used for MR against the target data, no solution is obtained. This indicates that the model needs extensive editing prior to MR, resulting in variable regions being trimmed and the model being adjusted and optimized in a unique fashion which turns out to be successful.

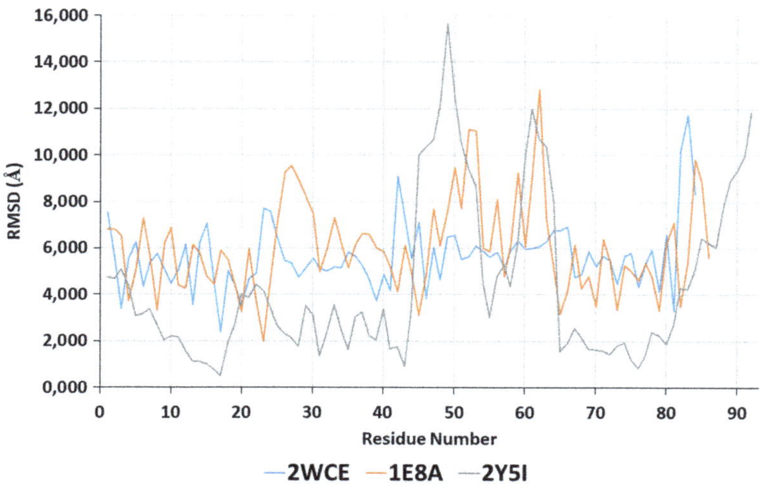

Fig. 2 C-alpha carbon superposition of human apo-S100Z versus human holo-S100A12 (orange), human apo-S100A12 (blue), zebrafish holo-S100Z (gray)

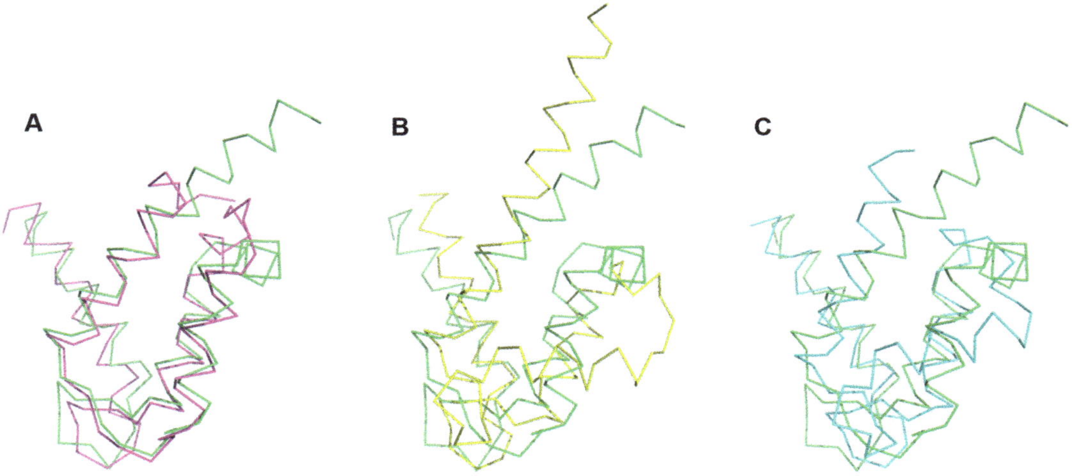

Fig. 3 Panel **a** backbone superposition of human apo-S100Z (green) versus human apo-S100A12 (purple). Panel **b** backbone superposition of human apo-S100Z (green) versus zebrafish holo-S100Z (yellow). Panel **c** backbone superposition of human apo-S100Z (green) versus human holo-S100A12 (light blue)

2.2 Calcium and PCA Analysis

The relationship between structural similarity and presence of calcium has been investigated through a more extensive comparison of the structural features of human apo-S100Z with all other S100 family member: this has been performed by the principal component analysis (PCA) of the angles among α-helices [23, 24, 31, 41, 42]. This method provides quantitative parameters to describe the open and closed conformations, and it has been used to sample the conformational space explored by the EF-hand domains in the S100 proteins. In each EF-hand domain, the six interhelical angles have been measured from the directions of the four α-helices by considering the eight residues immediately preceding and following each EF-hand loop [24].

These PCA data collected on several S100 proteins show that the hydrophobic surfaces, exposed on opposite sides of the homodimer when the proteins adopt the "open" conformation, are responsible for the interaction with their target proteins. In the apo state, most of the S100 proteins exhibit a closed conformation, and the conversion to the "open" state occurs after the binding with calcium(II) [21]. Only in few cases, the apo form is already in the "open" (i.e., for S100A10 [43] or "semi-open (i.e., for S100A14 [24]) conformation. It has also been observed that in few cases, both the apo and holo forms display a closed conformation (i.e., for S100A16 [23]). These three proteins are thus exceptions to the expected apo-holo PCA clustering.

The first principal component and second principal component values of the zebrafish holo-S100Z fall in the region defined by the open conformations of human S100s. The plot (Fig. 4) provides a visual representation of the extent to which the conformation of

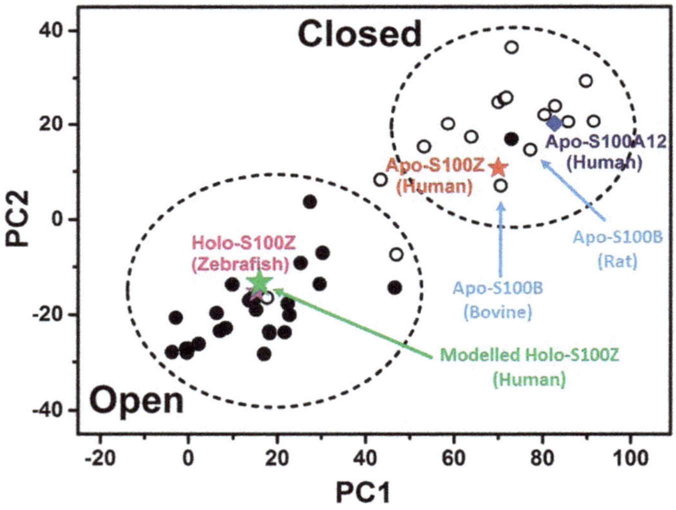

Fig. 4 Principal component analysis performed on the six interhelical angles of apo- and calcium(II)-loaded forms of S100 proteins. Apo and holo proteins are represented as open circles and filled circles, respectively. The purple star represents zebrafish holo-S100Z, the red star human apo-S100Z, the blue square human apo-S100A12, and the green star the modelled human holo-S100Z. The two open circles placed in the region of the protein in the open conformation correspond to the apo form of calbindin D9k (Cb) and S100A10 (A10). The filled circle placed in the region of the protein in the closed conformation corresponds to holo-S100A16 (A16). S100A14 (open circle between the open and closed regions) is in a "semi-open" conformation in physiological conditions. The two cyan arrows indicate the two structures that are closer to that of human apo-S100Z than that of human apo-S100A12

human S100Z is in the closed conformation as it is in general for apo-S100 proteins.

Still something meaningful has to be pointed out in Fig. 4: the two closest structures to human apo-S100Z appear to be bovine and rat S100B, respectively (both solved through solution NMR). Although the higher structural similarity of these two structures with respect to human apo-S100A12, they proved unsuccessful in solving the phase problem.

For this reason human S100Z can be considered as a further example of the continuum of conformational states occurring within the whole EF-hand protein superfamily. This variety of structures accounts for the different functions in which EF-hand proteins are involved and explains the difficulty of obtaining direct sequence–structure or sequence–function relationships. The clustering of human apo-S100Z in the closed conformation region is a further confirmation of the unsuitability of calcium-bound S100 structures to work as templates for apoproteins on MR.

At the same time, the large conformational space sampled by these proteins in both the apo and metalated forms stresses the importance of choosing the appropriate algorithm for model optimization and the pivotal role of the sequential exploration of the available models. All of the above is also supported by solution NMR studies which show a high flexibility of the members of S100 family [24, 44].

2.3 Calcium-Driven Formation of Supramolecular Assemblies

It is very well known that the binding of calcium (and of zinc in some cases) leads to conformational changes in S100s. These conformational changes are likely to promote the binding to other proteins. The basic quaternary structural motif of S100 proteins is a homodimer, but it is now well accepted that higher-order non--covalent oligomers are formed and that they play some functional role.

In order to shed light on this aspect, equilibrium analytical ultracentrifugation experiments have been carried out for several S100 proteins (S100A1, S100A2, S100A3, S100A4, S100A5, S100A6, S100A7, S100A8, S100A9, S100A10, S100A11, S100A12, S100A13, S100B, S100P, and S100Z) under reducing conditions in the absence and presence of calcium ions. It has been shown that the addition of calcium triggers the formation of tetrameric structures which could be even enforced in vivo where an additional effect of molecular crowding can be taken into account [28].

Human apo-S100Z shares the typical fold of all other apo-S100s. Figure 3 (panels a and b) shows the backbone superposition of human apo-S100A12 and zebrafish holo-S100Z with human apo-S100Z, respectively.

As expected, the r.m.s. deviation is significantly larger when apo and holo structures are superposed for each individual S100. This is due to the tertiary structure modification caused by the presence of calcium. This change is not just a mere structural modification, but it also influences another very important element which is the ability of the physiological dimer to form higher-order assemblies, like tetramers, for instance. Some members of the S100 family (S100A4, S100B, S100A8/S100A9, S100A12) have in fact been shown to form tetramers or higher-order multimers in the presence of calcium or zinc, and these supramolecular moieties have been supposed to be involved in some gain of function which could be the basis of several pathologies [44–46].

The comparison of the buried surface area (BSA) involved in potential tetramers formation between human apo-S100Z and zebrafish holo-S100Z (2Y5I) shows significant differences. In the first case, the BSA between the two dimers in the asymmetric unit is about 2800 $Å^2$; in the latter case BSA ranges from 3300 $Å^2$ to 3400 $Å^2$ since there are three dimers in the asymmetric unit. The buried surface area between the two monomers forming each dimer on the other side ranges from 2000 $Å^2$ to 2200 $Å^2$ in the case of the two dimers in the asymmetric unit of human apo-S100Z, whereas it ranges from 3700 $Å^2$ to 3900 $Å^2$ in the case of zebrafish holo-S100Z.

These data show clearly that the interaction between the monomers in each dimer is stronger in the case of the metalated protein and also that the interaction among dimers is stronger, supporting the role of oligomerization promoter which has been already claimed for calcium. Furthermore, it is not possible to identify superhelical multimers in the crystal in the case of human apo-S100Z (Fig. 5), as opposed to what can be done in the case of zebrafish holo-S100Z (Fig. 6), and this is indeed due to the difference in tertiary structure between the apo and holo form.

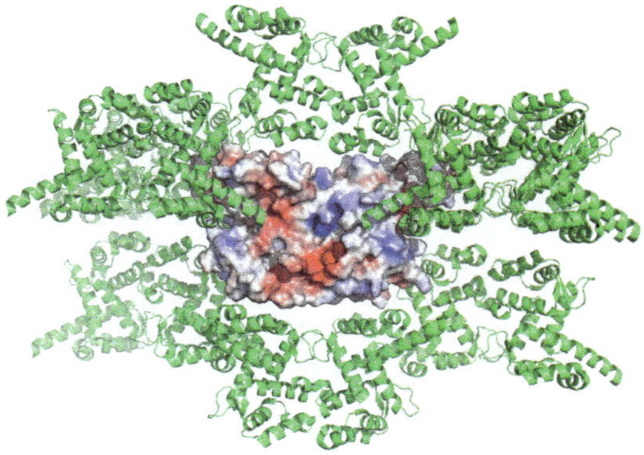

Fig. 5 Absence of superhelical arrangement of human apo-S100Z (5HYD) where no continuous structure is formed; the content of the asymmetric unit is shown as electrostatic potential-colored surface. The green ribbons represent the symmetry expansion of the content of the asymmetric unit

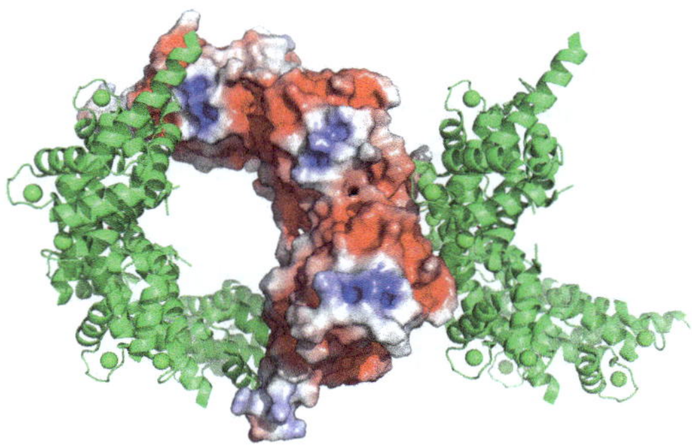

Fig. 6 Evident winding up of the zebrafish calcium-S100Z (2Y5I) molecules in the crystal to form a superhelical arrangement; the content of the asymmetric unit is shown as electrostatic potential-colored surface. The green ribbons and spheres represent the symmetry expansion of the content of the asymmetric unit and the calcium atoms, respectively

3 Conclusions

S100Z is reported in several review papers dealing with S100 protein family [3, 12, 15, 34]. However, the biological functions and the possible relevance in pathology of human S100Z are still unknown, and the structure of its apo form is one of the few that was still missing among all the members of the family (Table 1). There is barely no literature in fact about S100A17 and S100A18, whereas one paper identifies S100A19 in marsupials [47]. Taking into account the apo-holo structural varability, the abovementioned PCA unpredictability shown by S100A10, S100A14, and S100A16, and the intrinsic flexibility of the proteins demonstrated by solution studies, sequence homology considerations prove to be insufficient in predicting structural details of a new member on the basis of the structural features of already known members.

Here we compare the first crystal structure of the apo form of human S100Z with the members of the family. The availability of the structural details of this protein adds another piece of information to the puzzle of the S100 protein family and may provide new hints to reveal the role of these proteins in extracellular and intracellular processes.

Human apo-S100Z has been compared to the structure of zebrafish calcium-S100Z with which it shares 70% sequence identity and with human apo-S100A12 with which it shares 43% sequence identity and was the template for structure solution. The comparison with holo-S100Z shows that the surface of interaction between the monomers in the dimers decreases; likewise, the surface among the dimers is lower, with the consequent absence of superhelical arrangement of molecules in the crystals as observed in the case of some holo-S100s.

Table 1
List of the members of the S100 family whose structure has been solved (and whose structure still has to be solved) either by X-ray crystallography or NMR in the apo or holo state

Name	X-ray		NMR		PDB-ID	Reference
	Apo	Holo	Apo	Holo		
S100A1	N	Y	Y	Y	5K89	(56)
S100A2	Y	Y	N	N	4DUQ	(57)
S100A3	Y	N	N	N	1KSO	(58)
S100A4	N	Y	Y	Y	3C1V	(59)
S100A5	N	Y	Y	Y	4DIR	---
S100A6	Y	Y	Y	Y	1K96	(60)
S100A7	Y	Y	N	N	1PSR	(61)
S100A8	N	Y	N	N	5HLO	(62)
S100A8/S100A9	N	Y	Y	N	4GGF	(63)
S100A9	N	N	Y	N	5I8N	(64)
S100A10	Y	N	N	N	1A4P	(65)
S100A11	N	Y	Y	N	1QLS	(66)
S100A12	Y	Y	N	Y	2WCE	(51)
S100A13	N	Y	Y	Y	2EGD	(67)
S100A14	N	N	Y	N	2M0R	(44)
S100A15	N	Y	N	N	4AQI	(68)
S100A16	Y	N	Y	Y	3NXA	(24)
S100A17	N	N	N	N	---	---
S100A18	N	N	N	N	---	---
S100A19	N	N	N	N	---	---
S100B	Y	Y	Y	Y	4PE0	(69)
S100P	N	Y	Y	Y	1J55	(70)
S100Z	Y	Y	N	N	2Y5I	(54)

Acknowledgments

We thank the staff members of ID-29 beamline at ESRF (Grenoble, France) for assistance during data collection.

This work was supported by Instruct, part of the European Strategy Forum on Research Infrastructures (ESFRI), and supported by national member subscriptions. Specifically, we thank the EU ESFRI Instruct Core Centre CERM-Italy.

References

1. Marenholz I, Lovering RC, Heizmann CW (2006) An update of the S100 nomenclature. Biochim Biophys Acta 1763(11):1282–1283
2. Donato R (1986) S-100 proteins. Cell Calcium 7(3):123–145
3. Heizmann CW, Fritz G, Schafer BW (2002) S100 proteins: structure, functions and pathology. Front Biosci 7:d1356–d1368
4. Chazin WJ (2011) Relating form and function of EF-hand calcium binding proteins. Acc Chem Res 44(3):171–179
5. Schäfer BW, Heizmann CW (1996) The S100 family of EF-hand calcium-binding proteins: functions and pathology. Trends Biochem Sci 21(4):134–140
6. Kojetin DJ et al (2006) Structure, binding interface and hydrophobic transitions of Ca2+-loaded calbindin-D(28K). Nat Struct Mol Biol 13(7):641–647
7. Marenholz I, Heizmann CW, Fritz G (2004) S100 proteins in mouse and man: from evolution to function and pathology (including an update of the nomenclature). Biochem Biophys Res Commun 322(4):1111–1122
8. Anonymous (2001) Handbook on Metalloproteins Edited by Ivano Bertini (University of Florence), Astrid Sigel, and Helmut Sigel (University of Basel). Marcel Dekker: New York, Basel. 2001. xxx + 1182 pp (plus 16 color plates). $265.00. ISBN: 0-8247-0520-3. J Am Chem Soc 123 (50):12748–12748
9. Zimmer DB, Eubanks JO, Ramakrishnan D, Criscitiello MF (2013) Evolution of the S100 family of calcium sensor proteins. Cell Calcium 53(3):170–179
10. Reckenbeil J et al (2016) Cellular distribution and gene expression pattern of metastasin (S100A4), calgranulin A (S100A8), and calgranulin B (S100A9) in oral lesions as markers for molecular pathology. Cancer Investig 34 (6):246–254
11. Zimmer DB, Cornwall EH, Landar A, Song W (1995) The S100 protein family: history, function, and expression. Brain Res Bull 37 (4):417–429
12. Donato R (2003) Intracellular and extracellular roles of S100 proteins. Microsc Res Tech 60 (6):540–551
13. Donato R (1999) Functional roles of S100 proteins, calcium-binding proteins of the EF-hand type. Biochim Biophys Acta 1450 (3):191–231
14. Donato R et al (2012) Functions of S100 proteins. Curr Mol Med 13(1):24–57
15. Donato R (2001) S100: a multigenic family of calcium-modulated proteins of the EF-hand type with intracellular and extracellular functional roles. Int J Biochem Cell Biol 33 (7):637–668
16. Bresnick AR, Weber DJ, Zimmer DB (2015) S100 proteins in cancer. Nat Rev Cancer 15 (2):96–109
17. Moroz OV et al (2009) Both Ca2+ and Zn2+ are essential for S100A12 protein oligomerization and function. BMC Biochem 10:11
18. Arnesano F et al (2005) Structural interplay between calcium(II) and copper(II) binding to S100A13 protein. Angew Chem Int Ed Engl 44(39):6341–6344
19. Gilston BA, Skaar EP, Chazin WJ (2016) Binding of transition metals to S100 proteins. Sci China Life Sci 59(8):792–801
20. Xiao Y, Shaw GS, Konermann L (2017) Calcium-mediated control of S100 proteins: allosteric communication via an agitator/signal blocking mechanism. J Am Chem Soc 139 (33):11460–11470
21. Nelson MR, Chazin WJ (1998) An interaction-based analysis of calcium-induced conformational changes in Ca2+ sensor proteins. Protein Sci 7(2):270–282
22. Rety S et al (1999) The crystal structure of a complex of p11 with the annexin II N-terminal peptide. Nat Struct Biol 6(1):89–95
23. Babini E et al (2011) Structural characterization of human S100A16, a low-affinity calcium binder. J Biol Inorg Chem 16(2):243–256
24. Bertini I et al (2013) Solution structure and dynamics of human S100A14. J Biol Inorg Chem 18(2):183–194
25. Bhattacharya S, Large E, Heizmann CW, Hemmings B, Chazin WJ (2003) Structure of the Ca2+/S100B/NDR kinase peptide complex: insights into S100 target specificity and activation of the kinase. Biochemistry 42 (49):14416–14426
26. Okada M, Tokumitsu H, Kubota Y, Kobayashi R (2002) Interaction of S100 proteins with the antiallergic drugs, olopatadine, amlexanox, and cromolyn: identification of putative drug binding sites on S100A1 protein. Biochem Biophys Res Commun 292(4):1023–1030
27. Agamennone M et al (2010) Fragmenting the S100B-p53 interaction: combined virtual/biophysical screening approaches to identify ligands. ChemMedChem 5(3):428–435
28. Streicher WW, Lopez MM, Makhatadze GI (2010) Modulation of quaternary structure of

S100 proteins by calcium ions. Biophys Chem 151(3):181–186

29. Fritz G, Botelho HM, Morozova-Roche LA, Gomes CM (2010) Natural and amyloid self-assembly of S100 proteins: structural basis of functional diversity. FEBS J 277 (22):4578–4590
30. Botelho HM, Fritz G, Gomes CM (2012) Analysis of S100 oligomers and amyloids. Methods Mol Biol 849:373–386
31. Capozzi F, Luchinat C, Micheletti C, Pontiggia F (2007) Essential dynamics of helices provide a functional classification of EF-hand proteins. J Proteome Res 6(11):4245–4255
32. Calderone V, Fragai M, Gallo G, Luchinat C (2017) Solving the crystal structure of human calcium-free S100Z: the siege and conquer of one of the last S100 family strongholds. J Biol Inorg Chem 22(4):519–526
33. Raffat MA et al (2018) S100 proteins in oral squamous cell carcinoma. Clin Chim Acta 480:143–149
34. Moroz OV, Bronstein IB, Wilson KS (2011) The crystal structure of zebrafish S100Z: implications for calcium-promoted S100 protein oligomerisation. J Mol Biol 411(5):1072–1082
35. Calderone V, Fragai M, Gallo G, Luchinat C (2017) Solving the crystal structure of human calcium-free S100Z: the siege and conquer of one of the last S100 family strongholds. J Biol Inorg Chem 22:519–526
36. Calderone V, Fragai M, Luchinat C (2018) When molecular replacement has no trivial solution: the importance of model editing in human S100Z X-ray structure solution. Inorg Chim Acta 470:402–406
37. Otterbein LR, Kordowska J, Witte-Hoffmann-C, Wang CLA, Dominguez R (2002) Crystal structures of S100A6 in the Ca2+-free and Ca2+-bound states. Structure 10(4):557–567
38. Ishikawa K, Nakagawa A, Tanaka I, Suzuki M, Nishihira J (2000) The structure of human MRP8, a member of the S100 calcium-binding protein family, by MAD phasing at 1.9 Å resolution. Acta Crystallogr D Biol Crystallogr 56 (5):559–566
39. Koch M, Fritz G (2012) The structure of Ca2+-loaded S100A2 at 1.3-A resolution. FEBS J 279(10):17799–11810
40. Malashkevich VN et al (2008) Structure of Ca2+-bound S100A4 and its interaction with peptides derived from nonmuscle myosin-IIA. Biochemistry 47(18):5111–5126
41. Babini E et al (2005) Principal component analysis of the conformational freedom within the EF-hand superfamily. J Proteome Res 4 (6):1961–1971
42. Bertini I et al (2009) Solution structure and dynamics of S100A5 in the apo and Ca2+-bound states. J Biol Inorg Chem 14 (7):1097–1107
43. Leclerc E, Vetter SW (2015) The role of S100 proteins and their receptor RAGE in pancreatic cancer. Biochim Biophys Acta 1852 (12):2706–2711
44. Permyakov SE et al (2011) Intrinsic disorder in S100 proteins. Mol BioSyst 7(7):2164–2180
45. Korndörfer I, Brueckner F, Skerra A (2007) The crystal structure of the human (S100A8/S100A9)2 heterotetramer, calprotectin, illustrates how conformational changes of interacting alpha-helices can determine specific association of two EF-hand proteins. J Mol Biol 370(5):887–898
46. Moroz OV, Blagova EV, Wilkinson AJ, Wilson KS, Bronstein IB (2009) The crystal structures of human S100A12 in apo form and in complex with zinc: new insights into S100A12 oligomerisation. J Mol Biol 391(3):536–551
47. Kwek J et al (2013) Molecular evolution of a novel marsupial S100 protein (S100A19) which is expressed at specific stages of mammary gland and gut development. Mol Phylogenet Evol 69(1):4–16
48. Melville Z et al (2017) X-ray crystal structure of human calcium-bound S100A1. Acta Crystallogr F Struct Biol Commun 73 (Pt 4):215–221
49. Koch M, Fritz G (2012) The structure of Ca2+-loaded S100A2 at 1.3-A resolution. FEBS J 279(10):1799–1810
50. Mittl PR et al (2002) Metal-free MIRAS phasing: structure of apo-S100A3. Acta Crystallogr D Biol Crystallogr 58(Pt 8):1255–1261
51. Gingras AR et al (2008) Crystal structure of the Ca(2+)-form and Ca(2+)-binding kinetics of metastasis-associated protein, S100A4. FEBS Lett 582(12):1651–1656
52. Otterbein LR, Kordowska J, Witte-Hoffmann-C, Wang CL, Dominguez R (2002) Crystal structures of S100A6 in the Ca(2+)-free and Ca(2+)-bound states: the calcium sensor mechanism of S100 proteins revealed at atomic resolution. Structure 10(4):557–567
53. Brodersen DE et al (1998) EF-hands at atomic resolution: the structure of human psoriasin (S100A7) solved by MAD phasing. Structure 6(4):477–489
54. Lin H, Andersen GR, Yatime L (2016) Crystal structure of human S100A8 in complex with zinc and calcium. BMC Struct Biol 16(1):8
55. Damo SM et al (2013) Molecular basis for manganese sequestration by calprotectin and roles in the innate immune response to

invading bacterial pathogens. Proc Natl Acad Sci U S A 110(10):3841–3846

56. Chang CC et al (2016) Blocking the interaction between S100A9 and RAGE V domain using CHAPS molecule: a novel route to drug development against cell proliferation. Biochim Biophys Acta 1864(11):1558–1569
57. Rety S et al (2000) Structural basis of the Ca(2+)-dependent association between S100C (S100A11) and its target, the N-terminal part of annexin I. Structure 8(2):175–184
58. Imai FL, Nagata K, Yonezawa N, Nakano M, Tanokura M (2008) Structure of calcium-bound human S100A13 at pH 7.5 at 1.8 A resolution. Acta Crystallogr Sect F Struct Biol Cryst Commun 64(Pt 2):70–76
59. Murray JI et al (2012) Structural characterization of S100A15 reveals a novel zinc coordination site among S100 proteins and altered surface chemistry with functional implications for receptor binding. BMC Struct Biol 12:16
60. Cavalier MC et al (2014) Covalent small molecule inhibitors of Ca(2+)-bound S100B. Biochemistry 53(42):6628–6640
61. Zhang H et al (2003) The crystal structure at 2A resolution of the Ca2+ -binding protein S100P. J Mol Biol 325(4):785–794

Chapter 31

High Sensitive Quantitative Binding Assays Using a Nanoluciferase-Fused Probe for Analysis of ALG-2-Interacting Proteins

Wei Zhang, Rina Matsuo, Terunao Takahara, Hideki Shibata, and Masatoshi Maki

Abstract

Many non-catalytic cellular proteins exert biological functions by formation of stable or transient complexes with other proteins. Analysis of the signal-induced physical interactions is important to understand their physiological roles in cells. Here we describe a biochemical method for assessing the binding of ALG-2 (gene name, *PDCD6*) to its target proteins that are immunoprecipitated from cell lysates. Application of nanoluciferase (Nluc)-fused ALG-2 enables a rapid quantitative evaluation of Ca^{2+}-dependent interactions of target proteins with ALG-2 in vitro binding assays.

Key words Protein–protein interaction, ALG-2, Calcium-binding protein, EF-hand, NanoLuc, In vitro binding

1 Introduction

ALG-2, a 22-kDa intracellular penta-EF-hand Ca^{2+}-binding protein (gene name, *PDCD6*), plays regulatory roles in cell death, cell division, signal transduction, post-transcriptional regulation, and membrane trafficking through Ca^{2+}-dependent interactions with a variety of proteins [1–4]. Searching for new target proteins is required to gain a deeper insight into how proteins—lacking catalytic domains—work in cells. Recent developments in mass spectrometric proteomic analysis have enabled researchers to screen for interaction partners of interest. Several interactome databases are now open to the public. However, in order to exclude non-specific interactions, we have to apply different methods to confirm specific

Wei Zhang and Rina Matsuo contributed equally to this work.

Claus W. Heizmann (ed.), *Calcium-Binding Proteins of the EF-Hand Superfamily: From Basics to Medical Applications*, Methods in Molecular Biology, vol. 1929, https://doi.org/10.1007/978-1-4939-9030-6_31,

physical interactions. The in vitro binding assay—described here—is a good example to validate specific protein–protein interactions.

To detect specific interaction partners, antibodies are commonly used for co-immunoprecipitation of binding proteins, which are then analyzed by immunoblotting (Western blot analysis, WB). When specific antibodies are not available, tagging with epitopes or fusion with fluorescent proteins is also employed for detection of overexpressed interacting proteins in mammalian cells. However, WB is very time-consuming and needs multiple steps: (a) SDS-PAGE, (b) membrane transfer, (c) probing with a primary antibody, (d) probing with an enzyme-conjugated secondary antibody, and (e) detection of chemiluminescent signals from the enzyme-catalyzed substrate using a luminescent image analyzer. A wider linearity of the signals is essential for quantitative analysis, but inappropriate WB conditions often cause a saturation of detected signals and limit evaluation of the relative strengths of interactions.

An enzyme called luciferase catalyzes oxidation of photon-emitting substrates (luciferin) and produces bioluminescence. Because of the high sensitivity with a wide linearity of the luminescent signals, a variety of luciferases derived from different organisms have been used in the cell biology. The firefly luciferase (Fluc, 61 kDa) and the sea pansy *Renilla* luciferase (Rluc, 36 kDa) have served as reporter genes for transcriptional regulation studies. They are components of bioluminescence resonance energy transfer (BRET) imaging for the protein–protein interaction analysis. A 19-kDa catalytic subunit from the deep sea shrimp luciferase has been engineered to create a heat-stable, urea-resistant nanoluciferase (Nluc; NanoLuc) capable of a more efficient light emission using a novel substrate, furimazine [5]. Nluc produces glow-type luminescence (a signal half-life >2 h) with a specific activity ~150-fold greater than that of Fluc and Rluc. The superior characteristics of the Nluc system have broadened applications including protein-fragment complementation assays [6–8].

In this chapter we describe two methods using Nluc-fused ALG-2 to quantitatively analyze interaction of ALG-2 using Sec31A—an outer shell component of the coat protein complex II (COPII)—as a model ALG-2-binding protein [9–12] as well as a novel ALG-2-binding protein candidate SARAF [12, 13]. First, Nluc-ALG-2 is co-expressed with an epitope-tagged Sec31A Pro-rich region (Sec31A-p), and the amount of Nluc-ALG-2 - co-immunoprecipitated with Sec31A-p is quantified by measuring the luciferase activity. Second, the bacterially expressed recombinant Nluc-ALG-2 is used to bind to the immunoprecipitated partial fragment of SARAF on magnetic beads in vitro. The beads are washed, and then the remaining luciferase activity is measured. To normalize the immunoprecipitated proteins, the amounts of StrepTactin-conjugated alkaline phosphatase (AP) bound to the

Strep-tagged protein on the magnetic beads are measured. The ratio of Nluc activity to AP activity is expressed as relative luminescent units (RLU).

2 Materials

2.1 Nluc-ALG-2-Binding Assay by Co-immunoprecipitation

1. Expression vectors: pNluc (designated pNL1.1.CMV[Nluc/CMV], Promega); pNluc-ALG-2 (an Nluc sequence fused with 3× SGG-ALG-2; a GSSG-linker Nluc fusion vector available from Promega); pSGFP2-Sec31A partial fragment (790-1095 a.a.; designated protein, SGFP2-Sec31A-p); and expression vectors for candidate proteins to be analyzed (Fig. 1a).
2. Mammalian culture cells for DNA transfection: Grow human embryonic kidney (HEK) 293 cells (RIKEN BioResource Center Cell Bank; also available from American Type Culture Collection) (or shRNA-mediated ALG-2 knockdown HEK293

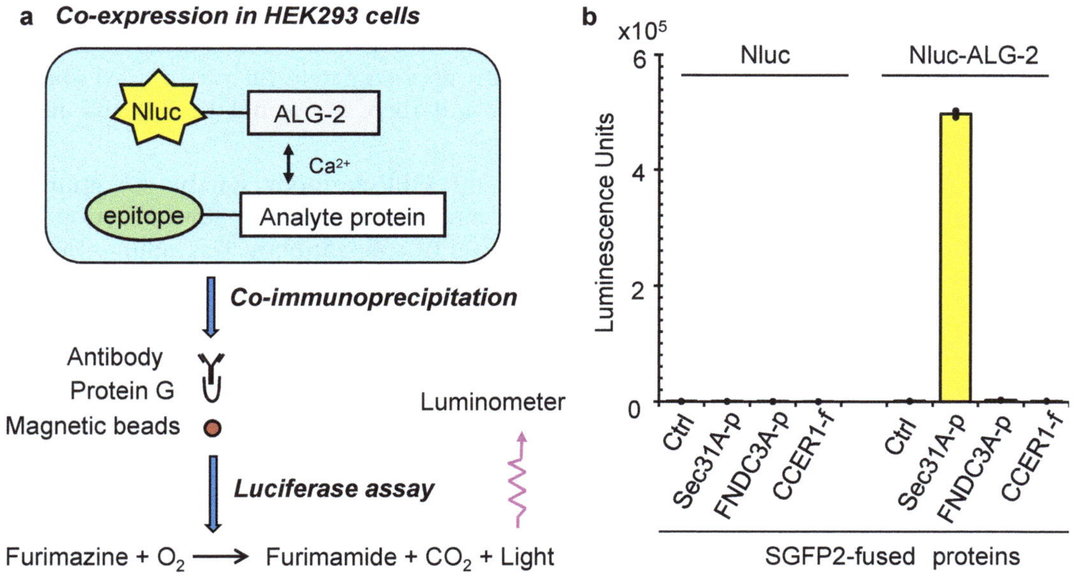

Fig. 1 Interaction analysis of co-expressed proteins. (**a**) Scheme of interaction analysis. Nluc-fused ALG-2 and epitope-tagged analyte proteins are co-expressed in HEK293 cells and interact in the presence of Ca^{2+}. Immunoprecipitated proteins are used for the Nluc luciferase assay using furimazine as a substrate, and the luminescent light is measured with a luminometer. (**b**) Example for Nluc assays. SGFP2 (negative control, Ctrl) or SGFP2-fused proteins (Sec31A-p, partial fragment of Sec31A, 790-1095 a.a.; FNDC3A-p, partial fragment of FNDC3A, 59-250 a.a.; CCER1-f, full-length CCER1) were co-expressed with either Nluc (negative control) or Nluc-ALG-2. Nluc and Nluc-ALG-2 were co-immunoprecipitated with SGFP2-fused proteins with an anti-GFP antibody in the presence of 100 μM $CaCl_2$. The amounts were measured by Nluc assays, expressed as luminescence units (LU, counts/s). Data obtained from triplicate assays are presented: dots, individual; bars, averaged LU

cells maintained in the presence of 1 μg/ml puromycin, *see* ref. 14 and **Note 1**) in Dulbecco's modified Eagle's medium (DMEM) supplemented with 0.584 mg/ml glutamine, 5% (or 10%, if necessary) fetal bovine serum (FBS), 100 units/ml penicillin, and 100 μg/ml streptomycin at 37°C under humidified air containing 5% CO_2.

3. Phosphate-buffered saline (PBS): 137 mM NaCl, 2.7 mM KCl, 8 mM Na_2HPO_4, and 1.5 mM KH_2PO_4 (pH 7.4), diluted from 10× PBS stock solution (stored at room temperature) with water. Prepare at the time of use or store at 4°C for less than 1 month.
4. Lysis buffer HKMN: 20 mM HEPES-NaOH (pH 7.5), 142.5 mM KCl, 1.5 mM $MgCl_2$, 0.2% (w/v) Nonidet P-40 (stored at 4°C), and supplemented with protease inhibitors. Just before use, add protease inhibitors to the following final concentrations using stock solutions stored at −20°C: 0.1 mM pefabloc (100 mM stock), 3 μg/ml leupeptin (3 mg/ml stock), 1 μM E-64 (1 mM stock), 1 μM pepstatin A (1 mM stock in ethanol), and 0.2 mM phenylmethylsulfonylfluoride (PMSF, 200 mM stock in methanol).
5. 200 mM ethylene glycol tetraacetic acid (EGTA): Dissolve 3.8 g of EGTA with approximately 20 ml of 1 M sodium hydroxide solution, and then adjust pH to 8.0 and fill up with water to 50 ml.
6. Antibodies: Rabbit anti-GFP antibody suitable for immunoprecipitation (antiserum was obtained by immunization with recombinant GST-EGFP, and antibody was affinity-purified using recombinant MBP-EGFP); mouse anti-GFP monoclonal antibody for Western blotting (Santa Cruz Biotechnology, clone B2); rabbit γ-globulins (Sigma-Aldrich) as control IgG.
7. Dynabeads Protein G: Equilibrate magnetic beads coated with protein G (Dynabeads™ Protein G, Thermo Fisher Scientific) in lysis buffer HKMN before use.
8. Nluc assay reagents: Passive lysis buffer (PLB), diluted from 5× PLB (Promega) with water; Nano-Glo® Luciferase Assay Reagent, diluted from 50× substrate (furimazine) solution with Nano-Glo® Luciferase Assay Buffer, both stored at −20°C and contained in the Nano-Glo® Luciferase Assay System (Promega). Warm up to room temperature before use.
9. Special equipment: Magnetic separation stands for multiple 1.5-ml and 0.5-ml tubes; a personal single-tube-type luminometer (ATTO luminescencer-MCA AB-2250 or equivalent); 75 × 12 mm clear polystyrene (PS) tubes (4.5 ml, Sarstedt, Nümbrecht, Germany).

2.2 Affinity Purification of Recombinant Nluc-ALG-2

1. Bacterial expression plasmid of Nluc-ALG-2: pET24d-Nluc-ALG-2, a codon-optimized synthetic Nluc-ALG-2 cDNA (DDBJ/GenBank/EMBL accession No. LC381846) synthesized and subcloned into the NcoI/EcoRI site of pET24d (Novagen/Millipore) by GenScript (Piscataway, NJ, USA).
2. *Escherichia coli* Rosetta (DE3): Rare codon tRNA-supplemented and proteases (*OmpT/Lon*)-deficient BL21 derivatives designed to enhance the expression of eukaryotic proteins in *E. coli* by the T7 RNA polymerase expression system (competent cells available from Novagen/Millipore).
3. ZB medium for overnight *E. coli* pre-culture: NZ Amine A 1.0 g, NaCl 0.5 g, water 100 ml. Autoclaved.
4. ZYG medium for expression culture: NZ Amine A 1.0 g, yeast extract 0.5 g, NaCl 0.5 g, water 100 ml, adjusted pH to ~7 with 0.1 N NaOH (~0.1 ml). After autoclaved and cooled, add 1 ml of autoclaved 40% (w/v) glucose to a final concentration of 0.4% (w/v).
5. Lysis buffer (buffer L): 10 mM Tris–HCl (pH 7.5), 50 mM NaCl, 0.05 mM EDTA, 5 mM 2-mercaptoethanol (2-ME), 2 mM PMSF, 5 mM benzamidine.
6. Column-binding buffer (buffer B): 10 mM Tris–HCl (pH 7.5), 100 μM $CaCl_2$, 5 mM 2-ME.
7. Elution buffer (buffer E): 10 mM Tris–HCl (pH 7.5), 5 mM 2-ME, 5 mM EGTA.
8. Stock solutions: 1 M isopropyl β-D-1-thiogalactopyranoside (IPTG); 30 mg/ml kanamycin sulfate; 0.5 M ethylenediaminetetraacetic acid (EDTA); 10% Triton X-100; 0.9% NaCl.
9. Affinity column: ALG-2-binding peptide-immobilized 1-ml HiTrap-NHS-activated column [15].

2.3 Nluc-ALG-2-Binding Assay Using Immunopurified Target Proteins

1. Lysis buffer HKN: 20 mM HEPES-NaOH (pH 7.5), 142.5 mM KCl, 0.2% (w/v) Nonidet P-40. Store at 4°C.
2. Gelatin stock solution (2% w/v): Dissolve 400 mg of gelatin in 20 ml of water by autoclaving. Store at 4°C.
3. Gelatin-pretreated Dynabeads Protein G: Wash the beads with lysis buffer HKN containing 2 mM EGTA and 0.1% gelatin three times.

2.4 StrepTactin-Conjugated Alkaline Phosphatase Assay

1. StrepTactin-AP solution: Immediately before use, dilute Precision Protein StrepTactin-AP Conjugate (0.5 mg/ml, BIO-RAD) at 1:5,000 with buffer HKN (*see* Subheading 2.3, **item 1**) containing 0.1% gelatin (final concentration, 0.1 μg/ml).

2. Assay buffer: 20 mM Tris–HCl and 1 mM $MgCl_2$ (pH 9.8), diluted from 10× assay buffer stock solution (stored at 4°C) with water.
3. Reaction buffer: CSPD chemiluminescent substrate 1:20 diluted with Reaction Buffer Diluent, both stored at 4°C and contained in the Phospha-Light™ SEAP Reporter Gene Assay System (Thermo Fisher Scientific). Warm up to room temperature before use.

2.5 Far-Western Assay Probed with Nluc-ALG-2

1. Tris-buffered saline (TBS): 10 mM Tris–HCl (pH 7.5) and 0.15 M NaCl. Prepare by diluting 10× TBS stock solution. Store at 4°C.
2. TBST: TBS containing 0.1% (w/v) Tween 20.
3. Blocking buffer: TBST containing 1% (w/v) gelatin and 0.02% NaN_3. Prepare TBS containing 1% (w/v) gelatin, dissolve by autoclaving, and then add Tween 20 and NaN_3 to 0.1% and 0.02%, respectively. Store at 4°C.
4. TBSTC: TBST containing 0.1 mM $CaCl_2$.
5. Binding buffer: TBSTC containing 0.1% gelatin. Prepare TBS containing 0.1% (w/v) gelatin, dissolve by autoclaving, and then add Tween 20 and $CaCl_2$ to 0.1% and 0.1 mM, respectively. Store at 4°C.
6. TBSC: TBS containing 0.1 mM $CaCl_2$.
7. CTZ-*h*: Dissolve 1 mg of coelenterazine-*h* [2-(4-Dehydroxy) coelenterazine, Wako Pure Chemicals; also available from Thermo Fisher Scientific/Invitrogen, Sigma-Aldrich, and others] with 1 ml of ethanol (final concentration, 1 mg/ml). Store at −20°C, protected from light.
8. Special equipment: LAS-3000 mini lumino-image analyzer (Fujifilm).

3 Methods

3.1 Nluc-ALG-2-Binding Assay by Co-immunoprecipitation

1. Seed 1 × 10^6 HEK293 cells in 60-mm dishes 1 day before transfection (*see* **Note 1**). Co-transfect cells with pNluc-ALG-2 and pSGFP2-Sec31A-p for a positive control by the conventional calcium phosphate precipitation method or liposome method (*see* ref. 16). For negative controls, transfect cells with pNluc and pSGFP2-Sec31A-p and/or with pNluc-ALG-2 and pSGFP2.
2. After 24 h post-transfection, harvest cells expressing proteins that are to be analyzed (*see* **Note 2**). Remove the medium with an aspirator, and wash cells in 3 ml of PBS per one 6-cm dish.

Add 1 ml of PBS to the dish, scrape cells with a rubber policeman, and transfer them to 1.5-ml microfuge tubes.

3. Spin down the cells at 5000 rpm (3900 × *g*) for 1 min. Discard the supernatants.
4. Add 0.6 ml of lysis buffer HKMN to the pelleted cells, mix, and leave for 30 min on ice (mix every 10 min).
5. Centrifuge the cell lysates at 14,000 rpm (18000 × *g*) for 10 min. Transfer the supernatants (0.5 ml) to empty 0.6-ml tubes.
6. Add 10 μl of 4 mM $CaCl_2$ (final concentration, approx. 100 μM) or 200 mM EGTA (final concentration, approx. 5 mM) when Ca^{2+} dependency needs to be investigated. Add 1 μg of antibody against GFP. Gently mix by rotating tubes at 4°C overnight.
7. Add 10 μl suspension of buffer HKMN-equilibrated Dynabeads Protein G to the antibody-treated cleared lysates. Continue rotation for 1 h at 4°C.
8. Set the tubes on a magnetic separation stand for 30 s, and collect magnetic beads at the side walls of the tubes. Remove supernatants (*see* **Note 3**). Wash the beads three times with 0.4 ml of buffer HKMN containing $CaCl_2$ or EGTA.
9. Suspend the washed beads with 50 μl of passive lysis buffer (PLB) (*see* **Note 4**).
10. Transfer 5 μl of the beads suspension into a 75 × 12 mm clear polystyrene tube suitable for a single-tube-type luminometer (*see* **Note 5**).
11. Add 25 μl of Nano-Glo® Luciferase Assay Reagent.
12. Vortex the reaction mixture gently, place the tube in the luminometer tube holder, and then measure luminescence at the following setting: delay, 10 s; integrated time of analysis, 10 s (Fig. 1b and *see* **Note 6**). Luminescence units are expressed as counts/s.

3.2 Purification of Recombinant Nluc-ALG-2

1. Preculture the *E. coli* transformant harboring pET24d-Nluc-ALG-2 in 5 ml of ZB broth containing kanamycin (30 μg/ml) in a test tube at 30°C overnight.
2. Transfer the preculture to the 200 ml ZYG medium containing kanamycin (30 μg/ml) in a 1-L flask, and incubate with shaking at 20°C until turbidity (A600) reaches 0.5 (approximately 10 h). Add 100 μl of 1 M IPTG (final concentration, 0.5 mM), and continue incubation by shaking at 20°C overnight.
3. Harvest cells by centrifugation at 5000 rpm (3900 × *g*) for 5 min at 4°C in a 250-ml centrifugation tube.

4. Suspend the pelleted cells with 40 ml of 0.9% NaCl, transfer to a 50-ml tube, and centrifuge at 3000 rpm (1600 × *g*) for 15 min at 4°C (*see* **Note 7**).
5. Suspend the harvested cells with 40 ml of buffer L, and sonicate at 20% duty cycle for 1 min. Repeat sonication five times.
6. Add 0.4 ml of 10% Triton X-100 (final concentration, 0.1%) to the cell lysate.
7. Centrifuge at 12,000 rpm (22500 × *g*) for 15 min at 4°C.
8. Save the supernatant and repeat centrifugation.
9. Filtrate the supernatant through a 0.45-μm pore-size cellulose acetate filter to remove particulates.
10. Measure the volume of the filtrate, and add 100 mM $CaCl_2$ to 150 μM (approximately surplus of 100 μM free Ca^{2+} over 50 μM EDTA contained in buffer L).
11. Apply the filtrate to the ALG-2-binding peptide-immobilized 1-ml affinity column using a peristaltic pump (*see* **Note 8**).
12. Wash the column with 10 ml each of buffer B containing 1 M NaCl and 50 mM NaCl, respectively.
13. Elute the bound Nluc-ALG-2 protein with buffer E, collect fractions (0.5 ml), and analyze the eluted proteins by SDS-PAGE for assessing the purity (Fig. 2a) and by measurement of Nluc activities (Fig. 2b).

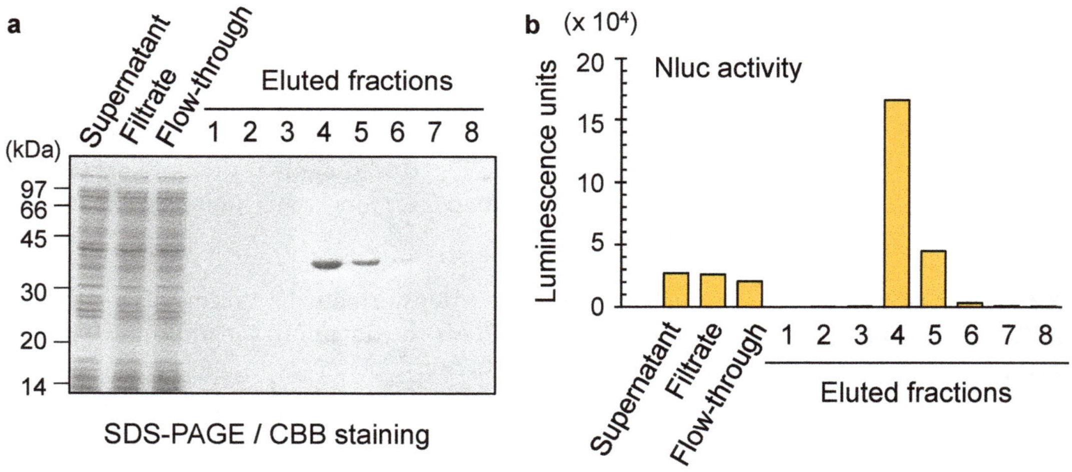

Fig. 2 Purification of recombinant Nluc-ALG-2 by affinity column chromatography. ALG-2 fused with Nluc was expressed in *Escherichia coli* and purified by an affinity column immobilizing an ALG-2-binding site (ABS) oligopeptide [15]. Nluc-ALG-2 was absorbed to the column in the presence of 100 μM Ca^{2+}, eluted with buffer E, and subjected to (**a**) SDS-PAGE followed by staining with Coomassie Brilliant Blue R-250 (CBB) and to (**b**) Nluc assays. Approximately 13% of the Nluc activities of the applied sample (filtrate) were recovered in the eluted fractions (*see* **Note 17**)

14. Collect peak fractions, and dialyze against 10 mM Tris–HCl (pH 7.5), 0.1 mM EDTA, 0.15 M NaCl (*see* **Note 9**).

3.3 In vitro Binding Assay Using Recombinant Nluc-ALG-2 and Immunopurified Target Proteins

1. Transfect HEK293 cells with expression plasmids for target proteins of choice (Fig. 3).
2. Proceed to Subheading 3.1, **steps 1–8** for immunoprecipitation, but use lysis buffer HKN (*see* Subheading 2.3, **item 1**) containing 2 mM EGTA and gelatin-pretreated Dynabeads Protein G (*see* Subheading 2.3, **item 2**). In addition, set a negative control by using control IgG for immunoprecipitation under the same condition.
3. After immunoprecipitation, suspend the washed beads with 100 μl of buffer HKN containing 2 mM EGTA and protease inhibitors (*see* Subheading 2.1, **item 4** and **Note 10**).
4. Take different aliquots (2.5 μl, 5 μl, 7.5 μl, and 10 μl) of beads suspension, and wash the beads three times with 0.5 ml of buffer HKN containing 10 μM $CaCl_2$ in 0.6-ml plastic tubes.

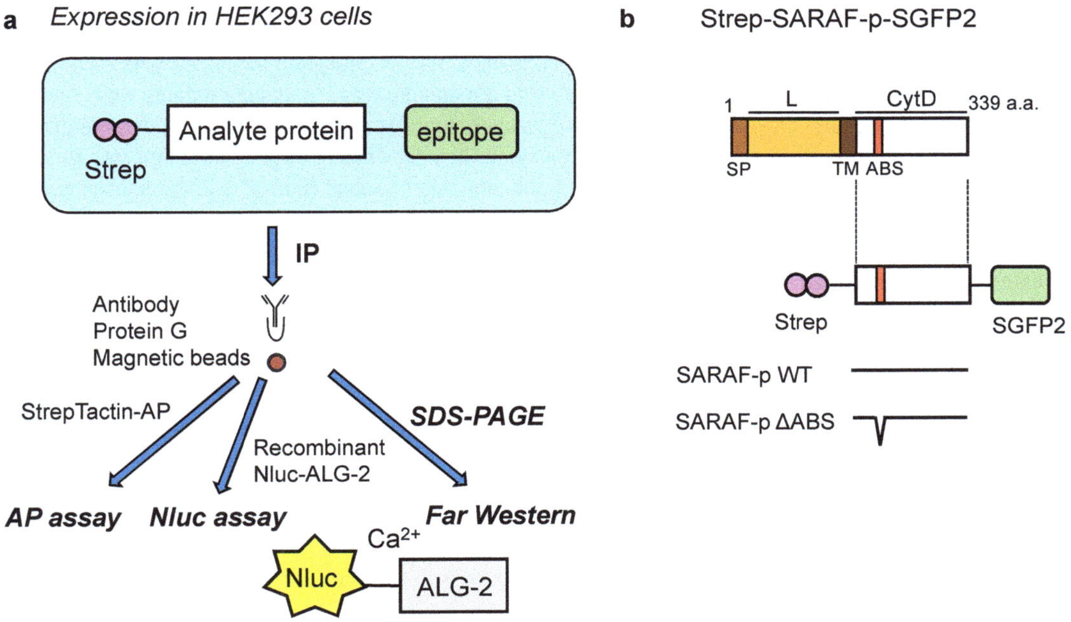

Fig. 3 Scheme of ALG-2-binding assays using recombinant Nluc-ALG-2. (**a**) Immunoprecipitated epitope-tagged proteins containing a Strep-tag are incubated with recombinant Nluc-ALG-2 in the presence of Ca^{2+}, washed, and subjected to luciferase assay for ALG-2-binding analysis. The relative amounts of immunoprecipitated proteins are quantified by the alkaline phosphatase (AP) assay using StrepTactin-AP. Direct binding of ALG-2 to analyte proteins is shown by Far-western blotting using Nluc-ALG-2 as a probe. (**b**) Diagram of an exemplar analyte protein used for Nluc-ALG-2-binding assays. SARAF is an ER-resident single-pass transmembrane protein. A partial fragment of SARAF (SARAF-p, 195-339 a.a.), corresponding to the cytosolic domain (CytD), is tagged with Strep and SGFP2 at its N-terminus and C-terminus, respectively. *SP* signal peptide, *L* luminal domain, *TM* transmembrane, *ABS* ALG-2-binding site, *WT* wild type, *ΔABS* ABS-deletion

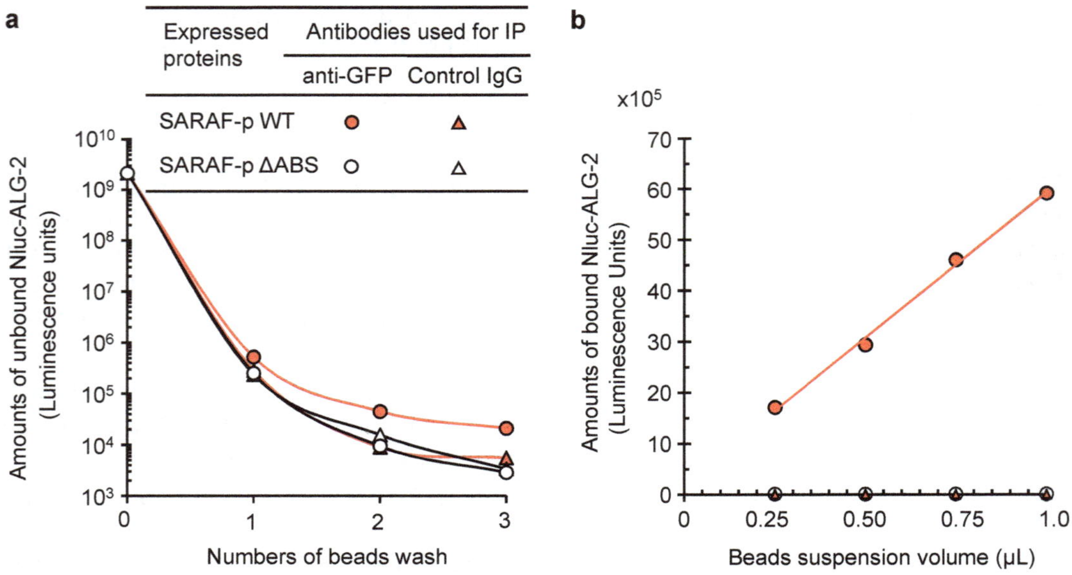

Fig. 4 Binding assays using recombinant Nluc-ALG-2. Strep-tagged and SGFP2-fused SARAF-p WT and an ALG-2-binding site (ABS, 223-229 a.a.) deleted mutant, SARAF-p-ΔABS (*see* Fig. 3b), were expressed in HEK293 cells and immunoprecipitated with anti-GFP or control IgG in buffer HKN containing 2 mM EGTA. Beads suspensions were washed with buffer HKN containing 10 μM $CaCl_2$ and incubated with recombinant Nluc-ALG-2 overnight. (**a**) Beads were collected by using a magnetic stand, and supernatants were saved for measurement of unbound Nluc-ALG-2. Beads were washed, and the numbers of washing steps required to remove non-specific binding were estimated by measuring Nluc activities of each supernatant (*see* **Note 18**). (**b**) Correlation between beads suspension volumes and amounts of bound Nluc-ALG-2. Luminescence units increased in direct proportion to the suspension volumes of the stock Dyanabeads Protein G, indicating that the amount of Nluc-ALG-2 used for the binding assay was sufficiently excessed over the amounts of immunoprecipitated SARAF-p

5. Re-suspend the collected beads with 0.2 ml of buffer HKN containing 10 μM $CaCl_2$ and 0.1% (w/v) gelatin (*see* **Note 11**).
6. Add approximately 50 ng of purified Nluc-ALG-2, and mix at 4°C for 2 h by constant agitation.
7. Collect beads and wash three times with 0.5 ml of buffer HKN containing 10 μM $CaCl_2$ (*see* **Note 12**).
8. Collect beads and re-suspend with 50 μl of PLB.
9. Proceed to Subheading 3.1, **steps 10–12** for measurement of luciferase activity (Fig. 4).

3.4 StrepTactin-Conjugated Alkaline Phosphatase Assay

1. Proceed to Subheading 3.3, steps **1–3** for beads suspension of immunopurified target proteins.
2. Take different aliquots (2.5 μl, 5 μl, 7.5 μl, 10 μl) of beads suspension, and wash the beads three times with 0.5 ml of buffer HKN in 0.6-ml plastic tubes.

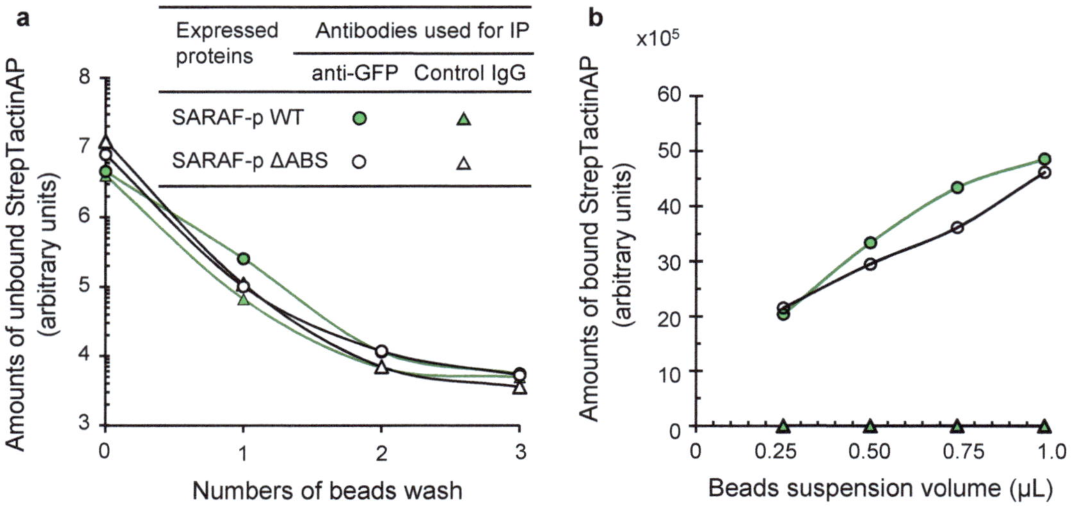

Fig. 5 Normalization of the amounts of immunoprecipitated proteins. Strep-tagged and SGFP2-fused proteins expressed in HEK293 cells were immunoprecipitated with anti-GFP as described in Fig. 4. Relative amounts of immunoprecipitated proteins were estimated by the StrepTactin-AP assay. Beads suspensions were incubated with StrepTactin-AP for 2 h. (**a**) The amounts of unbound StrepTactin-AP in the supernatants after each wash and (**b**) correlation between the suspension volumes of the stock Dynabeads Protein G and the amounts of bound StrepTactin-AP were analyzed (*see* **Note 19**)

3. Suspend the collected beads with 0.2 ml of buffer HKN containing 0.1% (w/v) gelatin and StrepTactin-AP (*see* Subheading 2.4, **item 1**).
4. Incubate the binding mixture at 4°C for 2 h by constant agitation.
5. Collect beads and wash three times with 0.5 ml of buffer HKN.
6. Collect beads and re-suspend with 200 μl of assay buffer (*see* Subheading 2.4, **item 2**).
7. Transfer 50 μl of the beads suspension into a 75 × 12 mm clear PS tube suitable for a single-tube-type luminometer, and incubate at room temperature for 5 min.
8. Add 25 μl of the reaction buffer (*see* Subheading 2.4, **item 3**).
9. Vortex the reaction mixture gently, and incubate at room temperature for 20 min (*see* **Note 13**).
10. Place the tube in the luminometer tube holder, and then measure luminescence at the following setting: delay, 0 s; integrated time of analysis, 3 s (Fig. 5).

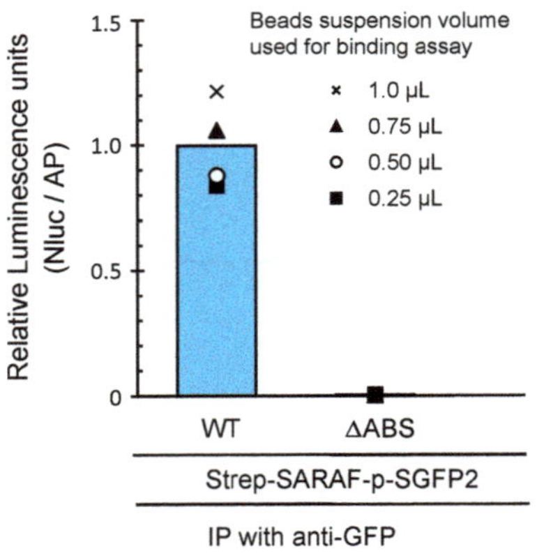

Fig. 6 Relative luminescence units of Nluc-ALG-2 binding. Relative amounts of Nluc-ALG-2 bound to immunoprecipitated Strep-SARAF-p-SGFP2 were normalized by calculating the ratio of Nluc activities shown in Fig. 4 to StrepTactin-AP activities shown in Fig. 5 and expressed as Nluc/AP (*see* **Note 20**)

11. To normalize the amounts of immunoprecipitated proteins, the ratio of Nluc activity (*see* Subheading 3.3, **step 9**) to AP activity (*see* Subheading 3.4, **step 10**) (Nluc/AP) is expressed as relative luminescent units (Fig. 6).

3.5 Far-Western Blot Analysis Probed with Nluc-ALG-2

1. Resolve the immunoprecipitated proteins (*see* Subheading 3.3, **step 3**) by SDS-PAGE, and electro-transfer the proteins from the gels to PVDF membranes (*see* **Note 14**).
2. Block the membrane by immersing the blots in blocking buffer (TBST containing 1% gelatin and 0.02% NaN_3) at 37°C for 1 h.
3. Rinse briefly twice with TBSTC at room temperature.
4. Incubate the membrane with TBSTC containing Nluc-ALG-2 (final concentration: 0.3 μg/ml) at 4°C overnight (*see* **Note 15**).
5. Wash the membrane twice by incubation with TBSTC at 37°C for 5 min.
6. Wash the membrane twice by incubation with TBSC at 37°C for 5 min.
7. Incubate the membrane with 20 μg/ml CTZ-*h* (*see* **Note 16**) diluted with TBSC for 3–5 s at room temperature, and detect chemiluminescence with a LAS-3000 mini lumino-image analyzer (Fujifilm) (Fig. 7).

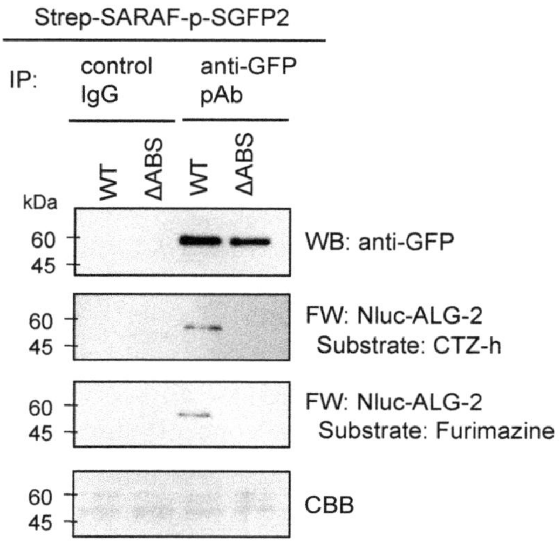

Fig. 7 Far-western blot analysis. The Strep- and SGFP2-fused SARAF fragment proteins expressed in HEK293 cells were immunoprecipitated with anti-GFP or control IgG. The immunoprecipitated proteins were resolved by SDS-PAGE (12.5% gel), followed by Western blotting (WB) with anti-GFP and Far-western blotting (FW) probed with recombinant Nluc-ALG-2 using CTZ-*h* or furimazine as a substrate. The membrane after FW (substrate, CTZ-h) was stained with CBB to check immunoprecipitated IgGs

4 Notes

1. When low amounts of endogenous ALG-2-interacting proteins are to be immunoprecipitated for the assay, use of ALG-2-deficient cells [14] is recommended. Binding efficiency of Nluc-ALG-2 may increase due to the lack of competition with ALG-2.
2. Harvested cell pellets can be temporally stored at −80°C. Nevertheless, immediate processing to immunoprecipitation is recommended in order to avoid degradation of the expressed proteins.
3. Save the supernatants to estimate immunoprecipitation efficiency of the antibody by testing the amount of the protein of interest remaining in the supernatants. Add more antibodies if the efficiency is poor. Because of the high sensitivity of the Nluc assay, low immunoprecipitation efficiency does not become a problem for this comparative binding analysis.
4. Samples need to be diluted with PLB when Nluc activities are too high.
5. When many samples need to be analyzed, a multi-well plate luminometer with an auto-injector may be useful.

6. The remaining PLB-suspended beads can be stored at 4°C or at −20°C for a later luciferase assay.
7. If not processed immediately, the harvested cells can be stored in a freezer at −80°C.
8. Before use, first regenerate the affinity column with 5 ml of 20 mM Tris–HCl (pH 7.5), 1 mM EDTA, 0.1 M NaCl, 6 M Guanidine HCl, and then equilibrate with buffer B containing 0.1% Triton X-100. After use, wash the column with 3 ml of 10 mM Tris–HCl (pH 7.5), 1 mM EDTA, 0.15 M NaCl, 0.02% NaN_3, and store at 4°C.
9. Store at 4°C. For a longtime storage, at −20°C, add glycerol (final concentration, 50%).
10. Take 25 μl or more volume of the beads suspension for Western blotting with anti-GFP and Far-western blotting with Nluc-ALG-2 (*see* Subheading 3.5 and Fig. 7).
11. Inclusion of 10 μM $CaCl_2$ in the binding buffer increases non-specific binding of the recombinant Nluc-ALG-2 to Dynabeads Protein G. This non-specific binding can be reduced by addition of 0.1% (w/v) gelatin to the binding buffer.
12. Washing three times is sufficient to remove non-specific binding (Fig. 4).
13. When luminescence of AP activity was measured at every 5 min after reaction start, stable luminescence was observed at about 20 min.
14. When furimazine is used as a substrate, nitrocellulose membrane is recommended to reduce background staining (see the instruction manual from Promega for a Nano-Glo HiBiT blotting system).
15. The protocol is comparable to Far-western using biotinylated ALG-2 as a probe [17], but fusion of NanoLuc with ALG-2 requires no secondary probes (streptavidin-horseradish peroxidase conjugate) resulting in shorter procedures.
16. CTZ-*h* has been used for nanoKAZ (identical with Nluc) [18]. Furimazine is available as a Nano-Glo® Luciferase Assay Substrate (Promega), but the concentration of the substrate solution is not given by the manufacturer. The substrate solution diluted 1:1000 with TBSC was used (Fig. 7).
17. Most of the activity remained in the flow-through fraction, partially because of degradation of the Nluc-ALG-2 protein.
18. Higher luminescence units in SARAF-p than SARAF-p ΔABS may be due to dissociation of Nluc-ALG-2 bound to SARAF-p during washing.

19. Although three washes seem sufficient, curbing of increase in the AP activities at larger beads suspension volumes suggests a better condition of using smaller beads volumes for the assay.
20. Greater values obtained at using larger beads suspension volumes are due to the deviation from direct proportion of StrepTactin-AP activities at larger beads suspension volumes (Fig. 5).

Acknowledgments

This work was supported by JSPS KAKENHI Grant Number 17H03803 (Grant-in-Aid for Scientific Research (B) to M.M.). We thank current and past members of the Laboratory of Molecular and Cellular Regulation for valuable suggestions and discussion. Wei Zhang and Rina Matsuo contributed equally to this work.

References

1. Takahara T, Inoue K, Arai Y, Kuwata K, Shibata H, Maki M (2017) The calcium-binding protein ALG-2 regulates protein secretion and trafficking via interactions with MISSL and MAP 1B proteins. J Biol Chem 292(41):17057–17072. https://doi.org/10.1074/jbc.M117.800201
2. Kanadome T, Shibata H, Kuwata K, Takahara T, Maki M (2017) The calcium-binding protein ALG-2 promotes endoplasmic reticulum exit site localization and polymerization of Trk-fused gene (TFG) protein. FEBS J 284(1):56–76. https://doi.org/10.1111/febs.13949
3. Ma J, Zhang X, Feng Y, Zhang H, Wang X, Zheng Y, Qiao W, Liu X (2016) Structural and functional study of apoptosis-linked gene-2 heme-binding protein 2 interactions in HIV-1 production. J Biol Chem 291 (52):26670–26685. https://doi.org/10.1074/jbc.M116.752444
4. Maki M, Takahara T, Shibata H (2016) Multifaceted roles of ALG-2 in Ca^{2+}-regulated membrane trafficking. Int J Mol Sci 17(9). pii: E1401. https://doi.org/10.3390/ijms17091401
5. Hall MP, Unch J, Binkowski BF, Valley MP, Butler BL, Wood MG, Otto P, Zimmerman K, Vidugiris G, Machleidt T, Robers MB, Benink HA, Eggers CT, Slater MR, Meisenheimer PL, Klaubert DH, Fan F, Encell LP, Wood KV (2012) Engineered luciferase reporter from a deep sea shrimp utilizing a novel imidazopyrazinone substrate. ACS Chem Biol 7 (11):1848–1857. https://doi.org/10.1021/cb3002478
6. Dixon AS, Schwinn MK, Hall MP, Zimmerman K, Otto P, Lubben TH, Butler BL, Binkowski BF, Machleidt T, Kirkland TA, Wood MG, Eggers CT, Encell LP, Wood KV (2016) NanoLuc complementation reporter optimized for accurate measurement of protein interactions in cells. ACS Chem Biol 11 (2):400–408. https://doi.org/10.1021/acschembio.5b00753
7. England CG, Ehlerding EB, Cai W (2016) NanoLuc: a small luciferase is brightening up the field of bioluminescence. Bioconjug Chem 27(5):1175–1187. https://doi.org/10.1021/acs.bioconjchem.6b00112
8. Schwinn MK, Machleidt T, Zimmerman K, Eggers CT, Dixon AS, Hurst R, Hall MP, Encell LP, Binkowski BF, Wood KV (2018) CRISPR-mediated tagging of endogenous proteins with a luminescent peptide. ACS Chem Biol 13(2):467–474. https://doi.org/10.1021/acschembio.7b00549
9. la Cour JM, Schindler AJ, Berchtold MW, Schekman R (2013) ALG-2 attenuates COPII budding in vitro and stabilizes the Sec23/Sec31A complex. PLoS One 8(9):e75309. https://doi.org/10.1371/journal.pone.0075309
10. Helm JR, Bentley M, Thorsen KD, Wang T, Foltz L, Oorschot V, Klumperman J, Hay JC (2014) Apoptosis-linked gene-2 (ALG-2)/Sec31 interactions regulate endoplasmic reticulum (ER)-to-Golgi transport: a potential

effector pathway for luminal calcium. J Biol Chem 289(34):23609–23628. https://doi.org/10.1074/jbc.M114.561829

11. Shibata H, Kanadome T, Sugiura H, Yokoyama T, Yamamuro M, Moss SE, Maki M (2015) A new role for annexin A11 in the early secretory pathway via stabilizing Sec31A protein at the endoplasmic reticulum exit sites (ERES). J Biol Chem 290(8):4981–4993. https://doi.org/10.1074/jbc.M114.592089
12. Takahashi T, Kojima K, Zhang W, Sasaki K, Ito M, Suzuki H, Kawasaki M, Wakatsuki S, Takahara T, Shibata H, Maki M (2015) Structural analysis of the complex between penta-EF-hand ALG-2 protein and Sec31A peptide reveals a novel target recognition mechanism of ALG-2. Int J Mol Sci 16(2):3677–3699. https://doi.org/10.3390/ijms16023677
13. Zhang W, Matsuo R, Achiha T, Teranishi N, Muramatsu A, Takahara T, Hideki Shibata H, Maki M (2017) Ca^{2+}-binding protein ALG-2 may function in modulating Ca^{2+} homeostasis by interacting with SARAF. In Paper presented at the 20th international symposium on calcium binding proteins and calcium function in health and disease (CaBP20), Awaji Yumebutai International Conference Center, Hyogo, Japan, 22–26 Oct 2017
14. Okumura M, Ichioka F, Kobayashi R, Suzuki H, Yoshida H, Shibata H, Maki M (2009) Penta-EF-hand protein ALG-2 functions as a Ca^{2+}-dependent adaptor that bridges Alix and TSG101. Biochem Biophys Res Commun 386(1):237–241. https://doi.org/10.1016/j.bbrc.2009.06.015
15. Shibata H, Suzuki H, Kakiuchi T, Inuzuka T, Yoshida H, Mizuno T, Maki M (2008) Identification of Alix-type and Non-Alix-type ALG-2-binding sites in human phospholipid scramblase 3: differential binding to an alternatively spliced isoform and amino acid-substituted mutants. J Biol Chem 283(15):9623–9632. https://doi.org/10.1074/jbc.M800717200
16. Green MR, Sambrook J (2012) Molecular cloning: a laboratory manual, 4th edn, Chap. 15, vol 2. Cold Spring Harbor Laboratory Press, New York
17. Osugi K, Shibata H, Maki M (2013) Biochemical and immunological detection of physical interactions between penta-EF-hand protein ALG-2 and its binding partners. Methods Mol Biol 963:187–200. https://doi.org/10.1007/978-1-62703-230-8_12
18. Inouye S, Sato J, Sahara-Miura Y, Yoshida S, Kurakata H, Hosoya T (2013) C6-Deoxy coelenterazine analogues as an efficient substrate for glow luminescence reaction of nanoKAZ: the mutated catalytic 19 kDa component of Oplophorus luciferase. Biochem Biophys Res Commun 437(3):23–28. https://doi.org/10.1016/j.bbrc.2013.06.026

Chapter 32

Calcium-Induced Protein Folding in Calumenin and Calmodulin

Marco Mazzorana and Thomas Lykke-Møller Sørensen

Abstract

Binding of calcium — and small molecules in general — often induce conformational changes in large molecules and complexes. The degree and type of change varies, but the resulting shift in specific affinities ultimately induces a physiological response. It is therefore important for our understanding of responses at the cellular level to define coupled changes at the molecular level.

Calumenin, a six-EF-hand calcium-binding protein localized in the endoplasmic reticulum, undergoes substantial calcium-induced rearrangement. We have demonstrated how calumenin changes from being unfolded in the absence of calcium to a compact trilobal fold in the presence of calcium (Mazzorana et al., PLoS One 11:e0151547, 2016).

Here, we describe protocols for the expression and purification of calumenin and calmodulin, another EF-hand protein modulated by calcium, along with protocols for biophysical techniques used to characterize calcium-induced changes to protein conformation. Analytical size-exclusion chromatography in the presence and absence of calcium provides an informed indication of any larger conformational movements. Circular dichroism spectroscopy reveals alterations to the secondary or tertiary structure, while small-angle X-ray scattering explores changes further providing low-resolution conformational details.

Surface plasmon resonance estimates binding kinetics and affinities completing the biophysical description of these events.

Key words Calcium-dependent folding, Calmodulin, Calumenin, EF-hand, Calcium signaling, SR-CD, SEC-MALS, SEC-SAXS, SPR

1 Introduction

Binding of ions and small molecules can bring about conformational changes in larger molecules which alters the molecular properties by exposing, shielding, or forming interaction sites, either modulating existing or creating new interactions [1]. Ultimately, the outcome of the modification is a change in the behavior of the macromolecule which either directly or indirectly leads to a shift of the cellular state and a physiological response. It is therefore

Claus W. Heizmann (ed.), *Calcium-Binding Proteins of the EF-Hand Superfamily: From Basics to Medical Applications*, Methods in Molecular Biology, vol. 1929, https://doi.org/10.1007/978-1-4939-9030-6_32,

important for our understanding of responses at the cellular level to define the coupled changes at the molecular level.

Calcium is a potent secondary messenger capable of initiating such diverse cellular processes as glycogenolysis, secretion of hormones and neurotransmitters, egg fertilization, cellular motility and growth, as well as contraction of muscles [2]. All these processes are brought about through calcium-induced conformational changes. The EF-hand domain is a fundamental structural element designed to transduce Ca^{2+} signal through rearrangements of helices around a Ca^{2+}-chelating loop. Here we present the calcium-induced changes in two proteins containing EF-hand domains.

Calumenin is a six-EF-hand calcium-binding protein localized in the endoplasmic reticulum. We have applied a range of biophysical techniques to characterize the calcium-dependent folding of this protein [3]. In the case of calumenin, these conformational changes are very substantial—from being unfolded in the absence of calcium to a compact trilobal fold in the presence of calcium. We have demonstrated this through methods that all contribute unique pieces of insight to the puzzle: analytical size-exclusion chromatography coupled with light-scattering (SEC–MALS) analysis revealed a significant calcium-dependent shift in apparent size with the apo-form of calumenin having an extremely high hydration radius for a protein of its size [4]; circular dichroism (CD) spectroscopy demonstrated how addition of calcium shifts the protein conformation from an unfolded state void of secondary structure elements to a mostly α-helical configuration [5]; surface plasmon resonance (SPR) confirmed calcium binding, specificity, and affinity [6]; and finally small-angle X-ray scattering (SAXS) described the shift to a more compact protein fold upon calcium binding in which secondary structure elements are arranged in an elongated shape [7].

More subtle changes can also be detected using the biophysical tools described here, and we have included data from measurement of calmodulin to demonstrate this. Calmodulin is the prototypical member of the EF-hand-containing protein superfamily also switching structure in response to variations in cytosolic Ca^{2+} concentration. In contrast to calumenin, calmodulin retains a core secondary structure in the absence of calcium [8, 9].

There is of course some overlap in the information gathered from these different approaches, and practical as well as scientific reasoning may decide which approach to take. Of key importance is that these techniques can all be used without requiring any direct protein modification or labeling. This means that a baseline understanding of relevant conformational changes induced by the presence of calcium in a physiological relevant concentration range can be established. Furthermore, recent technological developments also mean that these characterizations can now be completed with relatively small amounts of samples, and the same sample volume can to some extent be reused for the different techniques.

The insight gained from these experiments will guide further analysis with possible routes to pursue including isothermal titration calorimetry (ITC) providing a different approach to binding characterization, while techniques such as fluorescence resonance energy transfer (FRET) microscopy provide vital insight into the dynamics of the binding process and the associated conformational changes. Higher-resolution insight through macromolecular crystallography (MX), nuclear magnetic resonance (NMR), or electron microscopy (EM) will ultimately provide a structural framework to reveal the details of these conformational changes.

On a final note, although we have focused on calcium-induced conformational rearrangements, the approach taken in this study can easily be adapted to proteins where other ions or small molecules act as promoters of conformational change such as the silver-induced folding of SilE [10] or shed light on the mechanisms and dynamics of intrinsically disordered proteins in general [11].

2 Materials

2.1 Protein Expression

All solutions below are prepared with ultrapure deionized water (>18 MΩ cm at 25 °C) and chemicals provided from Sigma unless differently specified.

1. KpnI/HindIII linearized pOPIN-S and pOPIN-J plasmids (OPPF-UK).
2. Desalt quality oligonucleotides.
   (a) CaM-2FwJ: 5′-aagttctgtttcagggcccgGCTGACCAGCTGACTGAGGAGCAG.
   (b) CaM-149RvJ: 5′-atggtctagaaagctttaCTTTGCAGTCATCATCTGTACAAACTCTTC.
   (c) Calu-68FwS: 5′-gcgaacagatcggtggtGAGAGCAAGGAAAGGCTTGGAAAG.
   (d) Calu-315RvS: 5′-atggtctagaaagctttaGAACTCATCATGCCGTACTAAGGCCTCC.
3. Human cDNA libraries of synthetic genes encoding human calmodulin and calumenin.
4. *Sterile* (0.22 μm filtered) 20× NPS solution: 0.5 M $(NH_4)_2SO_4$, 1 M KH_2PO_4, 1 M g/L Na_2HPO_4 in water.
5. *Sterile* (autoclaved) 5052 solution: 250 g/L glycerol, 25 g/L D-glucose monohydrate, 100 g/L α-lactose in water.
6. *Sterile* (autoclaved) 10% glucose; 100 g/L D-glucose monohydrate in water.
7. SOC medium.
8. ZY medium (autoclaved): 5 g/L yeast extract, 10 g/L tryptone in water.

9. TB medium (autoclaved): 47.6 g/L TB modified medium in water.
10. LB-agar solid medium (autoclaved): 35.6 g/L LB broth with agar in water.
11. 50 mg/ml carbenicillin (Melford) in water, *sterile* (0.22 μm filtered).
12. 50 mg/ml kanamycin in water, *sterile* (0.22 μm filtered).
13. 25 mg/ml chloramphenicol in ethanol 100%, *sterile* (0.22 μm filtered).
14. Rosetta (DE3) competent *E. coli* cells (Merck).
15. Sterile pipettes (50 and 10 ml) and pipette tips (1 ml).
16. Autoclaved 1 ml glass cylinder (class B or higher).
17. Autoclaved 2.5 L Tunair™ shake flask.
18. Autoclaved 250 ml Erlenmeyer flasks.
19. Centrifuge tubes for harvesting pellet (compatible with type Beckman 8.1000 or equivalent).

2.2 Protein Purification

All buffers below are prepared with ultrapure deionized water (18 MΩ cm at 25 °C), filtered with 0.22 μm cutoff membrane, degassed and refrigerated to 4 °C prior to use. All reagents are from Sigma, purity >95% unless differently specified.

1. Lysis buffer: 25 mM Na-HEPES pH 7.5, 250 mM NaCl, 2.5 mM $CaCl_2$, 2.5 mM $MgCl_2$. Refrigerate to 4 °C, and add 10 μg/ml DNAseI and two tablets/100 ml complete mini-EDTA-free protease inhibitors (Roche) immediately prior to use.
2. Buffer A: 25 mM Na-HEPES pH 7.5, 250 mM NaCl, 2.5 mM $CaCl_2$, 25 mM imidazole-Cl (from 1 M stock solution at pH 7.5), 0.1% 1-thioglycerol.
3. Buffer B: 25 mM Na-HEPES pH 7.5, 250 mM NaCl, 2.5 mM $CaCl_2$, 250 mM imidazole-Cl (from 1 M stock solution at pH 7.5), 0.1% 1-thioglycerol.
4. Buffer C: 25 mM Na-HEPES pH 7.5, 250 mM NaCl, 2.5 mM $CaCl_2$, 0.1% 1-thioglycerol.
5. Buffer C*: 25 mM Na-HEPES pH 7.5, 250 mM NaCl, 1 mM EGTA (Fluka), 0.1% 1-thioglycerol.
6. 1 ml HisTrap HP column (GE healthcare).
7. Amicon Ultra-4 10 kDa cutoff filter devices (Merck).
8. Amicon Ultra-15 10 kDa cutoff filter devices (Merck).
9. His-tagged 3C protease [12].
10. His-tagged Ulp1/SUMO protease [13].

11. HiLoad 16/600 Superdex 75 pg (GE healthcare).
12. HiLoad 16/600 Superdex 200 pg (GE healthcare).
13. Liquid N_2.
14. NuPAGE 4–12% Bis–Tris gels (Thermo Fisher).
15. Novex™ NuPAGE™ MES SDS running buffer 20× (Thermo Fisher).
16. XCell SureLock™ Mini-Cell Electrophoresis System (Thermo Fisher) and power supply.
17. BenchMark™ Protein Ladder (Thermo Fisher).
18. 6× SDS-PAGE loading buffer (Thermo Fisher).
19. Simply blue safe stain (Thermo Fisher).
20. Cell disruptor (Constant Systems) or tip sonicator 6.
21. Beckman ultracentrifuge + rotor 45Ti + tubes.
22. FPLC system (GE Healthcare ÄKTA pure 25 or equivalent).
23. Nanodrop™ 2000 or equivalent.
24. Optional: Cell homogenizer (Stuart SHM2 or equivalent).
25. Magnetic stirrer.
26. Dialysis tube (22 mm, 3500 MWCO, Thermo Fisher).
27. 1 mm mesh strainer/sieve.

2.3 SEC–MALS

1. Superdex: Superdex 200 increase 10/300 GL (GE Healthcare).
2. HPLC system equipped with:
   (a) Inline MALS detector (Wyatt Technology Dawn Heleos II, or equivalent).
   (b) Inline refractometer (Wyatt Technology Optilab T-rEX, or equivalent).
   (c) UV–VIS photometer (for A_{280nm} measurements).
3. Buffer D: 25 mM Na-HEPES pH 7.5, 150 mM NaCl, 2.5 mM $CaCl_2$.
4. Buffer E: 25 mM Na-HEPES pH 7.5, 150 mM NaCl, 2.5 mM EGTA.
5. 100 μl Hamilton syringe.

2.4 Circular Dichroism

1. 5 ml HiTrap desalting column (GE Healthcare).
2. Cylindrical quartz cell (0.1 mm pathlength, Hellma Analytics cat. nr. 121-0.10-40).
3. Ultra-micro quartz cell (10 mm pathlength, Hellma Analytics cat. nr. 105-21-85-40).
4. Flat pipette tips (Mettler Toledo).

5. Ultrapure deionized water.
6. Cell wash solutions (concentrated HNO3, 100% ethanol).
7. HellmanexII (Hellma Analytics).
8. EGTA (Fluka).
9. Buffer F: 2.5 mM Na-HEPES pH 7.5, 25 mM NaCl, 5 mM EGTA, 0.5 μM $CaCl_2$.
10. Titrant T1: 50 mM $CaCl_2$ solution from serial dilutions of 1 M standard solution (Fluka).
11. Titrant T2: 1:10 dilution of titrant A in water.
12. SR–CD beamline (such as Diamond Light Source B23) or benchtop dichrograph (e.g., Chirascan Plus, Applied Photophysics).

2.5 SEC–SAXS

1. Shodex KW-402.5 column.
2. Buffer D: 25 mM Na-HEPES pH 7.5, 150 mM NaCl, 2.5 mM $CaCl_2$.
3. Buffer E: 25 mM Na-HEPES pH 7.5, 150 mM NaCl, 2.5 mM EGTA.

2.6 SPR Analysis

1. CM5 series S sensor chip (GE Healthcare).
2. Regeneration scouting kit (GE Healthcare).
3. Amine coupling kit (GE Healthcare).
4. 1 M solutions of $CaCl_2$, $SrCl_2$, $MgCl_2$, $MnCl_2$, $CoCl_2$.
5. Buffer G: HBS-N (GE Healthcare) + 5 mM $CaCl_2$.
6. Buffer H: HBS-N (GE Healthcare).
7. Biacore T200 (GE Healthcare) or equivalent.

3 Methods

3.1 Protein Expression

1. Following the protocol by Berrow et al. [14, 15], clone calmodulin 2–149 in vector pOPIN-J and calumenin 68–315 in vector pOPIN-S (*see* **Note 1**). To amplify full-length human calmodulin (Uniprot P0DP23, amino acids 2–149), use the primers CaM-2FwJ and CaM-149RvJ. Similarly the 6-EF-hand regions of human calumenin (Uniprot O43852, amino acids 68–315) are amplified using the primers Calu-68FwS and Calu-315RvS.
2. Purify and quantify the cloning product. Confirm the sequence is in frame and correct by DNA sequencing.
3. Incubate 50 μl chemically competent Rosetta (DE3) *E. coli* cells with 100 ng of each vector for 1 h in ice. Transform by heat shock 45 s at 42 °C, and place in ice for 2 min. Add 450 μl

SOC medium, and incubate at 37 °C for at least 1 h before plating in LB + 1.5% agar plates supplemented with the appropriate antibiotics (25 μg/ml chloramphenicol + 25 μg/ml kanamycin for pOPIN-S calumenin and + 25 μg/ml chloramphenicol + 50 μg/ml carbenicillin for pOPIN-J calmodulin). Incubate at 37 °C overnight.

4. Transfer a single colony to a 250 ml Erlenmeyer flask containing 50 ml of Terrific Broth medium supplemented with 500 μl 10% glucose and the same amount of antibiotics used in the LB-agar plates. Grow overnight at 37 °C under vigorous shaking (220 rpm).
5. Transfer 10 ml of the overnight culture in each of four Tunair flask, each containing 1 L of autoinduction medium produced according to Studier et al. [16]. In short, on the day, measure 930 ml sterile ZY medium into a sterile glass cylinder under flame or laminar flow. Supplement the medium with 50 ml NPS solution, 25 ml 5052 solution, and 1 ml 2 M $MgSO_4$. Add the appropriate antibiotics (as above), transfer in a sterile Tunair flask, and immediately inoculate with 10 ml overnight bacterial culture.
6. Grow the cell cultures 8 h at 37 °C, 220 rpm before decreasing the temperature to 18 °C, and then keep at 220 rpm for 60 h (*see* **Note 2**).
7. Harvest cells by centrifugation (5000 × *g* 10 min) using Beckman 8.1000 rotor or equivalent. Discard the supernatant, and transfer the pellet to 50 ml conical centrifuge tubes using a spatula.
8. Weigh the cell pellet, and proceed immediately to the purification step. Alternatively, label the tubes appropriately, flash freeze in liquid N_2, and store at −80 °C (leave the tube cap loose).

3.2 Protein Purification

1. Activate the refrigeration on the cell disruptor to allow it to cool to 10 °C or less while preparing the cell suspension.
2. For each gram of bacterial pellet, add 5 ml of ice-cold lysis buffer. Stir well using a magnetic stirrer, or use a cell homogenizer.
3. Strain the cell suspension through a sieve to avoid lumps and residues of plastic or other materials. Load in the cell disruptor, extrude at 25 kPsi, and collect the flow-through in ice. Repeat the operation two more times reflowing the cells (*see* **Note 3**).
4. Centrifuge at 200,000 × *g* 30 min using a fixed-angle rotor (Beckman 45 Ti or equivalent).
5. Collect the supernatant, and pass through a 0.45 μm syringe filter.

6. Load onto a 1 ml HisTrap HP column pre-equilibrated with buffer A at 1 ml/min using an FPLC system. Collect the flow-through fraction (FT) (*see* **Note 4**).
7. Wash extensively (at least 50 ml) with buffer A, and collect the wash fraction (W).
8. Using an FPLC system, elute the protein in three 20 ml steps of the following:
   - 10% buffer B + 90% buffer A.
   - 50% buffer B + 50% buffer A.
   - 100% buffer B.
9. Of each elution, collect 1 ml fractions, and label them accordingly (e.g., "10%—fraction 1"), or number them following the position in the fraction collector if using an automated FPLC system.
10. Load 10 μl of each of the desired fractions (+2 μl 6× sample buffer) and 10 μl benchmark molecular weight marker on each lane of a 10-well NuPAGE 4–12% gel. Run in MES buffer 1× at 200 V constant for 37 min. Remove the gel, and reveal the bands staining in SafeStain at room temperature for at least 5 min.
11. Look for the protein with an apparent molecular weight of 40 kDa for the His-GST-calmodulin$_{1-149}$ fusion and 50 kDa for His-SUMO-calumenin$_{68-315}$, and pool the corresponding fractions.
12. Load the pooled fractions in a 10 kDa cutoff Amicon Ultra-15 concentrator pre-equilibrated with buffer C, and reduce to 3 ml with 5-min cycles of centrifugation at 3000 × *g* alternated by thorough resuspension using a pipette tip.
13. Equilibrate the Superdex200 10/600 column (for pOPIN-J calmodulin) or the Superdex75 10/600 column (for pOPIN-S calumenin) with two column volumes of buffer C.
14. Load approximately 3 ml of the concentrated fractions containing the desired fusion protein in a 5 ml loop (save 50 μl for SDS-PAGE quality control), and inject in the appropriate Superdex size exclusion column at 0.5 ml/min. Perform an isocratic elution with buffer C. Collect 1 ml fractions between 0.7 CV and 1.5 CV. Monitor $A_{280\mathrm{nm}}$ during elution, and plot it against the elution volume.
15. From the two highest chromatogram peaks, select one fraction every three, and transfer 50 μl in wells of a PCR plate containing 10 μl sample buffer. Repeat the same for the pooled fractions loaded in gel filtration. Cover with plastic seal, and warm to 70 °C for 5 min.
16. Run the sample (and a MW marker reference) as before in 10-well NuPAGE 4–12% gels. This stage requires two gels,

which will need a decreased voltage for longer time. In the case of MES buffer, use 180 V constant for 45 min.

17. Stain the gels as before and select the purest fractions (i.e., desired protein band must account for >80% total proteins in the lane).
18. Pool them, and add the appropriate protease (3C protease for pOPINJ calmodulin, SUMO protease for pOPINS calumenin), in 1:20 molar ratio. Transfer the mixture in a dialysis tube, seal, and equilibrate overnight against 2 L of buffer C in a glass cylinder under mild agitation applied through a magnetic stirrer (50 rpm).
19. Perform a quality control of the cleavage on SDS-PAGE. Run 10 μl of one of the fractions pooled overnight and 10 μl of the sample after cleavage (50 μl + 10 μl 6× sample buffer).
20. If the intensity of the fusion protein band (at 40 or 50 kDa) is decreased in the treated sample to at least half of the original, proceed with the following step; otherwise, add a 1:20 molar ratio of the appropriate protease to the sample, and incubate for 2 h at room temperature.
21. Load the cleaved protein + protease mixture on a 1 ml HisTrap HP column pre-equilibrated with buffer C at 1 ml/min using a FPLC system. Wash with 10 ml buffer C, and collect both the flow-through and wash fractions containing cleaved untagged protein.
22. Concentrate the unbound fraction from the step above to approx. 3 ml using the procedure described previously.
23. Load 3 ml of the concentrated unbound fractions in a 5 ml loop, inject in the Superdex75 column at 0.5 ml/min, and perform an isocratic elution with buffer C. Collect 1 ml fractions between 0.7 CV and 1.5 CV. Monitor A_{280nm} during elution, and plot it against the elution volume.
24. Verify the purity of the eluted fractions loading 10 μl each on a SDS-PAGE, and pool those with purity >95%.
25. Concentrate to approx. 10 mg/ml using a 10 kDa cutoff AmiconUltra-4, and run as above at max 3000 × *g* in cycles of 5 min alternated with thorough resuspension with a pipette tip.
26. Verify the concentration of the protein using a Nanodrop Spectrophotometer (or similar) reading the A_{280nm} and applying the Beer–Lambert law using the following coefficients:
    (a) Calmodulin$_{2\text{–}149}$: MW = 16706 Da, ε_{280nm} = 2409 M−1 cm^{-1}.
    (b) Calumenin$_{68\text{–}315}$: MW = 29413 Da, ε_{280nm} = 50880 M^{-1} cm^{-1}.
27. Aliquot in PCR tubes 15 μl protein, flash freeze in Liquid N_2, and transfer in a labeled 50 ml Falcon tube for long-term storage at −80 °C.

3.3 Size-Exclusion Chromatography (SEC) Coupled with Multi-angle Light-Scattering (SEC–MALS) analysis

1. Connect a Superdex 200 10/300 increase column to an FPLC system equipped with a Dawn Heleos II MALS detector and Optilab T-rEX refractometer (Wyatt Technology).
2. Equilibrate the column with two volumes of buffer D passing the solution through a 100 μl loop (injection valve in inject position).
3. Maintain the flow rate constant to 0.7 ml/min until the refractive index value stabilizes.
4. Thaw one 10 mg/ml protein aliquot, and dilute the sample 1:5 with buffer C.
5. Fill a 100 μl Hamilton syringe with the diluted sample, and connect to the injection valve without loading in the loop.
6. Start the Astra control software (Wyatt Technology), and prepare to synchronize the chromatogram collection for 24 ml starting at the injection point.
7. Turn on the laser of the MALS detector, and wait for it to stabilize (approximately 5 min).
8. Switch the injection valve to load, and load the loop with the content of the glass syringe.
9. Start the injection and with it the collection of MALS, refractive index (Ri) data.
10. At the end of the Astra data recording, keep the flow constant to 0.7 ml/min (i.e., do not stop the FPLC). Save the data collection file with a meaningful name, and restart the same program if more samples need collecting in the same buffer.
11. Repeat the procedures above (with the exception of column equilibration) with all the samples that need to be measured in the presence of Ca^{2+}.
12. Before running the samples in the absence of Ca^{2+} (i.e., under EGTA chelation), pause the FPLC, wash the pump with buffer E, and flow two column volumes at 0.7 ml/min to re-equilibrate the column.
13. Once the column is equilibrated, load the sample and collect SEC/MALS data as in the procedure above.
14. To analyze the data, open the Astra software (each single file contains the raw data, and a series of tabs in the navigation menu indicates the steps of the analysis). The procedure in the following steps must be repeated for each run.
15. Look at the chromatograms for each of the 18 MALS sensors, and check that all chromatograms have the same shape, although with different scaling factor. If not, exclude from the analysis the odd sensor(s).

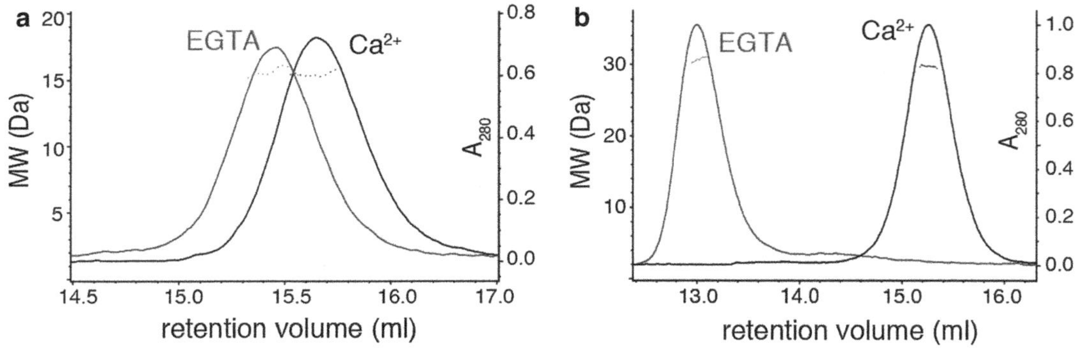

Fig. 1 Size-exclusion elution profile of calmodulin (**a**) and calumenin (**b**) coupled with the corresponding molecular-weight estimate from MALS (dotted lines in the middle of each peak). In black the curves obtained in the presence of Ca^{2+} in the buffer and in gray the equivalent measured in the presence of EGTA

16. Navigate to the baseline tab, and select two flat regions of the chromatogram containing the peaks to define the best baseline (black line). Apply this baseline to all sensors.
17. Define the region of the peaks (peaks tab), and plot for each curve (or multiple curves) the EASI graph. This plot shows the $A_{280\mathrm{nm}}$ signal from the FPLC detector as a function of the elution volume. For the region defined as the peak, the software plots the estimation of MW (derived from the MALS signal).
18. If within the peak, the MW estimate is approximately constant, the protein is monodisperse, and the value plotted provides an indication of its aggregation state (*see* **Note 5**).
19. Refer to the final report for the best estimate of the molecular weight.
20. Chromatograms can be exported as Encapsulated Postscript from Astra (*see* Fig. 1).

3.4 Circular Dichroism

This procedure can be performed at synchrotron CD instruments (SR–CD) such as the module B of Diamond Light Source's B23 beamline [17] or on benchtop CD photometers. The following instructions will indicate the differences between these setups.

1. Mount a 5 ml HiTrap desalting column on a FPLC system, and equilibrate with 25 ml buffer F at constant flow rate of 3 ml/min.
2. Thaw one aliquot of protein, and dilute it to 100 μl.
3. Inject the sample in the column, elute isocratically at 3 ml/min, and collect 0.5 ml fractions, monitoring the $A_{280\mathrm{nm}}$ and conductivity of solution.

4. Pool the fractions in the A_{280nm} peak for which the conductivity does not reach more than 5% of the maximum recorded.
5. Measure the concentration of protein using a Nanodrop as in the protein purification stage, and dilute in elution buffer to a final concentration of 5 μM.
6. For SR–CD experiments, load 150 μl of sample in a 0.1 mm cylindrical quartz cell using flat pipette tips, and avoid the formation of bubbles.

   For the benchtop setup, load 150 μl of sample in a 10 mm pathlength cuvette using regular pipette tips, and avoid the formation of bubbles.
7. Load the cuvette in the sample holder position of the chosen instrument, and collect four consecutive spectra in the far-UV region (185–260 nm) sampling in 1 nm step of the duration of 1 s each. CD data collection is performed at constant temperature of 20 °C.
8. Average the four spectra using the data collection software (OLIS software for B23) or any suitable statistical software package.
9. Export the values of absorbance to a master spreadsheet reporting in the first column the wavelength of each CD measurement and in the following columns the averaged values for the corresponding wavelength at different concentrations of calcium.
10. Using the software MaxChelator [18], calculate the amounts of 50 mM $CaCl_2$ or 5 mM $CaCl_2$ to add at each step to increase the free concentration of Ca^{2+} in a EGTA-buffered solution as in Table 1 (*see* **Note 6**).
11. At each step, add the requested amount (column 3, Table 1) to increase the total Ca^{2+} concentration to the desired level (column 4, Table 1), providing the free pCa (at which the spectrum has to be determined). Each addition of fresh Ca^{2+} solution is done using a flat pipette tip in the case of 0.1 mm cylindrical cell or a standard pipette tip for 10 mm cells. At each addition, the sample needs to be mixed thoroughly by pipetting, paying extreme attention to avoid the formation of bubbles (*see* **Note 7**).
12. Collect the spectra on the sample as above, and, after averaging, export the CD values in the master spreadsheet at the column of the corresponding pCa value.
13. At the end of each titration (or if the cell shows cloudy precipitates), clean the cuvette with 5 min wash in HNO_3 (under fume hood), followed by extensive rinse with water and ethanol and drying under N_2 flow.

Table 1
Ca^{2+} titration series

Initial pCa	Initial volume (μl)	Volume and type of titrant to add	$[Ca^{2+}]_{free}$ (mM)	Final pCa
8.36	150	1.60 μl T1	1.014	8
8	151.60	1.28 μl T1	1.436	7.8
7.8	152.88	1.50 μl T1	1.912	7.6
7.6	154.38	1.71 μl T1	2.440	7.4
7.4	156.09	1.71 μl T1	2.957	7.2
7.2	157.80	1.57 μl T1	3.410	7.0
7.0	159.37	1.21 μl T1	3.766	6.8
6.8	160.58	1.00 μl T1	4.038	6.6
6.6	161.58	0.64 μl T1	4.223	6.4
6.4	162.22	4.77 μl T2	4.352	6.2
6.2	166.99	3.14 μl T2	4.321	6.0
6.0	170.13	2.14 μl T2	4.306	5.8
5.8	172.27	1.14 μl T2	4.285	5.6
5.6	173.41	0.86 μl T2	4.282	5.4
5.4	174.27	0.64 μl T2	4.280	5.2
5.2	174.91	0.71 μl T2	4.283	5.0
5.0	175.62	3.56 μl T2	4.368	4.0
4.0	179.18	3.21 μl T1	5.190	3.0

Starting from 150 μl of buffer F (pCa = 8.36) and for additions of titrant solutions T1 and T2 to cover the range of pCa 8.0 ÷ 3.0

14. Plot the CD curves from the spreadsheet at various values of pCa, and analyze the differences between the extremes (*see* Fig. 2). Select the most variable indicator among the following (*see* **Note 8**):
    (a) If the wavelength of highest positive peak is stable, plot the CD value at this wavelength against pCa.
    (b) Otherwise choose one of the negative peaks, and plot the corresponding CD value as a function of pCa.
    (c) Alternatively choose any suitable indicator such as the total alpha-helical content calculated over the whole spectrum (using software packages such as CDPro) [19].
15. Export the values of the chosen indicator at the different values of pCa into data analysis software such as Graphpad Prism or Origin. Use the one-site binding model to calculate the kinetic values of Kd and the Hill function models to calculate the cooperativity coefficient H.

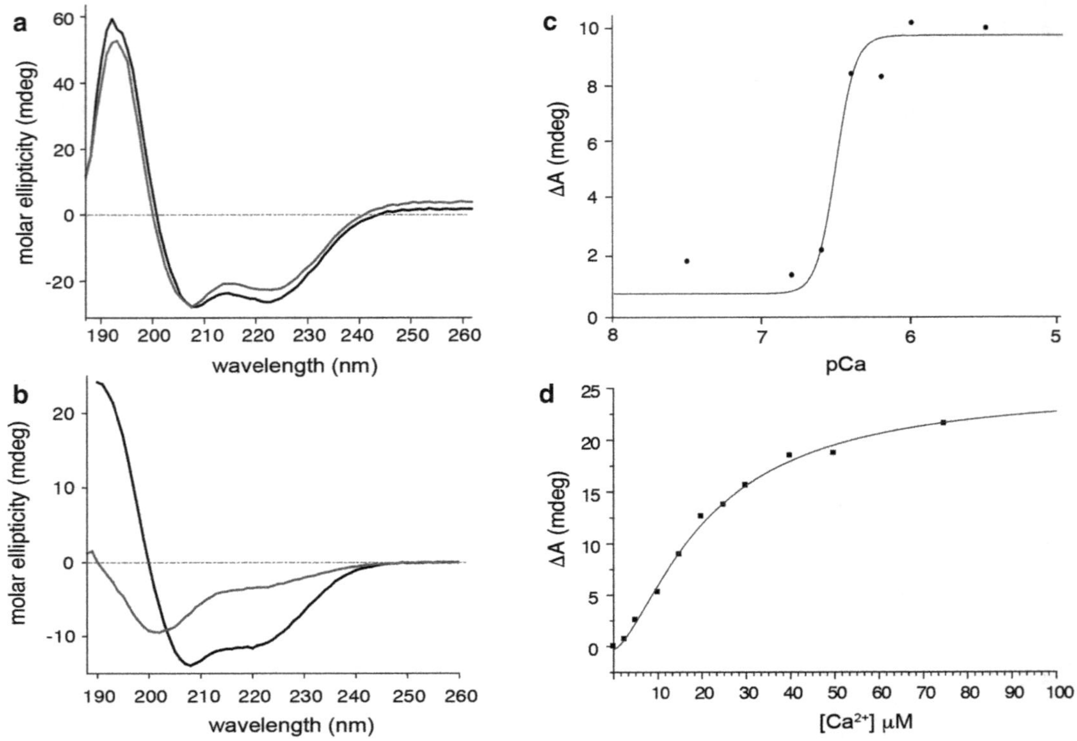

Fig. 2 CD spectra of calmodulin (**a**) and calumenin (**b**) in EGTA-buffered solution (gray) or in the presence of 1 mM free Ca^{2+} in solution (pCa = 3, black). The titration curve obtained by plotting the values of ellipticity at 191 nm for calmodulin (**c**) and 208 nm for calumenin (**d**) against pCa provides the Kd binding constants for Ca^{2+} to the two proteins

3.5 Size-Exclusion Chromatography-Coupled Small-Angle X-Ray Scattering (SEC–SAXS)

The following setup is optimized for small-angle X-ray scattering (SAXS) measurement at synchrotron instruments with an inline size-exclusion chromatography injection system such as Diamond Light Source's B21 beamline. The protocol is adaptable to static SAXS measurements.

1. Equilibrate the HPLC and Shodex size-exclusion column with 2 CV of buffer D.
2. Thaw 25 μl of concentrated protein and dilute 1:1 with buffer D.
3. Inject 30 μl of the diluted protein sample, and start an isocratic elution from the column in buffer D at a maximum flow rate of 0.04 ml/min.
4. Collect SAXS data (1 frame/s) referring to the beamline specification and staff for the optimal data collection parameters.
5. Perform buffer subtraction and radial averaging according to beamline specification (most beamlines provide a series of SAXS spectra as .dat radial intensity profile files).

6. Perform another injection of 30 µl pure buffer D, and collect the spectra as above [20]. This set of data will be used as reference (blank).
7. Repeat the operations above (i.e., equilibration of column, run for protein, run for blank) with buffer E (without Ca^{2+}).
8. Using the software ScÅtter, open all curves of a dataset (e.g., all curves obtained from the elution of protein in the presence of Ca^{2+}).
9. From the analysis tab, use the I(0):Rg plot to determine the images to be analyzed for the protein profile. I(0) profile would look similar to the $A_{280\mathrm{nm}}$ profile in the HPLC chromatogram, and the subset of images of the peak where Rg remains constant indicates a monodisperse population of protein.
10. Average the curves from the monodisperse peak (with the parameters chosen, 40 images can be used).
11. In ScÅtter load the data from the same frames used in the analysis of protein, and average them.
12. Subtract the averaged blank curve from the averaged samples to obtain the final curve to be analyzed by the software.
13. Following the most updated instructions at the website bioisis.net/tutorials, calculate the Guinier radius, Porod volume, and flexibility plot using the scattering curve obtained.
14. For real-space reconstruction, once the initial analysis is performed, and for all curves displaying a folded state of protein (i.e., not those such as the apo-calumenin whose profile is pure scattering), calculate the P(r) function to define the radial distribution of mass (*see* **Note 9** and Fig. 3).

3.6 Surface Plasmon Resonance (SPR) Analysis

All SPR experiments are performed at 25 °C, and each data point is repeated in quadruplicate, so that final values will result from the average of four measurements.

1. Prime the Biacore T200 instrument with buffer G. Meanwhile take a new CM5 chip, and allow to stand 15 min at room temperature.
2. Dock the sensor chip in the instrument following the manufacturer's instructions.
3. Dilute one aliquot of protein to 10 mM with buffer G.
4. Use the regeneration scouting kit to determine the optimal pH for the immobilization of the protein on the chip. Prepare buffers at different pH, and dilute the protein in the buffers to a final concentration of 5 mM. Following the manufacturer's instructions, load the protein samples at different pH on the chip, and measure the affinity level (the number of response units (RUs) at the end of the binding stage). Select the pH

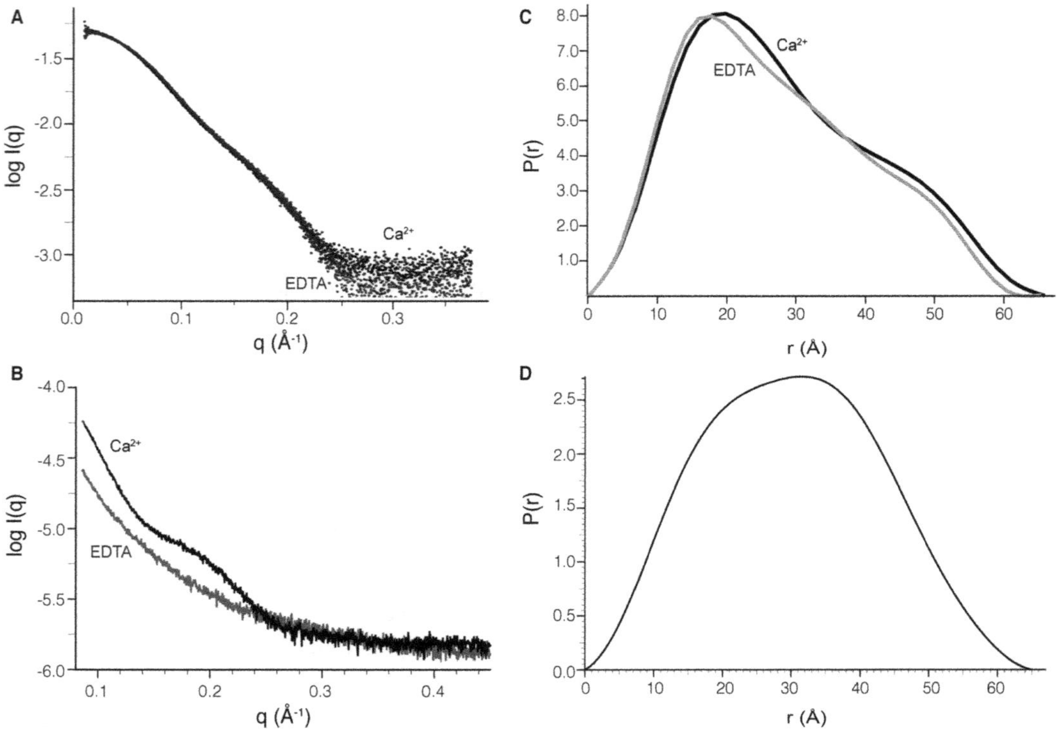

Fig. 3 SAXS profiles of calmodulin (**a**) and calumenin (**b**) in EGTA-buffered solution (gray) or in the presence of Ca^{2+} (black). Radial probability functions calculated for the two states of calmodulin (**c**) and for Ca^{2+}-bound calumenin (**d**)

giving the highest RU value for the immobilization state (*see* **Note 10**).

5. Proceed to the immobilization of the protein using the amine coupling kit according to the manufacturer's instruction. Use a dilute protein solution (5 mM) at the best pH identified in the previous stage, and immobilize the protein using circuit 2 of the sensor chip (circuit 1 will be used for the pure buffer as a reference). Measure the RUs as difference of response from channels 2 and 1. Ensure at least 2000 RUs are detected before proceeding to the amine-coupling immobilization phase (*see* **Note 11**).
6. Re-equilibrate the instrument using buffer H (*see* **Note 12**).
7. Design a program for metal binding using the following sequence:
   (a) Conditioning step in buffer H.
   (b) 3× startup washes in buffer H.
   (c) Metal binding step (60 s contact time, 30 μl/min, 30 s dissociation).

(d) Regeneration with 10 mM EGTA wash for 30 s.

(e) Re-equilibration with buffer H for 30 s.

(f) Repeat from metal binding step for all the metal concentrations to test.

8. Run the protocol for each of the following concentrations of Ca^{2+} in quadruplicate (0 μM, 15.6 μM, 31.25 μM, 62.5 μM, 125 μM, 250 μM, 500 μM, 1 mM).
9. Repeat the experiment with other divalent ions ($SrCl_2$, $MgCl_2$, $MnCl_2$, $CoCl_2$) (*see* **Note 13**).
10. Open the BiaEvaluation software, and select all curves of the experiment.
11. Perform an affinity analysis with a steady-state model for 1:1 binding (*see* **Note 14**). Briefly, from the top bar select Kinetics/ Affinity–Surface bound, evaluation in Single mode. Select curves resulting from the subtraction of the values of channel 1 (no protein) from channel 2 (protein immobilized) indicated as Fc = 2-1. Untick from the list curves to be discarded (e.g., if odd behavior or artifacts are spotted at visual inspection). Select Next to visualize blank subtracted chromatograms and then Affinity to show the response curve as function of Ca^{2+} concentration. Select Next, the desired model for binding, and fit the data by pressing Fit. A report will show the K_d values and statistics of fit. Repeat this operation for the other metals screened.

4 Notes

1. Alternatively any suitable *E. coli* expression vector can be used, but the cloning and expression strategy needs to be adapted accordingly. We found best results for the overexpression of these human genes in *Rosetta* (DE3) pLysS cells using autoinduction of the genes under lac promoter. In the case of calumenin in pOPIN-S (pET28 derivative), the protein could be harvested in large amounts also from leaky expression in Terrific Broth (TB) medium. Incidentally, the expression occurred also in static cultures, and we found the best results to occur in cells grown in TB to OD_{600nm} = 1.0 and left statically on the benchtop for 5 days before harvesting.
2. To minimize the impact on other activities, we found convenient to start the pre-culture on a Thursday night, so the large culture is left over the weekend when the demand for incubators is lower. Harvest is left for the beginning of the week allowing all the time for the whole purification.
3. In alternative to the cell disruptor, sonication is an efficient way of lysing bacteria. In this case, the slurry should be more

diluted (1:10 cell paste, lysis buffer ratio), and the procedure will need to be adapted to the machine specification. Best results are obtained by adding 10 μg/ml lysozyme and incubating 30 min in ice before the disruption. Usually 5 × 20 s pulses at medium power with 1-min pause in ice bath are sufficient for a good cell disruption. We found the best results to come from 10 min of pulsed disruption in a 50% ice-ethanol bath under constant agitation.

4. 1 ml HisTrap HP is sufficient for the purification of calumenin from 4 l of expression medium or that of calmodulin from 1 l of medium. Larger amounts of protein can be purified from higher quantities of bacteria by dividing the lysate in more than one affinity column; however, the quantity obtainable from this protocol yields 50–100 mg pure protein, which is sufficient for a large number of downstream characterization experiment such as those described in this work. Usually with this procedure, the protein is in excess of the resin capacity, so expect a large amount in the wash and 10% B elution fractions. However, the most concentrated and purest samples are visible in the 50% B and 100% B elution fractions.

5. Size-exclusion chromatography profiles of Ca^{2+}-binding proteins in the presence or absence of the ion can be very different. The Ca^{2+}-bound form of calmodulin has a lower retention volume than its apo-counterpart. Quite strikingly, the behavior is inverted for calumenin, which has a higher retention volume in the Ca^{2+}-bound state. Multi-angle light scattering (MALS) provides a shape-independent estimate of molecular weight which for both proteins does not differ in their apo- and ion-loaded form. This provides evidence that the binding of the ion alters the structure of the protein, without affecting its aggregation properties. The shift in the apparent molecular weight can be exploited during the purification stages, whereby in switching conformation, the desired protein has different retention volumes from contaminants of similar molecular weight. However, for the nature of these proteins, it is advisable to maintain them in their Ca^{2+}-bound state as this is more folded and resistant to degradation than the apo.

6. The experiments can be conducted also in the absence of a chelating agent with similar results. However, since in the presence of EGTA, the free concentration of Ca^{2+} is more finely tuneable, we suggest to use this method both in the CD titration and in SPR experiments.

7. As an alternative, one can prepare a larger stock of Ca^{2+}-binding protein, divide in aliquots of the same volume as the quartz cell, and add to each the desired amount of Ca^{2+} (chelated or free). Measurements can be performed every time on fresh

solutions prepared offline and loaded on the same cuvette, previously washed and decontaminated from precipitated sample from previous measurements. Despite requiring a larger amount of sample, this technique ensures that the protein measured at every step has not suffered radiation damage from prior irradiation (this could be a serious issue at synchrotrons where the intensity of the beam is much higher than for benchtop sources).

8. In the case of calumenin, since the folding process starts from a totally unfolded state, the highest variability peak did not have a fixed wavelength (i.e., 191 nm for the alpha helices such as in calmodulin), so it is necessary to spot another peak or use more complex indicators that provide an average variability of the various peaks. Alternatively, one can calculate the percentage of secondary structure and plot this value as a function of the increasing concentration of Ca^{2+}.
9. Despite the ease of use of programs such as ScÅtter and the ATSAS software suite, SAXS data analysis requires a high level of expertise. As this protocol provides generic guidance for the study of Ca-dependent protein folding, we remand the treatment of this part to more specialized publication and suggest anyone willing to undertake similar studies to contact beamline staff to discuss feasibility, plan preliminary experiments, and obtain guidance during data collection and, especially, data analysis.
10. For acidic proteins such as calmodulin$_{2-149}$ (pI = 4.09) and calumenin$_{68-315}$ (pI = 4.40), the pH range is extremely limited since the surface of the chip is positively charged at pH >3, but the protein is electrostatically attracted at pH < pI. For both proteins in this study, the optimal range is at pH 4.0 ÷ 4.5
11. As the SPR signal is proportional to the relative mass of the analyte that binds the ligand immobilized on the chip, when looking at ions or small molecules, the amount of protein captured needs to be sufficiently high. For proteins such as those described in this work, the maximum signal obtained upon binding is roughly 100 times lower than the immobilized RUs. Instruments such as Biacore T200 can work in this range, but lower immobilization rates will affect negatively signal/ noise ratios.
12. The proteins in the present study can be immobilized on a CM5 chip both in their Ca^{2+}-bound and in apo-form. In both cases, Ca^{2+}-dependent structural changes can be affected by the way proteins adhere and get tethered to the support upon amine coupling. For this reason, the maximum response to metal concentration is far from the ideal value described above. We tested both immobilization conditions: calumenin

immobilized in its Ca^{2+}-bound state responded slightly better than apo-immobilized protein (i.e., 50% and 40% of the ideal maximum response).

13. Although Ca^{2+} is the natural binding partner of such proteins, ions with similar charge/ionic radius ratios such as Sr^{2+}, Mg^{2+}, Mn^{2+}, and Co^{2+} can be studied. Calmodulin is known to partially restructure upon binding to Sr^{2+} [21]. The effect of other ions can be studied quickly and efficiently using SPR, minimizing the expense of protein sample. As in the CD experiments, these protocols can be easily and conveniently adapted to chelated metal solutions to provide more precise estimates of the binding constants.
14. Given the relatively low signal of these experiments, a number of artifacts (mostly buffer mismatch affecting refractive index) often impede a complete kinetic analysis (k_{on}, k_{off}). The affinity analysis, based on the equilibrium value, using a 1:1 model, is convenient and relatively straightforward. However, cooperativity is known to affect the binding of ions to these proteins, and a more detailed data analysis, such as the Hill plot described in the CD section, allows the description of more complex binding models even in the absence of a full kinetic investigation.

Acknowledgments

The authors would like to thank Rohanah Hussain, Tamas Javorfi, Robert Rambo, and Gemma Harris for expert help and advice.

References

1. Du X, Li Y, Xia YL et al (2016) Insights into protein-ligand interactions: mechanisms, models, and methods. Int J Mol Sci 17:144
2. Berridge MJ (1997) Elementary and global aspects of calcium signalling. J Physiol 499 (Pt 2):291–306
3. Mazzorana M, Hussain R, Sorensen T (2016) Ca-dependent folding of human calumenin. PLoS One 11:e0151547
4. Sahin E, Roberts CJ (2012) Size-exclusion chromatography with multi-angle light scattering for elucidating protein aggregation mechanisms. Methods Mol Biol 899:403–423
5. Scarlett G, Siligardi G, Kneale GG (2015) Circular dichroism for the analysis of protein-DNA interactions. Methods Mol Biol 1334:299–312
6. Drescher DG, Selvakumar D, Drescher MJ (2018) Analysis of protein interactions by surface Plasmon resonance. Adv Protein Chem Struct Biol 110:1–30
7. Rambo RP, Tainer JA (2010) Bridging the solution divide: comprehensive structural analyses of dynamic RNA, DNA, and protein assemblies by small-angle X-ray scattering. Curr Opin Struct Biol 20:128–137
8. Iida S, Potter JD (1986) Calcium binding to calmodulin. Cooperativity of the calcium-binding sites. J Biochem 99:1765–1772
9. Zhang M, Abrams C, Wang L et al (2012) Structural basis for calmodulin as a dynamic calcium sensor. Structure 20:911–923
10. Asiani KR, Williams H, Bird L et al (2016) SilE is an intrinsically disordered periplasmic "molecular sponge" involved in bacterial silver resistance. Mol Microbiol 101:731–742
11. Wright PE, Dyson HJ (2015) Intrinsically disordered proteins in cellular signalling and regulation. Nat Rev Mol Cell Biol 16:18–29

12. Ullah R, Shah MA, Tufail S et al (2016) Activity of the human rhinovirus 3C protease studied in various buffers, additives and detergents solutions for recombinant protein production. PLoS One 11:e0153436
13. Mossessova E, Lima CD (2000) Ulp1-SUMO crystal structure and genetic analysis reveal conserved interactions and a regulatory element essential for cell growth in yeast. Mol Cell 5:865–876
14. Berrow NS, Alderton D, Sainsbury S et al (2007) A versatile ligation-independent cloning method suitable for high-throughput expression screening applications. Nucleic Acids Res 35:e45
15. Berrow NS, Alderton D, Owens RJ (2009) The precise engineering of expression vectors using high-throughput In-Fusion PCR cloning. Methods Mol Biol 498:75–90
16. Studier FW (2005) Protein production by auto-induction in high density shaking cultures. Protein Expr Purif 41:207–234
17. Javorfi T, Hussain R, Myatt D et al (2010) Measuring circular dichroism in a capillary cell using the b23 synchrotron radiation CD beamline at diamond light source. Chirality 22 (Suppl 1):E149–E153
18. Bers DM, Patton CW, Nuccitelli R (1994) A practical guide to the preparation of Ca2+ buffers. Methods Cell Biol 40:3–29
19. Sreerama N, Woody RW (2000) Estimation of protein secondary structure from circular dichroism spectra: comparison of CONTIN, SELCON, and CDSSTR methods with an expanded reference set. Anal Biochem 287:252–260
20. Rambo RP, Tainer JA (2013) Accurate assessment of mass, models and resolution by small-angle scattering. Nature 496:477–481
21. Kursula P (2014) Crystallographic snapshots of initial steps in the collapse of the calmodulin central helix. Acta Crystallogr D Biol Crystallogr 70:24–30

Chapter 33

Measuring Calumenin Impact on ER-Calcium Depletion Using Transient Calumenin Overexpression and Silencing

Réginald Philippe and Olivier Mignen

Abstract

Calumenin is a secretory pathway protein regulating different endoplasmic reticulum (ER) proteins such as the sarco-endoplasmic reticulum calcium ATPase (SERCA) pumps. Combined with its diverse cellular distribution, its calcium-binding ability, and its interaction with proteins involved in calcium signaling, it is easy to speculate on future description of important roles of calumenin in calcium homeostasis in many cell types, as it was initially observed in muscle cells. In this chapter, we describe basic techniques to modulate calumenin expression and detect its impact on ER calcium content using classic transfection and Western blot techniques, as well as ER calcium measurement using microplate reader.

Key words Calumenin, ER calcium, Microplate reader, Transfection, Western blot, Ca^{2+} measurement

1 Introduction

Calumenin is a 48 kDa protein belonging to the CREC proteins together with endoplasmic reticulum (ER) Ca^{2+}-binding protein of 55 kDa (ERC-55), reticulocalbin, and Ca^{2+}-binding protein of 45 kDa (Cab45) [1, 2]. Calumenin is a calcium-binding protein with seven EF hands, including one pseudo-EF hand, and is mainly located in the ER but is also found in all the secretory pathway compartments and in the extracellular medium. Two isoforms were discovered followed by 13 new isoforms in 2014. Some of these isoforms reveals intriguing characteristics of calumenin as the lack of some EF-hand domains or new subcellular localization in the nucleus or cytosol [3].

The different localizations of calumenin isoforms indicate a wide range of cellular functions that still need to be identified. Calumenin was initially considered as a chaperone protein and later on proposed to be involved in protein folding and maturation [4–6].

Claus W. Heizmann (ed.), *Calcium-Binding Proteins of the EF-Hand Superfamily: From Basics to Medical Applications*, Methods in Molecular Biology, vol. 1929, https://doi.org/10.1007/978-1-4939-9030-6_33,

The best characterized functions of calumenin is the regulation of ER-located enzymes and its implication in the γ-carboxylation system [7, 8]. At the ER level, calumenin regulates ER stress by modulating the expression of some ER-stress proteins (like GRP78 or phosphorylated PERK) and proapoptotic (CHOP, p-JNK) or antiapoptotic (Bcl-2) proteins [4]. Calumenin is involved in calcification processes and in the regulation of γ-carboxylated glutamate (Gla residues) synthesis by the inhibition of γ-carboxylation [8–12]. Extracellular calumenin is also able to reorganize cytoskeleton and modulate cell cycle by decreasing septin 1 expression [13]. The implication of calumenin in extracellular matrix formation and calcification is probably not limited to thrombosis and atherosclerosis [12, 14, 15] but also to chondrocyte development, fracture resorption, or lung fibrosis [16–18].

Implication of calumenin was also described in cell migration and extracellular matrix production and maintenance. These assigned roles for calumenin may be directly linked to an altered expression in cancer cells and often associated with metastasis [2]. Its role in migration and invasion capacity of cancer cells could be related to cytoskeleton remodeling, filipod formation, ERK1/2 pathway regulation, and extracellular matrix protein fibulin-1 protection against metalloprotease degradation, as well as a role in cell cycle [3, 10, 13, 19–21].

Calumenin also regulates proteins implicated in calcium signaling such as the ryanodine receptor (RyR) and the sarco-endoplasmic reticulum calcium ATPase (SERCA), a calcium pump implicated in SR-Ca^{2+} refilling [6, 22]. As demonstrated by us and other groups, calumenin is clearly involved in calcium homeostasis and signaling regulation, probably through its capacity to bind calcium and to regulate ER calcium transport and buffer proteins [6, 22]. Others and we clearly need to keep deciphering the implication of calumenin in calcium homeostasis and its close association with different human pathologies.

Therefore, this chapter presents a method to evaluate calumenin expression by Western blot analysis and its impact on ER calcium signaling.

Determination of full-length calumenin expression by Western blot analysis is quite difficult mainly because of the relative specificity of available antibodies. Even if most antibodies are presented as specific for calumenin, the low reproducibility of detection and the lack of specificity can be a real constraint for the study of calumenin. Here we present a reliable, specific, and reproducible method to detect calumenin expression by Western blot. This method was validated by the modulation of calumenin expression using siRNA or transient overexpression. We also present a method to determine the impact of calumenin expression on ER-Ca^{2+} store signaling. Using the Fura-2 calcium sensible dye and a microplate reader, we explain how to rapidly and simply measure the modulation of

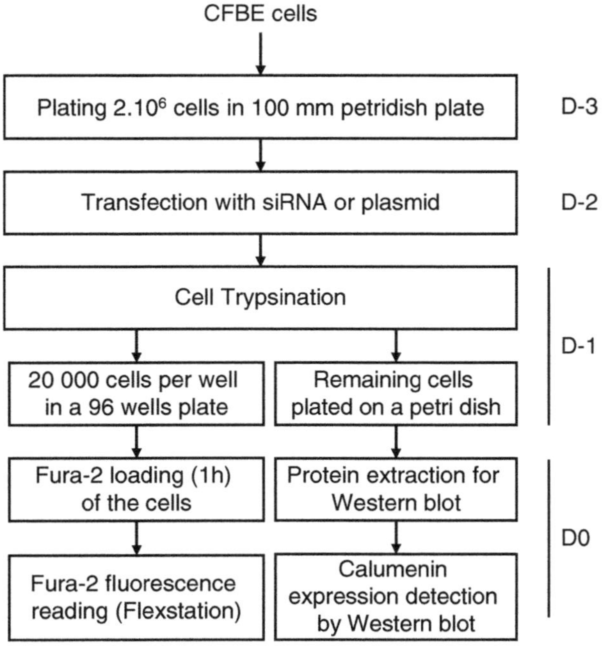

Fig. 1 Protocol summary for the evaluation of the impact of calumenin expression on ER-Ca^{2+} release. Cytosolic Ca^{2+} measurement using Fura-2 and Western blot for the detection of calumenin expression modulation may be performed from one single transfection experiment. Required amount of cells is seeded on a 96-well plate, and remaining cells are plated again on a petri dish for further protein extraction. D0 corresponds to the day of experiment as D-1 corresponds to 1 day before experiment (D-2 to 2 days before experiment, etc.)

ER-Ca^{2+} store release by calumenin (Fig. 1). This technique using a microplate reader is really efficient to obtain reproducible calcium signals and to test multiple experimental conditions.

2 Materials

2.1 Cells

1. Cystic fibrosis human bronchial epithelial (CFBE) cell lines.
2. Corning CellBIND surface flasks for tissue culture.
3. Petri dishes or 96-well plates coated with fibronectin (10 μg/ml) and BSA (100 μg/ml) in LHC basal medium at least one night before use.
4. DMEM completed with 10% SVF, penicillin, and streptomycin.

2.2 siRNAs and Plasmids

1. siRNA pool of 3 siRNAs of 19–25 nt directed against calumenin (siCalumenin) and a pool of scrambled siRNA that do not target specific cellular target (siControle). Both obtained from

SantaCruz (references, sc-60320 and sc-37007, respectively) (*see* **Note 1**).

2. N-terminal MycDDK-tagged calumenin (variant 1) transcript cDNA is inserted in pCMV-entry vector (empty vector). Both obtained from Origene (references, RC200589 and PS100001, respectively).

2.3 Transfection

1. Lipofectamine 2000® reagent (Invitrogen) for siRNA transfection (*see* **Note 2**).
2. Opti-MEM medium for mixing Lipofectamine 2000® and siRNA.
3. LipoD293™ DNA (Ver. II) (Tebu-bio) for plasmid transfection (*see* **Note 3**).

2.4 Western Blot

1. PBS is 137 mM NaCl, 2.7 mM KCl, 10 mM Na_2HPO_4, 1.8 mM KH_2PO_4, pH adjusted to 7.4.
2. TBS-Tween 20 (TBST) by addition of 0.1% (v/v) of Tween 20 in TBS buffer (150 mM NaCl, 50 mM Tris–HCl, pH adjusted to 7.4).
3. Lysis buffer: 20 mM Tris–HCl buffer (pH adjusted at 7.5), 150 mM NaCl, 1 mM EDTA, 1 mM EGTA, 1% (v/v) of Triton X-100, 2.5 mM Na^+ pyrophosphate, 1 mM glycerophosphate, 1 mM orthovanadate, 1 μg/ml of leupeptine. Stored at 4 °C.
4. Complete tablet of antiproteases added to 10 ml of lysis buffer just before use.
5. 5× loading buffer (for a 50 ml stock): 15 ml of 1 M Tris–HCl buffer (pH adjusted at 6.8), 5 g of SDS, 25 ml of glycerol, 10 ml of β-mercaptoethanol, few bromophenol blue for coloration. Aliquoted and stored at −20 °C.
6. Precast 10% polyacrylamide gels.
7. Membranes for Western blot transfer: Immobilon-P PVDF membranes with a pore size of 0.45 μm.
8. Blocking buffer: TBST with 5% dry milk.
9. Substrate for chemiluminescent horseradish peroxidase (HRP) reaction revelation: Luminata Forte (Millipore).
10. Chemiluminescent reaction acquisition on Chemi-Smart 5100 acquisition system (Vilber-Lourmat) with Chemi-Capt 5000 software.

2.5 Antibodies

1. Anti-calumenin primary antibody for Western blotting: rabbit antibody from Abcam (clone EPR9075) (*see* **Note 4**).
2. GAPDH used as a loading control, with an anti-GAPDH mouse antibody.

3. Hrp-coupled secondary antibodies: donkey anti-rabbit IgG antibody and donkey anti-mouse IgG antibody.

2.6 ER-Calcium Release Measurements

1. 96-Well black plates with clear bottom.
2. HEPES-buffered solutions for Ca^{2+} measurement:
   (a) 1.8 mM Ca^{2+}-containing solution: 137 mM NaCl, 4.7 mM KCl, 0.56 mM $MgCl_2$, 1 mM Na_2HPO_4, 10 mM HEPES, 1.8 mM of $CaCl_2$, 5.5 mM glucose; pH is adjusted to 7.4 with NaOH. Sterile filter and store at 4 °C.
   (b) Ca^{2+}-free solution: 137 mM NaCl, 4.7 mM KCl, 0.56 mM $MgCl_2$, 1 mM Na_2HPO_4, 10 mM HEPES, 100 μM EGTA, 5.5 mM glucose; pH is adjusted to 7.4 with NaOH. Sterile filter and store at 4 °C.
3. SERCA Ca^{2+} pump irreversible inhibitor thapsigargin, used to release ER-Ca^{2+} stores, dissolved in DMSO.
4. Commercial Fura-2QBT reagent containing a masking dye for extracellular Fura-2 used (Molecular Devices). Dye resuspended in 1.8 mM Ca^{2+} solution according to manufacturer volume indication and then aliquoted and stored at −20 °C.
5. ER-Ca^{2+} release measurements performed on a multiplate reader (Flexstation 3, Molecular Devices) (*see* **Note 5**).
6. SoftMax Pro software (Molecular Devices) used for experiment execution and data acquisition.
7. Microsoft Excel and GraphPad Prism used for Ca^{2+} measurement analysis.

3 Methods

3.1 Cell Transfection

3.1.1 Cell Preparation

Cells are seeded on 100 mm diameter petri dishes to obtain a cell confluency of approximatively 70–80% at day of transfection (*see* **Note 6**).

3.1.2 Calumenin Expression Knockdown Using siRNA

1. Cells are transfected 48 h before experiments with 500 pmol of the three anti-calumenin siRNA pool or control siRNA for a 100 mm petri dish.
2. Transfection is performed using Lipofectamine 2000® reagent according to manufacturer protocol. Briefly, 500 pmol of siRNA and 30 μl of Lipofectamine 2000® reagent are mixed in Opti-MEM medium and incubated for 20 min before addition to cells. Transfection medium should be changed to fresh medium 6 h after transfection.

3.1.3 Calumenin Overexpression Using Expression Plasmids

1. Cells are transfected with 5 μg of the pCMV-entry calumenin MycDDK expression vector or empty pCMV-entry vector.
2. Transfection is performed using LipoD293™ according to manufacturer protocol. Briefly, 5 μg of vector and 15 μl of LipoD293™ reagent are mixed in serum-free DMEM and incubated for 15 min before addition to cells. Medium is not changed.

3.1.4 Cell Seeding on 96-Well Plate

1. The day before the experiment (i.e., 24 h after transfection), transfected cells are trypsinized, centrifuged, and counted.
2. 20,000 cells are seeded in each well of a 96-well fibronectin-coated plate in fresh medium.
3. Remaining cells are plated again on 100 mm plates and harvested the next day for protein extraction and Western blot.

3.2 Evaluation of Calumenin Expression Modulation by Western Blot

Cell lysates are realized 48 h after transfection. All the following steps are performed on ice with ice-cold solutions.

3.2.1 Sample Preparation

1. Cells are rinsed twice in PBS and resuspended in PBS using a cell scraper.
2. Cells are pelleted by centrifugation at 300 × *g* for 5 min at 4 °C.
3. Pellets are then resuspended in the lysis buffer and lysed during at least 30 min (*see* **Note 7**).
4. Cellular debris are eliminated by a centrifugation at 16,000 × *g* during 12 min at 4 °C.
5. Supernatants are kept and stored at −80 °C if not used immediately.
6. Protein concentration is determined using Lowry method [23].

3.2.2 Western Blot

1. 50 μg of protein extracts is mixed with 5× Laemmli buffer containing SDS and β-mercaptoethanol and heated 15 min at 37 °C for protein denaturation.
2. Molecular-weight-dependent protein separation is performed in precast gel 10% polyacrylamide during 1 h at 200 V.
3. Protein transfer from the gel to a PVDF membrane is performed during 1 h 30 at 400 mA using a wet transfer system maintained at a cold temperature with ice.
4. Membranes are incubated for 1 h with blocking buffer to saturate nonspecific fixation sites.
5. PVDF membrane is incubated with diluted primary antibody directed against calumenin (1:1000 antibody dilution) or GAPDH (1:100,000 antibody dilution) overnight in the blocking buffer at 4 °C.

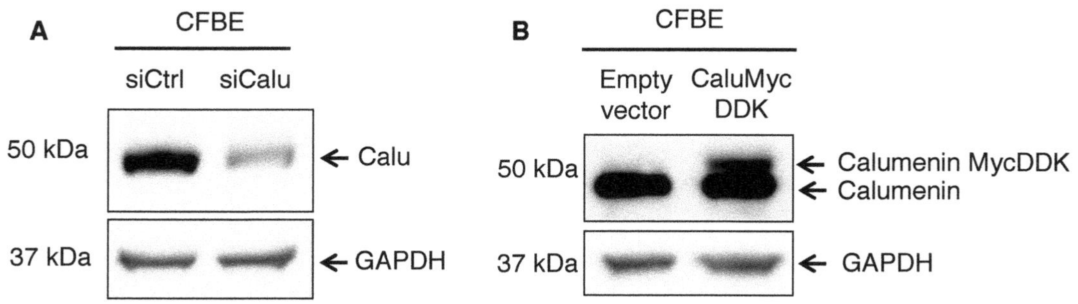

Fig. 2 (**a**) Typical Western blot of cell lysates from CFBE cells transfected with a siRNA directed against calumenin (siCalu) to knock down calumenin expression or CFBE cells transfected with a control siRNA (siCtrl). (**b**) Typical Western blot of cell lysates from cells overexpressing calumenin overexpression following a transient transfection with a MycDDk-calumenin plasmid. In both cases, calumenin is observed at the expected molecular weight of 48 kDa. The band observed at 50 kDa corresponds to MycDDK-tagged calumenin (4 kDa MycDDK peptide). Transfection with the siRNA directed against calumenin strongly decreases this protein expression (about 70%)

6. After four washes of 5 min in TBST, membranes are incubated for 1 h with secondary antibody diluted (1:10,000 dilution) in blocking buffer and then washed again four times during 5 min with the TBST solution.
7. Protein revelation is performed by measuring peroxidase activity following a 5-min incubation step with the Luminata Forte reagent. Luminescence is then measured using the Chemi-Smart 5100 acquisition system. Calumenin should have a size of about 48 kDa and calumenin MycDDK a size of 52 kDa (Fig. 2).
8. Images are analyzed to evaluate protein expression by densitometry using ImageJ software.

3.3 Thapsigargin-Mediated ER Calcium Depletion Measurement

3.3.1 Cell Loading with Fura-2

1. Cell culture medium is discarded from the well and replaced with the Fura-2 containing 1.8 mM Ca^{2+} solution.
2. Cells are incubated with 50 μl in the Fura-2 solution 1 h at 37 °C and in 5% CO_2 atmosphere.

3.3.2 Parameters Measured

1. F340: Fluorescence intensity measured at 510 nm when cells are excited in a 340 nM, corresponding to the Fura-2 dye quantity bound to Ca^{2+}.
2. F380: Fluorescence intensity measured at 510 nm when cells are excited in a 380 nM, corresponding to Fura-2 dye quantity unbound to Ca^{2+}.

3. F340/F380: Ratio of F340/F380 corresponding proportion of Fura-2 bound to Ca^{2+} compared to unbound dye (*see* **Note 8**).

3.3.3 FlexStation Drug Plate Preparation and Acquisition Programing

1. Thapsigargin is dissolved in a Ca^{2+}-free solution at 10 μM and added in one column of a 96-well drug plate.
2. FlexStation setup is done using the dedicated SoftMax Pro software as follows:
   (a) Selected reading mode on the FlexStation: FLEX mode.
   (b) Wavelengths: Excitation 1 (Ex1), 340 nm; Emission1 (Em1), 510 nm/Excitation2 (Ex2), 380 nm; and Emission2 (Em2), 510 nm.
   (c) Software is settled to calculate the ratio between fluorescence intensity of Em1 (F340) and fluorescence intensity of Em2 (F380). This ratio is called F340/F380.
   (d) Total acquisition time: 700 s (*see* **Note 9**).
   (e) Acquisition frequency: One acquisition every 4 s at each excitation wavelength (Ex1/Em1 and Ex2/Em2).
   (f) Injection time of thapsigargin: 100 s (*see* **Note 9**).
   (g) Injection volume: 20 μl of the thapsigargin solution in the selected wells.

3.3.4 ER-Calcium Depletion Measurement

1. After 1-h incubation of cells in Fura-2 solution, one end point acquisition is performed in order to evaluate resting Ca^{2+} concentration and dye loading by determination of the F340//F380 fluorescence ratio.
2. The Fura-2 solution is then removed, and cells are washed with the Ca^{2+}-free solution and recovered with 80 μl of the Ca^{2+}-free solution.
3. Acquisition is run in the "FLEX mode" for 700 s.
4. 20 μl of the thapsigargin solution is injected at 100 s after starting the acquisition (*see* **Note 9**) to obtain a final 2 μM thapsigargin concentration.
5. Acquisition should end (700 s) (*see* **Note 9**) when F340/F380 ratio nearly reaches its initial value.

3.3.5 Data Analysis

1. Data (F340, F380, and F340/F380) are imported on a spreadsheet.
2. Baseline value (F0) of the F340/F380 ratio is obtained by averaging ten values of this ratio measured just before thapsigargin injection (*see* **Note 10**).
3. F340/F380 is fitted to the baseline as follows: (F-F0)/F0 (Fig. 3).

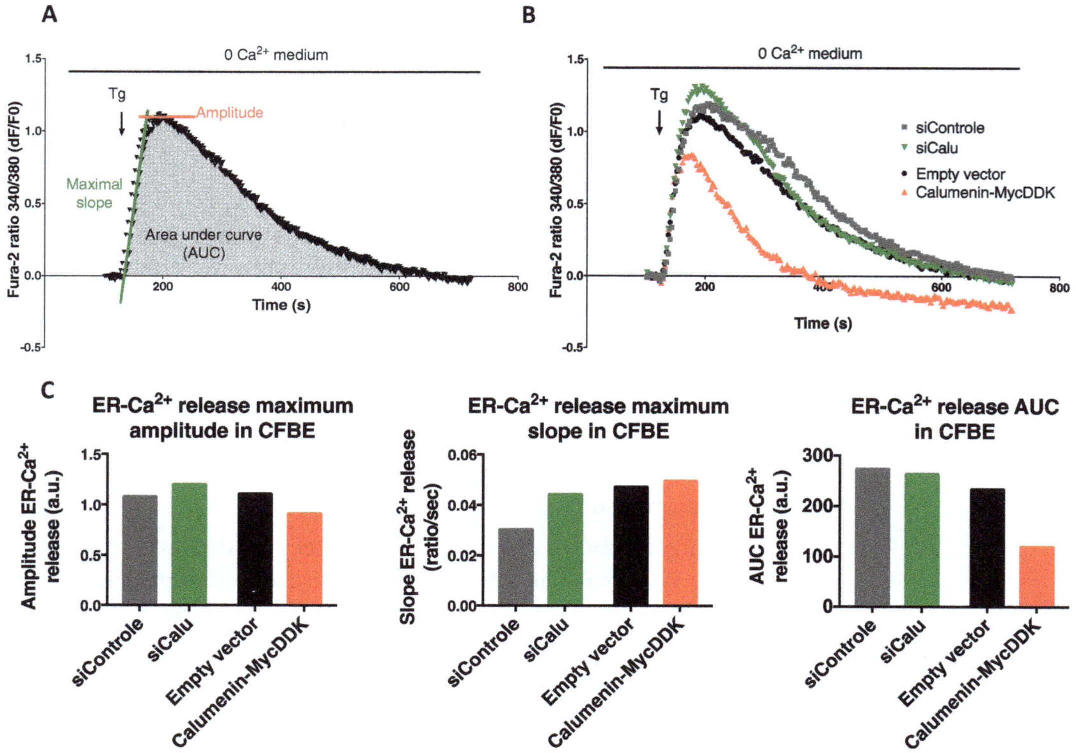

Fig. 3 Measurement of ER-Ca^{2+} store depletion in cells loaded with Fura-2 using thapsigargin (Tg), a SERCA pump inhibitor. (**a**) Representative curve of cytosolic Ca^{2+} concentration changes evaluated by measuring over time the normalized F340/F380 fluorescence ratio following the ER store depletion after Tg addition. Estimated parameters are the ER-Ca^{2+} release amplitude, corresponding to the maximum ratio value, and the maximal slope, corresponding to the maximum value of the slope of the Ca^{2+} raise following Tg addition. The area under the curve (AUC) is also calculated to estimate ER-Ca^{2+} release. (**b**) Representative traces obtained for cells transfected with control siRNA (in gray), a siRNA targeting calumenin (in green), an empty vector (in black), or a MycDDk calumenin vector (in red). (**c**) Histograms of corresponding values for the different parameters obtained from experiments presented in panel **b**

4. Maximum amplitude is determined by measuring the maximal value of the peak and represents the maximum cytosolic Ca^{2+} level observed during ER-Ca^{2+} store liberation (*see* **Note 11**).
5. Maximal slope is measured after determining the derivative of (F-F0)/F0 curve. The maximal value of this derivative (maximal slope) corresponds to the fastest leaking of Ca^{2+} stores from the stores (*see* **Note 11**).
6. Area under curve (AUC) can also be measured and is representative of the total Ca^{2+} release from the ER (*see* **Note 11**).

4 Notes

1. This siRNA pool was selected regarding its high and reproducible calumenin expression extinction power (around 80%) as observed by Western blot.
2. Other transfection reagents may be used, but Lipofectamine 2000® is preferentially used for siRNA transfection due to the high transfection efficiency and low cytotoxic effects observed in CFBE cells.
3. Other transfection reagents may also be selected depending on experimental cell type. LipoD293™ is used here to transfect CFBE cells. Many transfection reagents with different DNA/reagent ratios were tested with generally low transfection rates or high cytotoxicity. LipoD293™ reagent allows the best transfection results with a low cytotoxicity.
4. Many antibodies were tested to detect calumenin expression in the Western blot. Most of them were able to detect calumenin and its expression modulation but with very low reproducibility. This Abcam EPR9075 clone antibody gives the best results and reproducibility over time.
5. Measurement of calcium concentration changes may also be realized using an imaging, video-microscopy approach using Fura-2 or other calcium-sensitive dyes. Microplate reader was used here regarding the higher number of experimental conditions that can be quickly and simultaneously tested using a plate reader such as the FlexStation and also for the high level of reproducibility of measured calcium signals.
6. Typical seeding for CFBE cells is 2.10^6 cells 48 h before transfection or 4.10^6 cells 24 h before. These values should be adjusted considering cell type and growth speed.
7. An additional step of sonication may be added before final centrifugation, particularly for viscous samples.
8. Fura-2 is a ratiometric fluorescent dye, meaning there should be no impact of photobleaching and loss of fluorescence over time on the measured ratio.
9. Experimental timing for ER-Ca^{2+} kinetic evaluation is adjusted specifically to the cell type used. Timing of thapsigargin injection should be adjusted for each cell type to setup an injection when the fluorescence signal baseline in Ca^{2+}-free solution stabilizes, and total recording time should be long enough to obtain a return of fluorescence signal to baseline allowing a measurement of the area under curve if necessary.
10. F340/F380 ratio should be stable at this step.

11. An important point to note is that all amplitude, slope, and AUC of this Ca^{2+}-release curve are only representative of ER-Ca^{2+}. In fact, all these values also depend on activity of other Ca^{2+} transport actor's activities, like plasma membrane Ca^{2+} pumps (PMCA) or Ca^{2+} exchangers as well as mitochondrial buffering capacity.

References

1. Honoré B, Vorum H (2000) The CREC family, a novel family of multiple EF-hand, low--affinity Ca2+-binding proteins localised to the secretory pathway of mammalian cells. FEBS Lett 466:11–18
2. Honoré B (2009) The rapidly expanding CREC protein family: members, localization, function, and role in disease. Bioessays 31:262–277. https://doi.org/10.1002/bies.200800186
3. Feng H, Chen L, Wang Q et al (2013) Calumenin-15 facilitates filopodia formation by promoting TGF-b superfamily cytokine GDF-15 transcription. Cell Death Dis 4:e870. https://doi.org/10.1038/cddis.2013.403
4. Lee JH, Kwon EJ, Kim DH (2013) Calumenin has a role in the alleviation of ER stress in neonatal rat cardiomyocytes. Biochem Biophys Res Commun 439:327–332. https://doi.org/10.1016/j.bbrc.2013.08.087
5. Tripathi R, Benz N, Culleton B et al (2014) Biophysical characterisation of calumenin as a charged F508del-CFTR folding modulator. PLoS One 9:e104970. https://doi.org/10.1371/journal.pone.0104970
6. Philippe R, Antigny F, Buscaglia P et al (2017) Calumenin contributes to ER-Ca2+homeostasis in bronchial epithelial cells expressing WT and F508del mutated CFTR and to F508del-CFTR retention. Cell Calcium 62:47–59. https://doi.org/10.1016/j.ceca.2017.01.011
7. Wallin R, Hutson SM, Cain D et al (2001) A molecular mechanism for genetic warfarin resistance in the rat. FASEB J 15:2542–2544. https://doi.org/10.1096/fj.01-0337fje
8. Wajih N, Sane DC, Hutson SM, Wallin R (2004) The inhibitory effect of calumenin on the vitamin K-dependent gamma-carboxylation system: characterization of the system in normal and warfarin-resistant rats. J Biol Chem 279:25276–25283. https://doi.org/10.1074/jbc.M401645200
9. Hansen GAW, Vorum H, Jacobsen C, Honoré B (2009) Calumenin but not reticulocalbin forms a Ca2+-dependent complex with thrombospondin-1. A potential role in haemostasis and thrombosis. Mol Cell Biochem 320:25–33. https://doi.org/10.1007/s11010-008-9895-1
10. Hansen GAW, Ludvigsen M, Jacobsen C et al (2015) Fibulin-1C, C1 esterase inhibitor and glucose regulated protein 75 interact with the CREC proteins, calumenin and reticulocalbin. PLoS One 10:e0132283. https://doi.org/10.1371/journal.pone.0132283
11. Vorum H, Jacobsen C, Honoré B (2000) Calumenin interacts with serum amyloid P component. FEBS Lett 465:129–134. https://doi.org/10.1016/S0014-5793(99)01734-2
12. Jover E, Marín F, Quintana M et al (2015) CALU polymorphism A29809G affects calumenin availability involving vascular calcification. J Mol Cell Cardiol 82:218–227. https://doi.org/10.1016/j.yjmcc.2015.03.015
13. Østergaard M, Hansen GAW, Vorum H, Honoré B (2006) Proteomic profiling of fibroblasts reveals a modulating effect of extracellular calumenin on the organization of the actin cytoskeleton. Proteomics 6:3509–3519. https://doi.org/10.1002/pmic.200500686
14. Hernández-Romero D, Ruiz-Nodar JM, Marn F et al (2010) CALU A29809G polymorphism in coronary atherothrombosis: implications for coronary calcification and prognosis. Ann Med 42:439–446. https://doi.org/10.3109/07853890.2010.499131
15. Coppinger JA, Cagney G, Toomey S et al (2004) Characterization of the proteins released from activated platelets leads to localization of novel platelet proteins in human atherosclerotic lesions. Blood 103:2096–2104. https://doi.org/10.1182/blood-2003-08-2804
16. Wilson R, Norris EL, Brachvogel B et al (2012) Changes in the chondrocyte and extracellular matrix proteome during post-natal mouse cartilage development. Mol Cell Proteomics 11:M111.014159. https://doi.org/10.1074/mcp.M111.014159

17. Fietta A, Bardoni A, Salvini R et al (2006) Analysis of bronchoalveolar lavage fluid proteome from systemic sclerosis patients with or without functional, clinical and radiological signs of lung fibrosis. Arthritis Res Ther 8: R160. https://doi.org/10.1186/ar2067

18. Nakazawa T, Nakajima A, Seki N et al (2004) Gene expression of periostin in the early stage of fracture healing detected by cDNA microarray analysis. J Orthop Res 22:520–525. https://doi.org/10.1016/j.orthres.2003.10.007

19. Wang Q, Shen B, Chen L et al (2015) Extracellular calumenin suppresses ERK1/2 signaling and cell migration by protecting fibulin-1 from MMP-13-mediated proteolysis. Oncogene 34:1006–1018. https://doi.org/10.1038/onc.2014.52

20. Zheng P, Wang Q, Teng J, Chen J (2015) Calumenin and fibulin-1 on tumor metastasis: implications for pharmacology. Pharmacol Res 99:11–15. https://doi.org/10.1016/j.phrs.2015.05.001

21. Nagano K, Imai S, Zhao X et al (2015) Identification and evaluation of metastasis-related proteins, oxysterol binding protein-like 5 and calumenin, in lung tumors. Int J Oncol 47:195–203. https://doi.org/10.3892/ijo.2015.3000

22. Sahoo SK, Kim DH (2010) Characterization of calumenin in mouse heart. BMB Rep 43:158–163

23. Lowry OH, Rosebrough NJ, Farr AL, Randall RJ (1951) Protein measurement with the Folin-phenol reagent. J Biol Chem 193 (1):265–275. https://doi.org/10.1016/0304-3894(92)87011-4

Chapter 34

Secretagogin Purification and Quality Control Strategies for Biophysical and Cell Biological Studies

Anand Kumar Sharma, Radhika Khandelwal, and Yogendra Sharma

Abstract

Secretagogin (SCGN) has recently gained attention due to its modulatory effect on insulin/CRH secretion and function. However, a large pool of speculated SCGN functions remains unexplored. A major deficiency is the lack of knowledge about the biological functions of extracellular SCGN. We here describe convenient methods for the scalable production of His-tagged and untagged mouse SCGN. The protocol is optimized to remove endotoxins, and thus the protein is suited for biological applications such as cell culture treatment or animal injections. We also outline expedient methods to check the purity of SCGN preparation for biological applications.

Key words Secretagogin, Protein purification, Calcium sensor protein, His-tag purification, Chromatography, Circular dichroism

1 Introduction

Eighteen years ago, Wagner et al. reported the cloning of a novel calcium-binding protein enriched in pancreatic β-cells and named it as secretagogin (SCGN) [1]. Although the authors' reasoning of nomenclature remains ambiguous, today the name can be justified in two ways, i.e., (a) SCGN has a secretagogue-like activity in pancreatic β-cells and (b) SCGN is a secreted protein. Over the past decade, secretagogin (SCGN) gained attention due to its property to modulate secretion of two important hormones, namely, insulin [1–3] and CRH [4]. SCGN belongs to the class of six EF hand containing Ca^{2+}-binding proteins. Owing to large sequence similarity, SCGN is also known as Calbindin-like (CALBL) protein. Like Calbindin-D28k, SCGN acts as calcium buffer as well as calcium sensor protein [5]. In the pancreatic β-cells, SCGN enhances the second phase of glucose-stimulated

Anand Kumar Sharma and Radhika Khandelwal contributed equally in writing this manual.

Claus W. Heizmann (ed.), *Calcium-Binding Proteins of the EF-Hand Superfamily: From Basics to Medical Applications*, Methods in Molecular Biology, vol. 1929, https://doi.org/10.1007/978-1-4939-9030-6_34,

insulin secretion (GSIS) by inducing cytoskeleton remodeling to favor focal adhesion signaling [3]. In addition, SCGN's interaction with SNAP proteins [5] seems to be important in modulating CRH and insulin secretion and suggests a possible conservation of function (i.e., in modulating secretion of different hormones) in other SCGN-expressing neuroendocrine cells such as pituitary gland.

In the recent years, SCGN has been implicated in diabetes and has emerged as a potential biomarker or therapeutic lead against diabetes [6–8]. Besides these established intracellular functions, emerging studies suggest that even plasma/extracellular SCGN could be a functional protein as reflected in the altered plasma level with changing the glycemic status of the body [8] and an anticipated correlation of serum SCGN levels with insulin secretion [7; preprint manuscript on bioRxiv]. Nonetheless, the identification of the extracellular functions of SCGN remains unaccomplished.9

To study the unknown functions and biochemical basis of SCGN functioning and to decipher the functions of extracellular SCGN in cellular and animal models, establishing a convenient protocol for efficient and sterile preparation of bioactive and non-toxic recombinant SCGN is a precondition. In addition, to suit a specific purpose, one may prefer to prepare a suitably tagged protein with unaltered properties or an untagged protein. In this guide, we describe a comprehensive methodology for the heterologous overexpression and convenient purification of recombinant mammalian SCGN. We provide three different protocols for purification of bioactive mouse SCGN: (1) refolded tagged protein from inclusion bodies using His-tag affinity chromatography [adapted from Ref.], (2) soluble tagged protein using His-tag affinity chromatography [adapted from Ref. 10], and (3) untagged protein from soluble fraction using ion exchange chromatography followed by size-exclusion chromatography. All methodologies yield SCGN with similar biochemical characteristics and can be adapted to suit a specific purpose. Although the described protocol is for the purification of mouse SCGN, considering the large structural similarity between mouse and human SCGN (Fig. 1), a similar strategy can be adapted for the purification of human SCGN as well. We also describe rapid methods to check SCGN purity, biological applicability, and Ca^{2+}-binding features of SCGN.

2 Materials

2.1 Instruments

1. *Temperature-controlled orbital shaker*: An orbital shaker incubator with temperature and rotation speed control features is suitable such as Forma Orbital Shaker (Thermo Scientific) or Innova43 incubator Shaker Series (New Brunswick Scientific Co.).

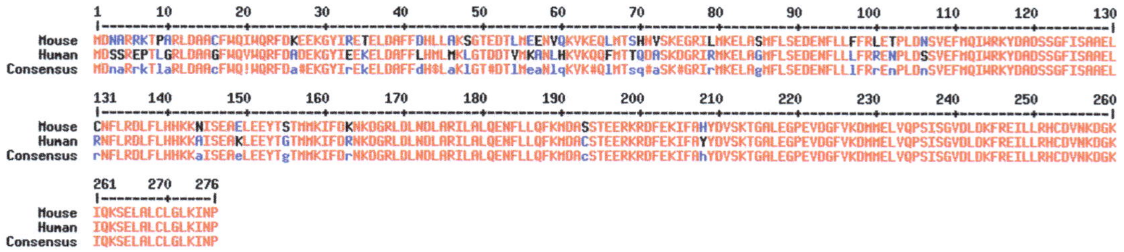

Fig. 1 Sequence alignment of mouse and human SCGN demonstrating conserved sequence similarity. Alignment was performed by using Multalign software [11]

2. *Sonicator*: We use Vibra-Cell VCX750 (Sonics & Materials Inc.) sonicator at 20% efficiency. The sonication efficiency can be adjusted accordingly if using a different instrument.
3. *Centrifuge*: Centrifuge (a floor-top model for large-scale preparation or a table-top model for small-scale preparation) with temperature and speed control feature is required. We use Avanti J-30I Centrifuge (Beckman Coulter) floor top model.
4. *SDS-PAGE gel casting apparatus*: A SDS-PAGE casting/running system (such as Bio-Rad Mini-PROTEAN Tetra cell) is required. Prepare 12% SDS-PAGE gel (follow Cold Spring Harbor Laboratory protocol) and keep TGS running buffer ready.
5. *FPLC workstation with the Superdex 75 prepacked column*: For high-resolution size-exclusion chromatography, an HPLC/FPLC system with UV/Quadtech detector (such as Bio-Rad BioLogic Duo-Flow) is used as final purification step. To perform size-exclusion, a prepacked Superdex 75 column (HiLoad 16/60, prep grade; GE Wipro) should be attached to FPLC workstation pump.
6. *Circular dichroism spectropolarimeter*: Our lab uses Jasco J-815 or ChiraScan CD spectrophotometers. Any other instrument with similar resolution and signal/noise ratio is appropriate to record CD spectra.
7. *Fluorescence spectroscopy*: High-sensitivity fluorimeter (such as Hitachi F-7000) is required.
8. *Cuvettes*: Quartz cuvettes for fluorescence (1 cm path length; 1 mL capacity), near-UV CD (1 cm path length; 3 mL capacity) and far-UV CD (0.2 mm path length; 60 μL capacity) are required.

2.2 Resin/Column Packing

1. *Ni-NTA gravity flow chromatography column:* We recommend using fast-flow, high-capacity resins such as GE Healthcare Ni-Sepharose 6 Fast Flow Ni-NTA beads. Pack 10 mL of slurry in a 20 mL capacity glass/polypropylene chromatography column (*see* **Note 1**).

2. *Q-Sepharose and Phenyl-Sepharose gravity flow chromatography column*: Pack 10 mL slurry of GE Healthcare High-Performance Q-Sepharose or Phenyl-Sepharose beads as described above (*see* **Note 2**).
3. *Chelex-100 resin*: Pack a Chelex-100 column with 10 mL of Bio-Rad Chelex-100 resin (50–100 mesh sodium form) slurry. Decalcify all buffers to be used for ion-dependent studies by passing through Chelex-100 column and store in plastic bottles.

2.3 BL21(DE3) Ultra-Competent Cells Transformed with pET21b Expression Vector with Mouse SCGN Coding Sequence

Clone mouse SCGN ORF into pET21b (with and without His-tag) bacterial expression vector, and confirm the insert by plasmid DNA sequencing. Transform the confirmed clone in a suitable expression host (such as BL21(DE3)) which is further used for large-scale recombinant protein purification (*see* **Note 3**).

2.4 Media

1. *LB broth*: To prepare 1 L of LB broth, add 10 g tryptone, 5 g yeast extract, and 10 g NaCl in 800 mL distilled water. After dissolving all components, make up the volume to 1000 mL by adding required volume of distilled water. Pour in a 2 L glass flask, and seal it tightly with a cotton plug. Autoclave and store at room temperature until needed.
2. *M9 minimal media*: To prepare M9 minimal media, make the following solutions:
   (a) 10× M9 salt solution—For 100 mL, add 12.8 g $Na_2HPO_4.7H_2O$, 3 g KH_2PO_4, 0.5 g NaCl.
   (b) 1 M $MgSO_4.7H_2O$—Add 12.3 g in 50 mL ddH_2O and autoclave.
   (c) 1 M $CaCl_2$—Dissolve 5.55 g in 50 mL ddH_2O and autoclave.
   (d) 20 % CAS amino acids—Dissolve 2 g in 10 mL ddH_2O, and autoclave.
   (e) 0.1% Vit. B1 or thiamine hydrochloride—Weigh 10 mg B1, dissolve in 10 mL ddH_2O, and autoclave.
   (f) Autoclave 850 mL ddH_2O in a 2-L flask sealed with a cotton plug.

Just before use, dissolve 1 g NH_4Cl in 10 mL water and prepare 20% glucose solution, and filter sterilize both the solutions.

M9 minimal media is prepared before adding the starter culture or primary culture. To prepare 1 L M9 minimal media, add 100 mL 10× M9 salts, 2 mL 1 M $MgSO_4$, 100 μL $CaCl_2$ (from 1 M stock), freshly prepared NH_4Cl, 2 mL 20% glucose solution, 2.5 mL CAS amino acids (from 20% stock), 500 μL 0.1% B1 solution, and 1 mL ampicillin (from 100 mg/mL stock) in 850 mL autoclaved ddH_2O. Mix the media, and then add 10 mL of primary culture.

2.5 Common Chemicals

1. *Ampicillin stock solution (100 mg/mL)*: Dissolve 500 mg of ampicillin sodium in 5 mL of autoclaved ddH_2O. Filter sterilize the solution with 0.22 μ filter membrane and use immediately (or store 1 mL aliquots at −30 °C for later use).
2. *1 M IPTG*: Dissolve 2.38 g of IPTG in 10 mL ddH_2O. Filter sterilize the solution with 0.22 μ filter membrane and dispense in 1 mL aliquots and store at −30 °C.
3. *1 M DTT*: Dissolve 1.54 g of DTT in 10 mL autoclaved ddH_2O and filter sterilized with 0.22 μ filter. Note: Use immediately or store 1 mL aliquots at −30 °C. It is suggested to not store DTT for longer durations.
4. *Chelex-purified water*: Pass freshly filtered ddH_2O through Chelex column and store at 4 °C in plastic bottles. Use of glass bottles is discouraged as it may leach contaminating Ca^{2+}.
5. *10× Chelex-purified buffer*: Mix 500 mL 10× solution (500 mM Tris pH 7.5, 1 M KCl) in freshly filtered ddH_2O (using 0.22 μ filter). Pass stock buffer through Chelex column, and store stock at 4 °C in plastic bottles. The same strategy should be used for the Chelex treatment of other solutions. Check pH before using a stored buffer.

2.6 Buffers

1. Buffer A—50 mM Tris; 100 mM KCl; 7 M urea; pH 7.5.
2. Buffer B—50 mM Tris; 100 mM KCl; pH 7.5.
3. Buffer C—50 mM Tris; 100 mM KCl; 2% Triton X-100; pH 7.5.
4. Buffer D—50 mM Tris; 100 mM KCl containing either 50 mM or 250 mM imidazole to set gradient; pH 7.5.
5. Buffer E—50 mM Tris; 100 mM KCl; 1 mM Ca^{2+}; pH 7.5.
6. Buffer F—50 mM Tris; 0.5 mM EDTA; pH 7.5.
7. Buffer G—50 mM Tris, pH 7.5.
8. Buffer H—50 mM Tris; 2% Triton X-100; pH 7.5.
9. Buffer I—50 mM Tris containing either 50 mM NaCl or 1 M NaCl; pH 7.5.

*Chelex-purified buffer—Prepare 10× buffer and pass through Chelex resin and store at 4 °C in sterile air-tight bottles. Before use, dilute to 1× using Chelex-purified water and filter sterilize.

3 Methods

3.1 His-Tagged SCGN Purification

Clone mouse SCGN cDNA in the pET21b vector to contain an in-frame C-terminal His_6 tag [9]. The overexpression of the recombinant protein in LB broth results in protein inclusion bodies. On-column refolding of solubilized inclusion bodies yields well-folded SCGN. It is the most convenient method for SCGN purification with pronounced yield. To purify SCGN from soluble fraction without denaturation, we describe a protocol for SCGN overexpression by growing bacterial culture in M9 minimal media. Soluble preparation yields a large amount of protein but requires an elaborate preparation of M9 media.

3.1.1 High-Yield Purification of His-Tagged SCGN from Inclusion Bodies

This protocol uses on-column refolding of SCGN aggregates solubilized in urea. Refolded protein is biochemically similar to the protein purified from soluble fraction and is suited for the large-scale purification of SCGN [9]. Purification from 1 L culture typically yields ~30–40 mg protein.

1. On day 0, inoculate BL21(DE3) cells carrying SCGN ORF plasmid in 10 mL of LB medium (containing 100 μg/mL ampicillin) and incubate overnight at 37 °C with shaking (220 rpm). This is termed as a primary culture.
2. Next morning, add 10 mL primary culture into 2-L-capacity flask containing 1 L autoclaved LB medium. Add 1 mL ampicillin (from 100 mg/mL stock) in 1 L medium. This is termed as the secondary culture.
3. Incubate secondary culture in an orbital shaker set at 220 rpm and maintained at 37 °C. Keep monitoring the growth of bacterial culture.
4. When the optical density (OD) at 600 nm reaches 0.6, induce the culture with 1 mM IPTG (*see* **Note 4**). Let bacterial culture grow for 7 h at 37 °C post-induction.
5. After 7 h of IPTG induction, pellet down the culture by centrifugation at 7000 RCF for 10 min (*see* **Note 5**).
6. Resuspend the bacterial cell pellet in Buffer A using Dounce homogenizer to make a homogenous suspension devoid of cell clumps. Add a pinch of deoxycholic acid to the lysate and mix well before proceeding for sonication.
7. Sonicate cell suspension for 40–60 min. Keep the falcon (containing cell suspension) on ice while sonicating. A "10 s on, 6 s off" sonic cycle yields good cell lysis without overheating. The adequacy of lysis is assessed by the drastic reduction in the viscosity of the suspension (*see* **Note 6**).
8. Centrifuge lysate for 1 h at 16,000 RCF. Separate supernatant immediately after the completion of centrifugation.

9. Load supernatant onto a pre-equilibrated Ni-NTA column. Equilibration is achieved by passing 100 mL of Buffer A through the column while the lysate is being centrifuged. Pass the lysate at a slow speed through the column 4–5 times.
10. Wash column with 100 mL of Buffer B (*see* **Note 7**).
11. Wash column with 100 mL Buffer C followed by washing with 200 mL of buffer B (*see* **Note 8**).
12. Elute the protein by setting a gradient with Buffer D containing either 50 mM or 250 mM imidazole using a gradient mixer. Collect 5 mL fractions in test tubes.
13. Check protein purity on 12% SDS-PAGE gel. Take 20 μL of protein eluents from alternate tubes in a 1.5 mL Eppendorf tube, and add an appropriate volume of bromophenol blue containing Laemmli SDS-loading buffer. Boil the samples and run on 12% SDS-PAGE gel with appropriate molecular weight marker.
14. *Optional size-exclusion chromatography purification*: A very high yield and purity are achieved by denaturing Ni-NTA chromatography described in this protocol (Fig. 2), and thus it is optional to perform size-exclusion chromatography (SEC). However, we recommend performing SEC if the set-up is available. To perform SEC purification, pool pure fractions and concentrate to 5 mL volume (or multiple of 5) using a 10 kDa cutoff membrane (or use Amicon-Ultra devices). Incubate the protein with 1 mM DTT at room temperature for 20 min (*see* **Note 9**). Inject protein sample on a Superdex 75 column pre-equilibrated with Buffer B supplemented with 1 mM DTT and monitor OD_{280} for protein peak. Based on OD_{280} signal, select protein-containing fractions and proceed to SDS-PAGE analysis as described in the previous step. We recommend using protein fractions having a single-band purity on SDS-PAGE visualized by Coomassie staining (Fig. 2d, fractions corresponding to lanes 5–8).

    If the protein is to be used for cell/animal experiments, stop at this step. Buffer exchange or dialyze with PBS and jump to ***step 17*** *(see* ***Note 10****). In case of calcium-dependent in vitro studies, proceed to subsequent steps.*
15. Pool all pure protein-containing fractions, and estimate tentative protein concentration by NanoDrop.
16. Add a 10× molar excess of EDTA (i.e., for 10 μM protein solution, add 100 μM EDTA) and incubate on ice for 30 min.
17. Perform buffer exchange with Chelex-purified Buffer B till the EDTA concentration reaches sub-picomolar range (*see* **Note 11**).

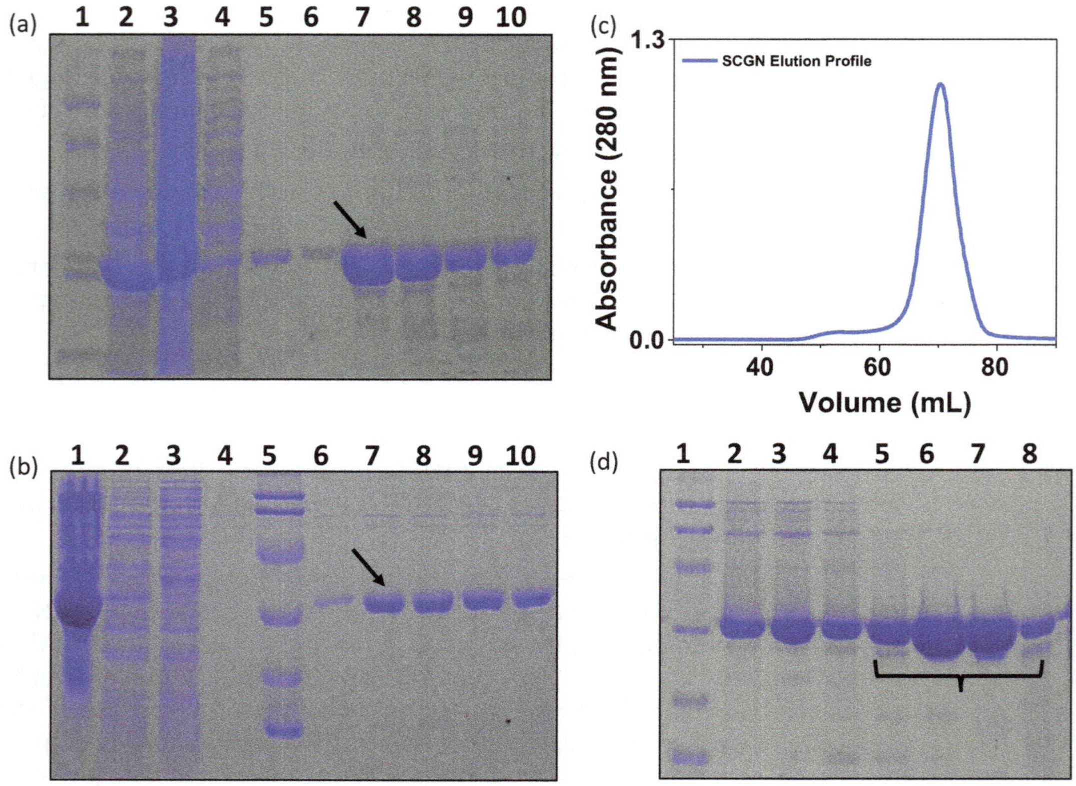

Fig. 2 Assessment of SCGN purity. (**a**) SDS-PAGE of eluents from Ni-NTA chromatography of solubilized inclusion bodies. Lane 1, low molecular weight marker; Lane 2, supernatant; Lane 3, flow through; Lane 4, wash; Lane 5–10, eluted fractions with increasing imidazole concentration (50–250 mM). The black arrow indicates the position of SCGN band. Note: The loss of protein seen in Lane 3 can be avoided by increasing the resin volume or decreasing the initial culture size. (**b**) SDS-PAGE of eluents from Ni-NTA chromatography of soluble fraction. Lane 1, supernatant; Lane 2, flow through; Lane 3, wash; Lane 4, eluent 1; Lane 5, low molecular weight marker; Lane 6–10, eluted fractions with increasing imidazole concentration (50 mM–250 mM). The black arrow indicates the position of SCGN band. (**c**) Preparative SEC chromatogram depicting the elution profile of SCGN after Ni-NTA chromatography. Note a prominent peak which corresponds to the monomeric SCGN and a fewer high molecular weight impurity. (**d**) SDS-PAGE gel depicting the purified SCGN fractions after SEC. The initial fractions contain SCGN as well as contaminating proteins (Lanes 2–4) corresponding to the early eluting population on SEC. The major peak in SEC contains homogenous SCGN without detectable contaminating proteins (Lanes 5–8)

18. Quantify protein concentration by using either BCA method or by taking the absorbance at 280 nm and dividing it by the molar extinction coefficient of the protein.
19. Protein is ready for experiments. For short-term usage, keep protein at 4 °C. For long-term storage, make small aliquots of 1–2 mg/mL protein concentration and store in −80 °C deep freezer. Avoid multiple freeze-thaw cycles and do not store highly concentrated protein.

3.1.2 High-Yield Purification of His-Tagged SCGN from Soluble Fraction

Purification from 1 L culture typically yields ~20–30 mg protein.

1. Inoculate transformed *E. coli* BL21(DE3) cells in 10 mL LB media for primary culture as described above.
2. Next morning, spin down the cell suspension at 5000 RCF for 5 min.
3. Prepare 1 L minimal media as described in the materials section.
4. Resuspend bacterial cell pellet in 10 mL minimal media.
5. Put secondary culture by adding resuspended bacterial cell pellet into 1 L minimal media.
6. Incubate the secondary culture in an orbital shaker (set at 220 rpm and maintained at 37 °C) till the optical density (OD) at 600 nm reaches 0.6–0.8.
7. At $OD_{600nm} = 0.6$, induce the culture with 1 mM IPTG, and incubate the induced culture for additional 10 h at 25 °C (*see* **Note 4**).
8. Pellet down the culture by centrifugation at 7000 RCF for 10 min (*see* **Note 5**).
9. Resuspend bacterial cell pellet in Buffer B supplemented with manufacturer-recommended concentration of protease inhibitor cocktail, DNase, and lysozyme and a pinch of deoxycholic acid. Homogenize the cell pellet properly using a Dounce homogenizer.
10. Sonicate and centrifuge the lysate as described in the previous protocol.
11. Load supernatant onto a pre-equilibrated Ni-NTA column. Equilibration is achieved by passing 100 mL of Buffer B through the column. Pass the lysate through the column 4–5 times.
12. Follow **steps 10–17** as described in Subheading 3.1.1 (*see* **Note 12**).

3.2 Purification of Untagged SCGN from the Soluble Fraction

Clone mouse SCGN cDNA in pET21b vector, and insert a stop codon after coding sequence (in the reverse primer). Transform *E. coli* BL21(DE3) with this plasmid, and maintain bacteria on ampicillin-containing LB agar plates.

1. On day 0, inoculate SCGN ORF plasmid-carrying BL21(DE3) cells in 10 mL of LB medium (containing 100 μg/mL ampicillin) and incubate overnight at 37 °C.
2. Next morning, put secondary culture as described in Subheading 3.1.1.
3. Incubate the secondary culture in an orbital shaker maintained at 37 °C with shaking at 220 rpm.

4. At $OD_{600nm} = 0.6$, induce culture with 0.8 mM IPTG and incubate at 25 °C for 10 h (*see* **Note 4**).
5. Pellet down grown culture by centrifugation at 7000 RCF for 10 min (*see* **Note 5**).
6. Resuspend bacterial cell pellet in Buffer E to make a homogenous suspension devoid of cell clumps. Add a pinch of deoxycholic acid, 1 mM PMSF, and 1 μg/mL DNase to the lysate.
7. Sonicate the cell suspension for 40–60 min. Keep the lysate on ice while sonicating. A "10 s on 6 s off" sonic cycle yields good cell lysis without overheating.
8. Centrifuge lysate for 1 h at 16,000 RCF. Separate supernatant immediately after the completion of centrifugation (*see* **Note 6**).
9. Load supernatant onto a Phenyl-Sepharose column pre-equilibrated with 100 mL Buffer E. Pass the lysate through the column at a slow speed for 4–5 times.
10. Wash column with 100 mL Buffer E.
11. Elute protein by passing Buffer F. Collect 5 mL fractions.
12. Check the purity of eluents on a 12% SDS-PAGE as described in **step 13**, Subheading 3.1.1.
13. Pool all the relatively pure fractions, and load the pooled protein fraction onto a pre-equilibrated Q-Sepharose column, and repeat this step 4–5 times (*see* **Note 13**). Equilibration of Q-Sepharose column is done by passing 100 mL of Buffer F.
14. Wash column with 100 mL of Buffer G.
15. Wash column with 100 mL Buffer H followed by 200 mL of Buffer G (*see* **Note 8**).
16. Elute protein using a gradient mixer with Buffer I containing either 50 mM NaCl or 1 M NaCl (i.e., 50 mM–1 M NaCl gradient). Collect 5 mL fractions in test tubes.
17. Follow **steps 12–18** from Subheading 3.1.1 (*see* **Note 12**).

3.3 Special Consideration for Cell Culture and Animal Treatment

It is expected that a protein purified from inclusion bodies using urea denaturation method (Subheading 3.1.1) will contain no impurities. In the case of SCGN, we observed a visually inappreciable amount of contaminating protein bands on SDS-PAGE suggesting absence of contaminating proteins. However, if mammalian cell lines (MIN6, HepG2, C2C12, and NIH-3T3 cell lines tested) were treated with the protein purified without Triton X-100 wash, to our surprise, we noticed considerable cell death within a few hours of treatment with this protein. In the case of purification from soluble fraction (Subheadings 3.1.2 and 3.2), the cytotoxicity is enhanced. Thus, it is recommended to wash the column-bound protein with 2% Triton-containing wash buffer during column chromatography. If the protein is to be used in cell culture or animal experiments, Triton X-100 wash should strictly be followed. It is recommended to store small aliquots of protein (2 mg/mL stock) in −80 °C freezer. Avoid freeze-thaw cycles and do not store at 4 °C for a longer duration.

3.4 Purity Check and Biological Usability

The first line of purity check for new preparation of SCGN is single-band homogeneity on SDS-PAGE. However, for the biological applications of protein (cell/animal treatment), this is misleading as such as preparation causes significant cell death if not washed with Triton X-100. Thus, it is necessary to perform cell cytotoxicity test with every batch of preparation. To do so, seed 20,000 MIN6 (or any other cell line) cells/well in a 96-well cell culture plate. After 24 h, treat cells with increasing concentrations of SCGN (100 nM to 1 μM) in serum-free DMEM. As a positive control treat one group with 50 nM insulin, treat another group with 2% Triton X-100 which would serve as negative control and one control group with no treatment. After 24 h of incubation, assess cell viability by MTT assay [7]. Insulin treatment results in increased cell viability, while Triton X-100 causes considerable cell death (0–10% cell viability). In our experience, SCGN treatment induces cell proliferation, and thus, if the protein preparation is pure, a dose-dependent increase in cell viability should be seen in the case of SCGN treatment (unpublished results from our lab).

3.5 Typical Properties of SCGN

The mouse SCGN purified using any of the abovementioned methodology yields spectroscopically similar protein. Ca^{2+} binding to SCGN induces large conformational rearrangement, which can be visualized by fluorescence spectroscopy or circular dichroism spectroscopy. Typical spectroscopic characteristics of recombinant mouse SCGN are discussed in the following text.

3.5.1 Tryptophan Fluorescence Spectroscopy

It is the most sensitive, inexpensive, time-saving, and confirmatory technique to check for the conformational state of the protein and to study ligand (such as Ca^{2+})-induced changes. To perform fluorescence spectroscopic analysis of SCGN:

1. Take 600 μL of 0.1 mg/mL (or 3 μM) protein in a 1 cm path length quartz cuvette.
2. Set a fluorimeter (such as F-7000 Spectrophotometer from Hitachi) on fluorescence emission wavelength scan mode.
3. Set excitation wavelength at 295 nm and emission range from 300 nm to 450 nm.
4. Record spectrum, and analyze the wavelength at which the maximum emission intensity is observed (known as λ_{max}).

A well-folded SCGN protein shows maximum emission at ~338–340 nm. A $\lambda_{max} > 450$ nm indicates unfolded (or misfolded)

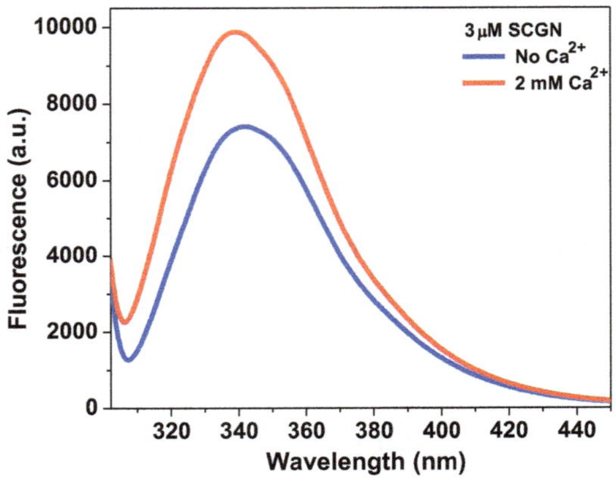

Fig. 3 Spectral changes monitored by intrinsic Trp fluorescence of Ca^{2+}-free SCGN (blue) and Ca^{2+}-saturated SCGN (red)

protein. Addition of Ca^{2+} to SCGN causes a blue shift in the λ_{max}. Ca^{2+} saturated SCGN shows a λ_{max} ~ 336 nm along with a considerable increase in the quantum yield (Fig. 3). Ca^{2+}-induced blue-shift coupled with an increase in the quantum yield is the hallmark of pure, Ca^{2+}-free native protein. The absence of these shifts suggests Ca^{2+} contamination or a non-native conformation of the protein.

3.5.2 Near-UV Circular Dichroism

Also known as tertiary CD, provides a qualitative fingerprint of the tertiary structure of a protein. To perform near-UV CD analysis of SCGN:

1. Take 1 mL of 1 mg/mL protein in a 10 mm path length quartz cuvette.
2. Set instrument (such as Jasco J-815) in wavelength spectrum scan mode.
3. Set wavelength range from 250 nm to 350 nm.
4. Record five replicates of the same sample.
5. Analyze spectrum while comparing ligand-free protein with ligand-bound protein.

SCGN tertiary fingerprint demonstrates well-defined aromatic bands between 260 and 300 nm with a prominent trough at 260–280 nm corresponding to abundant Phe (phenylalanine; F) signal. SCGN shows a huge conformational rearrangement upon Ca^{2+} binding. The apo fingerprint and the Ca^{2+}-induced rearrangements are very reproducible and thus can be used to study the ligand-induced conformational changes and even protein–ligand or protein–protein interactions (Fig. 4).

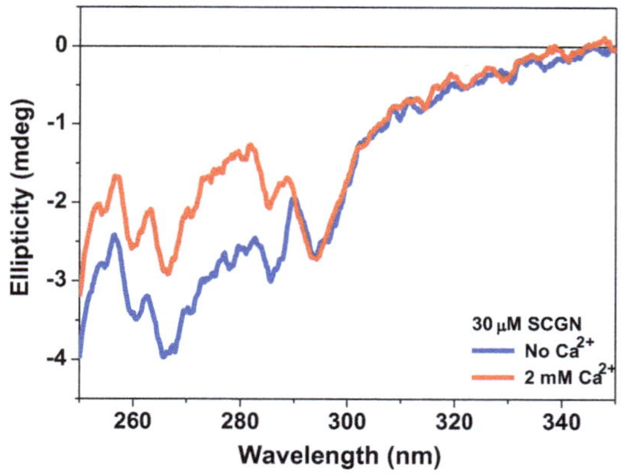

Fig. 4 Near-UV CD spectra of Ca^{2+}-free SCGN (blue) and Ca^{2+}-saturated SCGN (red). Addition of Ca^{2+} causes drastic readjustment of the tertiary structure fingerprints suggesting a conformational rearrangement in SCGN (*see* **Notes 14** and **15**)

3.5.3 Far-UV Circular Dichroism

Also known as secondary CD, provides a quantitative/qualitative information about the secondary structure content of a protein. To perform far-UV CD analysis of SCGN:

1. Take 60 μL of 1 mg/mL protein in a 0.2 mm path length quartz cuvette designed for CD measurements.
2. Set instrument (such as Jasco J-815) in wavelength spectrum scan mode.
3. Set wavelength range from 190 nm to 250 nm.
4. Record three replicates of the sample.
5. Analyze spectrum while comparing ligand-free protein with ligand-bound protein.

SCGN secondary CD spectrum shows a typical α-helical content with two prominent troughs at ~208 nm and ~222 nm. Titration of Ca^{2+} into SCGN exhibits no considerable secondary CD spectral rearrangement (Fig. 5).

4 Notes

1. To pack glass columns, use two-layered glass wool to stop resin loss (a number of convenient alternatives are available for ready-to-pack polypropylene columns with a prefixed stopper and resin bed, e.g., Econo-Pac columns from Bio-Rad). Wash with excessive double-distilled water (ddH_2O) to remove alcohol preservative. After every batch of the protein preparation, wash the column with 100 mL of 0.5 N NaOH followed by

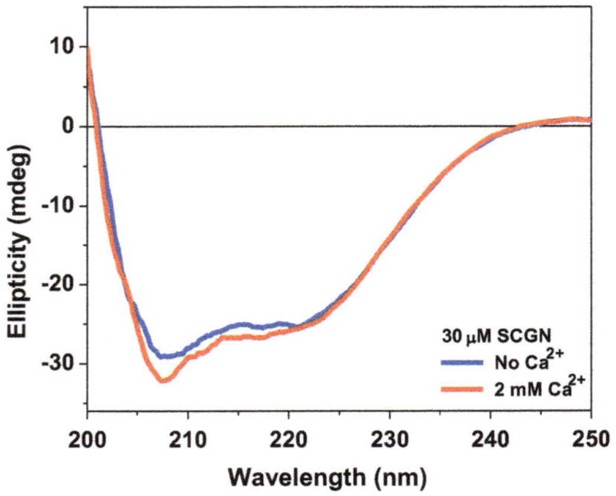

Fig. 5 Far-UV CD spectra of Ca^{2+}-free SCGN (blue) and Ca^{2+}-saturated SCGN (red). SCGN shows a typical all-alpha secondary structure

500 mL of ddH_2O. If the color of the beads fades, recharge column with $NiSO_4$ as per the manufacturer's protocol.

2. Follow manufacturer's wash/regeneration recommendation. Wash the resin extensively after every batch of the protein preparation.

3. For long-term usage, prepare glycerol stock by adding 300 μL of overnight-grown culture into 700 μL glycerol. Snap-freeze in liquid nitrogen and store in −80 °C deep freezer. Maintain BL21-transformed cells on LB agar nutrient plates with ampicillin and store in cold room. Re-streak cells on the freshly prepared plate every month or revive new cells from glycerol stock.

4. To reach OD = 0.6 (at 600 nm; i.e., $OD_{600} = 0.6$), secondary culture takes ~3–4 h depending on the primary culture growth and other variables. The permissible range of bacterial cell growth for induction with IPTG is $OD_{600} = 0.6$–0.8. In the case of culture in minimal media (Subheading 3.1b), $OD_{600} = 0.8$–1.0 is tolerated.

5. At this stage, the pellet can be transferred to a sterile 50 mL falcon and stored in −30 °C freezer until further use. The pellet can be stored for years with no loss of protein. Thus, it is recommended to store a backup pellet as spare.

6. Completion of cell lysis should be confirmed by taking a small aliquot of lysate and checking for the absence of protein band in the pellet fraction on 12% SDS-PAGE after 10 min' centrifugation at 15,000 RCF.

7. Washing the column with buffer containing no urea leads to on-column refolding of protein.

8. Washing the protein-bound column with 2% Triton X-100-containing buffer is necessary if the protein is to be

used for cell culture or animal experiments. Triton wash should be followed by wash with 200 mL triton-free wash buffer. Insufficient washing may lead to the retention of Triton X-100 which would be cytotoxic. If SEC purification is performed, the chances of triton contamination are minimized.

9. Secretagogin tends to form oligomers, which are further enhanced in the presence of Ca^{2+} [10]. Addition of DTT to the protein sample prior to SEC leads to dissociation of preformed oligomers to monomeric species, thus minimizing the loss of protein. Although the addition of DTT to the protein sample is optional, however, we strongly recommend it to get pure homogenous protein preparation.
10. In case the protein is to be used only for cell culture or animal experiments, the Superdex 75 column used for SEC could be equilibrated in PBS buffer only. This would eliminate the need for dialysis or buffer exchange.
11. Buffer exchange can be performed by dialysis or by centrifugal devices. The extent of buffer exchange should be sufficient to bring EDTA in sub-picomolar range; otherwise it will lead to misleading results in ion-binding studies.
12. It is recommended to perform size-exclusion chromatography for Subheadings 3.1.2 and 3.2 purification. Protein purified from soluble fraction usually contains impurities (both protein and nonprotein in nature) after Ni-NTA chromatography step. Pool the SEC elution fractions containing protein with no visible contaminating protein band. It also helps in removing the high concentrations of imidazole used to elute the protein in earlier step.
13. It is not possible to give Triton X-100 wash on a Phenyl-Sepharose column (Hydrophobic Interaction Chromatography, HIC, resin) because the detergent will interfere with the resin–protein hydrophobic interactions which would lead to untimely dissociation of protein from resin. Therefore, the fractions collected from HIC are then subjected to anion exchange chromatography which enables the removal of impurities as well as aids in the removal of endotoxins via Triton X-100 wash. A final step of SEC yields a pure homogenous protein preparation.
14. Ca^{2+}-induced increase in the Trp fluorescence quantum yield associated with a blue shift and rearrangement of near-UV CD spectra is hallmark of native Ca^{2+}-free SCGN. These features should be exploited to assess the Ca^{2+}-free preparation of SCGN and to validate that the particular batch of SCGN preparation contains native folded protein.
15. It is recommended to filter sterilize protein preparation before estimating the accurate concentration.

Acknowledgments

This work was supported partially by CSIR and DBT to YS. AKS was supported by the Senior Research Fellowship, University Grant Commission, and RK is the recipient of CSIR-GATE fellowship. Authors acknowledge occasional help from HC Amrutha during the preparation of this manuscript. Anand Kumar Sharma and Radhika Khandelwal contributed equally to this work.

References

1. Wagner L, Oliyarnyk O, Gartner W, Nowotny P, Groeger M, Kaserer K, Waldhausl W, Pasternack MS (2000) Cloning and expression of secretagogin, a novel neuroendocrine- and pancreatic islet of Langerhans-specific Ca2+-binding protein. J Biol Chem 275:24740–24751
2. Kobayashi M, Yamato E, Tanabe K, Tashiro F, Miyazaki S, Miyazaki J (2016) Functional analysis of novel candidate regulators of insulin secretion in the min6 mouse pancreatic β cell line. PLoS One 11:e0151927
3. Yang SY, Lee JJ, Lee JH, Lee K, Oh SH, Lim YM, Lee MS, Lee KJ (2016) Secretagogin impacts insulin secretion in pancreatic beta cells by regulating actin dynamics and focal adhesion. Biochem J 473:1791–1803
4. Romanov RA, Alpár A, Zhang MD, Zeisel A, Calas A, Landry M, Fuszard M, Shirran SL, Schnell R, Dobolyi A, Oláh M, Spence L, Mulder J, Martens H, Palkovits M, Uhlen M, Sitte HH, Botting CH, Wagner L, Linnarsson S, Hökfelt T, Harkany A (2014) A secretagogin locus of the mammalian hypothalamus controls stress hormone release. EMBO J 34:36–54
5. Rogstam A, Linse S, Lindqvist A, James P, Wagner L, Berggard T (2007) Binding of calcium ions and SNAP-25 to the hexa EF-hand protein secretagogin. Biochem J 401:353–363
6. Malenczyk K, Girach F, Szodorai E, Storm P, Segerstolpe Å, Tortoriello G, Schnell R, Mulder J, Romanov RA, Borók E, Piscitelli F, Di Marzo V, Szabó G, Sandberg R, Kubicek S, Lubec G, Hökfelt T, Wagner L, Groop L, Harkany T (2017) A TRPV1-to secretagogin regulatory axis controls pancreatic β-cell survival by modulating protein turnover. EMBO J 36 (14):2107–2125
7. Sharma AK, Khandelwal R, Chadalawada S, Ram NS, Raj A, Jerald MK, Sharma Y. SCGN administration prevents insulin resistance and diabetic complications in high-fat diet-fed animals. bioRxiv. 2017. https://doi.org/10.1101/189324
8. Hansson SF, Zhou AX, Vachet P, Eriksson JW, Pereira MJ, Skrtic S, Jongsma Wallin H, Ericsson-Dahlstrand A, Karlsson D, Ahnmark A, Sörhede Winzell M, Magnone MC, Davidsson P (2018) Secretagogin is increased in plasma from type 2 diabetes patients and potentially reflects stress and islet dysfunction. PLoS One 13(4):e0196601
9. Sharma AK, Khandelwal R, Sharma Y, Rajanikanth V (2015) Secretagogin, a hexa EF-hand calcium-binding protein: High level bacterial overexpression, one-step purification and properties. Protein Expr Purif 109:113–119
10. Khandelwal R, Sharma AK, Chadalawada S, Sharma Y (2017) Secretagogin is a redox-responsive Ca^{2+} sensor. Biochemistry 56 (2):411–420
11. Corpet F (1988) Multiple sequence alignment with hierarchical clustering. Nucl Acids Res 16 (22):10881–10890

Chapter 35

Tryptophan Scanning Mutagenesis of EF-Hand Motifs

Uday Kiran, Michael R. Kreutz, Yogendra Sharma, and Asima Chakraborty

Abstract

Ca^{2+} regulation in living systems occurs via specific structural alterations, subtle or drastic, in the Ca^{2+}-binding domains of sensor proteins. Sensor proteins perform designated nonredundant roles within the dense network of Ca^{2+}-binding proteins. A detailed understanding of the structural changes in calcium sensor proteins due to Ca^{2+} spikes that vary spatially, temporally, and in magnitude would provide better insights into the mechanism of Ca^{2+} sensing. This chapter describes a method to study various stages during apo to the holo transition of Ca^{2+}-binding proteins by Trp-mediated scanning of individual EF-hand motifs. We describe the applicability of this procedure to caldendrin, which is a neuronal Ca^{2+}-binding protein and to integrin-binding protein. Tryptophan mutants of full-length caldendrin were designed to reveal local structural changes in each EF-hand of the protein. This method, referred to as "EF-hand scanning tryptophan mutagenesis," not only allows the identification of canonical and noncanonical EF-hands using very low concentrations of protein but also enables visualization of the hierarchical filling of Ca^{2+} into the canonical EF-hands.

Key words Tryptophan scanning, EF-hand motif, CaBP, Caldendrin, Fluorescence

1 Introduction

Calcium (Ca^{2+}) is involved in crucial physiological events, and calcium homeostasis is essential for the viability of any living system. In multicellular eukaryotes, Ca^{2+} levels in different organelles are tightly controlled by a repertoire of Ca^{2+} signalling molecules that include Ca^{2+}-binding proteins or calcium sensors [1]. The transition of a calcium sensor or a Ca^{2+}-binding protein from its apo (Ca^{2+} free) to the holo (Ca^{2+} saturated) state or vice versa is a central component of calcium signalling, which, possibly under restricted intracellular Ca^{2+} supply, may be routed via a plethora of conformational states. Studying these routes, which vary in CaBPs, would enlighten how proteins function under the tight spatiotemporal regulation of intracellular Ca^{2+}. We know that the internal [Ca^{2+}] is maintained within a narrow range of 10^{-7}–10^{-8} M in nearly all living cells, while extracellular [Ca^{2+}] is

Claus W. Heizmann (ed.), *Calcium-Binding Proteins of the EF-Hand Superfamily: From Basics to Medical Applications*, Methods in Molecular Biology, vol. 1929, https://doi.org/10.1007/978-1-4939-9030-6_35,

in the 10^{-3} M range [2]. This situation is further complicated by the presence of a dense network of Ca^{2+}-binding proteins competing for limited available Ca^{2+} [3].

Structural methods provide atomic-level information about the coordination pattern around a metal-binding site. But we do not get any clues about the different conformational equilibrium states that a protein may adopt during the ion-binding process at different binding sites. The "conformational equilibrium state" here we mean that a protein exists in a particular conformation over a range of Ca^{2+} concentrations. These conformational states help in discriminating between interactions with multiple downstream targets. However, identification of the equilibrium state(s) during the apo-holo transition would require the dissection of the Ca^{2+}-binding event to individual EF-hand motifs. It is widely believed that the differences in the functions of Ca^{2+} sensors can be attributed primarily to how individual EF-hand motifs act and communicate with each other during the Ca^{2+}-binding/filling event [4].

Tryptophan (Trp) scanning mutagenesis is a sensitive technique in which a series of mutants are generated by replacing one residue at a time with Trp, making it possible to study local structural changes through fluorescence spectroscopy. It can be utilized as a routine technique to explore segments of proteins lacking Trp [5–7]. Trp as a reporter of Ca^{2+}-induced conformational changes of an individual EF-hand motif in CaBPs has been explored [8, 9], where EF-hand motif responded differently upon Ca^{2+} titration, revealing local effects of Ca^{2+}. Based on analogy with earlier Trp scanning studies, we have named our method "EF-hand scanning Trp mutagenesis." The past unfolding studies have shed light on the relative flexibilities of the EF-hand motifs, and the probable chronology of their folding in the Ca^{2+}-binding protein, caldendrin, as revealed by "EF-hand scanning Trp mutagenesis" has enabled us to explore the individual EF-hands both in the apo and holo forms.

In this chapter, Trp scanning of EF-hand motifs applied to study caldendrin is described because this protein has a unique bipartite structure, consisting of a flexible N-terminal acidic domain and a C-terminal domain containing four EF-hand motifs [10]. The flexible N-terminal region limits its detailed examination by NMR and X-ray crystallography; moreover, it is not amenable to most spectroscopic tools due to the lack of a Trp residue [10, 11]. For Trp scanning, we exploited the phenylalanine residues, which are placed proximal to the loop in each EF-hand motif. Any conformational strain induced by Ca^{2+} binding to an EF-hand motif would be sensed by this reporter Trp and recorded as a change in the fluorescence characteristics of each Trp mutant [8, 9]. Change in rigidity/flexibility of a motif caused by Ca^{2+} binding could be examined by chemical unfolding studies using Trp as a probe.

```
           1       10        20        30        40        50        60
           |--------+---------+---------+---------+---------+---------|
CaBP1             PSPSSPLPPARGRDGEERGL--SPALGLRGSLRARGRGDSVPAAASEADPFLH
CaBP2      MGNCAKRPWRRGPKDPLQWLGSPPRGSCPSPSSSPKEQGDPAPGVQGYSVLNSLVGPACI
CaBP5                                                        MQFPMGPACI
CaBP4                                                            MTEPWL
CaBP3                                        MLPAALQSHLVPVVFPAVSQQCLSSRR
Consensus  ............................................................

           61      70        80        90        100       110       120
           |--------+---------+---------+---------+---------+---------|
CaBP1      RLRPMLSSAFGQDRSLRPEEIEELREAFREFDKDKDGYINCR-DLGNCMRTMGYMPTEME
CaBP2      FLRPSIA-ATQLDRELRPEEIEELQVAFQEFDRDQDGYIGCR-ELGACMRTLGYMPTEME
CaBP5      FLRKGIA-EKQRERPLGQDEIEELREAFLEFDKDRDGFISCK-DLGNLMRTMGYMPTEME
CaBP4      ALGTSWTLPLQ-DRELGPEELDELQAAFEEFDTDRDGYISHR-ELGDCMRTLGYMPTEME
CaBP3      CRRGPWGTAWRGGVELAGHQSQDGPSSLLPVGAQPPYLLQHSFATSSCLSWLLLRSRCLR
Consensus  .lr..w..a.q..reL...#.##l..aflefd.#.dg.i.h...lg.c$rt$gympte$e

           121     130       140       150       160       170       180
           |--------+---------+---------+---------+---------+---------|
CaBP1      LIELSQQINMNLGGHVDFDDFVELMGPKLLAETADMIGVKELRDAFREFDTNGDGEISTS
CaBP2      LIEISQQIS---GGKVDFEDFVELMGPKLLAETADMIGVRELRDAFREFDTNGDGRISVG
CaBP5      LIELGQQIRMNLGGRVDFDDFVELMTPKLLAETAGMIGVQEMRDAFKEFDTNGDGEITLV
CaBP4      LLEVSQHIKMRMGGRVDFEEFVELIGPKLREETAHMLGVRELRIAFREFDRDRDGRITVA
CaBP3      AVTAGLEIRMNLGGRVDFDDFVELMTPKLLAETAGMIGVQEMRDAFKEFDTNGDGEITLA
Consensus  l.e.gq.IrmnlGGrVDF##FVELmtPKLlaETAgMiGVqE$RdAFkEFDt#gDGeItla

           181     190       200       210       220   226
           |--------+---------+---------+---------+-----|
CaBP1      ELREAMRKLLGHQVGHRDIEEIIRDVDLNGDGRVDFEEFVRMMSR
CaBP2      ELRAALKALLGERLSQREVDEILQDVDLNGDGLVDFEEFVRMMSR
CaBP5      ELQQAMQRLLGERLTPREISEVVREADVNGDGTVDFEEFVKMMSR
CaBP4      ELREAVPALLGEPLAGPELDEMLREVDLNGDGTVDFDEFVMMLSRH
CaBP3      ELQQAMQRLLGERLTPREISEVVREADVNGDGTVDFEEFVKMMSR
Consensus  ELqqAmqrLLGerltpr#isEvvr#aDvNGDGtVDF#EFVkM$SR.
```

Fig. 1 Sequence alignment of selected CaBPs showing conserved Phe residues. Different EF-hand Ca^{2+}-binding loops are underlined, and Phe residues location at either before or after the loop is highlighted with boxes. If Phe residue is not present near the loop sequence of 12 amino acids, another residue can be selected for replacing by Trp (*see* ref. 8; Leu was changed to Trp)

In case a protein already has a Trp residue, the problem can be overcome by mutating that Trp, followed by introducing new Trp at selected places. Introduction of Trp at the beginning or just after the end of the 12 amino acid loop region is preferable. In many CaBPs, there is a Phe residue naturally placed at these points, which could be replaced with Trp (Fig. 1). In this way, if a CaBP has four EF-hand motifs, then at least four Trp mutants can be generated (Fig. 2).

Once mutations are generated, the overexpression of mutants in *E. coli* is checked. Successful overexpression would indicate that a mutation is well tolerated. Lack of the overexpression of a Trp mutant would suggest that "homologous replacement" by Trp

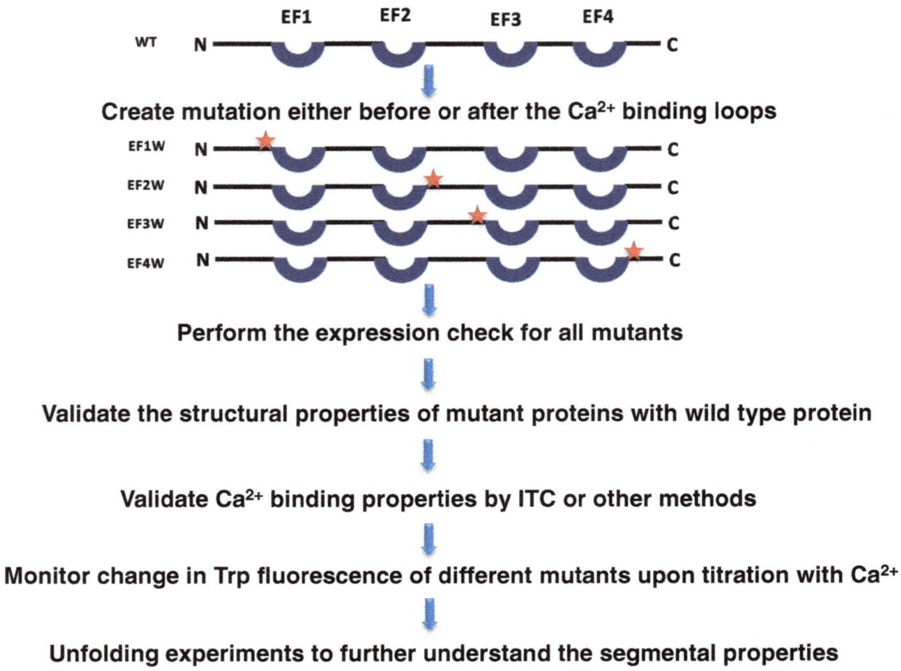

Fig. 2 A schematic presentation showing the steps required for performing "EF-hand Trp scanning." EF1 to EF4 represent four EF-hand motifs of a protein

residue could not be tolerated, implying a positional importance of that Phe residue. Steric hindrance is also a possibility. In such a scenario, the Trp location can be shifted to the other end of the loop. Perpendicular stacking of the indole ring of Trp with its edge on the Phe ring will not be a well-tolerated orientation for a stable protein conformation and would hinder or inhibit overexpression. We have encountered this phenomenon in the second EF-hand motif of caldendrin.

Before proceeding further, it should be made sure that a Trp mutation does not introduce undesirable perturbations in the overall conformation of the protein. Preservation of the conformation of mutants should be validated by comparing with wild-type protein. Contributions by Trp may or may not alter the near-UV CD spectrum, but it is worth comparing. Further validation can be performed by examining any mutant-induced change in surface hydrophobicity by employing the extrinsic fluorophore ANS as a probe. Now, we are all set for further analysis for monitoring changes in the fluorescence spectra by Ca^{2+} titration. Comparative plots of $[Ca^{2+}]$ with a change in fluorescence would provide feedback about the nature of Ca^{2+} filling to these EF-hand motifs (Fig. 3). Based on this simple plot, the EF-hand that accepts Ca^{2+} first can be identified (based on the response to Ca^{2+}). We add here that in a calcium vector protein, a plot of the normalized peak height of aromatic/amide protons by 1D NMR versus $[Ca^{2+}]$ was

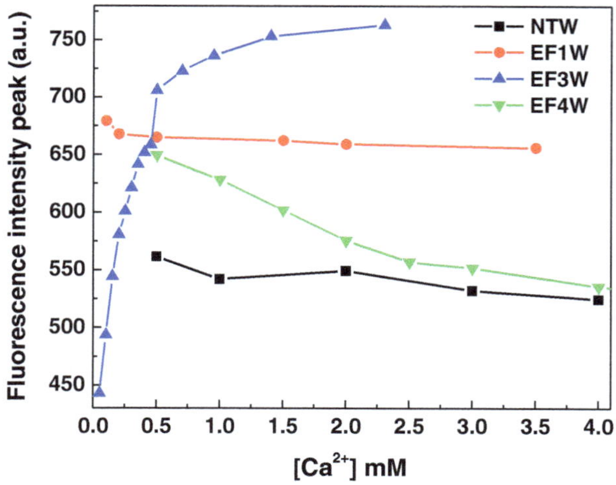

Fig. 3 The graph represents the change in Trp fluorescence intensity maximum of different mutants upon Ca^{2+} titrations. This is a sample representation of fluorescence intensity profile of four selected Trp mutants of caldendrin plotted against Ca^{2+} concentrations (NTW, black; EF1W, red; EF3W, blue; and EF4W, green). From the graph, it is clear that the EF3W responds to Ca^{2+} first, followed by EF4W, suggesting that it is this EF-hand motif which is filled first. It is also seen from the graph that EF4W responds to Ca^{2+} only after EF3W is almost saturated. Thus, the filling of EF3 also represents an equilibrium semi-filled state, different from either the apo or the holo form. Plotting the raw data of fluorescence intensity versus Ca^{2+} concentrations thus provides valuable information about the mode of Ca^{2+} filling

used to deduce the nature of Ca^{2+} filling to EF-hand motifs demonstrating sequential filling [12]. This plot is similar to the plot of fluorescence intensity of Trp mutants versus [Ca^{2+}] demonstrating similar phenomenon in caldendrin. Moreover, equilibrium unfolding studies would significantly highlight the comparative rigidity and role of Ca^{2+}. Here, data of all EF-hand motifs can be analyzed using various parameters, such as $C_{1/2}$ (the concentration of denaturant when 50% protein molecules are unfolded) and *m*-values (represents cooperativity of unfolding) of unfolding. These Trp mutants can be studied further for the analysis of ligand binding.

2 Materials

2.1 Site-Directed Mutagenesis

2.1.1 Reagents

1. Mutagenic primers to introduce Trp mutations at different selected positions of the gene (*see* **Notes 1** and **2**).
2. The *Pfu* enzyme with 3′ to 5′ exonuclease (proofreading) activity from QuikChange II site-directed mutagenesis kit.
3. 10× reaction buffer and dNTP mix provided with kit.
4. Nuclease-free water or autoclaved MQ water.
5. Template DNA.

2.1.2 Primers (See Notes 3–6)

NTW (F132W)

NTW-F: CGAGGCGGACCCGTGGCTCCACCAGCTG

NTW-R: CAGCTGGTGGAGCCACGGGTCCGCCTCG

EF1W (F165W)

EF1W-F: GAGGCCTTCCGGGAATGGGACAAGGACAAAGATGG

EF1W-R: CCATCTTTGTCCTTGTCCCATTCCCGGAAGGCCTC

EF3W (F242W)

EF3W-F: GACGCCTTTCGGGAGTGGGACACCAACGGTGAC

EF3W-R: GTCACCGTTGGTGTCCCACTCCCGAAAGGCGTC

EF4W (F292W)

EF4W-F: GGACTTCGAAGAGTGGGTCCGGATGATGTCTCGC

EF4W-R: GCGAGACATCATCCGGACCCACTCTTCGAAGTCC

2.1.3 Equipment

1. Calibrated micropipettes.
2. PCR tubes.
3. PCR machine with temperature gradient tool (Applied Biosciences or any other).

2.1.4 Agarose Gel Electrophoresis

2.1.5 Reagents

1. 50× stock TAE buffer: 2 M Tris (pH 8.0), 1 M glacial acetic acid, and 50 mM EDTA.
2. Dilute the 50× TAE to working concentration of (1×) with distilled water.
3. 1 kb DNA ladder to compare size of amplified product.
4. Commercially available 6× loading dye.
5. Agarose.
6. Ethidium bromide.

2.1.6 Equipment

1. Microwave oven.
2. Gel casting tray.
3. Well combs.
4. Electrophoresis apparatus with accessories.

2.1.7 DpnI Digestion

2.1.8 Reagents

1. DpnI enzyme.
2. 10× reaction buffer supplied with the enzyme.

2.1.9 Transformation

2.1.10 Reagents

1. DH5α competent cells (prepared by calcium chloride method).
2. BL21 (DE3) competent cells (prepared by calcium chloride method).
3. 100 mg/ml ampicillin.
4. LB-ampicillin (Amp) agar plates.

2.2 Protein Overexpression and Purification

2.2.1 Reagents

1. LB media (dissolve 10 g of NaCl, 10 g of peptone, and 5 g of yeast extract in distilled H_2O; make the volume to 1 l).
2. 100 mg/ml ampicillin (1 ml/L).
3. 1 M IPTG solution.
4. IMPACT™ Kit, NEB.
5. Chitin resin.
6. 1 M DTT.
7. 500 mM EDTA solution.
8. 50 mM Tris pH 8.5, 500 mM NaCl buffer prepared in MQ water (chitin buffer).
9. 50 mM Tris pH 7.5, 150 mM NaCl Chelex-treated buffer.
10. 50 mM Tris pH 7.5, 150 mM KCl Chelex-treated buffer.
11. Triton X-100 detergent.
12. Protease inhibitor cocktail.

2.2.2 Equipment

1. Autoclaved glass vials for primary small-scale cultures.
2. Gravity flow columns of 20–30 ml volume.
3. Ultrasonicator.
4. A high-speed centrifuge.

2.2.3 SDS-PAGE Electrophoresis

2.2.4 Reagents

1. 1.5 M Tris pH 8.8.
2. 1 M Tris pH 6.8.
3. 10% w/v SDS.
4. 10% w/v APS.
5. TEMED.
6. SDS-PAGE 4× loading dye.

7. Coomassie brilliant blue staining solution.
8. MQ water or destaining solution.

2.2.5 Equipment

1. Electrophoresis apparatus with accessories.

2.3 Circular Dichroism

2.3.1 Reagents

1. Purified protein samples.
2. 50 mM Tris pH 7.5, 150 mM NaCl Chelex-treated buffer.
3. 1 M $MgCl_2$ and $CaCl_2$ stocks.
4. Tris (2-carboxyethyl) phosphine (TCEP).

2.3.2 Equipment

1. Jasco J-815 spectropolarimeter.
2. Quartz cuvette of 0.1 cm and 1 cm path length for far-UV and near-UV spectra, respectively.
3. 1.5 ml Eppendorf tubes.

2.4 Isothermal Titration Calorimetry

2.4.1 Reagents

1. Purified protein samples.
2. 50 mM Tris pH 7.5, 150 mM KCl Chelex-purified buffer.
3. 1 M $MgCl_2$ and $CaCl_2$ stocks.

2.4.2 Equipment

1. MicroCal VP-ITC.
2. 1.5 ml Eppendorf tubes.
3. Hamilton syringe.
4. MicroCal ThermoVac (degas the samples).

2.5 Fluorescence Spectroscopy

2.5.1 Reagents

1. Purified protein samples.
2. 50 mM Tris pH 7.5, 150 mM KCl Chelex-treated buffer.
3. 1 M $MgCl_2$ and $CaCl_2$ stocks.
4. 2 mM ANS stock solution.
5. Tris (2-carboxyethyl) phosphine (TCEP).

2.5.2 Equipment

1. Hitachi F-7000 spectrofluorometer.
2. Quartz cuvette.

2.6 Equilibrium Unfolding

2.6.1 Reagents

1. 8 M GdmCl stock.
2. 50 mM Tris pH 7.5, 150 mM KCl Chelex-treated buffer.
3. 1 M $MgCl_2$ and $CaCl_2$ stocks.
4. 2 mM ANS stock solution.
5. Tris (2-carboxyethyl) phosphine.

2.6.2 Equipment

1. Refractometer.
2. Centrifuge.
3. 1.5 ml Eppendorf tubes.

4. Eppendorf tube stand.
5. Micropipettes and tips.

3 Methods

3.1 Modelling of Individual Trp Mutants

1. All Trp mutants should be modelled by using SWISS-MODEL program, with a suitable template (s-CaBP1 crystal structure was chosen as the template in case of caldendrin), to find out possible structural disturbances due to Trp insertion (*see* **Note 3**) [13].
2. To further rule out the hydrophobic environment of selected amino acid residues in proteins and Trp residues, modelled structures should be analyzed by solvent accessible surface area (SASA) calculation. Follow the below instructions to calculate SASA.
3. Load the structure in PyMOL.
4. Select the residue to calculate SASA.
5. Type as the following commends in the command line in PyMOL:
   (a) set dot_solvent.
   (b) set dot_solvent,1.
   (c) set dot_density,3.
   (d) get_area sele.
6. To rule out steric clashes in the modelled Trp mutants, use MolProbity software [14].
7. The Protein Interaction Calculator (PIC) software is an invaluable help for studying the different types of interactions between amino acids in wild-type and Trp mutant structures (*see* **Note 7**) [15].

3.2 Site-Directed Mutagenesis

3.2.1 Primer Design and Tm Calculation

The primers may be designed as per QuikChange II site-directed mutagenesis kit instructions and *Tm* of primers calculated by the formula given below:

For calculating *Tm*:

$$Tm = 81.5 + 0.41(\%GC) - (675/N) - \%\text{mismatch}$$

- N is the primer length in bases.
- Values for % GC and % mismatch are whole numbers.
- N does not include the bases which are being inserted or deleted.

3.2.2 DpnI Digestion

1. Amplified product should be treated with DpnI enzyme prior to transformation into DH5α cells.
2. DpnI treatment at 37 °C for 1 h results in digestion of methylated WT-plasmid (*see* **Note 8**).

3.2.3 Transformation

1. Keep the DH5α cells on ice for 10 min prior to addition of DpnI-digested PCR samples.
2. Add the digested samples to the cells and keep it on ice for 10 min.
3. Give heat shock to the cells at 42 °C for 90 s and place the cells immediately on ice for 10 min.
4. Add 900 μl of LB medium and keep the cells in a shaker incubator at 37 °C for 1 h.
5. Centrifuge the samples at low rpm (say 4000 rpm, ~1000 × *g*) for 5 min.
6. Remove the supernatant and retain up to 50 μl for resuspension.
7. Spread the cells with sterile spreader and incubate overnight at 37 °C.
8. Pick the well-grown colonies and allow them to grow overnight in 10 ml LB medium.
9. Isolate the plasmid for sequencing to confirm the mutation (*see* **Note 9**).
10. Transform the confirmed plasmid into BL21 (DE3) competent cells.

3.3 Protein Overexpression and Purification

1. BL21 (DE3) cells transformed with mutant positive plasmids should be checked for protein expression at low scale prior to large-scale protein expression and isolation.
2. In this protocol, all the mutants of caldendrin expressed well under similar conditions and were standardized against that of WT protein (0.2 mM IPTG at 37 °C for 4 h) (*see* **Note 10**).
3. Purify the mutants by chitin-intein affinity chromatography; the protein of interest would be expressed along with an intein tag that possesses chitin-binding domain, details of which is mentioned in the manufacturer's protocol provided with IMPACT™ kit, NEB.
4. Alternative methods of purification should be followed with respect to the choice of protein to be studied.
5. Lysis buffer composition includes chitin buffer with protease inhibitors, Triton X-100 detergent, and DNase I (*see* **Note 11**).
6. Sonicate for 40 min at 20% amplitude with 1 s ON and 1 s OFF pulse.

7. Centrifuge for 1 h at 12,000 rpm (~17000 × *g*).
8. Load the supernatant onto the column and wash with at least 10 column volumes (200 ml) of chitin buffer with 100 μM EDTA.
9. Activate column with the chitin buffer containing 50 mM DTT, and incubate it for 12 h (*see* **Note 12**).
10. Collect the eluates at 12-h interval until complete elution of protein from the column.
11. The eluted protein should be buffer-exchanged against Chelex-treated buffer till the concentration of added EDTA reduces to picomolar range.

3.4 Circular Dichroism

1. The near- and far-UV CD spectra recorded on a Jasco-810 spectropolarimeter.
2. The instrument should be purged with nitrogen gas for 30 min prior to starting the experiment, in order to avoid the formation of ozone due to the reaction of UV light with oxygen, which could be detrimental to the instrument.
3. Record the near-UV CD spectra for all proteins (1 mg/ml) over the wavelength range of 350–250 nm with a quartz cuvette of 0.5-cm path length.
4. At least 4–6 accumulations for each CD recording should be averaged to get the final spectrum.
5. Approximately 1 mg/ml protein in 50 mM Tris–HCl, pH 7.5, 150 mM NaCl, in presence of 5 mM TCEP is recommended for each recording.
6. A concentration of minimum 0.1 mg/ml is suggested for recording far-UV CD spectra between 200 and 250 nm with a quartz cuvette of 0.1-cm path length.

3.5 Isothermal Titration Calorimetry

1. Thermodynamics of Ca^{2+} binding in WT and mutant proteins should be performed by a MicroCal, VP-ITC, or any other method.
2. Take protein concentration of approximately 30–50 μM, prepared in 50 mM Tris, pH 7.5, 100 mM KCl buffer, may be appropriate in most cases (*see* **Notes 13** and **14**).
3. Standardize titration program that will titrate protein samples, saturated with 2 mM $MgCl_2$, with 5 mM $CaCl_2$.
4. Calculate association constant and other thermodynamic parameters using the provided Origin software.
5. The global dissociation constant K_d can be derived by the formula: $1/\sqrt{(k_{a1} * k_{a2})}$. ITC is optional for comparing the properties of mutant proteins with wild-type (*see* **Note 13**).

3.6 Fluorescence Spectroscopy

3.6.1 Intrinsic Trp Fluorescence

1. Trp fluorescence can be recorded in Hitachi F-7000 spectrofluorometer with a scan speed of 240 nm/min.
2. Trp is excited at 295 nm and emission can be recorded from 310 to 450 nm with 5 nm slit open on both ends.
3. Choose an optimal concentration of protein that gives appreciable fluorescence intensity (approximately 4 μM was optimal for caldendrin), and titrate the protein sample with Ca^{2+} by incubating protein for few minutes after each addition.
4. Record the fluorescence intensity values with each Ca^{2+} titration, and plot them against $[Ca^{2+}]$. This plot should be analyzed for comparing the response of different EF-hand motifs of the protein being studied, which will indicate the nature of Ca^{2+} filling to each site.

3.6.2 Extrinsic Fluorescence

1. To check the global surface hydrophobicity of the WT and mutants, ANS fluorescence should be done.
2. Add ANS to the protein samples (final concentration of ANS should be 10–100 μM) and incubate for 2 h before measurement.
3. ANS is generally excited at 370 nm wavelength and emission recorded from 400 to 600 nm.
4. Buffer blank (containing the same amount of ANS in the buffer) should be subtracted from all sample readings.
5. The wavelength corresponding to the peak (λ-max) and fluorescence intensity at the peak provide information about protein surface hydrophobicity and the environment where the ANS molecule is located.
6. Again, this is to check Trp mutations did not change the protein properties (*see* **Note 13**).

3.7 Equilibrium Unfolding

1. Prepare protein samples for unfolding experiments using guanidinium chloride (GdmCl) from 0 to 6 M, at 0.1 M interval between samples (*see* **Notes 15–18**).
2. All the samples should be analyzed by fluorescence, CD, and ANS experiments in identical conditions using optimal concentration of protein (0.1 mg/ml or ~4 μM concentration was sufficient for caldendrin).
3. In case of protein unfolding experiments, monitored by CD and ANS fluorescence, ellipticity and fluorescence intensity may be plotted, respectively, versus GdmCl concentration to visualize the trend of change.
4. The unfolding experimental data using Trp fluorescence may be fitted by two-state fitting using the following formula in GraphPad Prism 6. Data fitting should also be checked for

three or four states of unfolding; accordingly the equation will be modified as described earlier [16, 17]:

$$\Upsilon_{obs} = \frac{\{\Upsilon_N + \Upsilon_U \exp(-(\Delta G_U^o - m[D])/RT)\}}{\{1 + \exp(-(\Delta G_U^o - m[D])/RT)\}}$$

where Υ_N = fluorescence of the native state Υ_U = fluorescence of the unfolded state $\Delta G_U{}^\circ$ = standard free energy change of unfolding $[D]$ = concentration of the denaturant m = slope of the linear relationship between $\Delta G_U{}^\circ$ and $[D]$

The $\Delta G_U{}^\circ$ values, obtained from the fitting procedures, could be compared between different groups and conclusions may be drawn.

5. The unfolding transition profiles of different Trp mutants may or may not be similar which will indicate a difference in stabilities, representing comparative flexibilities of different motifs and how each responds to Ca^{2+}.

4 Notes

1. The Trp mutation should be located such that it is near to the EF-hand loop that binds Ca^{2+}; this will enable the introduced Trp residue to sense even minor structural changes.
2. It is suggested that either Tyr or Phe residues should be selected for replacing by Trp (provided they are proximal to the Ca^{2+}-binding loop), so that serious or adverse structural alterations may be avoided.
3. The length of the mutagenic primers should be between 25 and 40 bases for obtaining a successful PCR product with the desired mutation.
4. In case of unsuccessful amplification, despite several trials, primer length can be tweaked to attain the desired product.
5. The mutagenic site should be as close to the center of the primer as possible, with at least 15 bases on either side of the site.
6. The presence of G or C at the ends of primer would help in better annealing to the target site.
7. Crystal structure of the selected protein may be used as a template to model the mutants. Structural constraints or clashes could be analyzed with the help of t open access software like SWISS-MODEL, MolProbity, and PIC.
8. DpnI digestion (selectively digesting methylated DNA) is strongly recommended prior to transformation of the amplified product, in order to get rid of WT-DNA and hence minimize false-positive colonies.

9. Positive colonies should be validated by target gene sequencing.
10. Freshly prepared 1 M IPTG stock in autoclaved MQ water is preferable to solutions stored for a long time.
11. Addition of DNase and protease inhibitor to the lysis buffer and washing with nonionic detergent are highly recommended. This avoids protein degradation and nucleic acid and non-specific impurities in protein preparation, which could lead to noise in fluorescence spectroscopy.
12. The 1 M DTT solution should be prepared freshly. Prolonged storage at 4 °C results in loss of chemical activity.
13. Validation of structural and functional properties of mutant proteins is essential to make sure that the results are not due to mutation-induced structural changes.
14. Samples must be degassed before the ITC experiments to avoid the shift in baseline due to air bubbles.
15. In case of unfolding experiments, pipetting should be done extremely carefully and accurately, since erroneous preparation of samples for 60–70 data points may lead to erroneous results.
16. In this protocol, structural changes of the protein at different restricted Ca^{2+} concentrations have been demonstrated; therefore it is imperative to take serious precautions while preparing stock dilutions for the desired final concentration in a sample.
17. More data points are recommended for the initial range, from 0 to 2 M GdmCl, to enable identification of initial intermediate states in the unfolding pathway, if any.
18. The prepared chemical unfolding samples must be mix thoroughly by inverting the tubes 5–10 times to ensure uniform distribution of GdmCl. This should be followed by a short spin to settle the sample properly.

Acknowledgments

The work was supported by the CSIR fast-track SRA to AC, CSIR, DST, and DBT grants to YS and DFG (Kr1879/3-1) to MRK.

References

1. Berridge MJ, Bootman MD, Roderick HL (2003) Calcium: calcium signalling: dynamics, homeostasis and remodelling. Nat Rev Mol Cell Biol 4:517–529
2. Zhou Y, Xue S, Yang JJ (2013) Calciomics: integrative studies of Ca^{2+}-binding proteins and their interactomes in biological systems. Metallomics 5:29–42
3. Raghuram V, Sharma Y, Kreutz MR (2012) Ca^{2+} sensor proteins in dendritic spines: a race for Ca^{2+}. Front Mol Neurosci 5:61
4. Mäler L, Blankenship J, Rance M, Chazin WJ (2000) Site–site communication in the

EF-hand Ca^{2+}-binding protein calbindin D9k. Nat Struct Mol Biol 7:245–250

5. Wojciechowski D, Fischer M, Fahlke C (2015) Tryptophan scanning mutagenesis identifies the molecular determinants of distinct barttin functions. J Biol Chem 290:18732–18743
6. Deacon LJ, Billones H, Galyean AA, Donaldson T, Pennacchio A, Iozzino L, Dattelbaum JD (2014) Tryptophan-scanning mutagenesis of the ligand binding pocket in *Thermotoga maritima* arginine-binding protein. Biochimie 99:208–214
7. De Feo CJ, Mootien S, Unger VM (2010) Tryptophan scanning analysis of the membrane domain of CTR-copper transporters. J Membr Biol 234:113–123
8. Yamniuk AP, Silver DM, Anderson KL, Martin SR, Vogel HJ (2007) Domain stability and metal-induced folding of calcium-and integrin-binding protein 1. Biochemistry 46:7088–7098
9. Kiran U, Regur P, Kreutz MR, Sharma Y, Chakraborty A (2017) Intermotif communication induces hierarchical Ca^{2+} filling of Caldendrin. Biochemistry 56:2467–2476
10. Seidenbecher CI, Langnaese K, Sanmartí-Vila L, Boeckers TM, Smalla KH, Sabel BA, Garner CC, Gundelfinger ED, Kreutz MR (1998) Caldendrin, novel neuronal calcium-binding protein confined to the somatodendritic compartment. J Biol Chem 273:21324–21331
11. Reddy PP, Raghuram V, Hradsky J, Spilker C, Chakraborty A, Sharma Y, Kreutz MR (2014) Molecular dynamics of the neuronal EF-hand Ca^{2+}-sensor Caldendrin. PLoS One 9:e103186
12. Théret I, Baladi S, Cox JA, Sakamoto H, Craescu CT (2000) Sequential calcium binding to the regulatory domain of calcium vector protein reveals functional asymmetry and a novel mode of structural rearrangement. Biochemistry 39:7920–7926
13. Arnold K, Bordoli L, Kopp J, Schwede T (2006) The SWISS-MODEL workspace: a web-based environment for protein structure homology modelling. Bioinformatics 22:195–201
14. Chen VB, Arendall WB, Headd JJ, Keedy DA, Immormino RM, Kapral GJ, Richardson DC (2010) MolProbity: all-atom structure validation for macromolecular crystallography. Acta Crystallogr D Biol Crystallogr 66:12–21
15. Tina KG, Bhadra R, Srinivasan N (2007) PIC: protein interactions calculator. Nucleic Acids Res 35:473–476
16. Hung HC, Chen YH, Liu GY, Lee HJ, Chang GG (2003) Equilibrium protein folding-unfolding process involving multiple intermediates. Bull Math Biol 65:553–570
17. Muralidhar D, Jobby MK, Kannan K, Annapurna V, Chary KV, Jeromin A, Sharma Y (2005) Equilibrium unfolding of neuronal calcium sensor-1: N-terminal myristoylation influences unfolding and reduces the protein stiffening in the presence of calcium. J Biol Chem 280:15569–15578

Chapter 36

Mapping Calcium-Sensitive Regions in GCAPs by Site-Specific Fluorescence Labelling

Karl-Wilhelm Koch and Jens Christoffers

Abstract

Signal transduction processes that are under control of changes in cytoplasmic Ca^{2+}-concentration involve Ca^{2+}-sensor proteins, which often undergo pronounced conformational transitions triggered by Ca^{2+}. Consequences of conformational changes can be the structural rearrangement of single amino acids, exposition of small patches of several amino acids, or the movement of whole protein regions or domains. Furthermore, these conformational changes can lead to the exposure or movement of posttranslationally attached acyl groups. These processes could then control the function of target proteins, for example, by Ca^{2+}-dependent protein–protein interaction. Fluorescence spectroscopy allows for mapping these Ca^{2+}-sensitive regions but needs site-specific fluorescence labelling. We describe the application of a new group of diaminoterephthalate-derived fluorescence probes targeting either cysteines in guanylate cyclase-activating proteins, named GCAPs, or azide moieties in covalently attached acyl groups. By monitoring Ca^{2+}-dependent changes in fluorescence emission, we identify Ca^{2+}-sensitive protein regions in GCAPs and correlate conformational changes to protein function.

Key words Neuronal calcium sensor proteins, GCAP, Fluorescence, Myristoylation, Diaminoterephthalates

1 Introduction

Changes in intracellular concentration ($[Ca^{2+}]_i$) control a variety of physiological responses, which are mediated by Ca^{2+}-sensor proteins that regulate the specific function of a target protein. The prototype of a Ca^{2+}-sensor protein is calmodulin (CaM) that contains four so-called EF-hand Ca^{2+}-binding motifs consisting of a helix-loop-helix folding unit. CaM regulates many target proteins by adopting different three-dimensional structures, when a complex builds up under control of varying Ca^{2+}-concentrations [1]. Although CaM serves in many aspects as a benchmark for members of the protein superfamily of EF-hand Ca^{2+}-binding proteins, it differs in many molecular and regulatory aspects from other members.

Claus W. Heizmann (ed.), *Calcium-Binding Proteins of the EF-Hand Superfamily: From Basics to Medical Applications*, Methods in Molecular Biology, vol. 1929, https://doi.org/10.1007/978-1-4939-9030-6_36,

For example, neuronal Ca^{2+}-sensor (NCS) proteins regulate as a whole group a multitude of targets, but most of the single members have only a few or just one interacting partner out of a large diversity of different proteins [2, 3]. Further, NCS proteins have a rather similar compact and globular three-dimensional folding that differs from the conformational plasticity of CaM. Therefore, it is an unresolved issue how NCS proteins achieve target specificity.

Most NCS proteins harbor in the amino-terminal region consensus sites for myristoylation or palmitoylation. Acylation of proteins facilitates membrane association and contributes to specific cellular and subcellular distribution of NCS proteins. It can further modify the functional properties of proteins. Some myristoylated NCS proteins perform a Ca^{2+}-myristoyl switch, where the myristoyl group is buried in a hydrophobic cleft of the protein in the Ca^{2+}-free state and becomes exposed in the Ca^{2+}-bound state (Fig. 1). The protein dynamics of this process were deeply studied in the photoreceptor cell protein recoverin [4] but are less well understood in other NCS proteins [5]. Related to recoverin and also expressed in the rod and cone cells of vertebrate retinae is the NCS subgroup named guanylate cyclase-activating proteins (GCAPs), of which different paralogues and orthologues are known. Two GCAP forms are found in mice and bovine [6, 7], three in human [8], and six to eight in teleost fish [9–11]. GCAPs differ in their key properties, in particular in Ca^{2+}/Mg^{2+}-sensing and Ca^{2+}-dependent regulation of target guanylate cyclase activities. Differential Ca^{2+}-sensing of mammalian and zebrafish GCAPs accounts for the step-by-step response [12–15] to changing Ca^{2+}-levels leading to the following questions: What are the conformational changes in

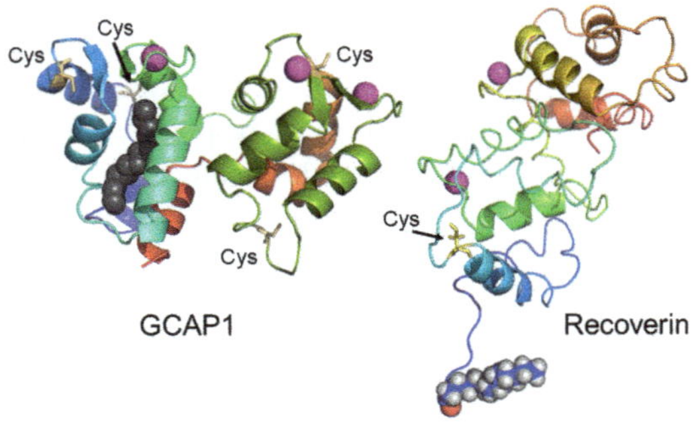

Fig. 1 Structural model of Ca^{2+}-bound GCAP1 (left part) and recoverin (right part). Tertiary structure of myristoylated GCAP1 created by PyMOL based on the crystal structure (PDB code: 2R2I) and that of recoverin based on the NMR structure (PDB code: 1JSA). Ca^{2+}-ions are symbolized in magenta spheres and cysteine residues are highlighted by yellow sticks. The myristoyl group is buried inside a cavity of GCAP1 (shown in gray spheres), but it is fully exposed in Ca^{2+}-saturated recoverin (myristoyl chain shown in blue and light gray)

related proteins that control on- and off-states of a particular Ca^{2+}-sensor allowing a distinct target regulation? Which regions in Ca^{2+}-sensor proteins determine their Ca^{2+}-sensitive properties and responsiveness? One experimental approach to address these questions are fluorescence studies using site-specific dye labelling and subsequent fluorescence spectroscopy. Cysteine residues are well suited for this approach, since they can selectively be modified by cysteine-reactive functional groups in dyes, mainly maleimide derivatives. The maleimide moiety is susceptible for conjugated addition of a thiol residue [16–18].

Another reactive functional group is cyclooctyne, which came recently into the focus of biological applications, since it can be ligated to biomolecules carrying azide moieties in a so-called copper-free click reaction [19–21], which is the uncatalyzed 1,3-dipolar cycloaddition of an alkyne and an organo azide furnishing a 1,2,3-triazole ring as ligating motif. Due to the pronounced ring strain of the cycloalkyne, this reaction proceeds smoothly and rapidly at ambient temperature without any copper catalyst. This is a newer development in the field of Click chemistry [22–24].

As the chromophore for carrying the functional unit, the maleimide or the cyclooctyne, we have chosen a 2,5-diaminoterephthalate (DAT) structural motif which is a powerful chromophore with remarkable fluorescence properties [25]. Containing two carboxylate and two amino functions, it defines a colored molecular scaffold which allows for orthogonal functionalization with up to four different functional units. In some cases, DATs serve as "turn-on" probes, because their fluorescence quantum yields are increased by orders of magnitude upon reaction with the molecular target, e.g., the conjugated addition of a thiol to a maleimide functional residue.

2 Materials

2.1 Heterologous Expression and Purification of GCAPs

1. Biological material, *Escherichia coli* BL21-Codon Plus (DE3) cells; plasmids containing GCAP-DNA (pET-11a/GCAP1 for expression of wild-type GCAP1 and pET21a/GCAP2 for expression of wild-type GCAP2); plasmid pBB131 (yeast *N*-myristoyltransferase).
2. Standard LB medium and dYT medium.
3. Antibiotics stock, 100 mg/mL ampicillin and 30 mg/mL kanamycin.
4. 2.5 mg/mL myristic acid in 50% ethanol.
5. 1 M isopropyl-β-D-thiogalactoside (IPTG).
6. Resuspension buffer: 50 mM Tris–HCl pH 8.0.
7. 1000 μg/mL DNAse.

8. 1 M DTT and 1 mM phenylmethanesulfonyl fluoride (PMSF).
9. Solubilization buffer, 6 M guanidinium hydrochloride, 1 mM DTT.
10. Dialysis buffer I, 150 mM NaCl, 20 mM Tris–HCl pH 8.0, 1 mM DTT.
11. Gel filtration buffer, 150 mM NaCl, 20 mM Tris–HCl, pH 7.5, 1 mM DTT.
12. Dilution buffer, 20 mM Tris–HCl pH 7.5, 1 mM DTT.
13. Stock solutions of 1 M $CaCl_2$ and 1 M EGTA.
14. Anion exchange buffer A, 50 mM Tris–HCl, pH 7.5, 50 mM NaCl, 1 mM DTT, EGTA, or $CaCl_2$ as indicated.
15. Anion exchange buffer B, 20 mM Tris pH 7.5, 1 M NaCl, 1 mM DTT, EGTA, or $CaCl_2$ as indicated.
16. Dialysis buffer II, 50 mM NH_4HCO_3.

2.2 Synthesis and Characterization of Fluorescent Dyes

The synthesis of both reactive probes **3** and **7** is routine synthetic organic chemistry and was published before in detail [26]. Both preparations started from DAT monomethyl ester **1**, which is actually accessed from succinic anhydride in five steps according to literature protocols [27]. The maleimide function is introduced by amidation of acid **1** and ammonium salt **2** (prepared by literature procedures, ref. [28]) with the aid of HATU [*O*-(7-azabenzotriazole-1-yl)-*N*,*N*,*N'*,*N'*-tetramethyluronium hexafluorophosphate] as coupling reagent and DIPEA (ethyldiisopropylamine) as base (step a). Compound **3** was purified by column chromatography and obtained in 85% chemical yield. As mentioned in the introduction, compound **3** behaves as a "turn-on probe," i.e., the fluorescence quantum yield increased by conjugated addition of a thiol from 0.1% to 8%. The synthesis of cyclooctyne derivative required three steps and started with the amidation of carboxylic acid **1** and mono-Boc ethylene diamine ($H_2NCH_2CH_2NHBoc$, prepared according to a literature protocol, ref. [29]) again with using HATU-DIPEA as coupling reagent (step b, 81% yield). The Boc-protective group was cleaved with trifluoroacetic acid (step c, 93%), and resulting amine **5** was converted to active carbonate **6** (commercially available from Sigma-Aldrich or Synaffix BV; HOSucc is a *N*-hydroxysuccinimide leaving group) to furnish cyclooctyne probe **7** (step d, 98% yield).

2.3 Site-Specific Labelling of GCAPs and Functional Test

1. Labelling buffer, 10 mM HEPES-KOH, pH 7.4 and 1 mM EGTA.
2. NAP-5 column for gel filtration.
3. Gel filtration buffer, 150 mM NaCl, 20 mM Tris–HCl, pH 7.5.
4. 50 µg/mL 12-azidododecanoic acid.

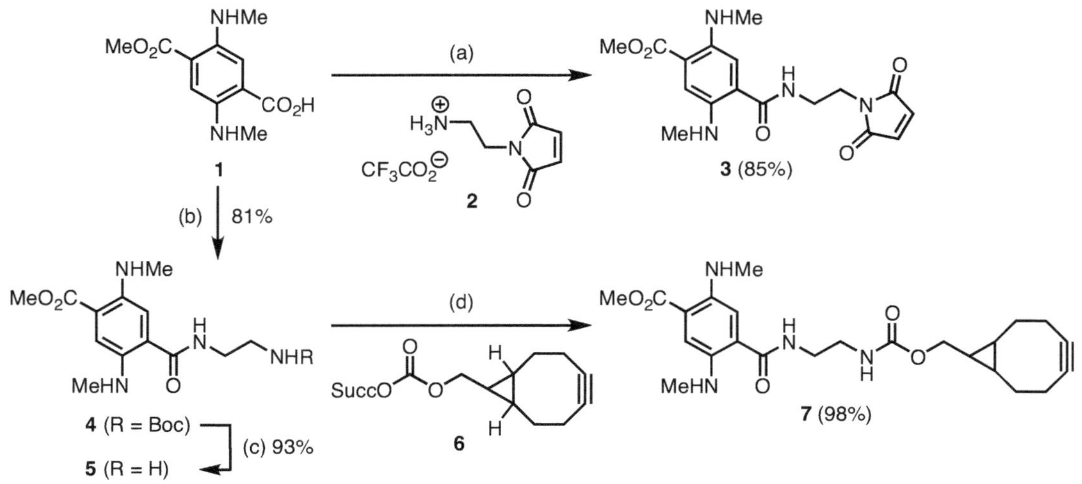

Scheme 1 Synthesis of maleimide (compound **3**) and cyclooctyne (compound **7**) functionalized probes from DAT monoester **1**. Precursors for reactive units are aminoethyl functionalized maleimide **2** (as salt with trifluoroacetic acid) and cyclooctyne building block **6**, which is an active carbonate of *N*-hydroxysuccinimide ("Succ"). For synthetic steps (**a**) to (**d**), see text above

5. Dissolving buffer, 10 mM HEPES-KOH, pH 7.4, 1 mM EGTA.
6. High-performance liquid chromatography column, Purospher C-18 column.
7. 0.1% trifluoroacetic acid (TFA) and 2 mM EGTA in H_2O.

2.4 Probing Conformational Changes

1. Fluorescence spectrophotometer, Photon Technology International (810/814 Photomultiplier Detection System or a similar device).
2. Stock solution of 1 M K_2H_2EGTA and 1 M CaH_2EGTA (*see* **Note 1**).
3. 1 M CsCl.

3 Methods

Site-directed mutagenesis of GCAP forms is based on standard cloning techniques, and well-established procedures are outlined in details in several publications including exchange of cysteine and other positions [30–32]. Mutagenesis is planned and conducted to have one cysteine residue per GCAP at different positions. The following procedures are described for bovine GCAP1 and GCAP2, of which WT forms have four and three cysteines, respectively. Native GCAPs are acylated, mainly myristoylated, at the amino-terminus. For expression in *Escherichia coli*, co-expression of yeast *N*-myristoyltransferase is necessary, when myristoylated variants are produced. If GCAPs have not a complete consensus sequence site for myristoylation by *N*-myristoyltransferase, an additional point mutation in the amino-terminal region is necessary to yield a high degree of myristoylation (e.g., D6S mutant of bovine GCAP1; *see* also ref. [33]).

3.1 Heterologous Expression and Purification of GCAPs

1. Overexpress GCAPs in *Escherichia coli* BL21-Codon Plus (DE3) cells. Therefore, transform competent *E. coli* cells with plasmids containing GCAP-DNA (pET-11a/GCAP1 for expression of wild-type GCAP1 and pET21a/GCAP2 for expression of wild-type GCAP2). Check amino acid sequences of GCAPs for a complete consensus site for myristoylation by yeast *N*-myristoyltransferase [34]. Use appropriate point mutants of GCAPs, if necessary. Cotransform cells with the plasmid pBB131 harboring the gene for a yeast *N*-myristoyltransferase (kindly provided by Dr. J. Gordon, Washington University School of Medicine, St. Louis, MO, USA) for producing myristoylated variants.
2. Thaw competent cells (100 μL) on ice, add plasmid DNA (1–10 ng), and incubate for 30 min on ice. Briefly increase temperature to 42 °C for 25 s and incubate on ice for 2 min. Add LB medium (400 μL) and grow cells for 1 h at 37 °C. Spread 50–500 μL on agar plates containing appropriate antibiotics (100 μg/mL ampicillin and 30 μg/mL kanamycin). Incubate overnight at 37 °C.
3. Supplement 5 mL of dYT medium with 100 μg/mL ampicillin and 25 μg/mL kanamycin, inoculate medium with single colonies of transformed cells, and grow at 37 °C overnight. Add 2 mL of an overnight culture to 500 mL of dYT medium containing 100 μg/mL ampicillin and 25 μg/mL kanamycin. Grow cells by intense shaking at 37 °C. Dissolve myristic acid in 50% ethanol and add to cell suspension at a final concentration of 100 μg/mL at an OD of 0.4 (expression of a myristoylated protein). Induce expression at an OD of 0.6 by adding 1 mM final concentration of isopropyl-β-D-thiogalactoside (IPTG). After 4 h, harvest cells by centrifugation (5000 × *g*, 10 min, 4 °C). Resuspend the pellet resulting from 1 L bacterial culture in 40 mL of 50 mM Tris–HCl pH 8.0.
4. Add 100 μg/mL lysozyme and 5 μg/mL DNAse to the bacteria suspension. Incubate at 30 °C for 30 min (water bath) and add 1 mM DTT and 0.1 mM PMSF. Centrifuge at 360,000 × *g* for 30 min at 4 °C. GCAPs are present in the soluble fraction and in inclusion bodies. For further purification from the soluble fraction, continue at **step** 7. Purification from inclusion bodies needs solubilization and subsequent dialysis to refold the protein.

5. Solubilize the pellet in 6 M guanidinium hydrochloride, 1 mM DTT (40–60 mL per 1 L of culture). Dissolve by stirring at room temperature. Dialyze against 5 L of dialysis buffer I for 5 h. Repeat dialysis (*see* **Note 2**).
6. Precipitate unsoluble material by centrifugation (360,000 × *g*, 30 min, 4 °C). Take the supernatant and add 67% $(NH_4)_2SO_4$ to precipitate proteins. Pellet by centrifugation at 60,000 × *g* for 30 min at 4 °C. The protein pellet can be stored at −20 °C for later use.
7. Dissolve the $(NH_4)_2SO_4$-pellet of precipitated GCAP in 2–3 mL of bidistilled water. Separate completely dissolved from non-dissolved protein by centrifugation (100,000 × *g*, 15 min, 4 °C). Continue with a size exclusion chromatography using a HiLoad 16/60 Superdex prep grade column. Equilibrate column with gel filtration buffer. Add 2 mM EGTA for the purification of GCAP1 or 2 mM $CaCl_2$ for the purification of GCAP2. Adjust a flow rate of 1 mL/min. Start chromatography by injection of the GCAP sample. Collect the eluted fraction and analyze fractions by SDS-PAGE.
8. Identify and combine the GCAP-containing fractions. Adjust to 50 mM NaCl by dilution with 20 mM Tris–HCl pH 7.5, 1 mM DTT. Equilibrate an anion exchange column (IEC, MonoQ or UnoQ) with anion exchange buffer A (with EGTA or $CaCl_2$ as indicated above). Apply the protein on the column at flow rate of 0.5 mL/min. First, wash the nonbound proteins from the column. Then increase the flow rate to 3 mL/min, and elute bound proteins by a gradient of 13 column volumes of 0–55% buffer B (EGTA or $CaCl_2$ as indicated). Collect the fractions and analyze by SDS-PAGE.
9. Combine GCAP-containing fractions and dialyze against 50 mM NH_4HCO_3. Repeat dialysis. Aliquot samples (0.5 or 1 mg) and lyophilize. Store at −80 °C until use.

3.2 Site-Specific Labelling of Cysteine Residues in GCAPs

Fluorescent dyes with a maleimide group for covalent coupling to cysteine residues are used for site-specific labelling of GCAP variants [26, 35, 36]. Ideally, proteins harbor one cysteine that is either present at the wild-type position or is engineered in a region of interest for structure/functional studies.

1. Dissolve lyophilized protein in dissolving buffer. Incubate with a fivefold excess of the thiol reactive dye (compound **3** in Scheme 1) for 2–3 h at room temperature. Add 2 mM DTT to terminate the reaction.
2. Equilibrate a NAP5 column with gel filtration buffer. Add 2 mM $MgCl_2$ to this buffer, if it is necessary for subsequent experiments. Remove excessive dye by size exclusion chromatography using the NAP5 column.

3. Determine the protein concentration by a Bradford assay. Measure the dye concentration by UV–VIS spectroscopy using molar extinction coefficients (*see* **Note 3**).

3.3 Site-Specific Labelling of an Acyl Residue Covalently Bound to GCAPs

In order to investigate the location or movement of a covalently bound acyl group (e.g., myristoyl chain in the case of GCAPs), we expressed the corresponding GCAP with an attached 12-azidododecanoic acid as a myristoyl substitute. Subsequent reaction with an alkyne-containing fluorescent dye results in a dye-labelled fatty acid incorporated into the protein [26].

1. Express acylated GCAP forms by cotransformation of *E. coli* cells with the plasmid pBB-131 containing the gene for yeast (*S. cerevisiae*) *N*-myristoyltransferase (*see* also Subheading 3.1 above). Supplement LB medium with 12-azidododecanoic acid (50 μg/mL) instead of myristic acid for incorporation of a fatty acid moiety with a functionalized azide group.
2. Purify GCAPs as described in Subheading 3.1.
3. Dissolve lyophilized protein in dissolving buffer (*see* **Note 4**). Incubate with a tenfold excess of the cyclooctyne-DAT conjugate (compound 7 in Scheme 1) for 2–3 h at room temperature (*see* **Note 5**).
4. Remove excessive dye by size exclusion chromatography using a NAP5 column and buffer conditions as described under Subheading 3.2, **step 2**.
5. Determine the degree of fluorescence labelling by high-performance liquid chromatography. Inject approx. 10 μg of labelled GCAP onto a Purospher C-18 column that is equilibrated in H_2O containing 0.1% TFA and 2 mM EGTA. Start elution using a linear gradient of acetonitrile (+0.1% TFA).
6. Monitor the elution profile from 220 nm to 500 nm to distinguish between labelled and non-labelled GCAP forms. Determine the relative amount, for example, by comparing the peak areas.

3.4 Functional Test of Modified GCAPs as Regulators of Guanylate Cyclase Activity

Dye labelling of GCAPs could change or even disturb their functional properties. The following tests are designed to detect possible interference of dye labelling with the regulation of membrane-bound photoreceptor-specific guanylate cyclases. GCAPs activate target guanylate cyclases at low Ca^{2+}-concentration undergoing a switch to a Mg^{2+}-bound state and inhibit targets at saturating high Ca^{2+}-concentration. Functional properties are tested by measuring guanylate cyclase activity in the presence of sufficient GCAP concentration (5–10 μM) and varying free Ca^{2+}-concentration. A second set of experiments is designed to measure the GCAP-dependent activation profile of a target GC at a fixed low Ca^{2+}-

concentration. A detailed protocol has been described in previous contributions and is not repeated here [12, 30, 35, 36].

3.5 Probing Conformational Changes

Linking a fluorescent dye to different protein regions allows distinguishing different conformational states of a protein. Labelled proteins can be probed by time-resolved fluorescence spectroscopy or steady-state fluorescence spectroscopy. We will focus on the application of steady-state fluorescence, but time-resolved fluorescence spectroscopy has been applied for studying protein dynamics bovine GCAP1 and GCAP2 [35, 36].

1. Use a commercially available fluorescence spectrophotometer, for example, from Photon Technology International (810/814 Photomultiplier Detection System).
2. Use 10 μM of labelled GCAP in Tris/NaCl buffer (*see* Subheading 3.2, **step 2**).
3. Adjust the measuring mode of the spectrophotometer according to the dye's properties. For example, the maleimide-functionalized dye (compound **3** in Scheme 1) attached to particular cysteine residues of GCAP mutants is excited at 425 nm; record the emission spectra from 540 to 650 nm. The cyclooctyne derivative (compound 7 in Scheme 1) attached to the *N*-terminal fatty acid of GCAP is excited at 469 nm, and emission spectra are recorded from 540 to 650 nm.
4. Ca^{2+}-dependent recordings of fluorescence emission intensities can be performed by monitoring changes in emission at different free Ca^{2+}-concentrations. Therefore, adjust the Ca^{2+}-concentration in the cuvette by a mixture of K_2H_2EGTA and CaH_2EGTA as described previously [37]. Try to cover a range of free Ca^{2+}-concentrations between 0.01 and 30 μM. Use exact stock solutions of 1 M (or higher) $CaCl_2$ and dilute accordingly in case you need to extend the titrations.
5. In case you test for Mg^{2+}-dependencies, record fluorescence emission in the presence of 100 μM EGTA.
6. For data recording and analysis, use the software package Felix32 (Photon Technology International). Determine the maximum fluorescence emission and plot as the function of free Ca^{2+}-concentration. Determine the Ca^{2+}-concentration at which the fluorescence emission is half-maximal.
7. For testing the solvent exposure of covalently attached dyes, use a fluorescence-quenching assay. Increase the concentration of CsCl (millimolar range) and record fluorescence emission. Compare the emission rates in the presence and absence of quencher. Less quenching is expected, when the fluorescent dye is more or less shielded from the solvent.

4 Notes

1. Tsien and Pozzan published an easy-to-follow procedure for the calculation and preparation of Ca-EGTA buffers [37]. The Ca^{2+}-chelator ethyleneglycol-bis(2-aminoethyl)ether-*N*,*N*,*N′*,*N′*-tetraacetic acid, EGTA abbreviated, is very often used, when buffers containing a very low Ca^{2+}-concentration are needed. Briefly, for preparing 25 mL 0.1 M K_2H_2EGTA, calculate with ca. two equivalent KOH per one equivalent EGTA (free acid).

   The stock solution of $K_2CaEGTA$ is prepared as follows: 0.96 g EGTA free acid were mixed in 15 mL H_2O_{bidest} with 0.238 g $CaCO_3$ and ca. 0.3 g KOH. When no longer CO_2 is released, the pH-value will be adjusted with 1 M KOH to ca. 7.4. Please make a note of the pH-value during the titration. Afterwards add 50 μL of 1 M $CaCl_2$-solution and the pH-value will be noted. The pH-value will reach the starting point by slow addition of KOH in small steps. Continue with this procedure until the addition of $CaCl_2$ leads to change of the pH, which is less than half of the change at the starting point. Rinse the pH-electrode with a small volume of H_2O_{bidest}, combine the rinsing water with the $K_2CaEGTA$-solution, and fill everything up to 25 mL. Occasionally appearing precipitates can be removed by filtration. Both stock solutions should have a pH $\approx$ 7.4.

2. Most GCAP variants can be solubilized from bacterial inclusion bodies by high concentrations of guanidinium hydrochloride (or urea), and subsequent refolding was in most cases successful. Subsequent guanylate cyclase assays [5, 35, 36] and/or circular dichroism spectroscopy can be used to test for correct of function and folding.

3. The molar absorbance coefficients of DAT dyes **3** and **7** in CH_2Cl_2 at λ_{max} = 454 nm (for both compounds) are $\varepsilon = 3388\ M^{-1}\ cm^{-1}$ (compound **3**) and $\varepsilon = 4571\ M^{-1}\ cm^{-1}$ (compound **7**). These data were determined from solutions in CH_2Cl_2 with defined concentrations.

4. GCAPs (and other NCS protein-like recoverin as well) tend to stick to glass surfaces leading to a lower protein concentration than expected from the protein determination before lyophilisation. Mild sonification can help to detach the protein from the surface. Precoating of the bare glass surface with another protein (e.g., serum albumin) is only recommended when the protein does not interfere with subsequent experiments with GCAPs.

5. The buffer for reaction of GCAPs with diaminoterephthalates can be varied to test for accessibility of functional groups in the

protein. For example, adding Ca^{2+} instead of EGTA or adjusting different salt concentrations or a different pH can trigger conformational changes, which might become detectable by a change in dye accessibility.

Acknowledgments

This work was supported by a grant from the Deutsche Forschungsgemeinschaft (GRK 1885).

References

1. Ikura M, Ames JB (2006) Genetic polymorphism and protein conformational plasticity in the calmodulin superfamily: two ways to promote multifunctionality. Proc Natl Acad Sci U S A 103:1159–1164
2. Philippov PP, Koch K-W (eds) (2006) Neuronal calcium sensor proteins. Nova Publishers, Hauppauge, NY
3. Burgoyne RD, Haynes LP (2014) Sense and specificity in neuronal calcium signalling. Biochim Biophys Acta 1853:1921–1932
4. Ames JB, Ishima R, Tanaka T et al (1997) Molecular mechanics of calcium-myristoyl switches. Nature 389:198–202
5. Sulmann S, Dell'Orco D, Marino V et al (2014) Conformational changes in calcium-sensor proteins under molecular crowding conditions. Chem Eur J 20:6756–6762
6. Dizhoor AM, Olshevskaya EV, Henzel WJ et al (1995) Cloning, sequencing, and expression of a 24-kDa Ca^{2+}-binding protein activating photoreceptor guanylyl cyclase. J Biol Chem 270:25200–25206
7. Howes K, Bronson JD, Dang YL et al (1998) Gene array and expression of mouse retina guanylate cyclase activating proteins 1 and 2. Invest Ophthalmol Vis Sci 39:867–875
8. Haeseleer F, Sokal I, Li N, Pettenati M et al (1999) Molecular characterization of a third member of the guanylyl cyclase-activating protein subfamily. J Biol Chem 274:6526–6535
9. Imanishi Y, Yang L, Sokal I et al (2004) Diversity of guanylate cyclase-activating proteins (GCAPS) in teleost fish: characterization of three novel GCAPs (GCAP4, GCAP5, GCAP7) from zebrafish (Danio rerio) and prediction of eight GCAPs (GCAP1-8) in pufferfish (Fugu rubripes). J Mol Evol 59:2204–2217
10. Takemoto N, Tachibanaki S, Kawamura S (2009) High cGMP synthetic activity in carp cones. Proc Natl Acad Sci U S A 106:11788–11793
11. Sulmann S, Vocke F, Scholten A et al (2015) Retina specific GCAPs in zebrafish acquire functional selectivity in Ca^{2+}-sensing by myristoylation and Mg^{2+}-binding. Sci Rep 5:11228
12. Hwang JY, Lange C, Helten A et al (2003) Regulatory modes of rod outer segment membrane guanylate cyclase differ in catalytic efficiency and Ca^{2+}-sensitivity. Eur J Biochem 270:3814–3821
13. Peshenko IV, Olshevskaya EV, Savchenko AB et al (2011) Enzymatic properties and regulation of the native isozymes of retinal membrane guanylyl cyclase (RetGC) from mouse photoreceptors. Biochemistry 50:5590–5600
14. Koch KW, Dell'Orco D (2013) A calcium relay mechanism in vertebrate phototransduction. ACS Chem Neurosci 4:909–917
15. Peshenko IV, Dizhoor AM (2004) Guanylyl cyclase-activating proteins (GCAPs) are Ca^{2+}/Mg^{2+} sensors: implications for photoreceptor guanylyl cyclase (RetGC) regulation in mammalian photoreceptors. J Biol Chem 279:16903–16906
16. Wu D, Cheung S, Devocelle M et al (2015) Synthesis and assessment of a maleimide functionalized BF_2 azadipyrromethene near-infrared fluorochrome. Chem Commun 51:16667–16670
17. Haimi P, Sikorskaite-Gudziuniene S, Baniulis D (2015) Application of multiplexed cysteine-labeled complex protein sample for 2D electrophoretic gel alignment. Proteomics 15:1777–1780
18. Dietz L, Bosque A, Pankert P et al (2009) Quantitative DY-maleimide-based proteomic 2-DE-labeling strategies using human skin proteins. Proteomics 9:4298–4308

19. Jewett JC, Bertozzi CR (2010) Cu-free click cycloaddition reactions in chemical biology. Chem Soc Rev 39:1272–1279
20. Bernard S, Audisio D, Riomet M et al (2017) Bioorthogonal click and release reaction of iminosydnones with cycloalkynes. Angew Chem Int Ed 56:15612–15616 Angew Chem 129: 15818–15822
21. Mamot A, Sikorski PJ, Warminski M et al (2017) Azido-functionalized 5′ cap analogues for the preparation of translationally active mRNAs suitable for fluorescent labeling in living cells. Angew Chem Int Ed 56:15628–15632 Angew Chem 129: 15834–15838
22. Tornøe CW, Christensen C, Medal M (2002) Peptidotriazoles on solid phase: [1,2,3]-triazoles by regiospecific copper(I)-catalyzed 1,3-dipolar cycloadditions of terminal alkynes to azides. J Org Chem 67:3057–3064
23. Rostovtsev VV, Green LG, Fokin VV et al (2002) A stepwise Huisgen cycloaddition process: copper(I)-catalyzed regioselective "ligation" of azides and terminal alkynes. Angew Chem Int Ed 41:2596–2599 Angew Chem 114: 2708–2711
24. Amblard F, Cho JH, Schinazi RF (2009) Cu (I)-catalyzed Huisgen azide-alkyne 1,3-dipolar cycloaddition reaction in nucleoside, nucleotide, and oligonucleotide chemistry. Chem Rev 109:4207–4220
25. Christoffers J (2018) Diaminoterephthalate fluorescence dyes – versatile tools for life sciences and materials science. Eur J Org Chem 2018:2366–2377
26. Sulmann S, Wallisch M, Scholten A et al (2016) Mapping calcium-sensitive regions in the neuronal calcium sensor GCAP2 by site-specific fluorescence labeling. Biochemistry 55:2567–2577
27. Freimuth L, Rozzi CA, Lienau C et al (2015) A diaminoterephthalate C_{60} dyad – a new material for opto-electronic applications. Synthesis 47:1325–1328
28. Antczak C, Bauvois B, Monneret C et al (2001) A new acivicin prodrug designed for tumor-targeted delivery. Bioorg Med Chem 9:2843–2848
29. Krapcho PA, Kuell CA (1990) Mono-protected diamines. N-tert-butoxycarbonyl-α,-ω-alkanediamines from α,ω-alkanediamines. Synth Commun 20:2559–2564
30. Hwang JY, Schlesinger R, Koch KW (2001) Calcium-dependent cysteine reactivities in the neuronal calcium sensor guanylate cyclase-activating protein 1. FEBS Lett 508:355–359
31. Sokal I, Li N, Klug CS, Filipek S et al (2001) Calcium-sensitive regions of GCAP1 as observed by chemical modifications, fluorescence, and EPR spectroscopy. J Biol Chem 276:43361–43373
32. Lim S, Peshenko IV, Dizhoor AM et al (2013) Structural insights for activation for retinal guanylate cyclase by GCAP1. PLoS One 8: e81822
33. Krylov DM, Niemi GA, Dizhoor AM et al (1999) Mapping sites in guanylyl cyclase activating protein-1 required for regulation of photoreceptor membrane guanylyl cyclases. J Biol Chem 274:10833–10839
34. Towler DA, Gordon JI, Adams SP et al (1988) The biology and enzymology of eukaryotic protein acylation. Annu Rev Biochem 57:69–99
35. Kollmann H, Becker SF, Shirdel J et al (2012) Probing the Ca^{2+} switch of the neuronal Ca^{2+} sensor GCAP2 by time-resolved fluorescence spectroscopy. ACS Chem Biol 7:1006–1014
36. Robin J, Brauer J, Sulmann S et al (2015) Differential nanosecond protein dynamics in homologous calcium sensors. ACS Chem Biol 10:2344–2352
37. Tsien R, Pozzan T (1989) Measurement of cytosolic free Ca^{2+} with Quin2. Methods Enzymol 172:230–262

Chapter 37

Quantification of Human Swiprosin-1/EFhd2 Expression on Protein and RNA Level

Barbara G. Fürnrohr and Dirk Mielenz

Abstract

Many Ca^{2+}-binding proteins are differentially regulated under pro-inflammatory conditions in different organs. Specific quantification of RNA and protein expression of those proteins demands validated protocols. Peripheral blood mononuclear cells (PBMC) can mirror an inflammatory status originating from several organs and can therefore be an important diagnostic tool. Swiprosin-1/EFhd2 (EFhd2) is a ~30 kDa Ca^{2+} and F-actin binding, cytoskeletal protein with two central EF hands and a C-terminal coiled-coil domain. Unbiased gene expression analyses and proteomics revealed that EFhd2 is regulated under pro-inflammatory conditions in several cell types and tissues. Here we describe validated protocols to quantify the expression of the human orthologue of Swiprosin-1/EFhd2 on RNA and protein level in PBMC. Both methods reveal that EFhd2 is stronger expressed in monocytes than in B cells of healthy donors. Thus, initial experiments relying on qPCR are likely to provide results with functional relevance. The higher expression of EFhd2 in monocytes could be related to monocyte migration under inflammatory conditions.

Key words Alcohol, Anxiety, B cell, Flow cytometry, Inflammation, Neurodegeneration, Neuroinflammation, Peripheral blood mononuclear cell, qPCR, Swiprosin-1/EFhd2

1 Introduction

Swiprosin-1/EFhd2 (EFhd2) is a ~30 kDa Ca^{2+}- and F-actin-binding protein with two central EF hands and a C-terminal coiled-coil domain [1, 2]. Functional studies have shown that EFhd2 is involved in axonal transport [1], actin remodeling [2], cell migration [2], humoral immunity [3], anxiety [4] and in ethanol drinking behavior [4]. Moreover, gene expression and proteomic data suggest that EFhd2 is differentially regulated in the brain and periphery in neurodegenerative and psychiatric diseases [4, 5] as well as under pro-inflammatory conditions [6–10]. Characterization of the expression pattern of EFhd2 may help to uncover specific disease mechanisms. Peripheral blood mononuclear cells (PBMC) reflect the inflammatory status and can even serve as a

Claus W. Heizmann (ed.), *Calcium-Binding Proteins of the EF-Hand Superfamily: From Basics to Medical Applications*, Methods in Molecular Biology, vol. 1929, https://doi.org/10.1007/978-1-4939-9030-6_37,

readout to monitor central nervous system-related inflammation, for instance, during neurodegeneration [9, 10]. To study the expression of human EFhd2, we established a qPCR protocol and a flow cytometric intracellular staining procedure of EFhd2 in human peripheral blood mononuclear cells (PBMC). We demonstrate that expression of EFhd2 can be quantified over several log scales by qPCR. Consistent with the RNA data, EFhd2 protein data reveal that human EFhd2 is stronger expressed in monocytes than in B cells in peripheral blood. These data support unbiased murine gene expression data (https://www.immgen.org/). The staining procedures can be expanded to assess EFhd2 expression in further PBMC subsets using cytometers with eight colors or more. Since qPCR and protein expression of EFhd2 correlate, qPCR analysis can be a valid readout to assess functional EFhd2 expression. However, there are also data about disease-related proteolytic processing of EFhd2 in rheumatoid arthritis [11]. The higher expression of EFhd2 in monocytes suggests an important functional role, which could be related to monocyte or cell migration under inflammatory conditions [12]. Having established protocols to assess EFhd2 expression, it will now be important to examine PBMC under various inflammatory conditions, such as arthritis, systemic lupus erythematodes, or neuroinflammation.

2 Materials

2.1 Equipment

1. Four-color flow cytometer.
2. Ultrasound device.
3. Incubators, centrifuges.
4. Reaction tubes, cell culture material.
5. qPCR machine (e.g., Applied Biosystems AB7300 Real-Time PCR System).
6. Cell counter or hemocytometer and microscope.
7. Photometer.
8. SDS-PAGE equipment.

2.2 Plasmids

1. pGEX-2T (27-4801-01) (Amersham Pharmacia Biotech).
2. pGEX-2T_EFhd2 [13] (the EcoRI fragment from pCR2.1 containing the EFhd2 coding sequence was cloned into the EcoRI site of PGEX-2T [14]; available upon request).
3. pCru5 and pCru5_EFhd2 (retroviral vectors containing IRES-GFP [15]; a fragment from pCR2.1 containing the EFhd2 coding sequence [15] was cloned into pCru5; available upon request).

2.3 PCR Primers

1. Human efhd2 primers: fwd 5′ AAGGACATGGAGAAGATGTTCAAGCAG (exon 1), rev 5′ AAGTTTCTCCATCATGAGTTTTAGCTCC (exon 2).
2. Human β-actin primers: fwd 5′ CTACGAGCTGCCTGACGG, rev 5′ GCCACAGGACTCCATGCCC.

2.4 Bacteria

1. Luria broth (LB) (prepared according to standard procedures [16]).
2. 100 mM isopropyl-β-D-thiogalactoside (IPTG); stored at −20 °C in 1 ml aliquots.
3. 100 mg/ml ampicillin (store in 1 ml aliquots at −20 °C).
4. Competent *E. coli* BL21 or Rosetta strains.

2.5 Cell Culture

1. Human HeLa breast cancer cells, 293HEK transformed human embryonic kidney cells, and Phoenix-Eco cells (ATCC).
2. DMEM medium supplemented with 10% heat-inactivated fetal bovine serum (FBS), 2 mM L-glutamine, 1 mM sodium pyruvate, and 100 μg/ml penicillin-streptomycin (D10 medium) (all from Gibco); grow cells at 37 °C, 7.5% CO_2, 95% humidity.
3. RPMI1640 medium supplemented with 10% heat-inactivated fetal bovine serum (FBS), 2 mM L-glutamine, 1 mM sodium pyruvate, 50 μM β-mercaptoethanol, and 100 μg/ml penicillin-streptomycin (R10 medium) (all from Gibco); grow cells at 37 °C, 5% CO_2, 95% humidity.

2.6 Purification of GST-Tagged Fusion Proteins from Bacteria

1. Lysis buffer and wash buffer: 0.5% Nonidet P40 (Igepal CA360), 20 mM Tris(hydroxy)aminomethane (Tris)/HCl pH 8.0, 140 mM NaCl, 1 mM dithiothreitol (DTT), 1 mM ethylenediaminetetraacetic acid (EDTA), 1 mM phenylmethylsulfonyl fluoride (PMSF).
2. Elution buffer: 100 mM reduced glutathione (GSH) in *200 mM* Tris/HCl pH 8.0, 140 mM NaCl, 1 mM DTT, 1 mM PMSF.
3. Dialysis buffer: 20 mM Tris/HCl pH 8.0, 140 mM NaCl, 1 mM DTT.
4. GSH agarose (GE Healthcare).

2.7 Transfection

1. 1 mg/ml polyethylenimine (stored in aliquots at −20 °C).
2. Opti-MEM medium (Gibco).

2.8 Antibodies and Staining Reagents

1. Monoclonal anti-EFhd2 antibodies or hybridoma supernatant [13] (available upon request) or commercial anti-EFhd2 antibody (to be tested).
2. Murine IgG1k antibody (M9269, Sigma).

3. Goat anti-mouse Fc-specific antibody coupled to Alexa647 or Cy3 (Jackson ImmunoResearch; distributed by Dianova, Germany).
4. Antihuman CD19 antibody (FITC conjugated; # 555412, Becton Dickinson Biosciences).
5. FACS buffer (2% FBS in PBS containing 0.02% sodium azide <toxic!>).
6. 4% paraformaldehyde in PBS, pH 7.0 <toxic!>.
7. 1% paraformaldehyde in PBS, pH 7.0 <toxic!>.
8. 0.1% Tween-20 in PBS.

2.9 Isolation of Primary Human Blood Cells [17]

1. Appropriate informed consent of blood donors (according to the principles outlined by the Helsinki Declaration); Ethics permit.
2. Monocyte Isolation Kit II (Miltenyi Biotec, Bergisch Gladbach, Germany) plus magnets.
3. B Cell Isolation Kit (Miltenyi Biotec, Bergisch Gladbach, Germany) plus magnets.
4. Lymphoflot density medium (Biotest, Dreieich, Germany; any other standard density medium such as Ficoll-Paque will do as well).
5. Syringes and tubes precoated with EDTA (BD Bioscience).
6. Prewarmed PBS containing 2 mM EDTA (20 °C).

2.10 Isolation, Activation, and Infection of Primary Mouse B Cells

1. Institutional permit to sacrifice mice.
2. 10 mg/ml lipopolysaccharide in PBS (store in 100 μl aliquots at −20 °C).
3. Erythrocyte lysis buffer (10 mM $KHCO_3$, 155 mM NH_4CL, 0.1 mM EDTA, pH 7.5).
4. Mouse B Cell Isolation Kit (EasySep; Stemcell Technologies, Canada).
5. 70 μM mesh, 5 ml syringe.
6. 1 mg/ml polybrene.

2.11 Quantitative PCR

1. SYBR Green Master Mix Applied Biosystems #4309155.
2. MQ H_2O.
3. 2 ng/μl cDNA.
4. 5 μM primer stock solutions.

3 Methods

3.1 Quantifying Human EFhd2 on RNA and Protein Level

The procedures describing intracellular EFhd2 protein staining have been published [13] (Fig. 1). Several primer combinations to quantify human EFhd2 expression have been assessed using the method shown in Fig. 2 but could not be validated (not shown). We describe here a protocol and a primer combination that has worked specifically (Fig. 2), demonstrating that human *efhd2* RNA expression can be quantified linearly over several log scales from cDNA obtained from tissue culture. This method can also be applied to assess *efhd2* expression in subsets of peripheral blood cells (Fig. 3).

3.2 Quantifying Human EFhd2 Expression on RNA Level

3.2.1 Preparation of RNA from Transfected HeLa and 293 Cells

1. Plate 1×10^6 cells per 6 well in D10 medium.
2. Next day: for every well use 2 μg DNA; then, bring the total volume to 31.7 μl with Opti-MEM medium (without antibiotics); add 7.6 μl PEI (1 μg/μl; use 4 × the amount of the total DNA amount used) and vortex lightly. Wait 5 min and add to well dropwise. Incubate at 37 °C and 7.5% CO_2. Replace with fresh prewarmed medium after 5 h.
3. Next day: harvest cells, prepare RNA according to standard procedures, and transcribe reversely [16].

3.2.2 Preparation of RNA from Human Blood Cells [17, 18]

1. Draw blood into EDTA tubes.
2. Separate leukocytes from erythrocytes and platelets by centrifugation through Lymphoflot or Ficoll-Paque: dilute blood with an appropriate amount of prewarmed PBS/1 mM EDTA. Underlay cell suspension with density medium. Centrifuge at 20 °C for 20 min at 2000 rpm in a standard cell culture centrifuge.
3. Harvest cells carefully with a 2 ml pipette. Wash cells with PBS once and count.
4. Isolate B cells or monocytes using the Monocyte Isolation Kit II or the B Cell Isolation Kit.
5. Prepare RNA according to standard procedures and transcribe reversely [16].

3.2.3 Quantitative SYBR Green PCR

Standard protocol: Mix 7.5 μl SYBR Green Master Mix, each 0.3 μl forward, and reverse primer (5 μM), 2.5 μl cDNA (2 ng/μl) per reaction. Denaturing, 15 min at 95 °C; annealing temperature, 60 °C; elongation, 60 s; number of cycles, 40. Size of the efhd2-specific PCR product: 99 bp.

The antibodies we developed to stain EFhd2 recognize both human and murine EFhd2 [13]. As a positive control for

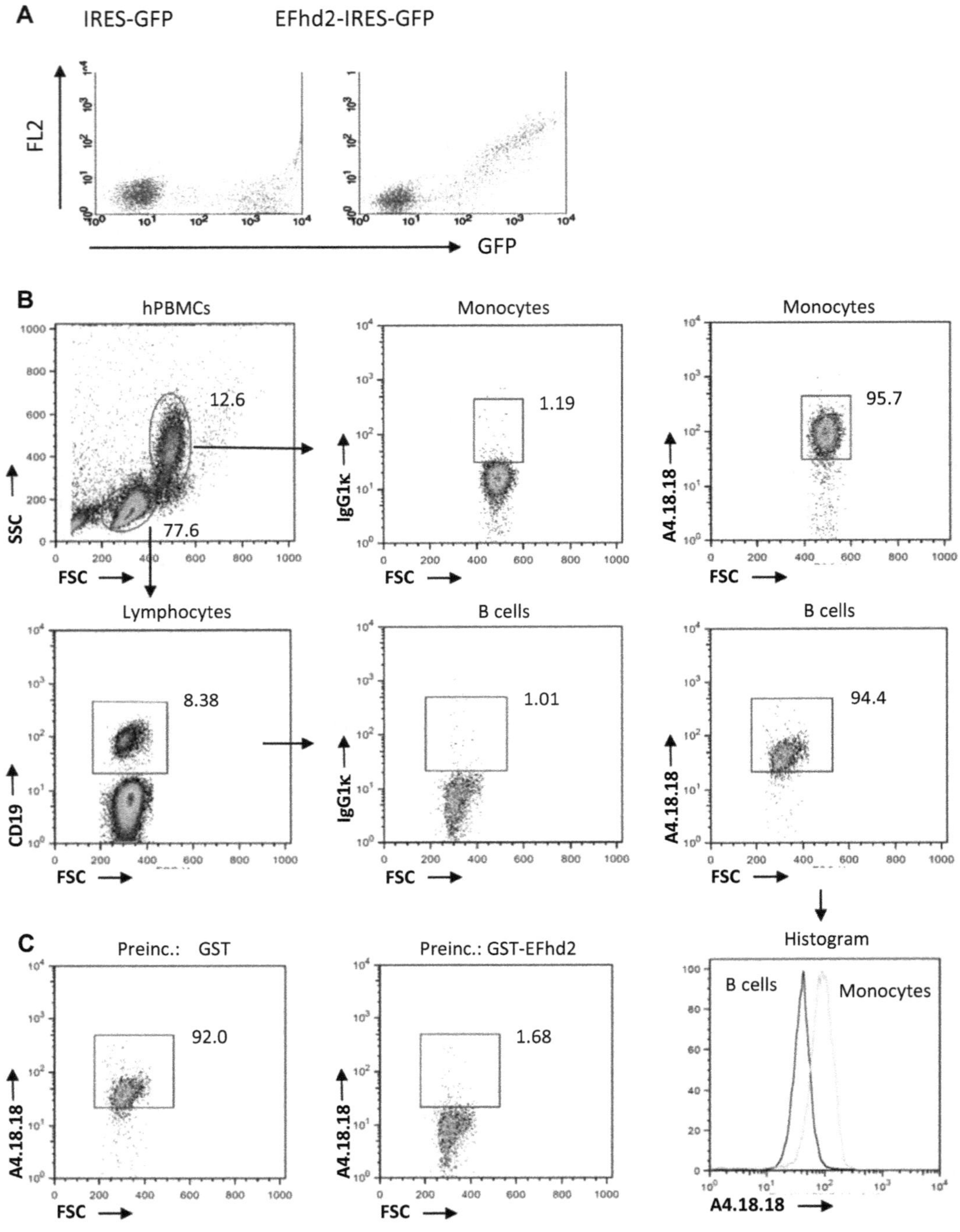

Fig. 1 Quantification of EFhd2 expression in infected cells and in human PBMC by flow cytometry. (**a**) Primary splenic B cells were stimulated with lipopolysaccharide and infected with a retrovirus encoding an IRES-GFP cassette or EFhd2Myc followed by an IRES-GFP cassette 48 h later cells were stained with anti-EFhd2 mAb A41818 followed by secondary Cy3-conjugated mouse Fcγ-specific Ab (FL2). (**b**) Human PBMCs were fixed, permeabilized, and stained for CD19 and EFhd2 (mAb A41818) as indicated. A murine IgG1κ antibody served as isotype control. Cells were analyzed in monocyte and lymphocyte gates defined by forward scatter (FSC)

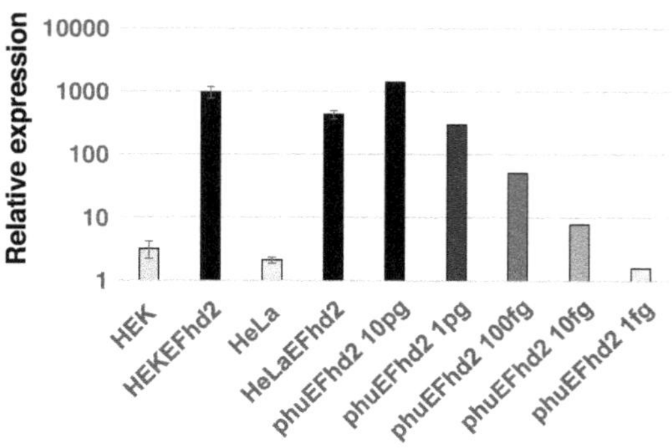

Fig. 2 Quantitative assessment of human EFhd2 expression over several log scales. 293HEK cells of HeLa cells were left untransfected or transfected with 2 μg of a plasmid encoding human EFhd2 in triplicate. RNA was prepared, reversely transcribed, and assessed for human EFhd2 expression with EFhd2-specific primers by qPCR. The ΔCt method was applied to normalize EFhd2 expression to human β-actin. The plasmid encoding human EFhd2 was subjected to qPCR with EFhd2-specific primers in tenfold serial dilutions as indicated in duplicate. Bars represent mean ± SEM (transfected cells) or mean (plasmid)

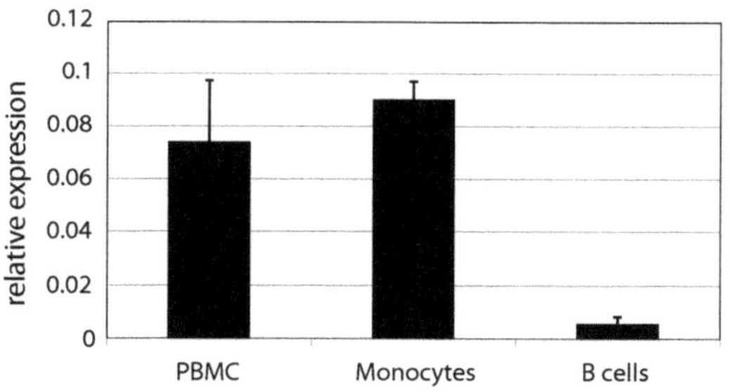

Fig. 3 Quantification of EFhd2 expression in human PBMC by qPCR. Human PBMCs were purified from EDTA blood obtained from healthy donors (HD) upon informed consent [17]. CD19+ B cells and CD14+ monocytes were further purified using the Miltenyi human B cell or monocyte isolation kit (1718). RNA was prepared, reversely transcribed, and analyzed using EFhd2- and actin-specific primers by qPCR $N = 7$ persons from each group. The RNA used in this experiment was previously prepared [17]. Data represent mean ± SD

Fig. 1 (continued) and side scatter (SSC). Numbers indicate % cells in the gates. (**c**) Prior to anti-EFhd2 staining of human PBMC, anti-EFhd2 mAb A41818 was incubated with a threefold molar amount of either GST or GST-EFhd2 fusion protein for 15 min. Shown are human B cells. Representative of two experiments. Numbers indicate % cells in the gates. This figure has been published before [13]. With kind approval © Mary Ann Liebert

3.3 Intracellular Staining of Human EFhd2 Protein

intracellular EFhd2 protein staining, we use therefore primary infected B cells ectopically expressing EFhd2-IRES-GFP (Fig. 1); as negative controls we use preincubation of the anti-EFhd2 mAb with the antigen (GST-EFhd2) or an isotype control antibody (murine IgG1κ).

3.3.1 Isolation of GST-EFhd2 Fusion Protein (According to Standard Procedures; 16)

1. Transform bacteria with pGEX-2T or pGEX-2T_EFhd2 according to standard procedures [16], and plate on LB_{Amp} agar plates.
2. Pick a colony and grow overnight in LB medium containing 100 μg/ml ampicillin.
3. Dilute 1:100 in prewarmed LB medium containing 100 μg/ml ampicillin.
4. After ~2 h check OD; add 1 mM IPTG from stock at an OD_{600} of ~0.6.
5. Grow for 2 h, centrifuge, wash once in ice-cold PBS, and lyse in 20 ml ice-cold lysis buffer/0.5 l culture.
6. Ultrasound for 2–5 min (70% power, 10 s intervals, on ice).
7. Centrifuge at 10,000 × g for 30 min.
8. Incubate supernatant with 2 ml of washed GSH agarose (per 20 ml of lysate) for 2 h at 4 °C on a rotating wheel.
9. Wash twice with 20 ml of lysis buffer, and elute with 3 × 1 ml elution buffer.
10. Analyze fractions by SDS-PAGE; dialyze, determine protein concentration, and store in aliquots at −70 °C.

3.3.2 Infection of Primary B Cells with Retroviral Plasmid Encoding EFhd2

Preparation of Retroviral Supernatants

1. Plate 1 × 10^6 cells per 6 well in D10 medium.
2. Next day: for every well use 2 μg DNA; then, bring the total volume to 31.7 μl with Opti-MEM medium (without antibiotics); add 7.6 μl PEI (1 μg/μl; use 4 × the amount of the total DNA amount used) and vortex lightly. Wait 5 min and add to well dropwise. Incubate at 37 °C and 7.5% CO_2. Replace with fresh prewarmed medium after 5 h.
3. 48–72 h later: harvest supernatant, filter through 0.45 μM, and store at −70 °C; avoid repeated thawing/freezing of the retroviral supernatants.

Infection of Primary B Cells

1. Sacrifice C57Bl/6 mouse according to institutional guidelines and remove the spleen.
2. Make single cell suspension using a 70 μm mesh and the piston of a 2 ml syringe in cold R10 medium.
3. Wash once in ice-cold PBS and add 5 ml of erythrocyte lysis buffer at room temperature.
4. Stop lysis with R10 medium, and wash with PBS.

5. Isolate B cells using the Mouse B Cell Isolation Kit, and determine the cell number.
6. Plate cells at 0.5×10^6/ml in R10 medium containing 10 μg/ml LPS.
7. 24 h later spin-infect 10^6 cells by placing in prewarmed retroviral supernatant including 4 μg/ml polybrene for 3 h at 30 °C and 3300 rpm in a standard cell culture centrifuge.
8. Wash and plate in prewarmed R10 medium containing 10 μg/ml LPS at 0.5×10^6 cells/ml.
9. 24–48 h later wash in PBS.
10. Fix 0.5×10^6 cells with 4% PFA/PBS for 15 min at RT, wash with PBS, and permeabilize with 0.1% Tween/PBS for 10 min at 37 °C.
11. Store at 4 °C in PBS (stable for some days).

3.3.3 Isolation, Staining, and Analysis of Human Peripheral Blood Monocytic Cells

1. Isolate human PBMCs from blood samples of volunteers as described above (*see* Subheading 3.2.2) [17, 18], but do not separate B cells and monocytes.
2. Resuspend cells in 2% FCS/PBS buffer (30 μl/10^7 cells).
3. Block FC receptors with human Ig (from Monocyte Isolation Kit II, Miltenyi Biotec, Bergisch Gladbach, Germany) (10 μl per 30 μl) for 10 min on 4 °C, and wash with 2 ml FACS buffer.
4. Fix 0.5×10^6 cells with 4% PFA/PBS for 15 min at RT, and wash and permeabilize with 0.1% Tween/PBS for 10 min at 37 °C.
5. Preincubate antibody/hybridoma supernatant with isomolar amounts of GST (5 μg/ml) or GST-Efhd2 fusion protein (10 μg/ml) (*see* Subheading 3.3.1) for 20 min on 4 °C (compare Fig. 1).
6. Stain PBMC and infected and fixed B cells (*see* Subheading 3.3.2) with 10 μg/ml A4.18.18 anti-EFhd2 antibody or supernatant [13] or murine IgG1κ (M9269, Sigma) as isotype control for 20 min (*see* Fig. 1).
7. Wash twice with 1 ml of FACS PBS.
8. Incubate with goat anti-mouse Alexa647-coupled IgG Fc-γ-specific antibody for 15 min on ice, and wash once with 1 ml of FACS PBS.
9. Fix samples with 1% PFA for 10 min at RT and wash with 1 ml of 0.1% Tween/PBS.
10. Stain cells with anti-human CD19 antibody coupled to FITC (1:200) for 10 min on ice.
11. Wash once in 1 ml of FACS buffer and analyze by flow cytometry (*see* Fig. 1).

4 Notes

1. We recommend as a positive control for anti-EFhd2 flow cytometric staining murine B cells overexpressing EFhd2. Any other cell line overexpressing EFhd2 works as well. The best positive control for human PBMC would be lentivirally infected primary blood cells since they exhibit similar forward/side scatter properties, thereby, facilitating proper gating during flow cytometry. Very good negative controls to test new anti-EFhd2 antibodies (https://www.genecards.org/cgi-bin/carddisp.pl?gene=EFHD2) are knockout tissue [1] or Crispr-Cas9 engineered knockout cell lines. Rabbit antibodies raised against an EFhd2 peptide did not work in flow cytometry in our hands (not shown).
2. To obtain PBMC we collaborated with medical doctors [17].

Acknowledgments

This work was supported by the Deutsche Forschungsgemeinschaft (DFG; TRR130, to D.M.)

References

1. Purohit P, Perez-Branguli F, Prots I, Borger E, Gunn-Moore F, Welzel O, Loy K, Wenzel EM, Grömer TW, Brachs S, Holzer M, Buslei R, Fritsch K, Regensburger M, Böhm KJ, Winner B, Mielenz D (2014) The Ca2+ sensor protein swiprosin-1/EFhd2 is present in neurites and involved in kinesin-mediated transport in neurons. PLoS One 9:e103976. https://doi.org/10.1371/journal.pone.0103976
2. Kwon MS, Park KR, Kim YD, Na BR, Kim HR, Choi HJ, Piragyte I, Jeon H, Chung KH, Song WK, Eom SH, Jun CD (2013) Swiprosin-1 is a novel actin bundling protein that regulates cell spreading and migration. PLoS One 8:e71626. https://doi.org/10.1371/journal.pone.0071626
3. Brachs S, Turqueti-Neves A, Stein M, Reimer D, Brachvogel B, Bösl M, Winkler T, Voehringer D, Jäck H-M, Mielenz D (2014) Swiprosin-1/EFhd2 limits germinal center responses and humoral type 2 immunity. Eur J Immunol 44:3206–3219. https://doi.org/10.1002/eji.201444479
4. Mielenz D, Gunn-Moore F (2016) Physiological and pathophysiological functions of Swiprosin-1/EFhd2 in the nervous system. Biochem J 473:2429–2437. https://doi.org/10.1042/BCJ20160168
5. Mielenz D, Reichel M, Jia T, Quinlan EB, Stöckl T, Mettang M, Zilske D, Kirmizi-Alsan-E, Schönberger P, Praetner M, Huber SE, Amato D, Schwarz M, Purohit P, Brachs S, Spranger J, Hess A, Büttner C, Ekici AB, Perez-Branguli F, Winner B, Rauschenberger V, Banaschewski T, Bokde ALW, Büchel C, Conrod PJ, Desrivières S, Flor H, Frouin V, Gallinat J, Garavan H, Gowland P, Heinz A, Martinot JL, Lemaitre H, Nees F, Paus T, Smolka MN, Schambony A, Bäuerle T, Eulenburg V, Alzheimer C, Lourdusamy A, Schumann G, Müller CP (2018) EFhd2/Swiprosin-1 is a common genetic determinator for sensation-seeking/low anxiety and alcohol addiction. Mol Psychiatry 23(5):1303–1319. https://doi.org/10.1038/mp.2017.63 Epub 2017 Apr 11
6. Wendeln A-C, Degenhardt K, Kaurani L, Gertig M, Ulas T, Jain G, Wagner J, Häsler LM, Wild K, Skodras A, Blank T, Staszewski O, Datta M, Centeno TP, Capece V, Islam MR, Kerimoglu C, Staufenbiel M, Schultze JL, Beyer M, Prinz M, Jucker M, Fischer A, Neher JJ (2018) Innate immune memory in the brain shapes neurological disease hallmarks. Nature

556:332–338. https://doi.org/10.1038/s41586-018-0023-4

7. Dotzlaw H, Schulz M, Eggert M, Neeck GA (2004) Pattern of protein expression in peripheral blood mononuclear cells distinguishes rheumatoid arthritis patients from healthy individuals. Biochim Biophys Acta 1696:121–129. https://doi.org/10.1016/j.bbapap.2003.09.015
8. Borger E, Herrmann A, Mann DA, Spires-Jones T, Gunn-Moore F (2014) The calcium-binding protein EFhd2 modulates synapse formation in vitro and is linked to human dementia. J Neuropathol Exp Neurol 73:1166–1182. https://doi.org/10.1097/NEN.0000000000000138
9. Comabella M, Cantó E, Nurtdinov R, Río J, Villar LM, Picón C, Castilló J, Fissolo N, Aymerich X, Auger C, Rovira A, Montalban X (2016) MRI phenotypes with high neurodegeneration are associated with peripheral blood B-cell changes. Hum Mol Genet 25:308–316. https://doi.org/10.1093/hmg/ddv473
10. Lang SC, Harre U, Purohit P, Dietel K, Kienhöfer D, Hahn J, Baum W, Herrmann M, Schett G, Mielenz D (2017) Neurodegeneration enhances the development of arthritis. J Immunol 198:2394–2402. https://doi.org/10.4049/jimmunol.1601472
11. Schulz M, Dotzlaw H, Mikkat S, Eggert M, Neeck G (2007) Proteomic analysis of peripheral blood mononuclear cells: selective protein processing observed in patients with rheumatoid arthritis. J Proteome Res 6:3752–3759. https://doi.org/10.1021/pr070285f
12. Zhang S, Tu Y, Sun YM, Li Y, Wang RM, Cao Y, Li L, Zhang LC, Wang ZB (2018) Swiprosin-1 deficiency impairs macrophage immune response of septic mice. JCI Insight 3. https://doi.org/10.1172/jci.insight.95396
13. Brachs S, Lang C, Buslei R, Purohit P, Fürnrohr B, Kalbacher H, Jäck HM, Mielenz D (2013) Monoclonal antibodies to discriminate the EF hand containing calcium binding adaptor proteins EFhd1 and EFhd2. Monoclon Antib Immunodiagn Immunother 32:237–245. https://doi.org/10.1089/mab.2013.0014
14. Hagen S, Brachs S, Kroczek C, Fürnrohr BG, Lang C, Mielenz D (2012) The B cell receptor-induced calcium flux involves a calcium mediated positive feedback loop. Cell Calcium 51:411–417. https://doi.org/10.1016/j.ceca.2012.01.004
15. Avramidou A, Kroczek C, Lang C, Schuh W, Jäck H-M, Mielenz D (2007) The novel adaptor protein Swiprosin-1 enhances BCR signals and contributes to BCR-induced apoptosis. Cell Death Differ 14:1936–1947. https://doi.org/10.1038/sj.cdd.4402206
16. Green MR, Sambrook J (2014) Molecular cloning, 3rd edn. Cold Spring Harbour Laboratory Press, Cold Spring Harbour
17. Fürnrohr BG, Rhodes B, Munoz LE, Weiß K, Vyse TJ, Schett G (2015) Osteoclast differentiation is impaired in a subgroup of SLE patients and correlates inversely with mycophenolate mofetil treatment. Int J Mol Sci 16:18825–18835. https://doi.org/10.3390/ijms160818825
18. Rhodes B, Fürnrohr BG, Roberts AL, Tzircotis G, Schett G, Spector TD, Vyse TJ (2012) The rs1143679 (R77H) lupus associated variant of ITGAM (CD11b) impairs complement receptor 3 mediated functions in human monocytes. Ann Rheum Dis 71:2028–2034. https://doi.org/10.1136/annrheumdis-2012-201390

Chapter 38

Three-Dimensional Reconstruction Imaging Method to Study the Function of EFHD2 in Invadopodia Formation

Yun Hyun Huh, Yu Ra Yeo, and Woo Keun Song

Abstract

Calcium and calcium-binding proteins play crucial roles in the regulation of actin dynamics, which contributes to cancer cell migration and invasion. In this chapter, we have focused on a three-dimensional imaging method to explore the pathophysiological function of EF-hand domain-containing protein D2 (EFHD2), a novel actin-binding protein. To overcome the limitations of two-dimensional imaging on substrate-coated cover glass for examination of invasive protrusions of cancer cells, we suggest three-dimensional reconstruction from optical *z*-sections of cells cultured on substrate-impregnated membrane filters of Transwell.

Key words Calcium, EFHD2, Actin, Invadopodia

1 Introduction

Calcium-mediated remodeling of the actin cytoskeleton makes essential contributions to the migration and invasion of cancer cells [1]. Precise control of the assembly and disassembly of the actin cytoskeleton is necessary for coordinated cell movements [2]. Calcium and calcium-binding proteins thus act as key regulators that exert multiple effects on actin dynamics.

EF-hand domain-containing protein D2 (EFHD2; also known as swiprosin-1) has recently been characterized as an actin-binding protein. EFHD2 promotes F-actin bundling in the presence of Ca^{2+}. This process is dependent on self-dimerization through both EF-hand motifs and a coiled-coil domain [3]. In particular, direct interactions between EFHD2 and F-actin promote the formation of motile protrusions, such as lamellipodia and filopodia [4], and ultimately enhance migration and invasion of B16F10 mouse melanoma cells [5].

Claus W. Heizmann (ed.), *Calcium-Binding Proteins of the EF-Hand Superfamily: From Basics to Medical Applications*, Methods in Molecular Biology, vol. 1929, https://doi.org/10.1007/978-1-4939-9030-6_38,

Cancer cell invasiveness was investigated earlier by visualizing extracellular matrix (ECM)-degrading activities of cancer cells on cover glass coated with substrates such as gelatin or collagen, which are major components of ECM [6]. However, this two-dimensional approach has limitations for visualization of the elaborate molecular structures of invadopodia due to the restricted protrusion depth of shallow gelatin substrates on glass covers [7, 8]. A three-dimensional approach using a polycarbonate filter coated with gelatin was subsequently applied for the study of podosomes of dendritic cells [9, 10]. In this chapter, we have discussed the development of a three-dimensional technique for examination of molecular structures of invadopodia induced by EFHD2 in invasive cancer cells using Transwell membrane impregnated with fluorescently labeled gelatin. Our protocol could be expanded to characterize the subcellular structures of invadopodia and membrane ruffles regulated by different calcium signaling pathways in cancer cell invasion.

2 Materials

Prepare all solutions in this protocol using ultrapure water generated by purifying deionized water to attain a sensitivity of 18 MΩ at 25 °C. The storage conditions for different solutions and reagents are variable.

2.1 Cell Culture

1. Cell preparation: Mouse melanoma cell line, B16F1: Myc-tagged EFHD2-overexpressing B16F1 (B16F1-myc-EFHD2) and control cells (B16F1-myc) [5].
2. Complete medium: Dulbecco's Modified Eagle's Medium (DMEM; Gibco-BRL, Grand Island, NY) supplemented with 10% (v/v) fetal bovine serum, 50 μg/mL streptomycin, and 50 units/mL penicillin.
3. Serum-free medium: DMEM.
4. Chemoattractant: Fibronectin, human plasma (Merck), 1 mg/mL stock concentration, 20 μg/mL working concentration (*see* **Note 1**).

2.2 Reagents for Labeling and Antibodies

1. FITC-conjugated gelatin (Thermo Fisher Scientific): 1 mg/mL (5×) stock solution, final working concentration, 0.2 mg/mL (*see* **Note 2**).
2. 6.5 mm Transwell with a 0.4 μm pore polycarbonate membrane cell culture insert (Merck) (*see* **Note 3**).

3. Primary antibodies: anti-cortactin Ab (Merck, 4F11) and anti-myc Ab (Merck, 9E10).
4. Fluorescent dye-labeled secondary antibodies: Alexa 555- or Alexa-647-conjugated anti-mouse IgG (Molecular Probes).
5. Alexa Fluor 647-phalloidin (Thermo Fisher Scientific) and phalloidin-tetramethylrhodamine B isothiocyanate (phalloidin-TRITC; Sigma).

2.3 Instrumentation and Software

1. Olympus laser scanning confocal microscope (FV1000).
2. FV10-MSASW software.

3 Methods

3.1 Preparation of Insert Filters of Transwell Impregnated with Fluorescently Labeled Gelatin

1. Prepare FITC-conjugated gelatin stock solution. Dissolve FITC-gelatin conjugates supplied in 5 mg unit sizes into 5 mL distilled water (dH_2O). Heat to 50 °C for 20 min to aid dissolution.
2. Dilute the gelatin conjugate solution by adding dH_2O to a final working concentration of 0.2 mg/mL.
3. Place the inserts of Transwell into 12-well plates and gently deposit 200 μL FITC-gelatin conjugate solution into each insert. Incubate the inserts with FITC-gelatin conjugate solution for 30 min at room temperature in the dark.
4. Briefly rinse the inserts twice with PBS and once with serum-free medium.

3.2 Confocal Imaging Analysis of Cell Invasion

1. One day before the experiment, plate cells into 60 mm cell culture dishes and grow in complete medium in an incubator at 37 °C and 5% CO_2.
2. On the day of experiment, suspend cancer cells in 1 mL trypsin/EDTA solution, and add 2 mL complete medium to inactivate trypsin. Collect cells via centrifugation at 1200 rpm for 3 min, and wash the cell pellet twice with serum-free medium.
3. Resuspend the cell pellet in serum-free medium and count the cells by Trypan blue staining.
4. Seed cells (5×10^3 cells/200 μL/insert) on FITC-gelatin-soaked inserts of Transwell.

1. Place 750 μL NIH3T3-conditioned medium in each lower chamber of the Transwell plate.
2. Incubate cells for 1–4 h at 37 °C and 5% CO_2 (*see* **Note 4**).

3. Transfer the inserts to vacant wells in 12-well plates and briefly rinse in PBS to remove unattached cells.
4. Fix the cells on gelatin-soaked inserts with 4% paraformaldehyde for 20 min at room temperature.
5. Wash the inserts three times with PBS for 2 min each time.
6. Permeabilize the cells on gelatin-soaked inserts with 0.1% Triton X-100 for 10 min at room temperature.
7. Wash the inserts three times with PBS for 2 min each time.
8. Add 1 mL of 1% BSA in PBS to cover the inserts and incubate cells at room temperature for 60 min.
9. Add the desired concentration of primary antibody diluted in 1 mL of 0.1% BSA in PBS [anti-cortactin (1:100) or anti-myc (1:250)] to cells on the gelatin-soaked inserts and incubate for 2 h at room temperature.
10. Remove the primary antibody solution and wash inserts three times with PBS for 2 min each time.
11. Add fluorescent dye-labeled secondary antibody diluted in 1 mL of 0.1% BSA in PBS (1:1000) to cells and incubate for 1 h at room temperature protected from light.
12. Wash the inserts three times with PBS for 2 min each time.
13. Add Alexa Fluor 647-phalloidin or phalloidin-TRITC diluted in 1 mL PBS (1:500) to cells, and incubate for 1 h at room temperature protected from light.
14. Wash the inserts three times with PBS for 2 min each time.
15. Cut the membrane from the insert filter with a knife.
16. Place the membrane face down onto a drop of anti-fading mounting medium on a slide glass (*see* **Note 5**). Add another drop of anti-fading mounting medium on the membrane and place a 18 mm glass cover on top. Avoid air bubbles and crumpling of the membrane (*see* **Note 6**).

3.3 Confocal Imaging Analysis of EFHD2 in Invadopodia Formation

1. Observe protrusions in the lumen of pores of polycarbonate filters by optical *z*-sectioning under a laser scanning confocal microscope (Fig. 1).
2. Obtain a three-dimensional reconstruction of the images captured at different focal planes. For *z*-stacks, the images are collected with the appropriate *x*, *y*, and *z* dimensions using FV10-MSASW software (Fig. 2).

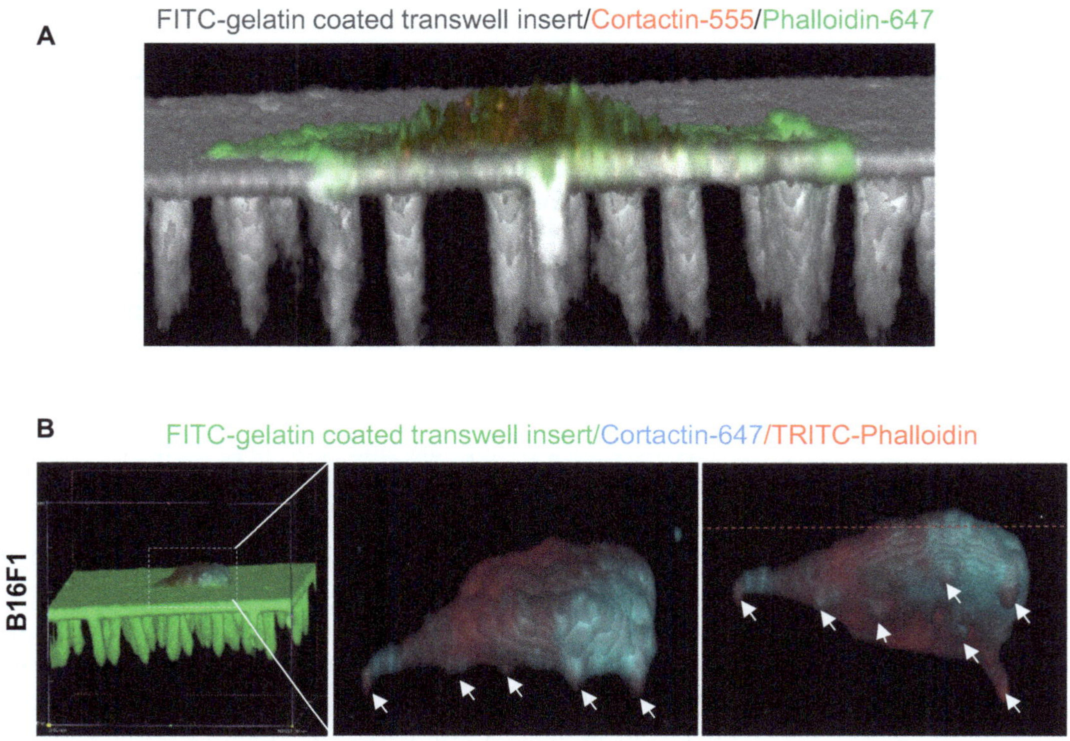

Fig. 1 Representative three-dimensional image of motile protrusions on a FITC-gelatin-impregnated filter. (**a**) B16F1 mouse melanoma cells were plated on polycarbonate membrane filters impregnated with FITC-gelatin (gray) for 2 h. Cortactin, a marker of invadopodia, was labeled with Alexa Fluor-555 (red) and F-actin stained with Alexa Fluor 647-phalloidin (green). The *z*-stack image from optical *z*-sectioning shows that the invadopodium protrudes into the lumen of pores of filters (*see* **Note 7**). (**b**) Three-dimensional reconstruction from the optical *z*-section image. The white arrows depict localization of cortactin and F-actin

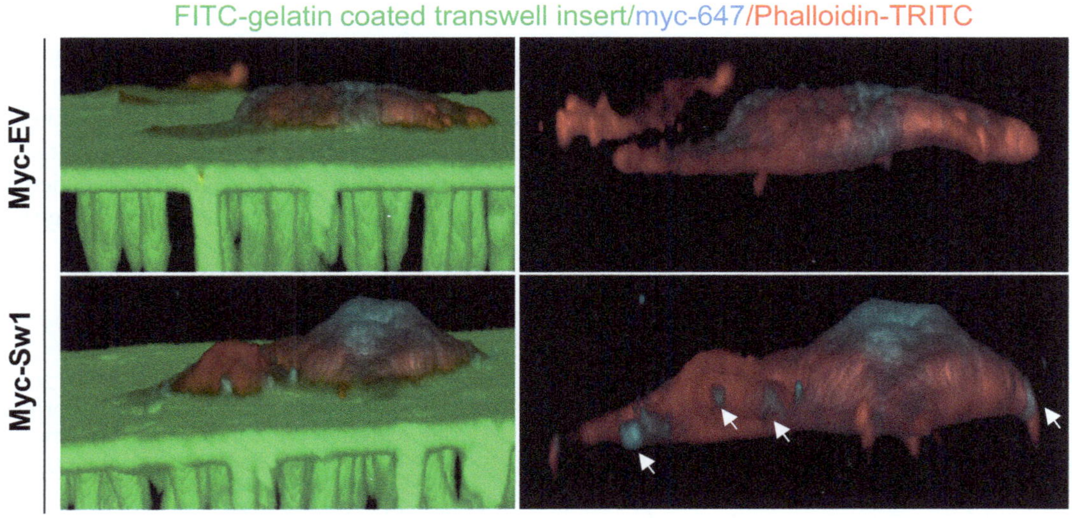

Fig. 2 Three-dimensional reconstruction images showing the involvement of EFHD2 in invadopodia formation. B16F1-myc-EFHD2 and control B16F1-myc cells were plated on FITC-gelatin-impregnated filters (green) for 2 h. Myc-tag was conjugated with Alexa Fluor-647 (blue) and F-actin stained with phalloidin-TRITC (red). The three-dimensional reconstruction images illustrate that EFHD2 overexpression promotes invadopodia formation

4 Notes

1. Conditioned media from fibroblasts can be used as the chemoattractant for migration and invasion of a number of cancer cell types.
2. We utilize gelatin conjugates lyophilized from 5 mL phosphate-buffered saline, pH 7.2 (PBS). For long-term storage, divide solutions into aliquots and freeze at −20 °C. Avoid repeated freezing and thawing and protect from light.
3. Inserts with 0.4 μm and 8.0 μm pores are generally used for detection of invadopodia and cell migration or invasion, respectively.
4. Cells on 0.2 mg/mL gelatin-coated inserts are incubated for 1–4 h to examine the invadopodia structure or 6–12 h to observe migration or invasion activity.
5. The direction of protrusions or cell movement can be easily analyzed by placing the membrane upside down for inverted confocal microscopy.
6. If the membrane is crumpled, the resulting images do not illustrate a flat membrane, and it is difficult to focus a microscope. Bubbles also disrupt the focal point.
7. Molecular structures of invadopodia may be determined by immunostaining of target molecules. Please refer to Huh et al. [5].

Acknowledgments

This research was supported by the Bio Imaging and Cell Logistics Research Center (NRF-2016R1A5A1007318), Brain Research Program NRF-2014M3C7A1046050, NRF-2016M3C7A1905475, and NRF-2016R1C1B1013237 to YHH through the National Research Foundation of Korea funded by the Ministry of Science, ICT and Future Planning. The Bio Imaging and Cell Logistics Research Center provided access to microscopes.

References

1. Greka A, Mundel P (2012) Calcium regulates podocyte actin dynamics. Semin Nephrol 32 (4):319–326. https://doi.org/10.1016/j.semnephrol.2012.06.003
2. Rottner K, Stradal TE (2011) Actin dynamics and turnover in cell motility. Curr Opin Cell Biol 23(5):569–578. https://doi.org/10.1016/j.ceb.2011.07.003
3. Kwon MS, Park KR, Kim YD et al (2013) Swiprosin-1 is a novel actin bundling protein that regulates cell spreading and migration. PLoS One 8(8):e71626. https://doi.org/10.1371/journal.pone.0071626
4. Huh YH, Kim SH, Chung KH et al (2013) Swiprosin-1 modulates actin dynamics by regulating the F-actin accessibility to cofilin. Cell

Mol Life Sci 70(24):4841–4854. https://doi.org/10.1007/s00018-013-1447-5

5. Huh YH, Oh S, Yeo YR et al (2015) Swiprosin-1 stimulates cancer invasion and metastasis by increasing the Rho family of GTPase signaling. Oncotarget 6(15):13060–13071. https://doi.org/10.18632/oncotarget.3637
6. Wang Z, Liang X, Cai M et al (2016) Analysis of invadopodia formation in breast cancer cells. Methods Mol Biol 1406:203–210. https://doi.org/10.1007/978-1-4939-3444-7_18
7. Linder S, Wiesner C, Himmel M (2011) Degrading devices: invadosomes in proteolytic cell invasion. Annu Rev Cell Dev Biol 27:85–211. https://doi.org/10.1146/annurev-cellbio-092910-154216
8. Murphy DA, Courtneidge SA (2011) The 'ins' and 'outs' of podosomes and invadopodia: characteristics, formation and function. Nat Rev Mol Cell Biol 12:413–426. https://doi.org/10.1038/nrm3141
9. Gawden-Bone C, Zhou Z, King E et al (2010) Dendritic cell podosomes are protrusive and invade the extracellular matrix using metalloproteinase MMP-14. J Cell Sci 123 (9):1427–1437. https://doi.org/10.1242/jcs.056515
10. Baranov M, Ter Beest M, Reinieren-Beeren I et al (2014) Podosomes of dendritic cells facilitate antigen sampling. J Cell Sci 127 (5):1052–1064. https://doi.org/10.1242/jcs.141226

Chapter 39

Characterization of Calcium-Binding Proteins from Parasitic Worms

Charlotte M. Thomas and David J. Timson

Abstract

Parasitic diseases caused by helminths (worms) represent a major burden on humanity with hundreds of millions of people infected worldwide. However, there are relatively few drugs to treat these diseases, and resistance is emerging to some of these. Therefore, there is a pressing need to characterize proteins from helminths as potential drug targets. Calcium signalling proteins represent attractive targets due to the vital nature of properly regulated calcium-mediated signalling and the presence of unusual calcium-binding proteins in helminths. Here we present methods to characterize these proteins in terms of their ion-binding properties, drug-binding properties, and oligomeric state, including a method to correct for the effects of non-spherical proteins in analytical gel filtration. In addition we present an overview of their recombinant expression and purification and methods to predict their structures.

Key words EF-hand, Native gel electrophoresis, Cross-linking, Protein modelling, Drug binding, Calcium ion, Divalent cation, Trematodes, *Schistosoma* spp., *Fasciola* spp., Neglected tropical disease

1 Introduction

The use of calcium ions in cellular signalling is widespread across the Kingdoms of life [1]. The widespread availability of the ion combined with its rapid solvation/desolvation properties enable its use and facilitate its binding (and unbinding) to protein targets [2]. These proteins generally act as transducers of the calcium signal, altering conformation and binding properties in response to interaction with the ion [3]. The most common interaction motif is the EF-hand which folds such that six functional groups are positioned to coordinate the calcium ion [4, 5]. These motifs typically occur in pairs and, on interaction with the ion, alter their conformation resulting in the exposure of more hydrophobic residues on the protein's surface [6]. The best-characterized calcium receptor protein is calmodulin which is present in all eukaryotes and many different cell types [7].

Claus W. Heizmann (ed.), *Calcium-Binding Proteins of the EF-Hand Superfamily: From Basics to Medical Applications*, Methods in Molecular Biology, vol. 1929, https://doi.org/10.1007/978-1-4939-9030-6_39,

The critical role of calcium ions in signalling pathways means that proteins involved in these processes are often considered as potential drug targets. Indeed, the antipsychotic drugs trifluoperazine (TFP) and chlorpromazine (CPZ) act partly through antagonism of calmodulin [8]. Calmodulin has also been proposed as a possible cancer drug target [9, 10]. Some antiparasitic drugs act through the disruption of calcium homeostasis. For example, praziquantel (PZQ) causes uncontrolled calcium influx, uncontrolled muscle contraction, and consequent paralysis in *Schistosoma* spp. and some other parasitic worms [11, 12]. Although the molecular target(s) of PZQ have yet to be definitively identified, it seems likely that it acts on components of calcium signalling pathways in the worm.

There remains a considerable deficit in the number of effective antiparasitic drugs. The majority of parasitic infections are considered neglected tropical diseases. Furthermore, resistance is known to some of the current, widely used drugs such as albendazole, artemisinin, chloroquine, ivermectin, melarsoprol, and pentamidine [13–18]. The disease burden from parasitic infections is significant with hundreds of millions of humans affected [19]. In addition to this direct effect on humanity, many farm animals and beasts of burden are also infected (or at risk of infection) [20]. There is, therefore, a need to investigate calcium signalling molecules from parasites. Such investigations help us understand the fundamental biochemistry of these organisms and may also enable us to identify novel drug targets. To date, the investigation of calcium signalling in invertebrates and protists has lagged substantially behind studies in mammals. However, the majority of human parasites are single-celled organisms such as *Plasmodium* spp. and *Trypanosoma* spp. or helminths (worms). Some of these species are known to have unusual calcium-binding proteins in addition to calmodulin [21]. In Trematoda (a class of parasitic flatworms), at least two of these occur. The 8 kDa family consists of two, low-affinity calcium-binding EF-hands [22, 23]. The second family consists of proteins which combine an N-terminal domain with two EF-hands and a C-terminal dynein light chain-like (DLC-like) domain [24]. There are typically several isoforms of these proteins in trematode species—for example, *Schistosoma mansoni* expresses at least 13 isoforms [25]. Interestingly, these isoforms have different biochemical properties. They bind different divalent cations (and a small number have been identified which appear to have no ion-binding activity), and they can be distinguished pharmacologically by their different drug-binding properties [26–34]. All proteins from this family whose oligomeric state has characterized to date dimerize [24, 35–37]. Here we present methods to characterize the ion- and drug-binding properties of calcium-binding proteins from parasites together with protocols to determine their oligomeric state and predict their structure. These methods are based on our studies of calmodulins and EF-hand/DLC-like proteins from *S. mansoni* and *Fasciola hepatica*. However, they could be readily applied to other groups of calcium-binding proteins from parasites.

2 Materials

2.1 General Materials, Equipment, and Buffers

Pipettes (at least: 1–20 μl, 20–200 μl, and 200–1000 μl).

Pipette tips.

Microcentrifuge tubes (1.5 ml).

Cuvettes and/or clear 96-well plates.

Gloves, laboratory coats, and eye protection.

Cell resuspension buffer (CRB): 50 mM Hepes-OH (pH 7.5), 150 mM sodium chloride, 10% (v/v) glycerol.

Buffer A: 50 mM Hepes-OH, pH 7.5; 500 mM sodium chloride; 10% (v/v) glycerol.

Buffer B: Buffer A, supplemented with 250 mM imidazole.

Dialysis buffer: 50 mM Hepes-OH (pH 7.5), 150 mM sodium chloride, 1 mM DTT, 10% (v/v) glycerol.

Hepes buffer (pH 7.4): 10 mM Hepes (pH 7.4).

Buffer R: 50 mM Tris–HCl, 17 mM Tris-base, 150 mM sodium chloride (pH 7.4; pH adjustment should not be necessary [38]).

SDS-PAGE running buffer: 25 mM Tris-base, 250 mM glycine, 3.5 mM SDS (pH 8.8; pH adjustment should not normally be necessary).

SDS-PAGE loading buffer: 120 mM Tris-base, 4% (w/v) SDS, 20% (v/v) glycerol, 1% (w/v) DTT, 0.1% (w/v) bromophenol blue (pH 6.8).

Continuous native-PAGE running buffer (pH 6.6): 25 mM histidine, 30 mM Mops.

Continuous native-PAGE running buffer (pH 7.4): 43 mM imidazole, 35 mM Hepes.

Continuous native-PAGE running buffer (pH 8.8): 25 mM Tris–HCl, 250 mM glycine.

Continuous native-PAGE running buffer (pH 9.4): 60 mM Tris-base, 40 mM CAPS.

Continuous native-PAGE loading buffer: 20% (v/v) native running buffer (pH to match running buffer), 20% (v/v) glycerol, 5% (w/v) bromophenol blue, 1% (w/v) DTT.

Coomassie blue stain: 20% (v/v) ethanol, 10% (v/v) acetic acid, 0.1% (w/v) Coomassie blue R350.

Destain: 0.75% (v/v) acetic acid, 0.5% (v/v) ethanol.

Acrylamide/bisacrylamide solution (e.g., 40%(v/v) 37.5:1 acrylamide/bisacrylamide solution from Sigma).

Ice.

Freezer (−70 or −80 °C).

Freezer (−20 °C).

Fridge (4 °C).

Dry block heaters, water baths, or incubators (37 and 95 °C).

Power pack—choose a model capable of delivering at least 20 mA constant current and 200 V constant voltage.

SDS-PAGE gel apparatus (choose a robust set as it will be used a lot; a gel size of approximately 10 cm by 10 cm and a thickness of 0.75–1.0 mm will be sufficient for the protocols described below).

2.2 Recombinant Expression and Purification

Source of cDNA encoding the gene. There are increasing numbers of resources which will supply these at no (or low) cost. For *Schistosoma mansoni*, the Schistosomiasis Resource Center (distributed by BEI Resources), NIAID, and NIH (Manassas, VA, USA; https://www.beiresources.org/) are very useful [39]. Alternatively, it is increasingly cost-effective to have coding sequences synthesized commercially and codon-optimized.

PCR consumables: primers (design as per instructions in the cloning kit); dNTP solutions; high-fidelity thermostable DNA polymerase (e.g., Phusion™ from Thermo Fisher).

PCR machine.

pET 46 Ek/LIC Cloning Kit (Merck) or equivalent.

DNA miniprep kits.

Petri dishes (sterile).

Shaking incubator (with fittings to shake cultures from 5 ml to 1 L).

Flasks for bacterial cultures (2.5 L flasks are recommended for 1 L cultures; smaller flasks or tubes can be used for the 5 ml overnight grow-ups).

Luria-Bertani (Miller) bacterial growth media: 1% (w/v) tryptone, 0.5% (w/v) yeast extract, 1% (w/v) sodium chloride (pH 7.0). Alternatively, can be purchased pre-made.

Ampicillin: 100 mg ml^{-1} in ethanol; store at −20 °C.

(*Optional*) Other bacterial growth media and antibiotics may be required if optimization of the expression conditions is required or alternative expression vectors and/or bacterial strains are used.

Cobalt affinity resin (e.g., His-Select from Sigma).

Glass or plastic columns for chromatography (e.g., Econo-Column 5 cm by 1 cm from Bio-Rad).

Sonicator (choose a model and tip which can process 20–50 ml samples).

Centrifuge, rotors, and tubes (choose a model and rotor which can process 250–500 ml bottles and 20–50 ml tubes).

Agar.

Escherichia coli strains HMS174(DE3), BL21(DE3) and Rosetta (DE3).

Ampicillin (other antibiotics may be required if alternative vectors or strains are used).

Autoclave.

Dialysis tubing. Either buy pre-treated tubing or treat by boiling for 5–10 min in 250 ml deionized water with two spatulas of sodium hydrogencarbonate and one spatula of EDTA dissolved. Wash in deionized water and store in 20%(v/v) ethanol at 4 °C.

Bradford reagent. For convenience, buy this pre-made.

Spectrophotometer.

2.3 Native Gel Electrophoresis

Native gel electrophoresis apparatus. We strongly recommend obtaining kit specifically for native gels. A gel size of approximately 10 cm by 10 cm and a thickness of 0.75–1.5 mm will be sufficient for the majority of the protocols described below. For higher resolution a larger gel may be required.

Ice box large enough to fit the gel apparatus inside comfortably.

Metal salt solutions—make up 100 mM stocks in CRB. Store at 4 °C.

Ethylene glycol-bis (2-aminoethylether)-*N*,*N*,*N*′,*N*′-tetraacetic acid (EGTA)—make up 100 mM stock in CRB. pH if required. Store at 4 °C.

Gel documentation system or good-quality digital camera.

(*Optional*) Software for densitometry analysis, e.g., myImageAnalysis from Thermo Fisher.

Peptides—obtain at least 95% purity. A yield of 1 mg will be enough for a set of native gel experiments. Unless there is a good experimental reason, order peptides with N-terminal acetylation and C-terminal amidation. Reconstitute at 1 mg ml^{-1} in sterile water and keep stocks frozen at −80 °C.

2.4 Differential Scanning Fluorimetry

Quantitative PCR (qPCR) machine, e.g., Rotor-Gene Q cycler (Qiagen).

Software for analyzing melting curves from qPCR machine.

Tubes (or plates) suitable for use in the qPCR machine.

SYPRO Orange (available at 5000× dilution from Sigma).

Drugs (e.g., praziquantel (PZQ), chlorpromazine (CPZ), N-(6-aminohexyl)-5-chloro-1-naphthalenesulfonamide hydrochloride (W7), trifluoperazine hydrochloride (TFP), and thiamylal (ThA)). Make up in 100%(v/v) DMSO and dilute into CRB to make working stocks. Store at −20 °C.

Software for statistical calculations, e.g., Excel (Microsoft) or Prism (GraphPad Software).

2.5 Fluorescence

Plate-reader fluorimeter (if possible, choose one with a refraction grating rather than filters; this will allow complete flexibility in the choice of wavelength).

Black, 96-well plates suitable for the fluorimeter.

Incubator (37 °C).

Aluminum foil.

Software for analyzing fluorescence data (normally provided with the fluorimeter, but it may be desirable to export raw data and analyze in programs like Prism or Excel).

2.6 Cross-Linking

Cross-linking reagent, *bis*(sulfosuccinimidyl)suberate (BS^3). Make up stock solutions in 5 mM sodium citrate. Dilute in CRB as required.

2.7 Limited Proteolysis

Protease, chymotrypsin. Make up stock solutions in CRB supplemented with 1 mM calcium chloride.

(*Optional*) Other proteases, e.g., trypsin, subtilisin, and thermolysin. Make up as for chymotrypsin.

2.8 Analytical Gel Filtration

Glass column approximately 1 cm in diameter and with sufficient volume to hold approximately 60 ml of resin.

Peristaltic pump.

Silicone tubing (various diameters).

Fraction collector.

Sephadex S300 (GE Healthcare or Sigma).

Glass rod.

Ethanol (20% (v/v).

Standard proteins (e.g., serum albumin, chymotrypsinogen, ribonuclease).

Blue Dextran 2000 (GE Healthcare).

Syringes (1 or 2 ml) and needles.

Software for linear regression, e.g., Excel or Prism.

(*Optional*) WinHydroPro software.

2.9 Structure Prediction

Computer with access to the Internet.

Software for visualizing .pdb files (e.g., PyMOL).

YASARA View software.

3 Methods

3.1 Recombinant Expression and Purification

1. Insert PCR-amplified cDNA into a suitable expression vector for use in *Escherichia coli*. We tend to use the ligation-independent cloning system from Merck (Novagen). The pET 46 Ek/LIC system enables the rapid insertion of the amplicon such that the resulting sequence incorporates bases which code for an N-terminal hexahistidine tag (MAHHHHHHVDDDDK). The manufacturers provide extensive protocol information, which should be followed carefully (*see* **Note 1**).
2. Verify the DNA sequence. The whole sequence should be checked to avoid errors introduced by PCR or in the cloning process.
3. Isolate purified plasmid using miniprep kits (*see* **Note 2**).
4. Transform the plasmid into a suitable *E. coli* expression strain such as HMS174(DE3) (*see* **Note 3**). Typically 0.5 μl of mini-prepped plasmid is sufficient to give a large number of colonies following transformation using the calcium chloride method.
5. Using aseptic technique, pick one colony and transfer it to 5 ml of sterile Luria-Bertani (LB) medium supplemented with 100 μg ml^{-1} ampicillin (*see* **Note 4**). Grow this culture overnight, shaking at approximately 200 rpm at 37 °C.
6. Dilute the overnight culture into 1 L of LB medium supplemented with 100 μg ml^{-1} ampicillin maintaining aseptic technique. Grow this culture for approximately 3 h, shaking at approximately 200 rpm at 37 °C. After this time, induce the culture using IPTG to a final concentration of 1.3 mM. At this stage, it is acceptable to add solid IPTG and not use strict aseptic technique. Grow the culture for a further 2–3 h (*see* **Note 5**).
7. Harvest by centrifugation at 4200 × *g* for 15 min. Pour off the supernatant. Resuspend the pellet thoroughly in approximately 20 ml of cell resuspension buffer. Freeze immediately at −80 °C. The cells can be kept for several weeks if required before moving to the next step.

8. Thaw the cells fully. This can be achieved in a reasonable time by placing the tube in a beaker of cold tap water. Do not use warm water or use a microwave oven to thaw the cells.
9. Disrupt the cells by sonication on ice. Typically three pulses of 100 W are required. Allow a gap of 30–60 s between each pulse as sonication tends to heat the cell suspension (*see* **Note 6**).
10. Immediately remove the cell debris by centrifugation at 20,000 × *g* for 20 min at 4 °C. Decant the supernatant to a clean container. Retain 20 μl of this cleared sonicate for SDS-PAGE analysis.
11. During the centrifugation step, prepare a cobalt-agarose affinity column. Shake the resin suspension well to disperse the resin. Add 1 ml to a clean column and allow the liquid to drain away. Wash the column with 20 ml buffer A. Resins are typically stored in ethanol solutions, and so the wash step is vital to remove any ethanol (*see* **Note 7**).
12. Apply the supernatant from 10 to the column. Allow it to pass through and collect the flow-through (wash 1). Wash the column with 20 ml buffer A and collect the flow-through (wash 2).
13. Elute the protein with three 2 ml washes with buffer B. Collect the three flow-throughs separately (elutions 1, 2, and 3). Keep these on ice.
14. Analyze the sonicate, washes, and elutions by SDS-PAGE. Identify which elution(s) contains the protein (*see* **Note 8**).
15. Dialyze protein-containing elution(s) against at least 2 l of dialysis buffer overnight at 4 °C.
16. Divide the protein solutions into aliquots of 20–100 μl and freeze at −80 °C. Determine the concentration of protein using Bradford's method (or suitable alternative) [40] (*see* **Note 9**).

3.2 Native Gel Electrophoresis for Ion Binding

1. Make a vertical slab 6% (w/v) polyacrylamide gel, pH 8.8. Volumes and dimensions will depend on the gel electrophoresis kit being used. A gel of approximately 10 cm by 10 cm and a thickness of 0.75–1.5 mm will normally provide sufficient resolution. Since some variation in conditions may be required (*see* **Note 11**), it is unlikely that native gels can be purchased pre-made. Mix the components of the native gel (*see* Subheading 2) and pour rapidly but smoothly into the gel mold. Insert the comb immediately and allow to set completely.
2. Prepare assay mixes (total volume 10 μl) using 10–40 μM protein. Add solutions of ions (1–2 mM) and EGTA (half the concentration of ions) as required. Typically we use 1 mM

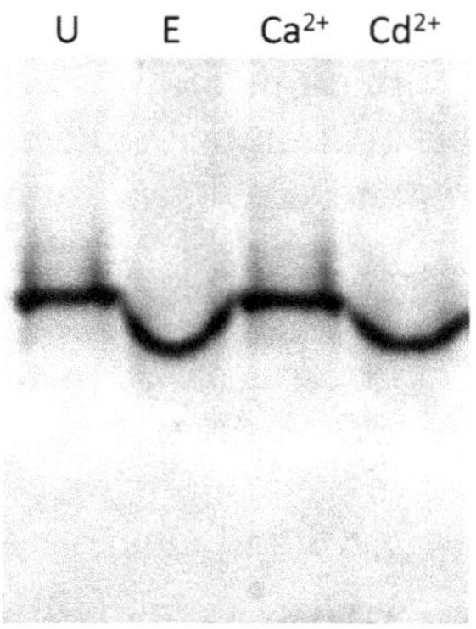

Fig. 1 Example of a gel shift assay. In this experiment, the *F. hepatica* protein FhCaBP2 (50 μM) was resolved on 10%(w/v) native continuous PAGE at pH 8.8. The first lane (U) is a sample of untreated protein. The second lane (E) is the protein treated with 1 mM EGTA. The last two lanes (Ca^{2+} and Cd^{2+}) are the protein treated with 1 mM EGTA and 2 mM metal ion salt. The reduced electrophoretic mobility in the presence of calcium ions, but not cadmium ions, provides evidence that this protein binds calcium but not cadmium ions. The untreated protein also had reduced mobility, suggesting that the recombinant protein was purified largely with calcium ions bound [30]

EGTA and 2 mM ion solution. Make up the volume with CRB (*see* **Note 10**).

3. Incubate at room temperature for 30 min. Then add 10 μl of native gel running buffer. Mix gently.
4. Electrophorese on a continuous, vertical slab 6% (w/v) polyacrylamide gel, pH 8.8, with a Tris-glycine (pH 8.8) running buffer at constant current (20 mA) for 1–4 h. For longer runs, cool the gel system by placing it in a polystyrene box and packing ice around it. Be careful not to let ice enter the gel system itself or get near the electrodes (*see* **Note 11**).
5. Stain with Coomassie blue dye (typically shaking at room temperature overnight) and destain thoroughly (shaking at room temperature for several hours). This may require several changes of destain.
6. Ion binding is indicated by a reduction in the protein's mobility when comparing EGTA-treated protein with protein in the presence of EGTA and a molar excess of ion. *See* Fig. 1 for an example (*see* **Note 12**).
7. Obtain a good-quality image of the gel using a gel documentation system or digital camera.

3.3 Native Gel Electrophoresis for Peptide Binding

1. Make a vertical slab 15% (w/v) polyacrylamide gel, pH 6.6. Volumes and dimensions will depend on the gel electrophoresis kit being used. A gel of approximately 10 cm by 10 cm and a

thickness of 0.75–1.5 mm will normally provide sufficient resolution. However, larger gels may be required, particularly if quantitative studies are required. Mix the components of the native gel (*see* Subheading 2) and pour rapidly but smoothly into the gel mold. Insert the comb immediately and allow to set completely.

2. Prepare assay mixes containing 10 μM protein and either EGTA (2 mM) or EGTA (2 mM)/calcium chloride (4 mM) in microcentrifuge tubes. Make up the volume to 9 μl with CRB and incubate at room temperature for 30 min.
3. Prepare peptide stock solutions in CRB such that the concentrations of the stocks are ten times those required in the assay (typical range is 5–160 μM final concentration in the assay). Check the approximate pH of these solutions using pH paper. Adjust to approximately 7.5 if required.
4. Add peptide solution (1 μl) and incubate at room temperature for a further 60 min. After this time add 10 μl of native gel loading buffer and mix gently (*see* **Note 13**).
5. Electrophorese the mixtures on the slab gel at 20 mA constant current until the blue dye runs off the end of the gel (*see* **Note 14**).
6. Stain with Coomassie blue dye (typically shaking at room temperature overnight) and destain thoroughly (shaking at room temperature for several hours). This may require several changes of destain.
7. Peptide binding is detected by reduced mobility compared to protein under identical conditions.
8. Obtain a good-quality image of the gel using a gel documentation system or digital camera.
9. (*Optional*) If a range of concentrations have been used, it may be possible to quantify the extent of the shifts using densitometry with myImageAnalysis software. Obtain an image of the gel and check for pixel saturation or uneven staining. Reimage the gel in the case of pixel saturation (noting that unsaturated images often look visually less appealing). Unevenly stained gels need to be repeated.
10. (*Optional*) Select the area around each band by eye. Correct for the different sizes of bands by dividing the intensity by the area. Select six random areas well away from the protein bands. For each of these, determine the intensity divided by the area. Calculate the mean of these six values to obtain a background value. Subtract this value from the values obtained for the bands.
11. (*Optional*) Ideally, values from at least three separate gels should be combined for each peptide concentration. Plot the

mean of these values (Υ) against peptide concentration. Using nonlinear curve fitting, fit the data to $\Upsilon = \Upsilon_{max}[\text{peptide}]/(K_{d,app}+[\text{peptide}])$, where Υ_{max} is the maximum possible value of Υ and $K_{d,app}$ is the apparent dissociation constant of the protein-peptide complex (*see* **Note 15**).

3.4 Differential Scanning Fluorimetry (DSF) for Ion and Drug Binding

1. Prepare assays (total volume 20 μl in PCR tubes as recommended by the manufacturer of the qPCR machine). Calculate the volume of CRB required to make up the volume and add this first. Add any additives (drugs, ions, etc.) second and protein third. (Protein concentrations are typically 2–10 μM; optimization may be required. Once optimized keep the same concentration for all experiments where comparison is required.) Add SYPRO Orange to a final concentration of 10× (manufacturer's concentration definition). Mix gently with a pipette and leave on ice for no more than 30 min. Each condition should be repeated in (at least) triplicate. If DMSO or similar solvents are used to dissolve additives, ensure that their concentration is constant between assays to enable fair comparison (*see* **Note 16**).
2. Place the tubes in a qPCR machine. We use a Rotor-Gene Q cycler but most other instruments can be used in this assay. Use the High Resolution Melt function setting the machine to heat from 30 to 95 °C in 5 °C increments every 4 s.
3. When the run has completed, the software will calculate a "melting temperature," T_m, from the first derivative of the denaturation curve (*see* **Note 17**).
4. Comparison of melting temperatures in the presence or absence of a drug or ion enables evidence to be gathered for or against interaction. Statistical analysis is recommended. Pairwise comparisons can be done using a t-test with Welch's correction (which does not assume equal standard deviation). Larger datasets should be analyzed by ANOVA with Tukey's post hoc test. A statistically significant difference in T_m suggests that the ligand binds to, and alters the stability of, the protein (*see* **Note 18**).
5. (*Optional*) Repeat the experiment with a range of ligand concentrations to obtain apparent dissociation constants. For each concentration calculate ΔT_m according to $\Delta T_m = T_m(\text{with ligand}) - T_m(\text{no ligand})$. Then plot ΔT_m against [ligand], and fit the data to $\Delta T_m = \Delta T_{m,max}[\text{ligand}]/(K_{d,app}+[\text{ligand}])$ where $\Delta T_{m,max}$ is the maximum value of ΔT_m (*see* **Note 19**).
6. If running several assays consecutively, ensure that the qPCR machine has fully cooled down to room temperature (which should be less than 30 °C) before starting the next assay.

3.5 Intrinsic Fluorescence for Ion and Drug Binding

1. Prepare assay mixes (in at least triplicate) and blanks in black 96-well plates. Mix buffer (10 mM Hepes-OH, pH 7.5), protein (10 μM) with buffer, and any additives (ions, drugs, etc.) in a total volume of 150 μl. Ions are typically added at 2 mM (with 1 mM EGTA to chelate any prebound calcium; *see* **Note 10**) and drugs at 100–250 μM. If DMSO is used, a solvent for the drugs ensure that no drug controls contain an equivalent concentration of DMSO. If a concentration range of drugs is to be used, it is helpful to make up a range of stocks each at ten times the concentration required in the assay. This way, an equal volume of drug solution can be added to each assay (reducing the chance of pipetting errors), and the final concentration of drug solvent will be constant in all assays.
2. If drugs are being included in the assays, make up a set of drug-only controls which contain the same components as the experimental assays, except for the protein. (Some drugs fluoresce in the same wavelength range as proteins.)
3. Incubate assay mixes at 37 °C for 30 min in the dark. If necessary, wrap the plate in aluminum foil to exclude light.
4. Measure the fluorescence in a plate-reader fluorimeter using an excitation wavelength (λ_{ex}) of 280 nm and an emission wavelength (λ_{em}) of 340 nm. Alternatively collect the emission spectrum between 330 and 400 nm (*see* **Note 20**).
5. Comparison of the fluorescence intensities in the presence and absence of drug or ion provides evidence for or against interaction. (NB, Check for fluorescence in the drug-only controls, if applicable. If significant fluorescence is detected here, it may be hard or impossible to interpret the data.) Since interaction may result in conformational changes which alter the environment of fluorescent side chains, the key spectral properties are the wavelength of maximum emission ($\lambda_{em,max}$) or the fluorescence intensity at one fixed wavelength (normally 340 nm). A statistical treatment is recommended (*see* **step 4** in Subheading 3.4) (*see* **Note 21**).

3.6 ANS Fluorescence for the Detection of Calcium-Dependent Conformational Changes

1. Prepare assays (in at least triplicate) in 96-well black plates. The final control assay mixes should contain protein (10 μM; some optimization may be required), EGTA (1 mM), and ANS (36 μM) in a total volume of 100 μl. Experimental assays will contain the same components plus 2 mM calcium chloride. Hepes-OH buffer (10 mM, pH 7.4) should be used to make up the volume. Add the buffer first and then the remaining components except the ANS. Mix by pipetting (*see* **Note 22**).
2. Incubate for 30 min at 37 °C. Then add the ANS, mix by pipetting, and incubate for a further 60 min at 37 °C in the

dark. Wrap the plate in aluminum foil to exclude light if necessary.

3. Measure the fluorescence in a plate-reader fluorimeter with an excitation wavelength of 350 nm and emission in the range 410–510 nm.
4. Conformational change is suggested by a significant increase in the fluorescence emission and a shift in the maximum emission wavelength to higher values.

3.7 Cross-Linking

1. Prepare assays in microcentrifuge tubes. Control assay mixes should contain protein (10–60 μM) in a total volume of 10 μl. The volume should be made up with CRB. Experimental assays should contain the same concentration of protein with either 1 mM EGTA or 1 mM EGTA/2 mM calcium chloride. Make up four tubes with each condition (*see* **Note 23**).
2. Incubate at 37 °C for 45 min and then add increasing concentrations of BS^3 (50–500 μM). Incubate at 37 °C for a further 60 min.
3. Add 10 μl of SDS-PAGE loading buffer, heat to 95 °C for 3 min, and analyze by 10%(w/v) SDS-PAGE. Stain fully (normally overnight) and destain until the gel background is clear. Obtain a good-quality image using a gel documentation system or digital camera.
4. Oligomerization is suggested by the appearance of additional bands at higher molecular masses. Comparison with molecular mass markers can be used to estimate sizes and, thus, the oligomeric state. For an example, *see* Fig. 2 (*see* **Note 24**).

3.8 Limited Proteolysis to Detect Conformational Changes on Drug Binding

1. Prepare assays in microcentrifuge tubes. Control assay mixes should contain protein (10–20 μM) and calcium chloride (800 μM) in a total volume of 9 μl. Experimental assays should contain the same components plus drug at 250 μM. Use CRB to make up the volume where required (*see* **Note 25**).
2. Incubate for 1 h at 37 °C. Then add chymotrypsin (800 nM) and incubate for a further 1 h at 37 °C.
3. Add 10 μl of SDS-PAGE loading buffer, heat to 95 °C for 3 min, and analyze by 10%(w/v) SDS-PAGE. Stain fully (normally overnight) and destain until the gel background is clear. Obtain a good-quality image using a gel documentation system or digital camera.
4. Observe the pattern of bands resulting from proteolysis. Changes in the pattern are likely to result from conformational changes resulting from drug binding (*see* **Note 26**).

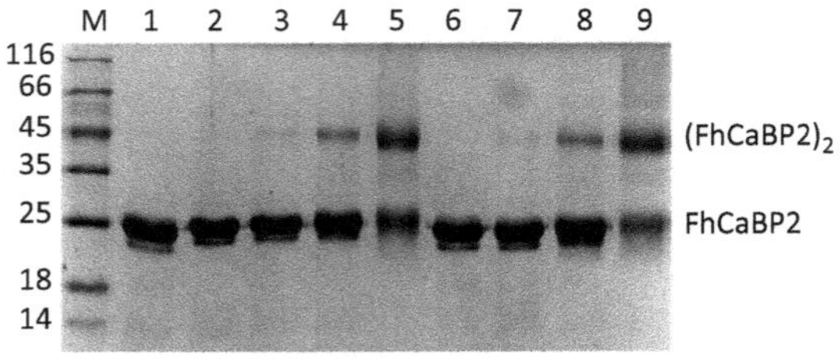

Fig. 2 Example of a cross-linking assay. In this experiment, the oligomeric state of FhCaBP2 (50 μM) was investigated by cross-linking. The reactions were analyzed by 15%(w/v) SDS-PAGE. Lane M is molecular mass markers (with masses, in kDa, to the left of the gel). Lane 1 is untreated protein (i.e., no cross-linker or other additives). Lanes 2–5 are protein in the presence of EGTA (2 mM) and 0, 8, 80, and 800 μM BS^3. Lanes 6–9 are protein in the presence of EGTA (2 mM)/$CaCl_2$ (4 mM) and 0, 8, 80, and 800 μM BS^3. The untreated protein resolves as a single main band (small amounts of impurities can also be detected). The same result is seen for the protein in the presence of EGTA or EGTA/calcium chloride. When cross-linker is added, an additional band is seen on the band (indicated at $(FhCaBP2)_2$ on the right-hand side of the gel). The intensity of this band increases with increasing cross-linker concentration. At the higher concentrations of cross-linker, some bands corresponding to higher molecular masses are also seen. These may correspond to higher-order oligomers or to aggregation. Although some proteins from this family alter their oligomeric state in response to calcium binding, FhCaBP2 forms dimers in the presence and the absence of calcium ions [26, 28, 30]. Based on this experiment, calcium ions appear have little effect on the oligomeric state of FhCaBP2

3.9 Analytical Gel Filtration

1. Pack a gel filtration column. Use approximately 65 ml of Sephacryl S-300 in a column of diameter approximately 1 cm. Agitate the bottle of resin well to disperse the resin evenly. Pour smoothly and steadily into the column. Allow the resin to settle and the liquid to run through the column. Pump slowly if required. When there is around 1 cm of liquid above the resin bed, add more resin. Continue to top up with resin until the resin bed is 1–2 cm from the top of the column. This will require several additions and may take 1–2 h. Do not allow layers to form. Before each addition, disrupt the top of the resin bed with a glass rod (*see* **Note 27**).
2. Measure the internal diameter of the column (and from this calculate the radius, r) and the height of the resin bed (h). Calculate the total volume of the column (V_t) using the equation $V_t = \pi r^2 h$.
3. The column should be stored in 20% (v/v) ethanol. If possible, avoid changing the temperature at which the column is stored. To store in 20% (v/v) ethanol, pump two to three times the total volume of the column of this solution through at a flow rate of 1 ml min^{-1}.

4. Equilibrate the column by pumping two to three times the total volume of the column of buffer R plus EGTA at a flow rate of 1 ml min^{-1}. (If the equilibration is from 20%(v/v) ethanol, it may be necessary to start with a slower flow rate as ethanol is harder to pump than buffer solutions.)
5. Estimate the void volume of the column (V_0). Apply a solution of blue dextran (250 μl at a concentration of 1 mg ml^{-1} in buffer R plus 2 mM EGTA) to the column and elute with buffer R plus 2 mM EGTA at a flow rate of 1 ml min^{-1}. Collect approximately 50 fractions (1 ml each) (*see* **Note 28**). Identify the fractions containing blue dextran by measuring the absorbance at 280 nm. The elution volume (which corresponds to V_0) can be determined by plotting a graph of absorbance against volume and estimating the volume at the maximum absorbance. Wash the column thoroughly (at least three times V_t of buffer R plus 2 mM EGTA).
6. Calibrate the column using standard proteins of known molecular mass (M) and, preferably, Stokes radius (R_S). Commonly used standards are albumin (67.0 kDa; $R_s = 2.55$ nm), chymotrypsinogen (25.0 kDa; $R_s = 2.09$ nm), and ribonuclease (13.7 kDa; $R_s = 1.64$ nm). Dissolve each protein to a final concentration of 50 μM in 250 μl of buffer R plus 2 mM EGTA. We recommend a separate run for each protein. Add the protein solution carefully to the top of the column; use a syringe to later it on top of the gel bed below the buffer layer. Quickly attach the pump. Elute with buffer R plus 2 mM EGTA at a flow rate of 1 ml min^{-1}. Collect approximately 50 fractions (1 ml each) (*see* **Note 28**). Identify the fractions containing protein by measuring the absorbance at 280 nm or by carrying out Bradford assays. The elution volume (V_e) can be determined by plotting a graph of absorbance against volume and estimating the volume at the maximum absorbance. Wash the column thoroughly (at least two times V_t of buffer R plus 2 mM EGTA). Repeat with the remaining standards (*see* **Note 29**).
7. Prepare a sample of protein (250 μl; at least 30 μM, preferably higher; supplemented with 2 mM EGTA). Incubate at room temperature for 20 min. Apply the sample to the column and chromatograph as for the standards (**step 6**). Record V_e.
8. Collect samples from protein-containing fractions. Verify the presence and identity of the protein by analyzing the samples by SDS-PAGE alongside a sample of the applied protein. Stain and destain the gel. Obtain a good-quality image.
9. For standards and the protein of interest, calculate the partition coefficient (K_{av}) according to the equation $K_{av} = (V_e - V_0)/(V_t - V_0)$. For the standards, plot K_{av} against the logarithm of

the molecular mass. The data should conform to a straight line with negative slope. Fit the data using linear regression (*see* **Note 30**). Use the equation of the line to estimate the molecular mass of the protein.

10. Using the protein sequence, estimate the monomeric molecular mass using ProtParam (https://web.expasy.org/protparam/) [41]. The experimental molecular mass should be (approximately) an integral multiple of the calculated monomeric molecular mass. This integer is the likely number of subunits per protein assembly under the conditions of the experiment (*see* **Note 31**).
11. Reequilibrate the column in buffer R plus 2 mM calcium chloride (use at least three times V_t). Repeat **steps 7–10** using a protein sample supplemented with 2 mM calcium chloride. This will enable the comparison of the solution molecular mass in the presence and absence of calcium ions (*see* **Note 32**).
12. (*Optional*) For the standards, plot K_{av} against R_s. Fit the data by linear regression, and use the equation of the line to estimate the Stokes radii for the protein in the presence and absence of calcium ions.
13. (*Optional*) Using a molecular model of the monomeric protein (see below), estimate the sedimentation coefficient ($S_{20,w}$) using WinHydroPro (which can be downloaded here: http://leonardo.inf.um.es/macromol/programs/hydropro/hydropro.htm) [42].
14. (*Optional*) Estimate the subunit stoichiometry (n) using the equation $nM = S_{20,w}N_A(6\pi\eta R_s)/(1 - \nu_2\rho)$, where M is the monomeric molecular mass (in daltons), N_A is Avogadro's number (6.023×10^{23} mol^{-1}), η is the viscosity of the solvent (0.01 g cm^{-1} s^{-1}), ν_2 is the partial specific volume (0.73 g cm^3 g^{-1}), and ρ is the density of the solvent (1.0 g cm^{-3}) [43] (*see* **Note 33**).

3.10 Structure Prediction

1. Enter the amino acid sequence into the appropriate box on the Phyre2 webserver (http://www.sbg.bio.ic.ac.uk/~phyre2/) [44]. Check to ensure that the complete sequence has been copy/pasted and no extraneous letters. The server will ignore numbers and spaces and so there is no need to remove these (*see* **Note 34**).
2. Choose a sensible name for the protein and enter this into the "optional Job description" box. If multiple structures are being predicted in one session, it is wise to record the descriptions and the proteins they relate to. Enter your email address in the appropriate box.

3. Choose the "intensive" modelling mode and click "Phyre search." Typical structure predictions take several hours in the intensive mode. If you leave the window open, it will display updates on progress and the final result, once available. A link is also sent to your email address.
4. The data output consists of a protein model (in .pdb format) and a rich source of information about the protein templates, etc. used to derive the model. Peruse the information and note the main templates which were used. Save the information locally as it is only retained on the server for 30 days.
5. (*Optional*) Submit your model to 3DLigandSite to predict which ions and small molecules might bind to it [45]. There is a click through link on the results page from Phyre2.
6. The quality of the model can be increased by computationally solvating it and minimizing the energy. Submit the model from Phyre2 to the YASARA webserver (http://www.yasara.org/minimizationserver.htm) [46]. Typical minimizations take several hours (but can take 1–2 days). Results are sent by email. Frustratingly, it is not possible to assign a name to jobs sent to the YASARA server, and they are returned numbered according to the order in which they were completed on the server. Note that this may not be the order you submitted them in if you are submitting multiple jobs.
7. YASARA returns results in .sce format. This format is largely used by the YASARA suite of programs. To convert it to .pdb format, download YASARA view (currently available free here: http://www.yasara.org/yasaradl.htm). Open the .sce file in YASARA view. Select File>save As>PDB. This brings up a box of options. Choose Sequence, End; Name, End; Belongs to or has, All. In the next window choose a filename and a location to save the file.
8. To view the resulting .pdb file, open it in PyMOL (www.pymol.org) or an alternative such as Discovery Studio Visualiser (free version available here for academic use: http://accelrys.com/resource-center/downloads/freeware/index.html) or Chimera (https://www.cgl.ucsf.edu/chimera/download.html) [47]. This will enable viewing of the model and the production of publication quality images.
9. The quality of the model should be checked. A straightforward check of the geometry can be done by generating a Ramachandran plot using servers such as RAMPAGE (http://mordred.bioc.cam.ac.uk/~rapper/rampage.php) [48]. The majority of residues should lie in the allowed areas of the plot. A large number of residues in the disallowed regions would throw doubt on the validity of the model. Various servers exist

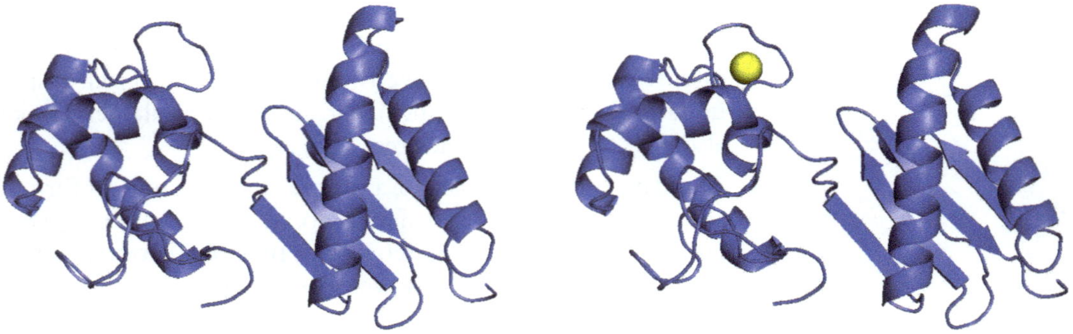

Fig. 3 Example of a molecular model. A model of the 22 kDa *F. hepatica* protein FhCaBP1 (FH22) was generated with Phyre2 and its energy minimized using YASARA (left). This model was used with the highest ranked calcium-binding template (the mouse Reps1 EH domain; PDB: 1FI6 [51]) to generate a calcium-bound form which was then energy minimized in YASARA (right). The calcium ion is shown as a sphere. The structures of the two forms of the protein are highly similar (rmsd = 0.433 Å over 2642 equivalent atoms) [27]

which assess the quality of protein models, for example, APOLLO [49].

10. To generate a calcium-bound model, identify the highest ranked template from Phyre2 which includes a calcium ion. Download this template protein's .pdb file from the Protein Data Bank (https://www.rcsb.org) [50].
11. Open both this file and the model in the same PyMOL window. Use the align command to overlap the two structures (e.g., align protein1, protein2). Take care to ensure the names are exactly correct. Check that the alignment has overlapped the EF-hand(s) so that the calcium ion(s) are sensibly positioned.
12. Use the sequence command to display the two protein sequences. Highlight the whole of the model sequence and the calcium ion(s) from the template. Save this as a new .pdb file (File> Save>Molecule; then pick "sel" and click ok). Open the new .pdb file in PyMOL to check that you have the model plus calcium(s) at approximately the expected positions.
13. Submit this file to YASARA for a further round of minimization. Once you have this minimized structure in .pdb format, you can compare the calcium and apo-forms. For an example, *see* Fig. 3 (*see* **Note 35**).

4 Notes

1. Those who study the Merck LIC system's instructions carefully will see that a considerable portion of the reaction mixes are wasted. Therefore, the volumes of reactions can be scaled down to save reagents. The T4 DNA polymerase is often depleted

before other reagents. It can be replaced by the same enzyme from reputable suppliers. Other cloning systems are available and can be used. Extensive protocols are available from the manufacturers and in the literature.

2. Many such kits are available on the market. Any reputable brand is suitable for this task. It is wise to maintain stocks of sequence verified plasmid in the laboratory to avoid having to reclone the sequence; this also enables the sharing of the expression vector with collaborators.
3. Overall, we have found that this strain is likely to result in high-level expression of the recombinant protein. However, if the yield is low, we suggest trying alternative strains such as BL21 (DE3) or Rosetta(DE3).
4. Ampicillin is required for selection of the recombinant plasmid. It is convenient to make a stock solution at 100 mg ml^{-1} in ethanol. This can be stored at −20 °C but does not freeze. If other vectors or bacterial strains are used, other antibiotics may be required instead of, or in addition to, ampicillin.
5. Many expression/purification protocols call for the careful monitoring of cell density by measurement of A_{600nm}. We do not routinely do this and find that the times indicated here normally give good yields. However, if the culture is slow-growing, we recommend waiting until $A_{600nm} = 0.6–0.8$ before adding IPTG. (As an approximation, $A_{600nm} = 0.6–0.8$ corresponds to a cell density in which, when the culture is held up to a light source, it is no longer possible to see your hand through.) If expression yields are low, we recommend changing culture times and temperatures. For example, growing the cells for a longer time (e.g., overnight) at a lower temperature (e.g., 16 °C) before induction often improves yield of "difficult" proteins.
6. Some investigators add protease inhibitors at this point. We tend to find these unnecessary for these proteins so long as **steps 8–11** are carried out as rapidly as possible and the cell suspension/extract is kept on ice. However, if degradation of the final protein product is observed, protease inhibitors may help. If so, we recommend adding them immediately prior to sonication.
7. Nickel-agarose can be used in place of cobalt-agarose. Both are equally effective, but cobalt-agarose tends to be cheaper. Of course, many sophisticated, automated protein purification systems exist, and there is no reason why they could not be used to purify these proteins. However, we have found that simple columns relying on gravity flow result in good yields of purified protein in several hours.

8. Any protocol for SDS-PAGE is likely to work. Choose the percentage acrylamide based on the molecular mass of the protein. Ten to twelve percent (w/v) generally works well for proteins in the 20–100 kDa range. Smaller proteins may require 12–18%(w/v) acrylamide. Acrylamide is highly neurotoxic and weighing of this chemical should be avoided: buy acrylamide/bisacrylamide solutions. Also note that the acrylamide/bisacrylamide ratio in these solutions will affect the resolution and effective molecular mass range of gels. The gel can also be used for troubleshooting if little or no protein is recovered. If the protein does not appear in the sonicate, it is worth investigating cell extracts before and after induction to determine if the protein is being made. If it being made but not released in the sonicate, then the protein may be insoluble in the cells. Try using reducing the temperature during the induction phase of the grow-up. If it is released but appears in the wash, then it may not be binding to the cobalt-agarose. Investigate alternative purification methods (e.g., phenylsepharose can often be used exploiting the calcium-dependent exposure of hydrophobic surfaces on the protein [6]). If the protein is in the sonicate and neither the washes nor the elutions, it may be binding very tightly to the resin. Try increasing the imidazole concentration or using EDTA as an eluant. In extreme cases, it may be necessary to elute with a mixture of imidazole and a strong chaotrope such as urea or guanidine hydrochloride. The chaotrope can be removed by extensive dialysis. In some cases it may be necessary to step the concentration of chaotrope down through successive dialyses. With some luck, the protein will refold as the chaotrope is removed according to Anfinsen's principle [52].
9. We tend to thaw each aliquot once and dispose of it at the end of a day's experiments. Refreezing is not recommended.
10. In the majority of cases, we have found that recombinant calcium-binding proteins tend to be purified largely in complex with calcium ions [26, 29]. Therefore, when investigating ions other than calcium, we add EGTA (a calcium-specific chelating agent) to remove calcium ions from the protein and a molar excess of the ion solution. To ensure identical experimental conditions, we also do this when investigating calcium ion binding (and in subsequent experiments, e.g., cross-linking). Note that EGTA is acidic and it may be necessary to adjust the pH of solutions of this chemical even if dissolving it in buffers. For some applications, EDTA can be used in places of EGTA, but care should be taken as this reagent chelates a much broader range of divalent metal ions.
11. It is hard to predict how proteins will behave in native PAGE. We find that a simple, continuous slab gel at pH 8.8 is a good

starting point and works for many proteins. However, it is likely that some trial and error may be required. It is possible to estimate the isoelectric point of proteins using applications like ProtParam in ExPASy [41]. However, these are estimates based on the sequence alone and cannot take into account the consequences of protein folding. Therefore, they should be used with caution. A protein will have no mobility in native PAGE at its isoelectric point. Above this point it will be negatively charged (and thus migrate to the positive electrode) and below it will be positively charged. Thus these estimates can be used to guide variations in the gel and/or buffer pH and which way round to set the electrodes. (Typically gel electrophoresis apparatuses are designed such as to detect proteins which migrate toward the positive electrode since this is what is required in SDS-PAGE. Reversing the connections at the power pack will detect any proteins which migrate to the negative electrode). It is possible that pH, running time, and direction may all have to be altered to result in good resolution. There is considerable literature on native gel electrophoresis, and this is a good source of alternative starting conditions (see, e.g., [53, 54]). We provide some examples of alternative buffers for continuous native PAGE in the Materials.

12. The difference in mobility is typically quite small. It is advisable to run several replicates to ensure that the shift is due to ion binding rather than experimental artifacts. Unfortunately native PAGE appears particularly susceptible to artifacts such as trailing at the edge and "smiling" of bands. Keeping a gel kit solely for native PAGE is advisable. Switching kit between SDS-PAGE and native PAGE is not recommended as residual SDS can be hard to eliminate entirely. Native gel kit should be kept clean and well-maintained to reduce artifacts. Note that while a genuine shift provides strong evidence for ion binding, the absence of shift does not equate to an inability to bind. It may be that any shift cannot be detected under the gel electrophoresis conditions or that binding does not result in sufficient conformational change to result in altered mobility.
13. Further complexity can be added to the experiment, for example, drugs which are hypothesized to antagonize the interaction.
14. Like with the ion-binding experiments, trial and error may be required to identify the ideal gel conditions for this experiment (*see* **Notes 11** and **12**). In carrying out a search of conditions, it is worth keeping in mind their physiological relevance. For example, good separation at very high or low pH may not reflect cellular conditions.

15. The curve should be rectangular hyperbola (similar to the classical Michaelis–Menten enzyme kinetics curve). For reasonable analysis, at higher values of [peptide] Υ should be approaching Υ_{max}, and $K_{d,app}$ should lie in the middle of the [peptide] values. Note that this is an ***apparent*** dissociation constant which is likely to differ considerably from the real one not least as the conditions of native gel electrophoresis differ substantially from those prevailing in the cell. Nevertheless, $K_{d,app}$ values can be useful in ranking affinities of complexes analyzed under identical conditions (e.g., *see* [55]).
16. SYPRO Orange is sold by Sigma-Aldrich at 5000× in DMSO. While it would be possible to determine the molar concentration by spectroscopy, a concentration of 10× in the assay generally works. The 5000× stock should be kept in at 4 °C in a dark bottle. We recommend mixing or vortexing the bottle before removing liquid from it to make diluted stocks. A fresh diluted stock is normally made for each assay.
17. This is, of course, not the temperature at which the protein changes from solid to liquid phase. It represents the temperature at which half the protein is unfolded due to thermal denaturation as estimated by the non-specific association of SYPRO Orange molecules with exposed hydrophobic patches [56].
18. Values of T_m can increase or decrease in response to binding. Increases are relatively easy to interpret: the ligand binds and the protein becomes more thermodynamically stable. This is a common situation in which proteins become slightly more compact and rigid on binding, and additional non-covalent interactions form in the binding site. It is unlikely that ligand binding to the native state results in reduced thermal stability. One possible explanation is that the ligand binds and stabilizes a partially folded state (note that proteins in solution exist in equilibrium between fully folded and partially folded forms). Like with native gel electrophoresis, while a change is T_m is good evidence for interaction, absence of a change does not prove lack of interaction (*see* **Note 12**).
19. The same "health warnings" apply to the apparent dissociation constants as for those obtained by native gel electrophoresis (*see* **Note 15**). Given that the thermal denaturation event is likely to be complex and cooperative, it is wise not to over-interpret these values. At most, they should be used to rank affinities.
20. Many protocols for protein fluorescence suggest using an excitation wavelength of 290 nm since this is close to the maximum absorbance of tryptophan side chains (the main fluorophore in most proteins). However, some calcium-binding proteins

(e.g., many calmodulins) lack tryptophan residues or only have one or two per molecule. Furthermore, most plate-reader fluorimeters are not able to excite with light of a single wavelength, and there is considerable "spillover." Therefore a good separation between λ_{ex} and λ_{em} is often required. However, if satisfactory results are not obtained with $\lambda_{ex} = 280$ nm, we recommend trying values up to 290 nm.

21. Unlike absorbance readings in spectrophotometry, fluorescence readings are arbitrary numbers. Take care to ensure that the values are not too low or too high, particularly if they exceed the linear range of the instrument. Since fluorescence values are arbitrary, they cannot be compared between instruments. Indeed caution should be exercised in comparing values obtained on the same machine on different days.
22. This protocol is designed to compare ANS fluorescence in the presence and absence of bound calcium ions and thus to detect calcium-dependent conformational changes. It can be readily adapted to measure conformational changes in response to other ions, drugs, etc.
23. This protocol is designed to compare the effects of the presence and absence of calcium ions on oligomeric state. It can be readily adapted to compare other conditions (e.g., the effect of drugs). We suggest using a range of cross-linker concentrations since the amount needed is somewhat dependent on the protein. If an optimum concentration is identified, future experiments could use just this.
24. Cross-linking depends on the juxtaposition of suitable reactive groups in the protein complex. Absence of additional bands does not provide convincing evidence for lack of interaction. If no interaction is observed, we suggest trying alternative cross-linkers such as glutaraldehyde or 1-Ethyl-3-(3-dimethylamino-propyl)carbodiimide (EDC). It is also possible to vary the incubation time and temperature.
25. 25 The activity of many proteases, including chymotrypsin, is affected by calcium ions [57]. Therefore this assay cannot be used to compare the effects of binding in the presence and absence of calcium ions. Indeed, we recommend making up stack solutions of chymotrypsin which contain calcium ions (*see* Subheading 2).
26. Ideally the experiment should be repeated with other proteases (e.g., trypsin and subtilisin; similar concentrations can be used) with different specificities. Confirmation of changes to the proteolysis pattern in the presence of the drug will strengthen the argument for interaction. It is also desirable to carry out control experiments using a protein for which there is no evidence of interaction with the drug. We tend to use

galactokinase or other metabolic enzymes. However, many proteins will be suitable. Avoid using serum albumin. This protein binds non-specifically to many drugs, especially more hydrophobic ones. These experiments rule out the possibility of an effect of the drug on the protease rather than the calcium-binding protein.

27. Alternatively purchase pre-prepared columns. In general, these need to be run on computer-controlled chromatography systems. These provide the advantage of being automated and having better run-to-run reproducibility. However, much greater care is required. For example, buffers generally need to be filtered, and the consequences of admitting even small amounts of air can be catastrophically expensive. Since the aim here is to determine the oligomeric state, we find that hand-made columns provide sufficient resolution to distinguish monomers, dimers, and tetramers. We do not filter buffers and the column can be repacked should it dry out or be contaminated.

28. It can be difficult using manual methods to set the flow rate to exactly 1 ml min^{-1}. A solution is to set it to approximately this value and then weigh the fractions to determine their volume (making the assumption that the density of the solutions is 1 g ml^{-1}).

29. Serum albumin forms dimers and so two peaks may be observed with this protein [58]. If so, the elution volume of the dimer peak can be used as an additional calibration point.

30. If the standards do not conform to a straight line, repeat using higher concentrations. If this fails, it may be necessary to repack the column.

31. Technically, gel filtration measures hydrodynamic volume and not mass. So this makes the assumption that proteins have approximately the same density (which is broadly true) and are approximately spherical (which is sometimes far from the truth). See optional steps for an approach to correct for this.

32. Strictly speaking **steps 5** and **6** (determination of V_0 and calibration) should also be repeated in the presence of calcium chloride. In practice the presence or absence of calcium ions makes little difference to these values in our experience.

33. This should result in a more reliable estimate of the stoichiometry. However, it is dependent on the quality of the molecular model (*see* **Note 35**).

34. This protocol describes the use of Phyre2, which we have used extensively. However, alternatives do exist notably SWISS-MODEL and MODELLER [59, 60]. The former is webserver based like Phyre2, and the latter requires a program to be

downloaded (which is currently free for academic users). The field is likely to continue advancing, and so new versions of these programs are likely to be released, and new competitors will emerge.

35. The minimization methods here will correct side-chain conformations and small issues with the backbone or the position of the calcium ion. They will not predict major conformational changes or secondary structure reorganizations. It is also critical to recognize that these structures are *models*. They are a prediction of the structure based on currently known, similar structures. They will not be perfect, especially in areas of "random coil" or regions which have no closely similar proteins in the Protein Data Bank. Phyre2 will estimate these regions using ab initio methods, but these are not always reliable [44, 61].

Acknowledgments

CMT thanks the Department of Employment and Learning Northern Ireland (DELNI, UK) for a PhD studentship.

References

1. Berridge MJ, Lipp P, Bootman MD (2000) The versatility and universality of calcium signalling. Nat Rev Mol Cell Biol 1:11–21
2. Megyes T, Grósz T, Radnai T, Bakó I, Pálinkás G (2004) Solvation of calcium ion in polar solvents: an X-ray diffraction and *ab initio* study. J Phys Chem A 108:7261–7271
3. Williams RJ (2006) The evolution of calcium biochemistry. Biochim Biophys Acta 1763:1139–1146
4. Gifford JL, Walsh MP, Vogel HJ (2007) Structures and metal-ion-binding properties of the Ca^{2+}-binding helix-loop-helix EF-hand motifs. Biochem J 405:199–221
5. Kawasaki H, Kretsinger RH (2017) Structural and functional diversity of EF-hand proteins: evolutionary perspectives. Protein Sci 26:1898–1920
6. Gopalakrishna R, Anderson WB (1982) Ca^{2+}-induced hydrophobic site on calmodulin: application for purification of calmodulin by phenyl-Sepharose affinity chromatography. Biochem Biophys Res Commun 104:830–836
7. Chin D, Means AR (2000) Calmodulin: a prototypical calcium sensor. Trends Cell Biol 10:322–328
8. Roufogalis BD, Minocherhomjee AM, Al-Jobore A (1983) Pharmacological antagonism of calmodulin. Can J Biochem Cell Biol 61:927–933
9. Hait WN (1987) Targeting calmodulin for the development of novel cancer chemotherapeutic agents. Anticancer Drug Des 2:139–149
10. Kang S, Hong J, Lee JM, Moon HE, Jeon B, Choi J, Yoon NA, Paek SH, Roh EJ, Lee CJ, Kang SS (2017) Trifluoperazine, a Well-Known Antipsychotic, Inhibits Glioblastoma Invasion by Binding to Calmodulin and Disinhibiting Calcium Release Channel IP3R. Mol Cancer Ther 16:217–227
11. Coles GC (1979) The effect of praziquantel on *Schistosoma mansoni*. J Helminthol 53:31–33
12. Vale N, Gouveia MJ, Rinaldi G, Brindley PJ, Gartner F, Correia da Costa JM (2017) Praziquantel for schistosomiasis: single-drug metabolism revisited, mode of action, and resistance. Antimicrob Agents Chemother 61:e02582–e02516
13. Furtado LF, de Paiva Bello AC, Rabelo EM (2016) Benzimidazole resistance in helminths: from problem to diagnosis. Acta Trop 162:95–102
14. Carmichael I, Visser R, Schneider D, Soll M (1987) *Haemonchus contortus* resistance to ivermectin. J S Afr Vet Assoc 58(2):93

15. Vieira LS, Berne ME, Cavalcante AC, Costa CA (1992) *Haemonchus contortus* resistance to ivermectin and netobimin in Brazilian sheep. Vet Parasitol 45(1–2):111–116
16. Blasco B, Leroy D, Fidock DA (2017) Antimalarial drug resistance: linking *Plasmodium falciparum* parasite biology to the clinic. Nat Med 23:917–928
17. Baker N, de Koning HP, Maser P, Horn D (2013) Drug resistance in African trypanosomiasis: the melarsoprol and pentamidine story. Trends Parasitol 29:110–118
18. Takala-Harrison S, Laufer MK (2015) Antimalarial drug resistance in Africa: key lessons for the future. Ann N Y Acad Sci 1342:62–67
19. Utzinger J, Becker SL, Knopp S, Blum J, Neumayr AL, Keiser J, Hatz CF (2012) Neglected tropical diseases: diagnosis, clinical management, treatment and control. Swiss Med Wkly 142:w13727
20. Murrell KD (1991) Economic losses resulting from food-borne parasitic zoonoses. Southeast Asian J Trop Med Public Health 22 (Suppl):377–381
21. Russell SL, Timson DJ (2014) Calcium binding proteins in the liver fluke, *Fasciola hepatica*. In: New developments in calcium signaling research. Nova Science Publishers, pp 89–104
22. Fraga H, Faria TQ, Pinto F, Almeida A, Brito RM, Damas AM (2010) FH8--a small EF-hand protein from *Fasciola hepatica*. FEBS J 277:5072–5085
23. Hu S, Law P, Lv Z, Wu Z, Fung MC (2008) Molecular characterization of a calcium-binding protein SjCa8 from *Schistosoma japonicum*. Parasitol Res 103:1047–1053
24. Thomas CM, Timson DJ (2016) A mysterious family of calcium-binding proteins from parasitic worms. Biochem Soc Trans 44:1005–1010
25. Fitzsimmons CM, Jones FM, Stearn A, Chalmers IW, Hoffmann KF, Wawrzyniak J, Wilson S, Kabatereine NB, Dunne DW (2012) The *Schistosoma mansoni* tegumental-allergen-like (TAL) protein family: influence of developmental expression on human IgE responses. PLoS Negl Trop Dis 6:e1593
26. Banford S, Drysdale O, Hoey EM, Trudgett A, Timson DJ (2013) FhCaBP3: a *Fasciola hepatica* calcium binding protein with EF-hand and dynein light chain domains. Biochimie 95:751–758
27. Cheung S, Thomas CM, Timson DJ (2016) FhCaBP1 (FH22): a *Fasciola hepatica* calcium-binding protein with EF-hand and dynein light chain domains. Exp Parasitol 170:109–115
28. Orr R, Kinkead R, Newman R, Anderson L, Hoey EM, Trudgett A, Timson DJ (2012) FhCaBP4: a *Fasciola hepatica* calcium-binding protein with EF-hand and dynein light chain domains. Parasitol Res 111:1707–1713
29. Thomas CM, Fitzsimmons CM, Dunne DW, Timson DJ (2015) Comparative biochemical analysis of three members of the *Schistosoma mansoni* TAL family: differences in ion and drug binding properties. Biochimie 108:40–47
30. Thomas CM, Timson DJ (2015) FhCaBP2: a *Fasciola hepatica* calcium-binding protein with EF-hand and dynein light chain domains. Parasitology 142:1375–1386
31. Subpipattana P, Grams R, Vichasri-Grams S (2012) Analysis of a calcium-binding EF-hand protein family in *Fasciola gigantica*. Exp Parasitol 130:364–373
32. Vichasri-Grams S, Subpipattana P, Sobhon P, Viyanant V, Grams R (2006) An analysis of the calcium-binding protein 1 of *Fasciola gigantica* with a comparison to its homologs in the phylum Platyhelminthes. Mol Biochem Parasitol 146:10–23
33. Kim YJ, Yoo WG, Lee MR, Kang JM, Na BK, Cho SH, Park MY, Ju JW (2017) Molecular and structural characterization of the tegumental 20.6-kDa protein in *Clonorchis sinensis* as a potential druggable target. Int J Mol Sci 18:557
34. Senawong G, Laha T, Loukas A, Brindley PJ, Sripa B (2012) Cloning, expression, and characterization of a novel *Opisthorchis viverrini* calcium-binding EF-hand protein. Parasitol Int 61:94–100
35. Nguyen TH, Thomas CM, Timson DJ, van Raaij MJ (2016) *Fasciola hepatica* calcium-binding protein FhCaBP2: structure of the dynein light chain-like domain. Parasitol Res 115:2879–2886
36. Jo CH, Son J, Kim S, Oda T, Kim J, Lee MR, Sato M, Kim HT, Unzai S, Park SY, Hwang KY (2017) Structural insights into a 20.8-kDa tegumental-allergen-like (TAL) protein from *Clonorchis sinensis*. Sci Rep 7:1764
37. Xu J, Ren Y, Xu X, Chen J, Li Y, Gan W, Zhang Z, Zhan H, Hu X (2014) *Schistosoma japonicum* tegumental protein 20.8, role in reproduction through its calcium binding ability. Parasitol Res 113:491–497
38. Durst RA, Staples BR (1972) Tris/tris-HCl: a standard buffer for use in the physiologic pH range. Clin Chem 18(3):206–208
39. Lewis FA, Liang YS, Raghavan N, Knight M (2008) The NIH-NIAID schistosomiasis resource center. PLoS Negl Trop Dis 2(7): e267

40. Bradford MM (1976) A rapid and sensitive method for the quantitation of microgram quantities of protein utilizing the principle of protein-dye binding. Anal Biochem 72:248–254
41. Gasteiger E, Hoogland C, Gattiker A, Duvaud S, Wilkins MR, Appel RD, Bairoch A (2005) Protein identification and analysis tools on the ExPASy server. In: Walker JM (ed) The proteomics protocols handbook. Humana Press, New York, pp 571–607
42. Ortega A, Amoros D, Garcia de la Torre J (2011) Prediction of hydrodynamic and other solution properties of rigid proteins from atomic- and residue-level models. Biophys J 101:892–898
43. Erickson HP (2009) Size and shape of protein molecules at the nanometer level determined by sedimentation, gel filtration, and electron microscopy. Biol Proced Online 11:32–51
44. Kelley LA, Mezulis S, Yates CM, Wass MN, Sternberg MJE (2015) The Phyre2 web portal for protein modeling, prediction and analysis. Nat Protoc 10:845–858
45. Wass MN, Kelley LA, Sternberg MJ (2010) 3DLigandSite: predicting ligand-binding sites using similar structures. Nucleic Acids Res 38: W469–W473
46. Krieger E, Joo K, Lee J, Lee J, Raman S, Thompson J, Tyka M, Baker D, Karplus K (2009) Improving physical realism, stereochemistry, and side-chain accuracy in homology modeling: four approaches that performed well in CASP8. Proteins 77(Suppl 9):114–122
47. Pettersen EF, Goddard TD, Huang CC, Couch GS, Greenblatt DM, Meng EC, Ferrin TE (2004) UCSF Chimera—a visualization system for exploratory research and analysis. J Comput Chem 25:1605–1612
48. Lovell SC, Davis IW, Arendall WB 3rd, de Bakker PI, Word JM, Prisant MG, Richardson JS, Richardson DC (2003) Structure validation by Cα geometry: ϕ,ψ and Cβ deviation. Proteins 50:437–450
49. Wang Z, Eickholt J, Cheng J (2011) APOLLO: a quality assessment service for single and multiple protein models. Bioinformatics 27:1715–1716
50. Berman HM, Westbrook J, Feng Z, Gilliland G, Bhat TN, Weissig H, Shindyalov IN, Bourne PE (2000) The protein data bank. Nucleic Acids Res 28:235–242
51. Kim S, Cullis DN, Feig LA, Baleja JD (2001) Solution structure of the Reps1 EH domain and characterization of its binding to NPF target sequences. Biochemistry 40 (23):6776–6785
52. Anfinsen CB (1973) Principles that govern the folding of protein chains. Science 181:223–230
53. Ornstein L, Davis BJ (1964) Disc electrophoresis-I: background and theory. Ann N Y Acad Sci 121:321–349
54. McLellan T (1982) Electrophoresis buffers for polyacrylamide gels at various pH. Anal Biochem 126:94–99
55. Atcheson E, Hamilton E, Pathmanathan S, Greer B, Harriott P, Timson DJ (2011) IQ-motif selectivity in human IQGAP2 and IQGAP3:binding of calmodulin and myosin essential light chain. Biosci Rep 31:371–379
56. Ericsson UB, Hallberg BM, Detitta GT, Dekker N, Nordlund P (2006) Thermofluor-based high-throughput stability optimization of proteins for structural studies. Anal Biochem 357:289–298
57. Wu FC, Laskowski M (1956) The effect of calcium on chymotrypsins alpha and B. Biochim Biophys Acta 19:110–115
58. Squire PG, Moser P, O'Konski CT (1968) The hydrodynamic properties of bovine serum albumin monomer and dimer. Biochemistry 7:4261–4272
59. Arnold K, Bordoli L, Kopp J, Schwede T (2006) The SWISS-MODEL workspace: a web-based environment for protein structure homology modelling. Bioinformatics 22:195–201
60. Webb B, Sali A (2017) Protein structure modeling with MODELLER. Methods Mol Biol 1654:39–54
61. Kelley LA, Sternberg MJ (2009) Protein structure prediction on the Web: a case study using the Phyre server. Nat Protoc 4:363–371

Part III

Biomarkers, Diagnostics and Laboratory Medicine

Chapter 40

High-Sensitivity Troponin Assays in Clinical Diagnostics of Acute Coronary Syndrome

Danielle Hof and Arnold von Eckardstein

Abstract

Nowadays, measurement of cardiac troponins (cTn) in patient plasma is central for diagnosis of patients with acute coronary syndrome (ACS). High-sensitivity (hs) immunoassays have been developed that can very precisely record slightly elevated and rising plasma concentrations of cTn very early after onset of clinical symptoms. Algorithms integrate measurements of hs-cTn at onset of clinical symptoms of acute myocardial infarction (AMI), and 1 or 3 h after onset, to rule-in and rule-out AMI patients. More and more point-of-care (POC) cTn assays conquer the diagnostic market, but thorough clinical validation studies are required before potential implementation of such POC tests into hospital settings. This review provides an overview of the technical aspects, as well as diagnostic and prognostic use of cardiac troponins in AMI patients and in the healthy population.

Key words Acute coronary syndrome, Acute myocardial infarction, Cardiac troponin, High-sensitivity assay, NSTEMI, STEMI, Laboratory medicine

1 Acute Coronary Syndrome

The acute coronary syndrome (ACS) embraces a range of conditions, including acute myocardial infarction (AMI) and unstable angina (UA) pectoris. AMI is defined by evidence of myocardial injury in a clinical setting consistent with myocardial ischemia [1] and is a major cause of death and disability worldwide. According to the 2017 statistics of the American Heart Association, every 40 s a US citizen suffers an AMI [2]. It is estimated that currently each year around 695,000 Americans will suffer a first AMI, and around 325,000 will suffer a recurrent event [2].

The most common cause of AMI is the rupture of one or more vulnerable atherosclerotic plaque(s) [3–5], with thrombus formation leading to complete or partial occlusion of the coronary artery, which subsequently leads to ischemia and myocardial necrosis. Patients with complete occlusion of the coronary artery generally

Claus W. Heizmann (ed.), *Calcium-Binding Proteins of the EF-Hand Superfamily: From Basics to Medical Applications*, Methods in Molecular Biology, vol. 1929, https://doi.org/10.1007/978-1-4939-9030-6_40,

present with ST-segment elevation myocardial infarction (STEMI), which is observed during electrocardiography (ECG). The rule-out or rule-in of AMI in ACS patients without ST-segment elevation depends on the measurement of cardiac troponins (cTn) in the blood as biomarkers of cardiac tissue damage. Patients with elevated cTn will be classified as non-ST-segment elevation MI (NSTEMI). Patients without ST-segment elevation and without biomarkers of cardiac damage will be diagnosed with unstable angina (UA). For the sake of early treatment, correct and fast triage of NSTEMI versus UA patients is crucial. In addition to the STEMI, NSTEMI, and UA categories, AMI can be clinically classified into five AMI types, based on clinical, pathological, and prognostic differences. These five categories, as described by Thygesen et al. in the consensus statement "Expert consensus document: third universal definition of myocardial infarction" [6], ensure consistency between AMI reporting and allow standardized treatment strategies for the different categories. The current five categories are demonstrated in Table 1. An update on this consensus document, the fourth universal definition of myocardial infarction, will be published in the upcoming months.

Clinical scores, patient history with clues for coronary artery disease (CAD), ECG, and measurement of cTn are combined for optimal and quick treatment, which is essential for patient outcome. Current international guidelines demonstrate that high-sensitivity hs-cTn have become gold standard biomarkers for cardiac tissue necrosis and are essential for early diagnosis of especially NSTEMI patients.

2 Troponins

The troponin complex was the first intracellular Ca^{2+}-sensor protein to be described [7, 8]. The troponin complex plays an important role in the regulation of striated skeletal and cardiac muscle contraction [9, 10]. It is found in thin filaments and exists of three subunits, namely, the calcium-sensing troponin C (TnC), the actomyosin ATPase inhibitory troponin I (TnI), and the tropomyosin-binding troponin T (TnT). Crystal structure analyses suggest that the troponin complex is highly flexible [11]. Being located on actin, the conformation of troponin and tropomyosin changes upon alterations in intracellular Ca2+ concentration, provoking a dynamic change in the interaction of actin with myosin [9–11].

2.1 Cardiac-Specific Isoforms of TnT and TnI, but Not TnC, Exist

TnT and TnI are both expressed in humans by three genes, encoding for slow skeletal, fast skeletal, and cardiac isoforms (see also the extensive review on the human troponin complex [12]). Due to alternative RNA splicing, even more isoforms of cardiac TnT (cTnT) exist. The cardiac isoforms of TnI and TnT are highly specific for cardiac tissues, but elevated cTnT has also been described, e.g., in

Table 1
Overview of universal classification of AMI

Type 1: spontaneous MI, related to atherosclerotic plaque rupture, ulceration, fissuring, erosion, or dissection
Type 2: MI secondary to ischemic imbalance, e.g., coronary endothelial dysfunction, coronary artery spasm, coronary embolism, arrhythmias, respiratory failure, and hypo-/hypertension
Type 3: MI resulting in death when biomarker values are unavailable, in case of sudden unexpected cardiac death with symptoms suggestive of MI, but when death occurred before blood samples were drawn, or before cTn could rise
Type 4a: MI related to percutaneous intervention, cTn are elevated > 5× ULN (when baseline values were normal), or a rise of > 20% if baseline values are elevated and are stable or falling. Clinical symptoms, ECG, and imaging are required
Type 4b: MI related to stent thrombosis, as detected by coronary angiography or autopsy and with a rise or fall of cTn values with at least one value above ULN
Type 5: MI related to coronary bypass grafting, cTn are elevated > 10× ULN (when baseline values were normal). ECG, angiography, and imaging findings are required

cTn cardiac troponin, *ECG* electrocardiography, *MI* myocardial infarct, *ULN* upper limit of normal
Adapted with permission from [6]

patients with neuromuscular diseases [13–15]. It has been postulated that this could be due to re-expression of fetal cTnT in diseased skeletal (i.e., non-cardiac) cells, although this hypothesis has been debated [16]. The most important posttranslational modification of cTnI and cTnT is phosphorylation by several protein kinases, which can affect both the structure and the function of the troponin complex [12].

The TnC protein is the actual calcium-sensor of the troponin complex, being a member of the EF-hand superfamily of calcium-binding proteins and containing four EF-hands. A cardiac-specific isoform of TnC is lacking, since TnC is expressed by only two genes, encoding "fast skeletal" and "slow skeletal" isoforms [12].

2.2 Release of cTnI and cTnT During AMI

Cardiac troponins are released from dying cardiomyocytes in two steps. Initially, the cytosolic pool (around 6–7% of cTn) is released quickly into the extracellular space [17]. The half-life of this cTn in the blood stream is approximately 2 h. Secondly, structure-bound cTn is released with a much longer half-life of about 20 h [18, 19]. Reperfusion therapy induces a washout of the cytosolic pool with an early peak of cTnI and cTnT. After this first increase, cTns are then slowly released and finally disappear; elevated levels can be detected up to 14 days after AMI [17].

In blood samples from patients suffering an AMI, proteomic approaches differentiated several proteoforms of cTnI and cTnT such as free, complexed, proteolysed, and posttranslationally modified forms [20–22]. A schematic representation of the release of the different forms of cTn during AMI is illustrated in Fig. 1.

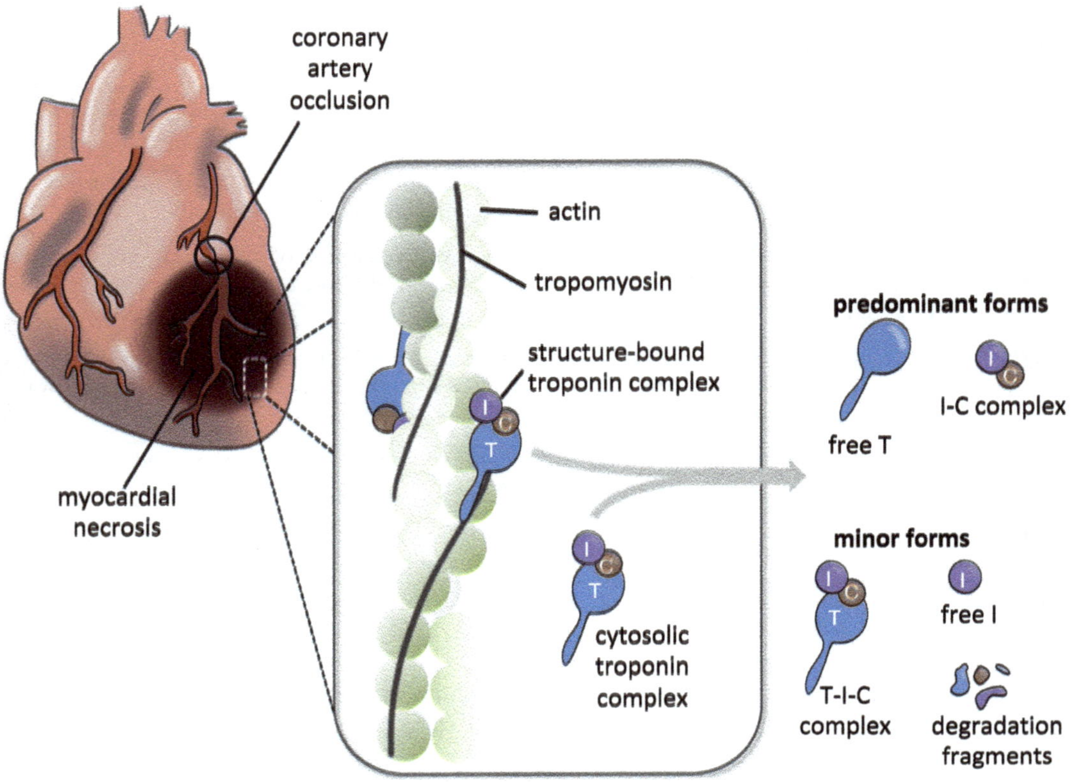

Fig. 1 Schematic representation of the troponin complex in cardiac tissue and release of major and minor cTn forms from cardiac necrotic cells. The majority of cardiac troponins are bound to the thin filaments. Only 6–7% of the complex is in the cytoplasm. Upon necrosis of cardiomyocytes, predominant forms found in plasma are free cTnT and the cTnI-cTnC complex

Due to their specific expression in cardiac tissue and release during cell death of cardiac cells, cTnI and cTnT isoforms—but not TnC—can be used as biomarkers of cardiomyocyte damage, which led to the development of highly specific and highly sensitive immunoassays for the clinical diagnostic detection of cTnI and cTnT in blood samples of patients after AMI.

3 Analytical Aspects of Cardiac Troponins

3.1 Cardiac Troponins are State-of-the-Art Biomarkers for ACS

Release of cTn correlates with infarct size, especially in STEMI patients [17], and measurement of hs-cTn has become a cornerstone for diagnostics of ACS, as reflected by evidence-based international guidelines of the European Society of Cardiology, the American Heart Association, and the American College of Cardiology [1, 23, 24]. The development of high-sensitive immunoassays for cTn has made the detection of the very low amounts of cTn—resulting from minor infarctions—possible and has considerably improved the diagnostics of AMI. Previously, the MB isoform of creatine kinase (CK-MB) has been used for many years as a marker of ACS. Today, after the introduction of cTn, CK-MB measurement is regarded as obsolete. Some hospitals still offer parallel measurement of cTn and CK-MB, but this does not provide incremental value to patient care [25]. On the contrary, it provides extra, unnecessary costs and may even in some cases deliver confusing results, when cTn is normal and CK-MB is elevated, not being related to ACS.

For the measurement of cTnT, only a few assays are available, because cTnT is patented by a single company, namely, Roche, Switzerland. For measurement of cTnI, however, many assays from different companies are available. Immunoassays for both cTnT and cTnI are categorized as high-sensitivity (hs) assays versus contemporary/sensitive assays versus not clinically acceptable tests [26]. This classification is primarily based on specifications for the functional assay sensitivity. High-sensitive cTn assays are able to measure the 99th percentile of cTn in a healthy population with an imprecision of 10% or less. The imprecision of contemporary/ sensitive cTn assays ranges between 10 and 20% at the 99th percentile of a healthy population. Assays with an imprecision of more than 20% at this decision point are not clinically acceptable. The Committee on Clinical Applications of Cardiac Bio-Markers of the International Federation of Clinical Chemistry provides overviews of hs and contemporary/sensitive assays to be used either as central laboratory or POC assays, including parameters such as the 99th percentile, coefficients of variation (CV) at different cTn concentrations, and epitopes recognized by the anti-cTn antibodies use [27–29]. Another overview including several additional tests can be found in a recent mini-review by Apple et al. [30].

3.2 High-Sensitivity cTn Assays

High-sensitivity cTn assays are defined to measure the 99th percentile of a healthy population with an analytical imprecision of $\leq$10% [31]. The 99th percentile is used as a diagnostic threshold for clinical decision-making in case of suspected AMI and is regarded as the upper limit of normal (ULN). The 99th percentile is method-dependent and, since hs-cTn assays are not standardized, has to be established for each assay individually [32]. Difficulties in standardization of cTn assays arise due to (1) the different (free, complexed and degraded) forms of cTn released during cardiac cell necrosis; (2) posttranslational modification of cTn, which can alter epitope recognition; (3) lack of suitable reference material; and (4) antibody design. Establishing correct 99th percentiles is likewise challenging. Harmonization on choosing the reference population characteristics (young, healthy, or age-matched, older healthy individuals), the method to examine the health condition of the chosen individuals, the number of individuals enrolled, and the method for the statistical calculation is needed. A further requirement of hs-cTn assays is that hs-cTn should be measurable above the limit of detection (LoD) in ≥50% of healthy individuals [33]. Results of hs-cTn assays should be reported in ng/L.

3.3 POC cTn Assays

By reducing the time needed for transportation and processing of the blood sample, point-of-care (POC) assays offer the advantage of having a rapid turnaround time, potentially reducing the time for patients to stay at the emergency department [34]. However, at present the majority of POC cTn assays have the disadvantage of a much lower sensitivity [24]. The majority of POC cTn assays [27, 30] are categorized as "clinically useful." Some are even "not clinically acceptable". Or manufacturers do not declare their performance [26, 33].

As will be described in paragraph 4, clinical algorithms for rule-in and rule-out of AMI, based on serial measurement of cTn at presentation and during the following hours, have been integrated in hospital and emergency setting. Most clinical studies of these algorithms with data on diagnostic sensitivity and specificity have been obtained using central laboratory immunoassay devices [24]. The diagnostic performance of POC cTn tests has not been as thoroughly investigated. In a recent study, five commonly used POC tests were compared to two hs-cTn tests for their clinical performance, using 90 patients with suspected NSTACS. Only one of the five POC tests, PATHFAST^{TM} (LSI Medience Corporation), was found to display comparable diagnostic performance to hs assays of central laboratories [35]. Thorough clinical studies on the performance of each individual POC tests are required to allow—if possible—correct and safe implementation of POC cTn assays in diagnostic algorithms in an emergency department setting.

3.4 cTn Can Be Elevated Under Other Circumstances than ACS

Although cTn are highly specific for cardiac tissue, it must be kept in mind that their blood concentration may rise due to other cardiac circumstances than AMI, such as myocarditis, heart failure, or arrhythmias. In addition, also many non-cardiac circumstances, such as pulmonary embolism, cardiotoxins, or severe head trauma, can lead to increases in cTn concentrations in blood [6, 24, 36]. Elevated cTn levels are also seen in healthy individuals after excessive physical exercise, for example, after marathon running. An extensive overview of such conditions is provided in Table 2. Since some conditions can clinically mimic AMI, it is essential to critically combine clinical findings, ECG, and laboratory measurements for correct diagnosis.

3.5 Interferences in cTn Assays

Generally, immunoassays are optimized to reduce analytical interferences [37]. However, as in any immunoassay, false-positive or false-negative hs-cTn results may still occur and must be kept in mind by physicians. Physical causes can be blood clots in the

Table 2
Overview of causes for cTn elevations above the 99th percentile

Cardiac	Non-cardiac
Tachy-/bradyarrhythmias	Aortic dissection
Hypertrophic cardiomyopathy	Septic shock
Takotsubo cardiomyopathy	Severe respiratory failure
Structural heart diseases (e.g., aortic stenosis)	Severe anemia
Coronary spasm	Severe pulmonary embolism or pulmonary hypertension
Coronary vasculitis (e.g., Kawasaki or Churg-Strauss syndrome)	Renal failure
Cardiac contusion	Stroke and subarachnoid hemorrhage
Cardiac surgery, ablation, pacing, coronary artery bypass surgery, and percutaneous coronary intervention	Amyloidosis, sarcoidosis, hemochromatosis, and scleroderma
Defibrillator shocks	Rhabdomyolysis
Cardiotoxic medication or poison (e.g., doxorubicin, 5-fluorouracil, Herceptin, carbon monoxide, snake venoms)	Extreme/endurance exercise (e.g., marathon running)
Heart failure	

Table 3
Overview of possible analytical interferences causing false-positive or false-negative cTn measurements

Heterophilic antibodies
Human anti-mouse antibodies (HAMA), human anti-animal antibodies (HAAA)
Rheumatoid factor
Troponin autoantibodies
Macrotroponin (troponin-autoantibody complexes)
Biotin treatment of the patient
High-dose hook effect
Cross-reactivity with endogenous molecules displaying similar epitopes
Strong hemolysis of the blood sample
Strong lipemia in the blood sample
Blood clots in the blood sample
Equipment malfunction

sample. Elevated biotin plasma levels due to treatment of patient can interfere with immunoassays using the biotin/streptavidin technology and cause false-low troponin levels. Interfering antibodies such as heterophilic or human anti-mouse antibodies are a cause of falsely elevated cTn levels [38, 39]. Autoantibodies against cTnI were shown to be present in approximately 10% of patients presenting with chest pain [40] and in similar amounts (cTnI or cTnT alone, or present in blood together) of healthy individuals [41]. An overview of possible interferences is presented in Table 3.

If the reported hs-cTn result is not in agreement with the clinical suspicion, and there is a hint for falsely elevated or falsely decreased results, a re-measurement of cTn in fresh samples should be performed to exclude any sample mix-up. If present in the first blood sample, interferences are highly likely to be stably present in the second blood sample at 1h, 3h, and 6h and during the following days and dynamic changes of cTn will probably not be observed. For further investigation of the interference, heterophilic blocking tubes can be used to remove interfering antibodies, dilutions can be made to observe linear dilution response, and cTn can be retested with an alternative immunoassay using different capture and detection antibodies that recognize different epitopes.

4 Clinical Applications of Cardiac Troponins

Only half of the patients presenting with acute chest pain at the emergency department will have cardiac disease: STEMI (5–10%), NSTEMI (15–20%), UA (10%), or other cardiac pathologies (15%), such as acute heart failure or cardiomyopathies. In the other 50% of patients with acute chest pain, the underlying cause can be, e.g., of pulmonary, vascular, gastrointestinal, and musculoskeletal origin [24]. On the other hand, especially NSTEMI patients may not present with chest pain or typical signs as radiation to neck, left arm, or lower jaw but can display atypical symptoms, such as shortness of breath, nausea, vomiting, fatigue, or syncope [1]. Indeed, a high proportion of AMI, estimated up to 40%, may remain clinically undiagnosed [42].

4.1 Diagnostics and Treatment of STEMI

STEMI is diagnosed upon the typical changes of the ECG, and measurement of cTn does only provide complimentary information (Fig. 2). In case of suspected ACS, a 12-lead ECG should be acquired and interpreted by an expert within 10 min [43]. Continued ECG monitoring with defibrillator capacity is recommended to detect life-threatening arrhythmias.

Preferably, patients with persistent ST-segment elevations and myocardial ischemia are treated within 12 h after onset of symptoms or within 2 h after diagnosis of STEMI. Treatment of choice is the so-called primary percutaneous coronary intervention (PCI), i.e., an emergency PCI with a stent (technique of first choice,

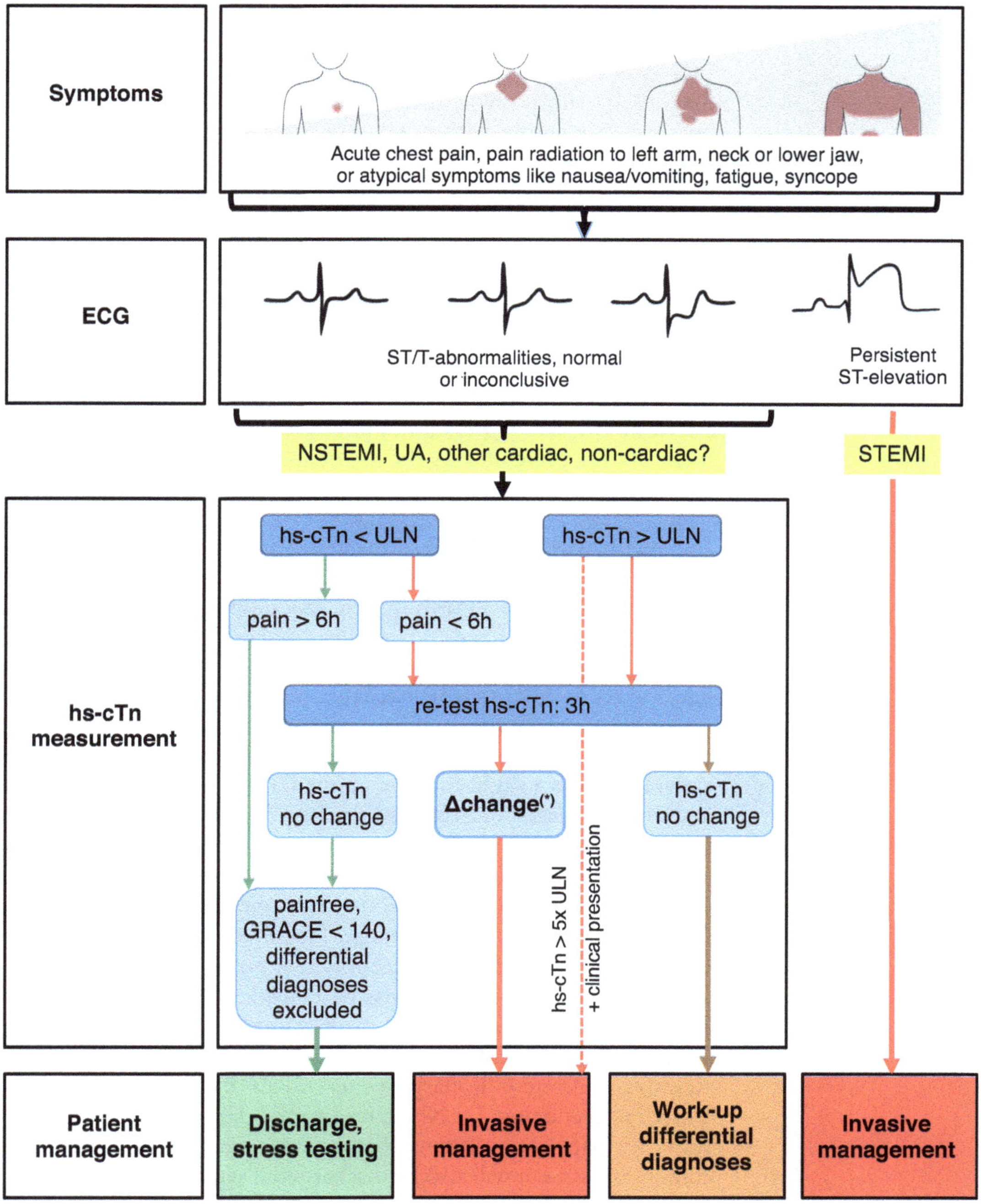

Fig. 2 Guideline for diagnostic triage and treatment strategies of patients with suspected AMI, based on ECG findings and 0h/3h-algorithm using hs-cTn. Asterisk: Δ change: absolute change in hs-cTn concentration between 0h and 3h is assay-dependent. *GRACE* Global Registry of Acute Coronary Event, *hs-cTn* high-sensitivity cardiac troponin, *NSTEMI* non-ST-segment elevation myocardial infarction, *STEMI* ST-segment elevation myocardial infarction, *UA* unstable angina pectoris, *ULN* upper limit or normal (i.e., 99th percentile of healthy population). Adapted with permission from [24]

especially using drug-eluting stents) or balloon (second choice) without previous fibrinolytic therapy [1, 44–46]. In case primary PCI is not timely possible (i.e., within 120 min after STEMI is diagnosed), fibrinolytic therapy must be initiated as quickly as possible, preferably within 10 min after diagnosis [1] provided that there are no contraindications, such as a known bleeding disorder or advanced liver diseases.

4.2 Diagnostics and Treatment of NSTEMI

Patients with ischemic cardiac symptoms will be classified as NSTEMI, if no ST-segment elevation is observed in ECG, but cTn plasma level is found to be above the 99th percentile (i.e., the ULN) and to increase or decrease upon follow-up measurements [24]. In some patients, ECG findings are normal, but in many patients ECG abnormalities are observed, such as transient ST elevation, ST depression, and T-wave changes [47]. Measurement of cTn, preferably by the use of an hs assay, is mandatory in all patients with suspected NSTEMI [24].

4.2.1 The 0h/3h-Algorithm

In patients with AMI cTn plasma levels rise quickly above the 99th percentile within 1 h. The thoroughly investigated and nowadays widely used 0h/3h-algorithm (Fig. 2) allows a quick rule-in and rule-out of NSTEMI patients based on hs-cTn levels [24, 48–52]. This algorithm is advocated by the ESC 2015 guideline for NSTEMI [24]. If hs-cTn is highly increased above ULN (>5 × ULN) at the time of first medical contact ($t = 0$h), combined with strong clinical evidence, intervention shall happen immediately. If there is a slight increase above ULN, another blood sample is drawn 3 h later and retested for hs-cTn. Dynamic changes, for which the absolute increase and decrease are defined per hs-cTn assay, diagnose NSTEMI as well, and patients are forwarded to invasive management. If initial hs-cTn is elevated above ULN, but there is no dynamic change in hs-cTn concentrations between the samplings at 0h and 3h, differential diagnoses (Table 2) or falsely elevated hs-cTn (Table 3) must be considered. If ACS symptoms had started 6 h or more before the first medical contact, the patient is pain-free, differential diagnoses are excluded, and the Global Registry of Acute Coronary Events (GRACE) score is <140 [53], a hs-cTn level below ULN will lead to the discharge of the patient. Low-risk patients as defined before who are asymptomatic upon clinical assessment will also be discharged if symptoms occurred less than 6 h ago, and hs-cTn concentrations are below ULN at baseline and do not increase during the next 3 h [53].

4.2.2 The 0h/1h-Algorithm

As the quicker alternative, the ESC 2015 guideline for NSTEMI [24] proposes a 0h/1h algorithm (Fig. 3). This less extensively investigated algorithm [54–59] can be used for few hs-cTn assays that are validated for this purpose. Assay-specific values for hs-cTn at 0h and defined increases between 0h and 1h are used for rule-in and safe rule-out of NSTEMI. Further testing for hs-cTn at 3h and

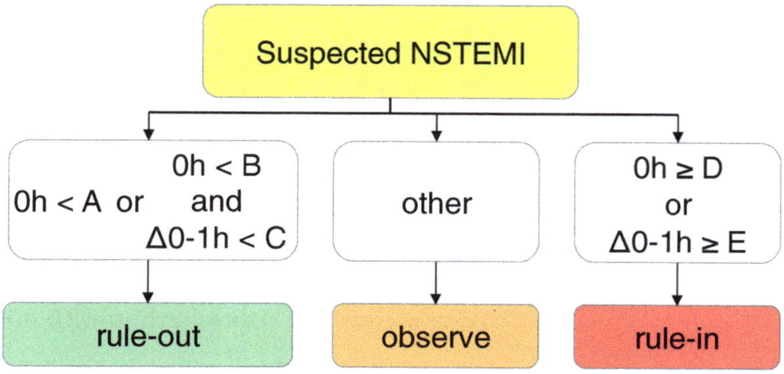

	A	B	C	D	E
hs-cTnT (Elecsys) (ng/l)	5	12	3	52	5
hs-cTnI (Architect) (ng/l)	2	5	2	52	6
hs-cTnI (Dimension Vista) (ng/l)	0.5	5	2	107	19

Fig. 3 Schematic overview of the 0h/1h-algorithm of NSTEMI patients with assay-specific absolute (A, B, and D) and delta changes (C and E), depicted in ng/L. Δ (delta) = increase between $t = 0$ (time of presentation) and $t = 1$ h. NSTEMI can be ruled out, if hs-cTn is very low at time of presentation (below A) (although only applicable if onset of chest pain is >3 h). NSTEMI can also be ruled out, if hs-cTn is below B and delta changes between 0h/1h is <C. NSTEMI can be ruled in if cTn is ≥D at $t = 0$ or if delta change between 0h/1h is ≥E. Adapted with permission from [24]

6h is recommended in non-conclusive cases. Such 0h/1h algorithms have been successfully implemented in several hospitals, allowing timely and safe dismissal of normal patients. If combined with clinical data and ECG findings, the negative predictive value for rule-out of AMI patients, using the 0h/1h- or the 0h/3h-strategy, is >98%. Positive predictive values have been reported to range between 66 and 78%. One has to be careful with patients that present very early (within 1 h after onset of symptoms) and wait for a re-measurement at 3h. Furthermore, in some patients, cTn are released very slowly, and if there is a high clinical suspicion of AMI or if chest pain persists, hs-cTn measurement should be continued.

4.2.3 Other Algorithms

Dozens of clinical studies describe alternatives to the 0h/3h- and 0h/1h-algorithms, using hs-cTn measurements at 0h, 1h, 2h, 3h, and 4h (see review [60]), all with a high negative predictive value (rule-out) but with positive predictive values (rule-in) varying from 19 to 87%.

4.2.4 Algorithms in Patients with Renal Disease

Diagnostics in patients with renal dysfunction (RD) are challenging, since they generally have elevated hs-cTn concentrations, even in the absence of acute myocardial ischemia. In addition, patients with RD have a higher incidence of AMI and may present more often with atypical clinical symptoms of AMI. Recent studies demonstrate that the diagnostic 0h/3h- and 0h/1h-algorithms can be applied to RD patients without adjustment for renal function, especially for rule-out with high sensitivities of >98%, comparable to patients without RD [61, 62]. Twerenbold et al. showed that the 0h/1h algorithm can be applied to RD patients with a rule-out sensitivity of 98.5%. Due to high pretest probability, efficacy of rule-out was reduced and allowed rule-out in 18% of RD patients, requiring further cTn re-measurement at later time points for the remaining RD patients. Specificity of rule-in was lower (84%) as compared to the control group with normal renal function (92%).

4.2.5 Treatment of NSTEMI Patients

NSTEMI patients should be monitored for arrhythmias and treated with antithrombotics and antianginal medication after stratification for bleeding risk. Further management of NSTEMI patients is based on response to antianginal medication and ischemic risk estimation by risk score assessment, e.g., by GRACE. Timing of invasive strategy depends on the risk situation of the patient [24]. For example, patients with hemodynamic instability or life-threatening arrhythmias are at very high risk, and invasive treatment should be done within 2 h. Early invasive strategy (within 24 h) is recommended in patients with at least one high-risk criterion, such as rise or fall of hs-cTn compatible with AMI or a GRACE score >140. For patients with intermediate risk, for example, because of diabetes mellitus or renal insufficiency, the maximal delay of 72 h can be allowed. If patients present with only one culprit lesion, PCI with drug-eluting stent is the first choice of intervention [24]. In patients with multivessel disease, choice for treatment by PCI or by coronary artery bypass graft has to be made for each patient individually.

Together with exercise-based rehabilitation, lifestyle interventions—such as smoking cessation, weight reduction, blood pressure control, and physical activities—are important for long-term prevention of secondary AMI events. In addition, antithrombotic therapy (aspirin, possibly combined with other antithrombotic drugs) and lipid-lowering therapy, with reduction of LDL cholesterol below 1.8 mmol/L, are recommended in NSTEMI, as well as STEMI, patients.

4.3 Risk Management and Prognostics of ACS Patients

Risk prediction of future adverse events in ACS patients is of utmost clinical importance. Several score systems are in use, among them the GRACE, the History, ECG, Age, Risk factors and Troponin (HEART), and the Thrombolyis in Myocardial Infarction (TIMI) risk scores. Mid-term performance of the three scores, i.e., their ability to predict a major adverse cardiac event (MACE) 6 weeks later, was compared in a study performed in nine Dutch hospitals and including 1748 patients with chest pain [63]. Receiver operating characteristics (ROC) curves with calculated areas under the curve (AUC) of the three scores clearly demonstrated that the HEART score had the highest overall discrimination of patients with and without MACE in the following 6 weeks (AUC under the ROC curve of 0.86), followed by the TIMI score (0.80) and GRACE (0.73). Gender is not commonly integrated as a risk factor in these scores, but a recent study demonstrated that risk prediction of the HEART score is safer in women than in men [64].

Traditional cardiac biomarkers such as CK-MB, myoglobin, or conventional cTn are already integral parts of the current GRACE, TIMI, and HEART risk scores. In several studies, their replacement by hs-cTnI and cTnT improved risk prediction in NSTACS patients but not STEMI patients. Other biomarkers, such as N-terminal pro-B-type natriuretic peptide (NT-proBNP) , high-sensitivity C-reactive protein, and copeptin to these scores, further improved mid- and long-term risk prediction of MACE in ACS [65–69].

Patients with heart failure usually have (slightly) elevated cTn level above the 99th percentile. In acute heart failure patients, serial hs-cTn measurements will only show minor hs-cTn changes [70]. More importantly, elevated cTn, and especially changes in cTn over time, have been strongly predictive of adverse events in patients with heart failure [70, 71].

4.4 Cardiac Troponins for Risk Prediction in the General Population

There is a strong medical need for accurate risk stratification and primary prevention of coronary heart disease (CHD) in asymptomatic individuals. Many research groups try to identify novel biomarkers for risk prediction and combine new [72] and already known biomarkers—such as hs-cTn—into risk prediction models. Examples are the Framingham [73] and PROCAM [74] risk scores as well as the ESC/EAS SCORE and AHA/ACC pooled cohort equation [75, 76] which all calculate the 10-year cardiovascular risk of asymptomatic individuals, by combining clinical (e.g., gender, smoking, age, blood pressure) and laboratory data (e.g., total cholesterol, HDL cholesterol).

Several large population studies, the largest including >74,000 individuals [77], have demonstrated that already slightly elevated hs-cTn is a strong and independent predictor of future AMI events, CHD death and secondary cardiovascular death, but not of non-cardiovascular death [77–81]. The proposed cutoff proposed by the large BiomarCaRe study is 6 ng/L for hs-cTnI [77], although it must be kept in mind that this is a specific cutoff for the Abbott Architect hs-cTnI assay and that other cutoffs for other assays may apply. Moreover, it was demonstrated that cTnI decreased during statin therapy [77, 79] and that decreasing cTnI was associated with better outcome, proposing that serial hs-cTn measurement could be a valuable surrogate biomarker for patient management and evaluation of therapeutic responses.

5 Conclusion

During the last decades, development of highly specific and highly sensitive cTn immunoassays and their integration into quick diagnostic rule-in and rule-out algorithms have revolutionized diagnostics of AMI, improved patient care, and reduced medical costs. Modern hs-cTn assays allow rule-out of NSTEMI within maximally 3 h after hospitalization of NSTACS patients compared to 6 h or more if contemporary/sensitive cTn assays or conventional biomarkers are used. However, it must always be kept in mind that cTn are specific for cardiac tissue damage rather than for AMI. Other diseases causing or enhancing damage of cardiomyocytes can also cause elevations of cTn blood levels. In addition, assay interferences may cause falsely positive or falsely negative results. Hence, cTn results should always be interpreted within the clinical context and ECG findings. More and more POC cTn tests, with different or even unknown analytical quality, are now available on the market. POC cTn tests yield very fast results, but as yet data on clinical utility are limited. Therefore physicians should use POC cTn results with precaution. Addition of hs-cTn to risk scores for prediction of MACE in patients having suffered an AMI improves risk prediction of mid- and long-term adverse events. Finally, hs-cTn levels slightly elevated above the limit of quantification are independent and strong predictors of long-term AMI and CAD events, hinting at further areas of diagnostic application for hs-cTn.

References

1. Ibanez B, James S, Agewall S et al (2018) 2017 ESC Guidelines for the management of acute myocardial infarction in patients presenting with ST-segment elevation. Eur Heart J 39:119–177. https://doi.org/10.1093/eurheartj/ehx393
2. Benjamin EJ, Blaha MJ, Chiuve SE (2017) Heart disease and stroke statistics—2017 update. A report from the American Heart Association. Circulation 135:e146–e603. https://doi.org/10.1161/CIR.0000000000000485
3. Muller JE, Tofler GH, Stone PH (1989) Circadian variation and triggers of onset of acute cardiovascular disease. Circulation 79:733–743. https://doi.org/10.1161/01.CIR.79.4.733
4. Davies MJ (2000) Coronary disease: the pathophysiology of acute coronary syndromes. Heart 83:361–366. https://doi.org/10.1136/heart.83.3.361
5. Bentzon JF, Otsuka F, Virmani R, Falk E (2014) Mechanisms of plaque formation and rupture. Circ Res 114:1852–1866. https://doi.org/10.1161/CIRCRESAHA.114.302721
6. Thygesen K, Alpert JS, White HD (2012) Expert consensus document. Third universal definition of myocardial infarction. Eur Heart J 33:2551–2567. https://doi.org/10.1093/eurheartj/ehs184
7. Ebashi S, Endo M (1968) Calcium ion and muscle contraction. Prog Biophys Mol Biol 18:123–183
8. Perry SV (2008) Background to the discovery of troponin and Setsuro Ebashi's contribution to our knowledge of the mechanism of relaxation in striated muscle. Biochem Biophys Res Commun 369:43–48. https://doi.org/10.1016/j.bbrc.2007.11.185
9. Gordon AM, Homsher E, Regnier M (2000) Regulation of contraction in striated muscle. Physiol Rev 80:853–924. https://doi.org/10.1152/physrev.2000.80.2.853
10. Manning EP, Tardiff JC, Schwartz SD (2011) A model of calcium activation of the cardiac

thin filament. Biochemistry (Mosc) 50:7405–7413. https://doi.org/10.1021/bi200506k

11. Takeda S (2005) Crystal structure of troponin and the molecular mechanism of muscle regulation. J Electron Microsc 54. (Supplement 1: i35–i41
12. Katrukha IA (2013) Human cardiac troponin complex. Structure and functions. Biochem Mosc 78:1447–1465. https://doi.org/10.1134/S0006297913130063
13. Jaffe AS, Vasile VC, Milone M et al (2011) Diseased skeletal muscle. A noncardiac source of increased circulating concentrations of cardiac troponin T. J Am Coll Cardiol 58:1819–1824. https://doi.org/10.1016/j.jacc.2011.08.026
14. Hof D, Jung HH, Bloch KE (2013) Troponin T elevation in amyotrophic lateral sclerosis without cardiac damage. Amyotroph Lateral Scler Frontotemporal Degener 14:75–77. https://doi.org/10.3109/17482968.2012.696659
15. Rittoo D, Jones A, Lecky B, Neithercut D (2014) Elevation of cardiac troponin T, but not cardiac troponin I, in patients with neuromuscular diseases. J Am Coll Cardiol 63:2411–2420. https://doi.org/10.1016/j.jacc.2014.03.027
16. Giannitsis E, Katus HA (2012) Challenging interpretation of elevated cardiac troponin T in complex case with rhabdomyolysis. J Am Coll Cardiol 60:1027–1028
17. Hallén J (2012) Troponin for the estimation of infarct size: what have we learned. Cardiology 121:204–212. https://doi.org/10.1159/000337113
18. Gerhardt W, Katus H, Ravkilde J et al (1991) S-Troponin T in suspected ischemic myocardial injury compared with mass and catalytic concentration of S-creatine kinase isoenzyme MB. Clin Chem 37:1405–1411
19. Bleier J, Vorderwinkler K-P, Falkensammer J et al (1998) Different intracellular compartmentations of cardiac troponins and myosin heavy chains: a casual connection to their different early release after myocardial damage. Clin Chem 44:1912–1918
20. Labugger R, Organ L, Collier C et al (2000) Extensive troponin I and T modification detected in serum from patients with acute myocardial infarction. Circulation 102:1221–1226. https://doi.org/10.1161/01.CIR.102.11.1221
21. Peronnet E, Becquart L, Poirier F et al (2006) SELDI-TOF MS analysis of the cardiac troponin I forms present in plasma from patients with myocardial infarction. Proteomics 6:6288–6299. https://doi.org/10.1002/pmic.200600158
22. Peronnet E, Becquart L, Martinez J et al (2007) Isoelectric point determination of cardiac troponin I forms present in plasma from patients with myocardial infarction. Clin Chim Acta 377:243–247. https://doi.org/10.1016/j.cca.2006.10.006
23. Amsterdam EA, Wenger NK, Brindis RG et al (2014) 2014 AHA/ACC Guideline for the management of patients with non–ST-elevation acute coronary syndromes. J Am Coll Cardiol 64:e139–e228. https://doi.org/10.1016/j.jacc.2014.09.017
24. Roffi M, Patrono C, Collet J-P et al (2016) 2015 ESC Guidelines for the management of acute coronary syndromes in patients presenting without persistent ST-segment elevation. Eur Heart J 37:267–315. https://doi.org/10.1093/eurheartj/ehv320
25. Alvin MD, Jaffe AS, Ziegelstein RC, Trost JC (2017) Eliminating creatine kinase–myocardial band testing in suspected acute coronary syndrome: a value-based quality improvement. JAMA Intern Med 177:1508. https://doi.org/10.1001/jamainternmed.2017.3597
26. Apple FS (2009) A new season for cardiac troponin assays: it's time to keep a scorecard. Clin Chem 55:1303–1306. https://doi.org/10.1373/clinchem.2009.128363
27. IFCC task force on clinical applications of cardiac bio-markers (2018) Point of care cardiac troponin I and T assay analytical characteristics designated by manufacturer (v060617). http://www.ifcc.org/media/477404/point-of-care-cardiac-troponin-i-and-t-assay-analytical-characteristics-designated-by-manufacturer-v07262018.pdf
28. IFCC task force on clinical applications of cardiac bio-markers (2018) Contemporary cardiac troponin I and T assay analytical characteristic designated by manufacturer. http://www.ifcc.org/media/477400/contemporary-cardiac-troponin-i-and-t-assay-analytical-characteristics-designated-by-manufacturer-v07262018.pdf
29. IFCC task force on clinical applications of cardiac bio-markers (2018) High sensitivity cardiac troponin I and T assay analytical characteristic designated by manufacturer. http://www.ifcc.org/media/477401/high-sensitivity-cardiac-troponin-i-and-t-assay-analytical-characteristics-designated-by-manufacturer-v07262018.pdf
30. Apple FS, Sandoval Y, Jaffe AS, Ordonez-Llanos J (2017) Cardiac troponin assays: guide to understanding analytical

characteristics and their impact on clinical care. Clin Chem 63:73–81. https://doi.org/10.1373/clinchem.2016.255109

31. Apple FS, Collinson PO (2012) Analytical characteristics of high-sensitivity cardiac troponin assays. Clin Chem 58:54–61. https://doi.org/10.1373/clinchem.2011.165795
32. Apple FS, Ler R, Murakami MM (2012) Determination of 19 cardiac troponin I and T assay 99th percentile values from a common presumably healthy population. Clin Chem 58:1574–1581. https://doi.org/10.1373/clinchem.2012.192716
33. Apple FS, Jaffe AS, Collinson P et al (2015) IFCC educational materials on selected analytical and clinical applications of high sensitivity cardiac troponin assays. Clin Biochem 48:201–203. https://doi.org/10.1016/j.clinbiochem.2014.08.021
34. Singer AJ, Taylor M, LeBlanc D et al (2018) Early point-of-care testing at triage reduces care time in stable adult emergency department patients. J Emerg Med 55(2):172–178. https://doi.org/10.1016/j.jemermed.2018.04.061
35. Suh D, Keller DI, Hof D et al (2018) Rule-out of non-ST elevation myocardial infarction by five point of care cardiac troponin assays according to the 0 h/3 h algorithm of the European Society of Cardiology. Clin Chem Lab Med 56:649–657. https://doi.org/10.1515/cclm-2017-0486
36. Arshed S, Luo HX, Zafar S et al (2015) Elevated troponin I in the absence of coronary artery disease: a case report with review of literature. J Clin Med Res 7:820–824. https://doi.org/10.14740/jocmr2280w
37. Sturgeon CM, Viljoen A (2011) Analytical error and interference in immunoassay: minimizing risk. Ann Clin Biochem 48:418–432. https://doi.org/10.1258/acb.2011.011073
38. Tate J, Ward G (2004) Interferences in Immunoassay. Clin Biochem Rev 25:105–120
39. Ward G, Simpson A, Boscato L, Hickman PE (2017) The investigation of interferences in immunoassay. Clin Biochem 50:1306–1311. https://doi.org/10.1016/j.clinbiochem.2017.08.015
40. Savukoski T, Ilva T, Lund J et al (2014) Autoantibody prevalence with an improved immunoassay for detecting cardiac troponin-specific autoantibodies. Clin Chem Lab Med 52:273–279. https://doi.org/10.1515/cclm-2013-0310
41. Adamczyk M, Brashear RJ, Mattingly PG (2010) Coprevalence of autoantibodies to cardiac troponin I and T in normal blood donors. Clin Chem 56:676–677. https://doi.org/10.1373/clinchem.2009.138099
42. de Torbal A, Boersma E, Kors JA et al (2006) Incidence of recognized and unrecognized myocardial infarction in men and women aged 55 and older: the Rotterdam Study. Eur Heart J 27:729–736. https://doi.org/10.1093/eurheartj/ehi707
43. Diercks DB, Peacock WF, Hiestand BC et al (2006) Frequency and consequences of recording an electrocardiogram >10 minutes after arrival in an emergency room in non–ST-segment elevation acute coronary syndromes (from the CRUSADE Initiative). Am J Cardiol 97:437–442. https://doi.org/10.1016/j.amjcard.2005.09.073
44. Zijlstra F, Hoorntje JCA, De Boer M-J et al (1999) Long-term benefit of primary angioplasty as compared with thrombolytic therapy for acute myocardial infarction. N Engl J Med 341:1413–1419
45. Andersen HR, Nielsen TT, Rasmussen K et al (2003) A comparison of coronary angioplasty with fibrinolytic therapy in acute myocardial infarction. N Engl J Med 349:733–742
46. Chan AW, Kornder J, Elliott H et al (2012) Improved survival sssociated with pre-hospital triage strategy in a large regional ST-segment elevation myocardial infarction program. JACC Cardiovasc Interv 5:1239–1246. https://doi.org/10.1016/j.jcin.2012.07.013
47. Savonitto S, Ardissino D, Granger CB et al (1999) Prognostic value of the admission electrocardiogram in acute coronary syndromes. JAMA 281:707–713. https://doi.org/10.1001/jama.281.8.707
48. Keller T, Zeller T, Ojeda F et al (2011) Serial changes in highly sensitive troponin I assay and early diagnosis of myocardial infarction. JAMA 306:2684–2693. https://doi.org/10.1001/jama.2011.1896
49. Rubini Giménez M, Hoeller R, Reichlin T et al (2013) Rapid rule out of acute myocardial infarction using undetectable levels of high-sensitivity cardiac troponin. Int J Cardiol 168:3896–3901. https://doi.org/10.1016/j.ijcard.2013.06.049
50. Thelin J, Melander O, Öhlin B (2015) Early rule-out of acute coronary syndrome using undetectable levels of high sensitivity troponin T. Eur Heart J Acute Cardiovasc Care 4:403–409. https://doi.org/10.1177/2048872614554107
51. Shah ASV, Anand A, Sandoval Y et al (2015) High-sensitivity cardiac troponin I at presentation in patients with suspected acute coronary syndrome: a cohort study. Lancet

386:2481–2488. https://doi.org/10.1016/S0140-6736(15)00391-8
52. Wildi K, Nelles B, Twerenbold R et al (2016) Safety and efficacy of the 0 h/3 h protocol for rapid rule out of myocardial infarction. Am Heart J 181:16–25. https://doi.org/10.1016/j.ahj.2016.07.013
53. Tang EW, Wong C-K, Herbison P (2007) Global Registry of Acute Coronary Events (GRACE) hospital discharge risk score accurately predicts long-term mortality post acute coronary syndrome. Am Heart J 153:29–35. https://doi.org/10.1016/j.ahj.2006.10.004
54. Reichlin T, Twerenbold R, Wildi K et al (2015) Prospective validation of a 1-hour algorithm to rule-out and rule-in acute myocardial infarction using a high-sensitivity cardiac troponin T assay. Can Med Assoc J 187:E243–E252. https://doi.org/10.1503/cmaj.141349
55. Rubini Giménez M, Twerenbold R, Jaeger C et al (2015) One-hour rule-in and rule-out of acute myocardial infarction using high-sensitivity cardiac troponin I. Am J Med 128:861–870. https://doi.org/10.1016/j.amjmed.2015.01.046
56. Jaeger C, Wildi K, Twerenbold R et al (2016) One-hour rule-in and rule-out of acute myocardial infarction using high-sensitivity cardiac troponin I. Am Heart J 171:92–102. https://doi.org/10.1016/j.ahj.2015.07.022
57. Mueller C, Giannitsis E, Christ M et al (2016) Multicenter evaluation of a 0-hour/1-hour algorithm in the diagnosis of myocardial infarction with high-sensitivity cardiac troponin T. Ann Emerg Med 68:76–87. https://doi.org/10.1016/j.annemergmed.2015.11.013
58. Neumann JT, Sörensen NA, Ojeda F et al (2017) Early diagnosis of acute myocardial infarction using high-sensitivity troponin I. PLoS One 12:e0174288. https://doi.org/10.1371/journal.pone.0174288
59. Boeddinghaus J, Nestelberger T, Twerenbold R et al (2017) Direct comparison of 4 very early rule-out strategies for acute myocardial infarction using high-sensitivity cardiac troponin I. Circulation 135:1597–1611. https://doi.org/10.1161/CIRCULATIONAHA.116.025661
60. Westermann D, Neumann JT, Sörensen NA, Blankenberg S (2017) High-sensitivity assays for troponin in patients with cardiac disease. Nat Rev Cardiol 14:472–483. https://doi.org/10.1038/nrcardio.2017.48
61. Twerenbold R, Badertscher P, Boeddinghaus J et al (2018) 0/1-Hour triage algorithm for myocardial infarction in patients with renal dysfunction. Circulation 137:436–451. https://doi.org/10.1161/CIRCULATIONAHA.117.028901
62. Gunsolus I, Sandoval Y, Smith SW et al (2018) Renal dysfunction influences the diagnostic and prognostic performance of high-sensitivity cardiac troponin I. J Am Soc Nephrol 29:636–643. https://doi.org/10.1681/ASN.2017030341
63. Poldervaart JM, Langedijk M, Backus BE et al (2017) Comparison of the GRACE, HEART and TIMI score to predict major adverse cardiac events in chest pain patients at the emergency department. Int J Cardiol 227:656–661. https://doi.org/10.1016/j.ijcard.2016.10.080
64. Bank IEM, de Hoog VC, de Kleijn DPV et al (2017) Sex-based differences in the performance of the HEART score in patients presenting to the emergency department with acute chest pain. J Am Heart Assoc 6:e005373. https://doi.org/10.1161/JAHA.116.005373
65. Bonaca M, Scirica B, Sabatine M et al (2010) Prospective evaluation of the prognostic implications of improved assay performance with a sensitive assay for cardiac troponin I. J Am Coll Cardiol 55:2118–2124. https://doi.org/10.1016/j.jacc.2010.01.044
66. Scirica BM, Sabatine MS, Jarolim P et al (2011) Assessment of multiple cardiac biomarkers in non-ST-segment elevation acute coronary syndromes: observations from the MERLIN-TIMI 36 Trial. Eur Heart J 32:697–705. https://doi.org/10.1093/eurheartj/ehq468
67. Widera C, Pencina MJ, Bobadilla M et al (2013) Incremental prognostic value of biomarkers beyond the GRACE (Global Registry of Acute Coronary Events) score and high-sensitivity cardiac troponin T in non-ST-elevation acute coronary syndrome. Clin Chem 59:1497–1505. https://doi.org/10.1373/clinchem.2013.206185
68. O'Malley RG, Bonaca MP, Scirica BM et al (2014) Prognostic performance of multiple biomarkers in patients with non–ST-segment elevation acute coronary syndrome. J Am Coll Cardiol 63:1644–1653. https://doi.org/10.1016/j.jacc.2013.12.034
69. Klingenberg R, Aghlmandi S, Räber L et al (2018) Improved risk stratification of patients with acute coronary syndromes using a combination of hsTnT, NT-proBNP and hsCRP with the GRACE score. Eur Heart J Acute Cardiovasc Care 7:129–138. https://doi.org/10.1177/2048872616684678
70. Magnussen C, Blankenberg S (2018) Biomarkers for heart failure: small molecules with high clinical relevance. J Intern Med 283:530–543. https://doi.org/10.1111/joim.12756

71. Januzzi JL, Filippatos G, Nieminen M, Gheorghiade M (2012) Troponin elevation in patients with heart failure: on behalf of the third Universal Definition of Myocardial Infarction Global Task Force: Heart Failure Section. Eur Heart J 33:2265–2271. https://doi.org/10.1093/eurheartj/ehs191
72. Wang J, Tan G-J, Han L-N et al (2017) Novel biomarkers for cardiovascular risk prediction. J Geriatr Cardiol 14:135–150. https://doi.org/10.11909/j.issn.1671-5411.2017.02.008
73. Gordon T, Castelli WP, Hjortlnd MC et al (1977) High density lipoprotein as a protective factor against coronary heart disease—the Framingham study. Am J Med 62:707–714
74. Assmann G, Schulte H, von Eckardstein A, Huang Y (1996) High-density lipoprotein cholesterol as a predictor of coronary heart disease risk. The PROCAM experience and pathphysiological implications for reverse cholesterol transport. Atherosclerosis 124(Suppl): S11–S20
75. Stone NJ, Robinson JG, Lichtenstein AH et al (2014) 2013 ACC/AHA guideline on the treatment of blood cholesterol to reduce atherosclerotic cardiovascular risk in adults. Circulation 129:S1–S45. https://doi.org/10.1161/01.cir.0000437738.63853.7a
76. Catapano AL, Graham I, De Backer G et al (2016) 2016 ESC/EAS guidelines for the management of dyslipidaemias. Eur Heart J 37:2999–3058. https://doi.org/10.1093/eurheartj/ehw272
77. Blankenberg S, Salomaa V, Makarova N et al (2016) Troponin I and cardiovascular risk prediction in the general population: the BiomarCaRE consortium. Eur Heart J 37:2428–2437. https://doi.org/10.1093/eurheartj/ehw172
78. Apple FS, Steffen LM, Pearce LA et al (2012) Increased cardiac troponin I as measured by a high-sensitivity assay is associated with high odds of cardiovascular death: the Minnesota Heart Survey. Clin Chem 58:930–935. https://doi.org/10.1373/clinchem.2011.179176
79. Ford I, Shah ASV, Zhang R et al (2016) High-sensitivity cardiac troponin, statin therapy, and risk of coronary heart disease. J Am Coll Cardiol 68:2719–2728. https://doi.org/10.1016/j.jacc.2016.10.020
80. Xiao W, Cao R, Liu Y et al (2017) Association of high-sensitivity cardiac troponin T with mortality and cardiovascular events in a community-based prospective study in Beijing. BMJ Open 7:e013431. https://doi.org/10.1136/bmjopen-2016-013431
81. Samman Tahhan A, Sandesara P, Hayek SS et al (2018) High-sensitivity troponin I levels and coronary artery disease severity, progression, and long-term outcomes. J Am Heart Assoc 7:e007914. https://doi.org/10.1161/JAHA.117.007914

Chapter 41

Targeted Mass Spectrometry of S100 Proteins

Juan Martínez-Aguilar and Mark P. Molloy

Abstract

The S100 protein family has attracted great interest in the field of biomarker research, and a growing number of studies reveal dysregulation of many of the 21 S100 protein isoforms in various human diseases. In cancer, S100 protein expression has been associated with tumor growth, progression, and response to treatment. Some S100 proteins are also considered candidate therapeutic targets. From an analytical perspective, multiplexed analysis of the family-wide S100 protein expression is challenging due to their relatively small size and high-sequence identity. Here we describe a mass spectrometry method using selected reaction monitoring which enables the targeted, multiplexed detection and quantitation of the entire S100 protein family in cell lines and tissue samples.

Key words S100 proteins, Ca^{2+}-binding, EF-hand, Biomarkers, Quantitative proteomics, Targeted mass spectrometry, Selected reaction monitoring

1 Introduction

Altered expression of specific members of the S100 protein family has been documented in several human diseases including inflammatory, neurological, cardiac, and cancer diseases [1–3]. S100 proteins have particularly been linked to tumourigenesis and cancer metastasis, and some of them have been proposed as therapeutic targets or biomarkers with diagnostic, prognostic, and predictive potential. Some examples include S100A2, S100A4, S100A6, S100A7, S100A8, S100A9, S100A11, S100A13, and S100P, which exhibit dysregulation in cancer of the breast, colon, lung, pancreas, prostate, stomach, or thyroid [4–14]. Overexpression of S100A4 in colorectal cancer and S100A6 in gastric cancer is associated with poor prognosis [6, 7], while low expression of S100A11 in bladder cancer is also linked to shorter survival [15]. S100A2 protein levels predict adjuvant therapy benefit in pancreatic cancer [16], and S100A13 overexpression is associated with chemotherapy resistance in melanoma [17]. S100B is widely valued as a clinically useful biomarker for melanoma [18].

Claus W. Heizmann (ed.), *Calcium-Binding Proteins of the EF-Hand Superfamily: From Basics to Medical Applications*, Methods in Molecular Biology, vol. 1929, https://doi.org/10.1007/978-1-4939-9030-6_41,

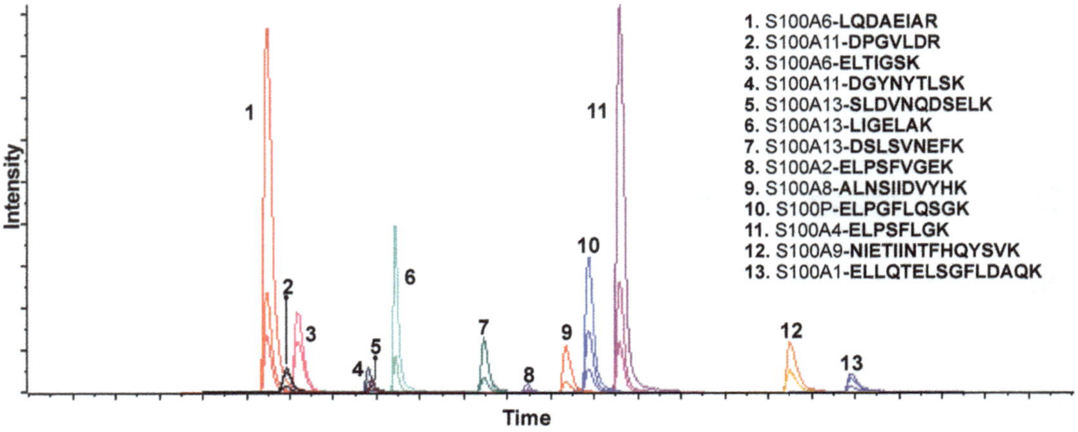

Fig. 1 An example of targeted detection of S100 peptides using SRM-MS

Evaluating the expression of S100 proteins is met with challenges due to their small size (9–13 kDa) and high-sequence identity that could present a problem for traditional immunoassays because of antigen cross-reactivity and costs related to antibody development. In addition, such methods generally lack the multiplexing capability needed to assess the abundance of the whole S100 protein panel at once. Here we describe an approach based on selected reaction monitoring mass spectrometry (SRM-MS) for the concurrent and specific quantitative analysis of the S100 protein family in cell lines and tumor tissue samples. The method takes advantage of the enrichment of S100 proteins using gel electrophoresis coupled with targeted MS. Proteins are recovered from cell/tissue lysates, enriched by gel electrophoresis and then proteolytically digested in situ. Recovered peptides are separated by reversed-phase chromatography in line with MS detection. Predetermined S100 peptide precursor and fragment ions are used for specificity, and ion areas are used for quantitative comparisons. An example of targeted detection of S100 peptides using SRM-MS is shown in Fig. 1.

2 Materials

Prepare all solutions using ultrapure water (purified deionized water, 18 MΩ-cm at 25 °C) and analytical grade reagents. All solutions should be stored at room temperature unless otherwise indicated.

2.1 Sample Preparation

1. Lysis buffer: 20 mM HEPES, 1% sodium deoxycholate (SDOC), 0.1% sodium dodecyl sulfate (SDS), 150 mM NaCl, 1 mM EDTA, pH 8, with complete protease inhibitor cocktail (Roche), added directly before use.

2. Bicinchoninic acid (BCA) protein assay kit (Pierce).
3. 96-well microplates for BCA assay.
4. Probe sonicator with pulsed mode.
5. Microplate absorbance spectrophotometer.
6. Incubator, set at 37 °C.
7. Ice bath.
8. Refrigerated centrifuge (4 °C) with rotor for 1.5 mL tubes.
9. 1.5 mL microcentrifuge tubes.

2.2 SDS-PAGE

1. Precast 4–12% Bis-Tris NuPAGE gels (8 × 8 cm, 1.0 mm thickness, 10-well). Store gels at 4–25 °C.
2. Lithium Dodecyl Sulfate (LDS) sample buffer, 4× (Invitrogen).
3. 500 mM dithiothreitol (DTT): Weigh 7.7.1 mg of DTT and dissolve it in 100 μL of water. Aliquots of this solution can be stored at −20 °C for at least 3 months.
4. MES SDS running buffer, 20× (Invitrogen): Prepare 1 L of 1× running buffer by adding 50 mL of 20× concentrate to 950 mL of water. Final concentrations are 50 mM MES, 50 mM Tris base, 0.1% SDS, 1 mM EDTA, pH 7.3.
5. Unstained protein molecular weight markers including the 10 kDa standard.
6. 1D gel electrophoresis unit (XCell SureLock Mini-Cell, Invitrogen) and power supply.
7. Heat block or microtube incubator.

2.3 Gel Fixing, Staining, and Gel Band Excision

1. Gel-fixing solution: 50% (v/v) ethanol/10% (v/v) acetic acid.
2. Background gel-destaining solution: 50% (v/v) methanol/10% (v/v) acetic acid.
3. Gel cubes destaining solution: 50% (v/v) acetonitrile (ACN) in 50 mM ammonium bicarbonate (AmBic, pH 8.0).
4. Coomassie stain: Dissolve 1.88 g of Coomassie brilliant blue G-250 in 250 mL of methanol for 1–2 h using a magnetic stir bar. Add 30 mL of 85% orthophosphoric acid and make up to 1 L with water.
5. Coomassie diluent: Dissolve 250 g of ammonium sulfate in 500 mL of water.
6. Platform rocker shaker.
7. Disposable scalpel, No. 10 blade.
8. Vacuum centrifuge (SpeedVac).
9. Microtube incubator/shaker.

2.4 In-Gel Digestion and Peptide Extraction

1. Reducing solution: 10 mM DTT in 100 mM AmBic.
2. Alkylating solution: 55 mM Iodoacetamide in 100 mM AmBic.
3. Washing solution: 50% (v/v) ACN in 50 mM AmBic.
4. Trypsin solution: 13 ng/μL of sequencing grade-modified trypsin (Promega) in 50 mM AmBic and 0.1% (w/v) sodium deoxycholate (SDOC). Suspend content of one trypsin vial (20 μg) in 1.5 mL of 50 mM AmBic/0.1% (w/v) SDOC.
5. Peptide extraction solution: 0.1% SDOC.
6. Formic acid, mass spectrometry grade.
7. Vacuum centrifuge (SpeedVac).
8. Microtube incubator/shaker.
9. Incubator, set at 37 °C.
10. Ultrasonic water bath.
11. 1.5 mL low-binding tubes.

2.5 Liquid Chromatography-Mass Spectrometry (LC-MS)

1. Peptide resuspension solution: 2% (v/v) mass spectrometry grade ACN/0.1% trifluoroacetic acid. Spike this solution with 18 fmol/μL of $^{13}C^{15}N$ stable isotope-labelled peptide ESDT-SYVSLK from human C-reactive protein (CRP), or alternative stable isotope-labelled control peptide.
2. Mobile phase A: 2% (v/v) mass spectrometry grade ACN/0.1% mass spectrometry grade formic acid.
3. Mobile phase B: 0.1% mass spectrometry grade formic acid in mass spectrometry grade ACN.
4. Waters symmetry C18 trapping column (180 μm × 20 mm, 5 μm particle size).
5. Waters BEH C18 nano UPLC column (100 μm × 100 mm, 1.7 μm particle size).
6. Waters nanoACQUITY UPLC system with column oven at 35 °C.
7. QTRAP 5500 mass spectrometer (AB Sciex).
8. LC-MS vials (e.g., waters certified clear glass 12 × 32 mm screw neck total recovery vials).
9. Ultrasonic water bath.
10. Mascot software search engine or AB Sciex ProteinPilot.

2.6 Data Analysis

1. MultiQuant software (AB Sciex)

3 Methods

3.1 Sample Preparation

1. Suspend cell or tumor samples in lysis buffer and put them in an ice bath (*see* **Note 1**).
2. Lyse samples on ice with a probe sonicator (three cycles of 20 s pulses and 30 s between pulses). *See* **Note 2**.
3. Centrifuge the lysates at 12,000 × *g* for 10 min at 4 °C and transfer the supernatant to a microcentrifuge tube.
4. Determine sample protein concentration using the BCA assay according to manufacturer's instructions.
5. For relative quantitative analysis of S100 proteins, prepare a reference pool sample by combining equal protein amounts from each sample to make up a total of 40 g × number of sample batches to be analyzed (*see* **Notes 3–5**). A sample batch is considered a group of up to 12 samples (including technical replicates) to be analyzed by SDS-PAGE, in-gel digestion, and LC-MS along with the reference pool sample analyzed in triplicate (15 samples in total).

3.2 SDS-PAGE

1. Mix 40 μg of sample protein with LDS sample buffer and 500 mM DTT. If using the maximum loading volume/well (25 μL), add 6.25 μL of LDS sample buffer and 2.5 μL of 500 mM DTT to 40 μg of sample (should be equal or less than 16.25 μL) and make up to 25 μL with water. Do the same for the reference sample.
2. Heat the sample mixture at 70 °C for 10 min in a microtube incubator.
3. Assemble the precast gel in the electrophoresis unit and fill the upper and lower chambers with MES SDS running buffer 1×.
4. Load unstained protein molecular weight markers into the first lane of the gel, and load the samples and pooled reference into the following lanes (*see* **Notes 6** and 7).
5. Run the gel electrophoresis at 200 V (constant voltage) for 25 min.

3.3 Gel Fixing, Staining, and Gel Band Excision

1. Remove the gel from the cassette and place it in a sealed container with gel-fixing solution.
2. Agitate the gel gently on a platform rocker for 1 h.
3. Replace the fixing solution with a 4:1 mixture of Coomassie stain and Coomassie diluent (e.g., for 50 mL, mix 40 mL of Coomassie stain with 10 mL of Coomassie diluent). Shake the gel for 1 h (*see* **Note 8**).

4. Destain the gel background briefly with 50% (v/v) methanol/10% (v/v) acetic acid. Rinse the gel with water and place it on a clean glass plate.
5. Using a clean disposable scalpel, excise the gel bands from around 7–14 kDa (*see* **Notes 9** and **10**). Cut closely to the stained area of each lane and dice each band into approximately 1 mm^3 cubes (1 mm depth is given by the gel thickness; *see* **Note 11**). Transfer the gel pieces to microcentrifuge tubes.
6. Cover the gel cubes with 50% (v/v) ACN in 50 mM ammonium bicarbonate (AmBic) and destain for 10 min with shaking. Remove the liquid and repeat this step until the gel pieces look clear.
7. Remove the liquid and cover the gel pieces with 100% ACN. Allow the gel to dehydrate for 5 min with shaking. The gel pieces should shrink and become white. Repeat step if necessary.
8. Remove supernatant and dry the gel pieces in a vacuum centrifuge for 5 min.

3.4 In-Gel Digestion and Peptide Extraction

1. Rehydrate gel pieces with 10 mM DTT in 100 mM AmBic and incubate at 56 °C for 1 h for protein reduction (*see* **Note 12**).
2. Remove supernatant and allow samples to cool down. Then, alkylate proteins with 55 mM iodoacetamide in 100 mM AmBic (enough to cover the gel cubes) at room temperature in the dark for 45 min.
3. Discard alkylating solution and wash gel pieces with 50% (v/v) ACN in 50 mM AmBic for 5 min with shaking.
4. Discard liquid and dehydrate with 100% ACN for 5 min in the microtube shaker.
5. Remove supernatant and dry the gel pieces in a vacuum centrifuge for 5 min.
6. Tryptic digestion: rehydrate dried samples with trypsin solution and keep on ice for 1 h. Typically 120 μL of trypsin solution are enough, but check if the complete volume is absorbed and add more solution to cover the gel pieces if necessary.
7. Remove residual trypsin solution and replace it with 50 mM AmBic/0.1% (w/v) SDOC.
8. Incubate samples overnight at 37 °C (*see* **Note 13**).
9. Transfer supernatant to a low-binding tube. Add enough peptide extraction solution (0.1 % SDOC) to cover the gel pieces and sonicate samples for 10 min. Pool supernatant with the first extract in the low-binding tube, and repeat the extraction twice with 0.1% SDOC solution, pooling the supernatants again.

10. Add 1 μL of formic acid to the combined peptide extract. Remove the precipitate by centrifugation at 14,000 × *g* for 10 min. Transfer the supernatant to a new low-binding tube.
11. Dry down the samples in a vacuum centrifuge at 35 °C (*see* **Note 14**).

3.5 LC-MS

1. Resuspend peptide samples in 10 μL of peptide resuspension solution (*see* **Note 15**) and sonicate for 10 min.
2. Centrifuge peptide resuspension at 14,000 × *g* for 10 min to remove microparticulates. Transfer the supernatant to LC-MS vials.
3. Load samples onto the trapping column installed in a Waters nanoACQUITY UPLC system using mobile phase A.
4. Separate peptides using a reversed-phase column (Waters BEH C18 nano UPLC column 100 μm × 100 mm, 1.7 μm particle size) at 35 °C and constant 0.4 μL/min. Start gradient with 3% mobile phase B for 1 min, 3–40% B in 30 min, and 40–85% B in 2 min holding for 3 min. Perform blank injections (mobile phase A) between each sample.
5. *Detection*: using a triple quadrupole system (e.g., QTRAP 5500, AB Sciex), use selected reaction monitoring (SRM)-triggered IDA MS/MS to detect and corroborate identity of S100 peptides present in the samples. Cover the full list of peptide sequences and fragment ions (transitions) shown in Table 1 in two separate acquisition methods using dwell times of 15 ms (*see* **Note 16**). Other parameters are set as follows: nano-ESI source operated in positive mode at 2.5 kV, interface heater temperature at 150 °C, curtain gas set to 20 psi, declustering potential at 70 V, and Q1/Q3 set at unit resolution. Search tandem mass spectra against the UniProt human database using Mascot search engine or AB Sciex ProteinPilot (*see* **Note 17**). Record identity of detected S100 peptides and their corresponding retention times (Rts).
6. *Acquisition of quantitative data*: set up a time-scheduled SRM method including all detected S100 peptide transitions and their recorded Rts. Use cycle time of 1 s and a transition window of 240 s.

3.6 Data Analysis

1. Integrate transition peak areas with MultiQuant software employing the MQ4 integration algorithm.
2. Peak assignment and removal of potential interferences. Peak identity assignment is guided by clear co-elution of transition signals and consistent Rts. Ensure accurate integration by manually inspecting the fragment ion peaks. Calculate the ratio between transitions for each S100 peptide across all samples and obtain the corresponding CV; remove transitions with CV>20% (*see* **Note 17**).

Table 1
***m/z* values of S100 peptide precursor (Q1) and fragment ions (Q3) for SRM analysis, reproduced with permission from [14]**

Q1 (*m/z*)	Q3 (*m/z*)	Protein.Peptide sequence.Fragment ion (transition)	CE	CEXP
846.45	778.41	S100A1.ELLQTELSGFLDAQK.+2y7	39.3	13
846.45	865.44	S100A1.ELLQTELSGFLDAQK.+2y8	39.3	13
846.45	1208.62	S100A1.ELLQTELSGFLDAQK.+2y11	39.3	13
503.27	763.40	S100A2.ELPSFVGEK.+2y7	22.7	12
503.27	666.35	S100A2.ELPSFVGEK.+2y6	30.7	11
503.27	579.31	S100A2.ELPSFVGEK.+2y5	30.7	9
675.84	750.38	S100A3.ELATWTPTEFR.+2y6	33.2	13
675.84	649.33	S100A3.ELATWTPTEFR.+2y5	33.2	13
675.84	936.46	S100A3.ELATWTPTEFR.+2y7	33.2	13
445.75	648.37	S100A4.ELPSFLGK.+2y6	21.5	19
445.75	551.32	S100A4.ELPSFLGK.+2y5	25.5	19
445.75	464.29	S100A4.ELPSFLGK.+2y4	25.5	16
455.22	693.36	S100A4.TDEAAFQK.+2y6	19.2	20
455.22	564.31	S100A4.TDEAAFQK.+2y5	19.2	20
455.22	493.28	S100A4.TDEAAFQK.+2y4	19.2	14
548.25	779.39	S100A5.NSDQEIDFK.+2y6	28.6	13
548.25	651.34	S100A5.NSDQEIDFK.+2y7	28.6	13
548.25	894.42	S100A5.NSDQEIDFK.+2y4	28.6	13
374.22	618.38	S100A6.ELTIGSK.+2y6	20.3	13
374.22	505.30	S100A6.ELTIGSK.+2y5	20.3	13
374.22	404.25	S100A6.ELTIGSK.+2y4	20.3	13
458.25	802.41	S100A6.LQDAEIAR.+2y7	25.4	13
458.25	674.35	S100A6.LQDAEIAR.+2y6	23.4	20
458.25	559.32	S100A6.LQDAEIAR.+2y5	31.4	14
628.81	708.36	S100A7.GTNYLADVFEK.+2y6	31.5	13
628.81	821.44	S100A7.GTNYLADVFEK.+2y7	31.5	13
628.81	984.50	S100A7.GTNYLADVFEK.+2y8	31.5	13
728.70	978.49	S100A7.IDFSEFLSLLGDIATDYHK.+3y17+2	37.2	13
728.70	904.96	S100A7.IDFSEFLSLLGDIATDYHK.+3y16+2+	37.2	13
728.70	1036.01	S100A7.IDFSEFLSLLGDIATDYHK.+3y18+2	37.2	13

(continued)

Table 1
(continued)

Q1 (*m/z*)	Q3 (*m/z*)	Protein.Peptide sequence.Fragment ion (transition)	CE	CEXP
718.70	781.91	S100A7A.IDFSEFLSLLGDIAADYHK.+3y14+2	36.7	13
718.70	889.95	S100A7A.IDFSEFLSLLGDIAADYHK.+3y16+2	36.7	13
718.70	963.49	S100A7A.IDFSEFLSLLGDIAADYHK.+3y17+2	36.7	13
426.57	423.22	S100A7A.GIHYLATVFEK.+3y3	20.8	13
426.57	522.29	S100A7A.GIHYLATVFEK.+3y4	20.8	13
426.57	623.34	S100A7A.GIHYLATVFEK.+3y5	20.8	13
721.32	526.25	S100A7L2.ENFPNFLSGCEK.+2y9+2	34.8	13
721.32	693.32	S100A7L2.ENFPNFLSGCEK.+2y6	34.8	13
721.32	1051.49	S100A7L2.ENFPNFLSGCEK.+2y9	34.8	13
449.19	606.25	S100A7L2.QYSGDDGR.+2y6	25.0	13
449.19	519.22	S100A7L2.QYSGDDGR.+2y5	25.0	13
449.19	462.19	S100A7L2.QYSGDDGR.+2y4	25.0	13
551.04	674.87	S100A7L2.VNYSEFLSLLGDITIDHHK.+4y12+2	27.6	13
551.04	574.81	S100A7L2.VNYSEFLSLLGDITIDHHK.+4y10+2	27.6	13
551.04	518.27	S100A7L2.VNYSEFLSLLGDITIDHHK.+4y9+2	27.6	13
636.85	774.41	S100A8.ALNSIIDVYHK.+2y6	31.8	13
636.85	974.53	S100A8.ALNSIIDVYHK.+2y8	31.8	13
636.85	887.50	S100A8.ALNSIIDVYHK.+2y7	31.8	13
711.36	965.45	S100A8.LLETECPQYIR.+2y7	34.5	13
711.36	1066.50	S100A8.LLETECPQYIR.+2y8	34.5	13
711.36	1196.53	S100A8.LLETECPQYIR.+2y9	34.5	13
411.71	694.36	S100A8.GADVWFK.+2y5	23.7	13
411.71	480.26	S100A8.GADVWFK.+2y3	23.7	13
411.71	579.33	S100A8.GADVWFK.+2y4	23.7	13
602.98	790.42	S100A9.NIETIINTFHQYSVK.+3y13+2	30.4	13
602.98	725.89	S100A9.NIETIINTFHQYSVK.+3y12+2	30.4	13
602.98	618.83	S100A9.NIETIINTFHQYSVK.+3y10+2	30.4	13
728.37	1148.56	S100A9.LGHPDTLNQGEFK.+2y10	35.1	13
728.37	1051.51	S100A9.LGHPDTLNQGEFK.+2y9	35.1	13
728.37	936.48	S100A9.LGHPDTLNQGEFK.+2y8	35.1	13
439.24	649.37	S100A9.DLQNFLK.+2y5	24.7	13

(continued)

Table 1 (continued)

Q1 (*m/z*)	Q3 (*m/z*)	Protein.Peptide sequence.Fragment ion (transition)	CE	CEXP
439.24	521.31	S100A9.DLQNFLK.+2y4	24.7	13
439.24	407.27	S100A9.DLQNFLK.+2y3	24.7	13
379.21	642.38	S100A10.DPLAVDK.+2y6	22.5	19
379.21	545.33	S100A10.DPLAVDK.+2y5	20.5	17
379.21	432.25	S100A10.DPLAVDK.+2y4	20.5	17
604.80	932.48	S100A10.EFPGFLENQK.+2y8	26.6	13
604.80	835.43	S100A10.EFPGFLENQK.+2y7	34.6	13
604.80	466.75	S100A10.EFPGFLENQK.+2y8+2	26.6	13
386.20	559.32	S100A11.DPGVLDR.+2y5	24.2	18
386.20	502.30	S100A11.DPGVLDR.+2y4	24.2	18
386.20	403.23	S100A11.DPGVLDR.+2y3	20.2	18
530.75	888.45	S100A11.DGYNYTLSK.+2y7	24.0	14
530.75	725.38	S100A11.DGYNYTLSK.+2y6	24.0	12
530.75	611.34	S100A11.DGYNYTLSK.+2y5	24.0	18
394.73	659.41	S100A12.ELANTIK.+2y6	23.1	13
394.73	475.29	S100A12.ELANTIK.+2y4	23.1	13
394.73	546.33	S100A12.ELANTIK.+2y5	23.1	13
452.73	448.28	S100A12.GHFDTLSK.+2y4	25.2	13
452.73	563.30	S100A12.GHFDTLSK.+2y5	25.2	13
452.73	710.37	S100A12.GHFDTLSK.+2y6	25.2	13
728.38	781.41	S100A12.LEEHLEGIVNIFHQYSVR.+3y13+2	37.2	13
728.38	837.95	S100A12.LEEHLEGIVNIFHQYSVR.+3y14+2	37.2	13
728.38	906.48	S100A12.LEEHLEGIVNIFHQYSVR.+3y15+2	37.2	13
519.76	836.45	S100A13.DSLSVNEFK.+2y7	23.6	13
519.76	723.37	S100A13.DSLSVNEFK.+2y6	23.6	20
519.76	636.34	S100A13.DSLSVNEFK.+2y5	23.6	20
372.24	630.38	S100A13.LIGELAK.+2y6	19.5	19
372.24	517.30	S100A13.LIGELAK.+2y5	17.5	14
372.24	460.28	S100A13.LIGELAK.+2y4	25.5	16
624.31	1047.50	S100A13.SLDVNQDSELK.+2y9	31.3	16
624.31	932.47	S100A13.SLDVNQDSELK.+2y8	31.3	16

(continued)

Table 1
(continued)

Q1 (*m/z*)	Q3 (*m/z*)	Protein.Peptide sequence.Fragment ion (transition)	CE	CEXP
624.31	833.40	S100A13.SLDVNQDSELK.+2y7	31.3	14
589.78	994.43	S100A14.IANLGSCNDSK.+2y9	30.1	13
589.78	880.38	S100A14.IANLGSCNDSK.+2y8	30.1	13
589.78	767.30	S100A14.IANLGSCNDSK.+2y7	30.1	13
625.82	1016.54	S100A14.SFWELIGEAAK.+2y9	29.0	17
625.82	830.46	S100A14.SFWELIGEAAK.+2y8	29.0	13
625.82	701.42	S100A14.SFWELIGEAAK.+2y7	32.0	12
834.36	1009.46	S100A14.SANAEDAQEFSDVER.+2y8	38.9	13
834.36	881.40	S100A14.SANAEDAQEFSDVER.+2y7	38.9	13
834.36	752.36	S100A14.SANAEDAQEFSDVER.+2y6	38.9	13
647.87	912.48	S100A16.AVIVLVENFYK.+2y7	28.0	14
647.87	799.40	S100A16.AVIVLVENFYK.+2y6	24.0	13
647.87	700.33	S100A16.AVIVLVENFYK.+2y5	26.0	12
683.35	1011.46	S100A16.LIQNLDANHDGR.+2y9	39.5	13
683.35	897.42	S100A16.LIQNLDANHDGR.+2y8	39.5	13
683.35	784.33	S100A16.LIQNLDANHDGR.+2y7	39.5	13
538.29	833.45	S100P.ELPGFLQSGK.+2y8	24.6	13
538.29	736.40	S100P.ELPGFLQSGK.+2y7	32.6	12
538.29	679.38	S100P.ELPGFLQSGK.+2y6	32.6	11
679.83	1108.55	S100P.YSGSEGSTQTLTK.+2y11	35.3	17
679.83	964.49	S100P.YSGSEGSTQTLTK.+2y9	35.3	16
679.83	835.45	S100P.YSGSEGSTQTLTK.+2y8	35.3	13
802.89	1007.50	S100G.GPNTLDDLFQELDK.+2y8	37.8	13
802.89	892.48	S100G.GPNTLDDLFQELDK.+2y7	37.8	13
802.89	1122.53	S100G.GPNTLDDLFQELDK.+2y9	37.8	13
686.42	557.37	S100G.LLIQAEFPSLLK.+2y5	33.6	13
686.42	904.51	S100G.LLIQAEFPSLLK.+2y8	33.6	13
686.42	1032.57	S100G.LLIQAEFPSLLK.+2y9	33.6	13
494.73	590.31	S100G.EGDPDQLSK.+2y5	26.7	13
494.73	687.37	S100G.EGDPDQLSK.+2y6	26.7	13
494.73	802.39	S100G.EGDPDQLSK.+2y7	26.7	13

(continued)

Table 1
(continued)

Q1 (*m/z*)	Q3 (*m/z*)	Protein.Peptide sequence.Fragment ion (transition)	CE	CEXP
898.94	1138.61	S100G.NGDGEVSFEEFQVLVK.+2y9	41.2	13
898.94	991.55	S100G.NGDGEVSFEEFQVLVK.+2y8	41.2	13
898.94	862.50	S100G.NGDGEVSFEEFQVLVK.+2y7	41.2	13
609.99	793.41	S100B.ELINNELSHFLEEIK.+3y13+2	41.8	13
609.99	736.87	S100B.ELINNELSHFLEEIK.+3y12+2	41.8	13
609.99	679.85	S100B.ELINNELSHFLEEIK.+3y11+2	41.8	13
423.71	460.28	S100B.EQEVVDK.+2y4	24.1	13
423.71	589.32	S100B.EQEVVDK.+2y5	24.1	13
423.71	717.38	S100B.EQEVVDK.+2y6	24.1	13
508.28	560.30	S100Z.IVQDLDANK.+2y5	27.2	13
508.28	675.33	S100Z.IVQDLDANK.+2y6	27.2	13
508.28	803.39	S100Z.IVQDLDANK.+2y7	27.2	13
416.72	474.29	S100Z.ETQLVDK.+2y4	23.9	13
416.72	602.35	S100Z.ETQLVDK.+2y5	23.9	13
416.72	703.40	S100Z.ETQLVDK.+2y6	23.9	13
627.81	911.43	S100Z.ELTEFLSCQK.+2y7	31.5	13
627.81	782.39	S100Z.ELTEFLSCQK.+2y6	31.5	13
627.81	1012.48	S100Z.ELTEFLSCQK.+2y8	31.5	13
568.28	919.50	CRP.ESDTSYVSLK.+2y8	27.0	13
568.28	703.50	CRP.ESDTSYVSLK.+2y6	27.0	13
568.28	616.50	CRP.ESDTSYVSLK.+2y5	27.0	13

CE Collision energy, *CEXP* Cell exit potential

3. Relative quantitation. For any sample, consider only peptide sequences where all transitions show signal-to-noise ratio S/N>10 and at least one transition above 5×10^4 counts; otherwise flag the sequence in that sample as not quantifiable. Sum transition peak areas of each S100 peptide and divide the value by the total peak area of the loading control (ESDTSYVSLK peptide). Divide this result by the total peak area of the same peptide in the reference pool sample belonging to the same batch as the sample of interest. Finally, obtain the average value between different peptides from the same protein (*see* **Notes 18** and **19**).

4 Notes

1. It is important to keep protein concentration high enough to not exceed the maximum loading volume/well on the electrophoresis gel, considering addition of LDS sample buffer and DTT. Recommended maximum volume on 10-well NuPAGE gels is 25 μL/well. 40 μg of sample protein will be loaded per well.
2. Hearing protection is recommended when using the sonicator. Minimize foaming by keeping the probe away from the air-liquid interface.
3. The reference pool sample helps control variations due to sample processing and changes in mass spectrometry system response with time. It is prepared and analyzed together with up to other 12 samples, including replicates. Although higher number of samples per batch could be possible, we have not tested this for label-free analyses. For tissue samples, we recommend including three technical replicates per biological replicate. In the case of cell lines, three biological replicates per cell line/condition are sufficient to assess coefficients of variation (CVs). The reference pool sample is meant to represent the average S100 protein composition across all samples, which should facilitate comparison of S100 peptide signals from each sample against this pooled reference. However, if it is expected that signal from one of the S100 peptides may get too diluted (below 5×10^4 counts), adjustment of protein amounts from corresponding samples is advisable. For tumor samples, where protein can be limited, a reference protein pool can be made up by mixing sample protein with protein from cell line(s) known to express the S100 proteins present in the samples (cell lines from the same tissue origin are the obvious choice). If the number of samples including replicates is lower than 12 and they will not be reanalyzed in the future, the reference pool sample may be omitted.
4. This protocol describes the relative label-free quantitation of S100 proteins. For quantitative purposes, it is possible to omit the use of a reference pool sample and a loading control in the peptide resuspension solution ($^{13}C^{15}N$ stable isotope-labelled peptide ESDTSYVSLK, Subheading 2.5) if stable isotope-labelled ("heavy") S100 peptide standards are added to the samples before tryptic digestion (e.g., SpikeTides TQL, JPT Peptide Technologies); *see* **step 8** in Subheading 3.4. Expected median coefficient of variation (CV) among technical replicates when using S100 "heavy" peptides is less than 6%, whereas label-free quantitation affords median CV around 20%. This

means it is also possible to work only with biological replicates when "heavy" labelled S100 peptide standards are available.

5. Before quantitative analysis, a screening step for the detection of S100 proteins is needed in order to know which S100 peptide sequences will be targeted for relative quantitation. To this end, follow the sample processing steps detailed from Subheading 3.2 until detection by LC-MS, but omit the pooled reference and load two replicates per sample. Two replicates per sample are needed since the targeted detection of the complete S100 protein family requires to run two methods (two S100 peptide transitions lists) for the same sample. Once S100 proteins have been identified, proceed with analysis of the entire sample panel.
6. 10 μL of Bio-Rad Precision Plus Protein™ unstained protein standards (#1610363) are recommended for Coomassie staining, but protein molecular weight markers from other vendors can be employed.
7. Allow samples to cool to room temperature before loading into the gel.
8. Protein molecular weight markers and sample protein bands can be seen after 1 h, but longer staining (overnight) can be performed for increased intensity of gel bands. In the second case, however, the destaining process will take longer.
9. Gel cutting should be carried out in a clean area to minimize keratin contamination.
10. 15 and 10 kDa molecular weight markers will aid in selecting the gel region. If there is no marker lower than 10 kDa, the recommendation is to cut below 10 kDa at least 1.5 times the length between the 10 and 15 kDa markers.
11. Do not cut the gel bands smaller than 1 mm^3 since it could lead to gel pieces being carried away during subsequent pipetting steps.
12. Approximately 120 μL of reducing solution is needed per sample, but make sure gel cubes are covered with solution after rehydration.
13. A dedicated benchtop incubator set at 37 °C is preferred for in-gel digestion since a microtube incubator (heat block or dry bath) may lack the temperature uniformity required along the length of the tubes during the incubation period.
14. Avoid over-drying. It is preferable to keep 1–2 μL of sample in the tube rather than excessive drying that may reduce peptide recovery. Dried samples may be stored at −80 °C for a few weeks, but it is better to analyze them soon after preparation.

15. $^{13}C^{15}N$ stable isotope-labelled peptide ESDTSYVSLK is used as loading control for LC runs. It helps normalize S100 peptide signals between runs. As loading control, it can be substituted for any other synthetic peptide known to be absent in the samples and to elute in or close to the mid-region of the gradient (for chromatographic reproducibility).
16. Dividing the list of peptide transitions listed in Table 1 into two methods is recommended to avoid long cycle times that may affect peptide detection.
17. Straightforward S100 peptide peak identification can be done by addition of stable isotope-labelled S100 peptides since transitions from both endogenous and "heavy" peptides should co-elute. Potential interferences can be detected by comparison of transition ratios, which should be very similar between "heavy" and endogenous peptides.
18. If stable isotope-labelled S100 peptides were spiked to the samples, only divide the sum of endogenous S100 peptide transition peak areas by the sum of transition areas from the corresponding "heavy" peptides. Obtain the average value between different peptides from the same protein.
19. Tumor tissue samples may have variable amounts of blood proteins. Because samples are loaded into the gel based on quantitated total protein values, this could lead to misrepresentation of relative S100 proteins abundance in the samples. To circumvent this issue, normalization of S100 protein levels against expression of reference proteins across samples is recommended. The latter can be determined following the protocol presented here applied to proteins in other regions of the gel. Some candidate reference proteins and their peptide sequences that could be useful for targeted quantitation by SRM can be found in [14]. For large-scale expression studies, the use of isotope-labelled peptides from reference proteins is advisable to streamline the protocol and gain precision in the determination of reference protein abundance. Synthetic "heavy" peptides (SpikeTides TQL, JPT Peptide Technologies) or metabolically labelled cell lines are equally suitable for this goal [14]. Additionally, reference proteins with stable expression across all samples could be identified by algorithms such as geNorm [19]. Reference proteins are not needed when working with cell lines.

Acknowledgments

MPM acknowledges support from the Cancer Institute NSW and Sydney Vital Translational Cancer Research Centre.

References

1. Marenholz I, Heizmann CW, Fritz G (2004) S100 proteins in mouse and man: from evolution to function and pathology (including an update of the nomenclature). Biochem Biophys Res Commun 322(4):1111–1122. https://doi.org/10.1016/j.bbrc.2004.07.096
2. Salama I, Malone PS, Mihaimeed F et al (2008) A review of the S100 proteins in cancer. Eur J Surg Oncol 34(4):357–364. https://doi.org/10.1016/j.ejso.2007.04.009
3. Chen H, Xu C, Jin Q et al (2014) S100 protein family in human cancer. Am J Cancer Res 4 (2):89–115
4. Gupta S, Hussain T, MacLennan GT et al (2003) Differential expression of S100A2 and S100A4 during progression of human prostate adenocarcinoma. J Clin Oncol 21(1):106–112. https://doi.org/10.1200/jco.2003.03.024
5. Cho YG, Kim CJ, Nam SW et al (2005) Overexpression of S100A4 is closely associated with progression of colorectal cancer. World J Gastroenterol 11(31):4852–4856. https://doi.org/10.3748/wjg.v11.i31.4852
6. Huang LY, Xu Y, Cai GX et al (2011) S100A4 over-expression underlies lymph node metastasis and poor prognosis in colorectal cancer. World J Gastroenterol 17(1):69–78. https://doi.org/10.3748/wjg.v17.i1.69
7. Wang XH, Zhang LH, Zhong XY et al (2010) S100A6 overexpression is associated with poor prognosis and is epigenetically up-regulated in gastric cancer. Am J Pathol 177(2):586–597. https://doi.org/10.2353/ajpath.2010.091217
8. De Petris L, Orre LM, Kanter L et al (2009) Tumor expression of S100A6 correlates with survival of patients with stage I non-small-cell lung cancer. Lung Cancer 63(3):410–417. https://doi.org/10.1016/j.lungcan.2008.06.003
9. Shekouh AR, Thompson CC, Prime W et al (2003) Application of laser capture microdissection combined with two-dimensional electrophoresis for the discovery of differentially regulated proteins in pancreatic ductal adenocarcinoma. Proteomics 3(10):1988–2001. https://doi.org/10.1002/pmic.200300466
10. Al-Haddad S, Zhang Z, Leygue E et al (1999) Psoriasin (S100A7) expression and invasive breast cancer. Am J Pathol 155 (6):2057–2066. https://doi.org/10.1016/s0002-9440(10)65524-1
11. Arai K, Takano S, Teratani T et al (2008) S100A8 and S100A9 overexpression is associated with poor pathological parameters in invasive ductal carcinoma of the breast. Curr Cancer Drug Targets 8(4):243–252. https://doi.org/10.2174/156800908784533445
12. Anania MC, Miranda C, Vizioli MG et al (2013) S100A11 overexpression contributes to the malignant phenotype of papillary thyroid carcinoma. J Clin Endocrinol Metab 98(10): E1591–E1600. https://doi.org/10.1210/jc.2013-1652
13. Arumugam T, Simeone DM, Van Golen K et al (2005) S100P promotes pancreatic cancer growth, survival, and invasion. Clin Cancer Res 11(15):5356–5364. https://doi.org/10.1158/1078-0432.Ccr-05-0092
14. Martinez-Aguilar J, Clifton-Bligh R, Molloy MP (2015) A multiplexed, targeted mass spectrometry assay of the S100 protein family uncovers the isoform-specific expression in thyroid tumours. BMC Cancer 15:199. https://doi.org/10.1186/s12885-015-1217-x
15. Memon AA, Sorensen BS, Meldgaard P et al (2005) Down-regulation of S100C is associated with bladder cancer progression and poor survival. Clin Cancer Res 11(2 Pt 1):606–611
16. Bachet JB, Marechal R, Demetter P et al (2013) S100A2 is a predictive biomarker of adjuvant therapy benefit in pancreatic adenocarcinoma. Eur J Cancer 49(12):2643–2653. https://doi.org/10.1016/j.ejca.2013.04.017
17. Azimi A, Pernemalm M, Frostvik Stolt M et al (2014) Proteomics analysis of melanoma metastases: association between S100A13 expression and chemotherapy resistance. Br J Cancer 110(10):2489–2495. https://doi.org/10.1038/bjc.2014.169
18. Gogas H, Eggermont AMM, Hauschild A et al (2009) Biomarkers in melanoma. Ann Oncol 20(suppl 6):vi8–vi13. https://doi.org/10.1093/annonc/mdp251
19. Vandesompele J, De Preter K, Pattyn F et al. (2002 Accurate normalization of real-time quantitative RT-PCR data by geometric averaging of multiple internal control genes. Genome Biol 3 (7):Research0034. doi: https://doi.org/10.1186/gb-2002-3-7-research0034

Chapter 42

Clinical Use of the Calcium-Binding S100B Protein, a Biomarker for Head Injury

Ramona Astrand and Johan Undén

Abstract

S100B is a calcium-binding protein most abundant in neuronal tissue. It is expressed in glial cells and Schwann cells and exerts both intra- and extracellular effects. Depending on the concentration, secreted S100B exerts either trophic or toxic effects. Its functions have been extensively studied but are still not fully understood. It can be measured in cerebrospinal fluid and in blood, and increased S100B level in blood can be seen after, e.g., traumatic brain injury, certain neurodegenerative disorders, and malignant melanoma. This chapter provides a short background of protein S100B, commercially available methods of analysis, and its clinical use, especially as a biomarker in minor head injury.

Key words S100B, EF hand, Calcium, Biomarker, Traumatic brain injury, Minor head injury, Malignant melanoma, Neurodegenerative diseases

1 Introduction

1.1 Biochemistry of S100 Proteins

The first member of the S100 family was isolated by Moore in 1965. It was given its name S100 due to solubility in 100% saturated ammonium sulfate at neutral pH. The protein S100 was initially purified from bovine brain and thought to be specific to neuronal tissue since it was mainly expressed in glial and Schwann cells [1]. Subsequent studies showed that the fraction contained two polypeptides, S100A1 and S100B; the two S100 proteins are most abundant in the nervous system [2].

The multigenic S100 family, consisting of more than 20 members, is characterized by the presence of two calcium-binding, EF-hand motifs (helix-loop-helix structure). S100A1–S100A16 proteins are coded by genes clustered at human chromosome 1q21. The other proteins S100B, S100P, S100Z, and S100G are coded by genes located at different chromosomes 21q22, 4p16, 5q14, and Xp22, respectively [3–5].

Claus W. Heizmann (ed.), *Calcium-Binding Proteins of the EF-Hand Superfamily: From Basics to Medical Applications*, Methods in Molecular Biology, vol. 1929, https://doi.org/10.1007/978-1-4939-9030-6_42,

The S100 proteins are small acidic proteins, with molecular weights between 9 and 13 kDa. They exist within cells as S100B homodimers with the exception of S100G (calbindin D9k) which is monomeric. In some cases, heterodimers can form, as in the case with S100A1/S100B, and even oligomers in some specific cells. Dimerization and oligomerization seem to be important for the biological activity, and upon calcium binding, conformational changes on the protein occur in order to be able to recognize its target proteins/receptors (e.g., RAGE) [6]. Several S100 proteins not only bind calcium but also zinc and copper.

1.2 Functional Roles of S100B

The S100 proteins relay signals from receptors on the cell surface to target molecules inside the cell, in the cytoplasm or nucleus. Extracellularly, several S100 proteins bind to RAGE (receptor for advanced glycation end products), which activates several intracellular signaling pathways [6]. In general, there are five major intracellular functions for S100 proteins: (1) regulations of phosphorylation mediated by protein kinases, (2) modulation of enzymatic activity, (3) maintenance of cell shape and motility, (4) influence of some signal transduction pathways, and (5) promotion of calcium homeostasis [7]. The function of S100B is, among others, regulation of calcium homeostasis and regulation of cell motility. Depending on the concentration, secreted S100B exerts trophic as well as toxic effects. At lower concentrations (nanomolar), S100B is thought to stimulate neuronal growth and enhancement of neuronal survival during development, while at higher concentrations (micromolar), S100B may have deleterious effects by increasing expression of the pro-inflammatory cytokine IL-6 and inducing apoptotic neuronal death [8].

1.3 Sources of S100B

Several of the S100 proteins are abundant in the central nervous system. S100B is expressed in astrocytes, oligodendrocytes, and Schwann cells, but it can also be found outside the nervous system in chondrocytes, adipocytes, and melanocytes [8, 9]. The extracranial concentration of S100B is much lower than that found in neuronal tissue [10].

S100B can be measured in both cerebrospinal fluid and in blood serum. The median concentration in blood among healthy adults is 0.05 μg/L and is independent of both age and gender [11]. There are still some uncertainties about the release mechanism of S100B into the blood. The most accepted theory is that of extracellular S100B passing into the blood due to a disrupted blood–brain barrier (BBB) [12] . Serum protein S100B is an early marker of the disruption of BBB after a blow to the head, though not always as a sign of neuronal damage. It is metabolized and excreted via the kidneys, and its half-life is estimated to 30–130 min [13, 14]. The stability of protein S100B in serum has been investigated by Raabe and colleagues, who found no effects of storage time and temperature prior to analysis [15]. S100B has also been found in other extracellular fluids, such as saliva and urine. The stability of urine S100B seems to be less stable [16]. Studies comparing jugular vein S100B to arterial concentrations have found contradicting results, one finding a significantly higher concentration in jugular blood [17] and the other with almost equivalent levels [18].

Reference levels for S100B in adults are both age and gender independent [11], and a clinical cutoff value ≤0.10 μg/L has been considered normal (within 95th percentile) [19] . Reference levels for children are slightly increased compared to adults [20].

2 Diagnostic Use of S100B

2.1 Head Injury

Traumatic brain injury is a significant cause of mortality and morbidity in adults and a leading cause of death in childhood. The diagnostic process includes clinical examination and, in more severe cases, neuroimaging, such as computed tomography of the head (CT) or magnetic resonance imaging (MRI). The majority of head injuries are so-called minor head injuries (85–90%), whereas only 5–10% are moderate to severe head injuries. The majority of the minor head injured are fully awake with a Glasgow Coma Scale (GCS) score of 15, without any loss of consciousness or amnesia post-injury, and, thus, have a very low risk of developing intracranial complications. These are according to the Scandinavian Neurotrauma Committee classified as minimal head injuries [21]. Those defined as being "mild head injuries" are the ones who appear fully or nearly fully awake (GCS 14–15) but have had or might have had a brief loss of consciousness and/or amnesia. The mild head injured have an approximately 1% risk of developing intracranial complications in need of neurosurgical intervention; less than 10% have pathological intracranial findings on CT [22].

2.2 Sensitivity and Specificity

Serum S100B increases almost immediately after a relevant brain injury, due to a brief breach of the BBB [12]. It has been discussed that since epidural hematomas may initially not be associated with an actual brain damage, the biomarker might not be increased. However, epidural hematomas reported in the literature have all shown elevated S100B >0.10 μg/L, although only slightly elevated [23]. Thus, S100B could be a marker of BBB disruption as well as brain cell damage [24]. The sensitivity of the protein is high, making S100B a better marker for identifying patients with a negligible risk for intracranial complications [25–29]. In a meta-analysis by Undén et al., the sensitivity for S100B (cutoff 0.10 μg/L and sample time within 3 h after minor head trauma) was found to be 98% for any type of intracranial pathology on CT. For clinically relevant findings/neurosurgical intervention, the sensitivity was 100% [30].

Protein S100B is not specific to the brain. Its presence is also found in low concentrations in adipose tissue, melanocytes in the skin, bone marrow, and heart muscle [31]. Undén et al. also showed that serum S100B is increased in patients with especially long bone fractures [32]; hence the extracerebral sources may be a limiting factor in the immediate assessment of multi-trauma patients [33]. S100B is not influenced by blood-alcohol levels [28, 34].

2.3 Minor Head Injury

Clinical use of S100B is mainly within the field of minor head injuries. Due to the high specificity and high negative predictive value for neurosurgically relevant intracranial complications, it has been estimated that S100B could reduce the frequency of head CT by 30% and still maintain patient safety. This was formerly proposed by Biberthaler et al. and further studied in a meta-analysis by Undén et al., strengthening the theory [30, 35]. Zongo et al. published a prospective study on the evaluation of S100B as a screening tool in minor head injury (GCS 13–15), which confirms the potential benefit of implementing S100B into clinical use [36]. A total of 1646 patients were included in the study. The cutoff value was chosen to be 0.12 μg/L, and blood was collected within 6 h after trauma. None of the patients were considered false negative. The sensitivity was 99.1%, and negative predictive value 99.7%. The potential percentage of CT scans that could be avoided was calculated to 19%. These studies only apply on adult patients.

The revised Scandinavian head injury guidelines from 2013 are the first head trauma guidelines to include a biomarker, protein S100B, as a means to rule out significant intracranial injury after head trauma [37]. The Scandinavian head injury guidelines are applicable to moderate, mild, and minimal head injuries; and the guidelines further divide mild head injuries into high, medium, and low risk. Serum S100B can be used for the mild, low-risk head injuries, which are the most abundant head injuries. The cutoff value for S100B sampled within 6 h of trauma is 0.10 μg/L. Thus, patients with serum S100B values <0.10 μg/L are considered not to have an intracranial injury and could therefore avoid a CT scan or admission. The guidelines, if used correctly, are potentially cost saving [38]. The guidelines have been validated, and are in clinical use in the Scandinavian countries [39]. Further validations and improvement of the guidelines are ongoing.

2.4 Severe Traumatic Brain Injury

Research on the potential use of biomarkers in severe head injury has primarily been directed at severity and outcome prediction. The clinical presentation of an unconscious patient, who is often sedated and/or with external mechanical ventilation, makes traditional clinical evaluation difficult. Initial level of consciousness (according to GCS), CT classifications (Marshall), pupil response, and events of intracranial hypertension or hypoperfusion/hypoxia are some of the traditional clinical tools for estimating clinical progress and outcome. However, a biochemical brain marker could be most valuable in the prediction of secondary complications in a neurointensive care patient.

Previous studies have found a correlation between serum S100B levels and clinical outcome according to Glasgow Outcome Scale (GOS) after severe head injury [40, 41]. Patients with poor outcome (vegetative state or death) have significantly higher S100B levels than patients with favorable to severely disabled outcome [42, 43]. One study found a 2.1-fold increase in serum S100B in those with unfavorable vs. favorable outcome [44]. In earlier studies, Raabe et al. found a strong association between a cutoff level of 2.5 μg/L (any time) and unfavorable outcome [40, 45]. Woertgen et al. found that serum levels above 2 μg/L (drawn within 6 h after trauma) predicted unfavorable outcome (positive predictive value 87%; negative predictive value, 77%) [46], while Nylén et al. reported a cutoff of 0.55 μg/L (at admission) for predicting unfavorable outcome 1 year post injury [47].

Böhmer et al. collected daily CSF samples in 20 patients with severe TBI and found that early elevations of S100B (up to 3 days) predicted deterioration to brain death [48]. Dimopoulou et al. reported elevated median serum S100B levels of 2.32 μg/L in those progressing to brain death, while survivors had significantly lower median values of 1.04 μg/L [49].

More recent studies have focused on the temporal profile and the prognostic ability of S100B after traumatic brain injury (TBI) [50]. A sudden increase in serum S100B levels could indicate a secondary neurological complication, such as progression of intracranial hematoma or cerebral infarction. Raabe et al. found that a secondary increase of 0.5 μg/L of daily drawn serum S100B correlated to secondary injury [51], and Undén et al. also noted that increased levels of S100B correlated to secondary neurological complications after TBI [52]. Thelin et al. also found that secondary increases as low as 0.05 μg/L beyond 48 h after trauma were strongly correlated to radiological deterioration and were significant predictors of unfavorable outcome [53].

Contrary to previous studies, some studies have shown no significant correlation between early CSF-S100B or serum S100B and GOS or other outcome scales [54] nor between dichotomized GOS [52, 55]. They conclude there is no clinically significant value of the marker as predictor of clinical outcome [55]. In a prospective, double-blind, randomized study, Olivecrona et al. investigated the prognostic value of biomarkers and concluded that serum S100B is a poor predictor of clinical outcome at 3 and 12 months [55].

2.5 Multi-trauma

One drawback of S100B is the presence of extracranial sources, especially bone marrow and adipose tissues [32, 33]. Pelinka et al. found that all patients studied with multiple organ trauma demonstrated raised S100B levels whether or not TBI was present, and they concluded that serum levels drawn during the first 24 h did not reliably predict clinical outcome in these patients [56]. Daily measurements were advocated. However, another study by da Rocha et al. did not find any correlation between higher S100B values and the presence of multi-trauma [42].

2.6 S100B Protein in Children

There are, by far, fewer studies performed on pediatric TBI and biomarkers, but similar to studies on adults, most studies have focused on the ability of S100B to predict outcome after TBI. Bechtel and colleagues described use of serum S100B as a screening tool to detect intracranial injuries in children admitted to the emergency department after closed head trauma. Although children with intracranial injuries had significantly higher S100B concentrations than children without intracranial injuries, the ability of S100B levels to detect such injuries was found to be poor [57]. Other studies have shown that biomarkers, such as glial fibrillary acidic protein (GFAP) and ubiquitin c-terminal hydrolase-L1 (UCH-L1), may have a higher diagnostic accuracy in outcome prediction and in detecting intracranial lesions than S100B [58, 59]. Child abuse and inflicted TBI are both serious and non-negligible. The presenting symptoms are diffuse, such as headache, abdominal pain, and excessive vomiting, and can be difficult to differentiate between other medical conditions, especially among infants. Neither CSF nor serum biomarkers have shown sufficient discriminative abilities to be used as a screening tool for inflicted TBI [60].

Castellani et al. recently investigated the correlation of S100B to CT results after mild TBI (GCS 13–15). Serum S100B concentrations were significantly higher in patients with abnormal CT (mean 0.64 μg/L) compared with patients with normal CT (mean 0.50 μg/L), and even though specificity was poor (0.42), the sensitivity and negative predictive value was 1.00. The authors concluded that normal S100B levels (<0.16 μg/L, Elecsys) safely rules out CT pathologies in minor head injured children [61]. Three independent studies have been performed to investigate the reference values of S100B in children and conclude that healthy children have increased S100B levels compared to adults and especially among infants [20, 62, 63].

S100B is not yet incorporated in any pediatric guidelines, or in routine clinical use, although there is a strong wish for this to reduce unnecessary CT scans and admissions. A recent meta-analysis investigated the prognostic value of S100B in pediatric mild traumatic brain injury and found an overall pooled sensitivity of 100%. The authors recommended that sampling should take place within 3 h of trauma and that cutoff levels should be based on pediatric reference ranges [64].

3 Neurodegenerative Diseases

The levels of S100B in blood and CSF have also been correlated to a number of other diseases, such as Alzheimer's disease (AD) [65], autoimmune disorders such as multiple sclerosis (MS) [66], psychiatric disorders such as schizophrenia [67], and acute cerebrovascular insults [68].

Alzheimer's disease (AD) is so far the most extensively studied neurodegenerative disorder coupled with S100B levels. S100B in CSF has previously been shown to be increased in mild to moderate stage of AD but nearly equal or decreased in severe AD compared to controls [69]. A more recent study reported positive correlations of the severity of dementia in AD and serum S100B levels [70]. This result differs from a previous study by Gruden et al. (2007), who reported lower levels in severe dementia compared to moderate and a peak of S100B immune response during moderate dementia [71].

The S100B findings in MS and schizophrenia support the theory that there are neurodegenerative mechanisms involved in the pathogenesis of these diseases. CSF and/or serum levels are increased in the active phase of MS [66, 72] and in the acute as well as chronic phase of schizophrenia [73, 74]. However, the clinical role of the S100B protein in these diseases is still quite poorly understood, and further studies are warranted.

4 Malignant Melanoma

Immunohistochemical procedures have improved the correct classification of tumors, although conventional histology is still considered the golden standard. Melanocytes, as opposed to non-melanocytic tumors, express S100B, which can be detected with monoclonal or polyclonal antibodies. Protein S100B has been used to improve melanoma staging; the more advanced tumor stage, the higher the blood levels. The sensitivity of serum S100B in patients with stage I and II malignant melanoma has been reported to be <15%, for stage III varying from 10 to 50%, and for stage IV 60–85% [9, 75]. S100B is so far considered as the most sensitive serum marker for detecting metastatic disease in malignant melanoma.

The prognostic factors for survival in malignant melanoma are tumor thickness and lymph node status. However, S100B has been shown to be a fairly good prognostic marker for survival [76]. Patients with low levels of S100B had on average 29.6 months longer survival than patients with elevated levels [77]. The next step and goal within this area is to find a potent inhibitor of S100B for the treatment of malignant melanoma [78].

Serum S100B is not specific to malignant melanoma. It is elevated in several malignant diseases, and increased levels can also be found in liver and renal diseases and in inflammatory and infectious diseases [79].

5 Diagnostic Tests

There are a number of commercially available immunoassays for serum S100B. The two most common are the immunoluminometric assay Liaison® Sangtec 100 (DiaSorin AB, Bromma, Sweden) and the electrochemiluminescence immunoassay Elecsys S100 (Roche Diagnostics, Mannheim, Germany). Both assays detect the heterodimers S100A1/B and the homodimer S100B protein.

Other assays commonly used in recent studies are the immunoluminometric assay LIA-mat® Sangtec 100 (DiaSorin AB, Bromma, Sweden), the ELISA Nexus DX™ S100 (SynX Pharma Inc., York, UK), the enzyme-linked immunosorbent assay (Nanogen), and the enzyme-linked immunoassay Elecsys® S100 (Roche, Mannheim, Germany).

CanAg S100A1/B EIA and CanAg S100BB EIA (Fujirebio Diagnostics AB, Gothenburg, Sweden) are so far the only assays to specifically analyze the S100A1B (focusing on the A1 dimer) and S100BB, respectively. Although commercially available, the analysis of S100BB in former studies has been limited. However, Nylén et al. investigated the relation of S100B, S100A1B, and S100BB in serum after severe TBI to outcome, with the hypothesis that S100BB dimer should be better related to outcome after severe TBI than either S100A1B or S100B, due to the greater abundancy of S100BB in the brain. They concluded that there most probably was of no advantage of analyzing S100BB instead of S100B alone [47].

There are a few studies performed to compare some of the commercially available assays. Muller et al. compared the Liaison® Sangtec 100 (DiaSorin AB) and the Elecsys® S100 (Roche) and found a mean difference of 0.14 μg/L between the measured values, where Liaison® Sangtec 100 was consistently higher than the Elecsys® S100 method. Similar results were found by Hallén et al., with a mean difference of 0.12 μg/L. In conclusion, differences between the analytical methods should be considered when interpreting results from different studies [16, 80, 81].

Some of the assays have also been used for measuring S100B in CSF, saliva, and other biological fluids, although their clinical implications (apart from CSF) have yet been quite limited.

6 Notes

The assay currently used for measuring S100B in the event of brain injuries, in the Scandinavian countries, is the electrochemiluminescence Elecsys S100 assay (Roche Diagnostics, Mannheim, Germany), mainly due to its availability. The majority of hospitals already have the laboratory setup and analyzers Elecsys 2010, Modular analytics E170, or Cobas, which are all compatible with the Elecsys S100 assay.

The Scandinavian guidelines for management of minimal, mild, and moderate head injuries in adults give a cutoff level of <0.10 μg/L for serum S100B [37], which is based on measurements made with the Elecsys S100 assay on Modular E170.

References

1. Moore BW (1965) A soluble protein characteristic of the nervous system. Biochem Biophys Res Commun 19:739–744
2. Isobe T, Okuyama T (1981) The amino-acid sequence of the alpha subunit in bovine brain S-100a protein. Eur J Biochem 116:79–86
3. Marenholz I, Heizmann CW, Fritz G (2004) S100 proteins in mouse and man: from evolution to function and pathology (including an update of the nomenclature). Biochem Biophys Res Commun 322:1111–1122
4. Marenholz I, Lovering RC, Heizmann CW (2006) An update of the S100 nomenclature. Biochim Biophys Acta 1763:1282–1283
5. Kizawa K et al (2011) S100 and S100 fused-type protein families in epidermal maturation with special focus on S100A3 in mammalian hair cuticles. Biochimie 93:2038–2047
6. Ostendorp T et al (2007) Structural and functional insights into RAGE activation by multimeric S100B. EMBO J 26:3868–3878
7. Santamaria-Kisiel L, Rintala-Dempsey AC, Shaw GS (2006) Calcium-dependent and -independent interactions of the S100 protein family. Biochem J 396:201–214
8. Donato R, Heizmann CW (2010) S100B protein in the nervous system and cardiovascular apparatus in normal and pathological conditions. Cardiovasc Psychiatry Neurol 2010:929712. https://doi.org/10.1155/2010/929712
9. Harpio R, Einarsson R (2004) S100 proteins as cancer biomarkers with focus on S100B in malignant melanoma. Clin Biochem 37:512–518
10. Haimoto H, Hosoda S, Kato K (1987) Differential distribution of immunoreactive S100-alpha and S100-beta proteins in normal nonnervous human tissues. Lab Investig 57:489–498
11. Wiesmann M et al (1998) Plasma S-100b protein concentration in healthy adults is age- and sex-independent. Clin Chem 44:1056–1058
12. Marchi N et al (2003) Peripheral markers of brain damage and blood-brain barrier dysfunction. Restor Neurol Neurosci 21:109–121
13. Jonsson H et al (2000) Elimination of S100B and renal function after cardiac surgery. J Cardiothorac Vasc Anesth 14:698–701
14. Blomquist S et al (1997) The appearance of S-100 protein in serum during and immediately after cardiopulmonary bypass surgery: a possible marker for cerebral injury. J Cardiothorac Vasc Anesth 11:699–703
15. Raabe A et al (2003) Measurements of serum S-100B protein: effects of storage time and temperature on pre-analytical stability. Clin Chem Lab Med 41:700–703
16. Hallen M et al (2008) A comparison of two different assays for determining S-100B in serum and urine. Clin Chem Lab Med 46:1025–1029
17. Raabe A et al (1998) Jugular venous and arterial concentrations of serum S-100B protein in patients with severe head injury: a pilot study. J Neurol Neurosurg Psychiatry 65:930–932
18. Kunihara T et al (2006) Arterio-jugular differences in serum S-100beta proteins in patients receiving selective cerebral perfusion. Surg Today 36:6–11
19. Muller K et al (2007) S100B serum level predicts computed tomography findings after minor head injury. J Trauma 62:1452–1456

20. Astrand R et al (2011) Reference values for venous and capillary S100B in children. Clin Chim Acta 412:2190–2193
21. Ingebrigtsen T, Romner B, Kock-Jensen C (2000) Scandinavian guidelines for initial management of minimal, mild, and moderate head injuries. The Scandinavian Neurotrauma Committee. J Trauma 48:760–766
22. af Geijerstam JL, Britton M (2003) Mild head injury - mortality and complication rate: meta-analysis of findings in a systematic literature review. Acta Neurochir 145:843–850 discussion 850
23. Unden J et al (2005) Serum S100B levels in patients with epidural haematomas. Br J Neurosurg 19:43–45
24. Unden J, Romner B (2009) A new objective method for CT triage after minor head injury—serum S100B. Scand J Clin Lab Invest 69:13–17
25. Biberthaler P et al (2001) Elevated serum levels of S-100B reflect the extent of brain injury in alcohol intoxicated patients after mild head trauma. Shock 16:97–101
26. Biberthaler P et al (2001) Evaluation of S-100b as a specific marker for neuronal damage due to minor head trauma. World J Surg 25:93–97
27. Ingebrigtsen T et al (2000) The clinical value of serum S-100 protein measurements in minor head injury: a Scandinavian multicentre study. Brain Inj 14:1047–1055
28. Mussack T et al (2002) Immediate S-100B and neuron-specific enolase plasma measurements for rapid evaluation of primary brain damage in alcohol-intoxicated, minor head-injured patients. Shock 18:395–400
29. Poli-de-Figueiredo LF et al (2006) Measurement of S-100B for risk classification of victims sustaining minor head injury—first pilot study in Brazil. Clinics (Sao Paulo) 61:41–46
30. Unden J, Romner B (2010) Can low serum levels of S100B predict normal CT findings after minor head injury in adults?: an evidence-based review and meta-analysis. J Head Trauma Rehabil 25:228–240
31. Kleine TO, Benes L, Zofel P (2003) Studies of the brain specificity of S100B and neuron-specific enolase (NSE) in blood serum of acute care patients. Brain Res Bull 61:265–279
32. Unden J et al (2005) Raised serum S100B levels after acute bone fractures without cerebral injury. J Trauma 58:59–61
33. Anderson RE et al (2001) High serum S100B levels for trauma patients without head injuries. Neurosurgery 48:1255–1258 discussion 1258–1260
34. Biberthaler P et al (2000) Influence of alcohol exposure on S-100b serum levels. Acta Neurochir Suppl 76:177–179
35. Biberthaler P et al (2006) Serum S-100B concentration provides additional information fot the indication of computed tomography in patients after minor head injury: a prospective multicenter study. Shock 25:446–453
36. Zongo D et al (2011) S100-B protein as a screening tool for the early assessment of minor head injury. Ann Emerg Med 59:209–218
37. Unden J et al (2013) Scandinavian guidelines for initial management of minimal, mild and moderate head injuries in adults: an evidence and consensus-based update. BMC Med 14:33. https://doi.org/10.1186/1741-7015-11-50
38. Calcagnile O, Anell A, Unden J (2016) The addition of S100B to guidelines for management of mild head injury is potentially cost saving. BMC Neurol 16:200. https://doi.org/10.1186/s12883-016-0723-z
39. Unden L et al (2015) Validation of the Scandinavian guidelines for initial management of minimal, mild and moderate traumatic brain injury in adults. BMC Med 13:292. https://doi.org/10.1186/s12916-015-0533-y
40. Raabe A, Grolms C, Seifert V (1999) Serum markers of brain damage and outcome prediction in patients after severe head injury. Br J Neurosurg 13:56–59
41. Nylen K et al (2006) Increased serum-GFAP in patients with severe traumatic brain injury is related to outcome. J Neurol Sci 240:85–91
42. da Rocha AB et al (2006) Role of serum S100B as a predictive marker of fatal outcome following isolated severe head injury or multitrauma in males. Clin Chem Lab Med 44:1234–1242
43. Wiesmann M et al (2010) Outcome prediction in traumatic brain injury: comparison of neurological status, CT findings, and blood levels of S100B and GFAP. Acta Neurol Scand 121:178–185
44. Vos PE et al (2010) GFAP and S100B are biomarkers of traumatic brain injury: an observational cohort study. Neurology 75:1786–1793
45. Raabe A et al (1999) Serum S-100B protein in severe head injury. Neurosurgery 45:477–483
46. Woertgen C et al (1999) Comparison of clinical, radiologic, and serum marker as prognostic factors after severe head injury. J Trauma 47:1126–1130
47. Nylen K et al (2008) Serum levels of S100B, S100A1B and S100BB are all related to outcome after severe traumatic brain injury. Acta Neurochir 150:221–227 discussion 227
48. Bohmer AE et al (2011) Neuron-specific enolase, S100B, and glial fibrillary acidic protein levels as outcome predictors in patients with

severe traumatic brain injury. Neurosurgery 68:1624–1631
49. Dimopoulou I et al (2003) Protein S-100b serum levels in trauma-induced brain death. Neurology 60:947–951
50. Rodriguez-Rodriguez A et al (2016) S100B and neuron-specific enolase as mortality predictors in patients with severe traumatic brain injury. Neurol Res 38:130–137
51. Raabe A et al (2004) S-100B protein as a serum marker of secondary neurological complications in neurocritical care patients. Neurol Res 26:440–445
52. Unden J et al (2007) Clinical significance of serum S100B levels in neurointensive care. Neurocrit Care 6:94–99
53. Thelin EP, Nelson DW, Bellander BM (2014) Secondary peaks of S100B in serum relate to subsequent radiological pathology in traumatic brain injury. Neurocrit Care 20:217–229
54. Bellander BM et al (2011) Secondary insults following traumatic brain injury enhance complement activation in the human brain and release of the tissue damage marker S100B. Acta Neurochir 153:90–100
55. Olivecrona M et al (2009) S-100B and neuron specific enolase are poor outcome predictors in severe traumatic brain injury treated by an intracranial pressure targeted therapy. J Neurol Neurosurg Psychiatry 80:1241–1247
56. Pelinka LE et al (2003) Serum S 100 B: a marker of brain damage in traumatic brain injury with and without multiple trauma. Shock 19:195–200
57. Bechtel K et al (2009) Relationship of serum S100B levels and intracranial injury in children with closed head trauma. Pediatrics 124: e697–e704
58. Mondello S et al (2016) Serum concentrations of ubiquitin C-terminal hydrolase-L1 and glial fibrillary acidic protein after pediatric traumatic brain injury. Sci Rep 6:28203. https://doi.org/10.1038/srep28203
59. Papa L et al (2015) In children and youth with mild and moderate traumatic brain injury GFAP out-performs S100beta in detecting traumatic intracranial lesions on CT. J Neurotrauma 33:58–64
60. Berger RP et al (2005) Serum neuron-specific enolase, S100B, and myelin basic protein concentrations after inflicted and noninflicted traumatic brain injury in children. J Neurosurg 103:61–68
61. Castellani C et al (2009) Neuroprotein s-100B—a useful parameter in paediatric patients with mild traumatic brain injury? Acta Paediatr 98:1607–1612
62. Castellani C et al (2008) Reference ranges for neuroprotein S-100B: from infants to adolescents. Clin Chem Lab Med 46:1296–1299
63. Simon-Pimmel J et al (2017) Reference ranges for serum S100B neuroprotein specific to infants under four months of age. Clin Biochem 50:1056–1060
64. Oris C et al (2018) The biomarker S100B and mild traumatic brain injury: a meta-analysis. Pediatrics 141. https://doi.org/10.1542/peds.2018-0037
65. Petzold A et al (2003) Cerebrospinal fluid S100B correlates with brain atrophy in Alzheimer's disease. Neurosci Lett 336:167–170
66. Rothermundt M et al (2003) S100B in brain damage and neurodegeneration. Microsc Res Tech 60:614–632
67. Schroeter ML et al (2009) Neuron-specific enolase is unaltered whereas S100B is elevated in serum of patients with schizophrenia—original research and meta-analysis. Psychiatry Res 167:66–72
68. Beer C et al (2010) Systemic markers of inflammation are independently associated with S100B concentration: results of an observational study in subjects with acute ischaemic stroke. J Neuroinflammation 7:71. https://doi.org/10.1186/1742-2094-7-71
69. Peskind ER et al (2001) Cerebrospinal fluid S100B is elevated in the earlier stages of Alzheimer's disease. Neurochem Int 39:409–413
70. Chaves ML et al (2010) Serum levels of S100B and NSE proteins in Alzheimer's disease patients. J Neuroinflammation 7:6. https://doi.org/10.1186/1742-2094-7-6
71. Gruden MA et al (2007) Differential neuroimmune markers to the onset of Alzheimer's disease neurodegeneration and dementia: autoantibodies to Abeta((25-35)) oligomers, S100b and neurotransmitters. J Neuroimmunol 186:181–192
72. Mitosek-Szewczyk K et al (2011) Some markers of neuronal damage in cerebrospinal fluid of multiple sclerosis patients in relapse. Folia Neuropathol 49:191–196
73. Rothermundt M, Ahn JN, Jorgens S (2009) S100B in schizophrenia: an update. Gen Physiol Biophys 28 Spec No Focus:F76–F81
74. Yelmo-Cruz S, Morera-Fumero AL, Abreu-Gonzalez P (2013) S100B and schizophrenia. Psychiatry Clin Neurosci 67:67–75
75. Hauschild A et al (1999) S100B protein detection in serum is a significant prognostic factor in metastatic melanoma. Oncology 56:338–344

76. Weide B et al (2012) Serum markers lactate dehydrogenase and S100B predict independently disease outcome in melanoma patients with distant metastasis. Br J Cancer 107:422–428
77. Hauschild A et al (1999) Prognostic significance of serum S100B detection compared with routine blood parameters in advanced metastatic melanoma patients. Melanoma Res 9:155–161
78. Hartman KG et al (2013) The evolution of S100B inhibitors for the treatment of malignant melanoma. Future Med Chem 5:97–109
79. Molina R et al (2002) S-100 protein serum levels in patients with benign and malignant diseases: false-positive results related to liver and renal function. Tumour Biol 23:39–44
80. Muller K et al (2006) Analysis of protein S-100B in serum: a methodological study. Clin Chem Lab Med 44:1111–1114
81. Alber B et al (2005) Multicenter evaluation of the analytical and clinical performance of the Elecsys S100 immunoassay in patients with malignant melanoma. Clin Chem Lab Med 43:557–563

Chapter 43

Serum S100B Levels in Melanoma

Anna Lisa Frauchiger, Reinhard Dummer, and Joanna Mangana

Abstract

Malignant melanoma is a cancer with increasing incidence worldwide with relevant socioeconomic impact. Despite progress in prevention and early detection, it is one of the most lethal forms of skin cancer. Therefore it is urgent need to identify suitable biomarkers in order to improve early diagnosis, precise staging, and prognosis, as well as for therapy selection and monitoring. In this book chapter, we are focusing on S100B and discuss its clinical relevance in melanoma.

Key words S100B, Calcium, EF hand, Melanoma, Staging, Follow-up, Therapy monitoring

1 Introduction

Malignant melanoma is a cancer arising from melanocytes. Its incidence is increasing worldwide with relevant socioeconomic impact [1, 2]. Fair skin population and patients with multiple dysplastic nevi or familiar dysplastic nevus syndrome are at greatest risk. Furthermore Australia, Northern Europe, and Northern America show the highest incidence of melanoma worldwide [3, 4]. Mortality rates are variable and influenced by geography, ethnicity, age, and sex [4–6]. Lately, mortality rates stabilized, especially in Australia, which is due to better prevention and effective drug therapies [5, 7].

Despite progress in prevention and early detection of melanoma, it is one of the most lethal forms of skin cancer and causes relevant health and economic burden on society [8, 9]. There are intercontinental variations, but in average, 5-year survival in primary cutaneous melanoma <1.5 mm in depth is about 90%. Five-year survival in mucosal and intraocular melanoma is about 70%. However, 5-year survival rates are 60–65% for locoregional metastases, decreasing to below 10% if widespread. Therefore, suitable biomarkers in order to improve early diagnosis, precise staging, and prognosis, as well as for therapy selection and monitoring, are

Claus W. Heizmann (ed.), *Calcium-Binding Proteins of the EF-Hand Superfamily: From Basics to Medical Applications*, Methods in Molecular Biology, vol. 1929, https://doi.org/10.1007/978-1-4939-9030-6_43,

urgently needed. Up to date, there are different serologic and histologic markers known like lactate dehydrogenase (LDH), S100 family of calcium-binding proteins, tyrosinase, cyclooxygenase-2, matrix metalloproteinases, and different stem and/or progenitor cell markers. Of clinical relevance are so far only LDH, tyrosinase, and the calcium-binding protein S100B.

Herein, we will focus on the S100B, a member of the EF-hand superfamily of Ca-binding proteins, and discuss its clinical impact in melanoma diagnosis and treatment.

2 Intra- and Extracellular Functions of S100B

In 1980, Gaynor et al. first reported that melanoma cells are able to secrete a soluble form of S100 protein [10].

S100 proteins belong to a multigene family of low-molecular-weight (9–13 kDa) calcium-modulated proteins being extracted from bovine brain tissue. The name S100 was given due to its solubility in 100% saturated ammonium sulfate solution at neutral pH [11]. The S100 gene family is the largest subfamily of calcium-binding proteins of EF-hand type. They are only expressed in vertebrates [12]. S100 proteins form either homodimeric or heterodimeric complexes with one another. There are 24 human S100 genes known; 19 of them (S100 proteins, group A) are located within chromosome 1q21. Other gene locations include S100A11P, located on chromosome 7q22-q3; S100B, which maps to chromosome 21q22; S100G, located on chromosome Xp22; S100P on chromosome 4p16; and S100Z on chromosome 5q13 [13, 14].

S100 proteins have either intracellular, extracellular, or both functions. Intracellular functions include the regulation of proliferation, differentiation, apoptosis, Ca^{2+} homeostasis, energy metabolism, inflammation, and migration/invasion through interactions with a variety of target proteins including enzymes, cytoskeletal subunits, receptors, transcription factors, and nucleic acids. Extracellular functions are of autocrine and paracrine mechanism, interacting with surface receptors, G-protein-coupled receptors, scavenger receptor, heparin sulfate proteoglycans, and N-glycans. Extracellular S100 proteins have regulatory activities on monocytes/macrophages/microglia, neutrophils, lymphocytes, mast cells, chondrocytes, muscle cells, neurons, astrocytes, Schwann cells, epithelial cells, myoblasts, and cardiomyocytes. They have a part in innate and adaptive immune responses, cell migration and chemotaxis, tissue development and repair, and leukocyte and tumor cell invasion [15].

S100B is expressed in astrocytes, certain neuronal populations, Schwann cells, melanocytes, chondrocytes, adipocytes, skeletal myofibers and associated satellite cells, certain dendritic cell and lymphocyte populations, and a few other cell types [16].

It is a stimulator of cell proliferation and migration and an inhibitor of apoptosis and differentiation [17–23].

In melanoma, it was shown that S100B binds directly to the p53 tumor suppressor protein [24–27], reduces p53 protein levels [27, 28], and inhibits wild-type p53 functions [27–29]. This leads to suppression of p53 tumor suppressor mechanism, resulting in uncontrolled tumor growth [30–32]. However, it still remains unclear, how S100B is shed into the blood by malignant melanocytes.

3 Serum S100B Immunoassays

Deregulation of S100 gene expression is observed in several human diseases suggesting either their valuable diagnostic capacity in wound healing, cardiomyopathy inflammation and cardiac infections, or prognostic significance in human tumors [13]. Elevated levels of S100B protein have been detected in malignant melanoma and less extent in thyroid and renal cell carcinoma [33].

There are several commercially available immunoassays for the detection of the S100B concentrations in serum (Table 1). S100B denotes a measurement of all S100 proteins, including at least one S100B monomer, i.e., the sum of the two dimers S100A1B and S100BB. Melanoma cells express the heterodimer S100A1B. Therefore central nervous damage (S100 BB) could give rise to increased serum S100 levels in melanoma patients [34].

The most widely used immunoassay in malignant melanoma measuring S100B is the Sangtec® 100 (Sangtec® 100 IRMA, Sangtec® 100 LIAISON, Sangtec® 100 ELISA) [35]. This is a two-step

Table 1
Commercially available S100B immunoassays

Assay name	Proteins detected	Manufacturer
CanAg S100A1B EIA	S100A1B	CanAg Diagnostics AB, Gothenburg, Sweden
CanAg S100BB EIA	S100BB	CanAg Diagnostics AB
Elecsys® S100	S100A1B, S100BB	Roche Diagnostics GmbH, Mannheim, Germany
Nexus Dx™ S-100 Elisa	S100BB, S100A1B	Syn X Pharma Inc., Toronto, Canada
Sangtec® 100 IRMA	S100A1B, S100BB	DiaSorin S.r.l., Saluggia, Italy
Sangtec® 100 ELISA	S100A1B, S100BB	DiaSorin S.r.l.
LIA-mat® Sangtec® 100	S100A1B, S100BB	DiaSorin S.r.l.
LIAISON® Sangtec® 100	S100A1B, S100BB	DiaSorin S.r.l.

incubation immunoassay based on three molecular antibodies (mABs): two catchers and one tracer [35]. It has a high specificity because all the mABs used bind exclusively to S100B as shown by real-time bio-specific interaction analyses being a reliable measurement of S100B in humans [36]. The assay was tested for possible cross-reactivity between S100B and members of S100A family with none being detected [36].

One study compared four S100B assays, two automated assays (LIAISON Sangtec 100 and Elecsys S100) and two manual ones (Sangtec 100 enzyme-linked immunosorbent assay and CanAg S100 EIA) with respect to clinical data, reference values, and correlation [37]. The inter-assay and intra-assay coefficients of variation were best for the automated assays. The overall sensitivity was highest with LIAISON Sangtec 100 being 47% for all stages of melanoma and increasing with disease stage (77% for stage IV).

4 S100B in Different Melanoma Stages

4.1 S100 in Early Melanoma

Melanomas are being classified according to TNM classification system that was recently revamped [38]. Stage 0 tumors are considered in situ, meaning they have not penetrated below the epidermis. Stage I tumors represent those that have invaded into the dermis, have low metastatic risk, and are divided in two substages depending on tumor thickness and presence or not of ulceration (stage IA and IB). Stage II melanomas are thicker than stage I, localized further into the dermis, and have a slightly higher chance of spreading. Stage III and IV melanomas represent advanced tumors that metastasized either locoregionally forming in-transit or lymph node metastases or to the distant skin, lymph nodes, or organs, respectively.

Tumor thickness in Breslow and ulceration are considered nowadays the most reliable factors for disease recurrence [38, 39]. Thick ulcerated melanomas show unfavorable prognosis with a 10-year survival being less than 80%. However, early detection of melanoma recurrence is crucial for optimal management of metastatic disease and prognosis improvement.

Since elevated serum S100B levels was detected in various patients with melanoma, it was suggested that S100B may be of value assessing the stage of disease [40].

The prognostic significance of S100B in the follow-up of early melanoma stages has been examined in a prospective study of 73 melanoma patients [41]. Only 1 out of 25 melanoma patients with primary melanoma (stage I/II) was measurable (0.30 μg/L). Additionally, S100 levels of the 130 control group comprised of healthy donors were not detectable in serum. Based on these results, the authors commented that S100B is not a reliable marker for early diagnosis or screening, though being helpful for treatment monitoring. The clinical significance of S100 was also investigated in 412 melanoma patients at different stages of disease [42]. In total, 1339 blood samples were available; 286 patients were in stage I/II. Disease progression in only 6/281 (2.1%)—initially S100 negative—stage I/II patients indicated a stage-dependent correlation and paralleled the previously mentioned study. Similarly in the study by Guo et al., S100B was rarely detected in stage I/II patients (1/80, 1.25%) [43].

Another prospective study of 411 melanoma patients during September 1997 and December 1998 evaluated the significance of S100 in screening for metastasis during regular follow-up [44]. No significant difference in detection of early metastasis in patients with low risk for tumor progression, i.e., primary melanoma and low tumor thickness (<1.5 mm), was determined. Nikolin et al. showed similar results with statistical differences in OS only in the high-risk IIC group meaning in patients with tumor thickness over 4 mm [45]. However, due to small number of tested patients, no safe conclusive conclusion can be made.

4.2 S100B in Stage III and IV Melanoma

There are different studies supporting the prognostic value of S100B in stage III and IV melanoma. The mechanism, how S100B is released into blood circulation, is not completely clear yet, but to date's evidence suggests that S100B in melanoma patients' blood is not a result of overexpression or increased release but a result of a loss of cell integrity and proteolytic degradation associated with apoptosis and cell death [46].

In a multivariate analysis by Balch et al., 2313 patients with AJCC stage III disease were evaluated prospectively. Preoperatively elevated S100B levels in patients with palpable nodes confirmed by PET or CT scan were associated with impaired disease-free survival (DFS). Furthermore, S100B seems to be a strong prognostic marker for lymphogenic metastasized melanoma [47, 48]. Damude et al. prospectively analyzed 107 patients with positive sentinel lymph node (SLNB) undergoing complete lymph node dissection (CLND). Preoperatively elevated S100B was the strongest predictor for lymph node metastases [49]. Another study by Kruijff et al. identified S100B as a marker for DFS in stage III patients with palpable lymph node metastases [47]. Wevers et al. confirmed these results in their cohort, where S100B levels in stage IIIB–C were also strongly associated with DFS [50]. In a Swiss study of 73 patients of all stages, S100B levels were elevated in 21.4% of stage III patients [41].

Several studies suggested S100B to be also a prognostic factor in the setting of stage IV disease [50–53]. In one study S100B levels higher than 1.6 μg/L were significantly associated with higher distant metastasis rates. In the same study, serum concentration of S100B correlated with Breslow thickness and tumor burden

[54]. S100B is highly specific; however, sensitivity is reported to be only around 60% [51, 55].

After the identification of mitogen-activated kinase pathway (MAPK) as driver in melanogenesis and BRAF as its most potent activator, a new era in melanoma therapy has been signaled [56, 57]. According to our own retrospective data of 162 advanced melanoma patients, BRAF wild-type patients with normal S100B levels experienced significantly higher overall survival (OS) compared to those with elevated S100B levels. In BRAF-mutant patients, no difference in OS was found, what we traced back to the effective BRAF inhibitor treatment in this group [58].

5 S100B for Treatment Response

Use of tumor markers for treatment monitoring is of greatest need, since changes in blood levels should allow conclusions on changes in tumor sizes reflecting therapy response. Serological measurement of S100B in melanoma has provided evidence on its efficacy in treatment monitoring [43, 59]. If there is a marked rise in serum S100B levels during treatment, progressive disease can be early detected and the vice versa; decreased S100B levels correspond to tumor regression or remission. The predictive value of S100B in treatment outcome of 64 stage IV melanoma patients during 77 treatment courses including chemotherapy and interferon alpha has been evaluated [59]. Seventy-eight percent of nonresponders demonstrated elevated S100B levels 4 weeks after treatment initiation, and 84% of those showed elevated levels by the end of treatment cycle.

Felix et al. reported S100B to be a useful marker to monitor response to chemo- and immunotherapy in stage IV melanoma [60]. In another study, S100B was investigated whether it is able to predict and monitor therapy success and tumor progression in vemurafenib-treated patients. During initial treatment, response of S100B levels was found; however, repeated measurements of S100B levels were not sufficient to detect relapse [61]. In a prospective study, S100B level during response to dabrafenib or vemurafenib treatment was of prognostic value. In this study, patients with high S100B levels showed shorter progression-free survival [62].

6 Conclusion

According to existing literature data, S100B has a high specificity between 93 and 100%, as in the majority of the studies control groups such as healthy individuals or patients with benign skin tumors, cutaneous autoimmune disease, dysplastic nevi, or inflammatory cutaneous diseases were consistently S100B negative. Howbeit, Bosserhoff et al. found elevated S100B levels in 4% of healthy patients, 16% of glioma, and 20% of sepsis patients [63]. Its sensitivity ranges between 30% and 62.5% in revealing a stage-dependent serum elevation. It is a useful tumor marker for monitoring of stage III and IV melanoma patients but less sensitive for screening or monitoring of stage I and II patients. Therefore, it has been implemented in Swiss and German guidelines for melanoma follow-up [64, 65]. In the metastatic session, a rise in S100B serum can indicate disease progression [41, 42]. On the other hand, few reports exist on S100B in early detection of melanoma recurrence [44, 66, 67]. In some cases at the time that S100B rises, the presence of advanced disease is also obvious; however in a minority elevated S100B can be the first indicator of recurrence [68].

Despite the fact that prognostic impact of S100B is debated, regular monitoring of this laboratory marker in the follow-up of stage III and stage IV patients and in the course of treatment effectiveness is highly recommended.

References

1. MacKie RM, Hauschild A, Eggermont AMM (2009) Epidemiology of invasive cutaneous melanoma. Ann Oncol 20:1–7. https://doi.org/10.1093/annonc/mdp252
2. Karimkhani C, Green AC, Nijsten T, Weinstock MA, Dellavalle RP, Naghavi M, Fitzmaurice C (2017) The global burden of melanoma: results from the Global Burden of Disease Study 2015. Br J Dermatol 177:134–140. https://doi.org/10.1111/bjd.15510
3. Kosary CL, Altekruse SF, Ruhl J, Lee R, Dickie L (2014) Clinical and prognostic factors for melanoma of the skin using SEER registries: collaborative stage data collection system, version 1 and version 2. Cancer 120(Suppl. 23):3807–3814. https://doi.org/10.1002/cncr.29050
4. Guy GP Jr et al (2015) Vital signs: melanoma incidence and mortality trends and projections – United States, 1982–2030 MMWR. Morb Mortal Wkly Rep 64:591–596
5. Whiteman DC, Green AC, Olsen CM (2016) The growing burden of invasive melanoma: projections of incidence rates and numbers of new cases in six susceptible populations through 2031. J Invest Dermatol 136:1161–1171. https://doi.org/10.1016/j.jid.2016.01.035
6. Sneyd MJ, Cox B (2013) A comparison of trends in melanoma mortality in New Zealand and Australia: the two countries with the highest melanoma incidence and mortality in the world. BMC Cancer 13:372. https://doi.org/10.1186/1471-2407-13-372
7. Azoury SC, Lange JR (2014) Epidemiology, risk factors, prevention, and early detection of melanoma. Surg Clin North Am 94:945–962, vii. https://doi.org/10.1016/j.suc.2014.07.013
8. Linos E, Swetter SM, Cockburn MG, Colditz GA, Clarke CA (2009) Increasing burden of melanoma in the United States. J Invest Dermatol 129:1666–1674. https://doi.org/10.1038/jid.2008.423
9. Nikolaou V, Stratigos AJ (2014) Emerging trends in the epidemiology of melanoma. Br J Dermatol 170:11–19. https://doi.org/10.1111/bjd.12492
10. Gaynor R, Irie R, Morton D, Herschman HR (1980) S100 protein is present in cultured human malignant melanomas. Nature 286:400–401
11. Moore BW (1965) A soluble protein characteristic of the nervous system. Biochem Biophys Res Commun 19:739–744
12. Donato R (2001) S100: a multigenic family of calcium-modulated proteins of the EF-hand type with intracellular and extracellular functional roles. Int J Biochem Cell Biol 33:637–668
13. Heizmann CW, Fritz G, Schafer BW (2002) S100 proteins: structure, functions and pathology. Front Biosci 7:d1356–d1368

14. Marenholz I, Heizmann CW, Fritz G (2004) S100 proteins in mouse and man: from evolution to function and pathology (including an update of the nomenclature). Biochem Biophys Res Commun 322:1111–1122. https://doi.org/10.1016/j.bbrc.2004.07.096
15. Donato R, Cannon BR, Sorci G, Riuzzi F, Hsu K, Weber DJ, Geczy CL (2013) Functions of S100 proteins. Curr Mol Med 13:24–57. https://doi.org/10.2174/156652413804486214
16. Donato R et al (2009) S100B's double life: intracellular regulator and extracellular signal. Biochim Biophys Acta 1793:1008–1022. https://doi.org/10.1016/j.bbamcr.2008.11.009
17. Hachem S, Laurenson AS, Hugnot JP, Legraverend C (2007) Expression of S100B during embryonic development of the mouse cerebellum. BMC Dev Biol 7:17. https://doi.org/10.1186/1471-213X-7-17
18. Saito T, Ikeda T, Nakamura K, Chung UI, Kawaguchi H (2007) S100A1 and S100B, transcriptional targets of SOX trio, inhibit terminal differentiation of chondrocytes. EMBO Rep 8:504–509. https://doi.org/10.1038/sj.embor.7400934
19. Mocellin S, Zavagno G, Nitti D (2008) The prognostic value of serum S100B in patients with cutaneous melanoma: a meta-analysis. Int J Cancer 123:2370–2376. https://doi.org/10.1002/ijc.23794
20. Brozzi F, Arcuri C, Giambanco I, Donato R (2009) S100B protein regulates astrocyte shape and migration via interaction with Src kinase: implications for astrocyte development, activation, and tumor growth. J Biol Chem 284:8797–8811. https://doi.org/10.1074/jbc.M805897200
21. Lin J, Yang Q, Wilder PT, Carrier F, Weber DJ (2010) The calcium-binding protein S100B down-regulates p53 and apoptosis in malignant melanoma. J Biol Chem 285:27487–27498. https://doi.org/10.1074/jbc.M110.155382
22. Tubaro C, Arcuri C, Giambanco I, Donato R (2010) S100B protein in myoblasts modulates myogenic differentiation via NF-kappaB-dependent inhibition of MyoD expression. J Cell Physiol 223:270–282. https://doi.org/10.1002/jcp.22035
23. Beccafico S, Riuzzi F, Puglielli C, Mancinelli R, Fulle S, Sorci G, Donato R (2011) Human muscle satellite cells show age-related differential expression of S100B protein and RAGE. Age (Dordr) 33:523–541. https://doi.org/10.1007/s11357-010-9197-x
24. Baudier J, Delphin C, Grunwald D, Khochbin S, Lawrence JJ (1992) Characterization of the tumor suppressor protein p53 as a protein kinase C substrate and a S100b-binding protein. Proc Natl Acad Sci U S A 89:11627–11631
25. Rustandi RR, Drohat AC, Baldisseri DM, Wilder PT, Weber DJ (1998) The Ca(2+)-dependent interaction of S100B(beta beta) with a peptide derived from p53. Biochemistry 37:1951–1960. https://doi.org/10.1021/bi972701n
26. Delphin C et al (1999) Calcium-dependent interaction of S100B with the C-terminal domain of the tumor suppressor p53. J Biol Chem 274:10539–10544
27. Lin J, Yang Q, Yan Z, Markowitz J, Wilder PT, Carrier F, Weber DJ (2004) Inhibiting S100B restores p53 levels in primary malignant melanoma cancer cells. J Biol Chem 279:34071–34077. https://doi.org/10.1074/jbc.M405419200
28. Lin J, Blake M, Tang C, Zimmer D, Rustandi RR, Weber DJ, Carrier F (2001) Inhibition of p53 transcriptional activity by the S100B calcium-binding protein. J Biol Chem 276:35037–35041. https://doi.org/10.1074/jbc.M104379200
29. Wilder PT et al (2006) Recognition of the tumor suppressor protein p53 and other protein targets by the calcium-binding protein S100B. Biochim Biophys Acta 1763:1284–1297. https://doi.org/10.1016/j.bbamcr.2006.08.024
30. Donato R (1991) Perspectives in S-100 protein biology. Cell Calcium 12:713–726
31. Wilder PT, Rustandi RR, Drohat AC, Weber DJ (1998) S100B(betabeta) inhibits the protein kinase C-dependent phosphorylation of a peptide derived from p53 in a Ca2+-dependent manner. Protein Sci 7:794–798. https://doi.org/10.1002/pro.5560070330
32. Zimmer DB, Cornwall EH, Landar A, Song W (1995) The S100 protein family: history, function, and expression. Brain Res Bull 37:417–429
33. Molina R, Navarro J, Filella X, Castel T, Ballesta AM (2002) S-100 protein serum levels in patients with benign and malignant diseases: false-positive results related to liver and renal function. Tumour Biol 23:39–44. https://doi.org/10.1159/000048687
34. Brochez L, Naeyaert JM (2000) Serological markers for melanoma. Br J Dermatol 143:256–268
35. Stigbrand T, Nyberg L, Ullen A, Haglid K, Sandstrom E, Brundell J (2000) A new specific

method for measuring S-100B in serum. Int J Biol Markers 15:33–40

36. Heizmann CW (2004) S100B protein in clinical diagnostics: assay specificity. Clin Chem 50:249–251. https://doi.org/10.1373/clinchem.2003.027367
37. Smit LH, Korse CM, Bonfrer JM (2005) Comparison of four different assays for determination of serum S-100B. Int J Biol Markers 20:34–42
38. Gershenwald JE et al (2017) Melanoma staging: evidence-based changes in the American Joint Committee on Cancer eighth edition cancer staging manual. CA Cancer J Clin 67:472–492. https://doi.org/10.3322/caac.21409
39. Balch CM et al (2009) Final version of 2009 AJCC melanoma staging and classification. J Clin Oncol 27:6199–6206. https://doi.org/10.1200/JCO.2009.23.4799
40. Fagnart OC, Sindic CJ, Laterre C (1988) Particle counting immunoassay of S100 protein in serum. Possible relevance in tumors and ischemic disorders of the central nervous system. Clin Chem 34:1387–1391
41. Henze G, Dummer R, Joller-Jemelka HI, Boni R, Burg G (1997) Serum S100 – a marker for disease monitoring in metastatic melanoma. Dermatology 194:208–212. https://doi.org/10.1159/000246103
42. Hauschild A, Engel G, Brenner W, Glaser R, Monig H, Henze E, Christophers E (1999) S100B protein detection in serum is a significant prognostic factor in metastatic melanoma. Oncology 56:338–344. https://doi.org/10.1159/000011989
43. Guo HB, Stoffel-Wagner B, Bierwirth T, Mezger J, Klingmuller D (1995) Clinical significance of serum S100 in metastatic malignant melanoma. Eur J Cancer 31A:1898–1902
44. Schlagenhauff B et al (2000) Significance of serum protein S100 levels in screening for melanoma metastasis: does protein S100 enable early detection of melanoma recurrence? Melanoma Res 10:451–459
45. Nikolin B, Djan I, Trifunovic J, Dugandzija T, Novkovic D, Djan V, Vucinic N (2016) MIA, S100 and LDH as important predictors of overall survival of patients with stage IIb and IIc melanoma. J BUON 21:691–697
46. Ghanem G et al (2001) On the release and half-life of S100B protein in the peripheral blood of melanoma patients. Int J Cancer 94:586–590
47. Kruijff S, Bastiaannet E, Kobold AC, van Ginkel RJ, Suurmeijer AJ, Hoekstra HJ (2009) S-100B concentrations predict disease-free survival in stage III melanoma patients. Ann Surg Oncol 16:3455–3462. https://doi.org/10.1245/s10434-009-0629-8
48. Balch CM et al (2010) Multivariate analysis of prognostic factors among 2,313 patients with stage III melanoma: comparison of nodal micrometastases versus macrometastases. J Clin Oncol 28:2452–2459. https://doi.org/10.1200/JCO.2009.27.1627
49. Damude S, Hoekstra HJ, Bastiaannet E, Muller Kobold AC, Kruijff S, Wevers KP (2016) The predictive power of serum S-100B for non-sentinel node positivity in melanoma patients. Eur J Surg Oncol 42:545–551. https://doi.org/10.1016/j.ejso.2015.12.010
50. Wevers KP, Kruijff S, Speijers MJ, Bastiaannet E, Muller Kobold AC, Hoekstra HJ (2013) S-100B: a stronger prognostic biomarker than LDH in stage IIIB-C melanoma. Ann Surg Oncol 20:2772–2779. https://doi.org/10.1245/s10434-013-2949-y
51. Banfalvi T, Gilde K, Gergye M, Boldizsar M, Kremmer T, Otto S (2003) Use of serum 5-S-CD and S-100B protein levels to monitor the clinical course of malignant melanoma. Eur J Cancer 39:164–169
52. Weide B et al (2012) Serum markers lactate dehydrogenase and S100B predict independently disease outcome in melanoma patients with distant metastasis. Br J Cancer 107:422–428. https://doi.org/10.1038/bjc.2012.306
53. Smit LH et al (2005) Normal values of serum S-100B predict prolonged survival for stage IV melanoma patients. Eur J Cancer 41:386–392. https://doi.org/10.1016/j.ejca.2004.10.019
54. Alegre E, Sammamed M, Fernandez-Landazuri S, Zubiri L, Gonzalez A (2015) Circulating biomarkers in malignant melanoma. Adv Clin Chem 69:47–89. https://doi.org/10.1016/bs.acc.2014.12.002
55. Wollina U, Karte K, Hipler UC, Knoll B, Kirsch K, Herold C (2000) Serum protein s100beta in patients with malignant melanoma detected by an immunoluminometric assay. J Cancer Res Clin Oncol 126:107–110
56. Montagut C, Settleman J (2009) Targeting the RAF-MEK-ERK pathway in cancer therapy. Cancer Lett 283:125–134. https://doi.org/10.1016/j.canlet.2009.01.022
57. Chapman PB et al (2011) Improved survival with vemurafenib in melanoma with BRAF V600E mutation. N Engl J Med 364:2507–2516. https://doi.org/10.1056/NEJMoa1103782
58. Frauchiger AL et al (2016) Prognostic relevance of lactate dehydrogenase and serum S100 levels in stage IV melanoma with known

BRAF mutation status. Br J Dermatol 174:823–830. https://doi.org/10.1111/bjd.14347

59. Hauschild A, Engel G, Brenner W, Glaser R, Monig H, Henze E, Christophers E (1999) Predictive value of serum S100B for monitoring patients with metastatic melanoma during chemotherapy and/or immunotherapy. Br J Dermatol 140:1065–1071
60. Felix J et al (2016) Relevance of serum biomarkers associated with melanoma during follow-up of anti-CTLA-4 immunotherapy. Int Immunopharmacol 40:466–473. https://doi.org/10.1016/j.intimp.2016.09.030
61. Abusaif S et al (2013) S100B and lactate dehydrogenase as response and progression markers during treatment with vemurafenib in patients with advanced melanoma. Melanoma Res 23:396–401. https://doi.org/10.1097/CMR.0b013e3283650741
62. Sanmamed MF et al (2014) Relevance of MIA and S100 serum tumor markers to monitor BRAF inhibitor therapy in metastatic melanoma patients. Clin Chim Acta 429:168–174. https://doi.org/10.1016/j.cca.2013.11.034
63. Bosserhoff AK et al (1997) Melanoma-inhibiting activity, a novel serum marker for progression of malignant melanoma. Cancer Res 57:3149–3153
64. Dummer R et al (2016) The updated Swiss guidelines 2016 for the treatment and follow-up of cutaneous melanoma. Swiss Med Wkly 146:w14279. https://doi.org/10.4414/smw.2016.14279
65. Garbe C et al (2016) Diagnosis and treatment of melanoma. European consensus-based interdisciplinary guideline – update 2016. Eur J Cancer 63:201–217. https://doi.org/10.1016/j.ejca.2016.05.005
66. Garbe C, Leiter U, Ellwanger U, Blaheta HJ, Meier F, Rassner G, Schittek B (2003) Diagnostic value and prognostic significance of protein S-100beta, melanoma-inhibitory activity, and tyrosinase/MART-1 reverse transcription-polymerase chain reaction in the follow-up of high-risk melanoma patients. Cancer 97:1737–1745. https://doi.org/10.1002/cncr.11250
67. Krahn G, Kaskel P, Sander S, Waizenhofer PJ, Wortmann S, Leiter U, Peter RU (2001) S100 beta is a more reliable tumor marker in peripheral blood for patients with newly occurred melanoma metastases compared with MIA, albumin and lactate-dehydrogenase. Anticancer Res 21:1311–1316
68. Beyeler M, Waldispuhl S, Strobel K, Joller-Jemelka HI, Burg G, Dummer R (2006) Detection of melanoma relapse: first comparative analysis on imaging techniques versus S100 protein. Dermatology 213:187–191. https://doi.org/10.1159/000095034

Chapter 44

The Ca^{2+}-Binding S100B Protein: An Important Diagnostic and Prognostic Neurobiomarker in Pediatric Laboratory Medicine

Diego Gazzolo, Francesca Pluchinotta, Giuseppe Lapergola, and Simone Franchini

Abstract

In recent decades a significant scientific effort has focused on projects regarding the use of neurobiomarkers in perinatal medicine with a view to understanding the mechanisms that interfere with physiological patterns of brain development and lead to ominous effects in several human diseases. Numerous potential neurobiomarkers have been proposed for use in monitoring high-risk fetuses and newborns, including markers of oxidative stress, neuroproteins, and vasoactive agents. Nonetheless, the use of these markers in clinical practice remains a matter of debate. Recently, the calcium-binding S100B protein has been proposed as being an ideal neurobiomarker, thanks to its simple availability and easy reproducibility, to the possibility of detecting it noninvasively in biological fluids with good reproducibility, and to the possibility of a longitudinal evaluation in relation to reference curves. The present chapter contains an overview of the most significant studies on the assessment of S100B in different biological fluids as a trophic factor and/or marker of brain damage in high-risk fetuses and newborns.

Key words S100B, Fetus, Newborn, Brain, IUGR, Perinatal asphyxia, Congenital heart disease, Biomarker, Calcium-binding protein, EF hand

1 Introduction

The brain is a complex network of different cell subsets that can not only be rewired but can also be structurally remodeled. The main consequence of neuroplasticity is the ability of several stimuli (daily experience, damage, recovery) to modulate the brain's anatomy and functions [1]. Following a brain insult, there begins a destructive cascade of biological events that continues over hours and days and that may worsen the patient's condition [1]. The main pathophysiological mechanism involves the cerebral autoregulation system (CAS): this is responsible for the maintenance of a constant cerebral blood flow (CBF) over a broad range of perfusion

Claus W. Heizmann (ed.), *Calcium-Binding Proteins of the EF-Hand Superfamily: From Basics to Medical Applications*, Methods in Molecular Biology, vol. 1929, https://doi.org/10.1007/978-1-4939-9030-6_44,

pressures [2–5] resulting from arteriolar vasoconstriction with increased perfusion pressure and vasodilation with decreased perfusion pressure [4, 6, 7].

The CAS is active in CNS of the later fetal and neonatal sheep-based model, as well as in that of the neonatal puppy and piglet [3, 5, 6]. The mechanisms underlying autoregulation are not entirely understood: historically it has been reported to be triggered by metabolic, myogenic, and neurogenic mechanisms; currently, most data suggest that CAS is mediated primarily by an interplay between endothelial-derived constricting and relaxing factors [8–10]. An adequate blood flow to the brain guarantees oxygen and glucose for tissue maturation and cellular functions, thereby maintaining the homeostasis. A reduced CBF causes a redistribution of supply toward posterior areas (cerebellum, basal ganglia, brain stem); however, an acute onset of brain hypoxia implies a generalized hypoperfusion with consequent brain damage [11]. Acute damage enhances the imbalance of cellular adenosine triphosphate synthesis and ion transport, leading to the increased release of excitatory neurotransmitters (glutamate) and, therefore, excitotoxicity. Moreover, membrane depolarization related to the inflow of sodium and calcium and the improper activation of nitric oxide synthase facilitates the generation of free radicals and lipid peroxidation [12]. The combination of these factors leads a vicious cycle that is responsible for cell apoptosis and necrosis (latent damage) [13]. The brain's neuroplasticity allows for a partial recovery during the first 30–60 min after an acute insult and the latent phase of injury begin (1–6 h). The oxidative stress and inflammation are partially recovered, while apoptotic cascade goes ahead [12]. A secondary phase of injury, characterized by cytotoxic edema, excitotoxicity, and secondary complete failure of mitochondrial activity, occurs among neonates with moderate-to-severe injury within approximately 6–15 h after the acute insult [14]. A tertiary phase of damage occurs during the months following the acute insult and involves late cell death, remodeling of the injured brain, and astrogliosis [15].

Thus, research studies on markers of brain damage both in humans and in specific animal models might help to understand the underlying mechanisms that disrupt this plasticity and provide new insight into several human diseases [16]. In recent years, several biomarkers have been proposed for monitoring high-risk fetuses and newborns. Among these, oxidative stress markers, neuroproteins, and vasoactive agents have been studied with promising scientific results [17]. However, their use in the clinical practice is still a matter of debate. The main controversial issues include assessment technique, measurement procedures, reproducibility, and costs. Among the specific characteristics that potential biomarkers should share are (1) simple measurement procedures with good reproducibility; (2) detection in different biological fluids (especially those fluids that can be collected noninvasively, such as urine, saliva); (3) possibility of longitudinal monitoring; (4) well-established use as an early and quantitative marker of brain lesions/damage; and (5) the availability of reference curves, in different biological fluids and in fetal-neonatal and pediatric populations [18].

Of the various biomarkers currently studied in perinatal medicine, the calcium-binding S100B protein seems to meet most of these criteria for an ideal biomarker. The possibility of assessing it in so-called noninvasive biological fluids is further evidence that fetal-neonatal longitudinal brain monitoring is becoming possible.

2 S100B in the Pre- and Perinatal Period

2.1 S100B and Central Nervous System (CNS)

S100B is expressed in several tissues, although it was regarded for more than a decade as specific to the CNS. In the CNS, the protein appears to be most abundant in glial cells and in a number of neuronal subpopulations (oligodendrocytes, Schwann cells, ependymal cells, enteric glial cells, retinal Muller cells) [19]. Cellular accumulation of S100B protein in the CNS is essentially completed within the third trimester of pregnancy, except in the cortex, where the protein appears later. The caudo-rostral pattern of S100B accumulation has been correlated with the biochemical, morphological, and electrophysiological maturation of the CNS [20].

2.2 Further Sources of S100B

In nonneural tissues S100B is expressed in melanocytes, Langerhans cells, chondrocytes, folliculostellate cells of the adenohypophysis, adrenal gland satellite cells, Leydig cells, adipocytes, skeletal muscle satellite cells, and interdigitating reticulum cells [21, 22]. Of the total amount of S100B found in different tissues during the postnatal period, the highest concentration is found in the brain (538,000 μg, 90.9%), followed by muscles (42,000 μg, 7.1%), adipocytes (10,500 μg, 1.77%), heart (1000 μg, 0.2%), and liver (200 μg, 0.03%) [23, 24]. Recently, S100B has been identified in placental tissues (villous and intermediate trophoblast cells) at different trimester of gestation and adipose tissue [22, 25]. The intensity of immunostaining increases according to gestational age (GA). In addition to S100B expression in the amnion, trophoblasts, and decidual cells of fetal membranes, the protein can be localized in endothelial cells of umbilical vessels at all GAs [25]. The possibility that the total amount of S100B detected in fetal and neonatal biological fluids may derive from extraplacental sources is still controversial and a matter of debate. On the one hand, studies of placenta immunohistochemistry have shown that feto-placental tissues contain S100B, which might therefore contribute significantly to the high levels found in fetal circulation in high-risk pregnancies [26]. On the other hand, no differences have been found in the localization or intensity of S100B staining in placental tissues between healthy pregnancies and those complicated by intrauterine growth restriction (IUGR) [25, 27]. The limited contribution of extra-cerebral sources of S100B to overall concentrations of this protein is borne out by the fact that the highest concentrations of S100B are found in the brain and by the limited data regarding extra-cerebral and placental sources [28].

3 S100B in Fetal-Neonatal Biological Fluids

S100B has been measured in several biological fluids such as cerebrospinal (CSF), blood, urine, amniotic fluid, saliva, and milk by a series of immunoassays (for review *see* Subheading 4).

3.1 Cerebrospinal Fluid

In the light of the earlier assumption that S100B was leaked primarily from damaged cells, the first fluid in which S100B was examined as a marker of active brain damage was CSF, owing to its anatomical proximity and biochemical peculiarities. Several studies of S100B measurements in CFS have been conducted in adults and children affected by neurodegenerative diseases, cerebral tumors, cerebral trauma, cerebrovascular diseases, and perinatal asphyxia (PA) [29–35].

In perinatal medicine, measurement of S100B in CSF has been used to monitor brain damage in infants complicated by prematurity, PA, posthemorrhagic ventricular dilatation, and congenital heart diseases (CHD) [35–37]. In this regard, the need to use age-matched reference values when evaluating S100B in CNS diseases has been emphasized. Although the data were promising, CSF was gradually abandoned in perinatal medicine, essentially on account of the stressful procedure involved in its collection, in other words, for ethical grounds [38].

3.2 Fetal-Neonatal Blood

Since the earliest observations in CSF, several studies have investigated the usefulness of measuring S100B in blood as an index of brain damage. The studies were based on the hypothesis that during active brain injury, at least some of the S100B released from the damaged tissue could spread through the blood–brain barrier (BBB) into the systemic circulation [39]. As for CSF, research regarding the detection of S100B in blood samples has been performed both in adults and in children [39–42].

In the perinatal period, S100B has been measured in cord and/or peripheral blood of healthy and IUGR fetuses and newborns [43, 44] and in those with a diagnosis of germinal matrix or intraventricular hemorrhage (GMH/IVH), PA, and CHD [45–47]. Finally, S100B levels in fetal-neonatal blood have been shown to be a useful marker for monitoring the effectiveness and side effects of antenatal therapeutic strategies such as glucocorticoids, nitric oxide donors, and antidepressive drugs [48, 49].

The longitudinal assessment of S100B in cord/peripheral blood has some limitations, the main one being the need of a central arterial catheter to collect cord blood and the related risks of higher infection rates and iatrogenic anemia [50]. Thus, as for CFS, the measurement of S100B in cord blood has been progressively abandoned in the late 1990s [50].

3.3 Maternal Blood

The measurement of S100B in maternal blood was based on previous observations in a sheep-based experimental model of PA in which S100B increased significantly both in maternal and in fetal blood soon after (within 3 h) the hypoxic insult. As no damage had been shown in the maternal brains of any of the animals used, the authors concluded that part of the higher maternal S100B levels could have had a fetal origin. The explanation was thought to lie in a putative passive/mediated S100B transport from the fetal to the maternal bloodstream [51, 52].

Results in experimental models were later confirmed in small for GA (SGA) and IUGR fetuses and, more recently, in healthy pregnancies between 33 and 36 weeks of GA at a stage when CNS development is at its highest peak [53–55].

3.4 Urine

Although studies of S100B concentrations as a pathologic marker in CSF and blood offered consistent and useful results, the potential for improving longitudinal monitoring using less invasive and painful procedures for the collection of biological fluids such as urine opened further avenues for research. Results provide evidence that S100B is (1) essentially eliminated by the kidney route (about 98%) [28, 56, 57]; (2) present in the bloodstream, where it has been shown in IVH infants [58]; (3) correlated with GA [59]; (4) similar to S100A1 in structure and molecular weight and exhibits the same patterns of concentration both in blood and urine [60]; and (5) absent in the kidney tissue [61]. Thus, as for cord and peripheral blood, increased S100B levels have been found in healthy infants and in those complicated by prematurity, IVH, IUGR, PA, and early postnatal death [58, 62–66].

Finally, S100B urine levels have been found to be a useful marker for monitoring the effectiveness or side effects of antenatal therapeutic strategies such as glucocorticoids, vasodilating agents, and antidepressive drugs [49, 67].

3.5 Amniotic Fluid

The amniotic fluid represents the standard source for the collection of S100B for CNS monitoring in the prenatal period. Bearing in mind that the collection of amniotic fluid requires an invasive procedure and that for ethical reasons repeated samplings are not safe for either mother or fetus, the usefulness of S100B is limited. Indeed, current guidelines restrict this procedure to the second trimester of pregnancy when amniotic fluid assessment is available for genetic investigation [68]. Results to date have shown significant changes in S100B levels in healthy fetuses, as well as in those complicated by CNS malformations (spina bifida, anencephalia, myelomeningocele), by maternal diseases (alcohol and/or cocaine addiction), and by unexplained intrauterine death [38, 69–74]. Interestingly, S100B levels in healthy fetuses correlated with GA and with ultrasound parameters expression of brain development (head circumference, biparietal diameter) [75]. Of note, comparisons in healthy twin pregnancies between S100B levels in mono-amniotic and diamniotic pregnancies found that the latter were half those of the former. The data suggested that each fetus releases a physiologically defined aliquot of the protein [74, 76].

3.6 Saliva

Saliva meets one of the main optimality criteria required for a biomarker with which to monitor the CNS in newborns: the possibility of a repeatable, simple, and noninvasive sampling procedure. The evidence that S100B detected in saliva is an indirect appraisal of brain and bloodstream concentrations of the protein is supported by (1) studies in human fetuses showing that members of the S100 protein family are present in different tissues of the salivary glands during ontogenesis; (2) the fact that different S100 proteins are expressed during different phases of pregnancy up to adulthood; (3) the absence of S100B expression in salivary glands in the third trimester of pregnancy, confirming that there is no bias in the assessment of saliva S100B due to potential extra-salivary sources [18, 77, 78]; and (4) the evidence that saliva and blood are isotonic due to a continuous two-way exchange of ions and proteins [77]. As for the other biological fluids described above, a reference curve for S100B has been provided, and significant changes in its levels have been found in infants complicated by PA [79, 80].

4 Methods

Several immunoassays have been conceived, such as a microcomplement fixation assay, RIA, a particle-counting immunoassay, a two-site IRMA (Sangtec 100, AD Sangtec Medical, Bromma, Sweden), an immunoluminometric assay (Lia-mat Sangtec 100, AB Sangtec Medical, Bromma, Sweden), and an ELISA (SynX Pharma, Toronto, Ontario, Canada). PCR has also been used to analyze S100B in blood and amniotic fluid [39, 81–84]. From 1984 onward, technological advances have improved the sensitivity of assays, whose threshold has fallen from 1.5 μg/L to 0.2 μg/L and, even more recently, using chemiluminescence (Liaison S100, Dietzenbach, Germany), to 0.02 μg/L [85].

5 Results and Discussion

5.1 S100B and Pre- and Perinatal Growth and Diseases

5.1.1 Trophic Role in Biological Fluids

Several studies have tried to clarify which factors may influence the expression of S100B in biological fluids. For example, reference curves have been shown to be GA-, gender-, and delivery modality-dependent [43, 55, 59, 75, 79, 86, 87]. Concentrations of S100B in third trimester cord blood collected in a population of healthy term and preterm pregnancies were significantly higher among preterm newborns, declining steadily near term and becoming undetectable or at the detection limit in the term group. The higher levels of S100B in the cord blood of preterm newborns may reflect the greater permeability of the BBB compared with term infants: a higher GA is correlated with a more advanced maturation of the brain and of the BBB and with a lower requirement of neurotrophic factors [88]. Similar data have been found in urine samples [59]. In particular, a significant increase in S100B levels has been found in the late preterm period (33–36 weeks of gestation) [87]. S100B has also been measured in saliva, where its pattern of concentration is in agreement with that previously detected in cord blood and urine [79]. More recently, high concentrations of S100B have been detected in maternal blood during healthy and SGA pregnancies, suggesting the possibility of monitoring fetal well-being through the maternal bloodstream [53, 54]. Because the fetal brain induces an overproduction of S100B, the protein has also been measured in the maternal bloodstream of healthy pregnancies, where its pattern of concentration was superimposable on that detected in other biological fluids [55].

5.2 S100B and Perinatal Diseases

S100B protein concentration patterns have also been investigated as a useful predictor of the main causes of perinatal mortality and morbidity (Table 1).

5.2.1 Prematurity and Intrauterine Growth Restriction (IUGR)

Fetal IUGR is defined as persistent suppression of genetic growth potential that occurs in response to a decrease in substrate supply. IUGR is associated with increased perinatal mortality and accounts for 40% of neurologically damaged children. In IUGR, there is evidence that fetal pre-exposure to adverse intrauterine conditions in terms of decreased oxygen and substrate supplies may play a causal role in perinatal mortality and CNS injury such as periventricular leukomalacia and IVH [94]. From a hemodynamic point of view, it is widely known that IUGR is characterized by uteroplacental blood flow impairment up to 50% of normal perfusion [95] causing an important reduction in the supply of oxygen and nutrients from the mother to the fetus. At this stage, the fetus activates a cardiovascular response characterized by a redistribution of cardiac output to maintain oxygen supply to the brain, heart, and adrenals to preserve their function at the expenses of visceral and peripheral

Table 1
S100B level cutoff in different biological fluids as predictors of short- and long-term neurological outcome

Biological fluids	Population	Outcomes	Cutoff	Sensitivity (%)	Specificity (%)	PPV (%)	NPV (%)	References
U	IUGR	IVH	0.70 μg/L	100	100	NA	NA	[58]
U	IUGR	1 year	7.37 MoM	95	99	NA	NA	[89]
MB	IUGR	IVH	0.72 μg/L	100	99	93	100	[53]
B	PA	HIE	8.50 μg/L	71	90	71	90	[62]
U	PA	1 year	0.28 μg/L	100	87	46	100	[90]
U	PA	HIE	0.41 μg/L	91	95	81	98	[91]
U	PA	HIE	9.69 ng/mL	100	98	100	92	[92]
S	PA	1 year	3.25 MoM	100	100	NA	NA	[80]
A	F	Death	2.50 MoM	100	100	100	100	[76]
U	PN	Death	12.93 MoM	100	98	79	100	[66]
U	PA	Death	1.00 μg/L	100	100	NA	NA	[65]
U	PA	Death	42.00 μg/L	91	95	83	97	[93]
U	PA-HT	Death	1.11 μg/L	100	60	NA	NA	[150]
U	PA-HT	Death	0.66 μg/L	100	60	NA	NA	[150]
B	CHD	Death	0.10 μg/L	100	64	NA	NA	[161]

PPV positive predictive value, *NPV* negative predictive value, *U* urine, *MB* maternal blood, *B* peripheral/cord blood, *IUGR* intrauterine growth retardation, *PA* perinatal asphyxia, *F* fetus, *PN* preterm newborn, *IVH* intraventricular hemorrhage, *HIE* hypoxic-ischemic encephalopathy, *S* saliva, *A* amniotic, *HT* hypothermia, *CHD* congenital heart disease, *NA* not available, *MoM* multiple of median

organs [96]. This phenomenon, called *brain-sparing effect* (BSE), is a well-known warning signal of fetal hypoxia and of unfavorable perinatal outcome [97–99]. At this stage, while CNS damage is at a subclinical stage, no standard clinical, laboratory, and ultrasound parameter is able to detect cases at risk for CNS damage. Thus, the possibility to measure S100B in different biological fluids, in order to detect cases at risk for perinatal morbidity at an early stage, is of the utmost importance. Following the gradual abandonment, for ethical reasons, of the CSF as a monitoring fluid, cord blood became the main biological fluid investigated. In a prospective study conducted in a cohort of 60 pregnant women carrying an IUGR fetus with ($n = 10$) or without BSE ($n = 10$) and in healthy controls ($n = 40$), standard clinical, laboratory, and ultrasound parameters were recorded, and S100B arterial cord blood levels were measured at birth. The results showed higher ($P < 0.05$) levels of S100B in IUGR BSE newborns than in the IUGR non-BSE and control groups. S100B correlated with Doppler parameters measured in the umbilical and middle cerebral arteries suggesting BSE [27]. Following this first observation, an Italian multicenter case-control study was conducted to evaluate longitudinal urinary S100B levels from first urination up to 72 h in preterm newborns complicated by IVH ($n = 18$) compared with age-matched healthy preterm infants ($n = 18$). All infants underwent standard laboratory, clinical, and ultrasound evaluation. The results showed higher ($P < 0.001$) urine S100B in the IVH infants at all monitoring time-points. Indeed, S100B correlated with the IVH grade, indicating the extent of the brain lesion. An S100B cutoff of 0.70 mg/L at first void gave sensitivity and specificity of 100% as a predictor of IVH [58]. On this light, Florio et al. conducted a case-control study evaluating longitudinal urinary S100B levels in preterm infants complicated by IUGR ($n = 42$) and in control infants ($n = 84$). Routine laboratory variables, neurological outcome at 7-day follow-up, ultrasound imaging, and urine S100B concentrations were determined at 5 time-points from birth to 7 days after birth. Higher ($P < 0.05$) urine S100B levels were found at all monitoring time-points in the IUGR newborns than in control infants, especially in those showing adverse short-term neurological outcome. At a cutoff of 7.37 multiple of medians (MoM) at first void, S100B achieved a sensitivity of 95% and a specificity of 99.1% as a single marker for predicting an adverse short-term neurological outcome [100].

More recently, it has been hypothesized that in fetuses complicated by chronic hypoxia and IUGR, part of the total amount of the protein released from the CNS could be transported from the fetal to the maternal bloodstream. To test this hypothesis, an Italian multicenter case-control study was conducted in 106 pregnancies complicated by IUGR, including a subgroup ($n = 26$) who developed IVH after birth, and 212 unaffected pregnancies matched for GA. Clinical, laboratory, and ultrasound parameters, Doppler patterns in the uteroplacental and fetal vessels, and maternal blood collection for S100B assessment were performed before birth. The results showed that in the maternal blood of IUGR infants developing IVH within 7 days from birth, S100B was significantly ($P < 0.001$) higher than in IUGR infants without IVH and in controls. S100B at the cutoff value of 0.72 mg/L achieved a sensitivity of 100% and a specificity of 99.3% as a single marker for predicting IVH in IUGR newborns, with positive predictive values (PPV) and negative predictive values (NPV) of 93% and 100%, respectively [53].

These results showed for the first time that longitudinal fetal well-being or distress monitoring is becoming possible through use of the most suitable noninvasive biological fluid: maternal blood.

Following this pioneering observation, Sannia et al. showed, in a cohort of healthy ($n = 80$) and SGA ($n = 80$) fetuses, that S100B transport from the fetal to the maternal bloodstream is gradient-dependent [67]. Bearing in mind that S100B in the perinatal period is GA-dependent, Serpero et al. provided a reference curve of the protein in the maternal blood during the second and third trimesters of pregnancy, based on a cohort of 1213 healthy pregnant women. The mechanism through which the protein is transported from the fetal to the maternal district (passive, active, mediated) is still controversial and a matter of debate. However, these results opened the way to further studies regarding the monitoring of high-risk pregnancies and new neurobiomarkers (NBs) for use in assessing maternal blood [55].

5.2.2 Perinatal Asphyxia (PA) and Hypoxic-Ischemic Encephalopathy (HIE)

PA and its complication HIE are an important cause of mortality and morbidity in full-term newborns, accounting for 25–28% of neurological handicaps [101, 102]. The post-asphyxia period is crucial because at a time when standard monitoring procedures and radiologic pictures can still be negative, brain damage may already be at a subclinical stage with the symptoms hidden by intensive care procedures (mechanical ventilation and sedation) [103, 104]. At present, blood pH, uric acid, and lactate levels together with cerebral ultrasound and continuous amplitude-integrated electroencephalographic recordings (aEEG) are the major diagnostic tools in monitoring HIE newborns. Unfortunately, they are unable to give an early prediction of brain damage in these patients [105, 106]. Recently, a series of studies reported that the assessment of S100B in different biological fluids can be of help in the early diagnosis of PA infants complicated by HIE, IVH, and adverse neurological outcome. In particular, Nagdyman et al. longitudinally monitored a panel of NBs including S100B in cord/peripheral blood in 29 asphyxiated and 20 control infants. At 2 h after birth, S100B significantly ($P < 0.05$) increased in infants with moderate or severe HIE. At the 2-h time-point (cutoff value of 8.5 mg/L), S100B reached a sensitivity of 71%, a specificity of 90%, a PPV of 71%, and an NPV of 90% [62]. The same authors followed this up with a monocentric study to investigate whether S100B could be a reliable tool for the prediction of adverse neurological outcomes at 20 months after birth. They recruited 29 asphyxiated and 20 control infants. The results showed that S100B and the panel of NBs previously studied did not correlate with long-term neurodevelopmental delay [107].

S100B has also been investigated in a multicenter case-control study of PA infants complicated by IVH ($n = 20$) and controls ($n = 80$). The results showed higher S100B blood levels in the first 12 h from birth in PA-IVH infants than in controls [108]. On the basis of the above findings and in light of the need to promote longitudinal S100B monitoring in biological fluids using procedures that are less stressful for PA infants, researchers explored the possibility to measure the protein in urine, as for preterm infants.

The first observation of longitudinal urine S100B assessments in PA infants was by Gazzolo et al. who conducted a prospective study in 38 full-term PA infants and 96 controls. S100B levels were significantly higher and correlated at different monitoring time-points in PA infants with abnormal neurological outcome at 1 year. An S100B cutoff of 0.28 mg/L at first urination showed a sensitivity of 100%, a specificity of 87.3%, a PPV of 46.2%, and an NPV of 100% for predicting short-term neurological handicap [109]. The same authors investigated S100B in urine as predictor of HIE in a cohort of 44 infants with PA-HIE and 68 controls. S100B levels were significantly ($P < 0.001$) higher in PA-HIE infants at all monitoring time-points. A S100B cutoff of 0.41 μg/L at first urination had sensitivity of 91.3%, specificity 94.6%, PPV 80.8%, and NPV 97.8% for predicting the development of HIE [110].

On the basis of these results, the assessment of S100B in CSF and blood has been progressively abandoned, and other S100 superfamily dimers such as A1B have been measured in the urine of PA-HIE infants. The goal was to promote S100A1B as a marker of multi-organ failure, which, together with HIE, is the most dramatic complication following a hypoxic insult. The A1B dimer is also concentrated in different extra-nervous tissues (e.g., white fat, skeletal muscle, heart, liver, spleen, and kidney) [38, 76, 111–113]. Bashir et al. conducted a prospective study in 105 full-term infants, 42 of whom were complicated by PA and HIE and 63 were healthy controls. They investigated whether longitudinal assessment of urine S100 dimers (A1B and BB) in PA newborns may constitute a useful tool for the early detection of cases at risk for severe HIE. The results showed that A1B and BB dimers were significantly higher, at all monitoring time-points, in PA infants complicated by severe HIE. An A1B dimer cutoff of 186.6 ng/mL at first void reached a sensitivity of 100%, a specificity of 96.3%, a PPV of 85%, and an NPV of 100%. Similarly, a BB dimer cutoff of 9.69 ng/mL at first urination reached a sensitivity of 100%, a specificity of 97.8%, a PPV of 100%, and an NPV of 92%. The authors opened the way to further studies of A1B and BB dimers as predictors of short-/long-term adverse neurological outcome [114].

Another issue that needs further elucidation regards the extra-neural origin of S100B, since no conclusive studies have elucidated whether or not S100B is expressed in kidney tissue. The question is of relevance, since S100B is mainly eliminated by the kidneys and PA is often associated with kidney failure. The possibility that S100B could be released in part by a kidney-mediated mechanism, thereby modifying the protein's reliability as a brain damage marker, therefore needs clarification. In this regard, Risso et al. conducted a case-control study in a cohort of healthy ($n = 432$) and PA newborns ($n = 32$) in which kidney function parameters (blood urea and creatinine concentrations and urine gravity) and urine S100B concentrations were assessed in the first hours after birth. Multiple logistic regression analysis models showed no significant correlations among S100B and kidney function parameters: S100B correlated only with the occurrence of PA. The authors concluded that an altered kidney function is not affecting the reliability of S100B in urine as a marker of CNS damage [61].

Pioneering observations of human fetuses have reported that members of the S100 protein family are present in different tissues of the salivary glands during ontogenesis (i.e., S100A, S100A1, S100A9), while S100B is known to be essentially absent from fetal salivary glands from 32 weeks of gestation onward and in adulthood [115]. Studies on the physiology of salivary glands have shown that systemic blood and saliva are isotonic and have the same concentrations of ions in the same ratios with a continuous exchange of ions and protein between blood and saliva. In particular, an increase or a decrease in the volume of saliva released is triggered by a systemic circulation-dependent mechanism mediated by ortho-/parasympathetic activation [77]. Following on from this observation, it was demonstrated that salivary glands do not produce S100B in the third trimester of pregnancy and that S100B concentrations detected in saliva are derived from systemic circulation. Thus, the possibility that S100B could be measured in saliva, using procedures that are less stressful than those used with other previously investigated biological fluids, has opened new possibilities for the assessment of this protein in perinatal medicine. More recently, the International "Cooperative Multitask against Brain Injury of Neonates" (CoMBINe) study investigated whether the measurement of S100B in saliva from PA-HIE newborns constituted a useful tool for the early detection of neurological abnormalities at 1-year follow-up. They conducted a cross-sectional study in 292 full-term newborns, 48 of whom suffered PA, and 244 healthy controls. S100B was longitudinally measured at ten time-points from birth up to 96 h after birth. Routine laboratory variables, neurological patterns, cerebral ultrasound, and magnetic resonance imaging (MRI) were also obtained. The results showed that S100B was significantly ($P < 0.001$) higher in PA-HIE infants with poor prognosis than in PA-HIE infants with good prognosis and in controls up to 48 h from birth. A significant correlation was observed between S100B and cerebral MRI scores for disease. An S100B cutoff >3.25 MoM at birth achieved sensitivity and specificity of 100% as a single marker for predicting the occurrence of poor neurological outcome [80].

5.2.3 S100B and Perinatal Ominous Outcome

It is widely known that prematurity and PA constitute the main causes leading to perinatal death in about 10–15% of high-risk newborns [66]. The early detection of fetuses and newborns at risk for ominous outcome is still a huge challenge for obstetricians, neonatologists, perinatologists, neuroscientists, and biochemists. The detection of elevated levels of S100B in utero and soon after birth has been shown to constitute the most reliable among a series of standard clinical, laboratory, and ultrasound parameters correlating with the occurrence of unexpected fetal (before 28th week of gestation) and/or early neonatal death (within the first 7 days from birth) [116]. In particular, Florio et al. conducted a case-control study in a cohort of 758 healthy women undergoing amniocentesis at mid-gestation, 12 of whom suffered a spontaneous intrauterine fetal death before 28 weeks, and 746 matched controls. In both groups S100B and α-fetoprotein were measured in amniotic fluid between the 15th and 18th weeks of GA. The concentration of S100B was corrected for GA by conversion to MoM of healthy controls of the same GA. The results showed that S100B, but not α-fetoprotein, was significantly higher ($P < 0.001$) in women who later suffered spontaneous fetal death than in controls. At a S100B cutoff of 2.5 MoM, sensitivity, specificity, and positive and negative predictive values as a diagnostic test for unexplained fetal death were 100% [76].

Following previous observations in amniotic fluid, Gazzolo et al. conducted a cross-sectional study collecting longitudinal urine samples in 165 preterm newborns, of whom 11 suffered early neonatal death, 121 displayed no overt neurologic syndrome, and 33 suffered neonatal hypoxia and IVH but not ominous outcome. Urine S100B concentrations were determined at four time-points and expressed as MoM. The results showed that at a stage when clinical, laboratory, and ultrasound imaging were still negative, urine S100B levels were already higher ($P < 0.01$) in infants who later died than in controls at first urination and increased progressively between the 24- and 96-h time-points. Statistical analysis, using logistic regression analysis among a series of standard clinical, laboratory, and ultrasound parameters, revealed a significant correlation between S100B and the occurrence of early neonatal death. At an S100B cutoff of 12.93 MoM at first void (within 2 h of birth), the protein reached a sensitivity of 100%, a specificity of 97.8%, a PPV of 78.6%, and an NPV of 100% for predicting an ominous outcome [66].

More recently, the possibility that S100B could be a valuable predictor of early neonatal death in infants complicated by PA and HIE has also been investigated. In a cross-sectional study, longitudinal urine S100B levels were measured in 132 term infants, 60 of whom suffered PA complicated by HIE with ($n = 12$) or without ($n = 48$) the occurrence of early neonatal death. Standard clinical, laboratory, and ultrasound variables and urine concentrations of S100B were longitudinally determined from first void to 96 h from birth. The results showed that S100B was significantly higher ($P < 0.001$) from the first void to the 96-h monitoring time-point in infants who later died. Statistical analysis using multiple logistic regression analysis showed a significant correlation between S100B concentrations and the occurrence of neonatal death, and at a cutoff of 1.0 μg/L, S100B had a sensitivity, specificity, and PPV and NPV of 100% for predicting neonatal death [65].

On the basis of previous observations by Bashir et al. [92], the pattern of concentration of S100A1B and BB dimers in the urine of PA infants complicated by early neonatal death was investigated. Risso et al. conducted an international multicenter case-control study in 38 PA full-term newborns (PA neonatal death, $n = 11$; PA survivor, $n = 27$) with 38 healthy infants matched for GA. Clinical, laboratory, ultrasound parameters and longitudinal urinary S100B assessment from first void to 72 h after birth were monitored. The results showed that A1B and BB dimers in PA non-survivor infants were significantly higher ($P < 0.001$) than in PA survivor and control groups. Multivariate analysis with the occurrence of early neonatal death as the dependent variable showed a significant correlation with A1B and BB dimers. An A1B dimer cutoff of >141 ng/L reached a sensitivity of 77.7%, a specificity of 93.1%, a PPV of 72.7%, and an NPV of 93%. A BB dimer cutoff >42 ng/L reached a sensitivity of 90.9%, a specificity of 94.9%, a PPV of 83.3%, and an NPV of 97.4%. Notably, when A1B and BB ROC analyses were compared, the authors found that the BB dimer was the most accurate predictor for the occurrence of early neonatal death: they concluded that the A1B dimer is a reliable predictor of multi-organ failure in PA infants, while the BB dimer is more brain-/ominous outcome-specific [117].

In view of these findings, an increased secretion of S100B in biological fluids could be regarded as an early sign of impaired fetal and neonatal programming. In this regard, it has been previously reported that high concentrations of extracellular S100B in the brain might be deleterious because the protein participates in apoptosis and neuronal death processes. Although in the case of in utero ominous outcome further investigations are needed to identify the "primum movens" responsible for fetal death, in preterm and PA infants, there is evidence that acute/chronic hypoxic insult is responsible for a continuous release of S100B from the damaged CNS, thus triggering apoptosis and neuronal death mechanisms. This possibility has been investigated in adults, and the measurement of S100B has been included among perimortem procedures [118, 119].

5.3 S100B and Perinatal Therapeutic Strategies

There is increasing awareness that the Food and Drug Administration (FDA) and the European Medicine Agency (EMA) strongly support initiatives for research into biomarkers with a view to encouraging the assessment of biomarkers in order to evaluate drug efficacy and side effects in clinical practice, thus promoting biomarker qualification programs. Longitudinal S100B monitoring has already been used to investigate several therapeutic strategies that are currently standard care in the perinatal period in terms of efficacy/side effects.

5.3.1 Antenatal Maternal Glucocorticoid Supplementation

Antenatal maternal glucocorticoid (GC) supplementation is widely used for the prevention of lung immaturity [120, 121], but data on its possible harmful effects on other organs including the CNS are still inconclusive and matter of debate [89, 122–127]. Although perinatal teratogenic effects are a hot topic for perinatologists, data on longitudinal monitoring in high-risk pregnancies and newborns are still an open field of investigation. However, some results regarding the assessment of S100B in biological fluids of fetuses and newborns whose mothers were antenatally treated with GC are available. In this regard, there is a consensus that the administration of GC has some effect on S100B levels in different biological fluids [67, 128]. An early case-control study compared the patterns of concentration of S100B in the cord blood and urine of 39 infants whose mothers were antenatally GC-treated with those of 39 infants whose mothers were not GC-treated. S100B was measured at birth, in the arterial umbilical cord, and in urine at five predetermined points from first urination to 120 h after birth. The results showed that S100B was significantly ($P < 0.05$) lower in GC-treated infants than in controls at all monitoring time-points [128]. Notably, lower S100B levels have been shown in GC-treated infants when compared to no GC-treated ones, complicated or not by IVH. Bearing in mind that S100B acts as a cytokine playing a trophic role at physiological concentrations, the finding of lower S100B levels in GC-treated infants than healthy controls offers additional support to the debated question on GC side effects on CNS development. Another open question regards whether or not S100B changes in biological fluids are GC dose-dependent. In this regard, Sannia et al. conducted a case-control study assessing S100B urine concentrations in 70 preterm infants whose mothers received a complete GC-course ($n = 16$), half-course ($n = 19$), and controls ($n = 35$). In all infants, S100B levels in urine were longitudinally measured from birth to 72 h from birth. Results showed that S100B levels in newborns of complete GC-course were significantly lower ($P < 0.01$) at all monitoring time-points than those whose mother received half-course treatment and controls [67]. No differences in neonatal outcomes such as the incidence of respiratory distress syndrome and the need of mechanical ventilation and the length of hospital staying have been found between infants treated by a complete GC-course and half GC-course. Results open the way to further studies aimed at investigating the most suitable low GC-course regimen dose.

5.3.2 Nitric Oxide Donors

In recent decades there has been mounting interest in the potential use during the perinatal period of vasodilating agents, especially in pregnancies complicated by placental insufficiency and IUGR. This has led to the proposal to administer nitric oxide (NO) donor supplementation in the third trimester of pregnancy in order to improve uteroplacental circulation in IUGR fetuses [129–132]. In an Italian multicenter study, the authors conducted a randomized controlled trial (RCT) in 48 IUGR pregnancies in which subjects received either a placebo ($n = 25$) or transdermal glyceryl trinitrate at a dose of 5 mg every 16 h daily until delivery ($n = 23$). The aim was to investigate whether or not NO administration could affect fetal CNS development/function. All infants were monitored by cord blood S100B assessment at birth and by longitudinal clinical, laboratory, and ultrasound recordings from birth to the seventh day after birth. The results showed that S100B was significantly ($P < 0.05$) higher in the placebo than in the NO-treated group. These data were also corroborated by improved neurological and respiratory outcomes in NO-treated infants than in the placebo group [48].

5.3.3 Antidepressants

There is growing evidence that pregnancy can constitute a risk factor for major depressive disorders [133]. The use of antidepressants such as selective serotonin reuptake inhibitors (SSRIs) during pregnancy is dramatically increasing, with up to 2–3% of pregnant women in Europe and 8–10% in the USA taking them [90]. This trend has evolved without any solid evidence on their safety or efficacy and without taking into account that when treating depression and anxiety, the expected benefits must outweigh the risks. Recourse to pharmacological therapy is justified only when other non-pharmacological treatments are insufficient. Conversely, attempted suicides and relapses of depression have been recorded when antidepressant treatment during pregnancy is inappropriately halted [91]. Pharmacological treatment is therefore uniformly recommended for pregnant women diagnosed with depression, despite the potential side effects on fetus and newborn [134, 135]. In humans, neural tube defects, cardiovascular malformations, persistent pulmonary hypertension, preterm birth, neonatal respiratory distress and withdrawal syndrome (i.e., seizures, hyperexcitability syndrome), and neurodevelopment abnormalities have been described as being the result of antidepressant treatment [92, 93, 136–139]. The overall risk of antidepressants being associated with poor neonatal adaptation is estimated at 30% for all antidepressant drugs, with the exception of paroxetine in which the risk is even higher [140]. In line with FDA and EMA recommendations, studies on NBs and the assessment of S100B in different biological fluids (i.e., maternal blood, amniotic, arterial and venous cord blood) have been conducted in pregnant women and in newborns whose mothers were antenatally treated with SSRIs.

Recently, an international multicenter observational study by Bellissima et al. recruited 306 pregnant women of whom 75 were SSRI-treated and 231 healthy controls. Standard clinical, laboratory, ultrasound parameters and longitudinal S100B assessment in maternal, fetal, and neonatal biological fluids were performed. Neurological follow-up was set at 7 days from birth. The results showed that S100B levels in maternal-fetal-neonatal fluids were significantly higher in the SSRI-treated group. Indeed, the highest S100B levels were observed in infants with an adverse neurological outcome. These findings opened the way to further investigations aimed at monitoring the effects of antidepressant drug treatment on NBs and short-and long-term neurological follow-up [141].

5.3.4 Perinatal Asphyxia and Hypothermia

Total body hypothermia (HT) is standard therapeutic strategy for PA-HIE in term infants [142, 143]. According to national and international guidelines, the admission of patients to treatment is defined by clinical, laboratory, and aEEG parameters. However, standard monitoring procedures can be silent or of no avail especially in infants requiring invasive interventions such as sedation and mechanical ventilation while in neonatal intensive care units (NICUs). In this regard, inclusion in the daily practice of NICUs of additional sentinel markers is eagerly awaited in order to identify as soon as possible which infants will benefit from cooling.

In a prospective longitudinal cohort study, Massaro et al. evaluated in 83 PA-HIE infants whether serum S100B protein and a panel of NBs measured during HT were predictive of neurodevelopmental outcome at 15 months. Clinical, laboratory, aEEG parameters and S100B were longitudinally recorded before, during, and after HT procedure. The results showed that increased ($P < 0.05$) S100B levels during HT were associated with adverse neurodevelopmental outcomes at 15 month [144]. Furthermore, Massaro et al. conducted an observational study to investigate whether S100B correlated with cerebral MRI patterns and short-term neurological follow-up (at 14 days after birth) in 75 HT-treated PA-HIE infants. Clinical, laboratory, aEEG parameters and S100B were longitudinally recorded before, during, and after HT procedure. The results showed that serum protein levels were higher ($P < 0.05$) at all monitoring time-points in patients with unfavorable outcomes and pathological MRI patterns [145]. Data were partly in agreement with a previous observation by Roka et al. [146], who found, in a cohort of 24 HT-treated ($n = 13$) or non-HT ($n = 11$) PA-HIE infants, elevated ($P < 0.05$) S100B levels in both the HT and non-HT groups at all time-points. Thus, S100B may be a useful prognostic marker in infants following PA, and it may be a useful biomarker of the effects of treatment in studies of additional neuroprotective therapies. Recently, Massaro et al. conducted an RCT to evaluate the potential neuroprotective effects of erythropoietin in a cohort of 50 HT-treated PA-HIE infants, using a panel of NBs including S100B. Clinical, laboratory, and NB parameters were longitudinally assessed before, during, and after HT procedure. The results showed increased ($P < 0.05$) NB and S100B levels in infants showing pathological cerebral MRI patterns. No significant effects of erythropoietin treatment on NB levels were detectable [147].

Therapeutic HT can be administered as selective head cooling (SHC) or whole-body cooling (WBC). However, data comparing the effectiveness of the two HT modalities are still controversial. In a prospective RCT pilot study, Celik et al. compared the efficacy of the two HT procedures using a panel of NBs including S100B in a cohort of 21 PA-HIE infants [148]. Clinical, laboratory, and NB parameters were longitudinally assessed before, during, and after HT procedure. Neurological follow-up was set at 1 year after birth. The results showed no differences in NBs between the two employed HT methods.

Finally, the advantage of the less invasive procedure of urine collection to detect S100B protein was investigated by Alshweki et al. who measured urinary S100B levels during HT treatment. The authors conducted a prospective observational study in 31 PA-HIE newborns of whom 13 had a normal neurological examination at discharge and 18 had ominous/adverse neurological outcome detected by MRI. Clinical, laboratory, and NB parameters were longitudinally assessed before, during, and after the HT procedure. The results showed higher ($P < 0.05$) S100B levels in infants with unfavorable outcome. An S100B cutoff on day 1 of 1.11 μg/L had a sensitivity of 100% and specificity of 60% for the prediction of neonatal death, while at a cutoff of 0.66 μg/L, the protein recorded a sensitivity of 83% and a specificity of 70% for the prediction of an abnormal neurological follow-up [149].

5.4 S100B and Brain Monitoring in Infants with CHD

In CHD infants the etiology of cerebral injury is both varied and complex. Brain damage can occur before, during, and after heart surgery [150]. Advances in cardiothoracic surgical and anesthetic techniques, cardiopulmonary bypass (CPB), and deep hypothermic circulatory arrest (DHCA) have led to a reduction in perioperative mortality and morbidity. However, neurodevelopmental consequences in survivors still constitute the main complications in the perioperative period. In particular, disturbances of cognition, behavior, attention deficit, hyperactivity disorder, executive planning, feeding, speech and language, socialization, and fine and gross motor coordination are the key components of the clinical repertoire that can complicate CHD infants [151–155]. This is because open-heart surgery represents a phase of programmed and intentional hypoxic-ischemic (H-I) injury, which is the price to pay in the treatment of or palliative therapy for CHD. In the past, brain injury was simply conceived as a complication of heart surgery and CPB. Today, postsurgical management is believed to have a relevant role in triggering and perpetuating brain damage. Intraoperative events such as CPB, DHCA, hypo-reperfusion stress, air emboli, and systemic inflammatory response are considered the most important causes of brain damage in CHD infants. For example, the length and the degree of cooling can be deleterious on children leading to cerebral anaerobic metabolism ATP depletion and energy failure, cause of membrane ionic pump failure, and depolarization [156]. Moreover, systemic inflammatory response activation (including cell activation due to contact with the bypass circuit, mechanical shear stress, hemodilution, hypothermia, and ischemia) enhances H-I reperfusion injury, activates the complement and coagulation systems, releases cytokines (such as interleukin [IL]-8), and leads to vascular resistance [157].

The occurrence of brain injury after H-I insult is an evolving process that starts during the acute insult and is followed by the reperfusion phase. As for non-CHD infants, the main pathogenetic mechanism underlying neurological damage due to hypoxemia/ischemia, or both, is the dramatic lowering of glucose and oxygen supplies.

In CHD infants, S100B has been assessed in order to monitor brain stress in different perioperative phases. In an observational study of 13 CHD infants, clinical, laboratory, and S100B protein assessments were recorded before, during, and after surgical procedure [42]. The concentrations of S100B were significantly ($P < 0.05$) higher at the end of CPB than either before or after the completion of surgery and were highest in the two patients who died in the early postoperative phase. Furthermore, S100B concentrations correlated with the duration of surgery and CPB.

These pioneering results were subsequently confirmed by several observational studies of CNS stress during different CPB phases such as cooling, rewarming, and DHCA [158–160]. In particular, Varrica et al. investigated the usefulness of S100B as a predictor of perioperative death. The authors conducted a case-control study in 88 CHD infants, 22 of whom died shortly after surgery and 66 of whom survived. Clinical, laboratory, and S100B parameters were recorded before, during, and after surgical procedure. The authors showed that S100B concentrations were significantly higher in the CHD death group at all monitoring time-points. A presurgery S100B cutoff of 0.1 μg/L reached a sensitivity of 100%, a specificity of 63.7%, a PPV of 47.8%, and an NPV of 100% [161]. Notwithstanding these promising results, the fate of S100B as a useful brain damage marker in the perioperative period of CHD infants is still controversial. In fact the protein, which was initially reported as a useful tool, was later largely abandoned for monitoring the brains of CHD adults and children [18–21]. This was because of the potential contribution of cardiac adipose tissue release during CPB phases to circulating levels of S100B [22–25]. Since adipocytes are a nonneural source of S100B, it was essential to assess the reliability of S100B as an NB during the CPB procedure. In this regard, Varrica et al. evaluated circulating S100B and adiponectin (an adipocyte-specific protein) in a prospective study on children who underwent surgical CHD repair [24]. They enrolled 26 CHD infants in whom clinical, laboratory, S100B, and adiponectin levels were recorded before, during, and after surgery. In line with earlier findings, S100B levels showed an increase during surgery, with a peak at the end of the CPB procedure and a steady concentration 24 h after the end of surgery. By contrast, adiponectin showed a flat trend with no significant differences throughout the entire study period. The authors concluded that S100B protein is not affected by release from nonneural tissue, as suggested by the absence of changes in circulating adiponectin concentrations [24]. More recently, perioperative S100B measurements were performed to monitor CPB management especially during the reoxygenation phase and to record differences in perioperative stress between cyanotic and non-cyanotic CHD infants. The results showed that hypoxia/hyperoxia mediated the release of S100B in the perioperative period in CHD infants. Cyanotic CHD infants are notably more prone to perioperative brain stress/damage. Further studies are needed regarding the usefulness of S100B to identify the "optimal" PaO_2 target in order to avoid the side effects associated with reoxygenation during CPB [160, 162].

6 Conclusion

In the last decade thanks to advances in laboratory technology and to the interrelationship among researchers from different disciplines, the usefulness of biochemical markers for early brain injury detection in high-risk fetuses, newborns, and children has been empowered. Experts from laboratory medicine, intensive care medicine, neonatology, pediatrics, obstetrics, neurosciences, and biochemistry have been involved as strongly supported by FDA and EMA. Therefore, NB assessment in different biological fluids has been investigated not only for CNS development and damage monitoring but, more recently, for new neuroprotective strategies validation. Among NBs, oxidative stress markers, neuroproteins, and vasoactive agents have been studied with promising results [17]. Conversely, their use in the clinical practice is still controversial and matter of debate. The main issues regard assessment technique, measurement procedures, reproducibility, costs, etc. However, according to FDA and EMA statements, a panel of criteria to be fulfilled has been agreed in order to include NBs in daily clinical practice [17]. The following items have been proposed for the choice of an ideal biomarker: (1) simple measurements with good reproducibility; (2) detection in different biological fluids especially those at low stress for sampling such as urine and saliva; (3) possibility of longitudinal monitoring; (4) well-established use as an early and quantitative marker of brain lesions/damage; and last but not least (5) the availability of reference curves, in different biological fluids and in fetal-neonatal and pediatric populations, mandatory for biomarker inclusion in clinical guidelines. At this stage, among several NBs investigated, S100B seems to meet most of these criteria for an ideal biomarker particularly those regarding assessment-measurement procedures and the possibility of detection in urine, saliva, and milk fluids [79, 86, 87, 163]. Another relevant point resides in the fact that S100B cost/benefit is lower when compared with any standard monitoring procedure currently used for CNS monitoring in infants and children. On this light, before S100B inclusion in the clinical practice, few items need to be fulfilled in agreement with FDA and EMA. The main point resides in S100B validation as diagnostic test through its correlation with a long-term neurological follow-up. Investigations in urine and saliva of preterm and term newborns are in progress, and preliminary results are promising. Another point regards protein assessment technique (i.e., ELISA, chemiluminescence, etc.) and related reference curves in different biological fluids.

In conclusion, S100B assessment in high-risk infants and children can represent, together with standard monitoring procedures, a golden standard in the early detection of CNS injury. Further multicenter investigations will corroborate the role of the protein in clinical guidelines.

References

1. Douglas-Escobar M, Weiss MD (2015) Hypoxic-ischemic encephalopathy – a review for the clinician. JAMA Pediatr 169 (4):397–403
2. Kuschinsky W, Wahl M (1978) Local chemical and neurogenic regulation of cerebral vascular resistance. Physiol Res 58:656
3. Lassen NA, Christensen MS (1976) Physiology of cerebral blood flow. Br J Anaesth 48:719–734
4. Paulson OB, Strandgaard S, Edvinsson L (1990) Cerebral autoregulation. Cerebrovasc Brain Metab Rev 2:161–192
5. Florence G, Seylaz J (1992) Rapid autoregulation of cerebral blood flow: a laser-Doppler flowmetry study. J Cereb Blood Flow Metab 12:674–680
6. MacKenzie ET, Strandgaard S, Graham DI et al (1976) Effects of acutely induced hypertension in cats on pial arteriolar caliber, local cerebral blood flow, and the blood-brain barrier. Circ Res 39:33–41
7. Mchedlishvili GI, Nikolaishvili LS, Antia RV (1976) Are the pial arterial responses dependent on the direct effect of intravascular pressure and extravascular and intravascular pO2, pCO2, and pH? Microvasc Res 10:298–311
8. Rubanyi GM, Botelho LH (1991) Endothelins. FASEB J 5:2713–2720
9. Faraci FM (1991) Role of endothelium-derived relaxing factor in cerebral circulation: large arteries vs. microcirculation. Am J Physiol 26:H1038–H1042
10. Greenberg DA, Chan J, Sampson HA (1992) Endothelins and the nervous system. Neurology 42:25–31
11. Harteman JC, Nikkels PG, Benders MJ et al (2013) Placental pathology in full-term infants with hypoxic-ischemic neonatal encephalopathy and association with

magnetic resonance imaging pattern of brain injury. J Pediatr 163(4):968–995.e2
12. Wassink G, Gunn ER, Drury PP et al (2014) The mechanisms and treatment of asphyxia encephalopathy. Front Neurosci 8:40. https://doi.org/10.3389/fnins.2014.00040
13. Ferriero DM (2004) Neonatal brain injury. N Engl J Med 351(19):1985–1995
14. Bennet L, Roelfsema V, Pathipati P et al (2006) Relationship between evolving epileptiform activity and delayed loss of mitochondrial activity after asphyxia measured by near-infrared spectroscopy in preterm fetal sheep. J Physiol 572(pt 1):141–154
15. Bennet L, Tan S, Van den Heuij L et al (2012) Cell therapy for neonatal hypoxia-ischemia and cerebral palsy. Ann Neurol 71 (5):589–600
16. Gazzolo D, Li Volti G, Gavilanes AWD et al (2015) Biomarkers of brain function and injury: biological and clinical significance. Biomed Res Int 2015:389023. https://doi.org/10.1155/2015/389023
17. Serpero LD, Bellissima V, Colivicchi M et al (2013) Next generation biomarkers for brain injury. J Matern Fetal Neonatal Med 26 (S2):44–49
18. Serpero LD, Pluchinotta F, Gazzolo D (2015) The clinical and diagnostic utility of S100B in preterm newborns. Clin Chim Acta 444:193–198
19. Yang Q, Hamberger A, Hyden H et al (1995) S100B has a neuronal localization in the rat hindbrain revealed by an antigen retrieval method. Brain Res 696:49–61
20. Zuckerman JE, Herschman HR, Levine L (1970) Appearance of a brain specific antigen (the S-100 protein) during human foetal development. J Neurochem 17:247–251
21. Haimoto H, Hosoda S, Kato K (1987) Differential distribution of immunoreactive S100α and S100β proteins in normal non-nervous human tissues. Lab Invest 57:489–498
22. Michetti F, Dell'Anna E, Tiberio G et al (1983) Immunochemical and immunocytochemical study of S100 protein in rat adipocytes. Brain Res 262:352–356
23. Jönsson H (2003) S100B and cardiac surgery: possibilities and limitations. Restor Neurol Neurosci 21(3-4):151–157
24. Varrica A, Satriano A, Frigiola A et al (2015) Circulating S100B and adiponectin in children who underwent open heart surgery and cardiopulmonary bypass. Biomed Res Int 2015:402642. https://doi.org/10.1155/2015/402642
25. Marinoni E, Di Iorio R, Gazzolo D et al (2002) Ontogenetic localization and distribution of S100h protein in human placental tissues. Obstet Gynecol 99:1093–1099
26. Wijnberger LD, Nikkels PG, van Dongen AJ et al (2002) Expression in the placenta of neuronal markers for perinatal brain damage. Pediatr Res 51:492–496
27. Gazzolo D, Marinoni E, di Iorio R et al (2002) Circulating S100beta protein is increased in intrauterine growth-retarded fetuses. Pediatr Res 51:215–219
28. Jönsson H, Johnsson P, Höglund P et al (2000) Elimination of S100B and renal function after cardiac surgery. J Cardiothorac Vasc Anesth 14:698–701
29. Petzold A, Eikelenboom M, Gveric D et al (2002) Markers for different glial cell responses in multiple sclerosis: clinical and pathological correlations. Brain 125 (Pt 7):1462–1473
30. Syeda T, Muhammad Hashim AS, Rizvi HA et al (2013) Serum S100B in patients with brain tumours undergoing craniotomy. J Coll Physicians Surg Pak 23(2):112–115
31. Chen DQ, Zhu LL (2005) Dynamic change of serum protein S100b and its clinical significance in patients with traumatic brain injury. Chin J Traumatol 8(4):245–224
32. Bustamante A, López-Cancio E, Pich S et al (2017) Blood biomarkers for the early diagnosis of stroke: the Stroke-Chip Study. Stroke 48(9):2419–2425
33. Berger RP, Pierce MC, Wisniewski SR et al (2002) Neuron-specific enolase and S100B in cerebrospinal fluid after severe traumatic brain injury in infants and children. Pediatrics 109: E31
34. Hardemark HG, Ericsson N, Kotwica Z et al (1989) S-100 protein and neuron-specific enolase in CSF after experimental traumatic or focal ischemic brain damage. J Neurosurg 71:727–731
35. Blennow M, Sävman K, Ilves P et al (2001) Brain-specific proteins in the cerebrospinal fluid of severely asphyxiated newborn infants. Acta Paediatr 90:1171–1175
36. Whitelaw A, Rosengren L, Blennow M (2001) Brain specific proteins in posthaemorrhagic ventricular dilatation. Arch Dis Child Fetal Neonatal Ed 84(2):F90–F91
37. Sellman M, Ivert T, Ronquist G et al (1992) Central nervous system damage during cardiac surgery assessed by 3 different biochemical markers in cerebrospinal fluid. Scand J Thorac Cardiovasc Surg 26:39–45

38. Michetti F, Gazzolo D (2002) S100B protein in biological fluids: a tool for perinatal medicine. Clin Chem 48(12):2097–2104
39. Persson L, Hardemark HG, Gustafsson J et al (1987) S-100 protein and neurospecific enolase in cerebrospinal fluid and serum: markers of cell damage in human central nervous system. Stroke 18:911–918
40. Bhattacharya K, Westaby S, Pillai R et al (1999) Serum S100B and hypothermic circulatory arrest in adults. Ann Thorac Surg 68 (4):1225–1229
41. Westaby S, Johnsson P, Parry AJ et al (1996) Serum S100 protein: a potential marker for cerebral events during cardiopulmonary bypass. Ann Thorac Surg 61:88–92
42. Gazzolo D, Vinesi P, Geloso MC et al (1998) S100 blood concentrations in children subjected to cardiopulmonary by-pass. Clin Chem 44:1058–1060
43. Gazzolo D, Vinesi P, Marinoni E et al (2000) S100B protein concentrations in cord blood: correlations with gestational age in term and preterm deliveries. Clin Chem 46:998–1000
44. Gazzolo D, Michetti F, Bruschettini M et al (2003) Pediatric concentrations of S100B protein in blood: age- and sex-related changes. Clin Chem 49:967–970
45. Gazzolo D, Vinesi P, Bartocci M et al (1999) Elevated S100 blood level as early indicators of intraventricular hemorrhage in preterm infants. Correlation with cerebral Doppler velocimetry. J Neurol Sci 170:32–35
46. Gazzolo D, Masetti P, Vinesi P et al (2002) S100B blood levels correlate with rewarming time and cerebral Doppler in pediatric open heart surgery. J Card Surg 17(4):279–284
47. Gazzolo D, Visser GHA, Lituania M et al (2002) S100B protein cord blood levels and fetal behavioural states of development: a study in normal and small for dates fetuses. J Maternal-Fetal Neonat Med 11:378–384
48. Gazzolo D, Bruschettini M, Di Iorio R et al (2002) Maternal nitric oxide supplementation decreases cord blood S100B in intrauterine growth-retarded fetuses. Clin Chem 48:647–650
49. Pawluski JL, Galea LA, Brain U et al (2009) Neonatal S100B protein levels after prenatal exposure to selective serotonin reuptake inhibitors. Pediatrics 124:e662–e670
50. Straus RG (1995) Neonatal anemia: pathophysiology and treatment. Immunol Invest 24:341–351
51. Garnier Y, Berger R, Alm S et al (2006) Systemic endotoxin administration results in increased S100B protein blood levels and periventricular brain white matter injury in the preterm fetal sheep. Eur J Obstet Gynecol Reprod Biol 124(1):15–22
52. Garnier Y, Frigiola A, Li Volti G et al (2009) Increased maternal/fetal blood S100B levels following systemic endotoxin administration and periventricular white matter injury in preterm fetal sheep. Reprod Sci 16:758–766
53. Gazzolo D, Marinoni E, Di Iorio R et al (2006) High maternal blood S100B concentrations in pregnancies complicated by intrauterine growth restriction and intraventricular hemorrhage. Clin Chem 52:819–826
54. Sannia A, Zimmermann LJ, Gavilanes AW et al (2011) S100B protein maternal and fetal bloodstreams gradient in healthy and small for gestational age pregnancies. Clin Chim Acta 412(15-16):1337–1340
55. Serpero LD, Bianchi V, Pluchinotta F et al (2017) S100B maternal blood levels are gestational age- and gender-dependent in healthy pregnancies. Clin Chem Lab Med 55 (11):1770–1776
56. Moore BW (1965) A soluble protein characteristic of the nervous system. Biochem Biophys Res Commun 19:739–744
57. Heizmann CW (1999) Ca2+-binding S100 proteins in the central nervous system. Neurochem Res 24:1097–1100
58. Gazzolo D, Bruschettini M, Lituania M et al (2001) Increased urinary S100B protein as an early indicator of intraventricular hemorrhage in preterm infants: correlation with the grade of hemorrhage. Clin Chem 47:1836–1838
59. Gazzolo D, Bruschettini M, Lituania M et al (2001) S100b protein concentrations in urine are correlated with gestational age in healthy preterm and term newborns. Clin Chem 47:1132–1133
60. Usai A, Kato K, Abe T et al (1989) S100ao protein in blood and urine during open-heart surgery. Clin Chem 35:1942–1944
61. Risso FM, Serpero LD, Zimmermann LJ et al (2012) Perinatal asphyxia: kidney failure does not affect S100B urine concentrations. Clin Chim Acta 413:150–153
62. Nagdyman N, Komen W, Ko HK et al (2001) Early biochemical indicators of hypoxic-ischemic encephalopathy after birth asphyxia. Pediatr Res 49:502–506
63. Liu L, Zheng CX, Peng SF et al (2010) Evaluation of urinary S100B protein level and lactate/creatinine ratio for early diagnosis and prognostic prediction of neonatal hypoxic-ischemic encephalopathy. Neonatology 97:41–44

64. Florio P, Marinoni E, Di Iorio R et al (2006) Urinary S100B protein concentrations are increased in intrauterine growth-retarded newborns. Pediatrics 118:e747–e754
65. Gazzolo D, Frigiola A, Bashir M et al (2009) Diagnostic accuracy of S100B urinary testing at birth in full-term asphyxiated newborns to predict neonatal death. PLoS One 4:e4298. https://doi.org/10.1371/journal.pone.0004298
66. Gazzolo D, Florio P, Ciotti S et al (2005) S100B protein in urine of preterm newborns with ominous outcome. Pediatr Res 58 (6):1170–1174
67. Sannia A, Risso FM, Serpero LD et al (2010) Antenatal glucocorticoid treatment affects preterm infants' S100B urine concentration in a dose-dependent manner. Clin Chim Acta 411:1539–1541
68. The American College of Obstetricians and Gynecologists – Women's Health Care Physicians (2016) Practice Bulletin No. 162 Summary: prenatal diagnostic testing for genetic disorders. Obstet Gynecol 127 (5):976–978
69. Anneren G, Esscher T, Larsson L et al (1988) S-100 protein and neuron-specific enolase in amniotic fluid as markers of abdominal wall and neural tube defects in the fetus. Prenat Diagn 8:323–328
70. Sindic CJ, Freund M, Van Regemorter N et al (1984) S-100 protein in amniotic fluid of anencephalic fetuses. Prenat Diagn 4:297–302
71. Eriksen JL, Gillespieb R, Druseb MJ (2002) Effects of ethanol and 5-HT1A agonists on astroglial S100B. Develop Brain Res 139:97–105
72. Clarke C, Clarke K, Muneyyirci J et al (1996) Prenatal cocaine delays astroglial maturation: immunodensitometry shows increased markers of immaturity (vimentin and GAP 43) and decreased proliferation and production of the growth factor S-100. Brain Res Dev Brain Res 91:268–273
73. Akbari HM, Whitaker-Azmitia PM, Azmitia E (1994) Prenatal cocaine exposure decreases the trophic factor S100b and induced microcephaly: reversal by postnatal 5-HT1A receptor agonist. Neurosci Lett 170:141–144
74. Michetti F, Gazzolo D (2003) S100B testing in pregnancy. Clin Chim Acta 335(1-2):1–7
75. Gazzolo D, Bruschettini M, Corvino V et al (2001) S100b protein concentrations in amniotic fluid correlate with gestational age and with cerebral ultrasound scanning results in healthy fetuses. Clin Chem 47:954–956
76. Florio P, Michetti F, Bruschettini M et al (2004) Amniotic fluid S100B protein in mid-gestation and intrauterine fetal death. Lancet 364:270–272
77. Humphrey SP, Williamson RT (2001) A review of saliva: normal composition, flow, and function. J Prosthet Dent 85:162–169
78. Gazzolo D, Michetti F (2010) Perinatal S100B protein assessment in human unconventional biological fluids: a minireview and new perspectives. Cardiovasc Psychiatry Neurol 2010:703563. https://doi.org/10.1155/2010/703563
79. Gazzolo D, Lituania M, Bruschettini M et al (2005) S100B protein levels in saliva: correlation with gestational age in normal term and preterm newborns. Clin Biochem 38:229–233
80. Gazzolo D, Pluchinotta F, Bashir M et al (2015) Neurological abnormalities in full-term asphyxiated newborns and salivary S100B testing: the "Cooperative Multitask against Brain Injury of Neonates" (CoMBINe) International Study. PLoS One 10(1): e0115194. https://doi.org/10.1371/journal.pone.0115194
81. Michetti F, Massaro A, Murazio M (1979) The nervous system-specific S100 antigen in cerebrospinal fluid of multiple sclerosis patients. Neurosci Lett 11:171–175
82. Michetti F, Massaro A, Russo G et al (1980) The S-100 antigen in cerebrospinal fluid as a possible index of cell injury in the nervous system. J Neurol Sci 44:259–263
83. Fagnart OC, Sindic CJ, Laterre C (1988) Particle counting immunoassay S100 protein in serum. Possible relevance in tumors and ischemic disorders of the central nervous system. Clin Chem 34:1387–1391
84. Yang YH, Kim IK, Oh SH et al (1998) Rapid prenatal diagnosis of trisomy 21 by polymerase chain reaction-associated analysis of small tandem repeats and S100B in chromosome 21. Fetal Diagn Ther 13:361366
85. Zhang H, Qi S, Rao J et al (2013) Development of a rapid and high-performance chemiluminescence immunoassay based on magnetic particles for protein S100B in human serum. Luminescence 28:927–932
86. Gazzolo D, Frulio R, Roletti A et al (2007) S100A1B and S100BB urine levels in preterm and term healthy newborns. Clin Chim Acta 384:186–187
87. Sannia A, Risso FM, Zimmermann LJ et al (2013) S100B urine concentrations in late preterm infants are gestational age and gender dependent. Clin Chim Acta 417:31–34

88. Gazzolo D, Vinesi P, Bartocci M et al (1999) Elevated S100 blood level as early an indicator of intraventricular hemorrhage in preterm infants. Correlation with cerebral Doppler velocimetry. J Neurol Sci 170:32–35
89. Sousa N, Madeira MD, Paula-Barbosa MM (1998) Effects of corticosteroids treatment and rehabilitation on the hippocampal formation of neonatal and adult rats. An unbiased stereological study. Brain Res 794:199–210
90. Lattimore KA, Donn SM, Kaciroti N et al (2005) Selective Serotonin Reuptake Inhibitor (SSRI) use during pregnancy and effects on the fetus and newborn: a meta-analysis. J Perinatol 25:595–604
91. Austin MP, Kildea S, Sullivan E (2007) Maternal mortality and psychiatric morbidity in the perinatal period: challenges and opportunities for prevention in the Australian setting. Med J Aust 186(7):364–367
92. Noorlander CW, Ververs FF, Nikkels PG et al (2008) Modulation of serotonin transporter function during fetal development causes dilated heart cardiomyopathy and lifelong behavioural abnormalities. PLoS One 3(7): e2782. https://doi.org/10.1371/journal.pone.0002782
93. Hemels ME, Einarson A, Koren G et al (2005) Antidepressant use during pregnancy and the rates of spontaneous abortions; a meta-analysis. Ann Pharmacother 39:803–809
94. Hagberg H, Mallard C (2000) Antenatal brain injury: aetiology and possibilities of prevention. Semin Neonatol 5:41–51
95. Bates JA, Evans JA, Mason G (1996) Differentiation of growth retarded from normally grown fetuses and prediction of intrauterine growth retardation using Doppler ultrasound. Br J Obstet Gynaecol 103:670–675
96. Behrman RE, Lees MH, Peterson EN et al (1970) Distribution of the circulation in the normal and asphyxiated fetal primate. Am J Obstet Gynecol 108:956–969
97. Levene MI, Fenton AC, Evans DH et al (1989) Severe birth asphyxia and abnormal cerebral blood-flow velocity. Dev Med Child Neurol 31:427–434
98. van Bel F, van de Bor M, Stijnen T et al (1987) Cerebral blood flow velocity pattern in healthy and asphyxiated newborns: a controlled study. Eur J Pediatr 146:461–467
99. Fellman V, Raivio KO (1997) Reperfusion injury as the mechanism of brain damage after perinatal asphyxia. Pediatr Res 41:599–606
100. Florio P, Marinoni E, Di Iorio R et al (2006) Urinary S100B protein concentrations are increased in intrauterine growth-retarded newborns. Pediatrics 118(3):e747–e754
101. Freeman JM, Nelson KB (1988) Intrapartum asphyxia and cerebral palsy. Pediatrics 82:240–249
102. Hagberg B, Hagberg G, Beckung E et al (2001) Changing panorama of cerebral palsy in Sweden: VII. Prevalence and origin in the birth year period 1991–1994. Acta Paediatr 90:271–277
103. Rennie JM, South M, Morely CJ (1987) Cerebral blood flow velocity variability in infants receiving assisted ventilation. Arch Dis Child 62:1247–1251
104. Ilves P, Talvik R, Talvik T (1998) Changes in Doppler ultrasonography in asphyxiated term infants with hypoxic-ischaemic encephalopathy. Acta Paediatr 87:680–684
105. Hellstrom-Westas L, Rosen I, Svenningsen NW (1995) Predictive value of early continuous amplitude EEG recordings on outcome after severe birth asphyxia in full term infants. Arch Dis Child 72:F34–F38
106. Huang CC, Wang ST, Chang YC et al (1999) Measurement of the urinary lactate:creatinine ratio for the early identification of newborn infants at risk from hypoxic-ischemic encephalopathy. N Engl J Med 341:328–335
107. Nagdyman N, Grimmer I, Scholz T et al (2003) Predictive value of brain-specific proteins in serum for neurodevelopmental outcome after birth asphyxia. Pediatr Res 54 (2):270–275
108. Gazzolo D, Di Iorio R, Marinoni E et al (2002) S100B protein is increased in asphyxiated term infants developing intraventricular hemorrhage. Crit Care Med 30 (6):1356–1360
109. Gazzolo D, Marinoni E, Di Iorio R et al (2003) Measurement of urinary S100B protein concentrations for the early identification of brain damage in asphyxiated full-term infants. Arch Pediatr Adolesc Med 157 (12):1163–1168
110. Gazzolo D, Marinoni E, Di Iorio R et al (2004) Urinary S100B protein measurements: a tool for the early identification of hypoxic-ischemic encephalopathy in asphyxiated full-term infants. Crit Care Med 32(1):131–136
111. Albuerne M, Mammola CL, Naves FJ et al (1998) Immunohistochemical localization of S100 proteins in dorsal root, sympathetic and enteric ganglia of several mammalian species,

including man. J Peripher Nerv Syst 3:243–253
112. Hachem S, Aguirre A, Vives V et al (2005) Spatial and temporal expression of S100B in cells of oligodendrocyte lineage. Glia 51:81–97
113. Haglid KG, Yang Q, Hamberger A et al (1997) S-100beta stimulates neurite outgrowth in the rat sciatic nerve grafted with acellular muscle transplants. Brain Res 753:196–201
114. Bashir M, Frigiola A, Iskander I et al (2009) Urinary S100A1B and S100BB to predict hypoxic ischemic encephalopathy at term. Front Biosci (Elite Ed) 1:560–567
115. Lee SK, Kim EC, Chi JG et al (1993) Immunohistochemical detection of S-100, S-100 alpha, S-100 beta proteins, glial fibrillary acidic protein, and neuron specific enolase in the prenatal and adult human salivary glands. Pathol Res Pract 189(9):1036–1043
116. Volpe JJ (1995) Intracranial hemorrhage: germinal matrix-intraventricular hemorrhage of the premature infant. In: Volpe JJ (ed) Neurology of the newborn. WB Saunders, Philadelphia, pp 403–463
117. Risso FM, Serpero LD, Zimmermann LJ et al (2013) Urine S100 BB and A1B dimers are valuable predictors of adverse outcome in full-term asphyxiated infants. Acta Paediatr 102 (10):e467–e472. https://doi.org/10.1111/apa.12343
118. Duda I, Krzych Ł, Jędrzejowska-Szypułka H et al (2017) Serum levels of the S100B protein and neuron-specific enolase are associated with mortality in critically ill patients. Acta Biochim Pol 64(4):647–652
119. Sieber M, Dreßler J, Franke H et al (2018) Post-mortem biochemistry of NSE and S100B: a supplemental tool for detecting a lethal traumatic brain injury? J Forensic Leg Med 55:65–73
120. Crowley PA (1995) Antenatal corticosteroid therapy: a meta-analysis of the randomized trial, 1972 to 1994. Am J Obstet Gynecol 173:3223–3235
121. O'Shea TM, Kothadia JM, Klinepeter KL et al (1999) Randomized placebo-controlled trial of a 42-day tapering course of dexamethasone to reduce the duration of ventilator dependency in very low birth weight infants. Pediatrics 104:15–21
122. Vidaeff AC, Mastrobattista JM (2001) Controversies in the use of antenatal steroids for fetal maturation. Semin Perinatol 25:385–396
123. Whitelaw A, Thoresen M (2000) Antenatal steroids and the developing brain. Arch Dis Child Fetal Neonatal Ed 83:154–157
124. Gilstrap LC, Christensen R, Clewell WH et al (1995) The effect of corticosteroids for fetal maturation on perinatal outcomes. National Institutes of Health Consensus Development Panel. JAMA 273:413–418
125. Mulder EJ, Derks JB, Visser GH (1997) Antenatal corticosteroid therapy and fetal behaviour: a randomized study of the effects of betamethasone and dexamethasone. Br J Obstet Gynecol 104:1239–1247
126. Huang WL, Beazley LD, Quinlivan JA et al (1999) Effect of corticosteroids on brain growth in fetal sheep. Obstet Gynecol 94:213–218
127. Uno H, Lohmiller L, Thiem C et al (1990) Brain damage induced by perinatal exposure to dexamethasone in fetal rhesus macaques. I. Hippocampus. Brain Res Dev Brain Res 53:157–167
128. Gazzolo D, Kornacka M, Bruschettini M et al (2003) Maternal glucocorticoid supplementation and S100B protein concentrations in cord blood and urine of preterm infants. Clin Chem 49(7):1215–1218
129. Lyall F, Greer IA, Young A et al (1996) Nitric oxide concentrations are increased in the feto-placental circulation in intrauterine growth restriction. Placenta 17:165–168
130. Di Iorio R, Marinoni E, Coacci F et al (1997) Amniotic fluid nitric oxide and uteroplacental blood flow in pregnancy complicated by intrauterine growth retardation. Br J Obstet Gynaecol 104:1134–1139
131. Cacciatore B, Halmesmaki E, Kaaja R et al (1998) Effects of transdermal nitroglycerin on impedance to flow in the uterine, umbilical, and fetal middle cerebral arteries in pregnancies complicated by preeclampsia and intrauterine growth retardation. Am J Obstet Gynecol 179:140–145
132. Grunewald C, Kublickas M, Carlström K et al (1995) Effects of nitroglycerin on the uterine and umbilical circulation in severe preeclampsia. Obstet Gynecol 86:600–604
133. Bennett HA, Einarson A, Taddio A et al (2004) Prevalence of depression during pregnancy: systematic review. Obstet Gynecol 103:698–709
134. Austin MP (2006) To treat or not to treat: maternal depression. SSRI use in pregnancy and adverse neonatal effects. Psychol Med 36 (12):1663–1670
135. Bellissima V, Ververs TF, Visser GH et al (2012) Selective serotonin reuptake

inhibitors in pregnancy. Curr Med Chem 19 (27):4554–4561
136. Moses-Kolko EL, Bogen D, Perel J et al (2005) Neonatal signs after late in utero exposure to serotonin reuptake inhibitors: literature review and implications for clinical applications. JAMA 293:2372–2383
137. Louik C, Lin AE, Werler MM et al (2007) First-trimester use of selective serotonin-reuptake inhibitors and the risk of birth defects. N Engl J Med 356:2675–2683
138. Latendresse G, Ruiz RJ (2011) Maternal corticotropin-releasing hormone and the use of selective serotonin reuptake inhibitors independently predict the occurrence of preterm birth. J Midwifery Womens Health 56:118–126
139. Kallen BA, Otterblad Olausson P (2007) Maternal use of selective serotonin reuptake inhibitors in early pregnancy and infant congenital malformations. Birth Defects Res A Clin Mol Teratol 79:301–308
140. Levinson-Castiel R, Merlob P, Linder N et al (2006) Neonatal abstinence syndrome after in utero exposure to selective serotonin reuptake inhibitors in term infants. Arch Pediatr Adolesc Med 160:173–176
141. Bellissima V, Vesser GHA, Tessa F (2015) Antenatal maternal antidepressants drugs affect S100B concentrations in fetal-maternal biological fluids. CNS Neurol Disord Drug Targets 14(1):49–54
142. Azzopardi DV, Strohm B, Edwards AD et al (2009) Moderate hypothermia to treat perinatal asphyxial encephalopathy. N Engl J Med 361:1349–1358
143. Edwards AD, Brocklehurst P, Gunn AJ et al (2010) Neurological outcomes at 18 months of age after moderate hypothermia for perinatal hypoxic ischaemic encephalopathy: synthesis and meta-analysis of trial data. BMJ 340: c363. https://doi.org/10.1136/bmj.c363
144. Massaro AN, Chang T, Baumgart S et al (2014) Biomarkers S100B and neuron-specific enolase predict outcome in hypothermia-treated encephalopathic newborns. Pediatr Crit Care Med 15 (7):615–622
145. Massaro AN, Chang T, Kadom N et al (2012) Biomarkers of brain injury in neonatal encephalopathy treated with hypothermia. J Pediatr 161(3):434–440
146. Roka A, Kelen D, Halasz J et al (2012) Serum S100B and neuron-specific enolase levels in normothermic and hypothermic infants after perinatal asphyxia. Acta Paediatr 101 (3):319–323
147. Massaro AN, Wu YW, Bammler TK et al (2018) Plasma biomarkers of brain injury in neonatal hypoxic-ischemic encephalopathy. J Pediatr 194:67–75
148. Çelik Y, Atıcı A, Gülaşı S et al (2015) The effects of selective head cooling versus whole-body cooling on some neural and inflammatory biomarkers: a randomized controlled pilot study. Ital J Pediatr 41:79
149. Alshweki A, Pérez-Muñuzuri A, López-Suárez O et al (2017) Relevance of urinary S100B protein levels as a short-term prognostic biomarker in asphyxiated infants treated with hypothermia. Medicine 96(44):e8453. https://doi.org/10.1097/MD.0000000000008453
150. Abella R, Varrica A, Satriano A et al (2015) Biochemical markers for brain injury monitoring in children with or without congenital heart diseases. CNS Neurol Disord Drug Targets 14(1):12–23
151. Newburger JW, Jonas JA, Wernovsky G (1993) A comparison of the perioperative neurologic effects of hypothermic circulatory arrest versus low-flow cardiopulmonary bypass in infant heart surgery. N Engl J Med 329:1057–1064
152. Bellinger D, Wypij D, duPlessis AJ et al (2003) Neurodevelopmental status at eight years in children with dextro-transposition of the great arteries: the Boston Circulatory Arrest Trial. J Thorac Cardiovasc Surg 126:1385–1396
153. Bellinger DC, Jonas RA, Rappaport LA et al (1995) Developmental and neurologic status of children after heart surgery with hypothermic circulatory arrest or low-flow cardiopulmonary bypass. N Engl J Med 332:549–555
154. Bellinger DC, Rappaport LA, Wypij D et al (1997) Patterns of developmental dysfunction after surgery during infancy to correct transposition of the great arteries. J Dev Behav Pediatr 18:75–83
155. Bellinger DC, Wernovsky G, Rappaport LA et al (1991) Cognitive development of children following early repair of transposition of the great arteries using deep hypothermic circulatory arrest. Pediatrics 87:701–707
156. Arrica M, Bissonnette B (2007) Therapeutic hypothermia. Semin Cardiothorac Vasc Anesth 11:6–15
157. Alcaraz AJ, Manzano L, Sancho L et al (2005) Different proinflammatory cytokine serum pattern in neonate patients undergoing open heart surgery. Relevance of IL-8. J Clin Immunol 25:238–245

158. Golej J, Trittenwein G (1999) Early detection of neurologic injury and issues of rehabilitation after pediatric cardiac extracorporeal membrane oxygenation. Artif Organs 23:1020–1025
159. Trakas E, Domnina Y, Panigrahy A et al (2017) Serum neuronal biomarkers in neonates with congenital heart disease undergoing cardiac surgery. Pediatr Neurol 72:56–61
160. Yuan SM (2014) S100 and S100β: biomarkers of cerebral damage in cardiac surgery with or without the use of cardiopulmonary bypass. Rev Bras Cir Cardiovasc 29 (4):630–641 + 40 192,193
161. Varrica A, Satriano A, Tettamanti G et al (2015) Predictors of ominous outcome in infants undergone to cardiac surgery and cardiopulmonary by-pass: S100B protein. CNS Neurol Disord Drug Targets 14:1
162. Varrica A, Satriano A, Gavilanes ADW et al (2017) S100B increases in cyanotic versus noncyanotic infants undergoing heart surgery and cardiopulmonary bypass (CPB). J Matern Fetal Neonatal Med 28:1–7
163. Gazzolo D, Bruschettini M, Lituania M et al (2004) Levels of S100B protein are higher in mature human milk than in colostrum and milk-formulae milks. Clin Nutr 23:23–26

Chapter 45

S100A7 in Psoriasis: Immunodetection and Activation by CRISPR technology

Mariagrazia Granata, Evangelia Skarmoutsou, Maria Clorinda Mazzarino, and Fabio D'Amico

Abstract

Psoriasis, an inflammatory autoimmune skin disease, is the result of a chronic interaction between hyperproliferative keratinocytes and infiltrating activated immune cells. The mechanisms underlying psoriasis pathogenesis remain largely unknown, although a combination of genetic and environmental factors plays an important role in its development. S100A7 is overexpressed in psoriasis, and there is growing evidence that S100A7 may be involved in the pathogenesis of psoriasis. Since the mechanisms underlying S100A7 regulation and function remain elusive, a better understanding of these mechanisms may be useful to uncover additional treatment approaches for psoriasis. Immunohistology provides invaluable tools for a better understanding of the pathogenetic mechanisms of psoriasis. Here, we describe basic methods for immunofluorescence and immunohistochemistry analysis of S100A7 expression in psoriatic patients as well as in S100A7 CRISPR-activated human keratinocyte cell line.

Key words S100A7, Psoriasis, Calcium, EF-hand, CRISPR, Immunofluorescence, Immunohistochemistry

1 Introduction

Psoriasis, a chronic inflammatory skin disease, is characterized by hyperproliferation of keratinocytes and immune cell activation and infiltration in the skin [1]. Under the influence of the psoriatic inflammatory milieu, keratinocyte produces several antimicrobial peptides and proteins, including members of the S100 protein family [2, 3]. Comprised of approximately 25 members of small acidic EF-hand-type calcium-binding proteins, S100 protein family is involved in a wide variety of intracellular and extracellular processes, including calcium homeostasis, regulation of cell proliferation, differentiation, and apoptosis. Moreover, S100s show to have chemotactic, angiogenic, and antimicrobial activities [4]. In particular, S100A7, also named psoriasin [5], may be overexpressed in

Claus W. Heizmann (ed.), *Calcium-Binding Proteins of the EF-Hand Superfamily: From Basics to Medical Applications*, Methods in Molecular Biology, vol. 1929, https://doi.org/10.1007/978-1-4939-9030-6_45,

response to inflammatory stimuli, and it has been associated with the infiltration of $CD4^+$ lymphocytes and neutrophils in the epidermis [6, 7]. S100A7 is overexpressed in several inflammatory diseases, including psoriasis, thus suggesting an important role of this protein in keratinocyte differentiation and pathogenesis of psoriasis [8, 9].

In psoriasis and in different psoriasis models, S100A7 production is induced by several cytokines, including IL-1α, IL-1β, IL-17A, IL-36, and TNF. S100A7 may activate many different intra- and extracellular signaling pathways which induce the upregulation of many pro-inflammatory cytokines involved in the pathogenesis of psoriasis [9].

In conclusion, S100A7 may have a pivotal role in amplifying and maintaining the inflammatory process in psoriatic skin through the presence of a S100A7-dependent network of different pro-inflammatory mediators. The mechanisms underlying S100A7 regulation and function remain incompletely elucidated. A better understanding of these mechanisms may be useful to uncover additional treatment approaches for psoriasis.

It is well established that immunohistochemistry has become a powerful and essential tool for research and diagnostic purposes. Many aspects of the mechanisms underlying the pathogenesis of diseases have been disclosed by immunohistology and related techniques. In particular, the application of histo- and cytochemical approaches provided, and is continuing to do so, an invaluable contribution to understanding the molecular processes involved in the pathogenesis of psoriasis.

In this chapter, we describe basic methods for S100A7 detection by immunofluorescence in normal human keratinocyte cell culture. In particular, a double immunofluorescence protocol for the simultaneous immunodetection of S100A7 and JAB1 (c-Jun activation domain-binding protein 1) is given. S100A7 contains a binding site for JAB1 [10]. The S100A7/JAB1 interaction has been shown to be involved in inflammatory diseases and cancer progression [11, 12].

Moreover, a detailed basic protocol of immunohistochemistry for the detection of S100A7 in archival formalin-fixed paraffin-embedded skin biopsies from psoriatic patients is given. Furthermore, we provide the description of a procedure of S100A7 activation in keratinocyte cell culture by CRISPR methodology, which may allow further analysis and functional assays.

Immunodetection protocols are optimized for the use of monoclonal antibody anti-S100A7 (sc-52948, Santa Cruz Biotechnology, USA).

2 Materials

2.1 S100A7 (and JAB1) Double Immunofluorescence Staining

1. Phosphate-buffered saline (PBS): dissolve 8 g of NaCl, 0.2 g of KCl, 1.44 g of Na_2HPO_4, and 0.24 g of KH_2PO_4 in 900 ml of distilled water. Adjust to pH 7.4. Add distilled water to a total volume of 1 l.
2. PBST: add 0.1% Tween 20 to PBS.
3. Blocking solution and antibody diluent: PBST + 1% bovine serum albumin (BSA).
4. 2% buffered formaldehyde: mix 20 ml of 10% formaldehyde (methanol-free) and 80 ml of PBS (*see* **Note 1**).
5. Permeabilization solution: PBS + 0.2% Triton X-100.
6. Primary antibodies: anti-S100A7 antibody (IgG_1 mouse monoclonal antibody, cat. no. sc-52948, Santa Cruz Biotechnology, USA). Anti-JAB1 antibody (IgG rabbit polyclonal antibody, cat. no. sc-9074, Santa Cruz Biotechnology).
7. Secondary antibodies: goat anti-rabbit IgG Alexa Fluor® 488-conjugated antibody (cat. no. A11034, Thermo Fisher Scientific, USA). Goat anti-mouse IgG Alexa Fluor® 546-conjugated antibody (cat. no. A11030, Thermo Fisher Scientific).
8. Antifade mounting medium containing 4′,6-diamidine-2′-phenylindole dihydrochloride (DAPI) for DNA staining.

2.2 S100A7 Immunohistochemical Staining

1. Silanized glass slides, coverslips.
2. Plastic microwaveable staining jars.
3. Hydrophobic wax pen.
4. Ethanol, xylene (alternatively, substitutes clearing agents).
5. Wash buffer: PBST (*see* Subheading 2.1).
6. Antibody diluent: *see* Subheading 2.1.
7. Antigen retrieval solution: 0.01 M sodium citrate buffer, pH 6. Dissolve 2.9 g of trisodium citrate dihydrate in 900 ml of distilled water. Adjust to pH 6.0 with 1 M HCl, and make up to 1 l. This solution may be stored at +4 °C for several months.
8. Primary antibody: monoclonal antibody anti-S100A7 (sc-52948, Santa Cruz Biotechnology).
9. Normal serum/secondary antibody: ImmPRESS™ HRP anti-mouse IgG (peroxidase) polymer detection kit, made in horse (MP-7402, Vector Laboratories, USA).
10. Peroxidase substrate: ImmPACT™ DAB (SK-4105, Vector Laboratories). Add one drop of chromogen to 1 ml of diluent. Mix well before use.
11. Hydrophobic mounting medium.

2.3 S100A7 CRISPR-Based Transfection

1. Six-well plate for cell cultures.
2. Normal growth medium antibiotic-free: minimal essential medium (MEM) with 10% fetal bovine serum and 1% L-glutamine (200 mM).
3. Bürker chamber.
4. Trypan Blue.
5. Psoriasin CRISPR Activation Plasmid (sc-418211-ACT, Santa Cruz Biotechnology).
6. Control CRISPR Activation Plasmid (sc-437275, Santa Cruz Biotechnology).
7. Plasmid transfection medium (sc-108062, Santa Cruz Biotechnology).
8. Transfection reagent (sc-395739, Santa Cruz Biotechnology).

3 Methods

3.1 S100A7 (and JAB1) Double Immunofluorescence Staining (Simultaneous and Indirect Method) in Normal Human Keratinocyte Cell Culture

1. After culturing cells directly on glass coverslips placed inside six-well plates, remove medium.
2. Wash in PBS three times, 1 min each.
3. Fix cells in 2% (para-)formaldehyde in PBS, pH 7.4 for 10 min at room temperature (r. t.).
4. Wash in cold PBS three times, 5 min each.
5. Permeabilize cells with PBS containing 0.2% Triton X-100, for 10 min.
6. Wash cells with PBS three times, 5 min each.
7. Block with blocking solution for 30 min at r. t.
8. Eliminate (*do not* wash) the blocking solution from the slides.
9. Incubate cells with the mixture of primary antibodies against S100A7 and JAB1 (use the antibody diluent) in a humidified chamber overnight at 4 °C. Dilute anti-S100A7 (1:100) and anti-JAB1 (1:50) antibodies by mixing 1 part of S100A7 antibody, 2 parts of JAB1 antibody, and 97 parts of antibody diluent.
10. Wash in PBS three times, 5 min each.
11. Incubate with the mixture of the two fluorochrome-conjugated secondary antibodies diluted in 1% BSA (1:300) for 1 h at r. t. in the dark.
12. Wash in PBS three times, 5 min each in the dark.
13. Rinse in distilled water.
14. Mount coverslips with a drop of aqueous mounting medium containing DAPI.

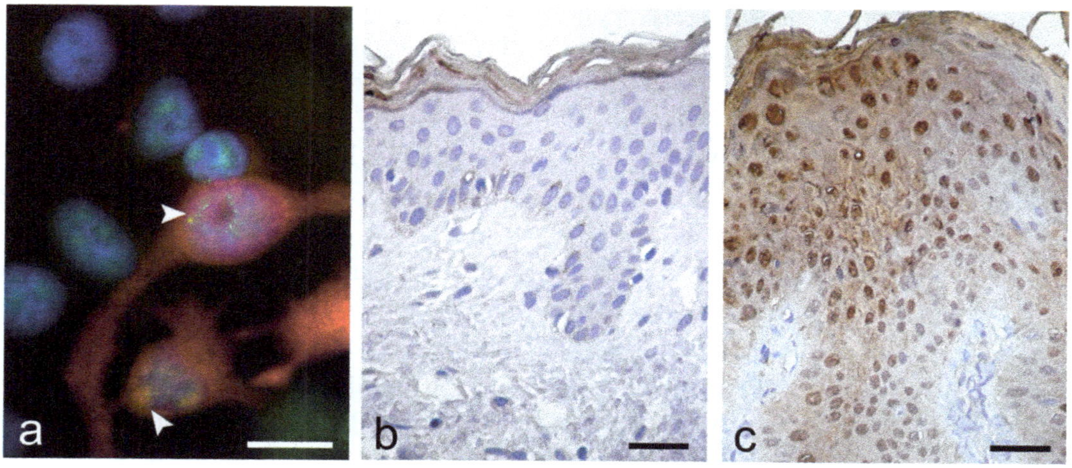

Fig. 1 Representative micrographs of immunofluorescence detection of S100A7 and JAB1 (**a**) and immunohistochemical localization of S100A7 in normal (**b**) and psoriatic skin (**c**). (**a**) S100A7 and JAB1 are immunolabeled with a double and indirect immunofluorescence staining method in normal human keratinocytes cell culture. S100A7, activated by the CRISPR-based approach, is localized in cytoplasm and nuclei (red), whereas JAB1 is mainly localized in cell nuclei (green). Some nuclear areas (arrowheads) are stained by the overlapping detections of both proteins (yellow), thus suggesting the occurrence of a S100A7/JAB1 interaction. Nuclei are colored by DAPI (blue). (**b**) S100A7 immunodetection is almost absent in skin sections from healthy donors. (**c**) S100A7 is immunolocalized in keratinocyte cytoplasm and nuclei (brown) of the epidermis from psoriatic patients. Nuclei are counterstained by hematoxylin (blue/violet) [bars (**a**) 15 μm; (**b**, **c**) 50 μm]

15. Seal coverslips with nail polish (optional).
16. Observe slides under a fluorescence microscope with appropriate filters. Positive fluorescent staining should appear as red (S100A7) or green (JAB1). Nuclei appear colored in blue by DAPI (Fig.1a).
17. In parallel, perform adequate control experiments (*see* **Note 2**).
18. Slides may be stored in the dark at +4 °C.

3.2 S100A7 Immunohistochemical Staining of Skin Biopsies

1. Cut sections 5 μm thick from formalin-fixed and paraffin-embedded skin tissue. Allow the slides to dry overnight in an oven at 37 °C. Dried slides may be stored at r. t. until ready for use.
2. Remove paraffin from sections by immersing slides in xylene (or alternative clearing agents), two changes, 5 min each (*see* **Note 3**).
3. Hydrate sections to water through a decreasing graded series of ethanol (100%, 100%, 95%, 70%, 3 min each).
4. Rinse in distilled water.

5. Place slides in the microwaveable staining jar with the antigen retrieval solution. Place the staining jar in a domestic microwave at 700 W for 10 min (*see* **Note 4**). After removing the staining jar, allow the slides to cool for 20 min at r. t.
6. Wash in distilled water, then in PBS twice, 5 min each.
7. Draw a circle around the sections with an appropriate hydrophobic pen.
8. Block non-specific antibody binding with horse nonimmune serum (20 min at r. t. in a humidified chamber) (*see* **Note 5**).
9. Drain off (*do not* wash) the blocking buffer from the slides.
10. Incubate tissue sections with diluted primary antibody (1:100 with the antibody diluent solution) in a humidified chamber (60 min at r. t.).
11. Wash sections with PBS twice, 5 min each.
12. Incubate sections with ImmPRESS^{TM} solution (30 min at r.t. in a humidified chamber) (*see* **Note 6**).
13. Wash sections with PBS twice, 5 min each.
14. Incubate sections in peroxidase substrate solution (10 min at r. t. in a humidified chamber). Avoid direct light during this step (*see* **Note 7**).
15. Wash in distilled water twice, 3 min each.
16. Counterstain sections with hematoxylin for 30 s.
17. Wash in tap water (10 min).
18. Dehydrate through three changes of ethanol (95%, 100%, and 100%, 3 min each).
19. Clear sections in the clearing agent (twice, 3 min each), and mount slides with a hydrophobic mounting medium.
20. Observe slides with a light microscope. Immunostaining should appear as a dark brown color. Nuclei are colored by hematoxylin counterstain in blue/purple.
21. In parallel, perform adequate control experiments (*see* **Note 2**).
22. Refer to Fig. 1b, c for representative results.

3.3 S100A7 CRISPR-mediated Activation in Normal Human Keratinocyte Cell Culture

Clustered regularly interspaced short palindromic repeats (CRISPR)-based technologies enable mammalian genome editing and have various applications [13]. In order to study the downstream effects of the upregulation of S100A7 in normal human keratinocyte cell culture, we routinely use Psoriasin CRISPR Activation Plasmid (sc-418211-ACT, Santa Cruz Biotechnology). This technique is specific for S100A7 gene and allows to upregulate its protein levels. A control CRISPR Activation Plasmid (sc-437275, Santa Cruz Biotechnology) is used as a negative control in order to evaluate the efficacy and specificity of the CRISPR Activation

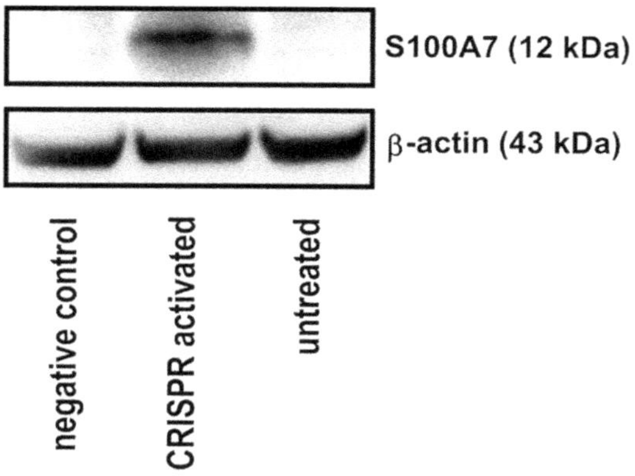

Fig. 2 Western Blot analyses of cell lysates after S100A7 CRISPR-mediated transfection. In particular, S100A7 and β-actin proteins are shown. S100A7 protein levels are increased only in the transfected lysates (central lane). Control CRISPR and untreated lysates do not show immunostaining for S100A7 (left and right lanes, respectively)

Plasmid-based transfection (*see* **Note 8**). An untreated six-well tissue culture plate is used as control.

We optimized the transfection protocol for a six-well tissue culture plate and for NCTC 2544 cell line.

1. In a six-well tissue culture plate, seed 2×10^5 cells in 3 ml of an antibiotic-free standard growth medium. Confluency must achieve 80% prior to transfection (*see* **Note 9**).
2. For each transfection, dilute 1 μg of plasmid DNA into 140 μl of plasmid transfection medium (sc-108062; *see* **Note 10**) to obtain the solution A. Incubate for 5 min at r. t.
3. Dilute 10 μl of transfection reagent (sc-395739) in 140 μl of plasmid transfection medium. Incubate for 5 min at r. t.
4. Add solution A directly to solution B. Vortex immediately the obtained solution, and let them incubate for 35 min at r. t.
5. Replace medium with fresh antibiotic-free normal growth medium.
6. Add 300 μl of the solution obtained from solution A and B per well, and gently mix up turning the plate. Incubate plates for 48 h at 37 °C in a cell culture incubator (5% CO_2).
7. Verify the transfection efficacy with real-time PCR.
8. Perform protein analysis with Western Blot 96 h after transfection. *See* Fig. 2.

4 Notes

1. It is strongly advisable to use depolymerized methanol-free paraformaldehyde (e.g., cat. no. 04018, Polysciences, USA). Alternatively, prepare a formaldehyde solution by adding 2 g of paraformaldehyde to 100 ml of PBS. Heat to 70 °C in a fume hood until complete solubilization. After allowing the solution to cool to r. t., adjust to pH 7.4 and filter. Formaldehyde is highly toxic. Refer to material safety data sheets for correct and safe handling.
2. Controls. Appropriate controls must be run in order to exclude possible false-positive results caused by background fluorescence and non-specific immunostaining. Use a fluorochrome-conjugated monoclonal, concentration and isotype-matched and not specific for the species under investigation. In addition, antibody specificity should be confirmed by the substitution of the primary antibody with nonimmune serum. Pre-absorption of primary antibody with an excess of purified immunogen is advisable. The above controls should yield negative reactions.
3. Xylene is a flammable liquid used in histology as a clearing agent. However, it may cause severe ill effects on health. We suggest to replace xylene with alternative clearing agents. There are several commercially available replacement products for xylene, which yield excellent results (i.e., Histo-Clear, National Diagnostics, USA; Histolemon, Carlo Erba, Italy).
4. Microwave incubation may cause a rapid evaporation of the antigen retrieval solution. Please check the level of the solution, and, if necessary, make up to the initial volume with distilled water.
5. There are many approaches for blocking unspecific antibody binding, such as 1–3% BSA in PBS or 1% gelatin in PBS. We routinely use ready-to-use 2.5% normal horse blocking serum (S-2012, Vector Laboratories, USA).
6. Routinely, we use HRP-polymer secondary antibodies in order to increase sensitivity and eliminate endogenous biotin background.
7. Peroxidase substrate contains diaminobenzidine (DAB), which is a suspected carcinogen. Use good laboratory procedures (fume hood, gloves, eye protection, lab coats). Dispose waste in accordance with local regulations. Furthermore, we highly recommend the use of pre-formulated DAB solutions in order to minimize health risks. We routinely use ImmPACT™ DAB peroxidase substrate, which contains the diluent and chromogen solutions. Optimal incubation time may vary. If inadequate

staining is observed, time should be determined by the investigator.

8. The control CRISPR Activation Plasmid contains 20 nucleotides of noncoding scrambled RNA sequence.
9. Optimal cell number for initial cell seeding and optimal time to reach the desired confluency depend on the cell line used. The confluency before transfection strongly influences the success of the reaction.
10. The transfection medium is a modification of Eagle's minimum essential medium, buffered with HEPES and sodium bicarbonate and supplemented with hypoxanthine, thymidine, sodium pyruvate, L-glutamine, growth factors, and phenol red. Furthermore, this medium contains reduced quantity of serum, and it must not be added with antibiotics.

Acknowledgment

This work was supported by intramural grants from the University of Catania, Italy.

References

1. Nestle FO, Kaplan DH, Barker J (2009) Psoriasis. N Engl J Med 361:496–509
2. Liang SC, Tan XY, Luxenberg DP, Karim R, Dunussi-Joannopoulos K, Collins M, Fouser LA (2006) Interleukin (IL)-22 and IL-17 are coexpressed by Th17 cells and cooperatively enhance expression of antimicrobial peptides. J Exp Med 203:2271–2279
3. Nakatsuji T, Gallo RL (2012) Antimicrobial peptides: old molecules with new ideas. J Invest Dermatol 132:887–895
4. Donato R, Cannon BR, Sorci G, Riuzzi F, Hsu K, Weber DJ, Geczy CL (2013) Functions of S100 proteins. Curr Mol Med 13:24–57
5. Marenholz I, Heizmann CW, Fritz G (2004) S100 proteins in mouse and man: from evolution to function and pathology (including an update of the nomenclature). Biochem Biophys Res Commun 322:1111–1122
6. Hoffmann HJ, Olsen E, Etzerodt M, Madsen P, Thøgersen HC, Kruse T, Celis JE (1994) Psoriasin binds calcium and is upregulated by calcium to levels that resemble those observed in normal skin. J Invest Dermatol 103:370–375
7. Di Nuzzo S, Sylva-Steenland RM, Koomen CW, de Rie MA, Das PK, Bos JD, Teunissen MB (2000) Exposure to UVB induces accumulation of LFA-1$^+$ T cells and enhanced expression of the chemokine psoriasin in normal human skin. Photochem Photobiol 72:374–382
8. Broome AM, Ryan D, Eckert RL (2003) S100 protein subcellular localization during epidermal differentiation and psoriasis. J Histochem Cytochem 51:675–685
9. D'Amico F, Skarmoutsou E, Granata M, Trovato C, Rossi GA, Mazzarino MC (2016) S100A7: a rAMPing up AMP molecule in psoriasis. Cytokine Growth Factor Rev 32:97–104
10. León R, Murray JI, Cragg G, Farnell B, West NR, Pace TC, Watson PH, Bohne C, Boulanger MJ, Hof F (2009) Identification and characterization of binding sites on S100A7, a participant in cancer and inflammation pathways. Biochemistry 48:10591–10600
11. Emberley ED, Niu Y, Curtis L, Troup S, Mandal SK, Myers JN, Gibson SB, Murphy LC, Watson PH (2005) The S100A7-c-Jun activation domain binding protein 1 pathway

enhances prosurvival pathways in breast cancer. Cancer Res 65:5696–5702

12. Webb M, Emberley ED, Lizardo M, Alowami S, Qing G, Alfia'ar A, Snell-Curtis LJ, Niu Y, Civetta A, Myal Y, Shiu R, Murphy LC, Watson PH (2005) Expression analysis of the mouse S100A7/psoriasin gene in skin inflammation and mammary tumorigenesis. BMC Cancer 5:17
13. Komor AC, Badran AH, Liu DR (2017) CRISPR-based technologies for the manipulation of eukaryotic genomes. Cell 168:20–36

Chapter 46

S100A8/A9 in Myocardial Infarction

Gopalkrishna Sreejit, Sunil Kiran Nooti, Baskaran Athmanathan, and Prabhakara Reddy Nagareddy

Abstract

S100A8/A9 represents a novel biomarker and therapeutic target in sterile inflammatory diseases. Among the various S100 proteins, S100A8 and S100A9 have been shown to be the most important of all the damage-associated molecular pattern (DAMP) proteins in sterile inflammatory conditions such as diabetes, cardiovascular disease, autoimmune disorders, etc. We present here methods to quantify S100A8/A9 expression in various tissues in mouse models of myocardial infarction (MI) using flow cytometry (FC), immunofluorescence, quantitative real-time polymerase chain reaction (q-RT-PCR), and enzyme-linked immunosorbent assays (ELISA).

Key words S100A8/A9, Myocardial infarction, Neutrophils, Mononuclear cells

1 Introduction

S100A8 (myeloid-related protein 8, MRP-8) and S100A9 (MRP-14) are members of the S100 calcium-binding family of proteins whose abundance in circulation is significantly associated with the prognosis of both acute coronary syndrome (ACS) [1, 2] and acute myocardial infarction (AMI) [3]. S100A8 and S100A9 constitute approximately 40% of the cytosolic protein fraction in circulating human blood neutrophils and 1% in monocytes [4] and are secreted by these cells at the site of inflammation, displaying antimicrobial activities against a variety of microorganisms [5]. The genes coding for S100A8/A9 in general show a parallel regulation of gene expression and can be expressed as homodimer or form a heterodimer of S100A8/A9 (called calprotectin), but they can also be elevated individually and act independently [6, 7]. S100A8 has a higher turnover than S100A9, and in the absence of S100A9 as in S100A9 knockout mice, S100A8 serum levels are almost undetectable [8, 9]. Gene deletion of S100A8 is embryonic lethal [10]; S100A9 KO mice have a normal phenotype although the

Claus W. Heizmann (ed.), *Calcium-Binding Proteins of the EF-Hand Superfamily: From Basics to Medical Applications*, Methods in Molecular Biology, vol. 1929, https://doi.org/10.1007/978-1-4939-9030-6_46,

neutrophils in these mice are less responsive to some chemoattractants [11, 12].

S100A8/A9 are associated with incidence and recurrence of acute cardiovascular events (CV) [2]. Elevated serum S100A8/A9 levels in acute myocardial infarction (AMI) patients have been found to correlate with circulating neutrophil count and the risk of subsequent MI [2, 13]. Following myocardial ischemia reperfusion injury in mice, S100A8/A9 is rapidly expressed and secreted by inflammatory cells and fibroblasts which then induce pro-inflammatory signaling, leukocyte infiltration, and cardiac dysfunction [14]. AMI causes an abrupt rise in leukocyte numbers in blood and myocardium. Neutrophils which are the first cells to infiltrate the infarcted myocardium release S100A8/A9 and other damage-associated molecular patterns (DAMPs) that bind to cognate pattern recognition receptors (PRRs) on the surface of various cell types to drive the expression of a large panel of inflammatory cytokines. These cytokines act to promote adhesive interactions between leukocytes and endothelial cells, resulting in the transmigration of pro-inflammatory Ly6Chi and anti-inflammatory Ly6Clo monocytes [15], resulting in tissue repair or progression to chronic inflammation. We present here methods, to quantify S100A8/A9 expression in the blood and heart tissue and in the infiltrating neutrophil and monocyte subsets of the heart following myocardial infarction in mice, using flow cytometry, immunofluorescence, q-RT-PCR, and ELISA.

2 Materials

2.1 Equipment

1. Surgical instruments for collecting mouse heart and blood: fine scissors, bone-cutting scissors, syringes, needles, forceps, and 40 μm filters/mesh, and heparinized capillary tubes for blood collection
2. Tissue culture supplies: pipettes, 50, 15, 1.5, and 0.5 ml polypropylene tubes, tissue culture dishes, and Petri dishes
3. Cellometer (Nexcelom K2 or equivalent) for cell counting
4. Water bath set to 37 °C
5. Refrigerated centrifuge and refrigerator
6. Flow cytometer equipped with appropriate lasers and detection filters
7. Microplate reader (BioTek)
8. Tissue dissociator (gentleMACS Octo Dissociator with C tubes)
9. Rotary Microtome
10. Bullet blender or equivalent
11. Real-time thermal cyclers

2.2 Reagents and Solutions

1. 50%, 70%, 90%, 95% and absolute ethanol
2. 10% formalin
3. Triton X-100
4. Xylenes
5. Donkey serum
6. 1× PBS
7. 1× HBSS
8. Chloroform
9. FC buffer (PBS, 1% FBS, 0.1% sodium azide)
10. Collagenase type II
11. RBC lysis buffer
12. Fixation and permeabilization solution (BD)
13. Perm/wash solution (BD)
14. Vectashield
15. Trizol reagent
16. RNA mini kit (Ambion Purelink)
17. RNaseZap
18. Antigen unmasking solution (Vectorlabs)
19. SuperScript IV VILO Master Mix (Thermo Fisher)
20. Fast SYBR Green Master Mix (Thermo Fisher)
21. Calprotectin ELISA kit (Alpco)
22. RNAqueous-Micro Total RNA Isolation Kit (Thermo Fisher)
23. Primers:

(a) S100A8 forward primer	5′ TCACCATGCCCTCTACAAGA 3′
(b) S100A8 reverse primer	5′ CCAATTCTCTGAACAAGTTTTCG 3′
(c) S100A9 forward primer	5′ TCAGACAAATGGTGGAAGCA 3′
(d) S100A9 reverse primer	5′ GTCCAGGTCCTCCATGATGT 3′
(e) GAPDH forward primer	5′ CGTCCCGTAGACAAAATGGT 3′
(f) GAPDH reverse primer	5′ TTGATGGCAACAATCTCCAC 3′

2.3 Antibodies for Flow Cytometry and Immunohisto-chemistry

1. CD45 APC/Cy7, CD115 BV711, GR1 PerCP Cy5.5 (BioLegend)
2. S100a9 AF647 (BD)
3. S100A8, Gr1 (Novus Biologicals)
4. AF 488 Donkey anti-rat (Thermo Fisher)
5. Cy3 Donkey anti-rabbit (Jackson ImmunoResearch)

2.4 Mice

C57BL/6J mice (Jackson Laboratory)

2.5 Software

Flow cytometry software (FlowJo) and plate reader software (Gen 5).

3 Methods

3.1 Harvesting Organs and Purified Cells from Mice

Myocardial infarction is induced in mice by permanent ligation of left anterior descending artery (LAD) (*see* **Note 1**). Sacrifice the sham and MI mice at the time of interest after MI to collect organs for analysis as follows.

3.1.1 Harvesting the Heart from Mice

1. Sacrifice a mouse, and quickly place the mouse on its back on a Styrofoam pad wrapped in foil, and spread and pin all the four limbs using 20-gauge needles.
2. Spray the chest and stomach with 70% alcohol, and make a vertical incision from the lower abdomen to the top of the thorax.
3. Open the abdominal wall below the ribcage, hold the sternum with tweezers, and make bilateral incisions through the rib cage toward the axilla.
4. Lift up and pin down the sternum near the mouse's ear to expose the thoracic cavity.
5. Cut the abdominal aorta below the heart.
6. Gently push a 24-gauge needle on a 10 ml syringe into the apex of the left ventricle, and slowly flush the blood off with 10 ml of PBS (*see* **Note 2**).
7. Harvest the heart by cutting the great vessels (ideally the heart should be beating at this point).
8. Transfer the heart to a Petri dish with ice-cold PBS, and trim away the left and right atria, excess vessels, and fat tissue, and gently dry the heart on a paper towel to remove excess liquid. Weigh the heart and put in ice-cold PBS.

3.1.2 Bone Marrow Isolation by Centrifugation

1. Remove the skin and muscle from the entire lower limb so that the bone is completely exposed, being careful not to cut the bone.
2. Remove the entire leg by cutting above the hip joint, being careful not to cut the top of the femur.
3. Clean the bones using Kim wipes, pull apart the femur and tibia, hold these bones with tweezers, and expose the bone marrow by making a cut where the bone marrow ends using a scissors (*see* **Note 3**).

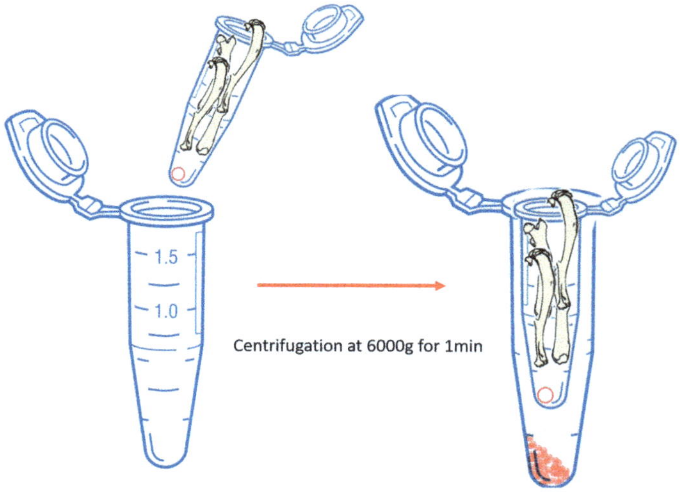

Fig. 1 Flushing out bone marrow cells by centrifugation

4. Puncture the base of 0.5 ml tube using 18-gauge needle (the needle tip can be flamed to make this easy) (Fig. 1).
5. Place all the four bones with the cut end toward the bottom of the tube, and add 200 μl of PBS to this tube.
6. Place 0.5 ml tube with the bone in a fresh 1.5 ml tube, and centrifuge at 6000 × *g* for 1 min.
7. Discard the 0.5 ml tube with the bones, resuspend the BM cells in the 1.5 ml tube and spin down again at 300 × *g* for 5 min, and collect the supernatant in a fresh tube and store at −80 °C (*see* **Note 4**).
8. Wash the flushed bone marrow cells with PBS, pass through a 40 μm cell strainer, and proceed with RBC lysis.

3.1.3 Blood Collection from Mouse

Retro-orbital bleeding can be used to collect blood before collecting the heart and bones. Collect the blood into EDTA-rinsed 1.5 ml Eppendorf tubes (final concentration of EDTA should be 5 mM). Gently mix the blood and anticoagulant; for plasma collection centrifuge for 10 min at 1500 × *g* at 4 °C (*do not* use brake to stop centrifuge). Carefully transfer the supernatant (plasma) to a fresh tube, and store at −20° or −80°C. Add RLT lysis buffer to the pellet for RNA extraction.

3.1.4 Digestion of the Heart for Single-Cell Preparation

1. Each mouse heart is divided into approximately 20 pieces.
2. Transfer these slices into gentleMACS C tube containing the disassociation mix (2 mg/ml collagenase type II in 1× HBSS (with calcium)).

3. Run program "me-neoheart-01" on gentleMACS Octo Dissociator©, and incubate sample for 15 min at 37 °C with gentle agitation.
4. Repeat the previous step two more times.
5. Prewet a 40 μm cell strainer on a 50 ml tube with 1 ml PBS.
6. Apply the cell suspension to the prewet strainer.
7. Rinse the cell strainer with 10 ml of PBS.
8. Centrifuge the filtered dissociate at 500 × *g*, 4 °C for 5 min.
9. Aspirate supernatant completely.
10. Proceed with red blood cell lysis step.

3.1.5 Red Blood Cell Lysis Protocol

1. Dilute the 10× lysing buffer concentrate 1:10 with distilled water.
2. Add 10 ml of 1× lysing buffer to the cell pellet and resuspend using 1 ml pipette; vortex lightly.
3. Incubate on ice for 5 min for heart and bone marrow cells; for blood incubate at room temperature for 15 min.
4. Add 10 ml of 1× PBS and centrifuge at 500 × *g* for 5 min.
5. Carefully aspirate the supernatant without disturbing the pellet.
6. Resuspend in 1ml FC buffer (1× PBS, 1% heat inactivated FBS and 0.1% sodium azide)
7. Transfer this to 1.5 ml centrifuge tube and spin again at 500 × *g* for 3 min.
8. Carefully aspirate the supernatant and resuspend again in FC buffer.
9. Resuspend pellet in 1 ml PBS, and do a cell count using AO/PI dye with Nexcelom K2 Cellometer.
10. Count the cells and proceed to FC staining protocol.

3.2 Staining Intracellular S100A9 for Flow Cytometry

1. Prepare cells for surface staining by aliquoting cells (approx. 1×10^6 cells) from the RBC lysis step to 1.5 ml centrifuge tubes. Spin down at 300 × *g* at 4°C for 3 min and resuspend in 100 μl of FC buffer.
2. Prepare cell suspension for single colors and unstained control (use 1×10^6 cells).
3. Prepare antibody staining master mix for surface staining accordingly by scaling up for the number of samples one is working with; always make the antibody master mix for a few extra samples. The composition for staining one sample is as follows:

CD45 APC/Cy7	1.25 µl
CD115 APC	0.30 µl
GR1 PerCp Cy5.5	1.25 µl

4. Add the antibody master mix to the sample tube, pulse vortex the mix, and incubate for at least 30 min at 2–8°C.
5. Wash the cells in 1 ml FC buffer. Pellet the cells by centrifugation at 300 × *g* for 3 min at 4 °C.
6. Resuspend cells in 200 µl of BD Cytofix/Cytoperm Buffer© for 20 minutes at 2–8 °C. Wash the cells with 1 ml of 1× BD Perm/Wash buffer, and spin and remove the supernatant (as described above).
7. Resuspend the cells in 100 µl of BD Perm/Wash Buffer containing intracellular antibody (S100A9 AF647 1.00 µl) (*see* **Note 5**).
8. Incubate for 20–30 min at 2–8 °C; wash with 1 ml of 1× BD Perm/Wash Buffer.
9. Resuspend the cells in 0.2 ml of flow buffer and acquire data.
10. Export the acquired data as FCS 3.0 files and analyze in FlowJo (Fig. 2).

3.3 S100A9 Staining Protocol for Paraffin-Embedded Heart Section

1. The heart from Step 8 is fixed by immersion in 10% formalin overnight.
2. Dehydrate tissue by:
   (a) Immersing the tissue in 70% ethanol for 60 min
   (b) Immersing the tissue in 90% ethanol for 45 min
   (c) Immersing the tissue in absolute ethanol 45 min and repeating this step three times
   (d) Immersing the tissue in xylene for 60 min and repeating this step three times
3. Embed the heart in paraffin at 58 °C
4. Cut 15-µm-thick tissue section using a rotary microtome.
5. Float the sections in a 40 °C water bath to flatten them out.
6. Pick up the sections out of the water bath using microscope slides, and store upright on a slide rack.
7. Place the slide rack into an oven at 37 °C and allow to dry overnight.

3.3.1 Staining Protocol

1. Deparaffinize and rehydrate by immersing the slides through the following solutions:
   (a) Immerse the slide in xylene for 5 min, and repeat this one more time.

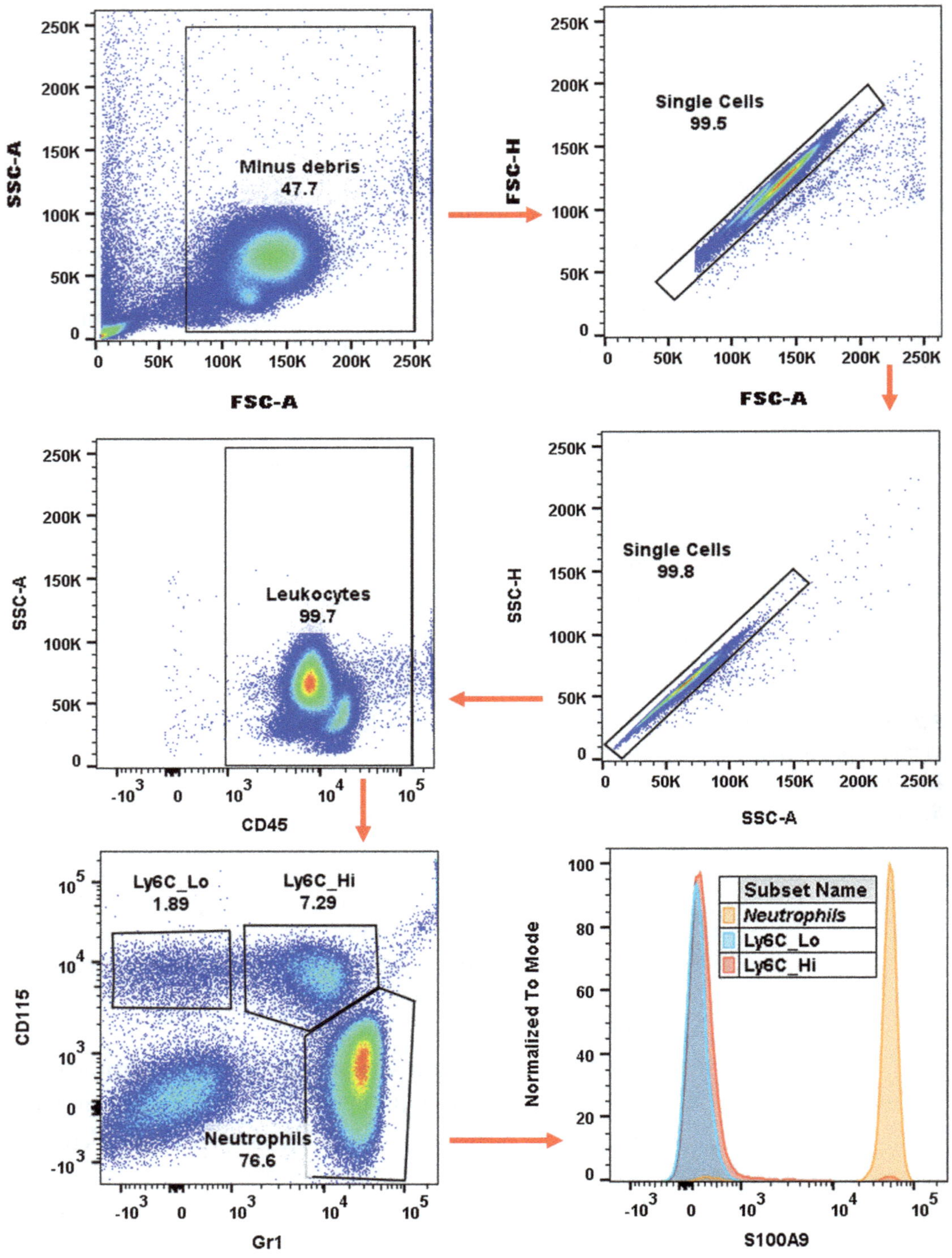

Fig. 2 Representative flow cytometry plots depicting blood leukocyte populations and S100A9 expression in different subsets

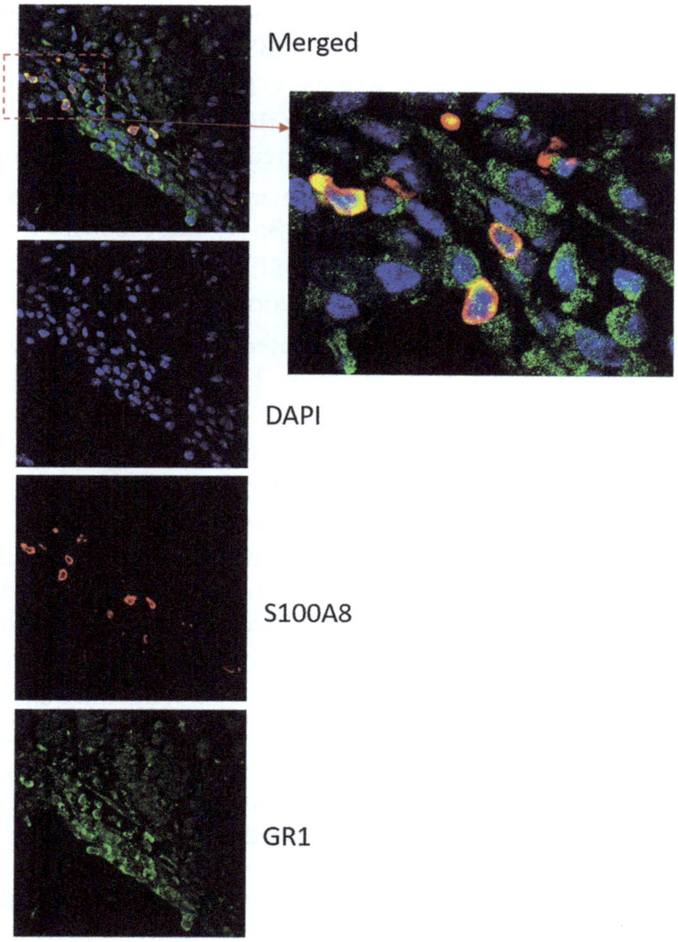

Fig. 3 S100A8 expression in neutrophils on MI heart slices as assessed by immunohistochemistry using anti-GR1 and anti-S100A8 antibody

(b) Immerse again in absolute ethanol for 5 min, and repeat once more.

(c) Immerse in 95% ethanol for 5 min.

(d) Immerse in 70% ethanol for 5 min.

(e) Immerse in 50% ethanol for 5 min.

(f) Rinse in dH_2O until ready for staining; the slide should not dry.

2. Antigen retrieval: the slides in a staining dish are placed in a preheated steamer containing diluted citrate buffer (1.9 ml in 200 ml dH_2O) at 95–100 °C for 40 min. Check the citrate buffer level, and add more if necessary. The slides are then allowed to cool to room temp for 20 min before proceeding with staining procedure.

3. Wash the sections twice for 10 min each with 1% animal serum in PBS-T (PBS with 0.4% Triton X-100). The species of the animal serum is dependent on the host of your secondary antibody.
4. Incubate the tissue sections with 5% serum in PBS-T for 30 min at room temperature to prevent nonspecific binding.
5. Add primary antibody (mouse S100A8) diluted (1:100) in 1% serum PBS-T, and incubate overnight at 4 °C or 90 min at room temperature.
6. Wash sections twice on a shaker with PBS for 5 min.
7. Add diluted (1:250) secondary antibody (Cy3 Donkey anti-rabbit) in 1% serum PBS-T, and incubate at room temperature for 1 h.
8. Add the second primary antibody (rat Gr1) diluted (1:100) in 1% serum PBS-T, and incubate overnight at 4 °C or 90 min at room temperature.
9. Add diluted (1:250) secondary antibody (AF 488 Donkey anti-rat) in 1% serum PBS-T, and incubate at room temperature for 1 h, and wash twice with PBS for 5 min.
10. Use Sudan Black (1 mg/ml in 70% ethanol, 30 min; wash three times using PBS) to reduce any autofluorescence.
11. Dry slides, add DAPI with mounting medium, mount cover-slip, and allow to dry for 5 min at room temperature. Slides can be stored at 4 °C in a dark slide box or slide book till they can be analyzed under a microscope (Fig. 3).

3.4 Real-Time Quantitative PCR of S100A8/A9 Transcripts

This protocol is designed for quantitative estimation of S100A8/S100A9 transcripts from heart tissues and purified cells but can be easily adapted to other tissues.

3.4.1 RNA Extraction from Heart Tissue and Purified Cells

For RNA extraction from heart tissue, use Trizol reagent with RNA purification kit (*see* **Note 6**). Before starting RNA isolation, spray and wipe the work area and pipettes with RNase decontamination solution, and only use RNAse-free certified pipette tips and tubes.

1. Add 1 ml Trizol per 100 mg of tissue in a safe lock tube having 100 mg of zirconium beads.
2. Use a bullet blender at max speed for 30 s and 30 s rest for three cycles.
3. Centrifuge the lysate for 5 min at 12,000 × *g* at 4° C, and transfer to a fresh tube.
4. Incubate the sample at 4° C for 5 min.
5. Add 0.2 ml of chloroform, vortex lightly and after incubating for 2–3 min at room temprature, centrifuge at 12,000 × *g* for 5 min.

6. Transfer the top colorless aqueous phase (approx. 300 μl) to a new tube containing equal volume of 70% ethanol and vortex thoroughly.
7. Transfer up to 700 μl of this sample to the spin cartridge, and centrifuge at 12,000 × *g* for 15 s at room temperature.
8. Discard the flow-through, and add 700 μl of Wash Buffer I to the spin cartridge. Centrifuge at 12,000 × *g* for 15 seconds at room temperature.
9. Place the spin cartridge with bound RNA into a new collection tube, add 500 μl Wash Buffer II (add ethanol as recommended in the kit) to the spin cartridge, and centrifuge at 12,000 × *g* for 15 s at room temperature.
10. Discard the flow-through and repeat the wash with Wash Buffer II.
11. Centrifuge the spin cartridge at 12,000 x *g* for 5 min to get rid of residual ethanol.
12. Insert the spin cartridge to fresh tube, add 30 μl of RNase-free water to the center of the spin cartridge, incubate at room temperature for 1 min, and spin down at 12,000 x *g* for 1 min (*see* **Note 7**).

When extracting RNA from sorted cells where the number of cells is limited (typically 50–100,000 cells per heart), use a kit that is designed for small cell numbers (*see* **Note 8**).

1. Resuspend purified cells in 100 μl lysis solution.
2. Add 50 μl of 100% ethanol and vortex briefly.
3. Load the lysate/ethanol mixture onto the Micro Filter Cartridge Assembly, and centrifuge for 10 s at maximum speed until the mixture has passed through the filter.
4. Empty the collection tube, and add 180 μl of wash solution I (with ethanol), and centrifuge to pass the solution through filter.
5. Wash the filter two times with wash solution 2/3 (with ethanol).
6. Discard the flow-through and centrifuge again to remove residual Wash Buffer.
7. Place the cartridge on a fresh tube, add 10 μl elution buffer heated to 75 °C, and after incubate at room temperature for 1 min. Elute the bound RNA with 10 μl of elution buffer heated to 75 °C.

3.4.2 Generation of cDNA for qPCR by Reverse Transcription

Thaw the VILO Master Mix and prepare the cDNA synthesis reaction mix as follows:

SuperScript IV VILO Master Mix	4 μl
RNA	1 pg–1 μg
Nuclease-free water	to 20 μl

Set up the cDNA synthesis reaction in a PCR machine with the following cycling conditions, and centrifuge and collect contents to the bottom of the tube before starting the reaction:

Anneal primers 25 °C for 10 min.

Reverse transcribe RNA 50 °C for 10 min.

Inactivate enzyme 85 °C for 5 min.

3.4.3 Quantitative Real-Time PCR

1. Write down the layout of the 96 or 364 well plate to avoid confusion during pipetting, and set up the reaction in duplicate.
2. Dilute the cDNA reaction mix such that it has an RNA concentration of 1 ng/μl, and use 5 μl of this diluted sample for quantitative real-time PCR (*see* **Note 9**).
3. This protocol is designed for use with Fast SYBR™ Green Master Mix, mix the fast SYBR© Green Master Mix thoroughly by swirling the bottle.
4. Calculate the amount of master mix required for the number of samples, and make the master mix for the gene of interest and for a housekeeping gene (make the master mix for few extra samples), minus the cDNA for a few extra tubes so as not to run out of master mix due to pipetting errors.
5. A typical reaction mix for one well is:

Master mix ($n = 1$)	
Master mix (SYBR Green)	10 μl
Forward primer (10 μm)	0.3 μl
Reverse primer (10 μm)	0.3 μl
Nuclease-free water	4.4 μl
cDNA	5.0 μl
Total volume	20 μl

6. Add 5 μl of cDNA first to the plate, and then add the master mix to get a final 20 μl reaction volume.
7. The thermocycler reaction protocol:

   Cycle 1: (1×)

   Step 1: 95 °C for 03:00

Cycle 2: (40×)

Step 1: 95 °C for 00:10

Step 2: 60 °C for 00:30

Data collection and real-time analysis enabled

8. After PCR is finished, remove the plate from the machine, export the Ct values, and analyze your data using delta-delta Ct method.

3.5 ELISA for the Detection of S100A8/S100A9

We use the calprotectin ELISA kit from ALPCO for detection of S100A8/S100A9 in serum. This assay uses the sandwich ELISA technique, where sample and the standards are added to wells of microplate coated with high-affinity anti-S100A8/S100A9 antibodies. Unbound antigen is washed out, and S100A8/S100A9 bound to the immobilized antibodies is detected by HRP-conjugated monoclonal anti-S100A8/S100A9 antibody. Tetramethylbenzidine is used as a substrate for peroxidase conjugated on detection Ab, and the color developed is proportional to the amount of S100A8/A9 bound. Color development is stopped by adding acid stop solution to terminate the reaction when the color changes from blue to yellow.

Before starting the assay:

A. Bring all reagents and samples to room temperature, and mix well before adding it to the wells.

B. Write down the layout of samples on the strip to avoid confusion, and also number the strips. The strips may drop off during washing, and labeling them would avoid confusion.

C. Make the following:

- Dilute the Wash Buffer concentrate by diluting 1:10.
- Reconstitute the lyophilized standards and controls by adding 500ul of ultrapure water; unused standard and control can be stored for 4 weeks at 2–8 °C.
- Reconstitute the lyophilized detection antibody in 450 μl diluted Wash Buffer, and allow to dissolve for 10 min at room temperature to insure complete reconstitution. Dilute the reconstituted detection antibody 1:401 in Wash Buffer. The reconstituted detection Ab can be stored at −20 °C for up to 4 weeks.
- Dilute the conjugate concentrate 1:101 in Wash Buffer. The diluted conjugate concentrate cannot be stored.

1. Wash each well five times by dispensing Wash Buffer into each well using a wash bottle; after each wash step, tap the plate forcefully on a stack of absorbent paper topped with Kim wipes to remove residual buffer.

2. Add 100 μl of standard, sample, controls into respective well.
3. Seal the strips tightly and incubate for 1 h at 37 °C on a horizontal shaker.
4. Discard the contents of each well, and tap on absorbent paper to get rid of residual sample. Wash each well by dispensing 250 μl of diluted Wash Buffer into each well. Repeat the washes five times, and after each wash step, remove residual buffer by tapping the plate on absorbent paper.
5. Add 100 μl diluted detection antibody into each well.
6. Cover the plate tightly, and incubate for 1 h at 37 °C on a horizontal shaker.
7. Discard the contents of each well. Wash each well by dispensing 250 μl of diluted Wash Buffer into each well. Repeat the washes five times, and after each wash step, remove residual buffer by tapping the plate on absorbent paper.
8. Add 100 μl diluted conjugate into each well.
9. Seal the strips tightly and incubate for 1 h at 37 °C on a horizontal shaker.
10. Discard the contents of each well. Wash each well by dispensing 250 μl of diluted Wash Buffer into each well. Repeat the washes five times, and after each wash step, remove residual buffer by tapping the plate on absorbent paper.
11. Add 100 μl of substrate into each well.
12. Incubate for 5–15 min at room temperature (15–30 °C) in the dark. Observe the color development and stop the reaction upon good differentiation.
13. Add 100 μl of stop solution into each well, and mix.
14. Determine absorption in an ELISA reader at 450 nm against 620 nm as reference. If the extinction of the highest standard exceeds the range of the photometer, absorption must be measured immediately at 405 nm against 620 nm as a reference.
15. Use the four-parameter algorithm on the ELISA reader software to calculate the result.

4 Notes

1. When MI is induced by LAD ligation, it is our experience that it is important to confirm MI in mice by echocardiography or other methods.
2. We find that flushing the left ventricle is sufficient and perfusion of the heart is not required. Hence make sure that the needle goes into the left ventricle only, and be careful not to

puncture the intraventricular septum and go into the right ventricle.

3. We normally flush out bone marrow out of the bones using centrifugation instead of crushing the bones or flushing the bone marrow out of the bones using PBS-filled syringe. We have compared the yield and viability of BM cells collected by centrifugation with other methods and found it can be comparable but easier.
4. This supernatant maybe analyzed for factors that promote hematopoiesis after MI. Another advantage of using the centrifugation method is the small volume in which BM cells were flushed out, which prevents dilution of any soluble factors present in the bone marrow for downstream analysis.
5. The protocols here which have used S100A9 can be adapted to S100A8 without any changes.
6. We find that using the lysis buffer provided with the kit for heart tissue does not yield good RNA in quantity or quality but using Trizol to homogenize heart tissue in combination with RNA kit gives better results.
7. To increase the concentration of eluted RNA, one can elute in 15 μl of water and do a second elution using the first eluate, but it should be noted that this will result in reduced total amount of RNA.
8. In our experience the regular RNA kits do not yield enough RNA with purified cells (due to the limited number of cells) from the heart, and we typically use RNAqueous-Micro Total RNA Isolation Kit, but similar micro kits could be used.
9. To prevent variation because of pipetting, avoid working with small volumes; we generally dilute samples so that the minimum volume needed to pipette is 5 μl. We also use SuperScript IV VILO Master Mix that alleviates the need to pipette multiple components of the master mix hence avoiding pipetting errors.

References

1. Altwegg LA, Neidhart M, Hersberger M et al (2007) Myeloid-related protein 8/14 complex is released by monocytes and granulocytes at the site of coronary occlusion: a novel, early, and sensitive marker of acute coronary syndromes. Eur Heart J 28:941–948
2. Morrow DA, Wang Y, Croce K et al (2008) Myeloid-related protein 8/14 and the risk of cardiovascular death or myocardial infarction after an acute coronary syndrome in the Pravastatin or Atorvastatin Evaluation and Infection Theraphy: Thrombolysis in Myocardial Infarction (PROVE IT-TIMI 22) trial. Am Heart J 155:49–55
3. Schaub N, Reichlin T, Meune C et al (2012) Markers of plaque instability in the early diagnosis and risk stratification of acute myocardial infarction. Clin Chem 58:246–256
4. Averill MM, Kerkhoff C, Bornfeldt KE (2012) S100A8 and S100A9 in cardiovascular biology and disease. Arterioscler Thromb Vasc Biol 32:223–229
5. Hsu K, Champaiboon C, Guenther BD et al (2009) Anti-infective protective properties of

S100 calgranulins. Antiinflamm Antiallergy Agents Med Chem 8:290–305

6. Foell D, Wittkowski H, Vogl T et al S100 proteins expressed in phagocytes: a novel group of damage-associated molecular pattern molecules. J Leukoc Biol 81:28–37
7. Geczy CL, Chung YM, Hiroshima Y (2014) Calgranulins may contribute vascular protection in atherogenesis. Circ J 78:271–280
8. van LPLEM, Blom AB, Schelbergen RFP et al (2012) Active involvement of alarmins S100A8 and S100A9 in the regulation of synovial activation and joint destruction during mouse and human osteoarthritis. Arthritis Rheum 64:1466–1476
9. Vogl T, Tenbrock K, Ludwig S et al (2007) Mrp8 and Mrp14 are endogenous activators of Toll-like receptor 4, promoting lethal, endotoxin-induced shock. Nat Med 13:1042
10. Passey RJ, Williams E, Lichanska AM et al (1999) A null mutation in the inflammation-associated S100 protein S100A8 causes early resorption of the mouse embryo. J Immunol 163:2209–2216
11. Hobbs JAR, May R, Tanousis K et al (2003) Myeloid cell function in MRP-14 (S100A9) null mice. Mol Cell Biol 23:2564–2576
12. Manitz M-P, Horst B, Seeliger S et al (2003) Loss of S100A9 (MRP14) results in reduced interleukin-8-induced CD11b surface expression, a polarized microfilament system, and diminished responsiveness to chemoattractants In vitro. Mol Cell Biol 23:1034–1043
13. Katashima T, Naruko T, Terasaki F et al (2010) Enhanced expression of the S100A8/A9 complex in acute myocardial infarction patients. Circ J 74:741–748
14. Volz HC, Laohachewin D, Seidel C et al (2012) S100A8/A9 aggravates post-ischemic heart failure through activation of RAGE-dependent NF-κB signaling. Basic Res Cardiol 107:250
15. Knorr M, Münzel T, Wenzel P (2014) Interplay of NK cells and monocytes in vascular inflammation and myocardial infarction. Front Physiol 5:295

Chapter 47

Enzyme-Linked Immunosorbent Assay to Measure S100A12 in Fecal Samples of Children and Adults

Steven T. Leach and Andrew S. Day

Abstract

In recent years there have been numerous attempts to establish reliable and consistent noninvasive biomarkers of gut inflammation. Here we describe the development of an enzyme-linked immunoassay for the measurement of the pro-inflammatory, calcium-binding protein S100A12 in fecal samples.

Key words ELISA, S100A12, Calcium, EF-hand, Fecal extraction, Inflammatory, Biomarker

1 Introduction

The assessment of inflammation in the gut typically involves undertaking an endoscopic procedure followed by the histological assessment of mucosal biopsies [1]. Although this is still considered the gold standard, it is invasive and costly, meaning that it is not feasible to repeat frequently. Serum markers, such as C-reactive protein and erythrocyte sedimentation rate, are easily available and inexpensive to measure; however, these provide inadequate sensitivity and specificity [2–4]. Much work has been undertaken in recent years to develop new ways to assess gut inflammation that might provide superior sensitivity and specificity.

The S100 family of proteins are calcium-binding proteins that have a variety of biological roles [5]. Several members of this family, especially S100A8, S100A9, and S100A12, have intra- and extracellular roles that include antimicrobial and anti-inflammatory activities [6]. S100A12 is expressed by various cell types, including neutrophils. In the setting of gut inflammation, it is present in high amounts in the inflamed mucosa and also able to be measured in serum and fecal samples [7, 8].

While measuring S100A12 in serum is shown to be helpful, it is thought that assessment of this protein in feces may provide more direct and specific indication of the current level of gut

Claus W. Heizmann (ed.), *Calcium-Binding Proteins of the EF-Hand Superfamily: From Basics to Medical Applications*, Methods in Molecular Biology, vol. 1929, https://doi.org/10.1007/978-1-4939-9030-6_47,

inflammation. This provided the rationale for earlier studies showing that fecal S100A12 levels were significantly greater in children with active inflammatory bowel disease (IBD), than in a control group of children without IBD [8]. Subsequent studies have shown that fecal S100A12 levels are increased in acute severe ulcerative colitis [9] and do not vary across children of various ages [10]. This report outlines the methodology utilized for the detection and measurement of S100A12 in fecal samples.

2 Materials

All solutions are prepared using ultrapure water (deionized water with resistance greater than 18 MΩ cm at 25 °C). Reagents should be stored at 4 °C (unless stated otherwise) although equilibrated to room temperature prior to use in the following procedures.

2.1 Fecal Extraction Materials

1. Fecal extraction buffer. 0.25 M Tris, 0.25 M citric acid, 2.5 M urea, 0.025 M $CaCl_2$ (*see* **Note 1**), 0.25% bovine serum albumin, 0.05% sodium azide, adjust pH to 8.0 using NaOH and filter through 0.22 μm filter (*see* **Note 2**).

2.2 ELISA Materials

1. Coating antibody. Rabbit polyclonal anti-human S100A12 (*see* **Note 3**).
2. Secondary antibody. Goat anti-human EN-RAGE polyclonal antibody, biotin conjugate (*see* **Note 4**).
3. S100A12 standard: Recombinant human EN-RAGE/S100A12 (*see* **Note 5**).
4. Streptavidin-horseradish peroxidase.
5. Substrate: TMB Substrate Kit.
6. Carbonate buffer. 0.1 M Na_2CO_3, 0.1 M $NaHCO_3$ (adjust the pH to 9.6, using Na_2CO_3 to decrease pH or $NaHCO_3$ to increase pH) (*see* **Note 6**).
7. Wash buffer. Phosphate buffered saline (*see* **Note** 7), 0.05% Tween 20 (*see* **Note 8**).
8. Blocking buffer. 1% Skim milk powder, phosphate buffered saline (*see* **Note** 7), 0.05% Tween 20 (*see* **Note 8**).
9. Diluent with fecal extraction buffer. Combine equal parts of fecal extraction buffer and blocking buffer (*see* **Note 8**).
10. Diluent without fecal extraction buffer. Combine equal parts of wash buffer and blocking buffer (*see* **Note 8**).
11. Stop solution. 2 M H_2SO_4 (*see* **Note 9**).

3 Methods

3.1 Fecal Extraction

1. Place plastic inoculation loop (*see* **Note 10**) in a 15 ml falcon tube (inoculation loop will act as an agitator to aid in homogenizing the fecal sample). Label tube and weigh to closest milligram (tube pre-weight).
2. Add approximately 100 mg of feces to the falcon tube with agitator (*see* **Note 11**).
3. Reweigh tube with feces (tube post-weight), and calculate feces weight using the equation: Feces weight (mg) = tube post-weight − tube pre-weight.
4. Add fecal extraction buffer at the ratio of 49 μl buffer per milligram of feces (*see* **Note 12**).
5. Vortex tube for at least 30 s, and then place on shaker table for 30 min of vigorous shaking (*see* **Note 13**).
6. Transfer 1.4 ml of homogenized feces to an Eppendorf tube, and centrifuge for 5 min at 13,000 × *g*.
7. Following centrifugation transfer 1.0 ml of clear supernatant to new Eppendorf tube (*see* **Note 14**), and if required store prior to assay (*see* **Note 15**).

3.2 ELISA

1. Coat each well of a 96-well ELISA plate (*see* **Note 16**) with 100 μl coating antibody (at 0.5 μg/ml in carbonate buffer). Cover plate with plastic wrap (*see* **Note 17**), and incubate overnight at 4 °C (*see* **Note 18**).
2. Aspirate coating solution and wash plate twice with wash solution. Tap on absorbent towel to dry wells.
3. Add 100 μl/well blocking buffer, and incubate for 2 h at room temperature.
4. Aspirate blocking buffer and wash plate twice with wash solution. Tap on absorbent towel to dry wells.
5. While plate is blocking, prepare the standards and samples for assay. Prepare S100A12 standard at 250 ng/ml, 125 ng/ml, 62.5 ng/ml, 31.25 ng/ml, 15.6 ng/ml, 7.8 ng/ml, 3.9 ng/ml (*see* **Note 19**). Prepare samples by diluting fecal extracts at the following suggested dilution factor 2, 20, and 200 (*see* **Note 20**).
6. Following blocking, add 100 μl/well of S100A12 standards and diluted samples in duplicate. Incubate for 1 h at room temperature.
7. Aspirate samples and standards, and wash plate three times with wash solution. Tap on absorbent towel to dry wells.

8. Add 100 μl/well of secondary antibody (0.5 μg/ml in diluent without fecal extraction buffer), and incubate for 1 h at room temp.
9. Aspirate secondary antibody, and wash plate three times with wash solution. Tap on absorbent towel to dry wells.
10. Add 100 μl/well of Streptavidin-horseradish peroxidase (diluted 1:1000 in diluent without fecal extraction buffer), and incubate at room temperature for 30 min.
11. Aspirate Streptavidin-horseradish peroxidase, and wash plate three times with wash solution. Tap on absorbent towel to dry wells.
12. Add 100 μl/well substrate, and leave for 1–5 min for color to develop (*see* **Note 21**).
13. Stop reaction with the addition of 50 μl/well stop solutions.
14. Read plate at 450 nm on ELISA plate reader within 30 min (*see* **Note 22**).

3.3 Calculations of Results

1. Average raw reads of duplicates for blank, standards, and samples should be used for calculation of results. Also, the average of the blank should be subtracted from the average of standard and sample duplicates for subsequent calculations.
2. Prepare a standard curve using the logistic 4-point (sigmoidal) curve, and calculate sample concentrations in ng/ml (Fig. 1) (*see* **Note 23**).
3. Use the following equation to calculate fecal S100A12 concentrations. Fecal S100A12 concentration (mg/kg) (*see* **Note 24**) = ELISA sample concentration (ng/ml) × Dilution factor × 0.05.

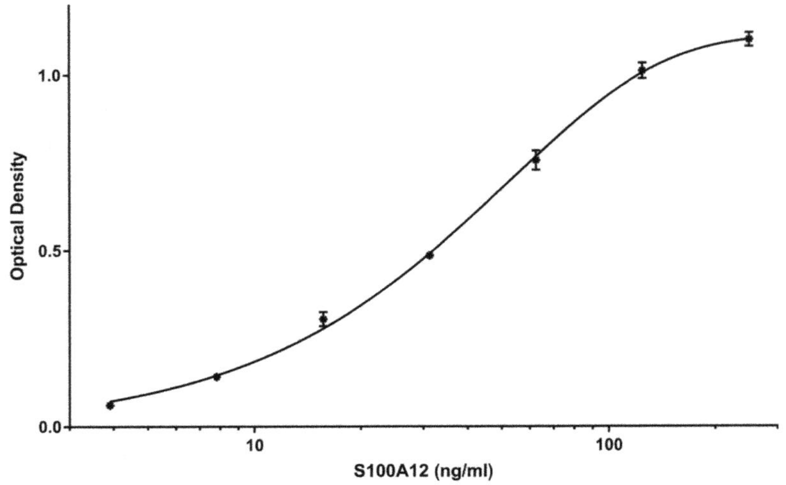

Fig. 1 Example of a standard curve with the logistic 4-point (sigmoidal) curve used to estimate sample concentrations

4 Notes

1. $CaCl_2$ alone can inhibit ELISAs. Diluent with fecal extraction buffer has been introduced into the procedure with the aim to ensure samples and standard are exposed to approximately equal concentrations of $CaCl_2$ and therefore minimize the effects of inhibition. If significant inhibition is observed, fecal extraction buffer without $CaCl_2$ can be tested; however this may alter S100A12 binding in the ELISA as S100A12 requires calcium for correct function.
2. Store at 4 °C for up to 12 months.
3. Stock solution of coating antibody at approximately 1 mg/ml in buffer containing serum albumin can be stored at 4 °C.
4. Stock solution of secondary antibody at approximately 1 mg/ml in buffer containing serum albumin can be stored at 4 °C.
5. Stock solution of recombinant human S100A12 at 2500 ng/ml can be aliquoted at 50 μl in buffer containing serum albumin and stored at −80 °C until required for assays.
6. Store at 4 °C for up to 12 months. If particulate matter is observed in buffer, discard and prepare a new buffer solution.
7. Recommend using commercially produced phosphate buffered saline to reduce variability introduced from bespoke batch preparations of phosphate buffered saline.
8. Make fresh buffer for each ELISA then discard after assay.
9. Store at room temperature.
10. Inoculation loops can be replaced by pipette tips or any other suitable agitator.
11. The minimum volume of fecal extract required is 1.4 ml with the maximum volume of the tube 15 ml. Therefore, fecal mass between 29 and 300 mg can be added to the tube. However, it is more practical to add between 50 and 250 mg feces to the tube to avoid inaccurate dilution at low mass and overfilling the tube at high mass.
12. As an example, for 100 mg of feces, add 4.9 ml of extraction buffer to produce a fecal extract at dilution factor of 50.
13. Shaking should be as vigorous as possible without risk of damage to tube or equipment. Tubes should be secured to the shaker table and agitated at greater than 100 revolutions per minute.
14. Avoid transferring any of the pellet.
15. Store the extract for not more than 48 h at 2–8 °C prior to assay. If longer-term storage is required, store at −20 °C. For assay of frozen extracts, allow to thaw slowly, preferably at

2–8 °C overnight, then allow extracts to equilibrate to room temperature before assay.

16. Recommend using Nunc-Immuno™ 96-microwell plates or similar.
17. Plastic wrap is required to avoid loss from evaporation. Plastic wrap can be either a food grade wrap or laboratory wrap such as Parafilm®.
18. As an alternative to overnight incubations, the plate can be incubated at room temperature for 2 h.
19. To prepare S100A12 standard, add 450 μl diluent with fecal extraction buffer to 50 μl of recombinant human S100A12 at 2500 ng/ml, to make 500 μl of 250 ng/ml standard. The subsequent standards can be prepared by twofold falling dilutions of this first standard. Prior to preparing samples and standards, tubes should be vortexed to ensure homogeneity of contents.
20. To prepare samples at dilution factor 2, add equal parts fecal extract and diluent without fecal extraction buffer. To prepare samples at dilution factor 20 and 200, add 1 part fecal extract, to 19 parts diluent with fecal extraction buffer and 199 parts diluent with fecal extraction buffer, respectively. These dilutions are suggested as they should cover the range of S100A12 concentrations that may be present in human feces.
21. Multiple factors affect the rate of color development making it very difficult to set the precise time required for the color reaction. The decision to stop the color reaction is based on experience and judgment. As a guide, the color reaction should be stopped when the most concentrated standard (250 ng/ml) reaches an optical density in the range of 1.5 to 2.5.
22. Preferably the plate should also be read at a second higher wavelength (e.g., 570 nm) with this read subtracted from the 450 nm wavelength read.
23. The logistic 4-point (sigmoidal) curve is most accurate in the middle (linear) part of the curve.
24. mg/kg is interchangeable with μg/g.

Acknowledgment

We would like to acknowledge the contribution of Professor Carolyn Geczy as this fecal assay is based on the detection of serum S100A12 assay developed in her laboratory.

References

1. IBD Working Group of the European Society for Paediatric Gastroenterology, Hepatology and Nutrition (2005) Inflammatory bowel disease in children and adolescents: recommendations for diagnosis—the Porto criteria. J Pediatr Gastroenterol Nutr 41:1–7
2. Turner D, Mack DR, Hyams J, LeLeiko N, Otley A, Markowitz J et al (2011) C-reactive protein (CRP), erythrocyte sedimentation rate (ESR) or both? A systematic evaluation in pediatric ulcerative colitis. J Crohns Colitis 5:423–429
3. Day AS, Hamilton D, Leach ST, Lemberg DA (2017) Inflammatory markers in children with newly diagnosed Inflammatory Bowel Disease. J Gastroenterol Hepatol Res 6:2329–2332
4. Mack DR, Langton C, Markowitz J, LeLeiko N, Griffiths A, Bousvaros A et al (2007) Laboratory values for children with newly diagnosed inflammatory bowel disease. Pediatrics 119:1113–1119
5. Leach ST, Mitchell HM, Geczy CL, Sherman PM, Day AS (2008) S100 calgranulin proteins S100A8, S100A9 and S100A12 are expressed in the inflamed gastric mucosa of *Helicobacter pylori* infected children. Can J Gastroenterol 22:461–464
6. Leach ST, Day AS (2006) S100 proteins in the pathogenesis and diagnosis of inflammatory bowel disease. Expert Rev Clin Immunol 2:471–480
7. de Jong NSH, Leach ST, Day AS (2006) Faecal S100A12: a novel non-invasive marker of gastrointestinal inflammation in children. Inflamm Bowel Dis 12:566–572
8. Sidler MA, Leach ST, Day AS (2008) Fecal S100A12 and fecal calprotectin as noninvasive markers for inflammatory bowel disease in children. Inflamm Bowel Dis 14:359–366
9. Turner D, Leach ST, Mack D, Uusoue K, Hyams J, Leleiko N et al (2010) Fecal calprotectin, lactoferrin, M2-pyruvate kinase, and S100A12 in severe acute ulcerative colitis: a prospective multicentre comparison of predicting outcomes and monitoring response. Gut 59:1207–1212
10. Day AS, Ehn M, Gearry RB, Lemberg DA, Leach ST (2013) Fecal S100A12 in healthy infants and children. Dis Markers 35:295–299

Chapter 48

S100 Proteins as Biomarkers in Risk Estimations for Malignant Transformation in Oral Lesions

Rainer Probstmeier, Dominik Kraus, Matthias Wenghoefer, and Jochen Winter

Abstract

Oncologic relevant members of S100 proteins are described as promising biomarkers in molecular pathology for risk estimation in oral neoplasia exhibiting different stages of malignancy: gingiva as healthy tissue, irritation fibroma as benign, leukoplakia as precancerous, and oral squamous cell carcinoma as malignant entity. Gene expression levels of S100A4 (metastasin), S100A7 (psoriasin), S100A8 (calgranulin A), and S100A9 (calgranulin B) were analyzed using quantitative RT-PCR. In addition, immunohistochemistry-based microscopy was used to examine cellular localization and distribution of these biomarkers in tissue sections. The results indicate that S100 proteins represent promising biomarkers for early-stage diagnosis in oral lesions. The inclusion of expression profiles and ratios for each entity even improves their diagnostic validity.

Key words S100 proteins, Calcium binding, EF-hand, Molecular pathology, Biomarkers, Oral lesions, Early-stage diagnosis

1 Introduction

S100 proteins are involved in various cellular processes like migration, differentiation, proliferation, apoptosis, energy metabolism, inflammation, host defense, and calcium homeostasis. They belong to an EF-hand containing calcium-binding subfamily of proteins that form homo- and heterodimeric complexes [1, 2]. S100 proteins bind extracellularly to various receptors, e.g., epidermal growth factor receptor (EGFR), toll-like receptor (TLR), or receptor for advanced glycation end products (RAGE), and can, thus, act in an autocrine as well as in a paracrine manner [2–4]. Intracellularly, S100 proteins have an impact on important tumor-associated molecules like p53, β-catenin, or NFκB. Due to their involvement

Rainer Probstmeier and Dominik Kraus contributed equally to this work.

Claus W. Heizmann (ed.), *Calcium-Binding Proteins of the EF-Hand Superfamily: From Basics to Medical Applications*, Methods in Molecular Biology, vol. 1929, https://doi.org/10.1007/978-1-4939-9030-6_48,

in a multitude of various cellular functions, there is strong evidence that dysregulation of S100 protein expression contributes to numerous diseases including cancer [2–5]. S100A4 (metastasin) is known to influence cell proliferation through interactions with EGFR ligands, p53, or RAGE. It has also been described to enhance a migratory phenotype and/or epithelial-mesenchymal transition (EMT), i.e., is involved in the formation of metastasis. Recently, S100A4 has already been shown to be associated with different oral lesions [2, 3, 6–9]. Increased gene expression of S100A7 (psoriasin) has been found in various epithelial malignancies, including head and neck squamous cell carcinomas, and has been associated with a poor prognosis. S100A7 plays a role in cell growth by affecting the β-catenin pathway. Furthermore, its gene expression is enhanced by inflammatory stimuli [2, 5]. S100A8 (calgranulin A) and S100A9 (calgranulin B) are often co-expressed. The proteins form a heterodimeric complex called calprotectin, which is present in epithelial cells mainly after stress, like wounding or inflammation. S100A8/A9 are tissue specifically involved in alterations of differentiation states of tumors. The protein complex binds to RAGE, thus inducing proliferative cell responses by establishing a pro-inflammatory tumor microenvironment. Recent studies could identify decreased expression of S100A8 and S100A9 to be associated with poor prognosis in oral carcinoma [4, 5, 10]. The objective of our studies was to analyze cellular localization and expression levels of the oncologic relevant members of the S100 family, S100A4, S100A7, S100A8, and S100A9, in the most frequent oral lesions representing different stages of malignancy in order to develop a putative diagnostic tool for the risk estimation of these types of oral neoplasms. Gingiva has been chosen as a member of healthy oral tissues, irritation fibroma as a benign form, leukoplakia as a precancerous form, and oral squamous cell carcinoma as a malignant cancer entity. Irritation fibromas are benign lesions which can undergo a malignant transformation in an inflammatory environment. Although oral leukoplakias are described as precancerous neoplasms, there is a high risk (38%) for a malignant transformation. Finally, oral squamous cell carcinomas are known to be preceded by premalignant lesions and associated with chronic inflammation [8, 9]. Hence, we think that analysis of these different members of oral neoplasms represents a convincing choice to develop a potent risk estimation tool for oral cancer early diagnosis.

2 Materials

Tissue samples should be accurately prepared avoiding unwanted surrounding healthy tissue that would otherwise falsify gene expression analysis. For immunohistochemistry analysis it might be an advantage to have some surrounded tissue parts still attached to the actual lesion to be able to compare the lesion with healthy tissue parts. In general, the small size of the samples to be used may hinder subsequent analysis. We normally use parts of the same tissue specimen for gene expression analysis and also immunohistochemistry examination. For this purpose, we separate the samples into two pieces and use one for RNA isolation and the other for immunohistochemistry.

2.1 Immunohistochemistry

1. Fixation: 4% buffered formalin solution
2. Embedding: Paraplast Plus (Roth, Germany); xylene (AppliChem, Germany); embedding automat VIP 2000/3000 (Vogel, Germany)
3. Preparation of tissue slices: microtome
4. Deparaffinization: Roti-Histol (Roth, Germany) as xylene replacement; ethanol; aqua dest
5. Blocking solution for endogenous peroxidases: 0.3% H_2O_2 in 100% methanol
6. Citrate buffer: 10 mM sodium citrate pH 6.00
7. Antibodies: final working concentration in 1% BSA/TBS
   (a) Anti-human S100A4 rabbit monoclonal (Cell Signaling): 1:400
   (b) Anti-human S100A7 rabbit polyclonal (Santa Cruz): 1:100
   (c) Anti-human S100A8 rabbit polyclonal (Santa Cruz): 1:50
   (d) Anti-human S100A9 rabbit polyclonal (Santa Cruz): 1:100
   (e) Anti-rabbit IgG goat polyclonal HRP conjugated (Dianova): 1:500
8. DAB (diaminobenzidine) staining: DAB solution (Thermo, Germany) 1:10 diluted with "Stable Peroxide Substrate Buffer" (Thermo, Germany)
9. Mayer's hematoxylin staining: stock solution (Merck, Germany) diluted 1:2 in aqua dest
10. DePex mounting: DePex stock solution (Serva, Germany)

2.2 Real-Time qRT-PCR

1. RNA extraction: RNeasy Protect Mini Kit (Qiagen, Germany).
2. RNA quantification: NanoDrop™ HND-1000 Spectrophotometer (NanoDrop Technologies, USA).
3. cDNA synthesis: iScript™ Select cDNA Synthesis Kit (Bio-Rad, Germany) with oligo(dT) primers.
4. Primers: all primers for PCR are from Metabion (Germany). Sequences are shown in Table 1.

Table 1
Genes, primer sequences, corresponding efficiencies, and annealing temperatures

Gene	Primer sequences (sense/antisense)	Efficiency	Annealing temperature (°C)
β-actin	5′-CATGGATGATGATATCGCCGCG-3′ 5′-ACATGATCTGGGTCATCTTCTCG-3′	1.84	69
GAPDH	5′-TGGTATCGTGGAAGGACTCA-3′ 5′-CCAGTAGAGGCAGGGATGAT-3′	1.93	67
S100A4	5′-ATGGCGTGCCCTCTGGAGAAGG-3′ 5′-TCATTTCTTCCTGGGCTGCTTATC-3′	2.08	69
S100A7	5′-ATGAGCAACACTCAAGCTGAGAGG-3′ 5′-TCACTGGCTGCCCCCGGAAC-3′	1.95	69
S100A8	5′-GACCGAGCTGGAGAAAGCCTTGA-3′ 5′-CCAGACGTCTGCACCCTTTTTCC-3′	1.92	62
S100A9	5′-AGCTGGAACGCAACATAGAGACCAT-3′ 5′-CCTCCATGATGTGTTCTATGACCT-3′	2.04	63

5. PCR mix: iQ™ SYBR® Green Supermix (Bio-Rad).
6. PCR cycler: iCycler® Thermal Cycler (Bio-Rad).

2.3 Statistics

One-way ANOVA and the post hoc Tukey's multiple comparison test are applied using a statistical software program (GraphPad Prism version 6 software, USA). P-values < 0.05 are considered to be statistically significant.

3 Methods

3.1 Immunohistochemistry

Tissue sampling is performed during routine surgical procedures followed by a protocol approved by the ethical board of the University of Bonn (Germany). The specimens used for immunohistochemistry are fixed as soon as possible after surgical treatment.

1. Sample fixation: tissue specimens are fixated in 4% PBS-buffered formalin solution for at least 24 h.
2. Embedding in paraffin (Paraplast Plus): tissue samples are embedded in paraffin using an embedding automat.
   (a) 60 min incubation in 70% ethanol at 40 °C (2×)
   (b) 60 min incubation in 80% ethanol at 40 °C
   (c) 60 min incubation in 96% ethanol at 40 °C (2×)
   (d) 75 min incubation in 100% ethanol at 40 °C (3×)
   (e) 60 min incubation in xylene at 40 °C
   (f) 30 min incubation in xylene at 40 °C

(g) 60 min incubation in Paraplast Plus at 60 °C (4×)

3. Preparation of tissue slices: tissue samples for immunohistochemistry are prepared by cutting the tissue sample containing paraffin block in 2 μm thick slices using a microtome. The slices are transferred onto object slides for later analysis.
4. Antibody staining:

   The object slides are put into glass chambers containing the different solutions necessary for rehydration.

   (a) 10 min incubation in Roti-Histol (2×)

   (b) 5 min incubation in 100% ethanol (2×)

   (c) 5 min incubation in 96% ethanol

   (d) 5 min incubation in 70% ethanol

   (e) 5 min incubation in aqua dest

   (f) 10 min incubation in TBS

   (g) 10 min incubation with H_2O_2 solution for blocking endoperoxidases

   (h) 10 min incubation in TBS

   (i) Heat treatment in citrate buffer (depending on the antibody): 5 min 95 °C boiling (microwave)/cooling at room temperature (3× for 5 min)

   (j) 10 min incubation in TBS

   (k) Overnight incubation with primary antibody at 4 °C in a humid chamber

   (l) 10 min incubation in TBS

   (m) 30 min incubation with secondary antibody at room temperature

   (n) 10 min incubation in TBS

   (o) DAB staining: 8 min at room temperature

   (p) 10 min incubation in TBS
5. Hemalaun staining: 2 s incubation with immediate blueing under running water.
6. Dehydration and mounting:

   (a) 30 s 70% ethanol

   (b) 30 s 96% ethanol

   (c) 2 min 100% ethanol (2×)

   (d) 2 min Roti-Histol (2×)
7. Mounting: add DePex on object slides and cover with cover slips.
8. Ready for microscopy (*see* example Fig. 1).

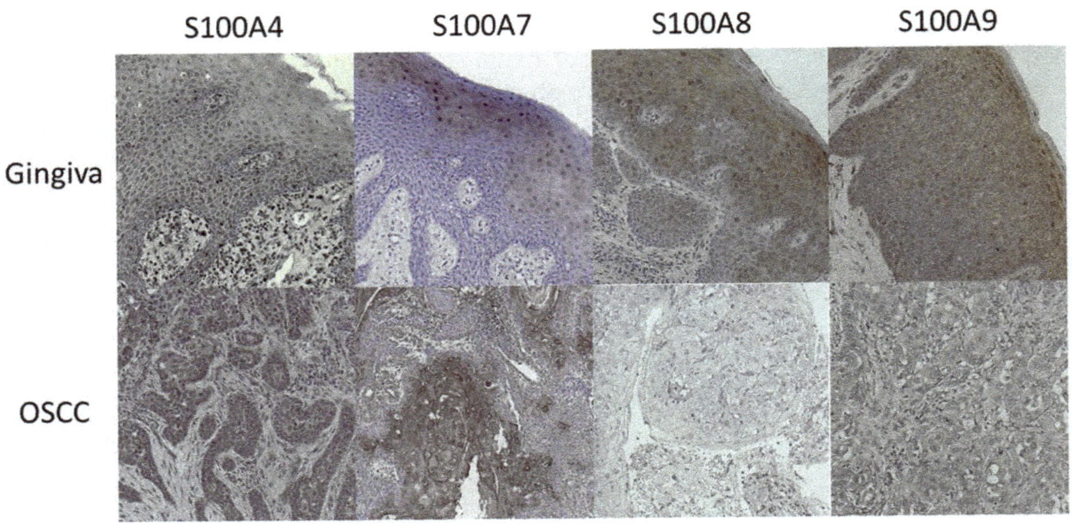

Fig. 1 Representative micrographs of S100A4, S100A7, S100A8, and S100A9 expression in healthy gingiva and malignant oral squamous cell carcinoma (OSCC)

3.2 Gene Expression Analysis Using Real-Time qRT-PCR

Relative differential gene expression is calculated using the method described by Pfaffl [11] with β-actin and glyceraldehyde phosphate dehydrogenase (GAPDH) serving as reference genes. Real-time PCR experiments are carried out according to "The MIQE guidelines: Minimum Information for Publication of Quantitative Real-Time PCR Experiments" [12]. Tissue samples are immediately stored frozen at −80 °C.

1. RNA extraction: RNA is isolated from tissue samples using the RNeasy Protect Mini Kit in a final volume of 100 µl (*see* **Note 1**).
2. RNA quantification: 1 µl of total RNA is quantified by measuring the absorbance ratio at wavelengths of 260/280 nm using the NanoDrop photometer (*see* **Note 1**).
3. cDNA synthesis: 1 µg of RNA is transcribed into cDNA using iScript™ Select cDNA Synthesis Kit with oligo(dT) primers according to the manufacturer's protocol. It is important to use oligo(dT) primers for first-strand synthesis to be sure to reversely transcribe even small RNA molecules (*see* **Note 2**). The final volume of the cDNA assay is 20 µl.
4. Real-time PCR: iQ™ SYBR® Green Supermix is a 2× mix. We normally use a final volume of 25 µl. Primers are added from a mix of the two specific primers (100 pmol/µl each) with a stock concentration of 50 pmol/µl for both primers. The PCR assay for one reaction is 12.5 µl Supermix, 11.375 µl aqua dest, 0.125 µl primer mix (0.5 pmol/µl), and 1.0 µl cDNA. For practical reasons we prepare a master mix of at least 100 µl Supermix, 100 µl aqua dest, and 1 µl primer stock. 1.0 µl of

Table 2
Relative gene expression of S100A4, S100A7, S100A8, and S100A9 compared to GAPDH and β-actin in gingiva, irritation fibromas, leukoplakias, and oral squamous cell carcinomas (OSCC) with $n = 15$ each

Entity	S100A4	S100A7	S100A8	S100A9
Gingiva	1.0	1.0	1.0	1.0
Irritation fibroma	1.2	11.0*	0.7	0.7
Leukoplakia	3.0*	25.0*	0.8	0.7
OSCC	3.0*	367.0*	0.3*	0.05*

Relative gene expression was standardized to the mean values for both reference genes
Healthy gingiva was set as "1.0." Significant differences to gingiva are depicted by asterisks (* for $p < 0.05$)

Table 3
Summary of differential expression states of S100A4, S100A7, S100A8, and S100A9 in benign, precancerous, and malignant lesions in comparison to healthy tissue

	Healthy	Benign	Precancerous	Malignant
S100A4	0	0	+	+
S100A7	0	+	+	++
S100A8	0	0	0	–
S100A9	0	0	0	––
S100A4/A7/A8/A9	0/0/0/0	0/+/0/0	+/+/0/0	+/++/–/––

Expression level compared to healthy state is shown for not different (0), for lower (–), and for higher range (+). Comparative analysis shows a characteristic pattern for each oral lesion entity being examined (last line)

cDNA is added to 24 μl of this master mix before starting the PCR under the following conditions: 5 min at 95 °C for denaturing, 50 cycles of 15 s at 95 °C, 15 s at annealing temperatures specific for primers (*see* Table 1), and 30 s at 72 °C for elongation (*see* **Notes 3** and **4**). Water instead of cDNA serves as negative control.

5. Quantification: Relative differential gene expression is analyzed using the method described by Pfaffl [11]. Calculations of molecule numbers are carried out by comparing Ct-values with standards of known molecule numbers (*see* **Notes 5** and **6**). ΔCt-values are determined by comparison to the mean from Ct-values of both housekeeping genes β-actin and GAPDH (*see* Table 2). To enhance the power of conclusion, it might help to directly compare the expression levels of all S100 genes beneath the lesions. With this method one can design a pattern which is characteristic for each entity (*see* Table 3).

4 Notes

1. It is necessary to check the RNA preparation for contamination with genomic DNA. In case genomic DNA is present, DNA digestion with RNAse-free DNase has to be performed prior to cDNA synthesis.
2. The lengths of S100A4/7/8/9 mRNAs are between 450 and 600 base pairs. For optimal quantification it is necessary that the process of reverse transcription catches as many mRNA molecules as possible during cDNA synthesis. Therefore, it is strongly recommended to use oligo(dT) primers which would guarantee the highest possible efficiency of cDNA molecule production.
3. It is highly recommended to put much effort into the design of the primers of choice. It is very important for the quality of differential gene expression analysis that primers are as specific as possible. We design our primers according to the "MIQE guidelines" [12]. Primers can be designed using "Primer-Blast" (https://www.ncbi.nlm.nih.gov). We routinely design primers with the highest annealing temperatures possible, to minimize unspecificity during the PCR reaction. It is recommended to double-check the primer sequences regarding their specificity by also using "Blast" (https://www.ncbi.nlm.nih.gov) for verification. Primers should only be used when one single specific PCR product of correct corresponding fragment length is amplified. The fragment should be sequenced to verify appropriate and specific amplification.
4. It is absolutely mandatory to verify the optimal annealing temperatures of all primer pairs to avoid/minimize unspecificity during PCR. We routinely use temperature gradients to optimize PCR conditions regarding the primers of choice.
5. To compare expression levels between different genes, it is necessary to determine the efficiency of primers regarding DNA amplification [11]. This factor of efficiency is identified through calculation of a regression graph based on different template input concentrations. Using the efficiencies it is possible to adjust the expression levels for direct comparison between multiple genes.
6. It might be important to know the minimum of detectable molecules. One can evaluate the Ct-value of the detection limit using samples containing known numbers of molecules. In our lab, samples generating Ct-values of below "35" are supposed to contain no target molecules anymore, i.e., Ct $<$ 35 is the limit of detection.

Acknowledgment

R. Probstmeier and D. Kraus contributed equally to this work.

References

1. Donato R (2001) S100: a multigenic family of calcium-modulated proteins of the EF-hand type with intracellular and extracellular functional roles. Int J Biochem Cell Biol 33:637–668
2. Donato R, Cannon BR, Sorci G, Riuzzi F, Hsu K, Weber DJ, Geczy CL (2013) Functions of S100 proteins. Curr Mol Med 13:24–57
3. Heizmann CW, Fritz G, Schäfer BW (2002) S100 proteins: structure, functions and pathology. Front Biosci 7:d1356–d1368
4. Donato R (2003) Intracellular and extracellular roles of S100 protein. Microsc Res Tech 60:540–551
5. Chen H, Xu C, Jin Q, Liu Z (2014) S100 protein family in human cancer. Am J Cancer Res 4:89–115
6. Wenghoefer M, Pantelis A, Najafi T, Deschner J, Allam JP, Novak N, Reich R, Martini M, Bergé S, Fischer HP, Jepsen S, Winter J (2010) Gene expression of oncogenes, antimicrobial peptides, and cytokines in the development of oral leukoplakia. Oral Surg Oral Med Oral Pathol Oral Radiol Endod 110:351–356
7. Winter J, Pantelis A, Allam JP, Novak N, Reich R, Martini M, Bergé S, Deschner J, Jepsen S, Wenghoefer M (2011) High α-defensin and S100A7 expression and missing DOC-1 down-regulation characterize irritation fibromas of the oral cavity and might counteract malignant transformation. J Craniofac Surg 22:100–104
8. Winter J, Pantelis A, Reich R, Jepsen S, Allam JP, Novak N, Wenghoefer M (2011) Risk estimation for a malignant transformation of oral lesions by S100A7 and Doc-1 gene expression. Cancer Investig 29:478–484
9. Reckenbeil J, Kraus D, Probstmeier R, Allam JP, Novak N, Frentzen M, Martini M, Wenghoefer M, Winter J (2016) Cellular distribution and gene expression pattern of metastasin (S100A4), Calgranulin A (S100A8), and Calgranulin B (S100A9) in oral lesions as markers for molecular pathology. Cancer Investig 34:246–254
10. Harris TM, Du P, Kawachi N, Belbin TJ, Wang Y, Schlecht NF, Ow TJ, Keller CE, Childs GJ, Smith RV, Angeletti RH, Prystowsky MB, Lim J (2015) Proteomic analysis of oral cavity squamous cell carcinoma specimens identifies patient outcome—associated proteins. Arch Pathol Lab Med 139:494–507
11. Pfaffl MW (2001) A new mathematical model for relative quantification in real-time RT-PCR. Nucleic Acids Res 29:e45
12. Bustin SA, Benes V, Garson JA, Hellemans J, Huggett J, Kubista M, Mueller R, Nolan T, Pfaffl MW, Shipley GL, Vandesompele J, Wittwer CT (2009) The MIQE guidelines: minimum information for publication of quantitative real-time PCR experiments. Clin Chem 55:611–622

Index

A

B

C

Claus W. Heizmann (ed.), *Calcium-Binding Proteins of the EF-Hand Superfamily: From Basics to Medical Applications*, Methods in Molecular Biology, vol. 1929, https://doi.org/10.1007/978-1-4939-9030-6,

D

N

O

P

Q

R

Lightning Source UK Ltd.
Milton Keynes UK
UKHW051305100219
336971UK00004B/52/P

9 781493 990290